AF700732

Soil Conditions and Plant Growth

Soil Conditions and Plant Growth

Edited by

Peter J. Gregory
Centre for Food Security
School of Agriculture, Policy & Development
University of Reading
Reading, UK
and
East Malling Research
New Road
East Malling
Kent, UK

Stephen Nortcliff
Soil Research Centre
Department of Geography and Environmental Science
University of Reading
Reading, UK

WILEY-BLACKWELL
A John Wiley & Sons, Ltd., Publication

This edition first published 2013 © 2013 by Blackwell Publishing Ltd.

Eleventh Edition published 1988 © 1988 Longman Group UK Limited (ed. A. White; co-published in the USA with John Wiley & Sons, Inc., New York). Tenth Edition published 1973 by Longman Scientific and Technical © 1973 E.W. Russell. Ninth Edition published 1961 by Longman Scientific and Technical © 1961 E.W. Russell. Eighth Edition published 1950 by Longmans, Green and Co. Rewritten and revised by E.W. Russell. Seventh Edition published 1937 by Longmans, Green and Co. Written by Sir E.J. Russell. Sixth Edition published 1932 by Longmans, Green and Co. In: *The Rothamsted Monographs on Agricultural Science* (ed. Sir E.J. Russell, DSc, FRS). Fifth Edition published 1927 by Longmans, Green and Co. In: *The Rothamsted Monographs on Agricultural Science* (ed. Sir E.J. Russell, DSc, FRS). Fourth Edition published 1921 by Longmans, Green and Co. In: *The Rothamsted Monographs on Agricultural Science* (ed. Dr E.J. Russell, DSc, FRS). Third Edition published 1917 by Longmans, Green and Co. In: *Monographs on Biochemistry* (eds. R.H.A. Plimmer, DSc, and F.G. Hopkins, MA, MB, DSc, FRS). Second Edition published 1915 by Longmans, Green and Co. In: *Monographs on Biochemistry* (eds. R.H.A. Plimmer, DSc, and F.G. Hopkins, MA, MB, DSc, FRS). First Edition published 1912 by Longmans, Green and Co. In: *Monographs on Biochemistry* (eds. R.H.A. Plimmer, DSc, and F.G. Hopkins, MA, MB, DSc, FRS).

Blackwell Publishing was acquired by John Wiley & Sons in February 2007. Blackwell's publishing program has been merged with Wiley's global Scientific, Technical and Medical business to form Wiley-Blackwell.

Registered Office
John Wiley & Sons, Ltd, The Atrium, Southern Gate, Chichester, West Sussex, PO19 8SQ, UK

Editorial Offices
9600 Garsington Road, Oxford, OX4 2DQ, UK
The Atrium, Southern Gate, Chichester, West Sussex, PO19 8SQ, UK
2121 State Avenue, Ames, Iowa 50014-8300, USA

For details of our global editorial offices, for customer services and for information about how to apply for permission to reuse the copyright material in this book please see our website at www.wiley.com/wiley-blackwell.

The right of the authors, to be identified as the authors, of this work has been asserted in accordance with the UK Copyright, Designs and Patents Act 1988.

All rights reserved. No part of this publication may be reproduced, stored in a retrieval system, or transmitted, in any form or by any means, electronic, mechanical, photocopying, recording or otherwise, except as permitted by the UK Copyright, Designs and Patents Act 1988, without the prior permission of the publisher.

Designations used by companies to distinguish their products are often claimed as trademarks. All brand names and product names used in this book are trade names, service marks, trademarks or registered trademarks of their respective owners. The publisher is not associated with any product or vendor mentioned in this book. This publication is designed to provide accurate and authoritative information in regard to the subject matter covered. It is sold on the understanding that the publisher is not engaged in rendering professional services. If professional advice or other expert assistance is required, the services of a competent professional should be sought.

Limit of Liability/Disclaimer of Warranty: While the publisher and author have used their best efforts in preparing this book, they make no representations or warranties with the respect to the accuracy or completeness of the contents of this book and specifically disclaim any implied warranties of merchantability or fitness for a particular purpose. It is sold on the understanding that the publisher is not engaged in rendering professional services and neither the publisher nor the author shall be liable for damages arising herefrom. If professional advice or other expert assistance is required, the services of a competent professional should be sought.

Library of Congress Cataloging-in-Publication Data

Soil conditions and plant growth / edited by Peter Gregory, Stephen Nortcliff.
p. cm.
Includes bibliographical references and index.
ISBN 978-1-4051-9770-0 (hardback : alk. paper) – ISBN 978-1-118-33731-8 (epdf/ebook) –
ISBN 978-1-118-33730-1 (emobi) – ISBN 978-1-118-33728-8 (epub) – ISBN 978-1-118-33729-5 (obook)
1. Soil science. 2. Crops and soils. 3. Plant-soil relationships. I. Gregory, P. J. II. Nortcliff, Stephen.
S591.S693 2012
631.4–dc23

2012024777

A catalogue record for this book is available from the British Library.

Wiley also publishes its books in a variety of electronic formats. Some content that appears in print may not be available in electronic books.

Cover image credits: small, top right photo: Peter J. Gregory; all remaining photos: Stephen Nortcliff
Cover design by Meaden Creative

Set in 10/12pt Times by SPi Publisher Services, Pondicherry, India

Contents

A colour plate section falls between pages 214 and 215

Contributors

Sayed Azam-Ali
Crops for the Future Research Centre
Selangor Darul Ehsan,
Malaysia

Richard D. Bardgett
Lancaster Environment Centre
Lancaster University
Lancaster, Lancashire, UK

A. Glyn Bengough
The James Hutton Institute
Invergowrie, Dundee, UK and
Division of Civil Engineering,
University of Dundee, Dundee, UK

Leo M. Condron
Agriculture and Life Sciences
Lincoln University
Canterbury, New Zealand

Erica Donner
Centre for Environmental Risk
Assessment and Remediation
University of South Australia
Mawson Lakes, South Australia, Australia

Gregorio Egea
Area of Agro-Forestry Engineering
School of Agricultural Engineering
University of Seville
Seville, Spain

Timothy S. George
The James Hutton Institute
Invergowrie, Dundee, UK

Keith W.T. Goulding
Department of Sustainable Soils and
Grassland Systems
Rothamsted Research
Harpenden, Hertfordshire, UK

The Late Duncan J. Greenwood FRS
University of Warwick
Wellesbourne, Warwick, UK

Edward G. Gregorich
Agriculture and Agri-Food Canada
Central Experimental Farm
Ottawa, Ontario, Canada

Peter J. Gregory
Centre for Food Security
School of Agriculture, Policy and
Development
University of Reading
Reading, UK
and
East Malling Research
East Malling, Kent, UK

Paul D. Hallett
The James Hutton Institute
Invergowrie, Dundee, UK

Phil M. Haygarth
Lancaster Environment Centre
Lancaster University
Lancaster, Lancashire, UK

Philippe Hinsinger
INRA
UMR Eco&Sols
Montpellier, France

Mark E. Hodson
Environment Department
University of York
Heslington, York, UK

David W. Hopkins
School of Life Sciences
Heriot-Watt University
Edinburgh, UK

Ken Killham
Honorary Fellow
The James Hutton Institute
Invergowrie, Dundee, UK

Daniel V. Murphy
Soil Biology Group
School of Earth and Environment
UWA Institute of Agriculture
The University of Western Australia
Crawley, Western Australia, Australia

Maria de Nobili
Dipartmento di Scienze Agrarie e Ambientali
Università degli studi di Udine
Udine, Italy

Stephen Nortcliff
Department of Geography and Environmental Science
Soil Research Centre
University of Reading
Reading, UK

David Powlson
Department of Sustainable Soils and Grassland Systems
Rothamsted Research
Harpenden, Hertfordshire, UK

Pete Smith
School of Biological Sciences
University of Aberdeen
Aberdeen, UK

Dominic Standing
School of Biological Sciences
University of Aberdeen
Aberdeen, UK

Elizabeth A. Stockdale
School of Agriculture
Food and Rural Development
Newcastle University
Newcastle-upon-Tyne, UK

Anne Verhoef
Department of Geography and Environmental Science
Soil Research Centre
University of Reading
Reading, UK

Philip J. White
The James Hutton Institute
Invergowrie, Dundee, UK

Preface

Since the last edition of *Russell's Soil Conditions and Plant Growth* in 1988, soil, plant and crop sciences have moved on considerably. For a long time during these 24 years, there was a diminishing interest in soil science as an underpinning element of crop production largely because the higher income countries of the world were food secure and, indeed, at times awash with surplus crops. Soil science became an environmental science with a much broader remit, and courses at universities changed to meet the training of this new wave of students. One consequence of these changes was that the traditional publisher of this textbook saw no demand for a new edition.

Recently, though, the mood has changed again, and both the rising awareness of global food insecurity and the need for soils and land to deliver simultaneously food and fibre and other ecosystem goods and services have focused attention on the requirement to better understand and manage the many interactions that occur between soils and plants. What better time for a new edition of *Soil Conditions and Plant Growth* to examine these interactions?

In preparing for this book, we decided at an early stage to let go of some of the content of the previous editions. There are now many introductory textbooks on soils and soil science; hence, we have quite deliberately excluded from this edition elements of pedology, chemistry, microbiology and soil survey that featured previously. Instead, we have tried to focus on what has always been the core feature of this book – the interactions between soils and plants. We have included accounts that detail how plants respond to soil properties but also how plants themselves are key agents in soil formation and modifiers of their environment. We have also chosen writers with some experience of how soils can be managed in both agricultural and ecological contexts to promote crop production but also to deliver high-quality water supplies, cope with the warming climate and all of the many other necessities of life that we have come to expect from our soils.

The contributors to this book would particularly like to thank (1) Dr Clare Benskin (Lancaster University, UK), Dr Tim George (James Hutton Institute, UK), Dr Alan Richardson (CSIRO, Australia) and Dr Ben Turner (Smithsonian Tropical Research Institute, Republic of Panama), who all contributed helpful material or input to preparing Chapter 5; (2) the financial support of The Scottish Government Rural and Environment Research and Analysis Directorate for Chapter 6, which is dedicated to the memory of Dr Duncan J. Greenwood (1932–2010); and (3) Dr Jos Raaijmakers and Dr Paolina Garbeva for providing the bacterial strain referred to in Figure 11.10.

Finally, we would like to thank acknowledge with gratitude the encouragement that we received in our careers from Professors Walter Russel, Dennis Greenland and Alan Wild and to Wiley-Blackwell for taking on the publication of this book. We also thank our many authors for so readily agreeing to write for us and for their enthusiasm in updating what, for many of us as students, was a soil science classic. This year, 2012, is the centenary of the first edition of Russell's book, and we hope that our readers will find the same inspiration as previous generations of students and researchers.

Peter J. Gregory
Stephen Nortcliff
March 2012

1 The historical development of studies on soil–plant interactions

Stephen Nortcliff[1] and Peter J. Gregory[2,3]

[1] *Department of Geography and Environmental Science, Soil Research Centre, University of Reading, Reading, UK*
[2] *Centre for Food Security, School of Agriculture, Policy and Development, University of Reading, Reading, UK*
[3] *East Malling Research, Kent, UK*

1.1 Introduction

How plants grow and how this growth varies through time and in response to changing conditions has been an interest of people for millennia. From the early cultivators to present-day gardeners, there has been a fascination in how a flourishing plant can be derived from a dry, apparently lifeless seed. Furthermore, there has been recognition that plant growth shows different patterns in response to weather conditions and that it varies from place to place. As global population continues to increase the need to understand the growth of plants and the role of soils in crop production becomes increasingly important. The demand for both food and biomass-derived energy from plants is increasing, so we must also seek to understand how to allocate land for multiple purposes. Soils must be used for these services and to obtain other essential services such as clean water and a diverse soil community of organisms.

Many early civilisations appear to have compiled information on plant growth and crop husbandry, and there was an extensive literature on agriculture developed during Roman times, which provided important guidance on crop growth and management for many centuries after the fall of the Roman Empire. The Roman literature was collected and condensed into one volume about the year 1309 by a senator of Bologna, Petrus de Crescentius (the book was made more widely available when published in 1471), whose book was one of the most popular treatises on agriculture of any time, being frequently copied, and in the early days of printing, passing through many editions. Many other agricultural books appeared in the fifteenth and early sixteenth centuries, notably in Italy and later in France. In some of these are found certain ingenious speculations that have been justified by later work. Such, for instance, is Palissy's remarkable statement in 1563:

> You will admit that when you bring dung into the field it is to return to the soil something that has been taken away…. When a plant is burned it is reduced to a salty ash called alcaly by apothecaries and philosophers…. Every sort of plant without exception contains some kind of salt. Have you not seen certain labourers when sowing a field with wheat for the second year

Soil Conditions and Plant Growth. Edited by Peter J. Gregory and Stephen Nortcliff.
© 2013 Blackwell Publishing Ltd. Published 2013 by Blackwell Publishing Ltd.

> in succession, burn the unused wheat straw which had been taken from the field? In the ashes will be found the salt that the straw took out of the soil; if this is put back the soil is improved. Being burnt on the ground it serves as manure because it returns to the soil those substances that had been taken away.

But while some of these speculations have been confirmed, many in other sources have not, and the beginnings of agricultural chemistry was to take place later when we had learnt the necessity for investigating possible relationships and pathways using experiments.

1.2 The search for the 'principle' of vegetation, 1630–1750

It was probably discovered at an early stage in agricultural development that manures, composts, dead animal bodies and parts of animals, such as blood, all increased the productivity of the land; and this was the basis of the ancient saying that 'corruption is the mother of vegetation'. Although there was empirical evidence for this linkage, the early investigators consistently ignored this ancient wisdom when they sought for the 'principle' of vegetation to account for the phenomena of soil fertility and plant growth. Thus, the great Francis Bacon, Lord Verulam, believed that water formed the 'principal nourishment' of plants, the purpose of the soil being to keep them upright and protect them from excessive cold or heat, though he also considered that each plant drew a 'particular juyce' from the soil for its sustenance, thereby impoverishing the soil for that particular plant and similar ones, but not necessarily for other plants. Similarly, van Helmont (1577–1644) regarded water as the sole nutrient for plants, and his interpretation of a carefully undertaken experiment in which he grew willows concluded that water was the principal requirement for plant growth (van Helmont, 1648).

Robert Boyle (1661) repeated the experiment with 'squash, a kind of Italian pompion' and obtained similar results. Boyle further distilled the plants and concluded, quite justifiably from his premises, that the products obtained, 'salt, spirit, earth, and even oil, may be produced out of water'. While these experiments were laudable, they ignored the part played by air, and in the van Helmont experiment there was a small reduction in the amount of soil present, which was ignored, although we now know this to be significant. In some respects, this might be taken as a guide for many of the future experiments undertaken in agriculture; if the hypotheses are wrong and other hypotheses are ignored, conclusions which may appear to be valid will often turn out to be incorrect because the alternatives have been ignored.

The primacy of water in plant growth was questioned by an experiment published by John Woodward in a fascinating paper (1699). Based on the experiments of van Helmont and of Boyle, he grew spearmint in water obtained from various sources and noted that all of these plants were supplied with an abundance of water so that all should have made equal growth had nothing more been needed. The amount of growth, however, increased with the impurity of the water (Table 1.1). He concluded:

> Vegetables are not formed of water, but of a certain peculiar terrestrial matter. It has been shown that there is a considerable quantity of this matter contained in rain, spring and river water, that the greatest part of the fluid mass that ascends up into plants does not settle there but passes through their pores and exhales up into the atmosphere: that a great part of the terrestrial matter, mixed with the water, passes up into the plant along with it, and that the plant is more or less

Table 1.1 Growth of spearmint using water from different sources.

Source of water	Mass (g) of plants when planted	Mass (g) of plants when harvested	Mass (g) gained in 7 days	Expense (g) of water (transpiration)	Ratio increase in mass:mass water used
Rain water	1.83	2.96	1.13	220.3	1:195
River Thames	1.81	3.50	1.69	161.5	1:95.6
Hyde Park Conduit	7.13	16.14	9.01	851.5	1:94.5
Hyde Park Conduit plus 105 g garden mould	5.96	24.36	18.40	968.8	1:52.7

Source: From Woodward (1699).

> augmented in proportion as the water contains a greater or less quantity of that matter; from all of which we may reasonably infer, that earth, and not water, is the matter that constitutes vegetables.

Taking account of the results in his experiment, he discussed the use of manures and the fertility of the soil from this point of view, attributing the well-known falling off in crop yield when plants are grown for successive years on unmanured land to the circumstance that:

> the vegetable matter that it at first abounded in being extracted from it by those successive crops, is most of it borne off.... The land may be brought to produce another series of the same vegetables, but not until it is supplied with a new fund of matter, of like sort with that it at first contained; which supply is made several ways, either by the ground's being fallow some time, until the rain has poured down a fresh stock upon it; or by the tiller's care in manuring it.

The best manures, he continued, are parts either of vegetables or of animals, which ultimately are derived from vegetables.

For a time there was little progress in relation to what plants needed in addition to water and how these needs might be met. Advances were, however, being made in agricultural practice. One of the most important was the introduction of the drill and the horse-hoe by Jethro Tull, an Oxford man of a strongly practical turn of mind, who insisted on the vital importance of getting the soil into a fine, crumbly state for plant growth. Tull (1731) was more than an inventor; he discussed in most picturesque language the sources of fertility in the soil. In his view, it was not the juices of the earth but the very minute particles of soil loosened by the action of moisture that constituted the 'proper pabulum' of plants. The pressure caused by the swelling of the growing roots forced these particles into the 'lacteal mouths' of the roots, where they entered the circulatory system. All plants lived on these particles, i.e. on the same kind of food; it was incorrect to assert, as some had done, that different kinds of plants fed as differently as horses and dogs, each taking its appropriate food and no other. Plants will take in anything that comes their way, good or bad. A rotation of crops is not a necessity, but only a convenience. Conversely, any soil will nourish any plant if the temperature and water supply are properly regulated. Hoeing increased the surface of the soil or the 'pasture of the plant' and also enabled the soil to better absorb the nutritious vapours condensed from the air. Dung acted in the same way, but was more costly and less efficient.

The position at the end of this period cannot better be summed up than in Tull's own words: 'It is agreed that all the following materials contribute in some manner to the increase of plants, but it is disputed which of them is that very increase or food: (1) nitre, (2) water, (3) air, (4) fire, (5) earth'.

1.3 The search for plant nutrients

1.3.1 The phlogistic period, 1750–1800

Great interest was taken in agriculture in the UK during the latter half of the eighteenth century. Many experiments were conducted, facts were accumulated, books written and societies formed for promoting agriculture. The Edinburgh Society, established in 1755 for the improvement of arts and manufactures, induced Francis Home 'to try how far chymistry will go in settling the principles of agriculture' (1757). The whole art of agriculture, he says, centres in one point: the nourishing of plants. Investigation of fertile soils showed that they contain oil, which is therefore a food of plants. But when a soil has been exhausted by cropping, it recovers its fertility on exposure to air, which therefore supplies another food. Home established pot experiments to ascertain the effect of various substances on plant growth. 'The more they [i.e. farmers] know of the effects of different bodies on plants, the greater chance they have to discover the nourishment of plants, at least this is the only road.' Saltpetre, Epsom salts, vitriolated tartar (i.e. potassium sulphate) all lead to increased plant growth, yet they are three distinct salts. Olive oil was also useful. It is thus clear that plant food is not one thing only, but several; he enumerates six: air, water, earth, salts of different kinds, oil and fire in a fixed state. As further proof he shows that 'all vegetables and vegetable juices afford those very principles, and no other, by all the chymical experiments which have yet been made on them with or without fire'.

Between 1770 and 1800, work was done on the effect of vegetation on air that was destined to revolutionise the ideas of the function of plants in the economy of nature, but its agricultural significance was not recognised until later. Joseph Priestley, knowing that the atmosphere becomes vitiated by animal respiration, combustion, putrefaction, etc., and realising that some natural purification must go on, or life would no longer be possible, was led to try the effect of sprigs of living mint on vitiated air (1775). He found that the mint made the air purer and concludes 'that plants, instead of affecting the air in the same manner with animal respiration, reverse the effects of breathing, and tend to keep the atmosphere pure and wholesome, when it is become noxious in consequence of animals either living, or breathing, or dying, and putrefying in it'. But he had not yet discovered oxygen and so could not give precision to his discovery; and when, later on, he did discover oxygen and learn how to estimate it, he unfortunately failed to confirm his earlier results because he overlooked a vital factor, the necessity for light. He was therefore unable to answer Scheele, who had insisted that plants, like animals, vitiate the air. It was Jan Ingen-Housz (1779) who reconciled both views and showed that purification goes on in light only, while vitiation takes place in the darkness. Ingen-Housz's conclusions might be summarised as follows: (1) light is necessary for this restoration (this we would now know as photosynthesis); (2) only the green parts of the plant actually perform restoration and (3) all living parts of the plant 'damage' the air (respire), but the extent of air restoration by a green plant far exceeds its damaging effect. Jean Senebier (1782) working in Geneva also concluded that the plant–atmosphere interactions were significant.

1.3.2 The period 1800–1860

The foundation of plant physiology

Progress to this point had been constrained by methodologies which were known and accepted. To gain new insights and further investigate some of the speculation, particularly in relation to the plant–atmosphere links, new methods were required. The work of Nicholas Theodore de Saussure (1804) in the early nineteenth century in establishing the broad principles of the quantitative experimental method, in some respects produced the paradigm shift which proved the basis for modern agricultural chemistry. The work of Boussingault, Liebig, Lawes and Gilbert drew on this new experimental method which still provides the basis of many investigations. De Saussure grew plants in air or in known mixtures of air and carbon dioxide, and measured the gas changes by eudiometric analysis and the changes in the plant by 'carbonisation'. He was thus able to demonstrate the central fact of plant respiration – the absorption of oxygen and the evolution of carbon dioxide – and further show the decomposition of carbon dioxide and evolution of oxygen in light. Carbon dioxide in small quantities was a vital necessity for plants, and they perished if it was artificially removed from the air. It furnished them not only with carbon, but also with some oxygen. Water is also decomposed and fixed by plants. On comparing the amount of dry matter gained from these sources with the amount of material that can enter through the roots, he concluded that even under the most favourable conditions the soil furnished only a very small part of the plant food. Small as it is, however, this part is indispensable: it supplies nitrogen which he described as an essential part of vegetation and, as he had shown, was not assimilated directly from the air, and also ash constituents, which he noted contributed to the solid parts of plants, just as with animals. Further, he showed that the root is not a mere filter allowing any and every liquid to enter the plant; it has a special action and takes in water more readily than dissolved matter, thus effecting a concentration of the solution surrounding it; different salts, also, were absorbed to different extents. Passing next to the composition of the plant ash, he showed that it was not constant, but varies with the nature of the soil and the age of the plant; it consists mainly, however, of alkalis and phosphates. All the constituents of the ash occur in humus. If a plant is grown from seed in water, there is no gain in ash: the amount found at the end of the plant's growth is the same as was present in the seed excepting for a relatively small amount falling on the plant as dust. Thus, he disposed finally of the idea that the plant *generated* potash.

While in retrospect we see the considerable insight into the basis of plant growth and plant nutrition presented by de Saussure, the ideas and approaches did not gain general acceptance. The two great books on agricultural chemistry then current still belonged to the old period. A. von Thaer and Humphry Davy did not realise the fundamental change in perspective introduced by de Saussure. Thaer published his *Grundsätze der rationellen Landwirtschaft* in 1809–1812, which was translated into English as late as 1844 by Cuthbert Johnson. In it he adopted the prevailing view that plants draw their carbon and other nutrients from the soil humus. Humphry Davy's book (1813) grew out of the lectures on agricultural chemistry which he gave annually at the Royal Institution between 1802 and 1812; it forms the last textbook of the older period. While no great advance was made by Davy himself, he carefully sifted the facts and hypotheses of previous writers and gives an account, which, although defective in places, represents the best accepted knowledge of the time, set out in the new chemical language. His great name gave the subject an importance it probably would not otherwise have had. He did not accept de Saussure's conclusion that plants obtain their carbon chiefly from the carbonic acid of the air: some plants, he says,

Table 1.2 Budgets of dry matter, carbon, hydrogen, oxygen nitrogen and mineral matter at Pechelbronn, Alsace.

	Weight in kg ha^{-1} of					
Crop	**Dry matter**	**Carbon**	**Hydrogen**	**Oxygen**	**Nitrogen**	**Mineral matter**
Beets	3172	1357.7	184	1367.7	53.9	199.8
Wheat	3006	1431.6	164.4	1214.9	31.3	163.8
Clover hay	4029	1909.7	201.5	1523.0	84.6	310.2
Wheat/turnips (catch crop)	4208	2004.2	230	1700.7	43.8	229.3
Oats	2347	1182.3	137.3	890.9	28.4	108.0
Total during rotation	17478	8192.7	956.5	7009.0	254.2	1065.5
Added in manure	10161	3637.6	426.8	2621.5	203.2	3271.9
Difference not accounted for taken from air, rain or soil	+7317	+455.1	+529.7	+4387.5	+51	−2206.4

Source: From Boussingault (1841).

appear to be supplied with carbon chiefly from this source, but in general he supposes the carbon to be taken in through the roots. Davy presented a list of sources of this carbon, but there was little supporting experimental evidence and subsequently the sources have been shown to be false, although Davy's reputation meant they persisted. His insistence on the importance of the physical properties of soils – their relationship to heat and to water – was more accurate and marks the beginning of soil physics. In mainland Europe, to an even greater extent than in Britain, it was held that plants drew their carbon and other nutrients from the soil humus.

The foundation of agricultural science

To this point, experiments had been conducted either in the laboratory or in small pots: around 1834, however, J. B. Boussingault, who was already known as an adventurous traveller in South America, began a series of field experiments on his farm at Pechelbronn in Alsace. These were the first of their kind: to Boussingault, therefore, belongs the honour of having introduced the method by which the new agricultural science was to be developed. He reintroduced the quantitative methods of de Saussure, weighed and analysed the manures used and the crop obtained, and at the end of the rotation drew up a balance sheet, showing how far the manures had satisfied the needs of the crop and how far other sources of supply – air, rain and soil – had been drawn upon. The results of one experiment are given in Table 1.2. At the end of the period, the soil had returned to its original state of productiveness, hence the dry matter, carbon, hydrogen and oxygen not accounted for by the manure must have been supplied by the air and rain, and not by the soil. On the other hand, the manure afforded more mineral matter than the crop took off, the balance remaining in the soil. Other things being equal, he argued that the best rotation is one which yields the greatest amount of organic matter over and above what is present in the manure.

The rotation had not impoverished the soil so he concluded that nitrogen may be taken directly in to the plant if the green parts are capable of fixing it. Boussingault's work covers the whole range of agriculture and deals with the composition of crops at different stages of their growth, with soils and with problems in animal nutrition. Unfortunately, the classic farm of Pechelbronn did not remain a centre for agricultural research, and the experiments came to an end in 1870.

During this period (1830–1840), Carl Sprengel was studying the ash constituents of plants, which he considered were probably essential to nutrition (1832). Schübler was working at soil physics, and a good deal of other work was quietly being done. No particularly important discoveries were being made, no controversies were going on, and no great amount of interest was taken in the subject.

But all this was changed in 1840 when Liebig's famous report to the British Association upon the state of organic chemistry, published as *Chemistry in Its Application to Agriculture and Physiology* in 1840 (Liebig, 1840), gave rise to the need to rethink the world of science. Liebig was highly critical of the plant physiologists of his day for their continued adhesion, in spite of accumulated evidence, to the view that plants derive their carbon from the soil and not from the carbonic acid of the air. 'All explanations of chemists must remain without fruit, and useless, because, even to the great leaders in physiology, carbonic acid, ammonia, acids and bases are sounds without meaning, words without sense, terms of an unknown language, which awake no thoughts and no associations.' Liebig stated that the experiments quoted by the physiologists in support of their view are all 'valueless for the decision of any question'. Liebig's ridicule did what neither de Saussure's nor Boussingault's logic had done: it finally killed the humus theory. Only the boldest would have ventured after this to assert that plants derive their carbon from any source other than carbon dioxide, and for a time carbon dioxide was considered to be the sole source of the carbon of plants. Hydrogen and oxygen came from water and nitrogen from ammonia. Certain mineral substances were essential: alkalis were needed for neutralisation of the acids made by plants in the course of their vital processes, phosphates for seed formation and potassium silicates for the development of grasses and cereals. The evidence lay in the composition of the ash: plants might absorb anything soluble from the soil, but they excreted from their roots whatever was non-essential. The fact of a substance being present was therefore sufficient proof of its necessity.

Plants, Liebig argued, have an inexhaustible supply of carbonic acid in the air. But time is saved in the early stages of plant growth if carbonic acid is being generated in the soil, for it enters the plant roots and affords extra nutrient over and above what the small leaves are taking in. Hence a supply of humus, which continuously yields carbonic acid, is advantageous. Further, the carbonic acid attacks and dissolves some of the alkali compounds of the soil and thus increases the mineral food supply. The true function of humus is to evolve carbonic acid.

Liebig further argued that the alkali compounds of the soil are not all equally soluble. A weathering process has to go on, which is facilitated by liming and cultivation, whereby the comparatively insoluble compounds are broken down to a more soluble state. The final solution is effected by acetic acid excreted by the plant roots, and the dissolved material now enters the plant.

Nitrogen is taken up as ammonia, which may come from the soil, from added manure, or from the air. In order that a soil may remain fertile, it is necessary and sufficient to return in the form of manure the mineral constituents and the nitrogen that have been taken away. When sufficient crop analyses have been made, it will be possible to draw up tables showing the farmer precisely what he must add in any particular case.

An artificial manure known as Liebig's patent manure was made up on these lines and placed on the market.

Liebig's book (1840) was meant to attract attention to the subject, and it did; it rapidly went through several editions, and as time went on Liebig developed his thesis and gave it a quantitative form: 'The crops on a field diminish or increase in exact proportion to the diminution or increase of the mineral substances conveyed to it in manure.' He further adds what afterwards became known as the Law of the Minimum, 'by the deficiency or absence of one necessary constituent, all the others being present, the soil is rendered barren for all those crops to the life of which that one constituent is indispensable'. These and other amplifications in the third edition, 1843, gave rise to much controversy. So much did Liebig insist, and quite rightly, on the necessity for alkalis and phosphates, and so impressed was he by the gain of nitrogen in meadow land supplied with alkalis and phosphates alone, and by the continued fertility of some of the fields of Virginia and Hungary and the meadows of Holland, that he began more and more to regard the atmosphere as the source of nitrogen for plants. Some of the passages of the first and second editions urging the necessity of ammoniacal manures were deleted from the third and later editions. 'If the soil be suitable, if it contain a sufficient quantity of alkalis, phosphates, and sulphates, nothing will be wanting. The plants will derive their ammonia from the atmosphere as they do carbonic acid', he writes in the *Farmer's Magazine*. Ash analysis led him to consider the turnip as one of the plants 'which contain the least amount of phosphates and therefore require the smallest quantity for their development'. These and other practical deductions were seized upon and shown to be erroneous by Lawes and Gilbert, who had for some years been conducting vegetation experiments. Lawes does not discuss the theory as such, but tests the deductions Liebig himself draws and finds them wrong. Further trouble was in store for Liebig; his patent manure when tried in practice *had failed*. This was unfortunate, and the impression in England at any rate was, in Philip Pusey's words: 'The mineral theory, too hastily adopted by Liebig, namely, that crops rise and fall in direct proportion to the quantity of mineral substances present in the soil, or to the addition or abstraction of these substances which are added in the manure, has received its death-blow from the experiments of Mr Lawes.'

And yet the failure of the patent manure was not entirely the fault of the theory, but only affords further proof of the numerous pitfalls of the subject. The manure was sound in that it contained potassium compounds and phosphates (it ought, of course, to have contained nitrogen compounds), but the compounds were rendered insoluble by fusion with lime and calcium phosphate so that the manure should not too readily wash out in the drainage water. Not until Way (1850) had shown that *soil precipitates soluble salts of ammonium, potassium and phosphates* was the futility of the fusion process discovered, did Liebig (1851) recognise the error he had made.

1.3.3 The second half of the nineteenth century

Meanwhile the great field experiments at Rothamsted had been started by Lawes and Gilbert in 1843. These experiments were conducted on the same general lines as those begun earlier by Boussingault, but they have the advantage that they still continue, having been on the same ground without alteration, except in occasional details, since 1852. The mass of information now accumulated is considerable and has become an invaluable source of data as we seek to understand aspects of sustainability and possible responses to environmental change (see, e.g. Leigh and Johnston, 1994). The experiments rapidly provided information and, by as early as 1855, the following points were clear:

1. Crops require phosphates and salts of the alkalis, but the composition of the ash does not afford reliable information as to the amounts of each constituent needed, for

example turnips require large amounts of phosphates, although only little is present in their ash.
2. Non-leguminous crops require a supply of some nitrogenous compounds, nitrates and ammonium salts being almost equally good. Without an adequate supply, no increases of growth are obtained, even when ash constituents are added. The amount of ammonia obtainable from the atmosphere is insufficient for the needs of crops. Leguminous crops behave abnormally.
3. Soil fertility may be maintained for some years at least by means of artificial manures.
4. The beneficial effect of fallowing lies in the increase brought about in the available nitrogen compounds in the soil.

Although many of Liebig's statements were shown to be wrong, the main outline of his theory as first enunciated stands. It is no detraction that de Saussure had earlier published a somewhat similar but less definite view of nutrition: Liebig had brought matters to a head and made people look at their cherished, but rarely examined, convictions. The effect of the stimulus he gave can hardly be over-estimated, and before he had finished, the essential facts of plant nutrition were settled and the lines were laid down along which scientific manuring was to be developed. The water cultures of Knop and other plant physiologists showed conclusively that potassium, magnesium, calcium, iron, phosphorus, along with sulphur, carbon, nitrogen, hydrogen and oxygen are all necessary for plant life. The list differs from Liebig's only in the addition of iron and the withdrawal of silica; but even silica, although not strictly essential for all plants, is advantageous for the nutrition of many cereals.

In two respects, however, the controversies continued for many years. Farmers were slow to believe that 'chemical manures' could ever do more than stimulate the crop and declared they must ultimately exhaust the ground. The Rothamsted plots falsified this assertion; manured year after year with the same substances and sown always with the same crops, they even now, after more than a 150 years of chemical manuring, continue to produce good crops, although secondary effects have sometimes set in. In France, the great missionary for artificial manures was Georges Ville, whose lectures were given at the experimental farm at Vincennes during 1867 and 1874–1875. He went even further than Lawes and Gilbert and maintained that artificial manures were not only more remunerative than dung but were the only way of keeping up fertility (Ville, 1879). In recommending mixtures of salts for manure, he was not guided by ash analysis but by field trials. For each crop, one of the four constituents, nitrogen compounds, phosphates, lime and potassium compounds (he did not consider it necessary to add any others to his manures), was found by trial to be more required than the others and was therefore called the 'dominant' constituent. For wheat he concluded that on his soil it required a good supply of nitrogen, less phosphate and still less potassium (Table 1.3).

Other experiments of the same kind showed that nitrogen was dominant for all cereals and beetroot, potassium for potatoes and vines, and phosphate for turnips and swedes. An excess of the dominant constituent was always added to the crop manure. The composition of the soil had to be taken into account, but soil analysis was at that time not good enough for this purpose. Instead, he drew up a simple scheme of plot trials to enable farmers to determine for themselves just what nutrient was lacking in their soil. His method was thus essentially empirical.

The second controversy dealt with the source of nitrogen in plants. Priestley had stated that a plant of *Epilobium hirsutum* placed in a small vessel absorbed during the course of the

Table 1.3 Yield data for wheat grown at Versailles, France.

Constituent added	Yield t ha^{-1}
Normal manure	2.98
Manure without lime	2.84
Manure without potash	2.14
Manure without phosphate	1.83
Manure without nitrogen	0.97
Soil without manure	0.83

Source: From Ville (1879).

month seven-eighths of the air present. De Saussure, however, denied that plants assimilated gaseous nitrogen. J. B. Boussingault's pot experiments showed that peas and clover could get nitrogen from the air while wheat could not, and his rotation experiments emphasised this distinction. While he did not make as much of this discovery as he might have done, he later fully realised its importance.

Liebig, as we have seen, maintained that ammonia, but not gaseous nitrogen, was taken up by plants, a view confirmed by Lawes et al. (1861) in the most rigid demonstration that had yet been attempted. A full summary of this work is provided in Lawes and Gilbert (1889). Plants of several natural orders, including the Leguminosae, were grown in surroundings free from ammonia or any other nitrogen compound. The soil was burnt to remove all traces of nitrogen compounds, while the plants were kept throughout the experiment under glass shades, but supplied with washed and purified air and with pure water. In spite of the ample supply of mineral food, the plants languished and died: the conclusion seemed irresistible that plants could not utilise gaseous nitrogen. For all non-leguminous crops, this conclusion agreed with the results of field trials. But there remained the very troublesome fact that leguminous crops required no nitrogenous manure, and yet they contained large quantities of nitrogen and also enriched the soil considerably in this element. Where then had the nitrogen come from? The amount of combined nitrogen brought down by the rain was found to be far too small to account for the result. For years experiments were carried on, but the problem remained unsolved. Looking back over the papers, one can see how very close some of the older investigators were to resolving the mystery: in particular, Lachmann carefully examined the structure of the nodules, which he associated with the nutrition of the plant, and showed that they contained 'vibrionenartige' organisms. His paper, however, was published in an obscure journal and attracted little attention (Lachmann, 1891). Atwater in 1881 and 1882 showed that peas acquired large quantities of nitrogen from the air and later suggested that they might 'favour the action of nitrogen-fixing organisms'. But he was too busily engaged to follow the matter up, and once again an investigation in agricultural chemistry had been brought to a standstill for want of new methods of attack.

1.4 The beginnings of soil microbiology

It had been a maxim with the older agricultural chemists that 'corruption is the mother of vegetation'. Animal and vegetable matter had long been known to decompose with the formation of nitrates: indeed nitre beds made up from such decaying matter were the recognised source of nitrates for the manufacture of gunpowder during the European wars of the seventeenth and eighteenth centuries. No satisfactory explanation of the process had been offered, although the discussion of rival hypotheses continued until 1860, but the conditions under which it worked were known and on the whole fairly accurately described.

No connection was at first observed between nitrate formation and soil productiveness. Liebig rather diverted attention from the possibility of tracing what now seems an obvious relationship by regarding ammonia as the essential nitrogenous plant nutrient, though he admitted the possible suitability of nitrates. Way (1850, 1852) came much nearer to the truth. He showed that nitrates were formed in soils to which nitrogenous fertilisers were added. Unfortunately he failed to realise the significance of this discovery. He was still obsessed with the idea that ammonia was essential to the plant, and he believed that ammonia, unlike other nitrogen compounds, could not change to nitrate in the soil but was absorbed by the soil. But he only narrowly missed making an important advance in the subject, for after pointing out that nitrates are comparable with ammonium salts as fertilisers he writes:

> Indeed the French chemists are going further, several of them now advocating the view that it is in the form of nitric acid that plants make use of compounds of nitrogen. With this view I do not myself at present concur: and it is sufficient here to admit that nitric acid in the form of nitrates has at least a very high value as a manure.

It was not until 10 years later, and as a result of work by plant physiologists, that the French view prevailed over Liebig's, and agricultural investigators recognised the importance of nitrates to the plant and of nitrification to soil fertility. It then became necessary to discover the cause of nitrification.

During the 1860s and 1870s, great advances were being made in bacteriology, and it was definitely established that bacteria bring about putrefaction, decomposition and other changes; it was therefore conceivable that they were the active agents in the soil and that the process of decomposition was not the purely chemical 'eremacausis' Liebig had postulated. Pasteur himself had expressed the opinion that nitrification was a bacterial process. The new knowledge was first brought to bear on agricultural problems by Schloesing and Müntz (1877, 1879, 1882) during a study of the purification of sewage water by land filters. A continuous stream of sewage was allowed to trickle down a column of sand and limestone so slowly that it took 8 days to pass. For the first 20 days, the ammonia in the sewage was not affected, then it began to be converted into nitrate; finally all the ammonia was converted during its passage through the column, and nitrates alone were found in the issuing liquid. Why, asked the authors, was there a delay of 20 days before nitrification began? If the process were simply chemical, oxidation should begin at once. They therefore examined the possibility of bacterial action and found that the process was entirely stopped by a little chloroform vapour, but could be started again after the chloroform was removed by adding a little turbid extract of dry soil. Nitrification was thus shown to be due to micro-organisms – 'organised ferments', to use their expression.

Warington (1878, 1879, 1884, 1891) had been investigating the nitrates in the Rothamsted soils, and at once applied the new discovery to soil processes. He showed that nitrification in the soil is stopped by chloroform and carbon disulphide; further, that solutions of ammonium salts could be nitrified by adding a trace of soil. By a careful series of experiments described in his four papers to the Chemical Society, he found that there were two stages in the process and two distinct organisms: the ammonia was first converted into nitrite and then to nitrate. But he failed altogether to obtain the organisms, in spite of some years of study,

Table 1.4 Relationships between nitrogen supply and plant growth.

Nitrogen in the calcium nitrate supplied per pot, g	None	0.056	0.112	0.168	0.224	0.336
Weight of oats obtained (grain and straw, g)	0.390	5.680	10.961	15.007	21.357	30.175
Weight of peas obtained (grain and straw, g)	3.093	2.137	7.725	5.619	8.186	11.352

Source: From Hellriegel and Wilfarth (1888).

by the gelatin methods then in vogue. However, Winogradsky (1890a–c) isolated these two groups of organisms, showing they were bacteria. He succeeded where Warington failed because he realised that carbon dioxide should be a sufficient source of carbon for them, so that they ought to grow on silica gel plates carefully freed from all organic matter; and it was on this medium that he isolated them in 1890.

Warington also established definitely the fact that nitrogen compounds rapidly change to nitrate in the soil, so that whatever compound is supplied as manure, plants get practically nothing but nitrate as food. This closed the long discussion as to the nitrogenous food of non-leguminous plants; in natural conditions, they take up nitrate only (or at any rate chiefly), because the activities of the nitrifying organisms leave them no option. The view that plants assimilate gaseous nitrogen has from time to time been revived, but it is not generally accepted.

The apparently hopeless problem of the nitrogen nutrition of leguminous plants was soon to be solved. In a striking series of experiments in sand cultures, Hellriegel and Wilfarth (1888) showed that the growth of non-leguminous plants, barley, oats, etc., was directly proportional to the amount of nitrate supplied – the duplicate pots agreeing satisfactorily – while in the case of leguminous plants no sort of relationship existed and duplicate pots failed to agree. After the seedling stage was passed, the leguminous plants grown without nitrate made no further progress for a time, then some of them started to grow and did well, while others failed. This stagnant period was not seen where nitrate was supplied. Results from two of their experiments are given in Table 1.4.

Analysis showed that the nitrogen contained in the oat crop and sand at the end of the experiment was always a little less than that originally supplied, but was distinctly greater in the case of peas; the gain in three cases amounted to 0.910, 1.242 and 0.789 g per pot, respectively. They drew two conclusions: (1) the peas took their nitrogen from the air and (2) the process of nitrogen assimilation was conditioned by some factor that did not come into their experiment except by chance. In trying to frame an explanation, they connected two facts that were already known. Berthelot had made experiments to show that certain micro-organisms in the soil can assimilate gaseous nitrogen. It was known to botanists that the nodules on the roots of Leguminosae contained bacteria. Hellriegel and Wilfarth, therefore, supposed that the bacteria in the nodules assimilated gaseous nitrogen, and then handed on some of the resulting nitrogenous compounds to the plant. This hypothesis was shown to be well founded by the following facts:

1. In the absence of nitrate, peas made only small growth and developed no nodules in sterilised sand; when calcium nitrate was added, they behaved like oats and barley, giving regular increases in crop for each increment of nitrate (the discordant results of Table 1.4 were obtained on unsterilised sand).

2. The peas grew well and developed nodules in sterilised sand watered with an extract of arable soil.
3. The peas sometimes did well and sometimes failed when grown without soil extract and without nitrate in *unsterilised* sand, which might or might not contain the necessary organisms. An extract that worked well for peas might be without effect on lupins or serradella. In other words, the organism is specific.

Hellriegel and Wilfarth read their paper and exhibited some of their plants at the Naturforscher-Versammlung at Berlin in 1886. Gilbert was present at the meeting, and on returning to Rothamsted repeated and confirmed the experiments. At a later date, Schloesing and Laurent (1892) showed that the weight of nitrogen absorbed from the air was approximately equal to the gain by the plant and the soil and thus finally clinched the argument. The organism was isolated by Beijerinck (1888a–c, 1989) and called *Bacillus radicicola*, but is now known as *Bradyrhizobium*.

Thus, another great controversy came to an end, and the discrepancy between the field trials and the laboratory experiments of Lawes, Gilbert and Pugh was cleared up. The laboratory experiments gave the result that leguminous plants, like non-leguminous plants, have themselves no power of assimilating gaseous nitrogen; this power belongs to the bacteria associated with them. This result was obtained because by excluding all traces of organic matter, and thereby ammonia, from the soil, the apparatus and the air, there was no chance of infection with the necessary bacteria and no assimilation could occur. In contrast, in the field trials the bacteria were active, and there was a gain of nitrogen.

The general conclusion that bacteria are the real makers of plant food in the soil, and are, therefore, essential to the growth of all plants, was developed by Wollny and Berthelot. It was supposed to be proved by Laurent's experiments. He grew buckwheat on humus, obtained from well-rotted dung, and found that plants grew well on the untreated humus, but only badly on the humus sterilised by heat. When, however, soil bacteria were added to the sterilised humus (by adding an aqueous extract of unsterilised soil), good growth took place. The experiment looks convincing, but is really unsound. When an organic-rich soil is heated, some substances are formed that are toxic to plants. The failure of the plants on the sterilised humus was, therefore, not due to absence of bacteria, but to the presence of a toxin.

1.5 The development and application of modern knowledge of soils

Our understanding of the physical, chemical and biological factors that control the fertility of soils has advanced greatly since the time of Gilbert and Lawes. The application of this knowledge has resulted in great increases in productivity. The most useful gains have been from new understanding of the storage and movement of water in soils, the value and valuation of reserves of plant nutrients, the physical, chemical and biological conditions in the rhizosphere, and from the role of cultivations in modern production systems. Both research and its application have been greatly aided by developments in the basic sciences, in optical and electronic instruments and in computers. Experimental work in the field with crops provides a practical assessment of advances in exploiting basic soil studies. In this context, the experimental designs initiated by R. A. Fisher proved to be invaluable by providing methods of solving problems in soil management and cropping systems which could not be attempted before. The development of statistical methods of assessing error in biological experiments was also a major advance. Field experimentation became a major technique in research on soil fertility and, because the precision of the results could be estimated, they were readily accepted by other scientists. Designs in which several factors were tested simultaneously – and their interactions measured – were an important advance. They opened the way to multidisciplinary research; for example, the effects of biological, physical and management factors on crop nutrition were then widely investigated.

During the twentieth century, agricultural research became recognised as important to national interest and attempts were made to bring some organisation and funding to research. For example, in the UK the Agricultural Research Council was established in 1931, and while it has changed its focus and remit, parts of the original aims of the Council are still within the much broader remit of the Biotechnology and Biological Sciences Research Council. The application of the research to develop agriculture was given to the newly established National Agricultural Advisory Service (NAAS) in 1946. This government-funded body later became the Agricultural Development and Advisory Service (ADAS) in 1971 and in 1997 was privatised. It now has a remit well beyond its original focus in agriculture as a provider of environmental solutions, rural development services and policy advice. This pattern of government-funded support for guidance in land management being progressively privatised is not unique to the UK and is now a common pattern in large parts of the western world.

Internationally, agricultural research to aid food production and agricultural development in less developed parts of the world has developed under the Consultative Group for International Agricultural Research (CGIAR). CGIAR has established research centres around the world which have in some cases a specific crop focus (e.g. the International Rice Research Institute, IRRI) and in others a regional agricultural focus (e.g. the International Centre for Research in the Dry Areas, ICARDA). These centres have provided a well-resourced focus for international collaboration on agricultural research issues.

1.5.1 Advances in soil science

Developments in our knowledge of soil science will be made clearer in the succeeding chapters of this book. Important advances were made last century resulting from the application of new methodologies, such as X-ray and spectrographic analyses, which led to our understanding of the crystalline structures of layered aluminosilicate minerals and information on the distribution of micronutrients.

Work on the forms of nutrient reserves in soils, their mobilities and availabilities, and the fate of nutrients applied in fertilisers continued throughout the last century. Work on phosphorus was advanced when the radioisotope ^{32}P became available. Other advances were made by applying thermodynamic concepts to the solubilities of nutrient ions. Understanding of cation relationships took a notable step forward when Schofield's Ratio Law was proposed in 1947; later advances came from the use of the quantity (Q) and intensity (I) factors. Concepts of nutrient ion mobility, first developed in the USA, were used in the mathematical modelling of processes of nutrient uptake by Nye and Tinker (1977) and subsequently by Barber (1995). These models of nutrient processes in soil have become the basis for management of nutrients in farming systems.

Much research has been done on the role of nitrogen in soil/crop systems. This is justifiable as, in many conditions, the supply of nitrogen has a greater effect on crop performance than do the supplies of other nutrients, so that on the world scale this nutrient dominates fertiliser markets. In past years, there was a period in which only about 30% of the nitrogen applied was, on average, taken up by the crops grown. The 70% that was not recovered represented a serious loss to farmers and was also the cause of environmental pollution – nitrate leached into waters used for public supply and nitrous oxide formed by denitrification contributed to greenhouse gas emissions and damage to the ozone layer of the lower part of the stratosphere. Improved technologies, including the use of the stable isotope ^{15}N, have made it possible to show in experiments that up to 90% or more of the nitrogen applied as fertiliser can be accounted for in uptake by the crop plus that nitrogen stored in the soil which may benefit future crops. Through improved nitrogen management, we are in sight of securing much higher efficiency of fertiliser nitrogen applied to crops, ranging from wheat in Europe to rice in Southeast Asia. The investigations described previously of the roles of micro-organisms in the fixation of nitrogen by leguminous crops have led to methods of preparing cultures of the organisms (*Rhizobium* spp.) which are specifically associated with particular legumes. These cultures have been made available to farmers for inoculating crops which are to be grown on soils where the appropriate species of *Bradyrhizobium* is lacking.

The management of other major nutrients, notably phosphorus and potassium but also calcium and magnesium, has been greatly aided by studies of the soils and crops in long-term experiments. The classical experiments at Rothamsted begun by Lawes in the nineteenth century have been invaluable in these studies; they have shown that reserves of phosphate and potassium accumulated in soils from fertiliser additions have considerable value in crop production. Long-term experiments also provide the best basis for relating soluble nutrients in soils to crop performance and to the need for fertilisers; they also lead to calculations of nutrient cycles which are essential for the efficient management of crop nutrition. Increasingly, the records provided by long-term experiments have provided both the underpinning knowledge to build models of how the soil system operates (for instance the Rothamsted Model of Soil Carbon, see, e.g., Jenkinson (1988) and Chapter 4 in this volume) and the data for the testing of model simulations.

The management of soil has been much improved as a result of scientific work. The cultivations necessary for good crop growth have been defined; minimum cultivation systems lessen the energy required and conserve soil structure and soil organic matter. Erosion of the cultivated layer of soil by water and by wind is a serious threat to efficient agriculture in many parts of the world. Studies of the mode and extent of losses of soil under practical conditions have led to recommendations for improved management of cropping systems which avoid these losses. Another serious hazard is the damage done to plant growth, and also to soil structure, by the salts which accumulate in saline soils. Salinity may occur naturally, or it may be the result of irrigating with unsuitable water. Investigations of this problem have led to definitions of the water quality that is required to remove soluble salts in drainage and to the use of gypsum for reclaiming saline soils. With good quality irrigation water and correctly managed drainage, salinity need not now be a problem for efficient crop production.

1.5.2 Soil surveys

The first proposals for the classification and mapping of soils were made by Russian workers in the middle of the nineteenth century. Surveys were made in Southeast England at the end of the nineteenth century, with other local surveys made early in the twentieth century (see, e.g., Kay, 1939). While the practice of soil survey has been widely undertaken in many parts of the world, Great Britain provides an example of many of the trends which have occurred. Soil mapping was undertaken principally in the period of 1945–1990 when maps on the scale of 1: 250 000 were produced across Great Britain and maps at scales of 1:63 360 and 1:25 000 were also produced for some areas. As part of the link between soil survey and agriculture, land capability maps were also derived. These maps use soil survey data, climate data and landscape data to show the limitations exhibited by areas of land for arable agricultural production. They had their origins in the USA and where an eight-class system was first introduced by USDA in 1961 (Klingebiel and Montgomery, 1961), with subsequent modifications for use in the national contexts; for example, Bibby and Mackney (1969) produced a modification of the USDA scheme for application in Great Britain with seven classes. In all systems, Class One land is land with no limitations and high potential productivity for arable crops. Class Two land has some minor limitations for arable agricultural production which limits the choice of crops and may restrict timing of cultivations (e.g. moderate or imperfect drainage and less than ideal rooting depth and moderate slopes). Class Three land exhibits moderate limitations which restrict the choice of crops but is still considered suitable for arable production with appropriate and good management. Class Four land is considered to have moderately severe limitations (e.g. poor drainage, shallow and/or very stony soils, etc.) which restrict choice of crops and need very careful management. Land in lower classes is not considered suitable for other than low productivity agricultural uses and non-agricultural uses such pasture, forestry and recreation. Most land capability systems identified subclasses of Classes Two to Four, where the major limitations to agricultural production were identified. These often included limitations due to drainage, erosion, wetness, climate and specific soil limitations. For specific discussion of land capability in relation to soil fertility, see Section 3.2.1.

Today in Great Britain, there is now no systematic mapping programme. The few soil surveys undertaken are for specific purposes, but databases containing soil information derived principally from earlier soil survey work and information gathered for specific tasks are available and can be interrogated.

One of the key contributions of soil surveys and soil classification was in the context of 'technology transfer'. Technology transfer involves combining information about soil properties gained from soil survey with the results of experimental work at that site to provide guidance on land use such as fertiliser recommendations; this knowledge is transferred to a distant site where the soil classification is the same. In addition to soil maps, advisers require computer-based information services giving the capability for cropping and physical, chemical and biological properties of the soil. Such information is becoming essential to modern management systems which aim to promote the productivity of soil. On a world scale, the most widely used tool for technology transfer is probably the *Soil Taxonomy* (1999) system developed in the USA. More recently, under the auspices of the International Union of Soil Science, the World Reference Base for Soil Resources (IUSS Working Group WRB, 2006) has been introduced. While soil surveys are still undertaken, there have been rapid changes in the manner in which soil data are recorded and managed. A recent significant shift has been the management and presentation of data using digital soil mapping.

The IUSS Working Group on Digital Soil Mapping (WG-DSM) defines 'Digital Soil Mapping' as 'the creation and the population of a geographically referenced soil databases generated at a given resolution using field and laboratory observation methods coupled with environmental data through quantitative relationships'. In addition, data may be added through inferred spatial and non-spatial relationships between soil and other environmental properties. The development of pedometrics (the application of statistical and mathematical methods for the study and understanding of soils, their distribution and development) has greatly enhanced the development of digital soil mapping.

DSM can rely upon, but is distinct from, traditional soil mapping or soil survey which involves the manual delineation of soil boundaries by field soil scientists. This digitised and georeferenced soil survey information does not become DSM until the Geographical Information System layer is used to derive other soil-related information within a GIS or similar information software application.

Digital Soil Mapping makes extensive use of previously collected soil survey data and progress has been considerable because of rapid development in computing and the increasing ease with which data are gathered and managed electronically.

1.5.3 Precision Farming

In recent years, as the costs of fertilisers and other agrochemicals have increased and there have been increasing controls on agriculture in terms of leaching of fertilisers, nutrients and pesticides to water courses and groundwater, there has been a shift towards precision agriculture or precision farming. Precision agriculture, or information-based management of agricultural production systems, has developed over the last two decades and is based on the recognition that soils and crops will vary within a field. Initially the focus was to adapt fertiliser distribution to varying conditions across a field, but increasingly with the increased use of new technologies such as global navigation systems and geographic information systems has resulted in a much wider set of applications. Initially these uses were extended to the effective and efficient use of other agrochemicals such as herbicides and pesticides and the correct timing and placement of irrigation water, but with time other practices have evolved such as automatic guidance of agricultural vehicles and implements, autonomous machinery and processes, and product traceability and software for the overall management of agricultural production systems. One consequence is that there is much more efficiency in the use of resources for crop production such as fertiliser, irrigation water, pesticides, etc. This is in sharp contrast to earlier practices where the same fertiliser or pesticide rate was applied to the whole field, often focusing on higher levels of application to address the demands of the poorer soils within that field, with a consequent over application in some parts of the field. This resulted in poor use of valuable resources and often resulted in environmental damage. While early approaches of precision agriculture were based on traditional soil survey maps and the expertise of the farmer, the availability of new technologies – such as global positioning systems, digital soil maps, remote and proximal soil and crop sensors, satellites or other aerial images – and information management tools such as geographical information systems to assess and understand variations has radically changed the practice of precision farming greatly increasing the precision of placement of the agrochemicals and other external resources. The increasing availability of digital soil maps and soil information now provides a wealth of information which is a further contribution to the efficient and effective management of resources. These various sources of information integrated with agricultural machinery, which is able to place agrochemicals and other crop needs to a high degree of resolution, has radically changed the amount of agrochemicals used in a field, introducing economic and environmental benefits as a result of better placement. Precision agriculture enables the efficient use of resources which provides cost savings and assists in the reduction of environmental damage. Gebbers and Adamchuk (2010) have briefly reviewed the considerable impact the use of precision agriculture has had on food production and food security.

Table 1.5 Fertiliser use in arable and grassland in Great Britain 1998–2010 (kg/ha).

	1998	1999	2000	2001	2002	2003	2004	2005	2006	2007	2008	2009	2010
Total N	126	125	123	116	117	113	110	109	107	105	95	97	102
Total K_2O	45	42	40	37	40	36	37	35	34	32	27	22	25
Total P_2O_5	35	32	32	29	31	28	28	27	25	24	20	15	19

Source: From DEFRA – http://www.defra.gov.uk/statistics/files/defra-stats-foodfarm-environ-fertiliserpractice-2010.pdf.

1.5.4 Fertilisers

The global use of fertilisers expanded greatly during the twentieth century and the first years of the twenty-first century, resulting in large increases in crop yield. According to Cooke (1982), global use of N, P and K fertilisers by 1913 was 1.4, 0.9 and 0.7 Mt, respectively. By 1998, the consumption had increased to 82.1, 14.2 and 18.0 (FAO, 2010). By 2007, the total global consumption of fertilisers was estimated at 179 Mt (FAO, 2010). These increases have resulted from the application by farmers' advisers of the research work which has identified the deficiencies of the main nutrients in soils and crops.

Both in the world as a whole, and in the UK, the amounts of nitrogen used dominate compared with applications of other nutrients; this nutrient is responsible for the major part of the cost of fertilisers to the world's farmers (in the UK about three-quarters of the total spent on fertilisers is for nitrogen). The extent to which farmers change their fertiliser practice on particular crops is shown by the *Surveys of Fertiliser Practice* which were initiated in the 1940s and still continue. Similar surveys are made in a few other countries. While fertiliser use during the twentieth century showed a steady increase in the amount used per hectare, data on the use of fertilisers in Great Britain between 1998 and 2010 show a pattern of a 'levelling off in use' and, in the most recent years, an overall reduction in the amounts of N, P and K applied per hectare of land (Table 1.5). In part this can be ascribed to our increasing knowledge about crop management and increased precision in fertiliser management, but the introduction of regulations enforcing nitrate-vulnerable zones have had a major effect (see DEFRA, 2011).

1.5.5 Ecosystem services

The publication of the Millennium Ecosystem Assessment (2005) brought the concept of ecosystem services to the fore. These services were broadly defined as 'provisioning services' such as food and water; 'regulating services' such as regulation of floods, droughts, land degradation and disease; 'supporting services' such as soil formation and nutrient cycling and 'cultural services' such as recreational, spiritual, religious and other non-material benefits.

Soils are widely recognised as an essential part of natural and agricultural ecosystems. Increasingly it is recognised that the role of soils is key to the provision of ecosystem services. These ecosystem services play key roles in the functioning of the ecosystems of which the soil is a part, in both natural and managed production systems. Soils provide beneficial services through the roles played in soil formation, nutrient cycling and primary production. In some respects, the regulatory roles of soils provide key services to other components of the system; yet these are frequently overlooked. These include buffering water fluxes between inputs in precipitation and outputs as run-off, throughflow and groundwater recharge; playing a key role in the carbon cycle through CO_2 emissions from respiration and carbon sequestration and the filtering and detoxification of harmful substances added to the soil. Soils, of course, provide cultural services in terms of supporting recreational space, support for buildings and infrastructure and indeed many human activities.

1.5.6 The needs for future research on soil–plant growth interactions

All agricultural production has its origin in plants which grow in soils. The ultimate potential yield is set by the genetic capacity of the crop which is grown and the solar radiation which it receives at the site. These potentials have been established by models based on the physiology of plants. Average yields throughout the world, as recorded by FAO, are below the recognised potentials because of a variety of constraints. Some constraints result from attacks by pests and diseases but many are associated with soil conditions. Chemical constraints may be due to shortages of nutrients or the presence of toxins; physical constraints arise from shortages or excesses of water or from mechanical impedance to root growth. Biological constraints arise where adverse conditions inhibit the survival and beneficial operation of micro-organisms which are key to the transformation and availability of nutrients. Scientific work to improve further agricultural production must therefore concentrate on the identification of constraints and the further development of inputs and/or management practices which overcome these constraints. The multifactorial field experiments which test these means of overcoming constraints must involve all the disciplines concerned with crop growth, and the objective must be to produce reliably the recognised potential yield at all sites.

In considering the nature of future experimental work, we should note the advice given by Sir Humphrey Davy in the 1821 edition of his book, for this is still relevant. He discussed the need to substitute sound and rational principles for 'vague, popular prejudices…. Nothing is more wanting in agriculture than experiments in which all the circumstances are minutely and scientifically detailed. The results of truly philosophical experiments in agricultural chemistry would be of more value in enlightening and benefiting the farmer than the greatest possible accumulation of imperfect trials conducted merely in the empirical spirit.'

The multidisciplinary experimentation in which soil science has a fundamental role requires adequate support from those responsible for agricultural improvement. It is therefore appropriate to quote from the writings of another giant of the nineteenth century: John Bennett Lawes concluded his paper published in 1850 to report the first work ever done on the effect of plant nutrients on water use efficiency with these words:

> The interest and progress of agriculture would be more surely and permanently served if its great patron Societies were to permit to their scientific officers a wider range of discretion, and more liberal means for the selection and carrying out of definite questions of research. Results of this kind promise, it is true, but little prospect of immediate and practical application, but by their aid the uncertain dictates, whether of common experience, theory, or speculation, may, ere long, be replaced by the unerring guidance of principles; and then alone can it reasonably be anticipated that miscellaneous and departmental analyses may find their true interpretation, and acquire a due and practical value.

References

Barber, S.A. (1995) *Soil Nutrient Bioavailability: A Mechanistic Approach*, 2nd edition. John Wiley & Sons, Inc., New York.

Beijerinck, M.W. (1888a) Die Bacterien der Papilionaceen-Knöllchen. *Botanische Zeitung*, **46**(46), 725–735.

Beijerinck, M.W. (1888b) Die Bacterien der Papilionaceen-Knöllchen. *Botanische Zeitung*, **46**(47), 741–750.

Beijerinck, M.W. (1888c) Die Bacterien der Papilionaceen-Knöllchen. *Botanische Zeitung*, **46**(48), 757–771.

Beijerinck, M.W. (1889) Künstliche Infection von Vicia Faba mit Bacillus radiixixola Ernährsungsbedingungen dieser Bacterie. *Botanische Zeitung*, **47**(52), 837–843.

Bibby, J.S. and Mackney, D. (1969) *Land Use Capability Classification*. Soil Survey Technical Monograph, No. 1, Soil Survey of Great Britain, Harpenden, UK.

Boussingault, J.B. (1841) De la discussion de la valeur relative des assolements par les résultats de l'analyse élementaire. *Annales de Chemie et de Physique*, **3**(1), 208-246.

Boyle, R.F. (1661) *The Sceptical Chemist*. J. Crooke, London.

Cooke, G.W. (1982) *Fertilizing for Maximum Yield*. Granada, London.

de Crescentius, P. (1471) *Ruralium commodorum libri duodecin*. Johan Schüssler, Augsburg, Germany.

Davy, H. (1813) *Elements of Agricultural Chemistry in a Course of Lecture*. Longmans, London.

DEFRA (2011) http://www.defra.gov.uk/food-farm/land-manage/nitrates-watercourses/nitrates/. Accessed on 9 October 2011.

FAO (2010) http://faostat.fao.org/site/575/default.aspx#ancor. Accessed on 10 December 2011.

Gebbers, R. and Adamchuk, V.I. (2010) Precision agriculture and food security. *Science*, **327**, 828–831.

Hellriegel, H. and Wilfarth, H. (1888) *Untersuchungen uber die Stickstoffnahrung der Gramineen und Leguminose*, 239 pp. Beilageheft zu der Zeitschrift des Vereins fur Rubenzucker-Industrie Deutschen Reichs. Kayssler & Co, Berlin, Germany.

Home, F. (1757) *Principles of Agriculture and Vegetation*. G. Hamilton and J. Balfour, Edinburgh, UK.

Ingen-Housz, J. (1779) *Experiments upon Vegetables, Discovering the Great Power of Purifying Common Air in Sunshine and of Injuring in the Shade and at Night*. P. Elmsly and H. Payne, London.

IUSS Working Group WRB (2006) *World Reference Base for Soil Resources 2006*. World Soil Resources Reports No. 10. FAO, Rome, Italy.

Jenkinson, D.S. (1988). Soil organic matter and its dynamics. In: *Soil Conditions and Plant Growth* (ed. A. Wild), 11th edn., pp. 564–607. Longmans Scientific & Technical, Harlow, UK.

Kay, F.F. (1939) *A Soil Survey of the Strawberry District of South Hampshire*. Faculty of Agriculture and Horticulture, University of Reading, Reading, UK.

Klingebiel, A.A. and Montgomery, P.H. (1961) *Land Capability Classification*. USDA Agricultural Handbook No. 210, Washington, DC.

Lachmann, J. (1891) Über Knollen and den Wurzelms der Leguminosen. *Bieder-Manns Zentrablat für Agrickulturchemie*, **20**, 837–854. (Originally published 1858 in *Landwirtschaftliche Mitteillungen Zeitschrift K. Lehranst. Vers. Sta. Popplesdorf* (Bonn), **37**.)

Lawes, J.B. and Gilbert, J.H. (1889) On the present position of the question of the sources of nitrogen of vegetation and some results and preliminary notice of new lines of investigation. *Philosophical Transactions of the Royal Society*, **180**, 1–107.

Lawes, J.B., Gilbert, J.H., and Pugh, E. (1861) On the sources of nitrogen of vegetation: Special reference to the question whether plants assimilate free or combined nitrogen. *Philosophical Transactions of the Royal Society of London*, **151**, 431–557.

Leigh, R.A. and Johnston, A.E. (1994) Long-term experiments in agricultural and ecological sciences. In *Proceedings of a Conference to Celebrate the 150th Anniversary of the Rothamsted Experimental Station*, Rothamsted, UK, 14–17 July 1993. CAB International, Wallingford, UK.

Liebig, J. (1840) *Chemistry in Its Application to Agriculture and Physiology*. British Association, London.

Liebig, J. (1851) *Familiar Letters on Chemistry and Its Relation to Commerce, Physiology and Agriculture*, 3rd edn. Chemical Society, London.

Millennium Ecosystem Assessment. (2005) *Ecosystem and Human Well-being – A Framework for Assessment*. Island Press, Washington, DC.

Nye, P.H. and Tinker, P.B. (1977) *Solute Movement in the Soil-Root System*. Blackwell, Oxford, UK.

Priestley, J. (1775) *Experiments and Observations on Different Kind of Air*. J. Johnson, London.
de Saussure, N.-T. (1804) *Recherches chimique sur la vegetation*. Nyon, Paris, France.
Schloesing, T. and Müntz, A. (1877) Sur la nitrification par les ferments organises. *Comptes rendus de l'Académie des Sciences*, **84**, 301–303.
Schloesing, T. and Müntz, A. (1879) Sur la nitrification par les ferments organises. *Comptes rendus de l'Académie des Sciences*, **86**, 892–895.
Schloesing, T. and Müntz, A. (1882) Sur la nitrification par les ferments organises. *Comptes rendus de l'Académie des Sciences*, **89**, 1074–1077.
Schloesing, T. and Laurent, E. (1892) Recherches sur la fixation de l'azote libre par les plantes. *Annals de l'Institut Pasteur*, **6**, 824–840.
Senebier, J. (1782) *Memoires Physico-Chemiques sur l'Influence de la lumière solaire por modifier les trios des être de la nature règne végétal*. Barthelemi Chirol, Geneva, Switzerland.
Soil Survey Staff. (1999) *Soil Taxonomy. A Basic System of Soil Classification for Making and Interpreting Soil Surveys*. Agricultural Handbook No. 436. Natural Resources Conservation Service, United States Department of Agriculture, Washington, DC.
Sprengel, C. (1832) *Chemie für Landwirthe, Forstmänner und Cameralisten*. Vandehoeck, Göttingen, Germany.
Tull, J. (1731) *Horse Hoeing Husbandry*. William Cobbett, London
Van Helmont, J.B. (1648) *Ortus medicinae*, pp.84–90. Complexium atque mistonium elementalium figmetnum, Amsterdam, the Netherlands.
Ville, G. (1879) *On Artificial Manures their Chemical Selection and Scientific Application to Agriculture* (translated by W. Crookes). Longmans, London.
Warington, R. (1878) On nitrification, *Journal of Chemical Society, Transaction*, **33**, 44–50.
Warington, R. (1879) On nitrification (Part II). *Journal of Chemical Society, Transactions*, **35**, 429–456.
Warington, R. (1884) On nitrification (Part III). *Journal of Chemical Society, Transactions*, **45**, 637–672.
Warington, R. (1891) On nitrification (Part IV). *Journal of Chemical Society, Transactions*, **59**, 484–520.
Way, J.T. (1850) On the power of soils to absorb manure. *Journal of the Royal Agricultural Society of England*, **11**, 313–379.
Way, J.T. (1852) On the power of soils to absorb manure. *Journal of the Royal Agricultural Society of England*, **13**, 123–143.
Winogradsky, S. (1890a) Recherches sur les organismes de la nitrification. *Annals de l'Institut Pasteur*, **4**(4), 213–231.
Winogradsky, S. (1890b) Recherches sur les organismes de la nitrification. *Annals de l'Institut Pasteur*, **4**(5), 257–275.
Winogradsky, S. (1890c) Recherches sur les organismes de la nitrification. *Annals de l'Institut Pasteur*, **4**(11), 760–771.
Woodward, J. (1699) Some thoughts and experiments concerning vegetation. *Philosophical Transactions of the Royal Society*, **20**, 193–227. http://www.defra.gov.uk/statistics/files/defra-stats-foodfarm-environ-fertiliserpractice-2010.pdf

2 Plant and crop science

Sayed Azam-Ali

Crops for the Future Research Centre
Selangor Darul Ehsan, Malaysia

2.1 Introductory principles of resource capture by crops

Communities of plants capture and convert environmental resources into carbon-based products. Their absolute capture and conversion of resources will determine their total productivity (biomass) and, where a proportion of this biomass is useful to human beings, their harvestable yield. Their relative capture and conversion of resources will determine the suitability of one crop over others at any location and in any growing season.

The principles of resource capture and conversion (hereafter, simply resource capture) can be considered as a framework for analysing the performance of crops and cropping systems. In this analysis, groups of plants (crops) act as vegetation 'factories' in which environmental resources are sequestered ('captured') and fixed ('converted') into chemical products. By using this physical analogy, we can apply concepts of 'work', 'energy' and 'efficiency' to the whole vegetation system and compare the performance of one vegetation system with another.

The resource capture framework allows quantitative decisions on how best we can manage the choice of crop (species, variety) and cropping system (architectural arrangement of one or more species) and the use of available resources naturally, from the environment, or applied, as irrigation, fertilisers, etc. For the framework to be applicable across environments, species and systems, it must be robust yet simple. The vegetation factory uses fuel (principally water, nutrients, solar radiation and carbon dioxide) from which it makes products (principally carbon-based compounds differentially composed of lipids, proteins and carbohydrates). The efficiency of the factory can be estimated from its productivity relative to the amount of 'fuel' it uses over the same period. When considering the efficiency of the crop factory, we should be aware that it must capture resources in two media, soil and air, and two dimensions, space and time. Hence, our resource capture framework requires an understanding of the soil and atmospheric processes that interact with a crop from the time it is sown or planted until such time that it is harvested.

Soil Conditions and Plant Growth. Edited by Peter J. Gregory and Stephen Nortcliff.
© 2013 Blackwell Publishing Ltd. Published 2013 by Blackwell Publishing Ltd.

2.1.1 The resource capture framework

Environmental resources required by plants must be sequestered from the atmosphere (principally carbon dioxide (CO_2), light and temperature) and the soil (principally water and nutrients from organic or inorganic matter and/or fertilisers). Of these resources, temperature and water (in the atmosphere and the soil) tend to limit the actual productivity of any particular crop below the potential set by its genetic characteristics and the availability of solar radiation and CO_2 at any geographical location. Irrespective of their composition, all green plants depend on their capture and use of these environmental resources which are finite in space and time and must be shared between individual plants for a crop community to progress through the processes of growth and development towards maturity and economic yield.

Radiant energy is only available for plant growth when it is absorbed in the leaf canopy in the wavelength range 0.4–0.7 μm (Squire, 1990; Daughtry *et al.*, 1992; Azam-Ali and Squire, 2002). This photosynthetically active radiation (PAR) or 'light' provides the high energy required for the process of photosynthesis. The final products of photosynthesis are carbon-based compounds (carbohydrates, lipids and proteins) which have different energy contents.

Longer wavelengths (above 3.0 μm), which are insufficiently energy-intense to drive photosynthesis, provide the thermal energy or 'heat' required to drive the process of development. In this way, plants progress through their life cycle, not in response to chronological time, but through the accumulation of thermal time or the amount of heat absorbed above a base temperature at which the development process begins.

In some plants, the process of flowering (a specialised example of development) requires the accumulation of a minimum ('short-day') or maximum ('long-day') daylength (photoperiod). Such daylength-sensitive plants require the additional photoperiod signal (0.66–0.73 μm) to switch from vegetative to reproductive development.

2.2 Growth and development

Irrespective of their other characteristics, the performance of all plants can be considered in terms of the twin but distinct processes of development and growth. In most arable crops, it is a specific fraction of growth (usually seeds or 'grain') that constitutes their economic yield. For example, cereals are harvested when the crop is physiologically mature and the final weight of grains (the reproductive fraction of the crop) is at a maximum. In other crops, it is not the reproductive or seed fraction but the vegetative fraction of growth that determines yield. For example, sugar is extracted from the stems of sugar cane and from the roots of sugar beet but not from the seeds of either species. Thus, except in plant breeding programmes, the production of seeds in sugar crops is an inconvenience because it curtails the process of sucrose accumulation.

2.2.1 Development

Development describes the life cycle of the crop from the start to the end of its duration. We can define development as 'progression of a plant through a series of discrete changes in the structure or number of individual organs' (Azam-Ali and Squire, 2002). For example, the germination of seeds, the initiation of leaves, roots and branches, the onset of flowering and the number of seeds produced are all developmental events within the duration of a crop. The duration of a specific developmental process can be measured by recording the time interval between two discrete events, e.g. the initiation of two successive leaves on a branch. The rate at which that process occurs is simply the reciprocal of its duration and can be expressed as $1/t$ where t is the chronological time for that event to occur. When a crop is adequately supplied with nutrients, the rate of each developmental event is usually determined by the prevailing temperature.

Thermal time

Because the progress of any developmental event depends both on temperature and time, the event can be expressed in terms of 'thermal time', measured in units called degree days (°C d). The germination of seeds is a good example of how temperature dominates a specific developmental event. For germination to occur, a seed must be viable, non-dormant and adequately supplied with water and oxygen. In addition, seeds of any particular variety require a minimum temperature (T_b) below which germination cannot occur. Above T_b, the germination rate increases linearly with temperature to an optimum (T_o) above which the rate declines linearly to a maximum temperature (T_m) at which germination stops (Squire, 1990).

In practice, the value of T_b is usually similar for different developmental processes within a crop variety, and thus the thermal time (Θ; °C d) for any developmental event can be generalised as:

$$\Theta = \sum_{i=1}^{i=n} (T - T_b)$$

where n is the number of days experienced by the plant at a mean daily temperature, T, above T_b. So, on any day, accumulated thermal time is simply the difference between the mean daily temperature and the base temperature of the crop, i.e. $T - T_b$. We can consider the values of T_b, T_o and T_m as cardinal temperatures for a particular crop variety.

Because crop species and their varieties have different cardinal temperatures, their accumulation of chronological time is a poor indicator of their rate of development. For example, on a day where T=20°C, Crop A with T_b=0°C will experience a development rate ($T - T_b$=20°C d), i.e. twice that of Crop B with T_b=10°C ($T - T_b$=10°C d). These differences can be scaled up to calculate the development rates of both crops over their whole duration (see Azam-Ali and Squire, 2002).

Because the concept of thermal time remains valid over different locations and environments, it can be used to normalise chronological durations of different crop varieties between sites and seasons. The same crop grown at a number of different sites and seasons may require very different amounts of chronological time but the same amount of thermal time to reach maturity.

Flowering

Apart from the accumulation of thermal time, in some crop cultivars two other signals, photoperiod and vernalisation, may be required for the switch from vegetative to reproductive development through the process of flowering. In most crops, there is usually an initial vegetative phase followed by a reproductive phase until crop maturity. Flowering marks the switch between these two phases. Different crops can be classified as determinate or indeterminate, depending on their reproductive habit and the point at which they are harvested. In some determinate crops, such as wheat, there is a sharp transition from vegetative to reproductive growth with the main axis being terminated by an inflorescence (flowering). In such crops, flowering and the subsequent development of seeds occur synchronously. Therefore, the whole crop reaches physiological maturity over a short period of time when all the seeds are ripe. In contrast, in indeterminate crops, such as groundnut, vegetative and reproductive phases occur concurrently because flowers are borne on a series of lateral shoots whilst the main axis continues to develop vegetatively. In such cases, seeds are produced throughout the reproductive phase and may be at different stages of maturity on the same plant.

In some crops, a signal is required for the switch from vegetative to reproductive activity via flowering. When this signal is provided by changes in daylength, the phenomenon is known as photoperiodism. Thus, crop varieties can be defined as photoperiod-sensitive or photoperiod-insensitive. The species that are sensitive to changes in daylength (or strictly photoperiod) can be further divided into those which respond to short days ('short-day' plants) and those which respond to long days ('long-day' plants). Species not affected by daylength are known as 'day-neutral' plants. Most tropical species are short-day plants, i.e. the vegetative period continues until the daylength declines below a critical minimum for a sequence of days before flowering can occur.

Some temperate species, such as winter-sown cereals, require an additional period of low temperature to induce flowering. This phenomenon is known as vernalisation. Table 2.1 summarises the photoperiod and vernalisation requirements of a range of major arable species.

The duration of a crop is the chronological time between when it is planted and harvested. Often, but not always, this will involve the sowing of seeds in soil and the subsequent harvesting of new seeds from the resultant plants. It is important to distinguish duration from the growing season of a crop which is that part of the year when atmospheric and

Table 2.1 The photoperiod and/or vernalisation requirements for flowering in a range of major crops.

		Photoperiod-sensitive			**Vernalisation requirement**
		Short	**Long**	**Neutral**	
Wheat	Spring		✓	✓	✓
	Winter		✓		✓
Rice		✓		✓	
Maize		✓		✓	
Barley	Spring		✓	✓	
	Winter		✓		✓
Sorghum		✓		✓	
Oat	Spring		✓	✓	
	Winter		✓	✓	✓
Millet		✓		✓	
Rye	Spring		✓	✓	
	Winter		✓	✓	✓
Soyabean		✓		✓	
Groundnut				✓	
Potato		✓		✓	
Cassava			✓		
Sweet potato		✓			
Sugar cane		✓			
Sugar beet			✓		✓

Source: From Azam-Ali and Squire (2002).

soil conditions permit the growth of a particular species. In general, the maximum yield of a crop species in a given environment is most likely to be obtained from a cultivar with a duration that is close to the length of the growing season.

In the humid tropics, crops can often be grown continuously because atmospheric and soil conditions rarely limit growth. Where modern short-duration varieties are cultivated, such as rice in South-East Asia, as many as three crops can be harvested annually. In contrast, in dry tropical regions, the length of the growing season is primarily determined by the period when the supply of water from rainfall and/or irrigation exceeds the demand from evaporation at the same location. In humid temperate regions, whilst the supply of water may exceed evaporative demand, the temperature at any location may set the beginning and end of the growing season where this falls below the value of T_b for any particular crop variety.

2.2.2 Growth

Within the duration of each developmental event, set by its thermal time (and, where appropriate, daylength and/or vernalisation requirement), the process of growth occurs. Plant growth can be defined as 'quantitative and irreversible changes in the length, area or weight of individual organs' (Azam-Ali and Squire, 2002). The area of individual leaves and the dry weights of leaves, roots, stems and grains are all examples of growth, whilst the number of leaves, roots, stems and grains are all examples of development.

Together, the duration and rates of different developmental and growth processes govern both the total productivity of a crop and the balance between the vegetative and reproductive fractions. Ultimately, growth depends on the division and enlargement of individual cells. However, at any particular time, active growth is restricted to certain embryonic regions of the plant called meristems. There are several different types of meristems. For example, axial organs, such as stems and roots, have apical meristems in which growth is restricted to the tip region. Apical meristems remain capable of growth over long periods (in perennial crops for many years) and are thus referred to as indeterminate meristems. In contrast, organs such as leaves, flowers and fruits grow only for a limited period before the whole organ reaches maturity. The growing regions of these organs are therefore termed determinate meristems.

Solar radiation and water: engines for crop growth

We have seen that between the developmental windows set by sowing and harvest, plants must accumulate biomass and partition this into different organs (roots, stems, leaves, flowers, seeds).

Development is controlled principally by the accumulation of temperature (thermal time) by plant organs. In some cases, daylength (photoperiod) and/or vernalisation may set the switch from vegetative (roots, shoots and leaves) to reproductive (seeds or grain) development through flowering.

Growth, on the other hand, is controlled principally by the accumulation of solar radiation, mainly by plant leaves, in which CO_2 is fixed through the process of photosynthesis into carbon-based compounds (carbohydrates, proteins and lipids).

Whilst the capture and conversion of solar radiation sets the potential productivity within the genetic limits of a species, the supply and availability of water to plant roots often determines the actual productivity of vegetation at any location. In such circumstances, we need to understand processes that govern both the above-ground capture and conversion of solar radiation and CO_2 by plant leaves and the below-ground capture of water and nutrients by plant roots. Together, these linked processes determine the potential and actual productivity of a crop system for any species and location. Later, we consider how a crop can be described as part of the 'soil-plant-atmosphere-continuum' (SPAC) that links the below-ground and the above-ground capture and conversion of resources by plant roots and shoots.

2.3 Solar radiation

The production of dry matter by crops is the consequence of leaves chemically fixing CO_2 by intercepting PAR, and the efficiency with which that intercepted PAR and CO_2 is used for dry matter production (Black and Ong, 2000). Crop 'canopies' represent an investment in dry matter, principally in photosynthetically active structures (e.g. leaves), that can intercept solar energy and CO_2 to produce new dry matter, some of which can be allocated to non-photosynthetically active structures (e.g. roots, stems and seeds) (Campbell and van Ever, 1994).

2.3.1 Radiation capture by crops

Because it is the surface that must intercept radiation, the size and longevity of a crop leaf canopy determine its rate and duration of dry matter accumulation. The size of the intercepting surface, or leaf area index (L), depends on the number of plants per unit area of ground (N_p), the number of leaves per plant (N_l) and the mean area of leaves per plant (A_s) (Azam-Ali and Squire, 2002):

$$L = N_p \times N_l \times A_s.$$

Irrespective of its total area, light must penetrate through the layers of a canopy for it to be effectively captured. In terms of light transmission, a crop can be considered as a number of horizontal layers (Azam-Ali and Squire, 2002). Thus, if radiation is measured at different levels in the canopy profile, each of $L = 1$, then the amount of light transmitted to any level is a function of the distribution of leaves above that level.

To drive photosynthesis, individual leaves within a crop stand must absorb or 'capture' the PAR component of radiation as it passes down through the canopy. Whilst the individual leaves of different species have different abilities to absorb, reflect and transmit radiation, at the crop scale these differences are less significant. For practical purposes, the most important factor is the fraction (f) of total radiation received by a crop that is intercepted before it hits the ground.

If we assume that a crop canopy is a homogenous block of vegetation in which leaves are randomly distributed without any row structure or clumping, the pattern of light transmission through the canopy obeys Beer's Law of exponential decay (Azam-Ali and Squire, 2002). In such circumstances, light transmission can be calculated using the Monsi–Saeki (1953) equation:

$$\frac{I}{I_o} = e^{-kL}$$

where

I_o is the irradiance above the crop canopy
I is the irradiance at a level within the canopy below a leaf area index L
k is the typical extinction coefficient of radiation for any particular species

Strictly, for light transmission to obey Beer's law, the leaves should be black, i.e. opaque to radiation. However, Goudriaan (1977) demonstrated Beer's Law to be a good approximation in many real crop canopies by plotting ln (I/I_o) against L, giving a straight line with a gradient, k which depends on the architecture of the crop. The estimated k relates to the fraction of radiation intercepted per unit leaf area index and can be used to compare the pattern of light interception between cultivars, species and environments (Hay and Walker, 1989).

Essentially, k is an estimate of the erectness or flatness of leaves in a crop canopy. Monteith (1981) reported k values of about 0.3 for crops with vertical leaves, about 0.6 for an intermediate leaf arrangement and 0.9 for crops with very horizontal leaves. For example, groundnut with intermediate leaf angle has k values in the range of 0.40–0.66 (Azam-Ali and Squire, 2002), whilst cereals, such as pearl millet, with erect leaves have k values ranging between 0.3 and 0.45 (Squire, 1990).

In practice, canopies are not always made up of uniformly distributed elements. For example, in row crops leaves are clearly clumped into lines of vegetation before complete canopy closure. At any time, the value of k depends on the size of the canopy and may increase as canopies expand and become more homogeneous and their leaves are more randomly oriented. It may also decrease as leaves become more vertically oriented as a canopy becomes denser with age, an effect enhanced at higher population density (Squire, 1990). However, k remains reasonably stable for a genotype over most of its life and across a wide range of conditions and can therefore be used to characterise differences between canopies with different architecture.

There is a functional link between f and L for crops with different canopy architectures. At any time, the value of f for a given crop canopy is determined by the actual value of L and the typical value of k. The relation between f and L can be expressed as:

$$f = 1 - \exp(-kL).$$

On any day, the radiation intercepted by a crop is the product of f and the irradiance, S (MJ m^{-2}), received that day. Over a growing season, seasonal intercepted radiation is therefore the sum of daily values of fS, i.e. ΣfS (MJ m^{-2}). Manipulating seasonal intercepted radiation is a critical feature of agronomic practices irrespective of species, location or environment. In practice (since we cannot manipulate S), successful crop management involves agronomic options that influence f and/or the crop duration. For example, directly or indirectly, sowing date, sowing rate, planting arrangement, irrigation and fertiliser applications all influence the rate of expansion, maximum value and seasonal total of f for any crop canopy. Similarly, agronomic options that influence crop duration, by selecting appropriate crop cultivars and sowing/harvest dates, can ensure that fractional interception best matches the seasonal pattern of irradiance at any site. In other words, it makes sense for growers to produce crops that achieve their maximum value of f in the brightest days of the growing season when most total light is available.

2.3.2 Radiation conversion by crops

Photosynthesis

Photosynthesis is a photochemical process that depends on the availability and interception of PAR by green leaves to drive biochemical pathways within them. The regulatory biochemistry of photosynthesis requires CO_2, temperature and the supply of water and mineral nutrients. Photosynthesis involves light-driven photochemical processes, enzymatic processes that need no light (so-called 'dark' respiration) and processes of gaseous diffusion into and out of leaves via apertures called stomata. The stomatal entry of CO_2 into a leaf depends on the concentration gradient between that in the ambient atmosphere and at the internal site of CO_2 fixation in the leaves. Through this common pathway, CO_2 and oxygen are exchanged between chloroplasts and the atmosphere (Larcher, 2003). Plants exert a physiological influence on this concentration gradient mainly through the opening and closing of stomata which regulate both the uptake of CO_2 and the loss of water from the plant (Azam-Ali and Squire, 2002; Larcher, 2003). Since leaves are the primary agents of crop photosynthesis, their efficiency of capture and use of PAR determine the overall productivity of a crop (Loomis and Connor, 1992).

The major pathway of photosynthesis is the conversion of atmospheric CO_2 and water into carbohydrates and oxygen, viz.:

$$6CO_2 + 6H_2O \xrightarrow{\text{Light}} C_6H_{12}O_6 + 6O_2.$$

Through the energy supplied by solar radiation, two energy-poor compounds are converted into two energy-rich compounds and, by this process, photosynthesis converts light energy into chemical energy (Azam-Ali and Squire, 2002).

Mechanisms of photosynthesis

All plant species exhibit one of three types of photosynthesis, viz. C_3, C_4 and Crassulacean Acid Metabolism (CAM) (Loomis and Connor, 1992; Larcher, 2003). The only significant crop species identified as using the CAM mechanism is pineapple; other CAM species include succulents such as cacti.

Most agricultural species employ the C_3 photosynthetic pathway in which CO_2 is fixed and the resulting molecules are processed in a series of enzyme-driven reactions known as the 'Calvin–Benson Cycle' (see Raghavendra, 2000). The cycle generates additional molecules to react with CO_2, the first products of which are molecules with three carbon atoms (hence 'C_3' photosynthesis), some of which are used as the first step in making sugars and starch.

C_4 photosynthesis is an adaptation of the Calvin–Benson cycle which uses a CO_2 concentrating system that, in effect, increases the availability of CO_2 at the active sites of photosynthesis and results in molecules with four carbon atoms. As a result, C_4 plants have faster rates of photosynthesis than C_3 plants and include crops such as maize, sorghum, millet and sugar cane which are amongst the most photosynthetically efficient species in the world.

Respiration

Respiration is important in the carbon budget of crops because it results in the loss of already fixed CO_2 from the plants through the process of growth. Respiration can be separated into that which results from growth of new tissues (photorespiration) and that which results from maintenance of existing tissues ('dark' respiration).

Since photorespiration acts on the CO_2 initially fixed by photosynthesis, its rate is closely linked to the CO_2 fixation rate. In C_3 crops, photorespiration increases with temperature, resulting in a reduction in the initial efficiency of light use by individual leaves at high temperatures. Because of their ability to concentrate CO_2 at the sites of photosynthesis, C_4 species do not undergo photorespiration – an advantage over C_3 species that is further enhanced at higher temperatures.

Irrespective of their photosynthesis system, all green plants undergo 'dark' respiration. Strictly, the term 'dark respiration' is a misnomer because the process continues in daylight at a comparable rate to that in the dark. However, because it must continue in daylight alongside photorespiration, the two components are difficult to distinguish except at night. During dark respiration, atmospheric O_2 is used by plants to convert carbohydrates into CO_2 and H_2O, with simultaneous liberation of energy (Azam-Ali and Squire, 2002). Plants use this energy to build more complex molecules from the initial products of photosynthesis which are used for the growth and maintenance of tissues. This explains why there is a higher energy cost for plants to produce unit mass of more complex lipids and proteins than simpler carbohydrates.

During the life of a crop, the relative contributions of the growth and maintenance components of respiration change with the age and weight of the crop. Maintenance respiration becomes more important later in the life of a crop because a larger amount of biomass requiring maintenance exists. Amthor (1989) reports that respiration can account for between 25% and 66% of the total assimilation of a crop.

Photosynthesis and light interception at leaf level

The photosynthetic response of individual leaves to their receipt of solar radiation and CO_2 is often described in terms of a 'light response curve' (Figure 2.1). In the light response curve, the initial photosynthesis (P_n) of an individual leaf responds to increasing irradiance (I). In very weak light, the relation for both C_3 and C_4 plant systems is linear because photosynthesis is limited almost exclusively by the absorption of light quanta (Loomis and Connor, 1992; Azam-Ali and Squire, 2002). This quantum efficiency is a measure of the amount of CO_2 absorbed per unit increase in irradiance. At $I=0$, the leaf evolves CO_2 by dark respiration at the rate R_d. At $I=I_c$, the light compensation point, $P_n=0$ because photosynthesis exactly balances any losses through dark respiration. Below the light compensation point, the leaf is actually losing more CO_2 through respiration than it is gaining through photosynthesis.

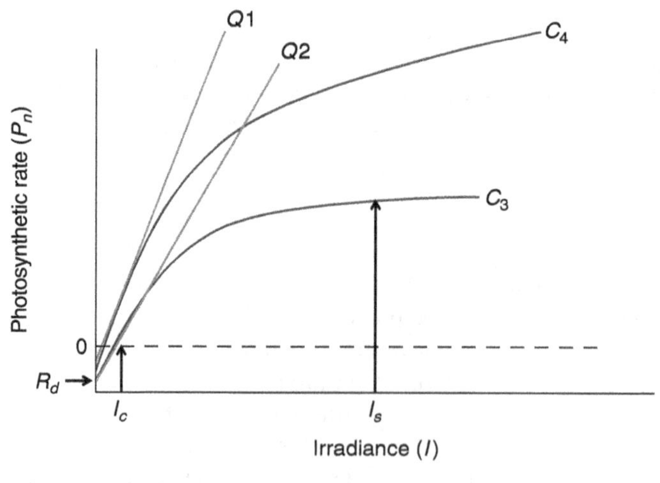

Figure 2.1 The idealised light response curve for a C_3 and C_4 leaf. Photosynthetic rate (P_n) is plotted against irradiance (I), and the response curve shows the light compensation point (I_c), saturating irradiance (I_s) and dark respiration (R_d) for a C_3 leaf. The initial, linear 'quantum efficiency' is shown for the C_4 (Q1) and C_3 (Q2) leaf types.

As C_4 plants do not undergo photorespiration, they have a lower light compensation point than C_3 plants. In C_3 species, as I increases above I_c, P_n increases in a curvilinear manner until it becomes light-saturated at a maximum net photosynthesis (P_{max}). In most C_3 species, P_n becomes light-saturated well below full sunlight and the maximum value of this saturating irradiance declines with leaf age and stresses, such as leaf water deficits (Azam-Ali and Squire, 2002). At saturating irradiance, photosynthesis is controlled by the rate at which CO_2 from the atmosphere is reduced to carbohydrates. In C_4 species, leaves rarely if ever reach P_{max} and, as a consequence, individual leaves of C_4 species continue to photosynthesise in strong light at which C_3 leaves are already photosaturated.

Photosynthesis and light interception at crop level

We have seen that the photosynthetic efficiency of single plant leaves can be measured by calculating their individual light response curves (Figure 2.1). However, light response curves for individual leaves are of limited value in crop situations where we need to scale up to the photosynthetic efficiency of the whole canopy. Monteith (1994) observed that the rate at which field crops accumulate dry matter is proportional to the rate at which their canopies absorb radiant energy. Figure 2.2 shows that, for any crop, the relation between total dry weight and the seasonal interception of solar radiation is often linear with a slope, ε_s (g MJ^{-1}), known as the dry matter/radiation ratio or often the 'radiation use efficiency' (RUE) (Azam-Ali and Squire, 2002).

The adjustment from the curvilinear responses at leaf level to the linear response to radiation at the canopy level can again be explained in terms of the Beer's Law analogy previously described for radiation capture. Effectively, the 'stacking' of leaves within a crop stand means that most operate at the linear part of the light response curve and therefore any additional radiation intercepted by each new leaf is proportionally converted into a photosynthetic product (Biscoe *et al*., 1975; Asseng and Hsiao, 2000).

Typically, C_3 species (most agricultural crops) have values of RUE (ε_s) of about 1.5 g MJ^{-1}, whilst C_4 crops have values closer to about 2.4 g MJ^{-1}. Whilst the value of ε_s for any cultivar remains fairly constant during much of the cropping season, it declines and eventu-

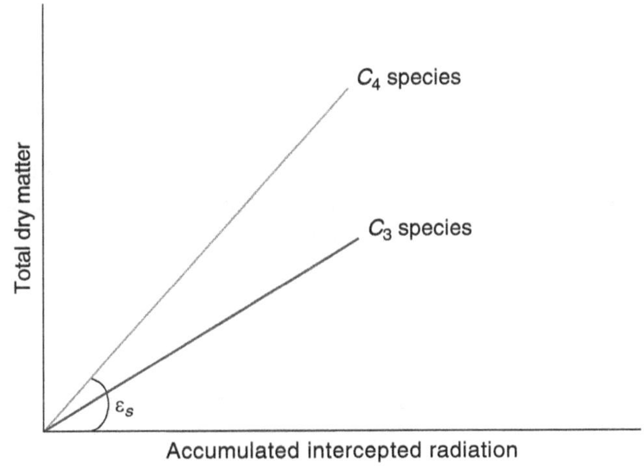

Figure 2.2 The idealised relation between total dry matter (including roots, leaves, stems and seeds) of a crop and accumulated intercepted radiation. The slope of the relation ε_s (g MJ^{-1}), known as the dry matter/radiation ratio, is shown for C_3 and C_4 crop species.

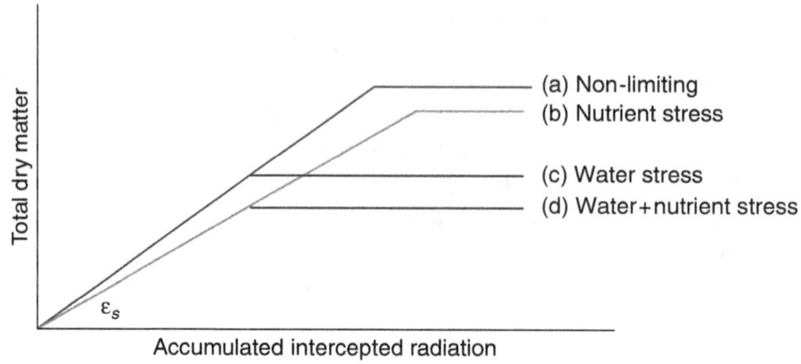

Figure 2.3 The effects of crop age and abiotic stresses on the dry matter/radiation ratio. Even without stress (line a), efficiency declines as the crop reaches maturity and leaves begin to senesce. The diagram shows progressive reductions in radiation use efficiency with (b) nutrient stress, (c) water stress and (d) nutrient and water stress.

ally levels off during reproductive growth, in response to drought and/or pests and diseases. Also, as it is a crude measure of biomass produced per unit of radiation intercepted, ε_s should be 'energy adjusted' to account for the differential energy contents of the carbohydrate, lipid and protein components of the crop (Penning de Vries *et al.*, 1989). Biomass composition also influences both the growth and maintenance respiration requirements of crops (Penning de Vries *et al.*, 1974; Amthor, 1989).

It is also important to note that ε_s represents the efficiency with which total biological productivity (including roots) is produced in relation to intercepted radiation. In most circumstances, only a fraction of total biomass is represented as yield. Therefore, to estimate the efficiency with which crop yields are produced, we need to know the 'harvest index' (grain yield/total biomass) of any particular crop.

From these considerations, we can now see that the biomass of a crop (*W*) can be related to its seasonal capture of radiation (ΣfS) and the efficiency with which it converts that captured radiation into biomass (ε_s), i.e.:

$$w = \varepsilon_s \, \Sigma \, f S.$$

Where crop roots are well supplied with water, the total radiation intercepted during the growing season and a knowledge of ε_s for any cultivar can often be used to estimate its productivity and yield. However, where drought and/or nutrient stress restrict the uptake of CO_2 through stomatal regulation, seasonal radiation interception is a poor indicator of productivity because ε_s does not remain constant (Figure 2.3).

2.4 Water

2.4.1 The soil-plant-atmosphere-continuum (SPAC)

Terrestrial plants must take up water from the soil through their roots, transport it upwards through roots, stems and leaves as a liquid from where it must be transferred into the atmosphere as vapour. The movement of water from the soil to the atmosphere can be described in terms of a SPAC. Water lost through transpiration drives the SPAC and must be replenished by further extraction from the soil by roots for the system to remain in equilibrium during the life cycle of a crop. Short-term water deficits result in symptoms of plant water stress which reduces plant growth rates by restricting CO_2 uptake with consequences for final crop yield. Long-term deficits between supply and demand may result in crop failure and plant death when water deficits prevent plants from maintaining metabolic functions and structural integrity.

To minimise water stress and avoid crop failure, growers must manage the water requirements of their crops to best match the availability of soil water to atmospheric demand and minimise water losses from the system that do not result in transpiration.

An understanding of the SPAC can help provide both a theoretical and practical basis for crop water management and a predictive basis to calculate the productivity of crops. In this section, we consider the individual components of the SPAC and how they contribute to the water use and productivity of crop plants.

2.4.2 Soil water

More comprehensive analyses of soil water characteristics appear in Chapter 9 and in Brady and Weil (1999). For our purposes, water in soils can be considered both in terms of its content, i.e. how much is there, and its energy status, i.e. how easy it is for plants to access.

The relation between soil water content and soil water energy

The amount of water available to plant roots is always less than either the actual soil water content (SWC) or its free energy status. Three important forces reduce the energy status of water in soil below that in its pure state. First, the particles that make up the soil exert an adhesive attraction for water within the soil matrix, i.e. a matric force that acts against its removal by plant roots. Second, the cohesive attraction of ions and other solutes in the soil water medium results in osmotic forces which reduce the water energy status below that of pure water. Third, gravitational forces impose a downward force on soil water because the energy status of water always declines with soil depth. Plant roots must remove water against the gravitational gradient which tends to move it away from the root zone. The difference between the free energy of soil water and that of pure water is known as soil water potential, ψ_s, and can be expressed as:

$$\Psi_s = \Psi_m + \Psi_s + \Psi_g$$

where

ψ_g is gravitational potential
ψ_m is matric potential
ψ_o is osmotic potential in a particular soil volume

Because actual water potential is always less than that of pure water, it is expressed as a negative value relative to zero (the energy status of pure water).

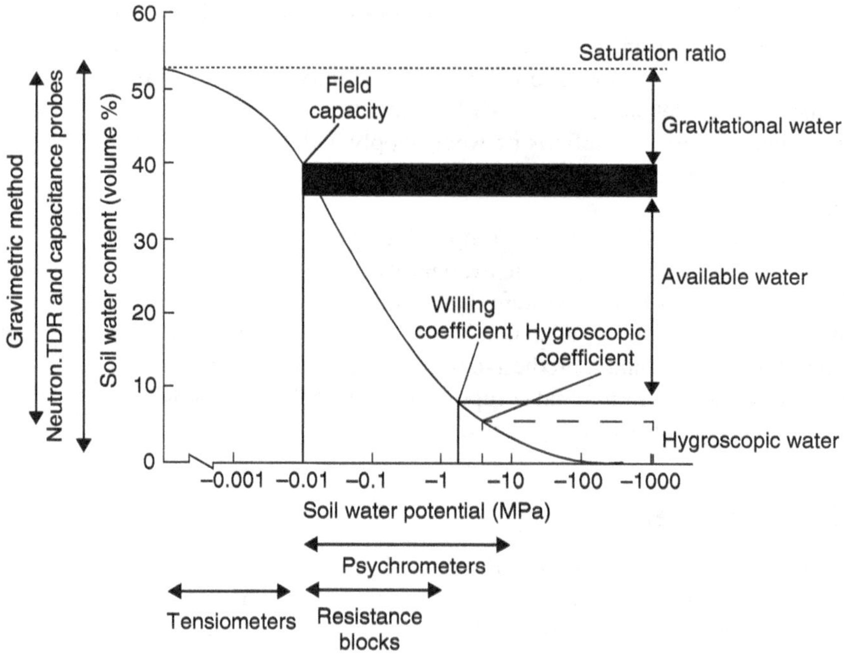

Figure 2.4 Soil water status and the instruments used to measure water content or water potential. (Part of the figure adapted from Brady and Weill, 1999. Refer to the text for terminologies on the figure.)

We now consider various factors that influence the volumetric SWC of soils (i.e. what percentage of the soil volume is made up of water) and its energy status, i.e. ψ_s. For different soil types (e.g. sandy, loamy and clay soils), there is a characteristic relationship between SWC and ψ_s known as the soil moisture release curve (SMRC).

Figure 2.4 shows the key components of the SMRC for a clay soil. Also marked on the diagram are the ranges over which instruments commonly used to measure SWC or soil water potential operate.

Saturation ratio (ϕ_s)

This represents the maximum amount of water that a soil can hold when water replaces all the available space in the soil volume. A soil that is saturated is at a water content equal to its saturation ratio (ϕ_s). Because of the various forces acting on it, especially gravity, ϕ_s is an unstable condition as water drains through the soil (deep drainage), runs across the soil surface (run-off) or evaporates from it (surface evaporation, E_s). In each case, water moves from a point of high energy status to a point of lower energy status.

When soils are at or close to saturation, they are considered to be waterlogged. This condition often occurs first at low spots in the field or when the drainage of the whole field is impaired by poor soil structure. Even well-drained soils can become waterlogged during a heavy rainstorm where infiltration rate is slower than rainfall (or irrigation) intensity. Soils compacted by ploughing or machinery can also encourage waterlogging by reducing soil porosity, i.e. the space available in the soil which can absorb water before saturation.

It is important to recognise that the infiltration of water into soil results in the simultaneous removal of air from the same soil volume. Soil aeration controls the level of O_2 and CO_2 in the soil volume. In turn, both gases influence the respiration of roots and soil micro-organisms. For soil respiration, O_2 must be supplied and CO_2 removed. For most crop plants, O_2 in the soil air must exceed 10% (Brady and Weil, 1999). Poor soil aeration occurs when water displaces gases and when the exchange of gases with the atmosphere is too slow, even if sufficient air space is available in the soil.

Field capacity

In a saturated soil, rapid drainage ceases as macroflow (saturated flow) slows because water in the macropores is increasingly replaced by air (unsaturated flow). At this stage, the soil is said to be at field capacity (FC). The matric potential of a soil at FC typically ranges between −0.01 and −0.03 MPa and is not a steady state because unsaturated flow continues, albeit more slowly than saturated flow.

Permanent wilting point

As its content continues to decline without being replenished by rainfall or irrigation, the water remaining in the soil volume becomes progressively more difficult for roots to extract. Plants eventually experience symptoms of water stress, most often exhibited by visible wilting of leaves, because roots are unable to replace water lost by evaporation. Plants which remain in this condition are said to be at permanent wilting point (PWP) which typically corresponds to a soil water potential of between about −1.0 and −2.0 MPa, with a mean of 0.15 MPa (Azam-Ali and Squire, 2002). The SWC at this stage within the range of plant roots is called the permanent wilting percentage. Water remaining in the soil is now too tightly held to be available to most plants as a liquid but is retained within the soil particles (matric potential), or bound to salts (osmotic potential).

Hygroscopic coefficient

Soil water below the PWP is so tightly held by soil colloids that it can only move as vapour. The moisture content in the soil at this stage is called the hygroscopic coefficient and is greater in colloid-rich clay soils than colloid-poor sandy soils.

Available water content

In terms of volumetric water content, the available water content (AWC) of a soil is often described as the difference between the FC of a soil and the PWP of the plants in it (Azam-Ali and Squire, 2002). However, we have seen that both FC and PWP are intrinsically arbitrary measures and depend on an unstable condition for an upper value (FC) and a plant-dependent lower value (PWP). In addition, AWC relates to differences in water content whilst plant roots act in response to water potential.

In terms of soil water potential, AWC can similarly be calculated as the difference between the value of ψ_s at FC and that at PWP. The presence of high levels of salts in the soil solution influences the availability of water to plants and hence the value of ψ_s. In effect, these osmotic forces mean that less water can be extracted than that in the soil at PWP point because of their effect on ψ_s rather than water content *per se*. In most humid soils, osmotic potential effects are negligible but they can be significant in soils of arid and semi-arid regions, especially those with high salt contents.

Knowledge of the SMRC for any soil provides a valuable basis to calculate how much water crops can extract at any location at any particular time. However, it ignores two important factors that influence the seasonal availability of soil water to a crop growing at that location. For this we require an understanding of the behaviour of the crop root system and the total available depth to which roots can descend. We consider the behaviour of root systems in more detail in Section 2.4.4. However, it is not simply the speed at which plant roots can descend down the soil profile but the depth of the soil profile itself. Soil depth is an important factor in many tropical soils, particularly shallow soils or those which have impenetrable layers of material within the soil profile. To compare different soil types, it is usual to present AWC in terms of millimetres of water per metre depth of soil; where the soil depth is known to be less than 1 m, the AWC should be adjusted accordingly.

2.4.3 Plant water

The relation between plant water content and plant water energy

As with soil, plant water can be considered in terms of energy and content. In metabolically active plants, cells contain 80–90% water. This water forms a liquid continuum from the surface of roots through the plant's tissues to the leaf surfaces from where it evaporates. Between its uptake from the soil and its loss to the atmosphere, plant water must also transport nutrients and metabolites to their sites of activity as well as contribute to plant structure by maintaining the turgidity of various organs.

Unlike soil water, plant water content must remain high for metabolic activities to continue. A decline of 20–25% below that at maximum hydration arrests most growth processes, and any further depletion in tissue water content may result in their death. New plant growth must depend either on tissues that have survived dehydration or on the production of new plant material. The ability of a plant to buffer its water status between atmospheric demand and soil supply determines its metabolic success and ultimately its survival.

Plant water potential

Water movement within plants occurs through gradients in plant water potential, ψ_p, from regions of high energy (abundant water) to those where the free energy is lower (less water). As with soils, the water potential within plant tissues is always less than that of pure water because of the influence of solutes on the osmotic potential (ψ'_o) and matric forces (ψ'_m) caused by the bonding of hydrogen ions to cell walls and proteins. Except in very tall trees, the effect of gravitational gradients is negligible. However, unlike soil water potential, plant cells must exert an additional outward hydrostatic or turgor pressure (ψ_h) to keep them turgid and functional both metabolically and structurally.

Plant water potential can be expressed as:

$$\Psi_p = \Psi_h + \Psi'_o + \Psi'_m.$$

For both soils and plants, water potentials are always less than those of pure water. Further, for water to move along the energy gradient between soil and atmosphere, ψ_p must always be more negative than ψ_s.

In practice, crops can only extract water from the soil in relation to the distribution of their active root systems and the water available locally to metabolically active roots. This means that root systems must often operate in soils where some is close to FC (between about 0.01 and 0.03 MPa) and some may be at or below PWP (between about −1.0 and 2.0 MPa). During daylight, water movement within the plant depends on the relative values of ψ_p across different parts of the plant, which in turn depends on water lost through stomata. At night, when stomata are closed, ψ_p returns to equilibrium with the soil (i.e. $\psi_p = \psi_s$).

Plant water content

Whilst water movement depends on energy (i.e. potential gradients), actual water content of tissues has a greater effect on plant metabolism and growth than any changes in ψ_p or ψ_h. Tissue water content can be expressed in various ways as water content per unit dry weight, per unit fresh weight and per unit weight of water at full hydration (i.e. the leaf relative water content; RWC). For our purposes, RWC is the most useful means of quantifying plant water content, since it is not influenced by changes in tissue dry weight.

The value of RWC can be estimated by measuring the fresh weight (W_2) of leaf samples and rehydrating these on water to obtain their saturated weight (W_3) (Azam-Ali and Squire, 2002). The discs are then oven dried to obtain their dry weights (W_1). RWC can then be calculated from:

$$\mathrm{RWC} = \frac{W_2 - W_1}{W_3 - W_1} \times 100 \tag{2.1}$$

2.4.4 Water capture systems

Root system size

Plants acquire water through their roots. Although there are numerous physical and physiological factors which control and regulate this process, ultimately, the total root length density (l_v; cm cm^3) and final depth (m) of the root system, together, play a critical role in its effectiveness in capturing water and nutrients. The final depth of the root system also allows us to calculate the potential volume of water that is available to the crop throughout its growing season. Because water moves down the soil profile, a deep rooting system is invaluable in meeting high transpiration demand when water availability is reduced. For example, Zhao *et al.* (2004) collated evidence that the mean rooting depth for desert vegetation may reach up to 13.4 m, with 31% of the total root biomass found below 0.3 m. They also reported that temperate forests have a maximum rooting depth of 3.7 m with 35% of the total biomass below 0.3 m and temperate grasslands have a maximum depth of 2.4 m with 17% of the total root biomass found below 0.3 m. In contrast, the effective rooting depth of annual crops rarely reaches 2 m and, in temperate crops well supplied with water, roots are largely restricted to the uppermost 30 cm or so of the soil.

The ability of a root system to take up water varies with factors such as root morphology and age as well as its total length. However, inflow rates are generally assumed to be a uniform function of l_v. Similarly, the distribution of a root system depends on many factors including soil type and depth as well as plant factors such as the rate of root growth and root system morphology. Gerwitz and Page (1974) concluded that a negative exponential function can usefully describe root system distribution.

So, if W_r (kg dry matter of roots per m^2 of ground) is the root mass and z (m) is the distance below the soil surface at $z=0$, then:

$$\beta W_r e^{-\beta z}$$

gives the root density per unit volume of soil at depth z, where β (m^{-1}) is a constant.

Integrating with respect to z between $z=0$ and z gives:

$$W_r(1-e^{-\beta z}).$$

Thus, $1 - e^{-\beta z}$ gives the fraction of root mass between the soil surface and depth z.

Root system speed

In addition to the final size of a root system, an important consideration in its success as a seasonal capturing system is the speed at which the root system extends into soil. This can be considered in terms of a root front velocity (mm d^{-1}) which depends both on the species in question and the physical characteristics of the soil.

2.4.5 Water movement systems

The movement of water through the plant components of the SPAC is often described as a transpirational stream. However, even when soil water supply to roots is unrestricted, diurnal changes also occur in the water potential of plant shoots because leaf surfaces must transpire at or close to the potential set by the atmosphere. The decrease in ψ_p of the shoot is caused by the hydraulic resistance, Z_p, to water movement in the transpirational stream. We can estimate the flux, F_p, of water through the plant system as:

$$F = \frac{\Delta\Psi_p}{Z_p} \tag{2.2}$$

where $\Delta\psi_p$ is the gradient of water potential across the plant. Often, a linear relationship (equivalent to Z_p) exists between F_p and $\Delta\psi_p$, because osmotic forces usually play only a small role in water movement and there is a steady state whereby water absorbed by the root system is equal to that subsequently transpired by the canopy. However, such conditions may not prevail where plant material includes young, actively growing tissues that require water for cell expansion. Also, many plant organs, such as leaves, stems and fruits, may act as 'capacitors' which change in size as a consequence of absorbing or losing water. Soil water deficits impose a restriction on supply of water that results in dehydration and loss of cellular water and an imbalance between the loss and gain of water by the whole plant. Ultimately, water must move through a system that connects roots, stems and leaves, that themselves contain water, as well as act as a connecting pathway between the sites of water uptake and water loss.

Root water uptake

Individual roots are made up of a cortex surrounding a central vascular cylinder (Zhao *et al.* (2004). Water enters the root and crosses laterally through a series of concentric layers of cells and then axially to the rest of the plant. In living root tissue, water is transported through both the apoplastic pathway, i.e. between cell walls, and the symplastic pathway, i.e. cell to cell transport (Steudle and Peterson, 1998).

Once it is within the plant, water moves passively along a water potential gradient which is driven by transpiration from the stomatal cavities in leaves. Transpiration reduces the water potential within the xylem vessels and effectively pulls water from the roots into the aerial tissues. When transpiration is reduced or absent (e.g. during the night or under drought conditions), solutes are actively pumped into the roots, thereby increasing the osmotic potential to provide a hydrostatic root pressure. Javot and Maurel (2002) showed that this positive upward pressure forces sap along the xylem. However, because the apoplastic pathway does not distinguish between water and solutes, it is unable to maintain the osmotic gradient for extended periods. The transport of water within the rooting system is therefore mainly driven by hydrostatic forces (Steudle, 1994).

Stem transport

Water moves from the roots up into the xylem (rigid, dead cells) which is surrounded by flexible, living cells. The connectivity of flow paths between the living tissues and xylem is limited and creates a resistance to the flow of water. As transpiration takes place, the living tissues (cambium and phloem) shrink as water flows from them into the xylem. When transpiration slows or stops, the potential gradient is reversed and water flows back into the living cells causing them to swell. This change in size of phloem and cambium is observed as trunk diameter fluctuations. When soil water availability is less than that lost in transpiration, shrinkage of the trunk dimension occurs in response to increased water potential gradient.

2.4.6 Water loss and conversion by crops

Evaporation

All atmospheric moisture originates from the earth's surface through the process of evaporation. This can occur from water or land surfaces as surface evaporation, or from vegetation as transpiration. In the agricultural context, surface evaporation from the soil (E_s) and transpiration (E_p) is collectively known as evapotranspiration, E_t, which is the total flow of water vapour from a cropped field into the atmosphere. It is important to recognise that direct evaporation and transpiration are the same physical process. In the case of E_s, water that evaporates must be replaced at the evaporating surface either through soil capillary action or replenishment by rain or irrigation. For E_p, transpired water must be replaced by upward water movement through the SPAC into the leaf surfaces from which further transpiration occurs.

Three atmospheric factors influence evaporation at any location. The first is the capacity of surrounding air to hold water vapour, i.e. the atmospheric demand for water which is a function of air temperature. The amount of water that can be absorbed by the atmosphere depends on its relative humidity at any temperature, i.e. how much water is already there relative to what it can hold at that temperature. The lower the relative humidity, the greater the potential for further evaporation at that temperature, since the atmosphere is able to absorb further water vapour until it is saturated.

The second factor is the energy or latent heat required to convert water from a liquid into vapour. This energy is mainly provided by solar radiation which, by facilitating this change in state, results in a cooling of the evaporating surface.

Thirdly, turbulence, usually measured as windspeed, is required to replace saturated air at the evaporating surface with unsaturated air from surrounding areas (Kowalik and Eckersten, 1984). For all evaporating surfaces, a boundary layer of saturated air exists between the surface and the turbulent atmosphere. Wind movements affect the stability and depth of this boundary layer and, therefore, evaporation occurs more quickly at high windspeeds that reduce the thickness of this layer than in relatively still air. For uniform surfaces, such as bare soil and water bodies, the boundary layer acts as a blanket of wet air above the location. In the case of crops, characteristics such as leaf angle and shape influence the depth and stability of the boundary layer which becomes progressively thicker down the canopy as windspeeds decrease.

Transpiration

Plants must absorb CO_2 from the atmosphere and use radiant energy absorbed by their leaves to drive the process of photosynthesis. However, only a very small proportion of the energy absorbed by leaves is used for photosynthesis. The rest tends to heat the crop canopy above the ambient temperature. To prevent canopy temperature increasing throughout the day, plants must dissipate the excess solar energy by transpiration, convection or radiation.

Transpiration is a special form of evaporation because it occurs not across the whole canopy surface but through stomatal apertures on leaf surfaces which can regulate the rate of water vapour loss. The average size and frequency of stomata and the total area of leaf surfaces determine how much transpiration can occur. When a crop covers the ground and its roots are well supplied with water, stomatal regulation is minimal (i.e. stomata are fully open) and direct evaporation from the soil surface is negligible. In these circumstances, evapotranspiration proceeds at a rate determined by the atmospheric conditions; transpiration is effectively the same as the reference evaporation (E_o) that would have occurred from an open water surface at the same location.

However, the transpirational component of E_t is rarely equivalent to E_0 because stomata exert a physiological influence on evaporation that depends on plant-related factors as well as atmospheric conditions. Physiological control of transpiration depends on the effectiveness of stomatal regulation by leaves and the total surface area (L) of the crop canopy. In practice, E_p is often limited by the supply of water to roots which in turn act on F_p and the flow of water through the SPAC. In arid climates, where air is hot and dry (large atmospheric demand) and water supply is limited (small soil water supply), stomata remain closed for part or most of the day and restrict seasonal transpiration (ΣE_p) to a finite proportion of E_o at the same location.

Conversion of resource (ε_w)

Plants must constantly balance the demands for CO_2 absorption to support growth and the need to restrict transpiration when soil water is limited. The link between CO_2 uptake and H_2O loss at the leaf level means that, at least in principle, seasonal transpiration can be used to estimate carbon assimilation at the crop scale. Figure 2.5 shows that, for any crop, the

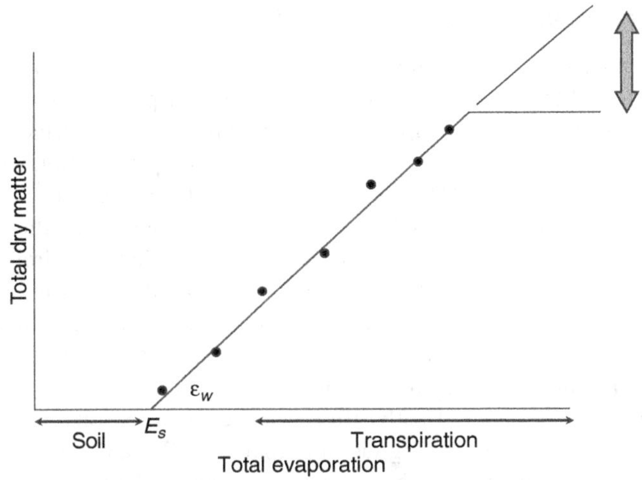

Figure 2.5 The relation between total dry matter (including roots, leaves, stems and seeds) of a crop and total evaporation. The slope of the relation ε_w (g kg^{-1}), known as the 'dry matter/transpired water', remains linear only when water lost directly from the soil surface is accounted for in the total evaporation. The arrow in the diagram illustrates how any extra transpiration through the crop results in a proportional increase in total dry matter.

relation between its accumulated dry weight and its transpiration over the same period is often linear with a slope, ε_w (g kg^{-1}), known as the dry matter/transpired water ratio (Squire, 1990). As with radiation, the relation between productivity and water use is often referred to as water use efficiency (WUE) of a crop (Azam-Ali and Squire, 2002). Hsiao *et al.* (2007) provide a comprehensive review of WUE and its improvement in agricultural systems.

Steduto (1996) and Steduto and Hsiao (2007) review the basis of the link between dry weight and transpiration. The closeness of this relation is probably because both dry matter accumulation and seasonal transpiration are closely related to the total solar energy intercepted by a crop (Monteith, 1981, 1986a). The value of ε_w appears to remain independent of nutrient uptake (Tanner and Sinclair, 1983) or water stress (Azam-Ali *et al.*, 1994). However, when the same crop is grown at different sites or seasons, the value of ε_w often varies between sites and/or seasons. There are two principal causes of this apparent variation in ε_w.

The contribution of roots to total productivity

When crops are well supplied with water, the root fraction of total dry weight is small and can often be ignored in calculations of ε_w. However, in arid climates and/or when soil moisture is limited, root weight may be a substantial component of total biomass. Because accurate measurements of root weight are rare, most calculations of ε_w are based on above-ground biomass. However, using total (i.e. including roots) or only above-ground biomass may lead to marked variations in published values of ε_w. For example, for three populations of groundnut growing on stored soil water in India, Azam-Ali *et al.* (1989) showed that the calculated value of ε_w increased from 1.57 to 3.00 g kg^{-1} when roots were included in measurements of crop dry weight.

The influence of atmospheric saturation deficit

For transpiration to occur, the sub-stomatal cavities of leaves must contain saturated air at the leaf temperature. Strictly, the driving gradient for transpiration depends on leaf temperature as this sets the saturation vapour concentration at the leaf surface, and the driving force for transpiration is therefore the difference between the saturation vapour concentration in the leaf and that in the air (*D′*). As a consequence, the value of ε_w is inversely related to *D′* during the season. When leaf and air temperatures are the same, the driving force for evaporation is the true saturation deficit of the air, i.e. *D*, and, in these circumstances, *D* sets the potential rate for transpiration. If leaf temperature is similar to air temperature, *D* is an acceptable approximation for the value of the driving gradient (i.e. *D′*). Similarly, although transpiration at any time is closely linked to the current saturation deficit, the seasonal mean value of *D* is often an acceptable approximation because changes in *D* during the season are usually small compared with differences between seasons (Monteith, 1986b).

When *D* is included as a normalising factor (Monteith, 1990), the value $\varepsilon_w D$ (g kPa kg^{-1}) appears to remain conservative throughout the season (Squire, 1990) and, unlike ε_w, remains fairly constant from site-to-site and year-to-year. The value $\varepsilon_w D$ is known as the transpiration equivalent, Ω_w, and can be treated as a quasi-constant for any particular crop (see Azam-Ali *et al.*, 1994; Azam-Ali and Squire, 2002).

However, there are a number of drawbacks to using *D* as a normalising factor for the relation between dry matter and transpiration. First, where canopy temperature is cooler than that of the air, the leaf vapour pressure can be substantially less than *D*. Similarly, under water stress leaf temperatures and the associated leaf vapour, pressure can be substantially greater than the mean value of *D*. In each case, *D* no longer serves as a true estimate of the driving gradient for transpiration because no account is made for the leaf vapour pressure.

Since accurate and frequent measurements of canopy temperatures are rare, the errors caused by variations in leaf to air temperature on the normalisation of ε_w are difficult to quantify and are strongly dependent on the type of weather data available (Allen *et al.*, 1998). A further complication is that normalisation by *D* becomes very sensitive at low values of *D* (Stockle *et al.*, 2003).

Steduto and Hsiao (2007) present the concept of crop biomass water productivity (WP_b) for a number of C_3 and C_4 crops. In their analysis, they compare the normalisation of the dry matter/transpiration relation for evaporative demand through two routes: via the transpiration gradient (i.e. the atmospheric vapour pressure deficit, *D*) and via a reference transpiration flux (i.e. the reference evaporation, E_o). They present evidence for numerous crops to show that the normalisation of ε_w by E_o is more robust than normalisation by *D* (Azam-Ali *et al.*, 1994; Clover *et al.*, 2001; Steduto and Albrizio, 2005) and provide theoretical and practical evidence for this improvement.

2.5 Integrating principles of resource capture by crops

2.5.1 Reassessing the solar and water engines for crop growth

The 'Resource Capture Framework' described in Section 2.1.1 and expounded in this chapter demonstrates that the biomass produced by a crop is a function of the amount of resources that it has captured during its life cycle and the efficiency with which it converts these resources into dry matter (Black and Ong, 2000; Azam-Ali and Squire, 2002). Where we know the fraction of biomass allocated to the reproductive or economic fraction (i.e. the harvest index), the framework allows us to quantify crop yield.

Crop productivity and yield can be estimated through a solar-driven or a water-driven engine by relating dry matter production to either the amount of radiation captured or the amount of water transpired over the period of growth (Azam-Ali *et al.*, 1994; Azam-Ali and Squire, 2002). Thus, the biomass of a crop (W) can be related to seasonal intercepted radiation (ΣfS) and dry matter/radiation coefficient (ε_s), i.e. $W = \varepsilon_s \Sigma fS$, or to seasonal transpiration (ΣE_p) and the transpiration equivalent (Ω_w), i.e. $W = \Omega_w \Sigma E_p$.

When plant roots are well supplied with water, stomata remain fully open for much of the day and allow transpiration and CO_2 uptake to continue at or close to the limits set by the environment. In such circumstances, the total productivity of a crop can be estimated using the solar-driven engine since the interception of solar energy determines productivity. However, water often limits the productivity of rain-fed crops in non-temperate climates (Squire, 1990; Zhang *et al.*, 2000; Lampurlanes *et al.*, 2001). Even in temperate areas, water stress can still cause a significant yield loss in many crops (Jones, 1992). As a result, the water-driven engine generally predominates, because crop growth is driven by radiation only when there is no water stress. Both the capture of solar radiation (i.e. f) and its conversion efficiency (i.e. ε_s) can be significantly reduced under drought (Collinson *et al.*, 1996, 1997; Craufurd and Wheeler, 1999).

Under drought, plants may modify their water extraction pattern from the soil, minimise water loss, reduce leaf area expansion and, in extreme cases, lose leaf area through abscission and/or senescence. All such modifications have implications on total crop productivity because each is linked with the stomatal regulation of photosynthesis.

In the semi-arid and arid tropics, rainfall is both erratic and low in some cases and most of the water received, as rainfall, is lost to run-off, drainage and evaporation – leaving only a small proportion for crop growth. Moisture stress (drought) is therefore a common phenomenon that crops in these regions have to experience during part of their growth periods, unless irrigation is applied and crops in such regions have developed mechanisms to withstand drought and ensure both survival and yield under soil moisture stress. The actual mechanisms employed vary widely both within and between species (Siddique *et al.*, 1990; Dardanelli *et al.*, 1997). There are three generally recognised strategies by which plants adapt themselves to drought conditions. These are classified as drought escape, drought tolerance and drought avoidance. Details of these strategies are discussed by Kramer and Bowyer (1995) and Jones (1992, 1993).

Generally, a crop with a high ε_w may give better yield than one with a low ε_w or enable it to use less water to achieve the same level of productivity. Variations in ε_w amongst species (Zhang *et al.*, 2000; Siddique *et al.*, 2001) and genotypes (Baigorri *et al.*, 1999) have been reported in grain legumes, and such differences can be used to select appropriate genotypes for drought-prone environments.

2.5.2 Agronomic management and the resource capture framework

Ultimately, the aim of a farmer is to achieve the maximum productivity of a crop based on its genetic potential and the resources available to it at any location. In *some* cases, where resources are insufficient from the natural environment (e.g. through rainfall and stored soil nutrients), they may be provided as additional inputs (e.g. through irrigation and

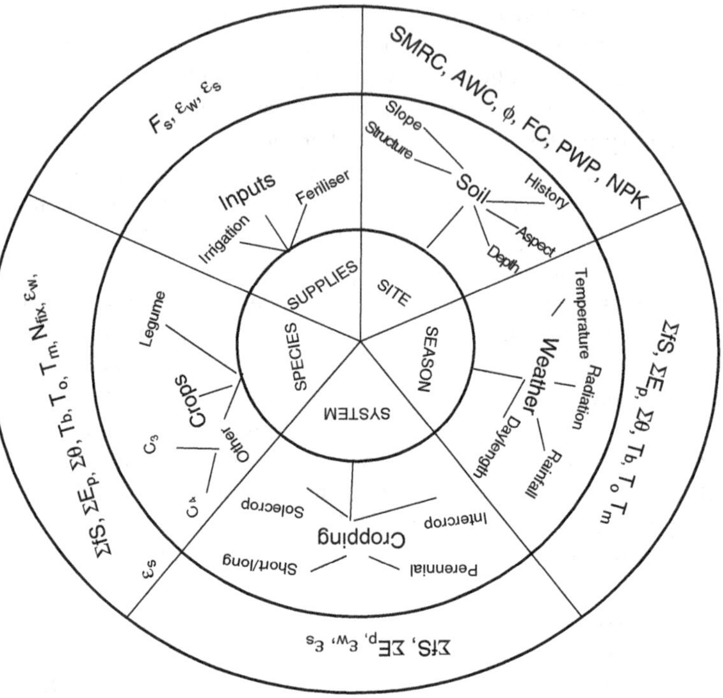

Figure 2.6 A 'Crop Circle' that links locational characteristics (site, season, system, species, supplies) with agronomic options (soil weather, cropping, crops and inputs) and specific resource capture terms. For details and definitions, see text.

fertilisers). Farmers must therefore make a series of agronomic decisions on how best to match crop requirements with available and/or applied resources throughout the life cycle of a crop.

Figure 2.6 attempts to summarise the key options available to a grower at any location in the form of a 'Crop Circle'. The factors are not definitive or necessarily chronological but provide a guide that links an inner circle of locational characteristics (site, season, system, species and supplies) in which a grower might assess available resources with a central circle of agronomic options which the grower might employ. An outer circle lists the various terms that have been presented in this chapter at the relevant position in relation to locational characteristics and agronomic options.

Site

A primary consideration in the cultivation of any crop is the existing conditions of the field site and location. The soil type and structure will influence the SMRC and moisture-related variables such as saturation ratio (ϕ_s), FC and PWP. These factors linked with soil depth provide an estimate of the total AWC at the location. Available macro (NPK) and micro nutrients can similarly be calculated at this location. Factors such as aspect and slope indicate the likely variations in solar radiation and warming during the growing season and the likelihood of run-off and run-on to the site.

Season

The growing season at any location sets the beginning and end of potential crop duration. The weather expected during a growing season sets the likely boundaries for crop growth and development in terms of total and seasonal accumulated solar radiation (ΣfS), accumulated thermal time ($\Sigma\theta$) and associated cardinal temperatures (T_b, T_o, T_m). The likely amounts and pattern of rainfall, linked with the site and soil characteristics, indicate the seasonal availability of water and likely seasonal transpiration (ΣE_p).

System

Here, system refers to the possible choice between a perennial or annual cycle of cropping and whether crops are to be grown as monocultures of a single species (sole crops) or as combinations of species (intercrops). Further considerations include the expected sowing and harvesting dates of individual species and the architectural arrangements of plants (sowing rate and row structure) that, together, set their pattern of resource capture (ΣfS; ΣE_p), demand for nutrients and overall biomass efficiencies (ε_s; ε_w) for contrasting cropping systems.

Species

A distinction within perennial and annual systems is the likely duration and, therefore, the start and end of the growth cycle of particular cultivars within the overall resources determined by the growing season. The choice(s) of crops and cultivars (C_3, C_4, legumes and non-legumes) against the likely availabilities of radiation and water will also influence the respective conversion efficiencies (ε_s; ε_w) of different species individually and collectively. These various permutations will influence the individual and collective capture of radiation and transpiration (ΣfS; ΣE_p), the accumulated thermal times ($\Sigma\theta$) and minimum and maximum temperatures for development (T_b; T_m) of each species and system and the respective demands for nutrients.

An important additional consideration within the choice of species is whether these include leguminous crops which are able to fix atmospheric nitrogen (N_{fix}) both for their own productivity and for that of successor or companion species.

Supplies

In addition to the resources naturally provided by the site and its environment, there may be supplementary resources that the grower can access to support crop production. Amongst others, such resources may include irrigation, which together with available water and nutrients and applied fertilisers will influence ΣfS and ΣE_p. Together, irrigation and fertilisers may also allow the crop to maintain photosynthesis at the genetic potential of the species (C_3 or C_4) and cultivars selected and thereby maintain ε_s.

2.6 Conclusion

The principles of resource capture provide a relatively simple yet robust framework for the analysis of crop growth, development and productivity in response to abiotic factors, principally solar radiation and water. For soil scientists and crop managers, this framework is particularly useful where it can guide in the application of water and/or nutrients so that the total biomass of a crop can reach the potential set by its genetic composition and its seasonal capture of solar energy. It is important to recognise that this framework refers to all the products of photosynthesis including roots, stems and leaves as well as seeds or grains. The concept of a 'soil-plant-atmosphere-continuum' provides a link between below- and above-ground capture and conversion systems in the optimal management of a range of crops in contrasting environments.

References

Allen, R.G., Pereira, L.S., Raes, D., and Smith, M. (1998) *Crop Evapotranspiration. Guidelines for Computing Crop Water Requirements*. Irrigation and drainage paper 56. Food and Agriculture Organisation, Rome, Italy.

Amthor, J.S. (1989) *Respiration and Crop Productivity*, Springer, New York.

Asseng, S. and Hsiao, T.C. (2000) Canopy CO_2 assimilation, energy balance and water use efficiency of an alfalfa crop before and after cutting. *Field Crops Research*, **67**, 191–206.

Azam Ali, S.N. and Squire, G.R. (2002) *Principles of Tropical Agronomy*, CABI, Wallingford, UK.

Azam-Ali, S.N., Simmonds, L.P., Nageswara Rao, R.C., and Williams, J.H. (1989) Population, growth and water use of groundnut maintained on stored water. III. Dry matter, water use and light interception. *Experimental Agriculture*, **25**, 77–86.

Azam-Ali, S.N., Crout, N.M.J., and Bradley, R.G. (1994) Perspectives in modelling resource capture by crops. In: *Resource Capture by Crops* (eds J.L. Monteith and M.H. Unsworth), pp. 125–148. Nottingham University Press, Nottingham, UK.

Baigorri, H., Antolin, M.C., and Sanchez-Diaz, M. (1999) Reproductive response of two morphologically different pea cultivars to drought. *European Journal of Agronomy*, **10**, 119–128.

Biscoe, P.V., Scott, R.K., and Monteith, J.L. (1975) Barley and its environment III. Carbon budget of the stand. *Journal of Applied Ecology* ,**12**, 269–293.

Black, C. and Ong, C.K. (2000) Utilisation of light and water in tropical agriculture. *Agricultural and Forest Meteorology*, **104**, 25–47.

Brady, N.C. and Weil, R.R. (1999) *The Nature and Properties of Soils*, 12th edn. Prentice-Hall, London.

Campbell, G.S. and van Ever, F.K. (1994) Light interception by plant canopies: efficiency and architecture. In: *Resource Capture by Crops* (eds J.L. Monteith and M.H. Unsworth), pp. 35–54. Nottingham University Press, Nottingham, UK.

Clover, G.R.G., Jaggard, K.W., Smith, H.G., and Azam-Ali, S.N. (2001) The use of radiation interception and transpiration to predict the yield of healthy, droughted and virus-infected sugar beet. *Journal of Agricultural Science*, **136**, 169–178.

Collinson, S.T., Azam-Ali, S.N., Chavula, K.M., and Hodson, D.A. (1996) Growth, development and yield of bambara groundnut (Vigna subterranea) in response to soil moisture. *Journal of Agriculture Science*, **126**, 307–318.

Collinson, S.T., Clawson, E.J., Azam-Ali, S.N., and Black, C.R. (1997) Effects of soil moisture deficits on the water relations of bambara groundnut (*Vigna subterranea* L. Verdc.). *Journal of Experimental Botany*, **48**, 877–884.

Craufurd, P.Q. and Wheeler, T.R. (1999) Effect of drought and plant density on radiation interception, radiation-use efficiency and partitioning of dry matter to seeds in cowpea. *Experimental Agriculture*, **35**, 309–325.

Dardanelli, J.L., Bachmeier, O.A., Sereno, R., and Gil, R. (1997) Rooting depth and soil water extraction patterns of different crops in a silty loam Haplustoll. *Field Crops Research*, **54**, 29–38.

Daughtry, C.S.T., Gallo, K.P., Goward, S.N., Prince, S.D., and Kustas, W.P. (1992) Spectral estimates of absorbed radiation and phytomass production in corn and soybean canopies. *Remote Sensing of Environment*, **39**, 141–152.

Gerwitz, A. and Page, E.R. (1974) An empirical mathematical model to describe plant root systems. *Journal of Applied Ecology*, **11**, 773–782.

Goudriaan, J. (1977) *Crop Micrometeorology: A simulation study*. Centre for Agricultural Publishing and Documentation, Wageningen, the Netherlands.

Hay, R.K.M. and Walker, A.J. (1989) *An Introduction to the Physiology of Crop Yield*, Longman Scientific, Harlow, UK.

Hsiao, T.C., Steduto, P., and Fereres, E. (2007) A systematic and quantitative approach to improve water use efficiency in agriculture. *Irrigation Science*, **25**, 209–231.

Javot, H. and Maurel, C. (2002) The role of aquaporins in root water uptake. *Annals of Botany*, **90**, 301–313.

Jones, H.G. (1992) *Plants and Microclimate: A Quantitative Approach to Environmental Plant Physiology*, 2nd edn. Cambridge University Press, Cambridge, UK.

Jones, H.G. (1993) Drought tolerance and water-use efficiency. In: *Water Deficits – Plant Responses from Cell to Community* (eds J.A.C. Smith and H. Griffiths), pp. 193–203. Bios Scientific Publishers, Oxford, UK.

Kowalik, P.J. and Eckersten, H. (1984) Water transfer from soil through plants to the atmosphere in willow energy forest. *Ecological Modelling*, **26**, 251–284.

Kramer, P.J. and Bowyer, J.S. (1995) *Water Relations of Plants and Soils*. Academic Press, London.

Lampurlanes, J., Angus, P., and Cantero-Martinez, C. (2001) Root growth, soil water content and yield of barley under different tillage systems on two soils in semiarid conditions. *Field Crops Research*, **69**, 27–40.

Larcher, W. (2003) *Physiological Plant Ecology. Ecophysiology and Stress Physiology of Functional Groups*, 4th edn. Springer-Verlag, Berlin, Germany.

Loomis, R.S. and Connor, D.J. (1992) *Crop Ecology: Productivity and Management in Agricultural Systems*. Cambridge University Press, Cambridge, UK.

Monsi, M. and Saeki, T. (1953) Uber den Lichtfaktor in den Pflanzengesellschaften und seine bedeutung fur die Stoffproduktion. *Japanese Journal of Botany*, **14**, 605–614.

Monteith, J.L. (1981) Does light limit crop production? In: *Physiological Processes Limiting Plant Productivity* (ed. C.B. Johnson), pp. 23–38. Butterworths, London.

Monteith, J.L. (1986a) Significance of the coupling between saturation vapour pressure deficit and rainfall in monsoon climates. *Experimental Agriculture*, **22**, 329–338.

Monteith, J.L. (1986b) How do crops manipulate water supply and demand? *Philosophical Transactions of the Royal Society, London*, **A316**, 245–259.

Monteith, J.L. (1990) Conservative behaviour in the responses of crops to water and light. In: *Theoretical Production Ecology: Reflections and Prospects* (eds R. Rabbinge, J. Goudriaan, H. Van Keulen, F.W.T. Penning de Vreis, and H.H. Van Laar), pp. 3–14. Pudoc, Wageningen, the Netherlands.

Monteith, J.L. (1994) Validation of the correlation between intercepted radiation and biomass. *Agricultural and Forest Meteorology*, **68**, 213–220.

Penning de Vries, F.W.T., Brunsting, A.H.M., and van Laar, H.H.I. (1974) Products, requirements and efficiency of biosynthesis a quantitative approach. *Journal of Theoretical Biology*, **45**, 339–377.

Penning de Vries, F.W.T., Jansen, D.M., ten Berge, H.F.M., and Bakema, A. (1989) *Simulation of Ecophysiological Processes of Growth in Several Annual Crops*, Pudoc, Wageningen, the Netherlands.

Raghavendra, A.S. (ed.) (2000) *Photosynthesis: A Comprehensive Treatise*. Cambridge University Press, Cambridge, UK.

Siddique, K.H.M., Belford, R.K., and Tennant, D. (1990) Root: shoot ratios of old and modern, tall and semi-dwarf wheats in a Mediterranean environment. *Plant and Soil*, **121**, 89–98.

Siddique, K.H.M., Regan, K.L., Tennant, D., and Thomson, B.D. (2001) Water use and water use efficiency of cool season grain legumes in low rainfall Mediterranean-type environments. *European Journal of Agronomy*, **15**, 267–280.

Steduto, P. (1996) Water use efficiency. In: *Sustainability of Irrigated Agriculture* (eds L.S. Pereira, R.A. Feddes, J.R. Gilley, and B. Lesaffre), NATO ASI Series E: Applied Sciences, pp. 193–209. Kluwer Academic Publishers, Dordrecht, the Netherlands.

Steduto, P. and Albrizio, R. (2005) Resource-use efficiency of field-grown sunflower, sorghum, wheat and chickpea II. Water use efficiency and comparison with radiation use efficiency. *Agricultural Forest Meteorology*, **130**, 269–281.

Steduto, P. and Hsiao, T.C. (2007) On the conservative behavior of biomass water productivity. *Irrigation Science*, **25**, 189–207.

Steudle, E. (1994) Water transport in roots. *Plant and Soil*, **167**, 79–90.

Steudle, E. and Peterson, C.A. (1998) How does water get through roots? *Journal of Experimental Botany*, **49**, 775–788.

Stockle, C.O., Donatelli, M., and Nelson, R. (2003) CropsSyst, a cropping systems simulation model. *European Journal of Agronomy*, **18**, 289–307.

Squire, G.R. (1990) *The Physiology of Tropical Crop Production*, CAB International, Wallingford, UK.

Tanner, C.B. and Sinclair, T.R. (1983) Efficient water use in crop production. In: *Limitations to Efficient Water Use* (eds H.M. Taylor, W.R. Jordan, and T.R. Sinclair), pp. 1–27. American Society of Agronomy, Madison, Wisconsin.

Zhang, H., Pala, M., Owesi, T., and Harris, H. (2000) Water use and water use efficiency of chickpea in a Mediterranean environment. *Australian Journal of Agricultural Research*, **51**, 295–304.

Zhao, C.X., Deng, X.P., Zhang, S.Q., Ye, S.Q., Steudle, E., and Shan, L. (2004) Advances in the studies on water uptake by plant roots. *Acta Botanica Sinica*, **46**, 505–514.

3 Soil fertility

Elizabeth A. Stockdale[1], Keith W.T. Goulding[2], Timothy S. George[3], and Daniel V. Murphy[4]

[1] School of Agriculture, Food and Rural Development, Newcastle University, Newcastle-upon-Tyne, UK
[2] Department of Sustainable Soils and Grassland Systems, Rothamsted Research, Harpenden, UK
[3] The James Hutton Institute, Dundee, UK
[4] Soil Biology Group, School of Earth and Environment, UWA Institute of Agriculture, The University of Western Australia, Crawley, Western Australia, Australia

3.1 Introduction

In *Soil Conditions and Plant Growth*, 9th Edition, E.W. Russell (1961) stated that 'a soil can only be fertile if it is a favourable environment for root growth' and that 'the root system … must also take up enough water and nutrients for the crop to give a good yield of whatever part is required'. The concept of 'soil fertility' is one way that soil scientists immediately engage the interest of gardeners and farmers; both hope that the scientists will answer one of their key questions: 'How can I manage this land in order to obtain the best and most profitable yield from it year after year?' All would agree that soil fertility has something to do with the capacity of a soil to support plant growth.

Soil fertility is a deceptively familiar term, but nonetheless an elusive one, where the disciplinary use of the term and its everyday use often do not agree (Patzel *et al.*, 2000). To what extent is soil fertility a fixed characteristic of the site itself and to what extent is it the result of management and hence under the control of land managers? Which properties of soil (and of the plants themselves) should be measured to give a robust assessment of soil fertility? If crop yields are maintained or increasing, then can we simultaneously speak of a decline in soil fertility? As different approaches are taken in answer to these and similar questions, it has been possible to identify only an 'observable mess of implicit, indirect and either over-stretched or over-restricted definitions of soil fertility' (Patzel *et al.*, 2000).

In the UK, the Agricultural Advisory Council (1970) recognised the range of definitions in use amongst agricultural scientists, advisors and farmers and hence adopted a broad definition to underpin a review of practice: 'we have regarded the qualities which make up a soil's fertility as those which allow crops to grow and produce optimum yields'. In this chapter, we will take this broad definition as our foundation and then briefly consider some of the frameworks within which soil fertility has been considered, highlighting the soil–plant interactions which lie at the heart of a full and integrated understanding of soil fertility. Finally, we will consider how advice on the management of soil fertility has developed and outline the challenges which remain for practical management of soil fertility in the future.

Soil Conditions and Plant Growth. Edited by Peter J. Gregory and Stephen Nortcliff.
© 2013 Blackwell Publishing Ltd. Published 2013 by Blackwell Publishing Ltd.

3.2 Concepts of soil fertility

3.2.1 Links between soil fertility and land capability and suitability assessment

Land use patterns and their spatial distributions at a global scale are driven largely by climate, giving broad agro-ecological zones. At regional/local scales, geology, in combination with climatic patterns, has a critical role together with associated variation in topography (gradient, aspect), hydrology, soil depth and erosion risk (FAO, 1996). To support effective and sustainable crop production, it is important to understand any permanent constraints to land use and crop yield which cannot be (economically) overcome by management and within which any farming system must be adapted. Land evaluation using a land capability classification approach was developed for farm planning in the USA by Klingebiel and Montgomery (1961) and has since been adapted and widely used across the globe. Permanent constraints (which may vary in importance in different climates/regions) are defined as limiting factors and used to assess land use capability. With this approach, it is the negative/limiting features of land which are taken into account to assign a land capability class, with the exception of location and accessibility. Few soil characteristics are routinely considered for land capability classification. The factors that are measured are those which lie beyond the control of the farmer: extremes of soil texture, soil depth and stoniness. Soil texture (the relative amounts of differently sized mineral fractions: sand silt and clay) and stoniness are inherent properties of the soil strongly related to characteristics of the parent material and cannot usually be easily or quickly changed. These inherent properties largely define the ranges within which management is able to control a number of soil properties, e.g. the formation and stability of soil structure, soil consistence/workability during tillage, infiltration capacity and depth and penetration of water and the capacity of soil to supply and retain nutrients. So, e.g., it is possible to give some general information on a range of soil characteristics based simply on the knowledge of its texture: a silty loam will tend to slake and cap at the surface; a sandy loam will hold less water and tend to become acid more quickly and its capacity to hold nutrients will be relatively low. However, even where all the permanent limiting factors are understood, land cannot be simply graded from best to worst on a single scale without taking its use into consideration (Gibbon, 1961).

The development of a Framework for Land Evaluation (FAO, 1976) was deliberately focussed on the assessment of land suitability in relation to particular defined land uses, i.e. 'Fitness for Purpose'. In these assessments, consideration of both permanent (inherent) and manageable constraints is included. As most suitability assessments have been carried out for various types of crop production, the factors used to define subclasses are commonly those associated with definitions of soil fertility (e.g. droughtiness, soil nutrient deficiencies, waterlogging, ease of cultivation). Definitions of suitability classes are expressed in terms of expected yields and/or the level of inputs required to achieve good yields. Measuring potential productivity along a continuous scale, rather than in discrete classes, is also possible through the use of arithmetic formulae combining key limiting factors within a number of parametric approaches to land/soil appraisal, e.g. Riquier *et al.* (1970) and Beatty *et al.* (1979).

Common to all these approaches is the identification of constraints to crop production (Figure 3.1) and the development of inputs and/or management practices which overcome these constraints, so that crop yield potential defined by genetic and climatic factors can be

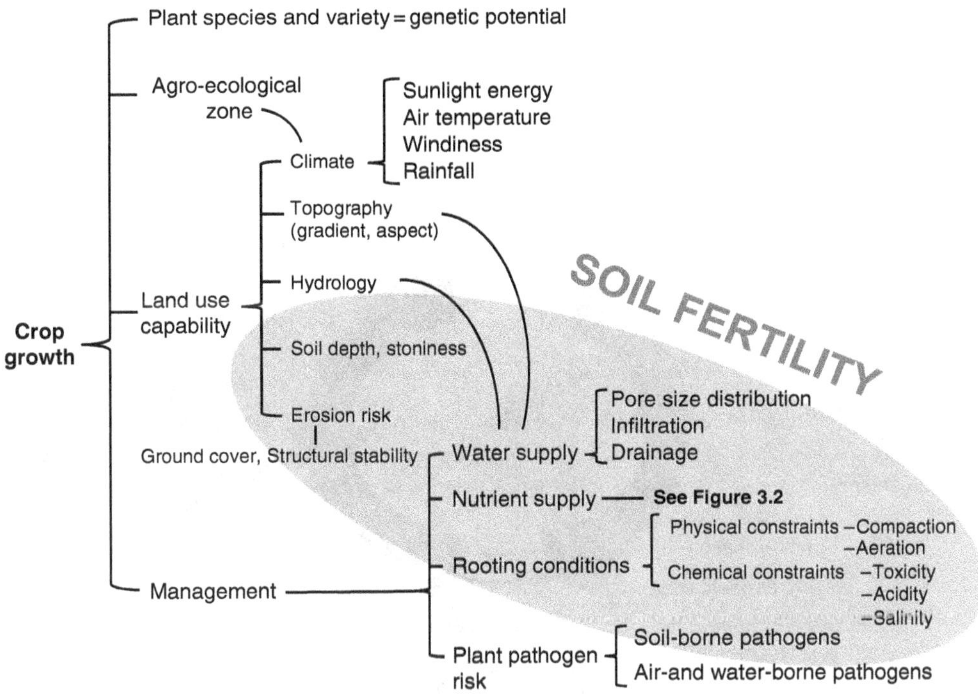

Figure 3.1 Potential constraints to crop growth include plant, climate and soil factors and their interactions. Soil fertility is defined by those soil factors which constrain crop growth. The key limiting factors in any location will vary, e.g. some locations are prone to salinity which then provides the major growth limiting factor, while other regions are mainly limited by acidity, drought, nutrient supply, etc. Liebig's law of the minimum suggests that crop growth will be constrained by the most limiting factor and that changes in other factors will have little impact on growth until they become most limiting. Where several constraints are high, but no single constraint is extreme, improvement of conditions in relation to either constraint will increase yield. Interactions also occur between the impacts of constraints on yield; e.g., with increasing salinity, drought stress is seen at higher levels of water availability. Crop variety selection can also be used to overcome some constraints as a result of the development of varieties with increased tolerance, e.g. aluminium toxicity, salinity.

achieved. Russell (1937) identified the main soil constraints to crop growth that can be controlled through management as water supply, nutrient supply, soil reaction (including acidity, salinity), erosion risk and depth of soil. Factors defining many of these constraints are routinely included in suitability assessments.

3.2.2 Equating soil fertility with the soil nutrient reserve

As described in Chapter 1, the role of the soil in supplying plant nutrients was at the heart of early controversy in agricultural science; developing, then patenting artificial manures provided much of the income stream for early research. As a result, the availability and use of fertilisers has expanded greatly on a global scale (see Section 1.5.4). Consequently during the twentieth century, soil fertility research often focussed on 'the study of all factors which affect the nutrition of agricultural crops' and 'how growth conditions for plants can be improved by manuring or fertilising' (de V Malherbe, 1964). This is not a small area of focus (Figure 3.2). Consequently for many agricultural scientists, the term 'soil fertility' developed a relatively

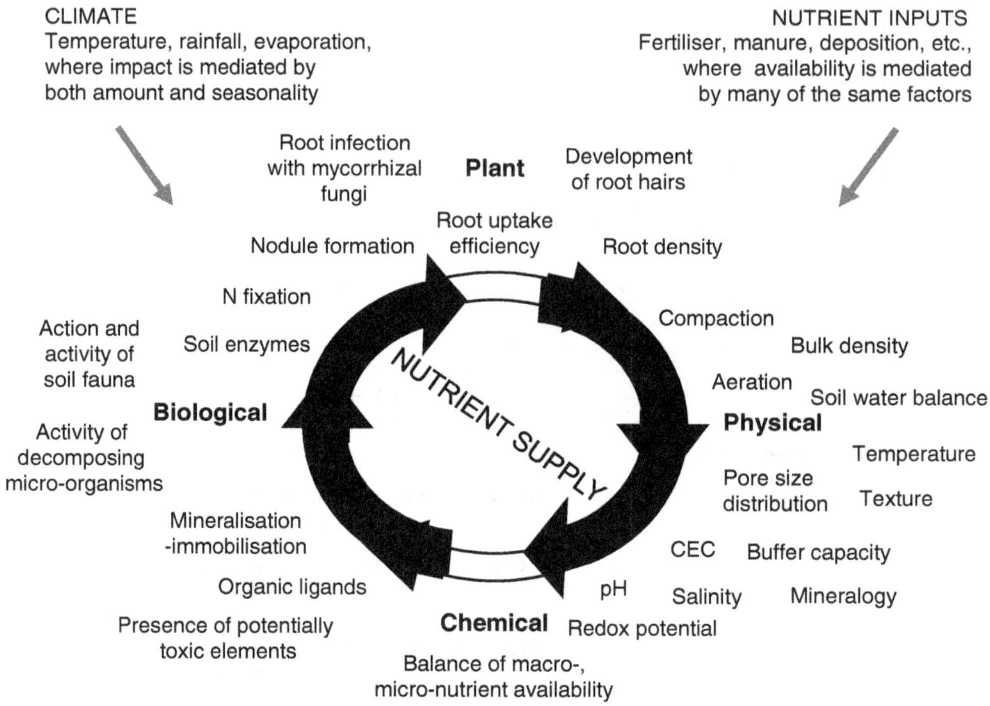

Figure 3.2 The interaction of plant/root, chemical, physical and biological factors in the soil with management (nutrient additions) and climate control nutrient supply to plants; presented as an example of how the main constraints to crop growth (Figure 3.1) can be considered to result from plant–soil interactions. Complex interlocking processes in the soil determine the amounts of essential plant nutrients present, their availability to crop plants, the transformations they undergo in soil, loss mechanisms, processes making them unavailable or less available to crop plants and the most appropriate means of replenishing them. Many agricultural and soil science textbooks in the twentieth century, most of them with the term 'soil fertility' in their title, have focussed solely on these issues and their practical management.

narrow meaning restricted to consideration of a soil's potential to supply nutrients for crop growth. This does not mean that the impact of other factors on yield, such as water availability or poor structure restricting root growth, were considered less important, but simply that these were considered separately to the management of soil fertility by fertiliser addition.

3.2.3 Understanding soil fertility in relation to concepts of 'soil quality'

At the end of the twentieth century, the concept of soil quality developed alongside an increasing awareness and valuation of the goods and services provided by ecosystems globally to support human life (Costanza *et al.*, 1997; Millenium Ecosystem Assessment, 2003) and the central role of the soil in the delivery of a range of key functions (Lal, 2001). Soil quality as defined by the Soil Science Society of America (Doran and Parkin, 1994) is 'the capacity of the soil, within land use and ecosystem boundaries to sustain biological productivity, maintain environmental quality and promote plant and animal health'. But the standard against which soil quality is to be compared is rarely agreed or well defined and, as the definition by Doran and Parkin indicates, it must be assessed in relation to the intended use of the soil: 'Fitness

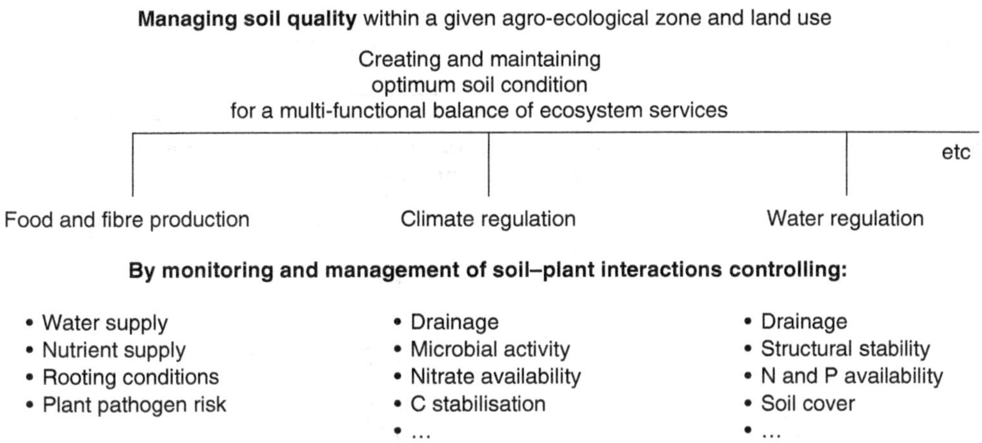

Figure 3.3 Maintaining optimum soil condition for food and fibre production (soil fertility) is one aspect of the management of soil quality. Soil quality frameworks need to take account of the requirement for soils to deliver a range of ecosystem services including provisioning (e.g. food, fibre, freshwater, genetic resources), regulating (e.g. climate, disease and water regulation), cultural (e.g. recreation, educational, cultural heritage) and supporting services (e.g. soil formation, nutrient cycling and primary production). The key ecosystem services required at any site and consequently the underpinning soil functions/characteristics required will vary according to location and societal demands.

for Purpose' again. The utility of an overarching concept of soil quality has been debated (Karlen *et al.*, 2001; Sojka *et al.*, 2003) but it now stands alongside definitions of acceptable measures of air and water quality in guiding environmental public policies. These issues are considered in more depth in Chapter 12.

It is clear, however, that embedded within all definitions of soil quality there is a recognition of the role of the soil in supporting food and fibre production, i.e. soil fertility as broadly defined (Figure 3.3). The relative emphasis given to aspects of soil fertility in the assessment of soil quality at any site will depend on the intended land use and the location; e.g., if the site is within a key catchment determining water supply for a city or next to a protected wildlife habitat, the role of the soil in supporting food and fibre production would not be a key priority. Many of the early examples of soil quality assessment cited in the literature focussed on crop productivity as they were carried out in agricultural areas (e.g. Romig *et al.*, 1995; Sinclair *et al.*, 1996; Stenberg, 1999) and are similar to the integrating approaches used for crop suitability assessments (see Section 3.2.1). Food and fibre production is recognised as a key soil function in all current definitions of soil quality. In this context, soil fertility might be regarded as the state of soil required for the effective performance of this function. Although some authors appear to use the terms interchangeably, soil fertility is not synonymous with soil quality and it should be considered as one aspect of a more integrated definition of soil quality.

3.2.4 Defining soil fertility in relation to ecosystems dynamics

The large diversity of soil organisms and their interactions (predator–prey and extensive omnivory) are the driving force behind many soil functions, including nitrogen (N) and phosphorus (P) mineralisation, structure formation and stabilisation, and consequently underpin

Table 3.1 Annual average biomass (kg C ha^{-1}) and simulated contribution to overall N mineralisation (kg N ha^{-1}) for the 0–10 cm layer of a winter wheat field under integrated management for key groups of soil organisms.

Soil organism group – by functional group if considered in the model	Average biomass (kg C ha^{-1})	Modelled annual N mineralisation (kg N ha^{-1})	% Reduction in amount of N mineralised if fauna group removed from the model
Bacteria	373.1	52.96	–
Fungi	21.0	4.53	–
Protozoa			
Amoebae	6.09	13.63	32.4
Flagellates	0.35	0.82	1.9
Nematodes			
Bactivorous nematodes	0.41	6.03	17.3
Fungivorous nematodes	0.02	0.05	1.2
Predatory nematodes	0.25	1.62	19.1
Enchytraeids	0.19	0.45	3.2
Earthworms	8.86	0.47	Not simulated

Data for small pools of soil microarthropods are not presented. The importance of predation in driving overall N mineralisation is shown by the relative increase in the estimated contribution made to mineralisation when it is estimated by the percentage reduction in the amount of N released through mineralisation as a result of exclusion of different groups of soil fauna.
Source: Data derived from de Ruiter *et al.* (1993).

plant growth (Coleman *et al.*, 1983; Wardle, 2002; Stockdale *et al.*, 2006). Consequently, the role of soils as the organising centres within terrestrial ecosystems has been emphasised by some authors (Coleman *et al.*, 1998). Russell (1937) clearly presented an integrated understanding of soil fertility, though he recognised that each specialist area has been tempted to claim the key to managing soil fertility for itself. Soil ecologists have not been exempt from this temptation: trophic interactions between soil organisms have been used as the sole basis of models of nutrient flows in soils; e.g., Brussaard *et al.* (1996) found that soil fauna can account overall for 30–40% of net N released into plant available forms (Table 3.1).

The agricultural intensification that occurred in the last half of the twentieth century in much of the world largely disregarded the biological interactions within the soil (Giller *et al.*, 1997). Despite this, Giller *et al.* (1997) found limited evidence for any significant loss of either biodiversity or capacity for ecosystem function in soil. Wardle *et al.* (1999) studied the impact of agricultural intensification over 7 years and also found little evidence that intensification has adverse consequences for soil biological activity. This is not to say that farm management practices do not affect below-ground biological interactions both directly and indirectly through physical and chemical changes (Stockdale *et al.*, 2006; Figure 3.4a, b): there is evidence that agricultural systems are associated with simplified food webs in the soil compared with semi-natural systems, and differences occur between agricultural systems that are predominantly pastoral or arable (van der Putten *et al.*, 2004).

Stockdale *et al.* (2006) placed an ecosystem-driven model at the heart of soil function. This conceptual model recognised that below-ground processes, which contribute to the delivery of important functions, result from the emergent soil ecosystem properties and processes arising from the interaction of soil habitats and their associated populations. An integrated understanding of soil fertility therefore requires a consideration of the interactions between roots, soil organisms and their habitat within the soil (Figure 3.5). The relative

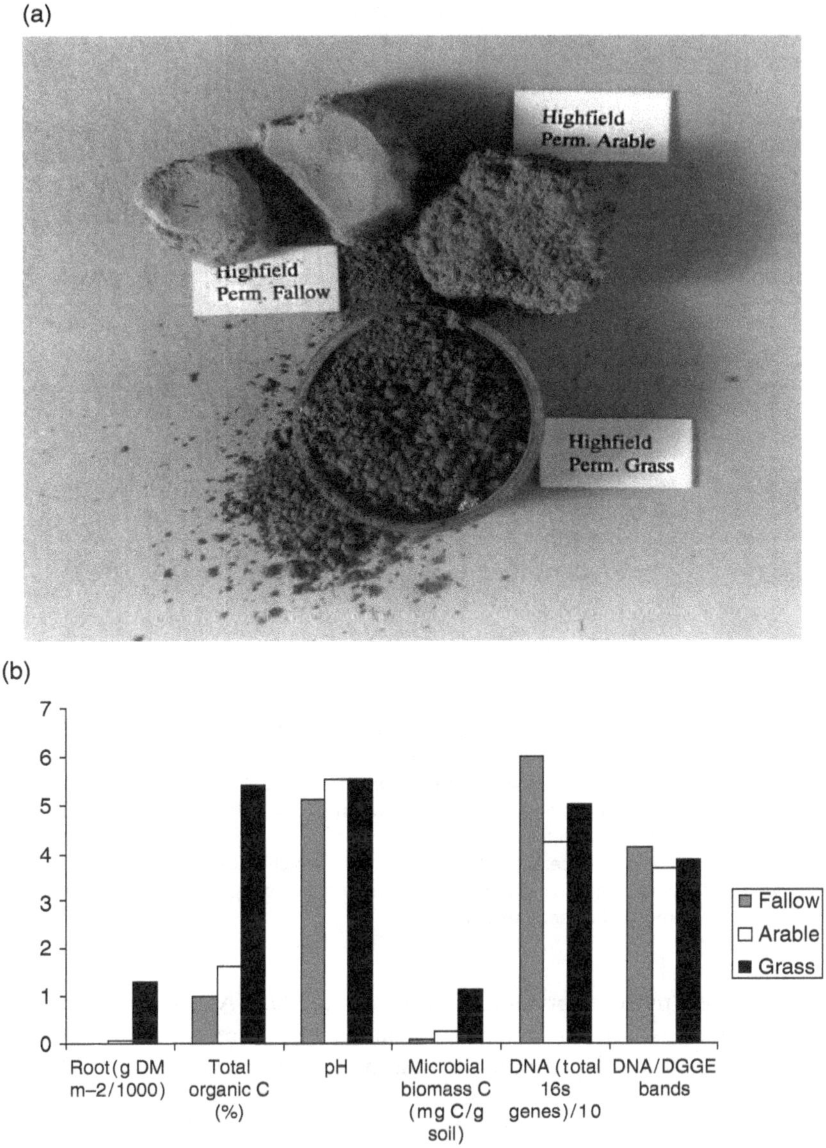

Figure 3.4 Data from the long-term Highfield experiment at Rothamsted, England, shows the impacts of long-term fallowing with intensive cultivation on (a) visible soil physical properties and (b) on some chemical and biological properties. The almost zero inputs of organic carbon via plants and tillage in the fallow plots has very significantly reduced soil organic carbon levels and soil microbial biomass but had no impact on pH in this well-buffered clay soil. Soil microbial diversity indicated by the number of bands in the DNA/DGGE is not reduced by fallowing; in fact, underlying soil microbial diversity is apparently revealed when the dominance of plant-associated bacteria is reduced. From Hirsch et al. (2009). With kind permission from Elsevier. For a colour version of this figure, please see Plate 3.1.

importance of abiotic factors (climate, management, inherent site constraints) in controlling the biological interactions and driving soil processes at a range of scales is still poorly understood (Bardgett, 2002). For soil fertility, increased understanding of the interactions driving soil water balance, nutrient cycling, rooting conditions and the regulation of soil-borne

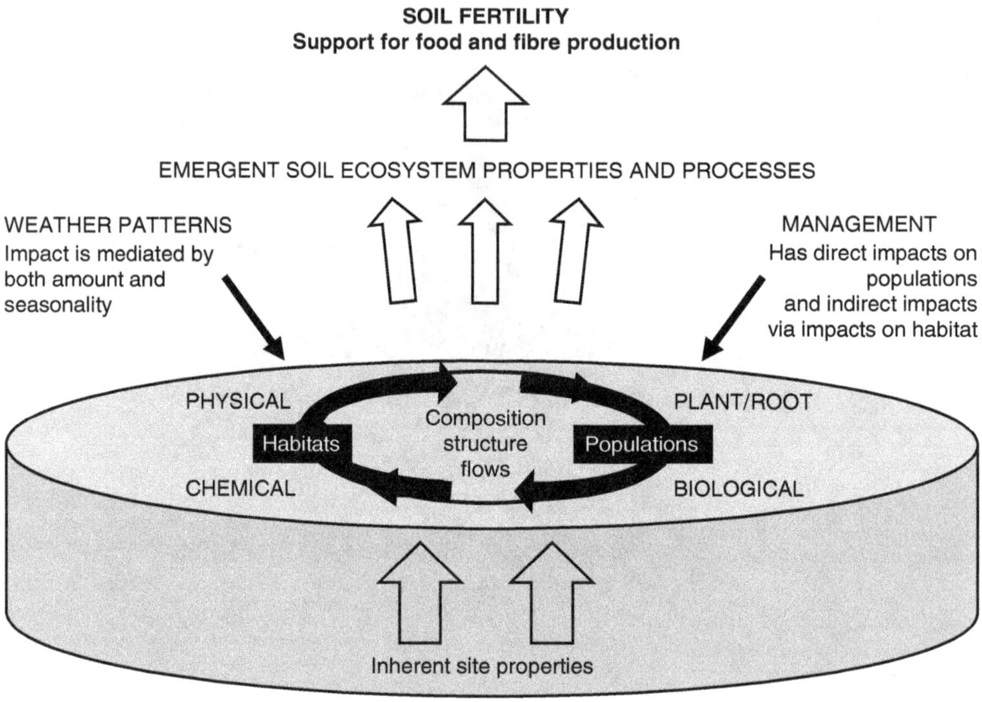

Figure 3.5 Soil ecosystem functioning is the result of physical, chemical and biological interactions between roots, soil organisms and their habitat within the soil. The interactions are driven by climate and management factors within the framework of permanent constraints set by the inherent site characteristics. The structure, composition and flows between soil populations and habitats determine the emergent properties together with the outcome and rate of soil processes observed at the soil scale. The outcome of these ecosystem processes is the soil functions which underpin all aspects of soil quality. For soil fertility, the interactions modifying nutrient and water availability, the environment for rooting and the activity of soil-borne pathogens are critical.

plant pathogens seem most important (Stockdale *et al.*, 2006). A mechanistic understanding of such interactions is needed if management practices are to be robustly evaluated (Brussaard, 1998) and developed in a way that protects the ecosystem functioning at the heart of soil fertility.

Sanchez (1994) considered that the underlying rationale of soil fertility management, i.e. every constraint should be overcome through external inputs, was outdated and highlighted the need to replace this with a second paradigm to underpin soil fertility research which increased focus on the use of endogenous processes and potentials within the soil-plant system. We recommend taking an integrated ecosystem approach to the definition of soil fertility and that research into soil–plant interactions, which ultimately allow crops to grow and produce optimum yields, will be the focus for the twenty-first century rather than separate studies of soil processes in the absence of plants and/or processes in the plant root in the absence of soils. As illustrated in Figure 3.2, a wide range of plant and soil properties interact to affect nutrient supply to and uptake by plants. The study of any of these factors without an understanding of the ecosystem context in which they take place may limit the application of increased understanding to the improvement of soil fertility management in practice. Similarly for soil-borne pathogens, van der Putten *et al.* (2009) highlight how modelling above- and below-ground plant–pathogen interactions could lead to new avenues for sustainable control of pests and diseases in crops, above- and below-ground pest and disease suppression, increased roles for arbuscular mycorrhizal (AM) fungi in plant nutrition and improved crop productivity.

3.3 Plant–soil interactions determining soil fertility

3.3.1 Impacts of soil on plant roots

Soil factors control plant growth largely through their impact on plant roots; in annual cropping systems, plant roots occupy relatively little of the soil volume (5–10%), whereas in grasslands it can be up to 40% of soil volume. The supply of water and nutrients to support plant growth critically depends on uptake from the soil across root membranes, so any factor that impedes plant root growth will impact water and nutrient uptake and apparent soil fertility. The rate of supply of nutrients to any root depends on the rates of mass flow and diffusion, but of critical overall importance for water and nutrient uptake is the distribution of roots within the soil, i.e. the total surface area over which uptake can occur. Root growth shows responses and interactions with a range of chemical, physical and biological factors in the soil (Hodge, 2006). In turn, many of these interact and plants are often able to compensate, so that restricted root growth in one part of the soil may be compensated for by increased growth in other regions that are less constrained.

The ability of the plant root to find space in which to extend through the soil is often an important factor limiting plant growth. The preparation of seedbeds with a good structure ('tilth') and consequent network of pores, holding the right balance of air and water appropriate to the seed to be sown, is an art underpinned by the science described in Chapter 8. The effects of soil compaction on root growth are also reviewed in Chapter 8; compact zones with high bulk density may restrict root penetration and radial growth of roots and so may reduce yield. The root cap perceives touch stimuli, and roots have also been shown to modify their gravitropic growth to avoid obstacles (Massa and Gilroy, 2003). Soil pore size and organisation together with water inputs to the soil determine soil water content (Chapter 9). The range in soil water content within which plant growth is not constrained by matric pressure, aeration or mechanical resistance (the least limiting water range; Da Silva *et al.*, 1994) is therefore one indicator of soil fertility.

The amount, form and distribution of nutrients or toxic elements in soil also influence the amount and distribution of root growth. Root branching may also be regulated by external nutrient concentrations with local stimulation of lateral root growth observed in zones of high N, P and potassium (K) availability (Drew, 1975; Forde and Lorenzo, 2001), though not all plants respond in the same way (Hodge *et al.*, 1998). Plants will usually selectively avoid rooting into zones of high toxicity. For example, in most plants, the first visible symptom of aluminium stress (often linked with acid conditions) is the inhibition of root elongation (Matsumoto, 2002); consequently rooting depth is often restricted by acid subsoils. Similar restrictions to rooting are seen where soils are saline or have boron toxicity. Under conditions of nutrient deficiency or toxicity, a range of specific physiological changes take place caused by multigenic interactions (Figure 3.6); e.g., over 1000 genes are differentially regulated under conditions of P starvation in Arabidopsis (Hammond *et al.*, 2003; Wu *et al.*, 2003; Morcuende *et al.*, 2007). Understanding the transcriptional regulation of these genes can allow the identification and selection of genotypes that are better suited to cope with conditions of low nutrient availability or elevated concentrations of potentially

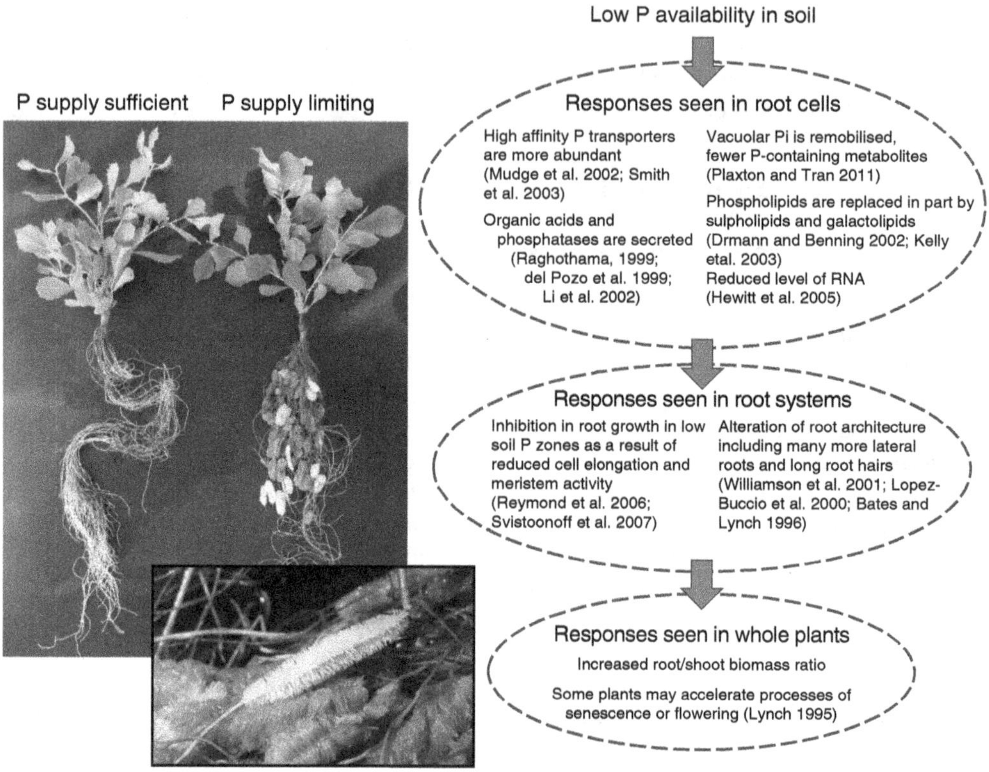

Figure 3.6 Complex and interlinked changes in plant physiology at root cell, root system and whole plant scales occur in response to low P availability in soil as the result of up-regulation of many genes. Photographs show *Hakea prostrata* R. Br. grown under sufficient and limiting phosphate supply. Cluster roots develop in some species as an adaptation for P acquisition in low P environments (details in lower image). Thanks to Stuart Pearse and Michael Shane from the University of Western Australia for access to greenhouse trials which provided the photographs. For a colour version of this figure, please see Plate 3.2.

toxic elements; it may also allow the use of transcriptional arrays to identify nutritional deficit or toxicity far in advance of any physical symptoms (Hammond *et al.*, 2004).

Direct damage to plant roots by root feeding and parasitic organisms can be considerable (particularly by beetle larvae and nematodes) and can therefore limit the extent to which roots exploit the soil volume. In addition, a wide range of soil micro-organisms that can cause plant disease through root invasion, in particular soil fungi, have been identified and their ecology described (for detailed descriptions see, e.g., Bruehl, 1987). Interactions between the plant species, soil organisms and environmental conditions determine infectivity with regard to the host crop (Smith *et al.*, 1999). Some soils have been shown to be disease suppressive (Menzies, 1959); a wide range of ecological mechanisms (including parasitism, direct and indirect antagonism) are now thought likely to contribute to this effect (Mazzola, 2002) and interactions with soil chemical and physical properties are also important (Duffy *et al.*, 1997).

3.3.2 Association and symbiosis between plant roots and soil organisms

The majority of plants are capable of forming mycorrhizal associations with various soil fungi (see Chapter 11). The most common is the AM association, which has been shown to have emerged very early in the evolution of land plants (Remy *et al.*, 1994). Developments in molecular and transcriptional analysis in recent decades mean we now understand a lot about the molecular responses of non-mycorrhizal plants to nutrient deficit, primarily due to the focus of transcriptional studies on the model plant *Arabidopsis*. Unfortunately, very little is known about such responses in mycorrhizal plants, which make up 90% of all land plant families; only very few crop species do not form such associations, among them *Brassicae* and Chenopodiaceae. Over 60 years ago, Sir Albert Howard suggested that 'the presence of an effective mycorrhizal symbiosis is essential to plant health' (Howard, 1943). The wide benefits of mutualistic relationships in an AM association to crop nutrition, health, stress resistance and the development of soil structure have since been confirmed and the impacts of management on their infectivity and effectiveness studied, e.g. Bethlenfalvy and Lindermann (1992), Harrier and Watson (2003) and Gosling *et al.* (2006). Most soils have indigenous populations of AM fungi, but more work is needed to understand whether and how these associations can be manipulated to support integrated soil fertility management approaches.

The nitrogen-fixing symbiosis that exists between plants of the Leguminosae (legumes) and the soil heterotrophic bacteria, known generally as rhizobia, has been understood for over 100 years (Atwater, 1884; Hellriegel and Willfarth, 1888). There are a number of other nitrogen-fixing associations between soil micro-organisms and plants, those involving cyanobacteria and those in which the microbial partner is an actinomycete, which are important in natural and in agricultural systems. Rhizobia are a phylogenetically disparate group of α- and β-proteobacteria; more than 70 species have been identified (Masson-Boivin *et al.*, 2009). Nodule formation in legumes occurs as a result of complex interactions between plant and microbial processes whose biological, biochemical and genetic aspects have been much studied (Sprent, 2001; Masson-Boivin *et al.*, 2009). The precise details of nodule formation are still not fully understood (Martinez *et al.*, 1990; Brewin, 1991; Povorov, 1998), although research into signalling and host plant cell responses during infection by rhizobia or arbuscular mycorrhizae (Oldroyd *et al.*, 2009) has suggested that it may be possible to induce colonisable nodules on plants other than legumes. Rhizobia–legume interactions have been shown to be highly specific (Long, 2001) and a wide variety of morphological variations have been observed. McInnes and Haq (2003) reported the impact of a wide range of environmental, plant (host and non-host) and soil (physical, chemical, biological) factors on the persistence and effectiveness of rhizobial populations in the nodulation of legumes. Even where nodulation is effective, the achievement of high rates of nitrogen fixation depends on a very complex mixture of interactions including the plant, the bacteria species, and climatic and soil factors (Peoples *et al.*, 2009). At the practical interface with agriculture, agronomists still struggle even to give broad ranges of N fixation associated with any particular species of crop or forage legumes, let alone give a site-specific estimate on which a rotational fertilisation plan can be based (Smil, 2001; Unkovich and Pate, 2001; Herridge *et al.*, 2008).

The development of molecular technologies is increasing the number of effective plant–microbial interactions that have been identified and understanding of their modes of interaction (Berg, 2009). Much of the work on these interactions to date has focussed on the understanding of the mechanism and possible isolation of the key organisms involved. Hayat *et al.* (2010) identify three main modes of action for plant growth promoting rhizobacteria (PGPR) species: (1) synthesising particular compounds for the plants, (2) facilitating the uptake of certain nutrients from the soil and (3) lessening the impact or protecting plants from diseases. A number of plant growth promoting fungi (PGPF) have also been isolated belonging to the genera *Trichoderma*, *Fusarium*, *Penicillium*, *Phoma*, etc. and have been reported to be beneficial to several crop plants, not only by promoting their growth but also by protecting them from disease (e.g. Shivanna *et al.*, 1996; Harman, 2011a). Trichoderma has been known as a fungal biocontrol agent for many plant diseases since the early twentieth century, although the mechanisms of action were not known. Much or most of the biocontrol activity of these fungi is now considered to be a result of their ability to induce systemic disease resistance, and effective strains also convey stress resistance to plants (Harman, 2011b). The basidiomycete *Piriformospora indica* has at least qualitatively similar effects on plants (Shoresh *et al.*, 2010), and hence it has been suggested that mixed species approaches might have greater impact. Ardanov *et al.* (2011) suggest that there may be an important difference between endophytic bacteria which are able to activate basal and inducible plant defence systems and biocontrol strains which have a plant growth promotion effect which is not correlated with induced resistance.

Plant growth promotion arising from interactions between soil organisms and plant roots is complex and often cannot be attributed to a single mechanism; it may result from the synergistic interactions of a number of species, not only soil bacteria or fungi (Richardson *et al.*, 2009). For example, many reports have documented the ability of soil micro-organisms (such as *Pseudomonas*, *Bacillus* and *Rhizobium* spp., actinomycetes and various fungi such as *Aspergillus* and *Penicillium* spp.) to solubilise insoluble mineral phosphate compounds (Richardson *et al.*, 2009). The most common mechanism used by micro-organisms for solubilising mineral phosphates in neutral and alkaline soils seems to be acidification and Ca-chelation via biosynthesis and release of a wide variety of organic acids (Whitelaw, 2000 and Chapters 5 and 11). However, attempts to improve mineral phosphate solubilisation by inoculation with isolated P-solubilising bacteria and/or fungi have not been particularly successful because of limitations such as poor ecological fitness, low metabolite production, variability in inoculant-delivery systems and inconsistent performance in field applications (Shenoy and Kalgudi, 2005; Richardson *et al.*, 2009).

3.3.3 Impacts of plants on soil

While soil properties have a role in controlling root distribution, plant roots also play a key role in soil structural development; roots are the most important agents creating structural pores in undisturbed soils. Roots expand by forcing a root tip into pores of 100–300 μm diameter. The young root then tends to compress soil around the pore by radial expansion, increasing the size of the pore occupied by the root but decreasing porosity and modifying pore size distribution in the rhizosphere; in the medium term, rooting tends to increase soil macroporosity through the provision of stabilised continuous channels (Angers and Caron, 1998). Differences between crops in their impact on soil structure have been noted, in part as a result of the fact that crops differ in the patterns and depth of rooting (Gregory, 2006; Chapter 10). While plant roots use pores of >100 μm diameter as points of entry, they can effectively empty pores down to 0.2 μm; this increases differences in the depth and patterns of soil wetting and drying.

The zone of soil in the immediate vicinity of the root is markedly different in its properties to the remainder of the soil as a result of the direct impacts of plants on the soil (Table 3.2). In 1904, Hiltner first used the term 'rhizosphere' for this interfacial zone between the plant roots and the bulk of soil, and this term is now used widely for the zone of soil surrounding the root within which the soil is directly affected by the presence of the root. The rhizosphere is a dynamic transit zone, with large fluxes of water and plant nutrients across it, and consequently small pool sizes for most nutrients, driven by patterns of plant demand and uptake (for more detail see Chapter 10).

Table 3.2 The rhizosphere has markedly different physical, chemical and biological characteristics than bulk soil which is not under the direct influence of plant roots.

Soil characteristic and observed change in the rhizosphere	Reference for paper where an example of these differences can be found
Water retention properties	
Smoother, less convoluted pores; larger pores on average	Whalley *et al.* (2005)
Nutrient availability	
Five- to tenfold accumulation or depletion depending on the balance between mass flow and uptake at the root surface	Marschner (1995)
pH	
Changes of 2–3 pH units (100–1000-fold difference in proton concentration); both increases and decreases in acidity of this magnitude are seen in the rhizosphere depending on plant species and management particularly the form of N taken up by the plant; uptake of N as NH_4 leads to proton excretion and is acidifying, NO_3 tends to increase pH	Neuman and Römheld (2002)
Redox conditions	
Reductase capacity can be increased up to 100-fold in the rhizosphere	Römheld (1990)
Labile organic matter	
Sugars and amino acids increased by 2–10 times in the rhizosphere as a result of root exudation	Uren (2000); Jones *et al.* (2004)
Microbial populations	
10- to 100-fold increase in microbial activity with particular stimulation of gram –ve bacteria in the rhizosphere but reduction in microbial diversity	Chineau *et al.* (2000); Kowalchuk *et al.* (2002)
Soil fauna	
Marked increase (around tenfold) in bacterial grazers (nematodes, protozoa, microarthropods)	Griffiths (1994); Bonkowski *et al.* (2000)

The amounts and composition of root exudates are the main driver for the large increase in size of the microbial biomass and change in the structure of communities of soil organisms found in the rhizosphere. For example, species-specific flavonoids have been identified as key signalling compounds in plant–micro-organism interactions (Shaw *et al.*, 2006). Distinct communities of bacteria have been found within the plant root and on the root surface (rhizoplane; Lupwayi *et al.*, 2004; Dennis *et al.*, 2009); around 4–10% of the root surface has been shown to be covered by soil micro-organisms. Some soil organisms stimulated in the rhizosphere are plant pathogens, and there is evidence that bacterial as well as fungal pathogens have co-evolved with plants showing a high degree of host specificity (Raaijmakers *et al.*, 2009). Plants are not the passive recipients of a pre-existing rhizosphere community that merely feed on root exudates: a series of interactions between plants and soil factors determines the structure and function of the associated rhizosphere community (Berg and Smalla, 2009).

Plants regulate and exclude elements and molecules at the root surface, manipulating the rhizosphere in a range of ways (changing pH and redox potential, hydraulic lift, exudation and secretion of organic acids and enzymes) in order to tolerate toxicities or deficiencies or manipulate microbial communities (Chapter 10). All such traits are controlled by specific genes or cascades of genes, some of which have been identified but many of which are still to be elucidated. Once understanding of the genetic control is gained, it may be possible to manipulate the rhizosphere directly by deploying crops and/or varieties in which the gene of interest is expressed, within appropriate crop rotations. Much potential exists to manipulate rhizosphere properties at a genetic scale, but such attempts need to be informed by an understanding of the controls imposed by the complex soil environment in which they need to be effective (George *et al.*, 2004, 2005).

3.4 Assessing soil fertility

Assessment of soil fertility takes place within the context of an individual farming system, where the crop choice and crop yield and quality goals are largely defined by constraints arising from societal policies, laws and financial incentives, economic costs and the aspirations, attitudes, education and training of the land manager. In the twentieth century, the assessment of soil fertility focussed on the assessment of the need for a range of external inputs that were used to overcome the identified soil constraints. Socio-economic factors also interact with the physical constraints arising from climate and location characteristics; together these constraints also determine which, if any, management strategies are available for soil fertility management. Towards the end of the twentieth century, increased awareness of the environmental impact of agriculture led to policies and legislation in a number of countries limiting the amount and/or timing of inputs, such as fertiliser and manure applications, and cultivations, such as ploughing grassland.

During the mid-late twentieth century, the use of fertilisers to overcome nutrient limitations to crop production expanded greatly, and fertiliser purchases are now a major part of the variable costs of production in most cash-cropping systems worldwide. Alongside these changes in management, developments in soil fertility assessment were focussed on the measurement of soil nutrient status to guide the development of recommendations for lime and fertiliser application to crops (Tisdale and Nelson, 1966; Cooke, 1982; Foth, 1997). Thousands of pot and field experiments have been conducted to evaluate soil fertility (i.e. given conditions *x*, how well does the plant grow, take up nutrients and water, produce grain/fruit) and to establish management recommendations for inputs of fertiliser, animal manures or lime (i.e. given conditions *x*, with an addition of *y*, a yield increase of *z* can be expected). However, in many developing countries, the adoption of such intensive input-focussed systems has not occurred to any great extent, in part because high yielding varieties, fertilisers and pesticides can be difficult to obtain due to poor distribution infrastructure and hence high costs. A range of renewable soil fertility replacement strategies have been proposed that are suitable for low-input systems (Ajayi *et al.*, 2007). For instance, significant attention has been given to the development of legume agroforestry (fertiliser tree) systems, the integration of green manure and dual purpose legumes and the careful use of animal manures often integrated into conservation farming approaches; the latter integrate reduced tillage, retention of crop residues, targeted planting and farm input use and the use of legumes in the rotation (Sanchez and Salinas, 1981; Palm *et al.*, 2001).

Increasingly in industrialised countries, farming systems are required to deliver broader societal goals, e.g. protection of wildlife habitat and/or seek alternative sustainable approaches, e.g. through certification to organic farming standards. Consequently, there has been an increased focus on practices and technologies that optimise inputs, make better use of existing soil nutrients and minimise the leakage of nutrients, especially N and P, from farms to the wider environment. This has seen farm nutrient management planning shift from a relatively small set of procedures designed to optimise fertiliser application rates for crop and animal production towards comprehensive whole-farm nutrient management approaches that consider a range of issues to deliver both farm productivity and environmental outcomes (Goulding *et al.*, 2008). In the twenty-first century, alongside predicted changes to climate, including the increasing frequency and intensity of extreme weather events, it seems almost certain that fossil-fuel-based inputs, such as N fertilisers and pesticides, as well as other non-renewable resources such as P fertilisers, will become increasingly expensive and scarce. These constraints are likely to necessitate the adoption of changed practices within intensive input-focussed farming systems (Kirschenmann, 2007). Consequently, the uptake of practices such as carefully planned crop rotations, maintenance of soil organic matter and the recycling of local organic 'wastes' is likely to increase in all farming systems (Hanson *et al.*, 2007) and be integrated with new approaches developed for lower input conditions. Within such systems, a broader integrated approach to the assessment of soil fertility that considers the integration of soil physical, biological and chemical aspects, together with inputs of water and nutrients from a range of sources, will be needed. It is therefore timely to review the assessment of soil fertility in this context.

3.4.1 Assessment of soil fertility and fertiliser requirements

The mere analysis of soil properties, however carefully selected, does not give a measure of soil fertility unless the selected parameters have been calibrated against plant growth. In yield response experiments, the aim is to describe the relationship between crop yield and increments of a nutrient or, in a factorial experiment, to measure the response to two or more treatments in combination. When carried out in the field, such an experiment can be used to assess the relative importance of limiting factors to crop yield at the site (Table 3.3) and may also allow the optimum soil fertility management strategy (cultivations, crop residue management strategies, inputs of organic matter, nutrients, water, lime) under the conditions of the experiment to be identified. The results for a single combination of crop, site and management factors are of very limited use alone. Consequently, farmers or their extension officers may need to carry out site-specific on-farm research to identify optimum system-adapted soil fertility management approaches. Additionally, by grouping results from similar soils and crops, it may be possible to draw out guidelines, whereby a set of measures of soil properties provide an index of soil fertility, which support the farmer to select from a limited range of soil fertility management strategies to obtain optimum economic benefit. For example, this approach has been used over many years to create and adapt the UK *Fertiliser Manual (RB209)* (Defra, 2010).

A number of steps are critically important in any system used to support the practical management of soil fertility:

1. Analysis of selected soil properties which have been shown to correlate with crop yield, nutrient uptake or quality.
2. Interpretation of the results to predict the degree of yield constraint as a result of expected deficiency of nutrient supply or intensity of some other constraint, e.g. compaction.
3. Development of recommendations for appropriate fertiliser, manure or other crop management practices that will enable yield optimisation.

The effective performance of any soil fertility assessment system also relies on the fact that the sample points measured in the field or presented for analysis are representative. More, deeper and perhaps more frequent, soil samples are needed for more mobile nutrients (e.g. nitrate, sulphate) or constraints (e.g. salinity). Hence, great care is taken as part of

Table 3.3 Bivariate regression of wheat grain yield with a range of weather, agronomic and soil physical, chemical and biological properties for 40 paddocks in each of two study areas (A and B) in Western Australia.

		Area		Coefficient[a]	P-value[b]	% variability
	Attribute	**A**	**B**			**explained[c]**
Climate	Rainfall (mm)	211[a]	206[a]	–	ns	3.7
Agronomy	N fertiliser (kg N ha^{-1})	20[a]	24[a]	0.02	0.055*	9.4
Physical	Clay content[d] (%)	11.0[a]	10.4[a]	0.08	0.062*	9.1
Chemical	Total carbon (tC ha^{-1})	9.0[a]	10.8[b]	–	ns	0.2
	pH ($CaCl_2$)	5.7[a]	5.6[a]	–	ns	0.4
	EC[d] (mS m^{-1})	80[a]	63[b]	–	ns	0.1
Biological	Labile C (kg C ha^{-1})	83[a]	118[b]	0.01	0.041**	10.5
	Microbial biomass C (kg C ha^{-1})	107[a]	183[b]	0.01	0.001***	30.3
	PMN (kg N ha^{-1})	7.0[a]	10.1[b]	0.14	0.003***	21.2

Average grain yield was 1.76 and 3.24 t ha^{-1} in areas A and B, respectively. Mean values of attributes that contributed significantly to the explanation of wheat grain yield are presented; attributes followed by the same letter are not significantly different between the two areas. Several soil attributes were identified that did not have a significant (direct) influence on grain yield; but may have had an indirect influence through their effect on the size of the microbial biomass. In this case, 66% of the variability in microbial biomass could be explained by clay content (log transformed data), pH and labile carbon.

[a]The coefficient can be interpreted as t ha^{-1} grain yield change per unit change in attribute.

[b]The variability explained has a maximum of 100% and is not additive between individual attributes. Using a multiple regression model that included all parameters listed in the table, we were able to explain 42% of the yield variability.

[c]Clay and EC data were assessed using robust regression analysis due to unusual data points. EC, electrical conductivity.

*Significant at $P<0.10$;

**Significant at $P<0.05$;

***Significant at $P<0.01$.

ns, not significant.

Source: Unpublished data provided by D.V. Murphy.

fertiliser recommendation systems to give clear guidance on the pattern, depth, timing and frequency of sampling (Sabbe and Marx, 1987; Defra, 2010); such recommendations are fitted to different land use types and perhaps even different cultivation methods; e.g., increased use of no-till approaches requires careful attention to the depth of samples taken as increased profile stratification is seen in no-till systems (e.g. Ismail *et al.*, 1994). Increased potential precision in fertiliser application, through the use of machinery guided by global positioning systems, has led to an increase in the use of nutrient mapping within fields on large, intensive farms. Such 'Precision Farming' needs to be underpinned by a good understanding of the spatial variability of the inherent soil characteristics such as texture and an appreciation of the many factors that control soil fertility and crop growth and yield. Farmers that use precision farming techniques value them for things such as savings in fertiliser costs (minimal overlapping during application). However, Anderson-Cook *et al.* (1999) showed no significant advantage in detailed grid sampling approaches compared to the use of stratified sampling approaches designed with an awareness of soil-type patterns to guide

P and K fertiliser recommendations for precision farming approaches, especially given the significantly increased costs associated with the collection and analysis of soil samples in most grid sampling approaches.

3.4.2 Role of plant analysis for soil fertility assessment

Where long-term assessment of soil fertility is required or where plant stress can be addressed by management within the growing season, plant analysis rather than soil testing may provide an appropriate indicator of soil fertility. Within-season testing of plant tissues at particular growth stages may permit fertiliser treatments to be used to address critical deficiencies which otherwise might limit yield (Walworth and Sumner, 1986). For example, Havlin *et al.* (1999) provide a clear summary of the use of plant tissue testing to guide in-season fertiliser for a wide range of crops within the USA. The availability of soil nutrients to a plant is determined both by the nutrient absorption capacity of the plant root and nutrient supply and retention characteristics of the soil. Consequently, post-harvest analysis of yield and nutrient concentrations in crops allows the evaluation of the integrated impact of both plant and soil processes and the efficacy of soil fertility management, given the actual yield, weather conditions and achieved management interventions (all of which may differ from what had been planned).

'Smart plants' have been developed under research conditions and it may become practical to use them in the field to diagnose nutrient deficiencies or growth limitations due to other constraints, e.g. acidity, in real time. Such plants are genetically engineered by transformation with constructs containing the promoter of a gene up-regulated specifically under the target condition (e.g. P deficiency) in an accessible tissue. For example, Arabidopsis bearing such a construct showed increased β-glucuronidase activity was linked to a green fluorescent protein signal molecule and was detectable, using confocal microscopy, when P was withdrawn from solution culture before growth was limited (Hammond *et al.*, 2003). Hence, such 'smart plants' could have the potential to monitor plant response to previously identified limiting constraints and allow targeted precision management within the growing season before the limitation becomes a problem.

3.4.3 Soil analysis

While plant nutrient concentrations are useful post hoc evaluation tools, they are of little use for the advanced planning of nutrient applications to annual crops. Soil analysis can be used to provide an indication of soil fertility before any crop is planted and hence support the planning of the season's management tactics as well as supporting long-term monitoring of soil fertility and the effectiveness of soil fertility management strategies. Which soil properties should be analysed to give a good indication of soil fertility? Put simply, those which indicate characteristics that need to be understood or modified to achieve optimum yield, namely rooting conditions, water supply, nutrient supply and risk from soil-borne pathogens. However, the identification of simple robust measures that correlate closely with nutrient uptake and crop yield for all climates, crops and soil types has proved to be problematic. In part, this is a result of the complex interaction of limiting factors and constraints to crop growth in the field (Figure 3.1).

Scientific selection of indicator measurements for soil fertility requires that they can be measured easily and economically, as well as accurately and precisely, across a wide range of soil types and soil conditions and that they are based on reliable and contemporary scientific knowledge (Brookes, 1993; Stenberg, 1999). The appropriateness of the indicators chosen for soil fertility assessment depends on how well they are understood with respect to the agricultural system of which they are part and, consequently, how well they can be used to guide on-farm management decisions (Lal, 2001). To date, crop responses to soil nutrient contents (and additions) and soil acidity have generally been studied more intensively than crop responses to physical condition, levels of soil organic matter and other biological indicators.

As described in Chapter 2, the uptake of nutrients by a plant is a result of both the root absorbing power and the nutrient concentration in the soil solution; both plant and soil factors are important. The key soil factors affecting nutrient cycling in soil and supply to plants are discussed in detail later (for N see Chapter 5). However, soil solution concentrations can be difficult to quantify accurately due to very small pool sizes. In addition, determination of the concentration of nutrients contained within the soil solution on one occasion has often proved to be very weakly predictive of plant growth across the season (e.g. Murphy *et al.*, 2009). This is because soil solution concentrations are dynamic, changing from day to day, and because the buffering capacity of the soil solid phases in relation to each nutrient is critically important in determining the rate at which nutrients in the soil solution can be replenished. For example, snapshot measurements of soil mineral N (NO_3^- and NH_4^+ in solution and NH_4^+ held on soil surfaces) can only give a partial prediction of soil N supply for crop growth, since the bulk of soil N is found in organic forms; as part of a long-term monitoring scheme, the average amount of mineral N measured in topsoil during autumn in the UK was only 76 kg N ha^{-1} compared to 7000 kg N ha^{-1} in organic matter (Shepherd *et al.*, 1996). The interlocking biological processes controlling N supply from soil organic matter (e.g. mineralisation, immobilisation, nitrification, denitrification) resulting from the interaction of soil organisms, physical, chemical and biological characteristics modified by interactions with plant roots, weather and management are still not fully understood. As such, seasonal patterns of N availability remain difficult to predict and thus include in fertiliser decision support systems.

The development of methods for assessing the size of the pool of available plant nutrients for the next season of growth has usually focussed on developing an extraction procedure which removes a 'labile' pool and which integrates a measure of pool size and rate of supply. However, no single method can simulate these complex dynamic processes and so measurements of 'available soil nutrients' are usually selected based on correlation with plant growth in otherwise non-limiting conditions. In most countries one or two tests are favoured; in many cases, the tests which are used are as much a matter of the history of their development as of rigorous comparative scientific evaluation. For example, the determination of K availability is usually directly linked to the determination of exchangeable K; a number of extractants are considered to be equally effective in displacing K from the exchange complex (Munson, 1985; Syers, 1998). In contrast, a large number of tests to assess available soil P are in use around the world. The extractants used vary in their strength and mode of action (Sibbesen, 1983). No single test is suitable for all soil types and few routinely used tests give information about the availability of organic forms of P in soil. Because of the multiplicity of methods, there is increasing interest in the development of on-the-go analyses using a range of sensing technologies. Several such approaches are now at proof-of-concept stage (as reviewed by Sinfield *et al.*, 2010) to provide information on soil physical, chemical and biological properties rapidly in the field to support immediate, spatially variable applications of lime and/or fertiliser. Spatial patterns of some soil properties can be determined by mid-infrared sensors (Janik *et al.*,

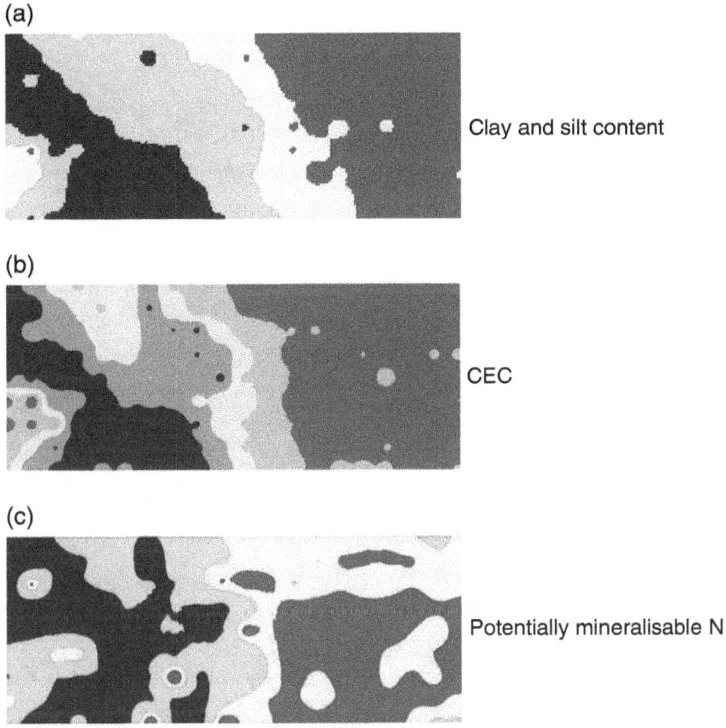

Figure 3.7 Spatial maps of (a) clay and silt content, (b) cation exchange capacity and (c) potentially mineralisable N (where the colours show quartiles ranges as an illustration) over a 10 ha area (200 × 500 m; 180 sampling points on a 25 × 25 m grid) where the properties were determined using mid infrared sensors. The spatial patterns were very strongly correlated with the same properties determined by traditional laboratory-based methods (r^2 = 0.91, 0.91 and 0.82, respectively). Unpublished data from the central wheatbelt, Western Australia, provided by D. V. Murphy. For a colour version of this figure, please see Plate 3.3.

1998; Figure 3.7a–c). This technology has the potential to be incorporated into rapid remote sensors that could be tractor mounted.

3.4.4 Interpretation of the results of soil analysis

The definition of justifiable target ranges or threshold values for soil nutrients and other properties relevant to soil fertility has also become a contentious issue in the development of interpretive frameworks to support soil quality monitoring (Arshad and Martin, 2002; Sparling *et al.*, 2003). Most of the detailed work carried out on relationships between soil fertility and crop growth has focussed on the relationship between yield and nutrient supply. There are data for relationships with other chemical properties e.g., acidity, and data are increasingly available for relationships between yield and physical properties (see Chapter 8). However, relatively few data are available to define suitability limits for soil biological properties for different crops, with the exception of some soil-borne diseases. Empirical relationships between yield and constraints usually show a curve of diminishing returns, so that as the constraint reaches non-limiting status the response to change decreases. Hence, absolute targets or thresholds are difficult to determine objectively.

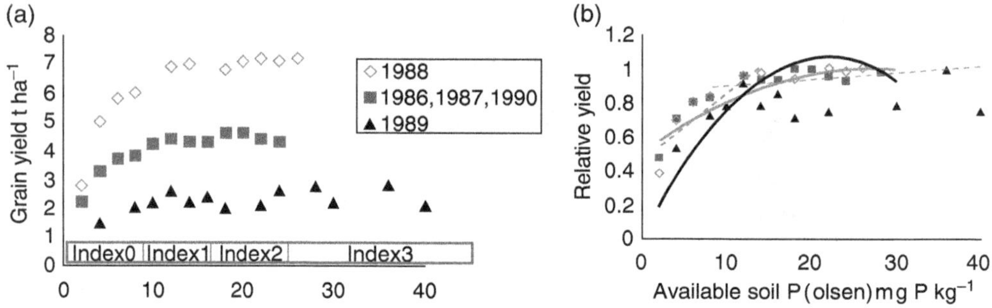

(c)

Fitted relationship (without constrained intercept) used from figure 3.8b	Available soil P (mg kg^{-1}) needed to achieve the target yield			
Operational relative yield	0.85	0.90	0.95	0.99
Curve	11.9	15.8	20.4	24.6
Two line fit (lines intersect at a relative yield = 0.898)	9.8	11.6	24.4	34.7

Figure 3.8 (a) Response of grain yield (t ha^{-1}) in spring barley to levels of available P (measured by bicarbonate extraction, Olsen) in a 5 year series of field experiments relating yields of spring barley to Olsen extractable P in the Hoos Barley Experiment at Rothamsted Research (silty clay loam soil) showing the critical ranges of soil P used to guide fertiliser recommendations, where Index 2 is considered the target range to provide sufficient soil P. (b) Response of grain yield for spring barley expressed as relative yield to normalise across years with available soil P. Response curves fitted to interpret the results and determine the optimal level of soil P using quadratic equations either with a constrained intercept = 0 (solid line) or a free intercept (dotted line) and with two straight lines with a free intercept (dotted line); these lines intersect at a relative yield of 0.898. In this case, the Olsen extract does not extract all plant available P; hence, it is not appropriate to constrain the responses to pass through the origin. (c) Soil available P values at a range of operational yield levels derived from the curve and straight line fits – operational yields of either 0.95 or 0.99 are commonly used to determine the optimum. Even small differences in the recommended level of available soil P can have large financial implications for the farmer; e.g., Defra (2010) indicates that increasing soil P by 10 mg kg^{-1} may need additional fertiliser applications totalling 850 kg ha^{-1} of triple superphosphate over a decade.

Sufficiency indices/non-limiting thresholds

Based on empirical observations of plant growth and yield responses to added nutrients, a common interpretation approach has been to categorise the range of nutrient availability into regions of low, adequate and high/excessive (Figure 3.8a). In the UK, this approach was used to develop the 'Index System' guidelines which provided an interpretation of soil test values with data from crop response experiments by Crowther and Yates (1941), and in the USA a similar approach was taken by Bray (1944, 1945, 1948). The use of relative yields in the analysis allows a better examination of the underlying relationship between soil supply and crop responses, even under different site and weather conditions (Evans, 1987). A range of curve-fitting approaches can then be used to support the interpretation of the data (Figure 3.8b). Because the Mitscherlich curve is asymptotic, the soil test value at a relative yield of either 0.99 or 0.95 is most often used to indicate the point at which there is no constraint to yield resulting from that nutrient under the typical management system in the field trials. The form of the response curve means that relative yields change slowly across a wide range of soil test values (Figure 3.8c), and interpretation of the data to give sufficiency levels to inform management depends on the underlying assumptions used. When errors associated with individual field experiments are taken into account, there is often no significant difference between fitting two intersecting straight lines to the response graph rather than a curve (Boyd *et al.*, 1976) as long as experiments are grouped by climate/soil/crop/management. The most appropriate interpretation approach also depends on the scientific understanding underpinning the development of the extraction methodology in the context of the soil-plant system; e.g., is the soil test effective in extracting all the available nutrient or does it extract only one part of plant available nutrient (hence, should the response curve be constrained to pass through the origin)? Hence decisions taken, when sufficiency ranges and thresholds are defined, are often more subjective than is realised. In the UK, expert groups (for P in 1962 and for K in 1963) were brought together to determine the most appropriate extraction methods for England and Wales and to identify soil test ranges linked to the likelihood of response to added fertilisers; the same ranges are still used in the current Fertiliser Manual (Defra, 2010; shown for P in Figure 3.8a) because no better, widely applicable method or framework for interpretation has been found. However, it is now important to consider whether, given the interactions of a number of constraints in the field and the increasing cost of inputs, it is still appropriate to set threshold values for all constraints at the highest relative yield levels with information derived from single factor response experiments. The same interpretative framework (sufficiency/non-limiting thresholds) can be used for calibration of multi-factor effects of soil constraints and management approaches – e.g., by using a response surface approach to support interpretation (Morton *et al.*, 2001; Bossa *et al.*, 2005).

The limiting factor/sufficiency approach can be applied to more than nutrients. For example, in Australia, recent developments in DNA-based assays, such as PredictaB® (South Australian Research and Development Institute), have enabled grain producers to determine inoculum levels of a broad range of soil-borne diseases before seeding and the data used to underpin decision support services for disease control such as the use of resistant varieties or a change in cropping patterns (Heap and McKay, 2009). Inoculum levels measured before sowing are usually correlated with root damage and crop biomass reduction and so can be interpreted as a disease risk rating. In most cases, high inoculum levels are expected to be associated with yield loss, but interactions during the growing season between weather conditions, crop nutrient status and soil type may confound this relationship (Heap and McKay, 2009).

Expert group approaches continue to be used to draw together a range of stakeholders with different focuses and expectations to define relationships for a wider range of soil properties and crop production systems and interpret them in terms of limiting or sufficiency ranges (Figure 3.9). Andrews *et al.* (2002) show how this was achieved for indicators of soil fertility in vegetable production systems in California and Lilburne *et al.* (2004) for indicators of soil quality in New Zealand. Participatory approaches with farmers are particularly important where the aim is to develop a scorecard for on-farm assessment and interpretation of soil properties affecting crop productivity (Romig *et al.*, 1995).

The sufficiency concept is the most common framework used for interpreting soil analysis for soil fertility. Its foundation is Liebig's law of the minimum. It is intuitively reasonable to assume that through management (irrigation, fertiliser, cultivation, etc.) the soil can be modified so that no limiting factors remain and weather conditions, together with crop genetic factors, become the primary limits to yield. Nonetheless, this interpretative approach has limitations for easily leached mobile nutrients (e.g. nitrate and sulphate) and in soils with very low cation exchange capacities that cannot retain sufficiently large

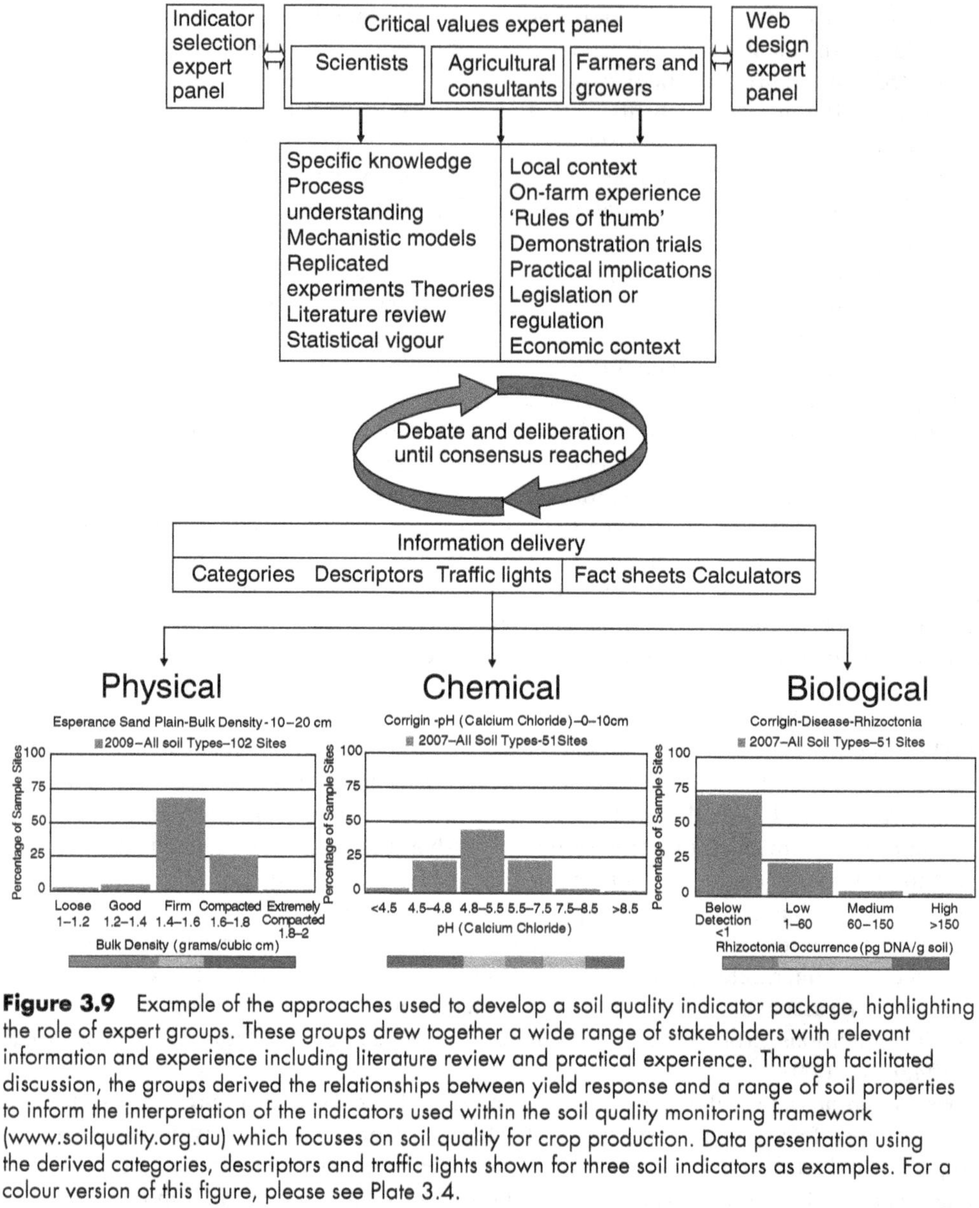

Figure 3.9 Example of the approaches used to develop a soil quality indicator package, highlighting the role of expert groups. These groups drew together a wide range of stakeholders with relevant information and experience including literature review and practical experience. Through facilitated discussion, the groups derived the relationships between yield response and a range of soil properties to inform the interpretation of the indicators used within the soil quality monitoring framework (www.soilquality.org.au) which focuses on soil quality for crop production. Data presentation using the derived categories, descriptors and traffic lights shown for three soil indicators as examples. For a colour version of this figure, please see Plate 3.4.

pools of cations such as K or magnesium (Mg). Interpretation also needs to take account of different crops; e.g., it is well known that cereals show yield limitation at a much lower level of soil test P than potatoes. Any sufficiency levels reported are inferred and not absolute and are consequently dependent both on the amount and quality of data available for analysis within key soil/crop/climate/management combinations and the approaches taken in data analysis (Figure 3.8). Changes in crop varieties and management practices may also mean that re-interpretation of sufficiency levels is required: e.g., a decline in the residual soil test P measured in row crops in Alabama was not accompanied by a decrease in yields; instead yields were found to have increased despite low soil P levels (Evans, 1987).

Wortmann *et al.* (2009) showed that the critical level of P for maize was affected by the previous crop with a critical level of 10 mg kg^{-1} (Bray extract) where maize followed soybean and 20 mg kg^{-1} where maize followed maize; they recommended that soil organic matter levels and pH be used together with soil test P values to improve soil fertility management decisions.

Base cation saturation ratios

Interpretation of soil tests on the basis of cation saturation is a common parallel approach to the use of sufficiency thresholds – especially in the USA. The basic cation exchange phenomena in soil which control the availability of calcium, magnesium and potassium and the impacts that the degree of saturation of one cation can have on the plant availability of itself and other cations are discussed in Chapter 6. Kopittke and Menzies (2007) highlight the historical development of the approach (often known as Albrecht analysis/interpretation), which resulted in the recommendation by Bear *et al.* (1945) that for an ideal soil 65% of the exchange complex should be occupied by Ca, 10% by Mg, 5% by K and 20% by hydrogen; it was considered that where this cation balance was achieved, chemical, physical and biological properties would all be in an appropriate equilibrium to support optimum plant growth. Review of the literature by Eckert (1987) and Kopittke and Menzies (2007) suggests that fairly wide variations in actual ratios are of little consequence for crop yield, quality or a wide range of soil physical and biological properties, as long as gross imbalances are not created. Consequently, sole focus in the interpretation of soil fertility on a target Ca/Mg ratio often leads to recommendations for management that are extremely expensive and have little impact on crop productivity or profitability.

Testing for mobile nutrients, e.g. nitrate

Measurement of soil mineral N (i.e. nitrate + ammonium, but often only nitrate because amounts of ammonium present are usually small) before planting or at a specific time during the crop growing season can be interpreted under certain conditions to give an indication of likely soil N supply. In climates without excessive overwinter leaching, the use of a pre-planting test for soil nitrate is widely used; in the western USA before 1982, 16 states were using such a test (Keeney, 1982). Values obtained for nitrate before planting were compared to the estimated crop N demand to obtain the likelihood of response to additional fertiliser. A pre-sidedress nitrate test has been developed for maize (Magdoff *et al.*, 1984), which allows a period of nitrate accumulation under field conditions before sampling takes place, and therefore also reflects the mineralisation potential under field conditions. Similar approaches are used in many countries in Europe and these were summarised by Mengel (1991). Measurement of soil mineral N can be combined directly with analysis of N already in the crop (i.e. that already supplied by the soil) to predict seasonal soil nitrogen supply and thus guide fertiliser requirements (Defra, 2010). However, research has shown that this single 'snapshot' in time is often very unrepresentative of soil nitrogen supply (Knight *et al.*, 2008; Murphy *et al.*, 2009). The UK Fertiliser Manual (Defra, 2010) consequently recommends that sampling and analysis of soil mineral N is used to give guidance '... where the supply of plant available nitrogen in the soil could be unusually large, and particularly where organic manures have been used regularly in recent years', not as a precise prediction tool.

3.5 Managing soil fertility

Where yield is constrained by soil factors, perhaps diagnosed by plant or soil analysis, the next step is to determine what action might be taken by the land manager to adjust their rotation to optimise productivity/profit given the constraint and/or to implement management strategies that mitigate the constraint. Management strategies might seek to improve performance over the medium to long term (e.g. to increase effective rooting depth by improving drainage, removal of pans by cultivation and use of deep-rooted crops within the rotation) or seek an immediate impact (though to improve soil depth), 'shattering the underlying rock or hard layers with explosives [is] not in general use' (Russell, 1937). Local knowledge and farmer understanding of soil fertility, as indicated by crop performance, weed species and diversity and observable soil properties have been used to select sites which are most likely to respond to improved soil management practices (Suarez *et al.*, 2001; Barrios and Trejo, 2003; Mairura *et al.*, 2008). Farmers' understanding of their own soils is strongly orientated towards easily observed characteristics that affect production, such as texture (heavy or light land for cultivation), waterlogging and drainage after heavy rain, nutrient retention (hungry soils) and the structural condition of the soil (tilth, heart). Such variations are often the reasons why field boundaries occur where they do, and the fields may have indicative names, e.g. Sweetlea, Hungry field, Starvall. Integration of local knowledge with technical information can provide a foundation from which to determine key local constraints or to identify which potential management practices could be used at each site (Barrios *et al.*, 2006). Careful targeting of proposed changes in soil management approaches to key crops and the selection of rotational systems to match geographic, market and social niches is needed to maximise effectiveness of any recommended interventions and to increase uptake by farmers (Ajayi *et al.*, 2007).

3.5.1 Development of recommendations for targeted inputs

Most of the research supporting soil fertility management has focussed on fertiliser and lime recommendations, as both of these practices seek to provide an economic return to farmers over the timescale of a crop rotation, if not more quickly. 'If the soil test falls below the sufficiency level, how much fertiliser should be applied?' is the primary question, but there are a wide range of further questions needed to support management decisions to inform choices about the type, timing and placement of fertiliser in practice. Fertiliser type often affects the availability of the nutrients added; so, e.g., there is a wide range in concentration and availability of P from diammonium phosphate, triple superphosphate, farmyard manure and phosphate rock (Table 3.4). Soil factors, not least pH and moisture content at application, also affect the availability of nutrients from fertiliser additions.

Often the same or a similar series of yield response experiments are used to determine optimum fertiliser application rates post hoc, as described in Section 3.2.1 for the determination of suitability ranges (Figure 3.8). However, for the development of fertiliser recommendations, economic considerations are central, with the determination of the optimum fertiliser application from a response curve determined in part by the ratio between the price of a unit of fertiliser nutrient and the saleable value of the crop; hence, actual rather than relative yields are used. Where environmental impacts are also an important consideration, the risk of nutrient loss (e.g. by leaching or denitrification) may also be taken into account alongside economic considerations. A wide range of response models can be

Table 3.4 Typical composition of some common fertilising materials; all organic fertilisers are very variable and consequently these data provide only an indication of likely composition.

	%			
	N	**P**	**K**	**Other notes**
Ammonium sulphate	21			24% S
Ammonium nitrate	32–34.5			
Urea	46			
Diammonium phosphate	21	23 (soluble)		
Triple superphosphate		20 (soluble)		14% Ca
Rock phosphate		12–16 (slowly soluble)		31–35% Ca
Potassium chloride			50	
Potassium sulphate			42	17% S
Cattle farmyard manure	0.6; 10% readily available	0.08 (soluble)	0.6 (soluble)	25% DM 0.1% S
Pig slurry	0.4%, 70% readily available	0.04 (soluble)	0.2 (soluble)	4% DM, 0.04% S
Poultry (broiler) litter	3%, 33% readily available	0.65 (soluble)	1.3 (soluble)	60% DM, 0.3% S
Greenwaste compost	0.75%, 3% readily available	0.06 (soluble)	0.4 (soluble)	60% DM, 0.05% S

Source: Data adapted from Defra (2010).

used but each is likely to provide a different estimate of the optimum fertiliser rates, and it is most important that a model is chosen so that it is appropriate to the data available. A good response model should (Cooke, 1982):

1. define both the slope of the rising part of the curve and the slope beyond the peak;
2. define accurately the yield near the optimum dressing and the dressing itself.

Actual yields and profitability are likely to have been affected by a wide range of factors other than those varied in the experiment, especially if the data are collated from a series of experiments carried out at different sites and in different years. Consequently, response curves are only valid as the basis of advice to farmers under the same conditions as used in the experiments. Although it is possible to compile tables of average responses of crops to fertilisers at various soil sufficiency levels, it is not clear that these provide anything other than very general guidelines to farmers.

For less mobile nutrients (e.g. P and K), fertiliser recommendation systems now commonly use the soil sufficiency level/range as a target value and develop a fertiliser policy which seeks to achieve/maintain this value (on a rotational basis), i.e. fertilising the soil. This approach often recommends applications of small amounts (maintenance dressings) of fertiliser, even where soil test values are above the sufficiency level, to replace the amounts likely to be removed by the crop to be grown. Olson *et al.* (1982) reported on a long-term comparison study of fertiliser recommendations based purely on the sufficiency approach (applications only where fertiliser response was likely) and those made by a number of commercial laboratories using a build/maintenance approach. No significant difference in crop yields were recorded, but fertiliser costs where a maintenance approach was used were on average twice those where fertiliser was applied on a sufficiency basis. Recommendations in the UK state that 'Once deficiency has occurred, a fresh application of P and K is most unlikely to be available for uptake by roots in time to benefit the crop being grown' (Defra, 2010). In addition, even in the low range below the soil sufficiency index, it has been observed that up to 40% of fields may not respond profitably to fertiliser application (Havlin *et al.*, 1999), in part due to nutrient supply from pools not accounted for in the soil test, or available nutrients below the sampling depth (Davis *et al.*, 1996), as well as the impact of other limiting factors in the year of observation. In farming systems with relatively stable crop rotations and production systems, it should be possible to couple plant and soil analysis for P and K to provide indicators of the effectiveness of management practices, rather than using such data to derive fertiliser recommendation *per se*. This type of farm-specific monitoring should allow modification of management underpinned by an understanding of the principles governing P and K availability (Chapters 5 and 6) rather than the application of average recommendations.

For mobile nutrients, such as N, fertiliser recommendations are usually based on yield potential; these are most often expressed as a farm/field-specific yield goal. This yield goal is used to determine the likely crop N demand in the season. A range of information can be used to guide the farmer's yield goal, including previous yields achieved at the site. The requirement to predict the likely weather, disease risk and other factors potentially impacting on crop growth is explicitly placed with the farmer or their advisor. In dryland agriculture, it is usually considered advantageous to set the grain yield goal above that of average yields to fully take advantage of the potential for above-average growing conditions, e.g. above-average rainfall (Johnson, 1991). However, Hergert (1987) highlighted the need to improve yield goal determination for fertiliser recommendations as 50% of farmers attained yields equivalent to only 80% of their yield goal. For N fertiliser, if the yield goal is too high, fertiliser recommendations will overestimate requirements, farmers will waste money and there is an increased risk of N loss to the environment. Development of new technological approaches that can be used during the season in the field (e.g. canopy reflectance, Raun *et al.*, 2001) may mean that it becomes increasingly possible to predict grain yield in-season and consequently adjust the fertiliser recommendations based on projected N demand.

Fertiliser N requirements are calculated from the crop requirement by adjusting for estimates of the soil N supply corrected for in-season losses:

$$(\text{Amount of N Fertiliser Needed}) = (\text{Crop N Requirement}) - (\text{Soil N Supply}) + (\text{N Losses})$$

and then adjusting for any nutrients to be supplied from manures, biosolids, etc. Improved fertiliser recommendations therefore depend in part on the ability to estimate N mineralisation and N losses (Mulvaney *et al.*, 2001; Zebarth *et al.*, 2009), and also amounts and rates of nutrients supplied from manures and other organic materials, e.g. animal manures, urban green waste composts, etc. Cooke (1982) outlined how the range of factors affecting crop response to nitrogen fertiliser had been integrated into an N index system. This used information on previous cropping and manuring to predict the likely response of sites to nitrogen fertiliser, i.e. providing an adjustment for soil N supply based on data collated from many years of field experiments (e.g. MAFF, 1996).

The development of models of soil nutrient dynamics (Shaffer *et al.*, 2001) may allow these to be integrated into decision support systems for fertiliser recommendations. This may be: (1) on a static basis, so that the model is run once to provide information on predicted soil N supply or losses, or (2) dynamically in which the model is run weekly or monthly throughout the growing season and the later applications of fertiliser are adjusted for model-predicted crop requirements, soil N supply and losses. Using such a system, fertiliser recommendations may be able to be revised to allow for the differences from average weather conditions or crop growth. However, even with such dynamic systems, it is not realistic to expect that fertiliser recommendations for the current crop will ever exactly match the optima measured at harvest due to the influence of the (largely) unpredictable weather on crop growth and soil processes. The use of models also provides the opportunity to predict the nitrogen response curve for a site; this information can be used to determine how sensitive crop yield is to small changes in rates or timings of fertiliser application. See an example in Figure 3.10.

3.5.2 Developing recommendations for integrated soil fertility management

There is an increasing use of broader integrated approaches to the assessment of soil fertility which consider the impact on plant growth of the emergent properties and processes arising from ecosystem-level interactions of plants/roots with physical, chemical and biological factors in the soil. It is therefore clear that soil fertility management cannot be restricted to advice on the seasonal management of nutrient, water inputs or even the use of green manures. Soil and water conservation measures, crop breeding and rotation design, tillage, drainage and pest management strategies together with farm landscape design (presence, composition and location of shelter belts, buffer strips, hedges) may all form important components in the integrated management of the below-ground ecosystem (Figure 3.11). In the same way that soil fertility is one aspect of a more integrated definition of soil quality (Section 3.2.3), integrated soil fertility management can be considered as one aspect of broader strategies for integrated natural resource management (Vanlauwe *et al.*, 2006).

Production practices such as the integration of reduced tillage, retention of crop residues, targeted planting and farm input use and the use of legumes in rotation (conservation farming approaches), legume agroforestry (fertiliser tree) systems, the integration of green manure and dual purpose legumes and the careful use of animal manures (Sanchez and Salinas, 1981; Palm *et al.*, 2001) are common where more integrated approaches to soil management are used. All combine inputs of organic matter with other changes to management systems. Because of the role of decomposition in providing energy to the soil ecosystem, regular inputs of organic matter can change the biomass, activity and diversity of soil organisms driving changes to a wide range of soil physical and chemical properties. Hence, responses to additions of organic materials to cropping systems are complex and interactive and rarely can be explained by a simple causal chain. In addition, there is increasing interest in the use of soil inoculants or plant sprays developed as plant growth stimulants or biopesticides (see Section 3.3.2). Rupela *et al.* (2006) reported findings of an experiment (established in 1999), which investigated the role of plant biomass (rice straw or farm waste combined with hedge clippings) as an input, together with applications of rock phosphate, biopesticides and reduced tillage in a rainfed cropping system on a Vertisol in the semi-arid topics (India). After a first year in which low-cost input/biologically based systems had lower yields (attributed to the lag phase in release of N and P from complex organic inputs), crop yields were close to those in optimally managed conventional plots, and net income to farmers was significantly higher due to the lower cost of inputs. N and P balances were positive and consequently there was no nutrient depletion; soil organic matter, microbial biomass C and

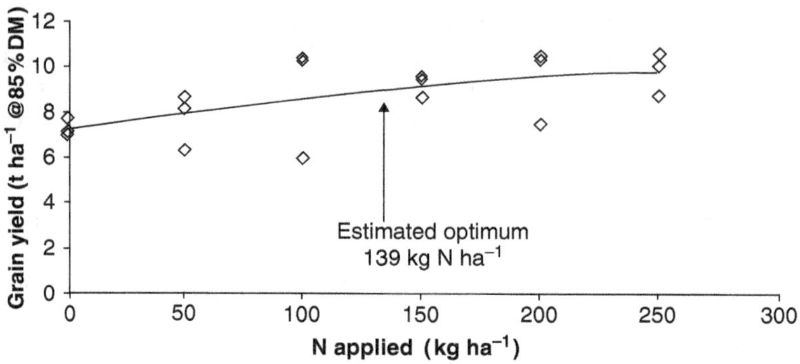

Method	Optimum*	RB209	SUNDIAL	
			Default weather and yield	Actual weather and yield
Recommendation	139	182	162	141
Calculated yield*	9.1	9.5	9.3	9.1
%difference from optimum yield		4	2	0

**Calculated from linear plus exponential curve fitted to trial data.*

Figure 3.10 Evaluation of a dynamic model (SUNDIAL) for fertiliser recommendation (Smith *et al.* 1996). In 1997, a project was initiated to establish a series of N response trials on working farms across the UK and to evaluate the performance of model recommendations across a range of typical crops and soils. Data are presented from one of these sites as an example. The site was a sandy clay loam soil in Suffolk, Eastern England. Winter oilseed rape (cv. Navajo) was harvested on 25 July 1998 and yielded 4.3 t ha^{-1} (@ 90% DM). Crop residues were chopped and incorporated and 40 t ha^{-1} of farmyard manure taken from the piggery was applied just before ploughing. Winter wheat (cv. Equinox) was drilled on 3 October 1998. The farmer was expecting a yield of 11–12 t ha^{-1}. The N response curve obtained in the field (three replicate plots of 6 N rates) was fairly flat and the replicate plots quite variable due to the N supplied by the pig manure. The response curve was fitted by a linear plus exponential relationship (Genstat) with an optimum estimated at 139 with a standard error of 17 kg N ha^{-1}. The actual yield obtained in the field was much lower than the farmer's target only reaching about 9.5 t ha^{-1}. The standard UK reference book *Fertiliser Recommendations for Agricultural and Horticultural Crops* (RB209; MAFF, 1996) recommended an application of 182 kg N ha^{-1}, while predictive use of the SUNDIAL model recommended 162 kg N ha^{-1}. The SUNDIAL recommendation was not significantly different from the measured optimum, while the RB209 recommendation was a significant overestimate. However, this over-application of N had little effect on the yield achieved, but may have increased the residual mineral N available for leaching after harvest. In plots that had received 250 kg N ha^{-1}, mineral N was increased very significantly above those plots which had received N applications close to the optimum. When SUNDIAL was used retrospectively, i.e. with the actual weather and the target yield set at the actual optimum achieved, the SUNDIAL recommendation almost exactly predicted the measured optimum. It is important that SUNDIAL achieves good retrospective predictions as this demonstrates that the model can be used reliably to carry over information from one year to the next without introducing significant errors.

N levels were increased in the treatments receiving additional C inputs. Such integrated system experiments are of considerable value in demonstrating how the principles underpinning integrated soil fertility management can be implemented in practice within inherent site constraints. Best practice may well be region and even site specific, e.g. the local availability of the biomass and biopesticides in the earlier example. However, such system management experiments are often the despair of scientists seeking to understand the

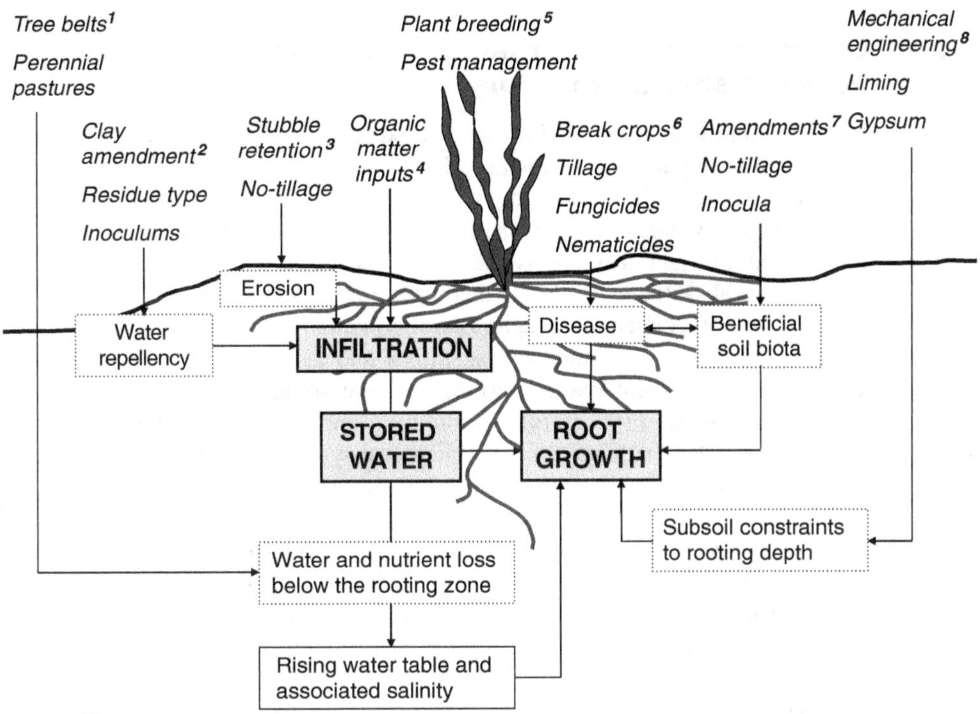

Figure 3.11 Developing integrated strategies for effective capture and utilisation of rainfall together with management of soil water is critical to achieve crop growth in dryland cropping systems (Tow *et al.*, 2012). In South Australia, a range of soil management approaches are integrated within farming systems so that the potential yields (defined by available water in any season) can be attained. These include (1) introduction of tree belts (Robinson *et al.*, 2006) and deep rooting perennial pastures to slow a rising (often saline) water table. (2) While wax degrading micro-organisms have been isolated (Roper, 2004), their success as inocula under field conditions has been limited. Instead clay amendment is an economically viable option (Hall *et al.*, 2010) to overcome water repellence where it can be sourced locally (e.g. from subsoil on the same farm). (3) No-tillage and stubble retention practices are now commonplace to manage erosion and ensure maintenance of soil carbon levels and associated soil biological function (Hoyle and Murphy, 2006). (4) Organic matter stabilises soil aggregates and improves water infiltration into soil by contributing to the development of a more porous soil structure (Hoyle *et al.*, 2012). (5) Plant breeding for improved water use efficiency (more grain per mm plant available water) and tolerance to toxicity (e.g. Al; Tang *et al.*, 2003a). Early vigour lines (Ludwig and Asseng, 2010), which enable roots to track early season water and nutrients in leaching soils, also have the benefit of growing the root tip through surface soil constraints (e.g. diseases) quicker. (6) Development of disease-suppressive soils has occurred through crop management and rotation (Wiseman *et al.*, 1996). (7) Amendments to soil can have either positive or negative effects on the soil biota (Bunemann *et al.*, 2006) which can impact on the level of micro-organisms involved in carbon and nutrient cycling. (8) Deep ripping to remove compaction and at same time including gypsum or lime (Tang *et al.*, 2003b) to depth to ameliorate subsoil constraints. Use of controlled traffic (Kingwell and Fuchsbichler, 2011) to stop vehicle movement over the majority of the soil surface and tractors with tank tracks instead of wheels to minimise compaction. For a colour version of this figure, please see Plate 3.5.

ecological interactions underpinning soil fertility and their links to individual component management practices. In general, in such systems, scientists have been very effective in cataloguing the impact of management practice changes on the root/soil system (chemical, physical and biological properties, e.g. Mäder *et al.* (2002) and Cookson *et al.* (2008)) but much less effective in developing management systems that achieve targeted changes.

3.5.3 Challenges for soil fertility management in the twenty-first century

As already noted, climate change coupled with increased costs and declines in the availability of many of the common inputs will necessitate changes to soil fertility management in the twenty-first century. Reduced tillage, legumes and other N-fixing plants as intercrops, green manures or as hedgerow trees and the selection of crops and varieties to promote rhizosphere-mediated nutrient availability have already been integrated into so-called 'sustainable' agricultural systems. It is likely that these and similar approaches will become a more common part of conventional agricultural systems as the twenty-first century progresses. These changes will require soil scientists and agronomists in the developed world to learn from their counterparts working in developing countries where farmer participatory research has been successful in promoting improved soil fertility, crop yield and farmer livelihoods. The challenge for all agricultural systems is to achieve sustainable intensification; this will require increased focus on the use of biological processes and potentials within the soil-plant system.

Simple single-factor response experiments (or even multifactorials), as commonly used for fertiliser trials, have little relevance in the evaluation of integrated soil fertility management approaches. Each input (e.g. compost, inoculant or new crop variety) or management practice (e.g. change in depth of tillage, incorporation of crop residues or green manures) has interacting impacts on multiple organisms and on the soil physical and chemical properties defining their habitats. Consequently, though there is a myriad of data cataloguing the impact of many of these practices on soil physical, chemical and biological practices, there is little systematic evaluation of these integrated management systems on the maintenance of soil fertility, as broadly defined, except on a case-by-case basis. Understanding how soil-plant systems respond to management and climate will require a more ecologically based approach to experimental design and analysis. Furthermore, experiments will need to be maintained for sufficient years to enable the resistance and resilience of systems to be determined and the underpinning soil–plant interactions to establish.

Integrated soil fertility management approaches often have at their heart the incorporation of organic inputs to the soil-plant system. However, if these systems are to become more widely adopted, consideration is needed of how the organic matter inputs required can be sustained over decades. The plant is the conduit bringing energy into the soil ecosystem; hence in natural ecosystems, net primary production is often strongly linked to the biomass and activity of the soil organisms. In many environments, effective capture of solar radiation is limited by water availability; in 'conventional' agricultural systems, plant breeding has focussed on increasing the biomass of the marketable component (e.g. grain, fruit) at the expense of crop residues, hence limiting or reducing inputs. A combination of rotational planning (exploiting the benefits of plant diversity in space and time) and plant breeding approaches (both to manipulate rhizosphere inputs, to increase amounts or manipulate quality of residues) will be needed to overcome this potential constraint for integrated fertility management approaches in the future.

The specialisation of soil science researchers has often led to their dissociation with the practical implementation of soil fertility; one might see the application of molecular methods to the analysis of soil microbial function as a particular example of this; function must be linked to process. The soil-plant ecosystem functions as a result of the complex interactions between soil organisms and their habitats, so the effective development of soil fertility management in the twenty-first century should be underpinned by collaborative interdisciplinary investigations emerging from dialogue amongst researchers, extension specialists and farmers/growers. Integrated soil fertility management should be at the heart of a global approach to improving the sustainability of agriculture, its impact on food security and the environmental impact of the industry. This will require the development of new plant genotypes, management systems, indicators and recommendations.

References

Agricultural Advisory Council of Great Britain. (1970) *Modern Farming and the Soil: Report of the Agricultural Advisory Council on Soil Structure and Soil Fertility*. Her Majesty's Stationery Office, London.

Ajayi, O.C., Akinnifesi, F.K., Sileshi, G., and Chakeredza, S. (2007) Adoption of renewable soil fertility replenishment technologies in the southern African region: lessons learned and the way forward. *Natural Resources Forum*, **31**, 306–317.

Anderson-Cook, C.M., Alley, M.M., Noble, R., and Khosla, R. (1999) Phosphorous and potassium fertilizer recommendation variability for two mid-Atlantic Coastal Plain fields. *Soil Science Society of America Journal*, **63**, 1740–1747.

Andrews, S.S., Karlen, D.L., and Mitchell, J.P. (2002) A comparison of soil quality indexing methods for vegetable production systems in Northern California. *Agriculture, Ecosystems and Environment*, **90**, 25–45.

Angers, D.A. and Caron, J. (1998) Plant-induced changes in soil structure: processes and feedbacks. *Biogeochemistry*, **42**, 55–72.

Ardanov, P., Ovcharenko, L., Zaets, I., Kozyrovska, N., and Pertila Anna, M. (2011) Endophytic bacteria enhancing growth and disease resistance of potato (*Solanum tuberosum L.*). *Biological Control*, **56**, 43–49.

Arshad, M.A. and Martin, S. (2002) Identifying critical limits for soil quality indicators in agro-ecosystems. *Agriculture Ecosystems and Environment*, **88**, 153–160.

Atwater, W.O. (1884) On the assimilation of atmospheric nitrogen by plants. *Report of the British Association for the Advancement of Science*, **54**, 685.

Bardgett, R.D. (2002) Causes and consequences of biological diversity in soil. *Zoology*, **105**, 367–374.

Barrios, E. and Trejo, M.T. (2003) Implications of local soil knowledge for integrated soil management in Latin America. *Geoderma*, **111**, 217–231.

Barrios, E., Delve, R.J., Bekunda, M., Mowo, J., Agunda, J., Ramisch, J., Trjo, M.T., and Thomas, R.J. (2006) Indicators of soil quality: a south-south development of a methodological guide for linking local and technical knowledge. *Geoderma*, **135**, 248–259.

Bear, F.E., Prince, A.L., and Malcolm J.L. (1945) The potassium needs of New Jersey soils. *New Jersey Agriculture Experimental Station Bulletin*, 721.

Beatty, M.T., Petersen, G.W., and Swindale, L.D. (eds) (1979) *Planning the Uses and Management of Land*, Agronomy Monograph 21, American Society of Agronomy, Madison, Wisconsin.

Berg, G. (2009) Plant-microbe interactions promoting plant growth and health: perspectives for controlled use of microorganisms in agriculture. *Applied Microbiology and Biotechnology*, **84**, 11–18.

Berg, G. and Smalla, K. (2009) Plant species and soil type cooperatively shape the structure and function of microbial communities in the rhizosphere. *FEMS Microbiology Ecology*, **68**, 1–13.

Bethlenfalvy, G.J. and Lindermann R.G. (eds) (1992) *Mycorrhizae in Sustainable Agriculture*. American Society of Agronomy Special Publication 54, Madison, Wisconsin.

Bonkowski, M., Cheng, W.X., Griffiths, B.S., Alphei, G., and Scheu, S. (2000) Microbial-faunal interactions in the rhizosphere and effects on plant growth. *European Journal of Soil Biology*, **36**, 135–147.

Bossa, J.R., Shannon, D.A., Adams, J.F., Mullins, G.L., and van Santen, E. (2005) Phosphorus, potassium and leucaena pruning interactions on corn in Haiti. *Nutrient Cycling in Agroecosystems*, **73**, 37–47.

Boyd, D.A., Lowsing, T.K.Y., and Needham, P. (1976) Nitrogen requirements of cereals. 1. Response curves. *Journal of Agricultural Science*, **87**, 149–162.

Bray, R.H. (1944) Soil–plant relationships I. The quantitative relation of exchangeable K to crop yields and to crop response to potash additions. *Soil Science*, **58**, 305–324.

Bray, R.H. (1945) Soil-plant relationships II. Balanced fertilizer use through soil test for K and P. *Soil Science*, **60**, 464–473.

Bray, R.H. (1948) Correlation of soil tests with crop response to fertilizers and with fertilizer requirement. In: *Diagnostic Techniques for Soils and Crops* (ed. H.B. Kitchen), pp. 53–86. America Potash Institute, Washington, D.C.

Brewin, N.J. (1991) Development of the legume root nodule. *Annual Review of Cell Biology*, **7**, 191–226.

Brookes, P.C. (1993) The potential of microbiological properties as indicators in soil pollution monitoring. In: *Soil Monitoring. Early Detection and Surveying of Soil Contamination and Degradation* (eds R. Schulin, A. Desaules, R. Webster, and B. von Steiger), pp. 229–254. Birkhäuser Verlag, Basel, Switzerland.

Bruehl, G.W. (1987) *Soil Borne Plant Pathogens*. Macmillan, New York.

Brussaard, L. (1998) Soil fauna, guilds, functional groups and ecosystem processes. *Applied Soil Ecology*, **9**, 123–135.

Brussaard, L., Bakker, J.-P., and Ollf, H. (1996) Biodiversity of soil biota and plants in abandoned arable fields and grasslands under restoration management. *Biodiversity and Conservation*, **5**, 211–221.

Bunemann, E.K., Schwenke, G.D., and Van Zwieten, L. (2006) Impact of agricultural inputs on soil organisms – a review. *Australian Journal of Soil Research*, **44**, 379–406.

Chineau, C.H., Morel, J.L., and Oudot, J. (2000) Biodegradation of fuel oil hydrocarbons in the rhizosphere of maize. *Journal of Environmental Quality*, **29**, 569–578.

Coleman, D.C., Reid, C.P.P., and Cole, C.V. (1983) Biological strategies of nutrient cycling in soil systems. *Advances in Ecological Research*, **13**, 1–55.

Coleman, D.C., Hendrix, P.F., and Odum, E.P. (1998) Ecosystem health: an overview. In: *Soil Chemistry and Ecosystem Health* (ed. P.H. Wang), pp. 1–20. Soil Science Society of America Special Publication 52, SSSA, Madison, Wisconsin.

Cooke, G.W. (1982) *Fertilizing for Maximum Yield*, Collins, London.

Cookson, W.R., Murphy, D.V., and Roper, M.M. (2008) Characterizing the relationships between soil organic matter components and microbial function and composition along a tillage disturbance gradient. *Soil Biology and Biochemistry*, **40**, 763–777.

Costanza, R., d'Arge, R., de Groot, R., Farber, S., Grasso, M., Hannon, B., Limburg, K., Naeem, S., O'Neill, R.V., Paruelo, J., Raskin, R.G., Sutton, P., and van den Belt, M. (1997) The value of the world's ecosystem services and natural capital. *Nature*, **387**, 253–260.

Crowther, E.M. and Yates, F. (1941) Fertiliser policy in wartime: the fertilizer requirements of arable crops. *Empire Journal of Experimental Agriculture*, **9**, 77–97.

Da Silva, A.P., Kay, B.D., and Perfect, E. (1994) Characterization of the least limiting water range of soils. *Soil Science Society of America Journal*, **58**, 1775–1781.

Davis, J.G., Walker, M.E., Parker, M.B., and Mullinix, B. (1996) Long-term phosphorus and potassium application to corn on coastal plain soils. *Journal of Production Agriculture*, **9**, 88–94.

Defra. (2010) *Fertiliser Manual (RB209)*. TSO, London.

Del Pozo, J.C., Allona, I., Rubio, V., Leyva, A., De la Pena, A., Aragoncillo, C., and Paz-Ares, J. (1999) Type 5 acid phosphatase gene from Arabidopsis thaliana is induced by phosphate starvation and by some other types of phosphate mobilising/oxidative stress conditions. *Plant Journal*, **19**, 579–589.

Dennis, P.G., Smith, S.J., Hirsch, P.R., Taylor, R., Valsami-Jones, E., and Miller, A.J. (2009) Linking rhizoplane pH and bacterial density at the microhabitat scale. *Journal of Microbiological Methods*, **76**, 101–104.

Doran, J.W. and Parkin, T.B. (1994) Defining and assessing soil quality. In: *Sustainable Environment* (ed. J.W. Doran), pp. 3–21, 53–64. Soil Science Society of America Special Publication 35, SSSA, Madison, Wisconsin.

Drew, M.C. (1975) Comparison of the effects of a localized supply of phopshate, nitrate, ammonium and potassium on the growth of the seminal root system, and the shoot, in barley. *New Phytologist*, **75**, 479–490.

Duffy, B.K., Ownely, B.H., and Weller, D.M. (1997) Soil chemical and physical properties associated with suppression of take-all of wheat by Trichodema koningii. *Phytopathology*, **87**, 1118–1124.

Eckert, D.J. (1987) Soil test interpretations: basic cation saturation ratios and sufficiency levels. In: *Soil Testing: Sampling, Correlation, Calibration, and Interpretation* (ed. J.R. Brown), pp. 53–64. Soil Science Society of America Special Publication 21, SSSA, Madison, Wisconsin.

Evans, C.E. (1987) Soil test calibration. In: *Soil Testing: Sampling, Correlation, Calibration, and Interpretation* (ed. J.R. Brown), pp. 23–30. Soil Science Society of America Special Publication 21, SSSA, Madison, Wisconsin.

Food and Agriculture Organisation, FAO. (1976) *A Framework for Land Evaluation*. FAO Soils Bulletin No. 32, Rome, Italy.

Food and Agriculture Organisation, FAO. (1996) *Agro-Ecological Zoning: Guidelines*. FAO Soils Bulletin 73. Accessed online at: www.fao.org/docrep/W2962E/W2962E00.htm on 27/7/11.

Forde, B. and Lorenzo, H. (2001) The nutritional control of root development. *Plant and Soil*, **232**, 51–68.

Foth, H.D. (1997) *Soil Fertility*, 2nd edn. CRC Lewis, Boca Raton, Florida.

George, T.S., Richardson, A.E., Hadobas, P., and Simpson, R.J. (2004) Characterisation of transgenic *Trifolium subterraneum* L. which expresses phyA and releases extracellular phytase: growth and P nutrition in laboratory media and soil. *Plant Cell and Environment*, **27**, 1351–1361.

George, T.S., Richardson, A.E., Smith, J.B., Hadobas, P.A., and Simpson, R.J. (2005) Limitations to the potential of transgenic *Trifolium subterraneum* L. plants that exude phytase, when grown in soils with a range of organic P content. *Plant and Soil*, **278**, 263–274.

Gibbon, F.R. (1961) Some misconceptions about what soil surveys can do. *Journal of Soil Science*, **12**, 96–100.

Giller, K.E., Beare, M.H., Lavelle, P., Izac, A.M.N., and Swift, M.J. (1997) Agricultural intensification, soil biodiversity and ecosystem function. *Applied Soil Ecology*, **6**, 3–16.

Gosling, P., Hodge, A., Goodlass, G., and Bending, G.D. (2006) Arbuscular mycorrhizal fungi and organic farming. *Agriculture, Ecosystems and Environment*, **113**, 17–35.

Goulding, K., Jarvis, S., and Whitmore, A. (2008) Optimizing nutrient management for farm systems. *Philosophical Transactions of the Royal Society B – Biological Sciences*, **363**, 667–680.

Gregory, P.J. (2006) Roots, rhizosphere and soil: the route to a better understanding of soil science? *European Journal of Soil Science*, **57**, 2–12.

Griffiths, B.S. (1994) Microbial feeding nematodes and protozoa in soil and their effects on microbial activity and nitrogen mineralisation in decomposition hotspots and the rhizosphere. *Plant and Soil*, **164**, 25–33.

Hall, D.J.M., Jones, H.R., Crabtree, W.L., and Daniels, T.L. (2010) Claying and deep ripping can increase crop yields and profits on water repellent sands with marginal fertility in southern Western Australia. *Australian Journal of Soil Research*, **48**, 178–187.

Hammond, J.P., Bennett, M.J., Bowen, H.C., Broadley, M.R., Eastwood, D.C., May, S.T., Rahn, C., Swarup, R., Woolaway, K.E., and White, P.J. (2003) Changes in gene expression in Arabidopsis shoots during phosphate starvation and the potential for developing smart plants. *Plant Physiology*, **132**, 578–596.

Hammond, J.P., Broadley, M.R., and White, P.J. (2004) Genetic responses to phosphorus deficiency. *Annals of Botany*, **94**, 323–332.

Hanson, J.D., Liebig, M.A., Merrill, S.D., Tanaka, D.L., Krupinsky, J.M., and Stott, D.E. (2007) Dynamic cropping systems: increasing adaptability amid an uncertain future. *Agronomy Journal*, **99**, 939–943.

Harman, G.E. (2011a) Multifunctional fungal plant symbionts: new tools to enhance plant growth and productivity. *New Phytologist*, **190**, 647–649.

Harman, G.E. (2011b) Trichoderma-not just for biocontrol anymore. *Phytoparasitica*, **39**, 103–108.

Harrier, L.A. and Watson, C.A. (2003) The role of arbuscular mycorrhizal fungi in sustainable cropping systems. *Advances in Agronomy*, **79**, 185–225.

Havlin, J.L., Beaton, J.D., Tisdale, S.L., and Nelson, W.L. (1999) *Soil Fertility and Fertilizers*, 6th edn. Prentice-Hall, Upper Saddle River, New Jersey.

Hayat, R., Ali, S., Amara, U., Khalid, R., and Ahmed, I. (2010) Soil beneficial bacteria and their role in plant growth promotion. *Annals of Microbiology*, **60**, 579–598.

Heap, J.W. and McKay, A.C. (2009) Managing soil-borne crop diseases using precision agriculture in Australia. *Crop and Pasture Science*, **60**, 824–833.

Hellriegel, H. and Willfarth, H. (1888) *Untersuchungen über die Stickstoff-Nahrung der Gramineen und Leguminosen. Beilageheft zu der Zeitschrift des Vereins für die Rübenzucker-Industrie des Deutschen Reiches, Buchdruckerei der "Post"*. Kayssler and Co, Berlin, Germany.

Hergert, G.W. (1987) Status of residual nitrate nitrogen soil tests in the United States of America. In: *Soil Testing: Sampling, Correlation, Calibration, and Interpretation* (ed. J.R. Brown), pp. 73–88. Soil Science Society of America Special Publication 21, SSSA, Madison, Wisconsin.

Herridge, D., Peoples, M., and Boddey, R. (2008) Global inputs of biological nitrogen fixation in agricultural systems. *Plant and Soil*, **311**, 1–18.

Hodge, A. (2006) Plastic plants and patchy soils. *Journal of Experimental Botany*, **57**, 401–411.

Hodge, A., Stewart, J., Robinson, D., Griffiths, B.S., and Fitter, A.H. (1998) Root proliferation, soil fauna and plant nitrogen capture from nutrient rich patches in soil. *New Phytologist*, **139**, 479–494.

Howard, A. (1943) *An Agricultural Testament*. Oxford University Press, London.

Hoyle, F.C. and Murphy, D.V. (2006) Seasonal changes in microbial function and diversity associated with stubble retention versus burning. *Australian Journal of Soil Research*, **44**, 407–423.

Hoyle, F.C., Baldock, J.A., and Murphy, D.V. (2012) Soil organic carbon – role in Australian farming systems. In: *Rainfed Farming Systems* (eds P, Tow, I. Cooper, I. Patridge, and C. Birch), pp. 333–361. Springer, Berlin, Germany.

Ismail, I., Blevins, P.L., and Frye, W.W. (1994) Long-term no tillage effects on soil properties and continuous corn yields. *Soil Science Society of America Journal*, **58**, 193–198.

Janik, L.J., Merry, R.H., and Skjemstad, J.O. (1998) Can mid infrared diffuse reflectance analysis replace soil extractions? *Australian Journal of Experimental Agriculture*, **38**, 681–696.

Johnson, G.V. (1991) General model for predicting crop response to fertilizer. *Agronomy Journal*, **83**, 367–373.

Jones, D.L., Hodge, A., and Kuzyakov, Y. (2004) Plant and mycorrhizal regulation of rhizodeposition. *New Phytologist*, **163**, 459–480.

Karlen, D.L., Andrews, S.S., and Doran, J.W. (2001) Soil quality: current concepts and applications. *Advances in Agronomy*, **74**, 1–40.

Keeney, D.R. (1982) Nitrogen availability indices. In: *Methods of Soil Analysis. Part 2* (eds A.L. Page, R.H. Miller, and D.R. Keeney), pp. 711–734. American Society of Agronomy, Madison, Wisconsin.

Kingwell, R. and Fuchsbichler, A. (2011) The whole-farm benefits of controlled traffic farming: an Australian appraisal. *Agricultural Systems*, **104**, 513–521.

Kirschenmann, F.L. (2007) Potential for a new generation of biodiversity in agroecosystems of the future. *Agronomy Journal*, **99**, 373–376.

Klingebiel, A.A. and Montgomery, P.H. (1961) *Land Capability Classification*, USDA Agriculture Handbook 21. US Government Printing Office, Washington, D.C.

Knight, S.M., Macdonald, A.J., Glendining, M.J., Whitmore, A.P., Dailey, A.G., Goulding, K.W.T., Sinclair, A.H., and Rees, R.M. (2008) *Better Estimation of Soil Nitrogen Use Efficiency by Cereals and Oilseed Rape*, HGCA Research Review 68, Home Grown Cereals Authority, London. Accessed online at: www.hgca.com/publications/documents/cropresearch/RR68_Final_Research_Review.pdf

Kopittke, P.M. and Menzies, N.W. (2007) A review of the use of the Basic Cation Saturation ratio and the "ideal" soil. *Soil Science Society of America Journal*, **71**, 259–265.

Kowalchuk, G.A., Buma, D.S., de Boer, W., Klinkhamer, P.G.L., and van Veen, J.A. (2002) Effects of above-ground plant species composition and diversity on the diversity of soil-borne micro-organisms. *Antonie van Leeuwenhoek International Journal of General and Molecular Microbiology*, **81**, 1–4.

Lal, R. (2001) Managing world soils for food security and environmental quality. *Advances in Agronomy*, **74**, 155–192.

Lilburne, L., Sparling, G., and Schipper, L. (2004) Soil quality monitoring in New Zealand: development of an interpretative framework. *Agriculture, Ecosystems and Environment*, **104**, 535–544.

Long, S.R. (2001) Genes and signals in Rhizobia-legume symbiosis. *Plant Physiology*, **125**, 69–72.

Ludwig, F. and Asseng, S. (2010) Potential benefits of early vigor and changes in phenology in wheat to adapt to warmer and drier climates. *Agricultural Systems*, **103**, 127–136.

Lupwayi, N.Z., Clayton, G.W., Hanson, K.G., Rice, W.A., and Biederbeck, V.O. (2004) Populations and functional diversity of bacteria associated with barley, wheat and canola roots. *Canadian Journal of Soil Science*, **84**, 245–254.

Mäder, P., Fliessbach, A., Dubois, D., Gunst, L., Fried, P., and Niggli, U. (2002) Soil fertility and biodiversity in organic farming. *Science*, **296**, 1694–1697.

Magdoff, F., Ross, D., and Amadon, J. (1984) A soil test for nitrogen availability to corn. *Soil Science Society of America Journal*, **48**, 1301–1304.

Mairura, F.S., Mugendi, D.N., Mwnaje, J.I., Ramisch, J.J., Mbugua, P.K., and Chianu, J.N. (2008) Scientific evaluation of smallholder land use knowledge in Central Kenya. *Land Degradation and Development*, **19**, 77–90.

de V Malherbe, I. (1964) *Soil fertility*, 5th English edn. Oxford University Press, Cape Town, South Africa.

Marschner, H. (1995) *Mineral Nutrition of Higher Plants*. Academic Press, London.

Martinez, E., Romero, D., and Palacios, R. (1990) The Rhizobium genome. *Critical Reviews in Plant Sciences*, **9**, 59–93.

Massa, G.D. and Gilroy, S. (2003) Touch modulates gravity sensing to regulate the primary roots of Arabidopsis thaliana. *The Plant Journal*, **33**, 435–445.

Masson-Boivin, C., Giraud, E., Perret, X., and Batut, J. (2009) Establishing nitrogen fixing symbiosis with legumes: how many rhizobium recipes? *Trends in Microbiology*, **17**, 458–466.

Matsumoto, H. (2002) Plant roots under aluminium stress: toxicity and tolerance. In: *Plant Roots the Hidden Half* (eds Y. Waisel, A. Eshel, and U. Kafkafi), 3rd edn, pp 821–838. Marcel Dekker, New York.

Mazzola, M. (2002) Mechanisms of natural soil suppressiveness to soil borne diseases. *Antonie van Leeuwenhoek International Journal of General and Molecular Microbiology*, **81**, 557–564.

McInnes, A. and Haq, K. (2003). Contributions of Rhizobia to soil nitrogen fertility. In: *Soil Biological Fertility: A Key to Sustainable Land Use in Agriculture* (eds L.K. Abbott and D.V. Murphy), pp. 99–128. Kluwer, Dordrecht, the Netherlands.

Mengel, K. (1991) Available nitrogen in soils and its determination by the 'Nmin' method and by electroultrafiltration (EUF). *Fertilizer Research*, **28**, 251–262.

Menzies, J.D. (1959) Occurrence and transfer of a biological factor in soil that suppresses potato scab. *Phytopathology*, **49**, 648–652.

Millenium Ecosystem Assessment. (2003) *Ecosystems and Human Well-Being. A Framework for Assessment*. Island Press, Washington, D.Columbia.

Ministry of Agriculture Fisheries and Food, MAFF. (1996) *Fertiliser Recommendations for Agricultural and Horticultural Crops*, 6th edn Reference Book 209. HMSO, London.

Morcuende, R., Bari, R., Gibon, Y., Zheng, W.M., Pant, B.D., Bläsing, O., Usadel, B., Czechowski, T., Udvardi, M.K., Stitt, M., and Scheible, W.R. (2007) Genome-wide reprogramming of metabolism and regulatory networks of Arabidopsis in response to phosphorus. *Plant Cell Environment*, **30**, 85–112.

Morton, J.D., Smith, L.C., Dodds, K.G., and Catto, W.D. (2001) Balanced and adequate potassium and phosphorus nutrition of pasture. *New Zealand Journal of Agricultural Research*, **44**, 269–277.

Mulvaney, R.L., Khan, S.A., Hoeft, R.G., and Brown, H.M. (2001) A soil organic nitrogen fraction that reduces the need for nitrogen fertilisation. *Soil Science Society of America Journal*, **65**, 1164–1172.

Munson, R.D. (ed) (1985) *Potassium in Agriculture*. American Society of Agronomy Inc., Crop Science Society of America Inc., Soil Science Society of America Inc., Madison, Wisconsin.

Murphy, D.V., Osman, M., Russell, C.A., Darmawanto, S., and Hoyle, F.C. (2009) Potentially mineralisable nitrogen: relationship to crop production and spatial mapping using infra-red reflectance spectroscopy. *Australian Journal of Soil Research*, **47**, 737–741.

Neumann, G. and Römheld, V. (2002) Root-induced changes in the availability of nutrients in the rhizosphere. In: *Plant Roots the Hidden Half* (eds Y. Waisel, A. Eshel, and U. Kafkafi), 3rd edn, pp. 617–649. Marcel Dekker, New York.

Oldroyd, G.E.D., Harrison, M.J., and Paszkowski, U. (2009) Reprogramming plant cells for endosymbiosis. *Science*, **324**, 753–754.

Olson, R.A., Frank, K.D., Grabouski, P.H., Rehm, G.W., and Knudsen, D. (1982) Economic and agronomic impacts of varied philosophies of soil testing. *Agronomy Journal*, **74**, 492–499.

Palm, C.A., Gachengo, C.N., Delve, R.J., Cadisch, G., and Giller, K.E. (2001) Organic inputs for soil fertility management in tropical agroecosystems: application of an organic resource database. *Agriculture, Ecosystems and Environment*, **83**, 27–42.

Patzel, N., Sticher, H., and Karlen, D.L. (2000) Soil fertility – phenomenon and concept. *Journal of Plant Nutrition and Soil Science*, **163**, 129–142.

Peoples, M.B., Brockwell, J., Herridge, D.F., Rochester, I.J., Alves, B.J.R., Urquiagas, S., Boddy, R.M., Dakora, F.D., Bhattarai, S., Maskey, S.L., Sampet, C., Rerkasem, B., Khaf, D.F., Hauggard-Nielsen, H., and Jensen, E.S. (2009) The contribution of nitrogen fixing crops to the productivity of agricultural systems. *Symbiosis* **48**, 1–17.

Povorov, N.A. (1998) Coevolution of rhizobia with legumes: facts and hypotheses. *Symbiosis*, **24**, 337–368.

van der Putten, W.H., de Ruiter, P.C., Bezemer, T.M., Wassen, J.A., and Wolters, V. (2004) Trophic interactions in a changing world. *Basic and Applied Ecology*, **5**, 487–494.

van der Putten, W.H., Bardgett, R.D., de Ruiter, P.C., Hol, W.H.G., Meyer, K.M., Bezemer, T.M., Bradford, M.A., Christensen, S., Eppinga, M.B., Fukami, T., Hemerik, L., Molofsky, J., Schädler, M., Scherber, C., Strauss, S.Y., Vos, M., and Wardle, D.A. (2009) Empirical and theoretical challenges in aboveground-belowground ecology. *Oecologia*, **161**, 1–14.

Raaijmakers, J.M., Paulitz, C.T., Steinberg, C., Alabouvette, C., and Moënne-Loccoz, Y. (2009) The rhizosphere: a playground and a battlefield for soil borne pathogens and beneficial microorganisms. *Plant and Soil*, **321**, 341–361.

Raun, W.R., Solie, J.B., Johnson, G.V., Stone, M.L, Lukina, E.V., Thomason, W.R., and Schepers, J.S. (2001) In season prediction of potential grain yield in winter whet using canopy reflectance. *Agronomy Journal*, **93**, 131–138.

Remy, W., Taylor, T.N., Haas, H., and Kerp, H. (1994) Four hundred million year-old vesicular arbuscular mycorrhizal. *Proceedings of the National Academy of Sciences of the United States of America*, **91**, 11841–11843.

Richardson, A.E., Barea, J.-M., McNeill, A.M., and Prigent-Combaret, C. (2009) Acquisition of phosphorus and nitrogen in the rhizosphere and plant growth promotion by microorganisms. *Plant and Soil*, **321**, 305–339.

Riquier, J., Bramao, D.L., and Cornet, J.P. (1970) *A New System of Soil Appraisal in Terms of Actual and Potential Productivity*. FAO Soil Resources, Development and Conservation Service, Land and Water Development Division, FAO, Rome, Italy.

Robinson, N., Harper, R.J., and Smettem, K.R.J. (2006) Soil water depletion by Eucalyptus spp. integrated into dryland agricultural systems. *Plant and Soil*, **286**, 141–151.

Romig, D.E., Garlynd, M.J., Harris, R.F., and McSweeney, K. (1995) How farmers assess soil health and quality. *Journal of Soil and Water Conservation*, **50**, 225–232.

Römheld, V. (1990) The soil-root interface in relation to mineral nutrition. *Symbiosis*, **9**, 19–27.

Roper, M.M. (2004) The isolation and characterisation of bacteria with the potential to degrade waxes that cause water repellency in sandy soils. *Australian Journal of Soil Research*, **42**, 427–434.

de Ruiter, P.L., Moore, J.C., Zwart, K.B., Bouwan, L.A., Hassink, J., Bloem, J., Devos, J.A., Matinissen, J.C.Y., Didden, W.A.M., Lebbink, G., and Brussaard, L. (1993) Simulation of nitrogen mineralisation in the below ground food webs of 2 winter wheat fields. *Journal of Applied Ecology*, **30**, 95–106.

Rupela, O.P., Gowda, C.L.L., Wani, S.P., and Bee, H. (2006) Evaluation of crop production systems based on locally available biological inputs. In: *Biological Approaches to Sustainable Soil Systems* (eds N. Uphoff, A.S. Ball, C. Palm, E. Fernandes, J. Pretty, H. Herren, P. Sanchez, O. Husson, N. Sanginga, M. Laing, and J. Thies), pp. 501–530. CRC Taylor & Francis, Boca Raton, Florida.

Russell, E.J. (1937) *Soil Conditions and Plant Growth*, 7th edn. Longmans, Green and Co., London.

Russell, E.W. (1961) *Soil Conditions and Plant Growth*, 9th edn. Longmans Green, London.

Sabbe, W.E. and Marx, D.B. (1987) Soil sampling: spatial and temporal variability. In: *Soil Testing: Sampling, Correlation, Calibration and Interpretation* (ed. J.R. Brown), pp. 1–14. Soil Science Society of America Special Publication 21, SSSA, Madison, Wisconsin.

Sanchez, P.A. (1994) Tropical soil fertility research: towards the second paradigm. In: *Transactions of the 15th World Congress of Soil Science*, Acapulco, Mexico. Mexican Society of Soil Science, Chapingo, Mexico, pp. 65–88.

Sanchez, P.A. and Salinas, J.G. (1981) Low-input technology for managing oxisols and ultisols in tropical America. *Advances in Agronomy*, **34**, 279–406.

Shaffer, M.J., Ma, L., and Hansen, S. (eds) (2001) *Modeling Carbon and Nitrogen Dynamics for Soil Management*. Lewis Publishers, Boca Raton, Florida.

Shaw, L.J., Morris, P., and Hooker, J.E. (2006) Perception and modification of plant flavonoid signals by rhizosphere microorganisms. *Environmental Microbiology*, **8**, 1867–1880.

Shenoy, W. and Kalgudi, G.M. (2005) Enhancing plant phosphorus use efficiency for sustainable cropping. *Biotechnology Advances*, **23**, 7–8.

Shepherd, M.A., Stockdale, E.A., Powlson, D.S., and Jarvis, S.C. (1996) The influence of organic N mineralization on the management of agricultural systems in the UK. *Soil Use and Management*, **12**, 76–85.

Shivanna, M.B., Meera, M.S., and Hyakumachi, M. (1996) Role of root colonization ability of plant growth promoting fungi in the suppression of take-all and common root rot of wheat. *Crop Protection*, **15**, 497–504.

Shoresh, M., Mastouri, F., and Harman, G.E. (2010) Induced systemic resistance and plant responses to fungal biocontrol agents. *Annual Review of Phytopathology*, **48**, 21–43.

Sibbesen, E. (1983) Phosphate soil tests and their suitability to assess the phosphorus status of soil. *Journal of the Science of Food and Agriculture*, **34**, 1368–1374.

Sinclair, H.R., Waltman, W.J., Waltman, S.W., Terpstra, H.P., and Reed-Margetan, D. (1996) *Soil Ratings for Plant Growth (SRPG)*. USDA-NRCS, National Soil Survey Centre, Lincoln, Nebraska.

Sinfield, J.V., Fagerman, D., and Colic, O. (2010) Evaluation of sensing technologies for on-the-go detection of macro-nutrients in cultivated soils. *Computers and Electronics in Agriculture*, **70**, 1–18.

Smil, V. (2001) *Enriching the Earth*. MIT Press, Cambridge, Massachusetts.

Smith, K.P., Handelsman, J., and Godman, R.M. (1999) Genetic basis in plants for interactions with disease suppressive bacteria. *Proceedings of the National Academy of Science USA*, **9**, 4786–4790.

Smith, F.W., Mudge, S.R., Rae, A.L., and Glassop, D. (2003) Phosphate transport in plants. *Plant and Soil*, **248**, 71–83.

Sojka, R.E., Upchurch, D.R., and Borlaug, N.E. (2003) Quality soil management or soil quality management: performance versus semantics. *Advances in Agronomy*, **79**, 1–68.

Sparling, G., Parfitt, R.L., Hewitt, A.E., and Schipper, L.A. (2003) Three approaches to define desired soil organic matter contents. *Journal of Environmental Quality* **32**, 760–766.

Sprent, J.I. (2001) *Nodulation in Legumes*. Royal Botanic Gardens, Kew, London.

Stenberg, B. (1999) Monitoring soil quality of arable land: microbiological indicators. *Acta Agriculturae Scandinavica*, **49**, 1–24.

Stockdale, E.A., Watson, C.A., Black, H.I.J., and Philipps, L. (2006) *Do Farm Management Practices Alter Below-Ground Biodiversity and Ecosystem Function? Implications for Sustainable Land Management.* JNCC Report No. 354, Joint Nature Conservation Committee, Peterborough, UK.

Suarez, S.A., de la Fuente, E.B., Ghersa, C.M., and Rolando, J.C.L. (2001) Weed community as an indicator of summer crop yield and site quality. *Agronomy Journal*, **93**, 524–530.

Syers, J.K. (1998) Soil and plant potassium in agriculture. *Fertiliser Society Proceedings No.*, **411**, 32 pp.

Tang, C., Asseng, S., Diatloff, E., and Rengel, Z. (2003a) Modelling yield losses of aluminium-resistant and aluminium-sensitive wheat due to subsurface soil acidity: effects of rainfall, liming and nitrogen application. *Plant and Soil*, **254**, 349–360.

Tang, C., Rengel, Z., Diatloff, E., and Gazey, C. (2003b) Responses of wheat and barley to liming on a sandy soil with subsoil acidity. *Field Crops Research*, **80**, 235–244.

Tisdale, S.L. and Nelson, W.L. (1966) *Soil Fertility and Fertilizers*, 2nd edn. Macmillan, New York.

Tow, P., Cooper, I., Partridge, I., and Birch, C. (eds) (2012) *Rainfed Farming Systems*. Springer, Berlin, Germany.

Unkovich, M.J. and Pate, J.S. (2001) An appraisal of recent field measurements of symbiotic N_2 fixation by annual legumes. *Field Crops Research*, **65**, 211–228.

Uren, N.C. (2000) Types, amounts and possible functions of compounds released into the rhizosphere by soil grown plants. In: *The Rhizosphere: Biochemistry and Organic Substances at the Soil Interface* (eds R. Pinton, Z. Varani, and P. Nannipieri), pp. 19–40. Marcel Dekker, New York.

Vanlauwe, B., Ramisch, J.J., and Sanginga, N. (2006) Integrated soil fertility management in Africa: from knowledge to implementation. In: *Biological Approaches to Sustainable Soil Systems* (eds N. Uphoff, A.S. Ball, C. Palm, E. Fernandes, J. Pretty, H. Herren, P. Sanchez, O. Husson, N. Sanginga, M. Laing, and J. Thies), pp. 257–272. CRC Taylor & Francis, Boca Raton, Florida.

Walworth, J.L. and Sumner, M.E. (1986) Foliar diagnosis—a review. In: *Advances in Plant Nutrition*, Vol. **III** (ed. P.B. Tinker), pp 193–241. Elsevier, New York.

Wardle, D.A. (2002) *Communities and Ecosystems: Linking the Aboveground and below Ground Components*. Monographs in Population Biology 34, Princeton University Press, Princeton, New Jersey.

Wardle, D.A., Nicholson, K.S., Bonner, K.I., and Yeates, G.W. (1999) Effects of agricultural intensification on soil -associated arthropod population dynamics, community structure, diversity and temporal variability over a seven year period. *Soil Biology and Biochemistry*, **31**, 1691–1706.

Whalley, W.R., Riseley, B., Leeds-Harrison, P.B., Bird, N.R.A., Leech, P.K., and Adderley, W.P. (2005) Structural differences between bulk and rhizosphere soil. *European Journal of Soil Science*, **56**, 353–360.

Whitelaw, M. (2000) Growth promotion of plants inoculated with phosphate-solubilizing fungi. *Advances in Agronomy*, **69**, 99–151.

Wiseman, B.M., Neate, S.M., Keller, K.O., and Smith, S.E. (1996) Suppression of Rhizoctonia solani anastomosis group 8 in Australia and its biological nature. *Soil Biology and Biochemistry*, **28**, 727–732.

Wortmann, C.S., Dobermann, A.R., Ferguson, R.B., Hergert, G.W., Shapiro, C.A., Tarkalson, D.D., and Walters, D.T. (2009) High yield corn response to applied phosphorus, potassium and sulphur in Nebraska. *Agronomy Journal*, **101**, 546–555.

Wu, P., Ma, L., Hou, X., Wang, M., Wu, Y., Liu, F., and Deng, X.W. (2003) Phosphate starvation triggers distinct alterations if genome expression in Arabidopsis roots and leaves. *Plant Physiology*, **132**, 1260–1271.

Zebarth, B.J., Drury, C.F., Tremblay, N., and Cambouris, A.N. (2009) Opportunities for improved fertiliser nitrogen management in production of arable crops in eastern Canada: a review. *Canadian Journal of Soil Science*, **89**, 113–132.

4 Soil organic matter

David Powlson[1], Pete Smith[2], and Maria De Nobili[3]

[1]*Department of Sustainable Soils and Grassland Systems, Rothamsted Research, Harpenden, Hertfordshire, UK*
[2]*School of Biological Sciences, University of Aberdeen, Aberdeen, UK*
[3]*Dipartmento di Scienze Agrarie e Ambientali, Università degli studi di Udine, Udine, Italy*

4.1 Introduction

It is the presence of organic matter that makes the difference between a collection of mineral particles and a functioning soil with a recognisable physical structure that provides a suitable medium for plant growth. Soils often contain only a small quantity of organic matter (typically a few per cent by mass in agricultural soils), yet this has a major influence on soil properties. For example, organic matter is a key factor determining the interactions between mineral particles and thus the formation of aggregates of varying size and stability and the distribution of spaces (pores) between the solid particles which may be filled with either air or water. This physical structure, relying in large part on the quantity and type of organic matter, determines whether or not a soil retains water, whether it becomes flooded or is well drained, whether it becomes easily compacted or retains a structure conducive to root growth. Organic matter also provides substrate for soil organisms and, often, anchorage points for bacteria or fungi. Interactions between organic matter, organisms and mineral particles form the basis for nutrient transformations on which all plant growth depends. Organic matter is a reservoir of many plant nutrients (such as N, P and S) and also provides a substantial part of the soil's cation exchange capacity (CEC), which is important for the retention of many nutrients. It is also the repository of organic carbon (C) in soil which, on a global scale, is the largest terrestrial stock of carbon after carbonate rocks. Thus, the influence of management practices on the quantity of organic C in soil is vitally important both for the maintenance of soil functions, and its influence on plant growth, and for its impact on global carbon cycling and climate change.

Even a small quantity of organic matter can have a profound influence on soil properties. Soil in arable cropping in a temperate climatic region may typically contain 1–5% organic carbon, yet small variations within this range have a large effect on soil properties.

Soil Conditions and Plant Growth. Edited by Peter J. Gregory and Stephen Nortcliff.
© 2013 Blackwell Publishing Ltd. Published 2013 by Blackwell Publishing Ltd.

4.1.1 Terminology – organic matter or organic carbon?

It is customary to refer to the total amount of organic matter (OM) in soil, in all forms, as soil organic matter (SOM). But chemical analysis normally gives a value for C content. It is common to use a conversion factor of 1.724 to convert a measured organic C value to organic matter, implying that organic matter contains 58% C, though this is based on work conducted in the nineteenth century. Values up to 2.0 for surface soils and 2.5 for subsoils have been published (Nelson and Sommers, 1982). A more recent review of data (Pribyl, 2010) shows that the value is variable but an average conversion factor of 2 (i.e. 50% C in organic matter) is more accurate in almost all cases. However, it is now strongly recommended not to quote OM values but use organic C, usually termed soil organic carbon (SOC).

When soil samples are analysed for organic C content, the result is conventionally expressed in units such as %, mg C kg^{-1} soil or μg C g^{-1} soil. These are units of *concentration*, though they are often incorrectly referred to as C *content*. To express C as a quantity or content rather than concentration, the concentration is multiplied by the mass of soil to a given depth, often determined from measurement of bulk density. SOC content can then be expressed in units such as Mg C ha^{-1} or g C m^{-2} to the defined depth. When expressed in such units, the quantity of C in 1 ha of soil can correctly be termed a C content or stock. Alternatively, the value may be multiplied by an area to give a C stock in an area of interest such as Europe, often in units of Tg or Pg. A term sometimes used in such studies is 'C density', which is synonymous with 'C stock per unit area'. Figure 4.1 shows an example of a C density map for the Amazonian region of Brazil (Cerri et al., 2007). In this region there is at least a fivefold range between areas with the lowest and highest C density. From the viewpoint of maintaining SOC stocks (as discussed later in relation to climate change mitigation), such maps can be useful as a tool for policy makers in determining appropriate land management practices for each area. For areas with a large C density, there is a priority to maintain them in their current state in order to avoid C loss and thus prevent the large C stock becoming a source of CO_2. For areas with a small C density, there may be potential to accumulate SOC through altered management or land use, thus creating a sink and helping to mitigate climate change. However, in many cases, a small C density reflects a small potential for SOC accumulation, either because of soil type (e.g. sandy soils which have less capacity to stabilise C than soils with a greater clay content) or through limited plant growth resulting from climatic factors such as low rainfall.

4.1.2 Analysis of SOC

It is now most common to analyse soil for C by high temperature combustion of a sample in the presence of oxygen; various commercial instruments are available for this, using a small finely ground soil sample. All C in the sample is converted to CO_2 and its quantity measured by a gas chromatographic detector, infrared analysis or (less commonly) gravimetrically. If the soil contains carbonate, the C contained in this will also be converted to CO_2, so the quantity derived is *total* soil C. To determine *organic* C, a separate analysis for *carbonate C* is made and this quantity subtracted from the total C value from combustion. Carbonate C is usually determined by treatment of a soil sample with dilute acid and measurement of the volume or mass of CO_2 evolved.

Before the advent of combustion instruments, it was usual to determine organic C by chemical oxidation by heating a soil sample with a mixture of potassium dichromate, sulphuric acid and, in some methods, phosphoric acid. Either the evolved CO_2 was determined gravimetrically or the amount of unused dichromate oxidising agent determined,

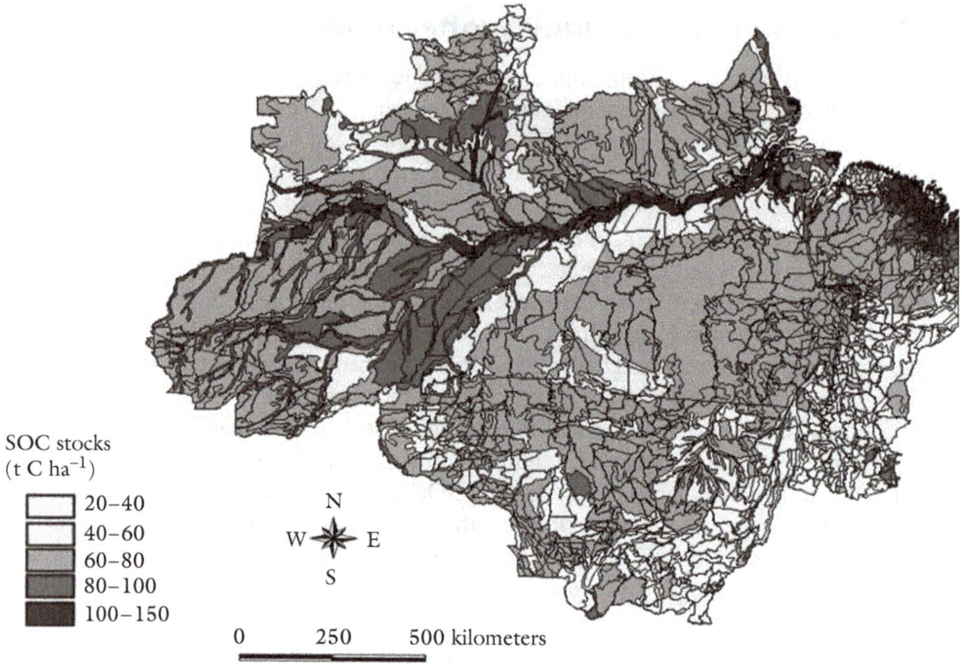

Figure 4.1 Map of soil organic carbon stocks (carbon density) in the Brazilian Amazon. Reprinted from Cerri et al. (2007). With kind permission from Elsevier. For a colour version of this figure, please see Plate 4.1.

often by titration against ferrous ammonium sulphate which was oxidised. Many variations on this principle have been published and used; see Nelson and Sommers (1982) and Kalembasa and Jenkinson (1973) for details of methods and discussion of their merits and limitations. One variation, known as the Walkley–Black method, relies on the heat naturally generated from the reaction between organic matter and the oxidising agent. In well-equipped laboratories, this method would rarely be used now, but in less well-endowed laboratories it is still in use and one sees reference to it in the literature.

An approximate method of analysis, suitable for soils with a high organic C content and where very large numbers of samples require analysis, is *loss on ignition*. In this method a sample is burned and the loss of mass determined. The sample is either dried first (so that water does not contribute to the mass loss) or a correction is made for the small water content of the soil which is normally air-dried prior to analysis. The assumption is that the loss of weight is due entirely to the destruction of organic matter, with the mineral components remaining.

4.2 Factors influencing SOM content

4.2.1 Inputs and outputs of organic carbon

The quantity of organic matter (or organic C) in soil is a balance between inputs and outputs (Figure 4.2). In agricultural soils, the main organic C inputs are plant roots and associated material (sloughed off roots cells, root exudates) and above ground plant parts (stems, fallen leaves, any other non-harvested plant parts) plus any manures or similar materials applied. Minor inputs might include organic matter in dust deposited at a site. The main output of C

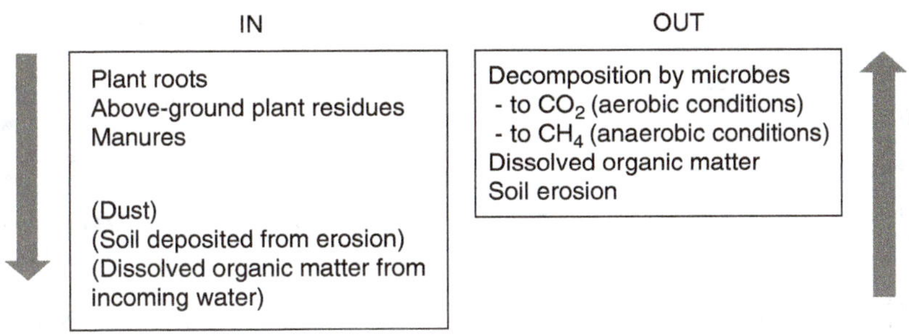

Figure 4.2 Inputs and outputs of C to and from soils.

is decomposition by the soil biological population: under aerobic conditions, C from decomposition is evolved to the atmosphere as CO_2 but under anaerobic conditions, such as in flooded soils growing paddy rice or natural or man-made wetlands, it is methane, CH_4. The rate and extent of decomposition is determined by environmental factors, especially temperature and soil moisture content. Decomposition will almost always be faster at higher temperature than lower assuming sufficient water is available for microbial activity. Very dry or very wet conditions will normally slow decomposition. Because of the generally increased rate of decomposition under warmer conditions, it is usual for soils in tropical regions to have a lower SOC content than those in cooler climates but under comparable land use.

4.2.2 Impact of management practices

Management practices, through their impacts on C inputs and decomposition rates, have a major impact on SOC content but changes often occur slowly – though faster under warmer climates. In this section several examples are presented to illustrate some of the main impacts and trends observed.

Figure 4.3 illustrates several aspects of management impacts on the SOC content of an agricultural soil in a temperate region. The data show SOC content in several treatments of the long-term Broadbalk Experiment at Rothamsted Research, UK, and also a simulation of the data using the RothC model (see Section 4.3). By the year 2005, the SOC content in the treatment given a high rate of farmyard manure (FYM) every year since the start of the experiment in 1843 was almost three times that in treatments given neither manure nor inorganic fertilisers. But the increase occurred slowly and the tendency of SOC to move asymptotically towards a new equilibrium value is clear. The annual rate of SOC increase was much greater in the early years: output from the RothC simulation shows that the average annual increase during the first 20 years was 1.0 t C ha^{-1} but during the last 20 years was only 0.2 t C ha^{-1}. Consequently about half of the final observed increase occurred within the first 30–40 years of manure application. During the period 1926–1966, all plots on Broadbalk were bare-fallowed one year in five, to control weeds, and no manure was applied in the fallow years. Thus, organic C inputs were reduced in these years due to absence of manure and absence of inputs from crop roots and stubble. The small decrease in SOC during the fallow years gives rise to the 'saw-tooth' shape in the simulations. The overall effect during this 40 year period was a small decline in SOC content in all treatments but is most clearly seen in the FYM treatment. After fallowing ceased (because weeds were controlled by herbicides), the increasing trend in the FYM treatment resumed but at a slower rate as the soil tended towards a new equilibrium value. If the soil had not had fallow years during this 40 year period, the new equilibrium value would presumably have been attained earlier and the

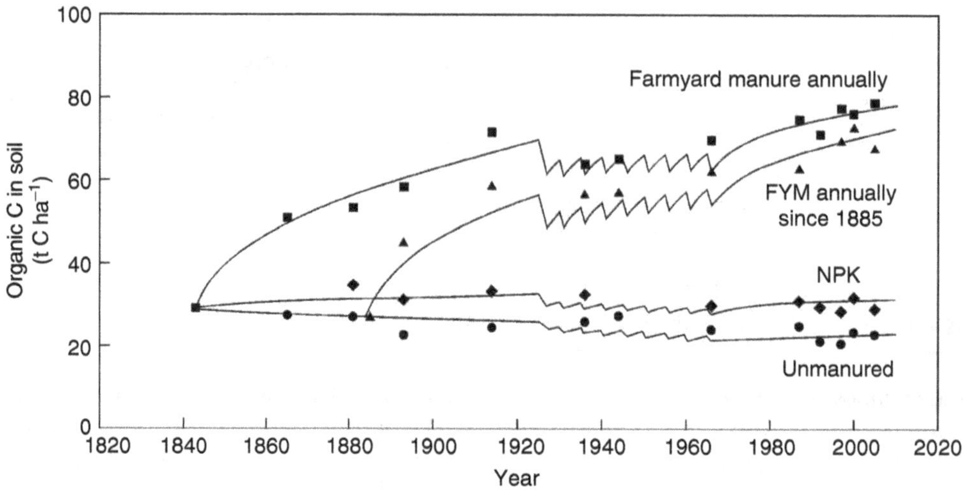

Figure 4.3 Changes in SOC content (0–23 cm) in selected treatments of the Broadbalk Wheat Experiments at Rothamsted Research, UK. Soils analysed using modern combustion methods on either recent samples or those from the archive. Values corrected for changes in soil bulk density. Treatments are farmyard manure (FYM) applied annually either since 1843 (■) or 1885 (▲) or inorganic fertilisers only (including N, P and K at 144, 35 and 90 kg ha^{-1}) annually (◆) and the unfertilised control (●). Points are measured values and lines are simulations using the RothC model. From Powlson et al. (2012). With kind permission from Elsevier.

annual rate of SOC increase in the later years would have been even less. The same trend can be seen in the later FYM treatment that started in 1885.

A surprising result in Figure 4.3 is the fact that SOC in the control treatment with no manure or fertiliser inputs did not decline but remained stable for the duration of the experiment. This is thought to be because the site had been in arable cropping for many centuries; the field is documented as arable at least 200 years before the start of the experiment and is thought to have been initially cleared of natural vegetation at least 1500 years earlier. So the high SOC content under native forest presumably declined during this long period and reached a new low equilibrium value that is maintained by the small crop inputs in the unfertilised and unmanured treatment. In the treatment receiving inorganic fertilisers, crop yields were much higher than in the control (similar to yields in the FYM treatment), so the organic inputs to soil from crop roots and stubble were also greater. This small increase in the annual input of organic C led to a small increase in SOC compared to the control treatment. This is a general trend seen in many long-term studies. Ladha et al. (2011) reviewed data on this point from 135 studies at 114 long-term sites worldwide. On average, application of inorganic N (often with P and K) led to the SOC and soil organic nitrogen (SON) contents of soils being 8% and 12% greater than in the corresponding controls. However, in many cases there was a general decline of SOC and SON over time in all treatments at the sites, and inorganic fertiliser applications did not reverse this; rather the decrease was slightly less where inorganic fertilisers were applied. These observations at so many sites contradict the claim that is sometimes made that inorganic fertilisers, especially N, cause a loss of SOM.

Figure 4.4 (redrawn from Gollany et al., 2011) shows data from a site where natural vegetation was cleared more recently than at Broadbalk, namely, the Morrow Plots in Illinois, USA. This is the oldest agricultural experiment in North America. The native prairie grassland was cleared in 1876 and a range of arable crop rotations and manure or fertiliser treatments established, though the data shown only starts in about 1905. It can be seen that

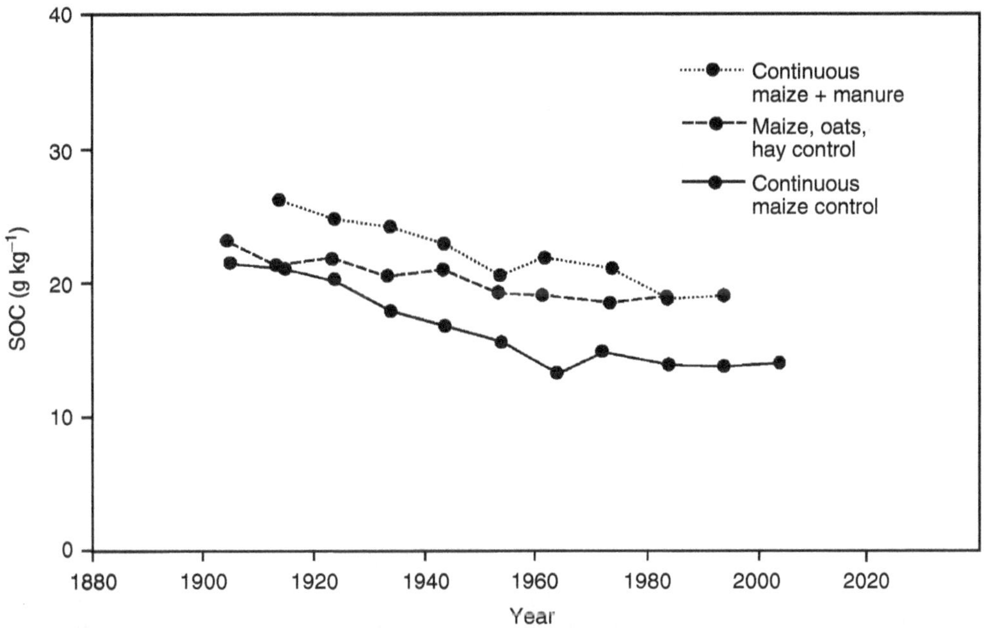

Figure 4.4 Changes in SOC content (0–15 cm) in selected treatments of the Morrow Plots, Illinois, USA. The experiment was started in 1876 after the site had been ploughed out of native prairie. Data shown from 1905. Redrawn from Gollany et al. (2011). With permission, copyright American Society of Agronomy.

SOC declined until about 1960 in all treatments and then approximately stabilised. This decline is attributed to (a) smaller inputs of organic C as arable crops replaced perennial grasses and (b) soil disturbance due to ploughing.

SOC in the continuous maize treatment (with no manure input) declined the most, from about 22 t C ha^{-1} in 1905 to 15 t C ha^{-1} after 70 years. Where manure was applied to continuous maize, SOC remained higher than in the control, but still declined throughout the period shown. In the maize–oats–hay rotation (with no manure), the rate of decline in the first half of the twentieth century was slightly lower and SOC content stabilised at about 5 t C ha^{-1} higher than under continuous maize. This is presumed to be due to the larger inputs of organic C in the year when hay is grown; inputs from oats are probably also greater than from maize due to the narrower row spacing. In all cases, maize stover (stubble) was removed until about 1960 but returned thereafter: this increase in C input will also have contributed to the slowing of the SOC decline that started about this time. If stover had continued to be removed, it is likely that SOC in both rotations would have stabilised at a slightly lower value. Where manure was applied to the continuous maize rotation, SOC continued to decline compared to the initial value but remained higher than in the treatment with no manure input. In this respect, the results are analogous to the effect of manure in the Broadbalk Experiments (Figure 4.3), though in the Morrow Plots the increase from manure was relative not absolute.

Figure 4.5 shows changes in SOC when tropical forest was cleared and converted to pasture (Cerri et al., 2003). The data were obtained through the use of a chronosequence rather than at specific experimental site. In this case, through historical records and interviews with land owners, it was possible to identify areas of pasture of different age. To make valid comparisons it was also necessary to establish that the soil type and management practices were very similar at all sites. Such an approach is sometimes termed a 'space for time' comparison. Figure 4.5

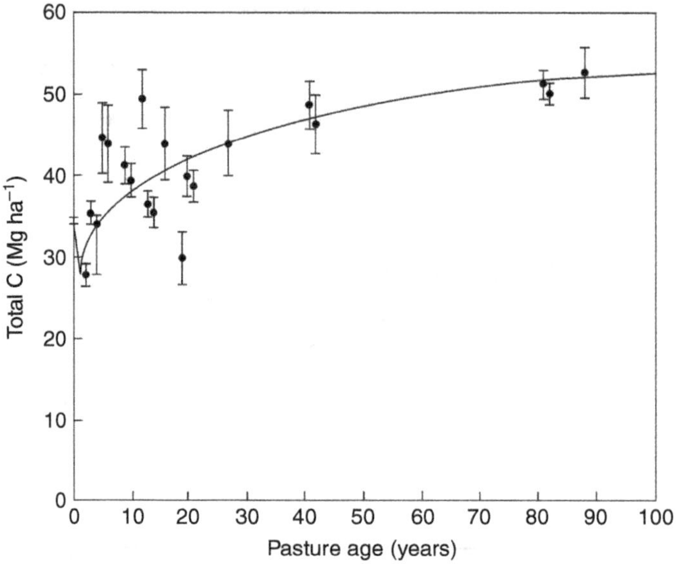

Figure 4.5 Changes in SOC (0–30 cm) in the Nova Vida Ranch chronosequence, Brazil. Native tropical forest was cleared and land converted to grazed pasture. SOC was measured under forest and under pastures of different ages, between 2 and 88 years after clearing. Sites had very similar soil types and pasture management regimes. Points are measured data and the line is a simulation using RothC. From Cerri et al. (2003). With permission, copyright Soil Science Society of America.

shows an initial rapid decline in SOC content after clearing forest but, somewhat surprisingly, this is followed by an increase and after 90 years SOC is at a higher value than in the original forest soil. Two factors are relevant here. Under well-managed grazed pasture in a tropical climate, with perennial grasses and some legumes, inputs of organic C to the soil can be higher than under forest. It is also observed that, under tropical forests, a larger proportion of the total C stock is above ground, in the trees, than is the case for temperate forests. Thus SOC content under tropical forests is often surprisingly small. Even though the long-term result of clearing forest and establishing pasture was a small increase in SOC stock (by about 20 tC ha^{-1} in this case), there was still a major loss of C from the system. It was estimated that 100–200 tC ha^{-1} were lost from the trees at the time of clearing.

Figure 4.6 shows an example of two contrasting land management changes in a temperate climate at two experimental sites at Rothamsted Research, UK (Johnston et al., 2009). In one case (the Highfield Experiment), a section of an area that had been in grass for over 100 years, and contained >60 t SOC ha^{-1} to a depth of 23 cm, was ploughed and converted to arable cropping. Over the next 20–30 years, its SOC content declined rapidly. Within 50 years it had lost one third of its original SOC and had reached the same content as that of a soil that had been under arable cropping for a period of well over 100 years. Figure 4.6 also shows that the section of grassland that continued under grass increased in SOC. This is because the intensity of management increased compared to that prior to the start of the experiment, including increased fertiliser N applications. At the arable site (the Fosters Experiment), the area that continued under arable cropping remained at an almost constant SOC content through the 50 years shown in Figure 4.6. The section of the arable field sown to grass and managed in the same way as the grass

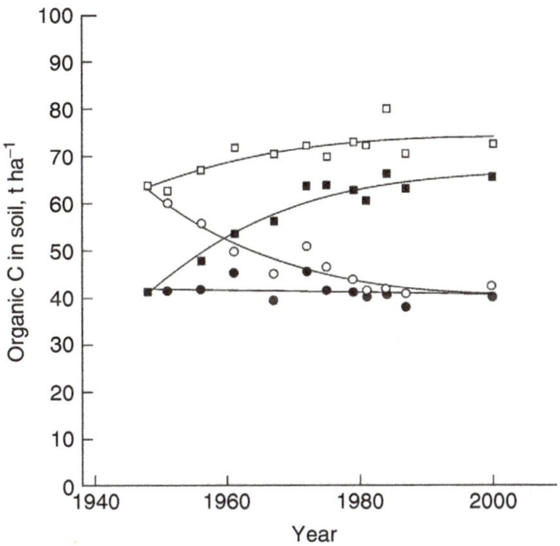

Figure 4.6 Changes in SOC (0–23 cm) in a silty clay loam soil in two ley-arable experiments at Rothamsted, UK, 1949–2002. Highfield Experiment: old grassland soil remaining in grass □ or ploughed and kept in arable cropping ○. Fosters Experiment: old arable soil remaining in arable ● or sown to grass and kept in grass ■. Reprinted from Johnston et al. (2009). With kind permission from Elsevier.

area of Highfield increased in SOC from 40 to 65 t ha^{-1} within 50 years but was still below the value in the continuing grass section of Highfield.

Figure 4.7 illustrates the importance of soil texture in determining SOC content. It shows the changes in SOC content over a period of almost 100 years at a site with a sandy soil containing only 7% clay (Johnston et al., 2009). The topsoil contained about 40 t C ha^{-1} at the start of the experiment, the relatively high value probably attributable to previous periods under grass which was ploughed up a few years before the start of the experiment. But SOC decreased to less than half the initial value during 100 years of arable cropping; even the management involving a four-course rotation receiving animal manure failed to significantly slow this decline. This is because clay and silt size particles, having a very large surface area, are so central to the stabilisation of organic matter in soil. Soils dominated by sand have much less potential to retain organic matter.

4.3 Modelling SOM dynamics

Models of SOM dynamics are simply mathematical mechanisms for expressing our conceptual understanding of how SOM decomposes and how carbon inputs and environmental factors affect SOM transformations. As with all models, SOM models are a simplification of reality, and there are a number of ways in which decomposition can be conceptualised. The vast majority of SOM models (and there are well over 30 globally) are process-based, describing SOM dynamics through first-order rate kinetics in one or more conceptual pools of carbon. A few models have been developed, which treat organic matter decomposition, and the subsequent change in the quality of the organic matter, as a continuum, with each fresh cohort of litter being tracked separately. Another small subset of models follows the

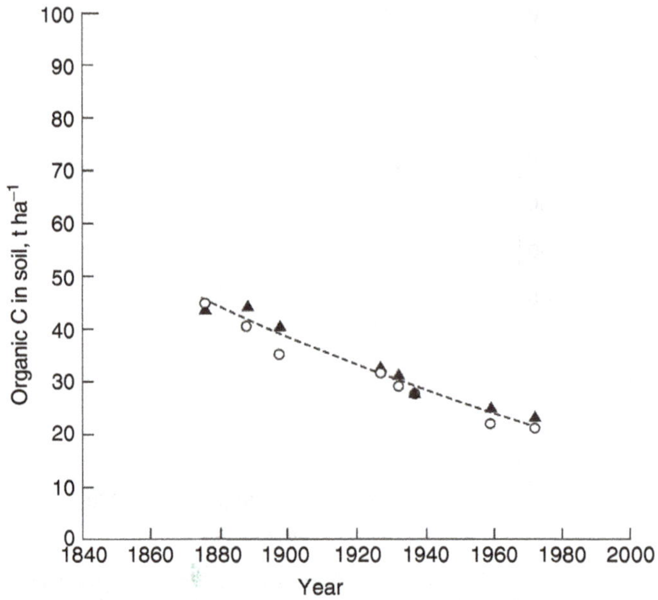

Figure 4.7 Decline in SOC content in a sandy soil (7% clay) over a period of 100 years. Growing a four-course rotation and adding manure (▲) did not halt the decline compared to continuous cereals given inorganic fertilisers only (○). Reprinted from Johnston et al. (2009). With kind permission from Elsevier.

flows of carbon and other nutrients and energy through the trophic web. These model types are described in more detail in Section 4.3.1.

4.3.1 Types of SOM model

Process-based models

Most models are process-based, i.e. they focus on the processes mediating the movement and transformations of matter or energy and usually assume first-order rate kinetics (Paustian, 1994). Early models simulated SOC as one homogeneous compartment (Jenny, 1941). A few years later, two-compartment models were proposed to reflect increased understanding (Beek and Frissel, 1973), and as computers became more accessible, multi-compartment models were developed (McGill, 1996; Molina and Smith, 1998). Of the 33 SOC models currently represented within the Global Change and Terrestrial Ecosystems (GCTE) Soil Organic Matter Network (SOMNET) database (Smith et al., 2001), 30 are multi-compartment, process-based models. Each compartment or SOC pool within a model is characterised by its position in the model's structure and its decay rate. Decay rates are usually expressed by first-order kinetics with respect to the concentration (C) of the pool: $dC/dt = -kC$, where t is the time. The rate constant k of first-order kinetics is related to the time required to reduce by half the concentration of the pool *when there is no input*. The pool's half-life [$h = (\ln 2)/k$], or its turnover time ($\tau = 1/k$) are sometimes used instead of k to characterise a pool's dynamics: the lower the decay rate constant, the higher the half-life, the turnover time and the stability of the organic pool (Molina and Smith, 1998).

The flows of *C* within most models represent a sequence of carbon going from plant and animal debris to the microbial biomass, then to soil organic pools of increasing stability. Some models also use feedback loops to account for catabolic and anabolic processes and microbial successions. The output flow from an organic pool is usually split. It is directed to a microbial biomass pool, another organic pool, and, under aerobic conditions, to CO_2. This split simulates the simultaneous anabolic and catabolic activities and the growth of a microbial population feeding on any given substrate. Two parameters are required to quantify the split flow. They are often defined by a microbial (utilisation) efficiency and stabilisation (humification) factor which control the flow of decayed C to the biomass and humus pools, respectively. The sum of the efficiency and humification factors must be inferior to unity to account for the release of CO_2. Examples of the structure of two commonly used SOM models are shown in Figure 4.8. A recent advance for allowing multi-pool models to be tested rigorously against data comes from fractionation techniques that yield fractions that are equivalent to measurable pools, so model simulations of the size of individual pools can be tested experimentally in addition to simulation of total SOC (Smith et al., 2002). Zimmermann et al. (2007) proposed an experimental fractionation scheme that yields measurable fractions that approximate to the conceptual pools in the RothC model, providing new opportunities for model initialisation and for testing the performance of models, component by component.

Cohort models describing decomposition as a continuum

Another approach to modelling SOC turnover is to treat each fresh addition of plant debris into the soil as a cohort (McGill, 1996). Such models consider one SOM pool that decays with a feedback loop into itself. Q-SOIL (Bosatta and Ågren, 1995), for example, is represented by a single rate equation. The SOC pool is divided into an infinite number of components, each characterised by its 'quality' with respect to degradability as well as impact on the physiology of the decomposer organisms. The rate equation for the model Q-SOIL represents the dynamics of each SOC component of quality q and is quality-dependent. Exact solutions to the rate equations are obtained analytically (Bosatta and Ågren, 1994).

Food-web models

Another type of model simulates C and N transfers through a food web of soil organisms, and such models explicitly account for different trophic levels or functional groups of biota in the soil. Figure 4.9 shows the structure of two food-web models (Hunt et al., 1987; de Ruiter et al., 1993). Some models have been developed, which combine an explicit description of the soil biota with a process-based approach (McGill et al., 1981). Food-web models require a detailed knowledge of the biology of the system to be simulated and are usually parameterised for application at specific sites.

4.3.2 Factors affecting decomposition in SOC models

Rate 'constants' (k) are constant for a given set of biotic and abiotic conditions. For non-optimum environmental circumstances, the simplest way to modify the maximum value of k is by multiplication by a reduction factor μ – ranging from 0 to 1. Environmental factors

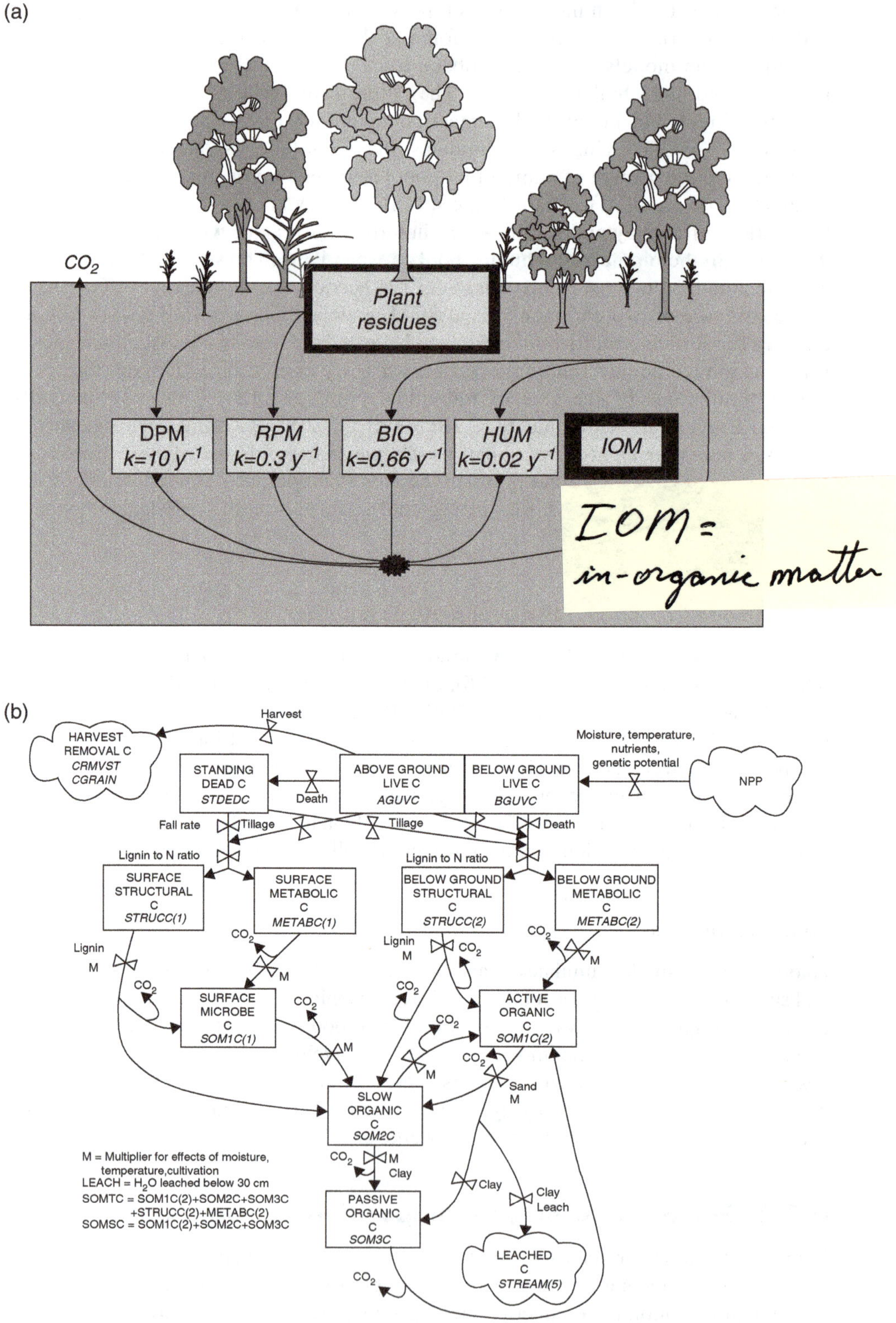

Figure 4.8 The structure of two commonly used SOM models. (a) RothC, (b) CENTURY.

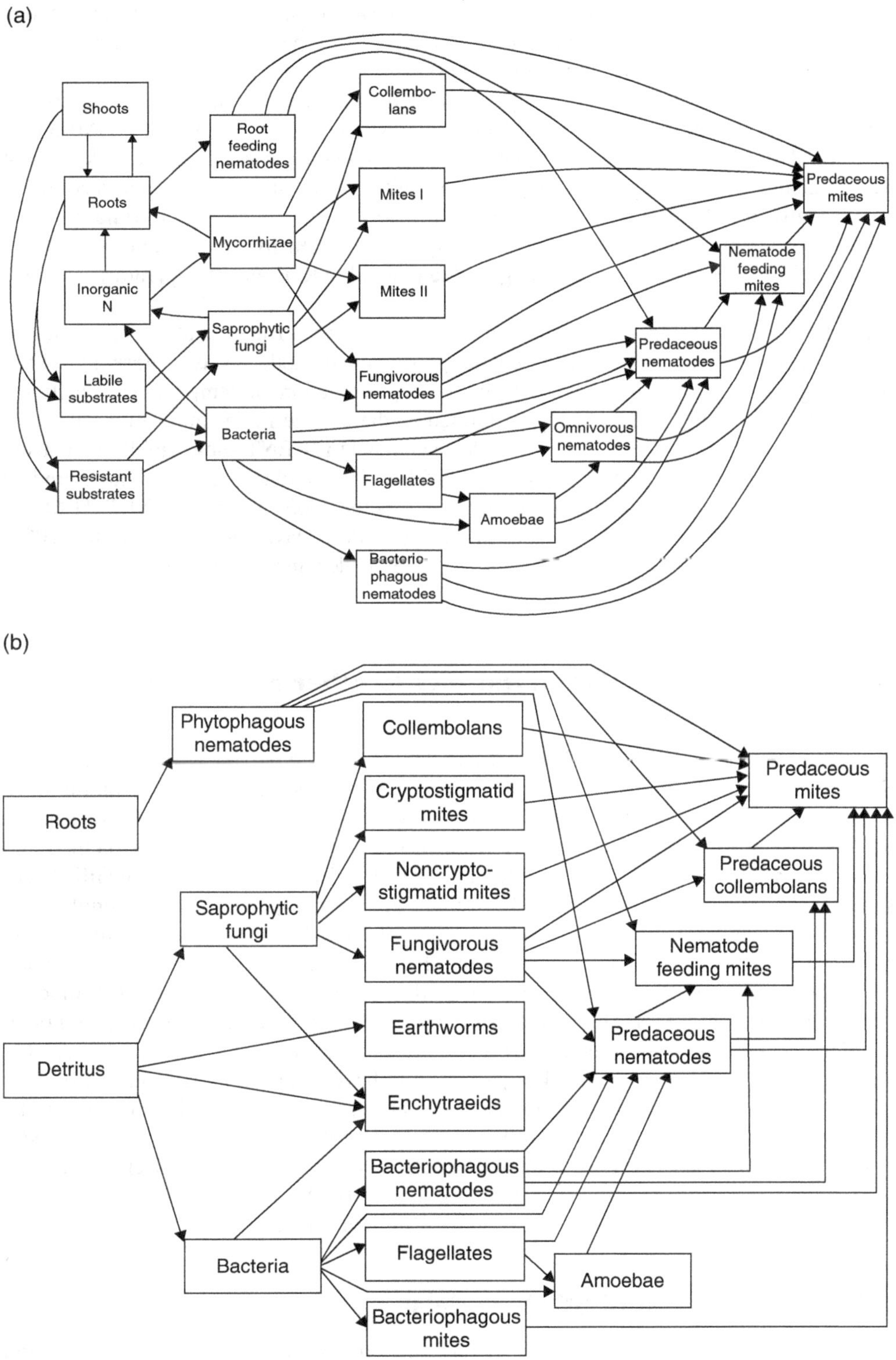

Figure 4.9 The structure of two commonly used food-web models. (a) Hunt et al. (1987), (b) de Ruiter et al. (1993). Reproduced with kind permission from Springer Science and Business Media.

considered by SOC models include temperature, soil water content, pH, nitrogen availability, oxygen concentration in soil air space, clay content, CEC, type of crop/plant cover and type or intensity of tillage (Molina and Smith, 1998).

Many studies show the effect of temperature on microbially mediated transformations in soil, either expressed as a reduction factor at lower temperatures or by the Arrhenius equation. However, the assumption that SOC decomposition was temperature-dependent was challenged by a study suggesting that old SOM in forest soils does not decompose more rapidly in soils from warmer climates than in soils from colder regions (Giardina and Ryan, 2000). This conclusion has been strongly challenged by other studies (e.g. Fang et al., 2005), with Knorr et al. (2005) suggesting that old SOC is more temperature-sensitive than young SOC.

Models represent temperature sensitivity in different ways. Burke et al. (2003) reviewed how temperature sensitivity in different biogeochemical models was represented. They showed great variation among models in the representation of temperature sensitivity, represented by the Q_{10} response. Q_{10} is defined as the change in decomposition rate for a change in temperature of 10°C. Some models such as DAYCENT show rapidly increasing Q_{10} at lower temperatures, while others such as CENTURY and RothC have a more gradual increase in Q_{10} at lower temperatures. Others, such as the 2003 versions of pNET, TEM and Biome-BGC, have a fixed Q_{10} of 2 at all temperatures. Davidson and Janssens (2006) reviewed the subject of temperature sensitivity of SOC decomposition, and it is considered further in Section 4.9.3.

4.3.3 Coupling of SOM dynamics to other ecosystem processes

Some models such as RothC (Coleman and Jenkinson, 1996) and ICBM (Andrén and Kätterer, 1997) only simulate SOC transformation, with no simulation of carbon inputs to the soil. Instead, values for C inputs have to be supplied by the user, based either on estimation or derived from measurements, or from fitting to initial SOC levels (Coleman and Jenkinson, 1996). Other SOC models (e.g. CENTURY, EPIC and DNDC; McGill, 1996) simulate the whole ecosystem: these have plant growth components which simulate the growth of plants and direct carbon (through roots, exudates, debris, etc.) to the soil components of the model. Models containing the greatest feedbacks between climate and soil C turnover are the coupled climate carbon cycle models (termed C4 models; Friedlingstein et al., 2006). These models not only have coupling between soil C and plant C inputs, but plants and soils also respond to, and feedback to, the climate via the atmospheric CO_2 concentration. C4 models tend to have a simpler representation of soils than dedicated soil or ecosystem models (Jones et al., 2005). Though C4 models can be used to examine climate impacts on soil C in a fully coupled way (Jones et al., 2005), ecosystem models can also simulate climate impacts (without climate feedback) through use of climate scenario driving data (e.g. Muller et al., 2007). Dedicated soil models can also be used, by deriving C inputs from other models or data sources (e.g. Smith et al., 2006).

4.3.4 Application of SOM models

SOM models are used for a variety of purposes, over a variety of timescales and spatial scales and have become an integral part of SOM research. SOM models are commonly used to test our understanding of SOM dynamics in given environments or at given sites. Perhaps the most common type of study involves applying models to explain changes in measured SOM dynamics at a given site. Such studies can also be used to test and compare the performance of the models (e.g. Smith et al., 1997).

Numerous model applications are now using eddy covariance data to simulate diurnal, seasonal and inter-annual carbon dynamics, particularly from the soil's perspective for examining factors such as the contribution of soil heterotrophic respiration to total net ecosystem exchange (e.g. Sus et al., 2010). At the other extreme, SOM modules in larger global models have been used to examine long-term temperature sensitivity using data over millennia. For example, Lenton and Huntingford (2003) constructed a simple model of vegetation and soil and tested various temperature sensitivities of, among other things, plant and soil respiration. They compared the terrestrial carbon sink observed since the Last Glacial Maximum to test which figures in the range of temperature sensitivities of soil respiration found in the literature could explain the observations and found that the upper limit of reported sensitivities (Q_{10}=3.63) under-predicted observed carbon storage. Such studies, using earth system observations/reconstructions over many thousands of years, allow long-term soil carbon responses to be tested, facilitating greater confidence in predictions of future responses to climate change.

SOM models remain valuable at the scale of a single field or land use unit in order to simulate and explain SOM dynamics in response to management practices at a specific site. But they are increasingly applied at a global scale (e.g. Jenkinson et al., 1991), often using detailed spatial data now available from remote sensing and other data collation exercises. Global Climate Models have moved from simple single-pool representations of SOM to having fully coupled SOM models integrated into them. RothC, for example, has been integrated into the Hadley Centre Climate Model and, when run globally, changed the behaviour of the model by dampening the response of the soil to climate fluctuations (Jones et al., 2005).

4.3.5 Confidence in SOM models: uncertainties and areas for further development

To test the validity of any model, it is necessary to compare model simulations with measured data. This was done in a comprehensive comparison of 9 models of SOC turnover, tested against 12 measured data sets from long-term experiments representing different land use, soil type, climate zone and land management (Smith et al., 1997). Through such evaluation, the applicability and limitations of models can be assessed. Where models can simulate SOC trends across a range of environmental conditions, they can be applied to future scenarios to project likely changes in SOC (Smith et al., 2008).

Areas of uncertainty requiring further research include the inability of current SOM models to represent potentially important ecological phenomena, including 'priming' of soil C decomposition in response to fresh organic C additions, particularly in subsoil (Fontaine et al., 2007; but see Jenkinson and Coleman, 2008; Jenkinson et al., 2008), the 'gadkil' effect and non-equilibrium dynamics (Schimel and Weintraub, 2003; Fontaine and Barot, 2005; Neill and Gignoux, 2006; Wutzler and Reichstein, 2007) and assumptions about the temperature sensitivity of different soil carbon pools (Davidson and Janssens, 2006). Other challenges include understanding the nature of impacts of land management change on the stability of different pools of soil carbon (e.g. the physical protection of organic matter – Denef et al., 2007) and the inclusion of microbial feedbacks of decomposition that are not captured by first-order assumptions (e.g. soil methanogenesis and methanotrophy).

4.4 Chemical structure of SOM as elucidated using new forms of spectroscopy – advances and pitfalls

What exactly do we mean by soil organic matter?

The term soil organic matter (SOM) is generally used to indicate the total of all organic components present in soil, including living organisms. Other definitions recall that suggested earlier by Waksman (1936) for 'humus', which excludes undecomposed or partially decomposed plant material and living organisms (Stevenson, 1994). At first sight this latter definition may seem more rational but the practical problem is that SOM defined in this manner would not coincide with any analysable fraction. This is because soil organisms (the soil microbial biomass) and microscopic plant and root debris cannot be reliably separated from the soil and are currently analysed as part of total organic C. The old word 'humus' itself is poorly defined and its meaning has not been completely agreed. It is generally agreed by soil chemists that, from a chemical point of view, SOM can be divided into non-humic and humic substances (Stevenson, 1994). The non-humic materials comprise organic substances that have defined chemical structures, such as carbohydrates, proteins, hydrocarbons, alcohols, aldehydes, resins and amino acids as well as aliphatic and aromatic acids. Humic substances are a largely heterogeneous mixture comprised of yellow to black-coloured polyphenolic polycarboxylic acids exhibiting a wide range of molecular weights. Although their functional groups and reactivity may be described in great detail, their actual chemical structure is not yet clearly defined. In fact, it is likely that there is no universal structure that can be defined in a rigorous chemical sense.

4.4.1 Structural information from nuclear magnetic resonance (NMR) spectroscopy

The development of modern solid state ^{13}C NMR techniques and instrumentation has led to substantial progress in SOM studies. The availability of high field magnets, from 600 to 1000 MHz, now permits the study of whole soil samples and allows the characterisation of SOM in situ, avoiding the need to extract organic materials and separate them from inorganic components. The extraction step, to provide a liquid sample, was unavoidable before the development of solid state ^{13}C NMR. But it has always been the most troublesome, for two reasons. With mild extractants, the recovery of SOC is low, so it is uncertain whether or not the material obtained is representative of the total. To obtain higher recoveries, solutions such as strong alkalis have been used but it is likely that these cause chemical alteration through hydrolysis reactions.

The main limitation of solid state NMR still remains the fact that solid state spectra cannot be recorded directly on soils rich in iron oxides because of interference from paramagnetic materials, notably Fe. Pre-treatment of soil with hydrofluoric acid is sometimes used to remove mineral components, including those containing Fe. However, removal of Fe is not always complete and the treatment may alter SOM chemical composition, thus nullifying the advantage of applying NMR spectroscopy to solid soil rather than a liquid extract. Even so, solid state NMR studies have substantially changed our perception of SOM composition: the most important change being the decreased importance of aromatic structure compared to aliphatic and the greater significance given to hydrophobic interactions (Mahieu et al., 1999; Ahmad et al., 2006). As a consequence new concepts have been developed to explain the complex chemical behaviour and intrinsic recalcitrance of SOM. One such concept is the hypothesis that a substantial fraction of SOM can be regarded as supramolecular aggregations based on hydrophobic interactions (Simpson et al., 2001) – this is discussed later.

Another limitation of ^{13}C NMR spectroscopy is poor sensitivity, in part because NMR only records C atoms having the relatively rare ^{13}C nucleus, comprising only 1.1% of the C in natural materials. Sensitivity is increased by a technique termed 'cross-polarisation' (CP) in which energy from adjacent ^{1}H nuclei is used to boost the signal from ^{13}C nuclei, but this can lead to overestimation of the quantity of C in alkyl groups (that are generally abundant in H atoms) and underestimation of C in aromatic groups that generally have fewer H atoms in comparison to C.

Solid state ^{13}C spectra are normally obtained using a combination of cross-polarisation and 'magic angle spinning' (MAS), the latter to overcome complications that otherwise make it impossible to obtain spectra from solid samples. This approach is termed ^{13}C CPMAS NMR. A comprehensive analysis of ^{13}C CPMAS NMR data from more than 300 soils, representing a wide range of SOM contents, climates and management practices (Mahieu et al., 1999), showed the remarkable similarities of functional group abundance. The largest fraction of soil organic C was within O-alkyl groups (on average 41–46% of total visible C), followed by alkyls (25%), aromatics (20%) and finally carbonyls. Grassland soils were the poorest in alkyls and richest in carbonyls, with the opposite for forest soils. The proportion of O-alkyls increased with C content, reaching up to 60% in high C soils. The general similarity of SOM structure between soils is not surprising. In spite of the great

Solid state nuclear magnetic resonance spectroscopy of SOM: limits and advancements

Solid state NMR spectroscopy, although still far from being the ultimate analytical tool, has proved to be very powerful. Yet only 30% of the soil organic C may be visible by NMR, due to various interferences and overall lack of sensitivity. Comparison of spectra recorded on soil samples before and after treatment with hydrofluoric acid (used to destroy clay minerals and remove Fe which interferes with NMR spectra) suggests that the distribution of organic C in solid state spectra is reasonably representative of the total. However, results obtained using more sophisticated techniques lead to the opposite conclusion. For example, aromatic and carbonyl C signals appear more intense in so-called 'Bloch decay' spectra, suggesting that they are underestimated in the more usual forms of ^{13}C NMR spectroscopy. It is possible that much of the C that is invisible in solid state CPMAS spectra could correspond to charcoal or black C originated from biomass burning or wild fires, though this remains to be seen.

Until the introduction of cross polarisation magic angle spinning (CPMAS), information gained by NMR was purely qualitative – thus these techniques represented a major move towards quantitation. However, as mentioned in Section 4.4.1, CP leads to disproportionate enhancement of C atoms located close to H atoms within molecules, thus exaggerating the abundance of alkyl chains at the expense of aromatic rings. A further inaccuracy is the formation of artefacts termed 'spinning side bands' that overlap with genuine signals, so appropriate measures must be taken in order to achieve the 'total suppression of side bands' (^{13}CPMAS TOSS NMR).

chemical complexity and range of input sources, above- and below-ground plant residues which constitute the main C inputs to soils comprise for the most part (>90%) a relatively narrow mixture of organic compounds: 50–60% polysaccharides (cellulose, hemicelluloses, starch, pectin), 15–20% lignin and 5% proteins. The abundance of the other compounds (tannins, waxes, lipids) is much more variable, but always limited to a few per cent.

In addition to organic entities synthesised within the soil as a result of biological processing of inputs, chemical species directly derived from plants can also be detected in soil. This is because some processes such as the action of tillage or fauna foster contact of plant residues with clay minerals and oxides (Wolters, 2000), resulting in physical stabilisation making these potential organic substrates unavailable to decomposers (Sohi et al., 2001; Six et al., 2002).

In natural ecosystems, above ground plant residues are deposited at the soil surface and enter the soil by the mixing action of invertebrates such as earthworms, insects and arthropods at various stages of their life cycle. Under temperate climatic conditions, the activity of earthworms is so crucial for the formation of stable soil aggregates that it is included in diagnostic criteria for humus forms and soil classifications (Zanella et al., 2011). Soil ingested by detritivores undergoes many alterations, including physical realignment of clay particles, which incorporate undigested fragments and favour the formation of organo-mineral complexes (Bossuyt et al., 2005). Aggregates formed by soil mixing and tillage can contain discrete particles of freshly incorporated plant residue (termed particulate organic matter, POM) and protect them from decomposition (Gregorich and Ellert, 1993). POM and the so-called light fraction (LF) are operationally defined fractions of SOM that are frequently used in soil C turnover studies. POM is isolated by sieving (size separates between 53 and 2000 μm) and makes up about 6% of soil C in cultivated soils and about 18% C in grassland and forest soils. Smaller fragments of POM, which make up the LF (isolated by density separation, suspending soil in a heavy liquid, density >1.9 g cm^{-3}), often comprise about 28% of soil C in forest soils and sometimes up to 20% in arable and grassland soils (Gregorich et al., 2006). So, in total, physically protected and non-protected chemically unaltered plant materials may sometimes comprise as much as 30–40% of SOM. On this basis, we can estimate that at least 5–8% of soil C is made up of polysaccharides of direct plant origin.

Cellulose, the most abundant natural polymer, is composed of approximately 8×10^3 D-glucopyranose residues linked by β 1→4 glycosidic bonds. About 36 of these glycan chains are held together by hydrogen bonds to form a cellulose microfibril, which binds to hemicellulosic polysaccharides, such as xylans, mannans and galactans, which have a lower degree of polymerisation and more branching than cellulose (Kogel-Knaber, 2002). Besides hemicelluloses, cellulose fibrils are closely associated with lignin in secondary cell walls of plants. Other common polysaccharides in plants are starch and fructans, but being much more easily degradable their presence in soil is transient.

Solid state ^{13}C CPMAS NMR allows monitoring of changes which occur during decomposition of plant materials in soil. It is observed that the proportion of alkyl C increases compared to O-alkyl C as decomposition proceeds as a result of polysaccharides being decomposed (Baldock et al., 1997). Figure 4.10 (from Webster et al., 2001) shows the ratio (alkyl C)/(O-alkyl C) in relation to the proportion of soil C in the microbial biomass in a set of soils from diverse environments and land uses. This shows a significant negative correlation between the (alkyl C)/(O-alkyl C) ratio and soil microbial biomass C content (expressed as a fraction of total organic C), consistent with the hypothesis that an increased proportion of alkyl C represents a greater degree of decomposition. These authors proposed

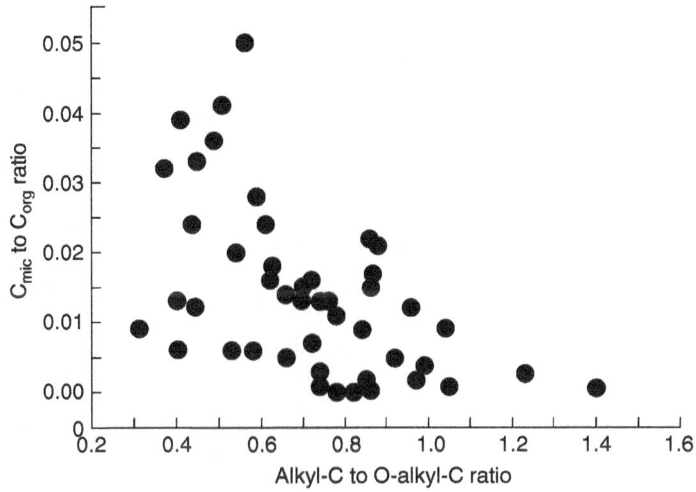

Figure 4.10 Relationship between the biomass C to total C (C_{mic} to C_{org}) ratio and alkyl-C to O-alkyl-C ratio derived from ^{13}C CP MAS NMR spectroscopy for 47 soils. The set includes mineral, organic and urban soils. Reproduced with permission from Webster et al. (2001).

the (alkyl C)/(O-alkyl C) ratio as an index of resource quality of soil C for the soil microbial biomass. This ratio is also negatively correlated with the C/N ratio of soil fractions, consistent with fractions having a larger proportion of polysaccharides being low in N.

Polysaccharides in SOM are not derived exclusively from plants but are also synthesised during decomposition of plant residues by soil microorganisms. These polysaccharides provide structural components for microbial cell walls, energy storage and extracellular mucigels which help bacteria to adhere to particle surfaces and provide protection from environmental stress. All these forms of polysaccharides accumulate during decomposition, partially masking the disappearance of carbohydrates from decomposing plant residues. Bacterial cell walls, particularly those of Gram-negative bacteria, contain murein, a carbohydrate backbone of alternating units of *N*-acetyl muramic acid and *N*-acetyl glucosamine cross-linked with oligopeptides. Fungal cell walls are mostly made of a different polysaccharide: chitin (*N*-acetyl-D-glucosamine in β-(1-4)-glycosidic bonds), which is also a structural component of the exoskeleton of insects. Microbially derived polysaccharides in soil cannot be separated from those of plant origin by physical isolation or selective hydrolysis. Their accumulation or disappearance can only be detected by analysing the composition of the products of hydrolysis with respect to the relative ratios of arabinose (A) and xylose (X), which are considered to be mostly originated from plants, whereas mannose (M) and galactose (G) in soil are mainly of microbial origin (Chantigny et al., 2000). When the ratio of (M + G) to (A + X) is lower than 0.5, the contribution is considered mainly from plant materials, whereas ratios higher than 2 are an indication of a microbial origin (Oades and Waters, 1991).

The most abundant polymers in plant residues, after cellulose, are lignins. Lignins are polydisperse polymers which are synthesised by random radical coupling reactions of phenylpropane units (Figure 4.11), namely, *p*-coumaryl, coniferyl and sinapyl alcohols (monolignols). This mode of synthesis produces irregular three-dimensional macromolecules with no extended sequences of regularly repeating units. Lignins are generally characterised by a high abundance of *p*-hydroxyphenyl propane (coumaryl alcohol), guaiacyl

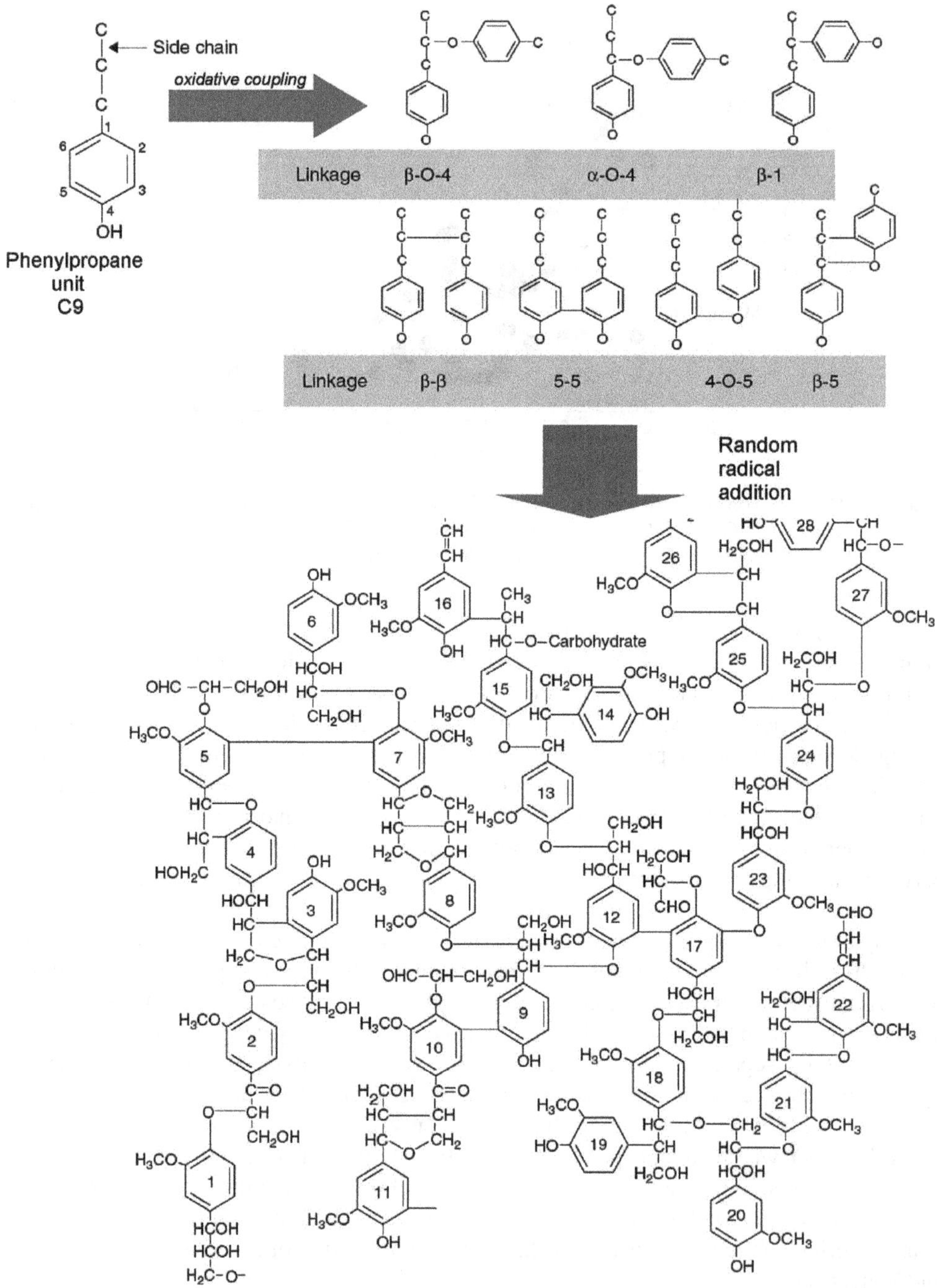

Figure 4.11 The formation of lignin from monolignol (phenylpropane) units dimerising by oxidative radical-mediated coupling leading to several possible types of linkage. The dimers randomly react together to produce a three-dimensional network of aromatic rings linked by an array of inter-unit linkages.

propane (coniferyl alcohol) and syringyl propane (sinapyl alcohol) units, derived from each of the three primary monolignols and by the distribution of inter-unit linkages in the polymer (e.g. β-aryl ether or β-O-4, phenylcoumaran or – β5, resinol or β-β, biphenyl or 5-5, diphenyl ether or 4-O-5). The few initial dimerisation reactions also produce hydroxycinnamyl alcohol end groups from structures such as non-cyclic α-aryl ethers and various other structures. Lignins of gymnosperms, angiosperms and grasses differ in their monolignol composition. Grass lignin is composed of equal proportions of the three primary monolignols, whereas that of gymnosperms (softwood lignin) derives in large part only from guayacyl propane monomers and that of angiosperms (hardwood lignin) from equal proportions of guayacyl and syringyl propane.

The irregularity and the aromaticity of its structure, together with the relatively large variety of linkages, makes lignin a recalcitrant and therefore a poorly degradable substrate for microorganisms. Only white rot fungi are able to completely mineralise lignin, and its decomposition in soil is probably carried out by a complex microbial community. The first step of lignin degradation, which can only occur under aerobic conditions, is demethylation of methoxyl groups, followed by oxidation of side chains. This leads to enrichment of the product in acidic functional groups (COOH and phenolic OH). Oxidation to quinones of orthohydroxy-benzene moieties resulting from demethylation favours condensation with NH_3 and amino-compounds. Lignin, however, is also depolymerised by fungi, which release dilignol components that are further transformed into coniferyl alcohol, coniferaldehyde, ferulic acid, syringaldehyde, siringic acid, vanillin, vanillic acid, etc. Lignin is therefore a source of polyphenols that in the soil environment are readily oxidised to quinones, which can eventually condense with amino-compounds.

Secondary components of plant and animal residues include a huge variety of complex organic molecules as well as relatively simple compounds, ranging from chlorophylls, sphingolipids, tannins, sterols, vitamins and nucleic acids to fatty acids and simple alkanes. Among these, a group of structurally different molecules, sharing a sufficient structural degree of hydrophobicity which makes them insoluble in water but extractable by non-polar organic solvents, are generally referred to as lipids (Bull et al., 2000). This empirically defined class of compounds is not unequivocally defined, as obviously much of its composition depends on the type of solvent used (e.g. hexane, acetone). It constitutes only from 4% to 8% of soil organic C but strongly influences the physical properties of soil aggregates (e.g. wettability, stability) and participates in many soil processes such as humus formation and clay dispersion. From a qualitative point of view, soil lipid extracts reflect the overlying vegetation and soil use. For example, the lipid extracts of the grazed grassland area of the Broadbalk Wilderness at Rothamsted are dominated by hexacosanol and other alkanol analogues (C_{18}–C_{34}); an adjacent area with herb vegetation is similar (Bull et al., 2000). Wax esters are extracted in much smaller quantities from these soils, indicating possible loss through hydrolysis or oxidation, and show an increased proportion of the lower simpler homologues. Root material, by contrast, is a predominant source of aliphatic dicarboxylic acids and of long-chain ($>C_{20}$) *n*-alkylcarboxylic acids, which derive from suberin.

4.4.2 Organic forms of nitrogen

^{15}N NMR and other spectroscopic methods of analysis show amide N to make up some 85% of the organic N in soil with a further 10% in amino acids, amino sugars, chitin, nucleic acids and heterocyclic N-containing aromatic rings. This dominance of amide N could be the result of an oversimplification in the interpretation of spectra, different spectral sensitivities to N in different chemical forms or by the overall poor sensitivity of ^{15}N NMR, which depends upon the low ^{15}N natural abundance (0.366%). However, the dominance of N in amide groups seems clear and is consistent with N contained in proteins. Proteins enter soil as a component of plant material and are also synthesised within the cells of soil microorganisms. They are located at different sites within cells, for example in cytoplasm or at different locations in cell walls, but are also released into soil solution. In fact, in order to break down plant residues and avail themselves of the energy and essential elements contained in natural macromolecules, microorganisms produce a range of extracellular enzymes which are released outside the membrane and then diffuse through soil solution to their specific substrates and hydrolyse them into small simpler units that can be utilised by the decomposing organisms (Lynd et al., 2002). However, proteins are among the most labile among natural macromolecules and, when added to soil under experimental conditions, normally persist for no more than 2 or 3 days, so it is surprising to find evidence of their dominant position within SOM. It is thought that this unexpected persistence is a result of stabilisation processes that include chemical entrapment by association with humic substances and physical adsorption on clay surfaces.

Total extractable protein concentration in soil is often around 1–2 mg protein C g^{-1} C (McClougherty and Linkins, 1988). Different proteins can fulfil different functions in soil. For example, glomalin is a glycoprotein that is a major component of the cell walls of the hyphae of mycorrhizal fungi. It remains in soil after hyphal death and is thought to have an important role in stabilising aggregate structure. It can be extracted from both native and agricultural soils with citrate solutions and is relatively resistant to decomposition because of its strong tendency to bind with multi-valent cations, particularly iron and aluminium. Glomalin is also partially hydrophobic: this may increase its stability and also confer stability on other molecules with which it interacts. Glomalin-related soil protein is operationally defined by its extraction method and cannot be regarded as a discrete chemical substance. Despite much discussion in recent years of its possible significance, its concentration in soil has probably been considerably overestimated (Gillespie et al., 2011).

Proteins that are exocellular enzymes, whether excreted from living soil organisms or remnants after cell death, can be stabilised in soil through association with other soil components such as sorption on clay surfaces or chemical or physical interactions with other organic materials. This gives rise to significant abiotic enzyme activity in soil that has been extensively measured and plays an important role in soil biological activities (Nannipieri, 2006; Kemmitt et al., 2008).

4.4.3 Black carbon

Black carbon or biochar is defined as residues derived from the combustion of vegetation, either through natural processes such as wildfires or human burning of vegetation. The term pyrogenic carbon is also used. If combustion is complete, the product comprises elemental C, or charcoal, but more commonly it is the result of incomplete combustion such that some chemical structures from the original plant material remains though the O/C ratio is low (pure charcoal would contain no oxygen). Black carbon can also include combustion condensates such as soot, derived from the burning of fossil fuel and reaching the soil through atmospheric deposition. It is thought that black carbon may comprise a significant proportion of the highly resistant fraction of soil C; some estimates are that it is up to 30% of SOM in agricultural soils in the USA and >45% of some German soils (Knicker, 2009), though whether these large values are generally true is not known. The mean residence time of black C in the environment has been assumed to exceed 1000 years. However, it is not inert and slow oxidation in aerobic soils occurs, either biotically or abiotically, leading to surface formation of carboxylic functional groups that can contribute to soil CEC (Knicker, 2009). Smaller fragments are preferentially found inside micro-aggregates, so that physical stabilisation probably concurs to increase persistence and retard oxidation processes (Brodowski et al., 2006).

There has been much recent discussion regarding the occurrence of black carbon in soils (Sohi et al., 2010) and its potential value as an additive to soils. Most evidence for beneficial effects comes from sites in the Amazon region with highly weathered soils where it appears that biochar was applied to soil at least 500 years ago and the impacts on soil properties persist. For example, there is circumstantial evidence from archaeological sites that the presence of biochar confers stability on other forms or organic C in soil (Liang et al., 2010) and that its modern application leads to the retention of nutrients in soil in high rainfall tropical environments (Major et al., 2010). However, there is as yet no unequivocal evidence that its addition to temperate soils leads to major improvements in soil functioning or plant growth (Powlson et al., 2011c) – research on this topic continues.

4.4.4 Modern concepts

The new insights provided by modern spectroscopic techniques have led scientists to rethink the nature of SOM and the process of humification, drawing attention towards classes of organic chemicals such as lipids and, for different reasons, charcoal that were previously considered minor constituents. In particular, the so-called supramolecular theory (see Schaumann, 2006) describes humic substances as associations of relatively small molecules arranged in micellar aggregations held together by hydrophobic interactions. The structures are hypothesised to contain some regions derived from relatively unaltered remnants of plant or microbial tissue interconnected with more degraded or newly synthesised small molecules. This theory contrasts with the classical view of SOM as comprising extremely large polymeric molecules. It is still a matter of debate as to which view of SOM is a more accurate representation – there is experimental evidence in support of both but neither completely accounts for the range of complex properties exhibited by SOM. See Swift (1999) for a critical review of evidence for both views, with a conclusion that the supramolecular theory, postulating an aggregation of relatively small molecules, is unlikely to explain the majority of SOM structure.

Another new theory, based on spectral evidence and improved knowledge of biotic processes, has recently highlighted the possibility that a class of recalcitrant secondary metabolites known as polyketides may be a significant component of the stabilised C pool that makes up the great majority of SOM residing in the smallest size fractions within soil (clay size particles and smaller) where the age of C is >1000 years (Schnitzer and Monreal, 2011). The polyketides comprise a very diverse family of natural products characterised by more than two carbonyl groups connected by single intervening carbon atoms, with diverse biological activities and pharmacological properties including antibiotics and natural insecticides. They are also very stable against biological decomposition. In addition to the stability of this class of compounds, it is hypothesised that the skeleton formed by polyketides can trap and bind other organics such as carbohydrates, peptides and lipids as well as inorganics such as clay minerals and hydrous oxides. This mechanism would confer additional stability on the trapped organic molecules, at least in part by holding them within a hydrophobic structure.

4.5 Organo-mineral interactions

Organic components in soil contain both hydrophobic and hydrophilic groups. Alkyl chains and aromatic rings are hydrophobic and thus have a strong tendency to avoid interaction with water, while hydroxyl, phenoxy and carboxyl groups are hydrophilic. Both classes of group can lead to an association between organic matter with minerals or amorphous oxides in soil that increase the stability of organic matter and also have profound impacts on soil physical properties. The main mechanism by which organic matter binds to minerals is through the formation of so-called cationic bridges (see Figure 4.12a). Clay particles have negative charge at their surface, which is balanced by association with cations such as the alkaline earth elements (most commonly Ca^{2+} or Mg^{2+}) or polycations such as Fe^{3+} and Al^{3+} derived from soil solution. These cations in turn are surrounded by water molecules which can be displaced by anionic groups (such as dissociated carboxyl or phenolic groups) on organic molecules. When organic entities bind to mineral surfaces, their stability against decomposition is increased because access to them by microorganisms or exocellular enzymes is limited. This stabilisation has been demonstrated by measuring the rate of decomposition of added organic materials to clays that are saturated with either Ca^{2+} or Na^{+} ions (Baldock and Oades, 1989). Lower evolution of CO_2 is observed with Ca^{2+} because this divalent ion favours such stabilisation through formation of bridges between organic matter and clays, whereas monovalent Na^{+} does not. Stabilisation through polyvalent cations such as Al^{3+} and Fe^{3+}, which can form strong coordination complexes with SOM, is even greater (Blaser et al., 1997). The decomposition of dissolved organic matter (DOM) extracted from a soil, then adsorbed on clays, was only 15–30% of the rate measured when the material was in solution (Schwezig et al., 2003). Sorption increased the mean residence time of the most stable DOM sample from 28 years to 91 years. The increase in stability due to sorption is greatest for labile DOM high in carbohydrates, and relatively small for stable DOM high in aromatic and complex molecules.

Figure 4.12b and c shows variations of organo-mineral interactions that are thought to occur. If the quantity of organic matter present is larger than in Figure 4.12a, crowding at the mineral surface can cause a different orientation of organic molecules (Figure 4.12b) such that the surface exposed to soil solution becomes more hydrophilic than in Figure 4.12a. If the quantity of organic matter becomes still larger, layers of organic molecules can build up via cationic bridges. This causes the surface to become increasingly hydrophobic (Figure 4.12c) and more like the situation in Figure 4.12a. These differences can have profound impacts on the surface properties of mineral particles influencing aggregate formation and wettability, both of which influence organic matter stabilisation (Bachmann et al., 2008).

A variation of the adsorption processes described earlier in this section is a process termed ligand exchange, which is particularly prevalent with iron and aluminium oxides, especially in tropical soils. In this process, a hydroxyl group on the mineral surface is displaced by an OH or oxygen atom belonging to an organic molecule. This group binds to an Al or Fe atom and becomes part of the mineral's structure, while the original OH group exits, becoming part of a water molecule or a solvated OH^{-} group. The organic molecule connected to the attached oxygen-containing group is now firmly linked to the mineral structure. The process is highly irreversible and results in permanent sequestration, so that even organic molecules that would otherwise be labile become unavailable for decomposition by microorganisms.

Another mechanism leading to binding between organic and mineral particles is through the hydrophobic sections within organic matter. These groups, and organic xenobiotics such as organic pollutants entering soil, can escape interactions with water either by binding with

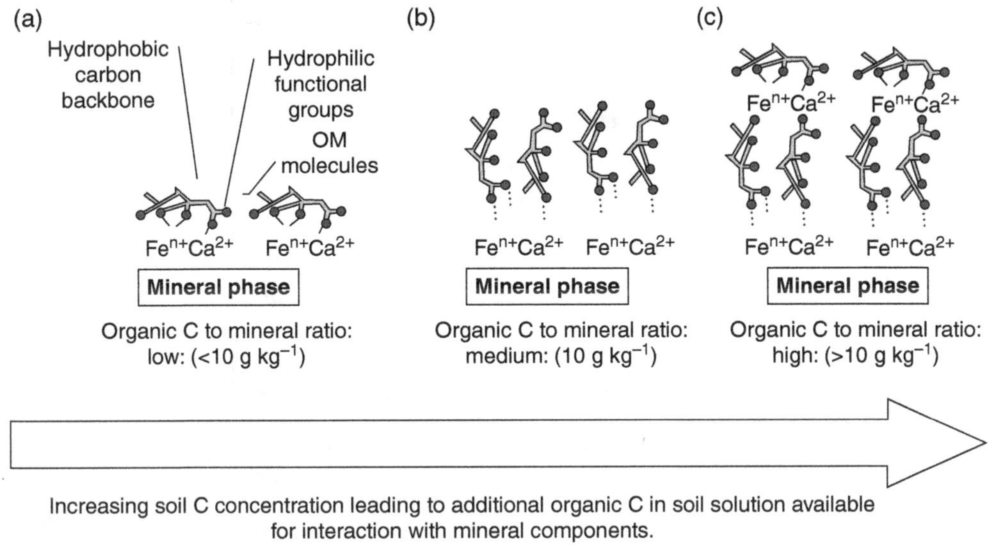

Figure 4.12 Different types of associations between organic molecules and mineral surfaces at (a) low, (b) medium and (c) high organic carbon:mineral ratios. Redrawn from Bachmann et al. (2008). With kind permission from John Wiley & Sons.

hydrophobic sites on other organic molecules or by binding to solid surfaces such as clay particles and aluminium or iron oxides. The bonding between hydrophobic groups and a mineral surface is through van der Waals forces which are weak. However, the process is favoured on the basis of thermodynamics. When hydrophobic groups are present in the aqueous phase, water molecules cannot form a solvation layer around the hydrophobic structure but instead, in an effort to avoid contact, tend to form an ordered cage-like structure in the vicinity and consequently lose freedom of movement. This represents a substantial decrease in entropy. When the organic molecule is bound to a mineral surface, the energy involved in forming the cage structure of water molecules is released and entropy increased. It is these thermodynamic factors that tend to drive the binding of hydrophobic regions of organic matter to mineral surfaces, even if the actual binding forces are weak.

The coating of clays and iron or aluminium oxides by organic molecules modifies their surface properties and eventually reduces the surface area accessible for other reactions: Figure 4.13 shows some examples of this (Kaiser and Guggenberger, 2003). In the experiments illustrated, organic matter bound to the different mineral fractions was destroyed (by oxidation with sodium hypochlorite) and then the specific surface area (SSA) of the particles determined by measuring adsorption of a monolayer of N_2 gas using standard methodology. The particles were then treated with a solution containing organic matter that had been extracted from an organic soil horizon using water and the SSA measured again. Figure 4.13 shows that the initial quantity of sorbed organic matter caused the largest decrease in SSA, and this was interpreted as organic matter being preferentially sorbed on surfaces at the mouths of micropores and thus blocking further potential sorption sites within these pores. This provides a mechanism whereby a small increase in the organic matter content in a soil may have a disproportionately large effect on the soil's physical properties. Preferential binding at reactive sites also favours surface accumulation of organic matter in patches rather than in a continuous coating, even on an initially uncovered mineral surface.

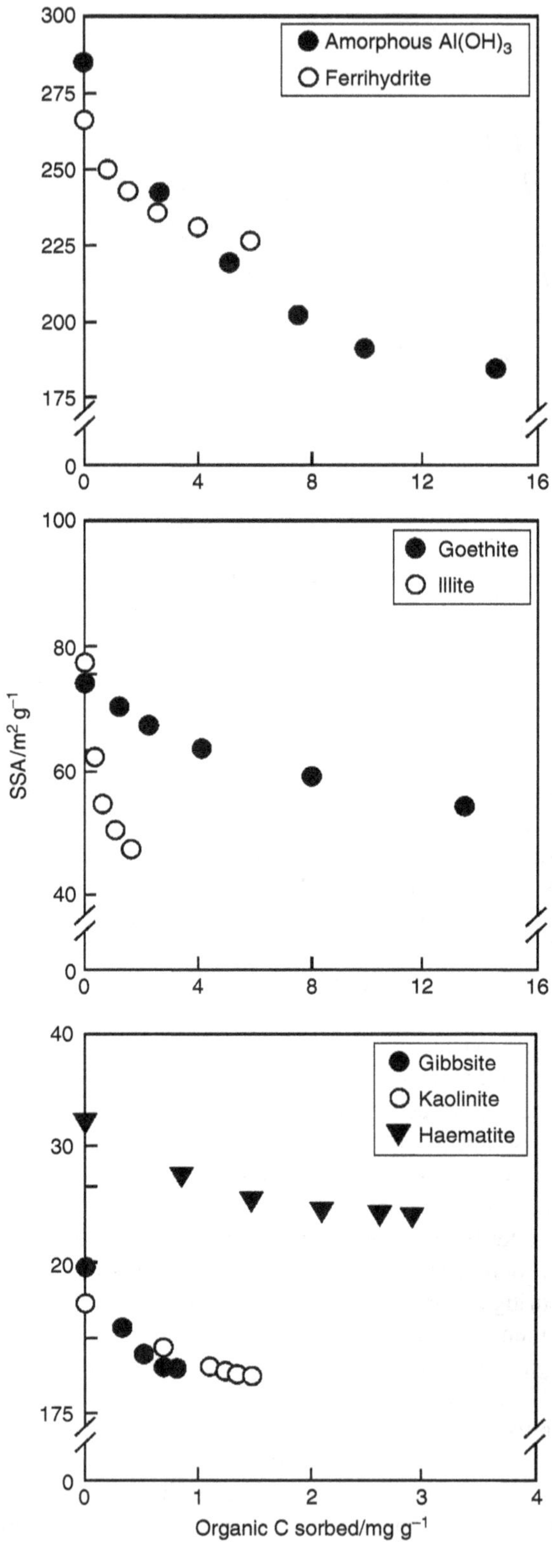

Figure 4.13 Effect of increasing amounts of sorbed organic C on the specific surface area (SSA) of seven mineral phases. Reprinted from Kaiser and Geggenberger (2003). With kind permission from John Wiley & Sons.

4.6 Fractionation methods for studying SOM

4.6.1 Fractionation based on solubility

An approach that has been used for many years is based on extraction of soil with alkali (e.g. 0.1 M NaOH) followed by acidification of the extract with H_2SO_4, though many variations exist. Conventionally organic matter that is precipitated after acidification is termed 'humic acid' while that which remains in solution is termed 'fulvic acid'. The material that is not extracted by alkali is termed 'humin'. These fractions have no specific significance in terms of chemical structure. This point was stated succinctly by Jenkinson (1973) in an earlier edition of this book who stated that they '... cannot be defined in precise chemical terms, despite the efforts of generations of chemists. They are complex mixtures, no two humic molecules being exactly alike, and cannot be crystallized or otherwise separated into classes of homogeneous molecules'. In other words, these fractions are 'operationally defined' and based on solubility at different pH values. There is also considerable likelihood that the extractant solutions will cause chemical changes and it was shown over 60 years ago that humic colloids in alkaline solution take up atmospheric oxygen, causing chemical change and an increase in their pH. Despite the clear limitations of this approach, one still sees reference to it in the soil science literature. In some cases, alkali extraction is simply used as a means of obtaining a liquid that can be subjected to various analyses that cannot be applied to solids; this was discussed in Section 4.4.1 in relation to obtaining material for liquid state NMR spectroscopy.

4.6.2 Physical fractionation approaches

Measuring the distribution of particles of different sizes (soil texture) is achieved by the total dispersion of soils into primary particles (usually using ultrasonics) followed by sedimentation. This is normally done after destroying organic matter using H_2O_2. But if dispersion is done without organic matter destruction, it is possible to determine the location of organic matter in relation to the different sizes. A useful way of expressing results is to use the concept of an enrichment factor, E_C. E_C is defined as the ratio of organic C concentration (e.g. expressed as mg g^{-1} or %) in a given fraction to that in whole soil. A value >1 indicated that the fraction is enriched in C compared to whole soil while a value <1 indicated depletion compared to whole soil. Table 4.1 (taken from Christensen, 1996) shows some typical results from soils from four continents. Particles in the clay and fine silt size categories are always enriched in organic C compared to whole soil: E_C values of about 2.0 or higher are common. Coarser particles, such as coarse silt (50–2 µm), are depleted in C, typically with E_C values of about 0.1. Sand size fractions are also typically low in C with E_C values <0.1, except in soils containing much macro-organic matter (such as fresh plant fragments) as these can appear in the sand-size fraction.

Figure 4.14 (also from Christensen, 1996) demonstrates the importance of clay and the finer silt particles for the retention of organic C in soil. For example, in a soil containing only 3% clay this fraction contains almost 50% of the total organic C in the soil. In the examples shown, with clay content ranging from 3% to 47%, the sand fraction contains a maximum 10% of total soil organic C, and usually much less. In soils with very low clay content, the silt fraction becomes highly significant for C retention: it contains about 40% of total soil C in the 3% clay soil. But as clay content increases the role of silt particles diminishes.

Table 4.1 Organic carbon enrichment factors, E_C, in different silt and clay size classes isolated by ultrasonic dispersion from predominantly cultivated soils.

Location (sampling depth)	USDA soil taxonomy	%C	Carbon enrichment, E_C								
			Silt (µm)						Clay (µm)		
			50–20	50–5	50–2	20–5	20–2	5–2	2–0.2	<2	<0.2
USA (0–25 cm)	Typic Argiudoll	2.18	0.2				1.1		2.2		2.3
Canada (Ap horizon)	Typic Argiboroll	1.71		0.6				3.8			
Canada (Ap horizon)	Udic Haploboroll	3.32		0.7				3.0			
Canada (A horizon)	Typic Cryoboroll 1.3	2.41		0.6				2.5			
Canada (Ap horizon)	Typic Haploboroll	2.50	0.1			1.1		2.4	2.6		2.1
Canada (Ap horizon)	Boralfic Cryoboroll	4.50	0.2			1.0		1.6	1.4		1.4
Canada (Ap horizon)	Aridic Natiboroll	2.10	0.1			1.2		2.9	2.8		2.4
Canada (0–30 cm)	Aridic Haploboroll	6.29	0.3			1.0		1.5	1.1		1.0
Australia (1–10 cm)	Aridic Calciaquo	5.35	0.2			1.2		1.3	0.9		0.5
USA (0–20 cm)	Typic Hapludalf	1.51	0.2				1.3		2.8		3.8
USA (0–10 cm)	Udollic Ochraqualf	1.13	0.1				0.6		2.9		2.6
Australia (0–10 cm)	Vertic Rhodoxeralf	1.04	0.2			0.8		1.9	2.6		3.0
Denmark (0–20 cm)	Typic Hapludalf	1.8	0.1				2.0			3.3	
Germany (Ap horizon)	Typic Eutrochrept	2.46					1.0			2.6	
Canada (0–15 cm)	Aquic Eutrochrept	2.31			0.6					2.7	
Australia (0–10 cm)	Chromic Pelloxererts	1.14			2.1				1.6		1.0
Brazil (0–20 cm)	Typic Haplorthox	1.41	0.7				2.2			1.5	
Denmark (0–21 cm)	Orthic Haplohumod	1.73	0.3				7.0			11.1	

The size class showing the greatest enrichment within a soil is underlined. E_C is defined as (C concentration in size fraction)/(C concentration in whole soil).
Source: From Christensen (1996).

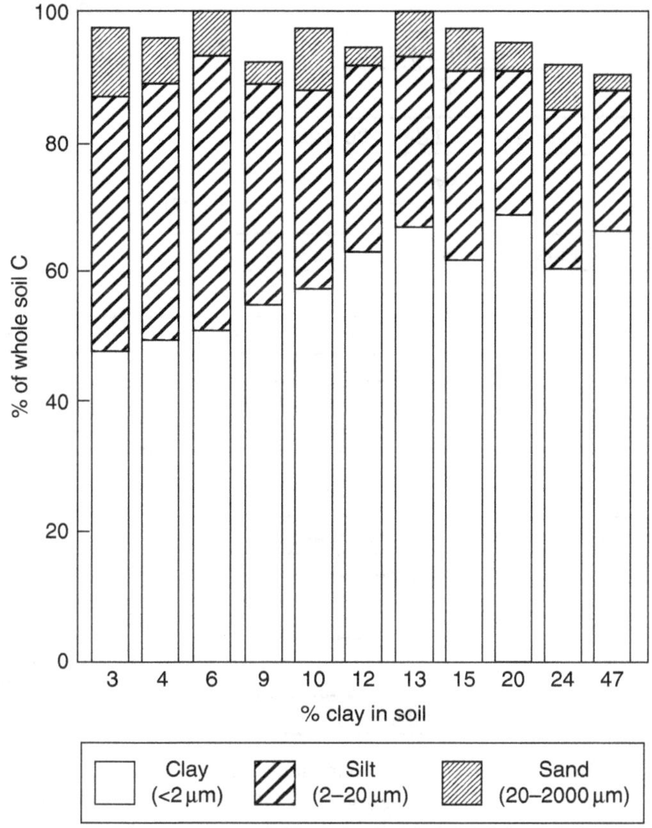

Figure 4.14 Distribution of whole soil organic C between clay (<2 µm), silt (2–20 µm) and sand (20–2000 µm) from Ap horizons of Danish arable soils. From Christensen (1996). With permission from CRC Press.

In addition to association of organic C with primary mineral particles, in recent years increased emphasis has been placed on the role of secondary structures within soil in stabilising organic C. For example, Killham et al. (1993) concluded that organic matter located within small pores was partially protected from decomposition by microorganisms. Many studies reviewed by Six et al. (2002) have emphasised the role of aggregates. It was concluded that macro-aggregates (defined as those >250 µm) provide little protection but micro-aggregates (<250 µm) offer a considerable degree of protection. Evidence comes, in part, from experiments in which aggregates of different size are crushed or disrupted in other ways and the effect on C mineralisation is measured. Disrupting macro-aggregates causes only a small increase in mineralisation rate, but with micro-aggregates it was three to four times greater (Bossuyt et al., 2002). Other evidence comes from experiments in which fresh substrates are added to soil; for example, Angers et al. (1997) incubated wheat straw labelled with ^{13}C and ^{15}N with soil under field conditions and found that the labelled C retained in soil was predominantly held within micro-aggregates. Micro-aggregates conferring stability on organic C can either be free in soil or contained within macro-aggregates. Although it is micro-aggregates that have the greatest influence on C stabilisation, it is thought that their formation is favoured if they are held within stable macro-aggregates – so soil-management practices that favour the stability of macro-aggregates indirectly favour C

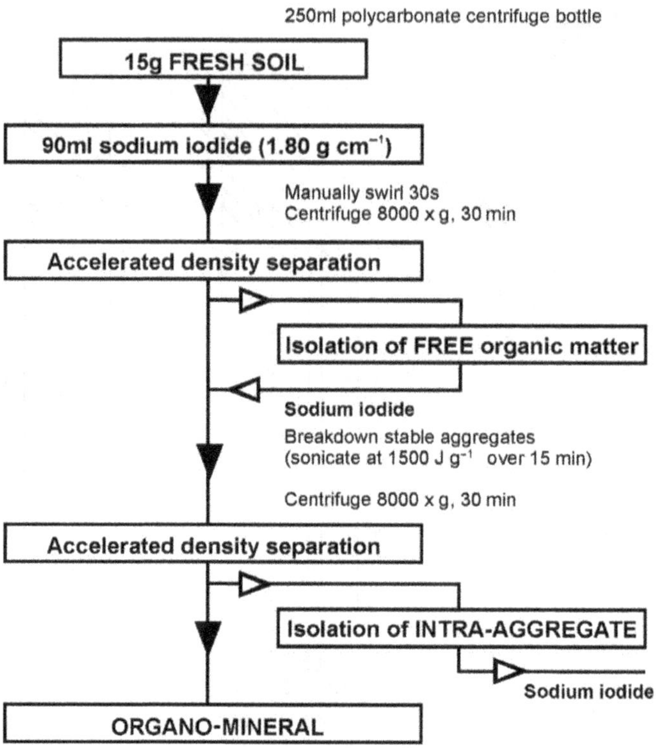

Figure 4.15 Physical fractionation scheme of Sohi et al. (2001). Soil is divided into free light fraction organic matter (FR-SOM), intra-aggregate light fraction organic matter (IA-SOM) and organo-mineral organic matter. The organo-mineral fraction can be further subdivided based on particle size. From Sohi et al. (2001). With permission, copyright Soil Science Society of America.

stabilisation within the micro-aggregates. This may be a significant factor in the accumulation of organic C near the soil surface where minimum tillage is practised as the formation and stabilisation of macro-aggregates is favoured under this practice compared to conventional tillage that causes greater disruption (Six et al., 2002; Denef et al., 2007). There is also evidence that the formation of micro-aggregates occluded within macro-aggregates, and the resulting protection of organic C against decomposition, is favoured by the activity of earthworms. Some of the organic matter stabilised through the influence of micro-aggregates is relatively undecomposed material derived from plants termed particulate organic matter (POM); indeed it seems likely that discrete POM particles form nuclei around which mineral particles accumulate and confer stability through the range of mechanisms discussed in Section 4.5.

Many physical fractionation schemes include a step in which soil is dispersed in a liquid of high density and a SOM fraction termed 'light fraction' floats to the surface. This fraction is predominantly composed of relatively fresh plant debris that has undergone only a small degree of decomposition, although it can also contain some charcoal or other forms of black carbon or biochar. The fractionation scheme published by Sohi et al. (2001) uses a very concentrated solution of sodium iodide (NaI) with a density of 1.8 g cm^{-3}. Soil is first dispersed with minimal disruption such that only macro-aggregates are broken and the floating material is termed 'free OM (FR-SOM)' or free light fraction (Figure 4.15). After its removal, the soil is again dispersed in NaI, but this time ultrasonic dispersion is used to break all aggregates including the more stable micro-aggregates, and it is found that additional organic matter is released and floats – this is termed 'intra-aggregate OM (IA-SOM)' and comprises discrete organic matter particles (sometimes termed 'particulate organic matter' or POM) that were previously entrapped within stable aggregates. A final fraction is the material that sinks in the high-density liquid and is mainly inorganic minerals with its associated organic matter, termed the organo-mineral fraction; this may be separated into size classes to give clay, silt and sand size fractions. Spectroscopic analyses (^{13}C NMR and a version of infrared spectroscopy) show that the intra-aggregate fraction contains a smaller proportion of C in O-alkyl groups and a larger proportion in alkyl groups compared to free OM fraction; i.e. IA-SOM had a smaller O-alkyl/alkyl ratio. This indicates that it has undergone greater alteration in structure due to decomposition. Other spectroscopic studies confirmed this conclusion and were consistent with the IA-SOM fraction containing a larger amount of microbial metabolites (Poirier et al., 2005). This difference in chemical properties of the two LFs (FR and IA) was also seen when the fractionation scheme was applied to soils from eight long-term experiments in widely differing environments including temperate regions in Europe and North America, an arid tropical site in Syria and two sites growing flooded rice (Sohi et al., 2005). It was therefore concluded that this difference was a general property of soils and gives further strength to the idea that physical location within soil structure is a significant factor in determining the stabilisation of organic matter in soil, in addition to chemical structure conferring an inherent resistance to decomposition.

4.7 The living part of SOM – the soil microbial biomass

Microorganisms and soil animals form a part of SOM and can be regarded as one specific fraction with extremely important properties. A common approach is to attempt the isolation of a single species or group from soil and then study its behaviour under laboratory conditions. An alternative is to regard the entire population as an entity: an analogy is the study of a whole forest rather than an individual tree. The following quotation (from Stockdale and Brookes, 2006) eloquently puts the case for the 'whole population' approach:

> Studies of single soil organisms, while useful in specialized cases, e.g. Rhizobia and mycorrhizae, do not yield information on the functioning of the soil ecosystem. This is because most important soil processes, e.g. carbon and nitrogen mineralization, depend upon interactions between entire suites of organisms, many of which still await identification and most of which remain unculturable. For many purposes, treating the soil microbial community as a single, undifferentiated unit, the soil microbial biomass (defined as all soil organisms $<5000\,\text{mm}^3$ volume), has much to commend it.

In addition to the activities of Rhizobia and mycorrhizae, there are additional soil processes that are of vital importance to soil functioning and plant growth and that are mediated by a single microbial species or small group. These include nitrification and denitrification, and in these cases it is logical to study the specific groups involved. The application of molecular methods in soil microbial ecology is a powerful approach that is bringing new insights, either on the activity of a defined group or species or on overall biodiversity in soil. But the case for the 'whole population' approach is extremely strong for many investigations.

During the last 30–40 years, a suite of techniques have been developed that permit the measurement and study of soil microorganisms as a whole unit, termed the soil microbial biomass. These methods are based on the finding that chloroform kills soil organisms, making their cell contents readily decomposable, and soluble in aqueous extractants, but does not alter the decomposability or extractability of non-living organic matter. In the original method soil was fumigated with chloroform and, after chloroform removal, the soil inoculated with a little fresh soil (to reintroduce living organisms) and CO_2 evolution measured for 10 days under standard conditions. Evidence was accumulated to demonstrate that the additional CO_2 evolved, compared to that from unfumigated soil, is derived from the decomposition of killed organisms by the reintroduced population (Jenkinson and Powlson, 1976a). By using a suitable conversion factor, the quantity of C held in the cells of the original population can be calculated (Jenkinson and Powlson, 1976b). This method, now termed fumigation-incubation (FI) has been largely superseded by the fumigation-extraction (FE) method in which the additional organic C extractable in 0.5 M K_2SO_4 is used as the measure of microbial biomass C (Vance et al., 1987; Wu et al., 1990).

Variations on the FE method now exist for measuring the quantities of N, P and S held in the microbial biomass – see Stockdale and Brookes (2006). Powlson (1994) also reviews the range of methods, their benefits and limitations, and the many pitfalls in applying them inappropriately and various ways in which they have been misused.

It is commonly found that the quantity of C in the microbial biomass is in the range of 2–4% of total organic C in soil. This translates to a surprisingly large quantity when considered at the field scale. In an old grassland soil, it has been estimated that the quantity of microbial biomass in topsoil is approximately equal to the mass of 100 sheep per ha! And the quantities of crop nutrients held in the microbial biomass are of significance to crop nutrition, especially because the rate of turnover of this fraction is faster than that of SOM as a whole (Kouno et al., 2002). For example, the quantity of P estimated to pass through the microbial biomass annually in a set of grassland soils was similar to or greater than annual P uptake by grass (Brookes et al., 1984).

A valuable application of biomass C measurements is to detect trends in total SOM content before they become measurable directly. For example, at two experimental sites in Denmark where the straw of spring barley had been either burned or incorporated for 18 years, no significant difference in soil total SOC or organic N content could be measured. But measurements of biomass C and N showed increases of 40% (Powlson et al., 1987). This is because the microbial biomass has a much higher turnover rate than total organic C or N and so, after a change in management practice, the biomass moves towards its new equilibrium value far quicker than total C or N. Powlson et al. (2011a) show other examples of biomass measurements acting as an 'early warning' of changes in total SOC before they become measureable by classical methods.

4.8 SOM and soil physical conditions and implications for plant growth

Soil physical properties are determined to a very great extent by soil texture but it has been well known for centuries that they are strongly modified by organic matter. For example, SOM increases water holding capacity, facilitates the formation of stable aggregate, increases formation of stable pore spaces that improve aeration and make root penetration

Table 4.2 Effect of soil organic C concentration on specific draught in the Broadbalk Experiment at Rothamsted, UK.

Long-term treatment	SOC %	Specific plough draught, *S* kPa
Nil	0.84	88
Farmyard manure	2.80	75
Inorganic fertilisers	1.08	77

Source: From Watts et al. (2006). Copyright Blackwell Publishing Ltd.

easier. Johnston and Dawson (2010) give numerous examples. A more recent realisation is that small changes in SOM content can have a disproportionately large impact on certain soil physical properties, almost certainly through the role of specific fractions within the totality of SOM (Powlson et al, 2011b). Some examples are given here to illustrate this.

Watts et al. (2006) used a specially designed ploughing apparatus to measure the energy required to pull the plough through soil on all the plots of the Broadbalk experiment at Rothamsted, UK. The measured factor was specific plough draught (S) defined as the force per cross-sectional area of worked soil in units of kPa. A larger value of S indicates that greater force is required to pull the implement through the soil; for ease of working and decreased fuel requirement, a low value of S is favoured. Because of spatial variability in the soil, clay content in topsoil varied between 19% and 39% between the different plots and this factor had the greatest impact on S. But organic matter content had a strong modifying effect. Table 4.2 shows data for three treatments after correcting for variation in clay content. The treatment-receiving FYM contained over three times the amount of SOC than the Nil plot receiving no manure or fertiliser. This very considerable increase in SOC led to a 15% decrease in specific draught, S, compared to the Nil treatment, reflecting the improved soil structure. The treatment-receiving inorganic fertilisers contained only 29% more SOC than the Nil treatment, yet the value of S was decreased almost as much as in the FYM treatment. Thus, a relatively small change in total SOC concentration had a disproportionately large impact on the energy required to pull an implement through the soil. In practical terms, this is significant because it is very difficult to increase SOC by a large amount in normal agricultural practice (see Section 4.2). This finding shows that even small changes can be beneficial and therefore worthwhile.

A similar finding was shown for the effect of SOC on aggregate stability and water infiltration. Blair et al. (1995) developed a technique to determine the so-called labile fraction of SOC (termed C_L) and soil organic N based on oxidation with dilute potassium permanganate. This reagent is relatively weak and only oxidises about 10% of total SOC, sometimes termed the readily oxidisable fraction. It is likely that this fraction comprises the soil microbial biomass (typically about 2% of total SOC in agricultural soils) plus microbial metabolites. It has been shown that the size of this labile fraction is correlated with certain soil physical characteristics, in particular aggregate stability (measured as mean weight diameter, MWD, after wet sieving) and water infiltration rate (indicated by the value of unsaturated hydraulic conductivity, K_{unsat}). The method was applied to plots in long-term experiments in Australia, Germany and the UK that covered a range of SOC contents due to past management practices including manure or fertiliser applications. As expected, soils at each site with higher SOC content showed greater aggregate stability and infiltration rate with a rough correlation between total SOC and MWD or K_{unsat}. However, the correlations between these measurements and labile C (as opposed to total SOC) were much closer. For example, at the UK site (Blair et al. 2006) the regression equations relating MWD to total SOC (C_T) or labile C (C_L), respectively, were:

$$\mathrm{MWD} = 0.12\mathrm{C_T} - 0.41 \quad \mathrm{r^2} = 0.43\,(\text{significant at P} < 0.05)$$

$$\mathrm{MWD} = 0.63\mathrm{C_L} - 0.12 \quad \mathrm{r^2} = 0.70\,(\text{significant at P} < 0.01)$$

As with the data on plough draught, this suggests that specific fractions with total SOC are particularly significant in influencing soil physical properties and that relatively small changes in total SOC content can have a disproportionately large impact. Some practical conclusions from this finding are:

1. Management practices that lead to only small increases in SOC can be worthwhile from the viewpoint of soil physical properties that influence management.
2. Even small decreases in total SOC resulting from soil management may have a greater negative impact on soil properties than might be deduced from the small magnitude of the change.

Whether or not improvements in soil physical conditions translate into improved crop growth depends on the soil type (whether or not it is inherently well structured) and the crop and environmental conditions. Clearly identifying such effects can also be difficult because organic matter in soil provides additional nutrients to crops, whether directly (e.g. from nutrients in added manure) or indirectly via mineralisation of the additional organic matter accumulated as a result of an ongoing management practice.

Figure 4.16 shows an example of additional organic matter in soil contributing to increased yields of arable crops. The data is from an experiment on sandy soil in the UK (Johnston and Poulton, 2005) where winter wheat or spring barley was grown following either (a) a long run of arable crops, (b) a 3-year grass ley or (c) a 3-year ley comprising a grass/clover mixture. Where the arable crops followed the ploughing up of the grass/clover mixture, yields with no added N fertiliser were almost 2 t ha^{-1} greater than following arable cropping. SOC would have been slightly greater following the grass/clover ley compared to long-term

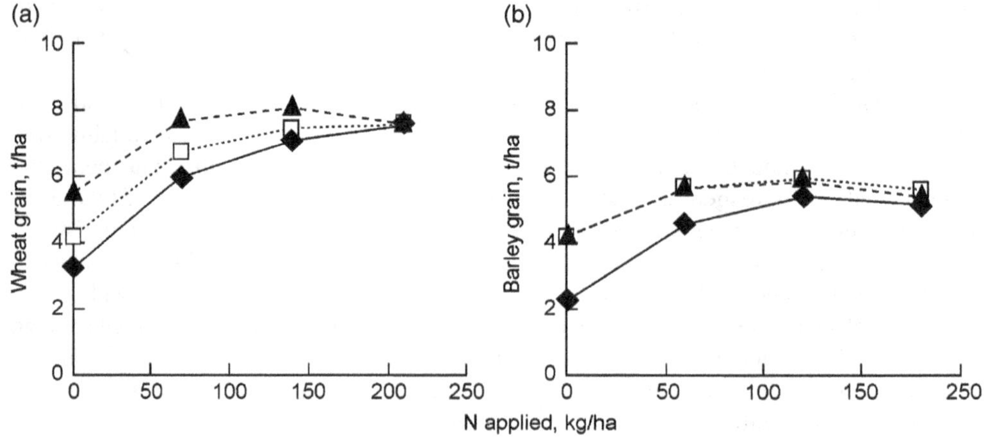

Figure 4.16 Yields of (a) winter wheat and (b) spring barley in the Woburn Ley-Arable Experiment on a sandy soil in southeast UK. Graphs show yields of arable test crops (t ha^{-1}) following three different previous cropping sequences: 3 years arable followed by 2 years arable test crops, ◆; 3 years grass ley + N followed by 2 years arable test crops, □; 3 years grass/clover ley followed by 2 years test crops, ▲. From Johnston and Poulton (2005). Reproduced with kind permission of the International Fertilizer Society.

arable cropping, but, more importantly, the soil would have contained additional nitrate and readily mineralisable N as a result of the legume. Yields of wheat or barley following grass/clover were increased further if additional N fertiliser was added. When additional N fertiliser was applied to the wheat or barley grown following long-term arable cropping, yields increased to levels comparable to those following grass/clover. Thus, in this example, all yield effects can be explained on the basis of N supply.

In the Broadbalk Wheat Experiment at Rothamsted, UK, yields of winter wheat achieved with FYM can be matched by applications of inorganic fertilisers, provided sufficient N is applied. Thus, there is little evidence of the additional organic matter in the FYM treatment having an impact on crop growth other than through nutrient supply, despite the improved soil physical structure. By contrast, in the neighbouring long-term experiment with spring barley grown each year (the Hoosfield Experiment), there is evidence of a crop benefit from increased SOM content (Johnston et al., 2009) – though the actual mechanism is still an open question. Figure 4.17 shows grain yields of spring barley in two main treatments of the experiment in four periods during which different crop varieties have been grown. With both main treatments (FYM or inorganic fertilisers, termed the PK treatment) four N treatments (including zero N) are superimposed. In the earliest period shown (1976–1979) when an old variety of spring barley was grown, grain yield in the FYM treatment could be matched in the PK treatment provided sufficient N fertiliser was applied (panel A). In later periods,

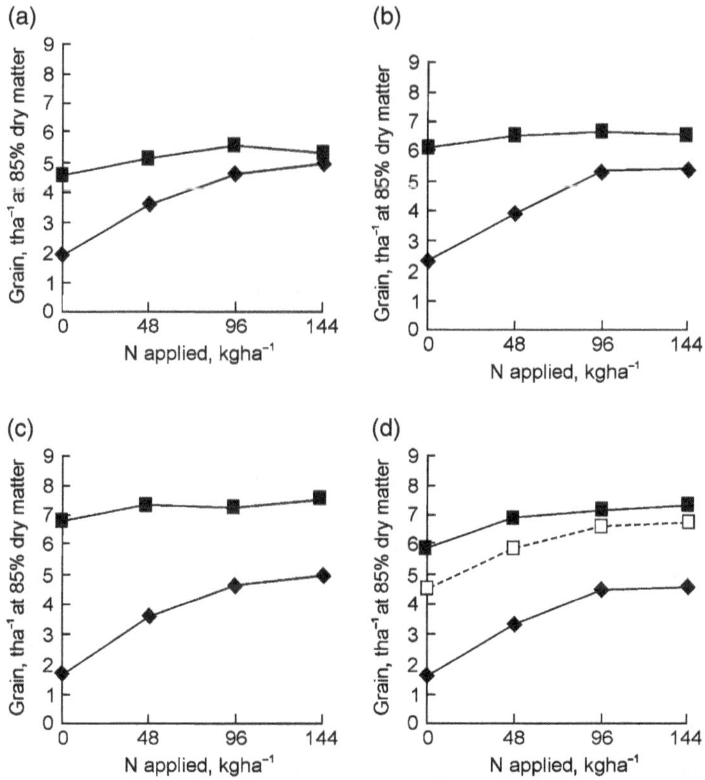

Figure 4.17 Yields of spring barley (t ha^{-1}) in the Hoosefield Barley Experiment at Rothamsted, UK. Shows changing response of different crop varieties to additional organic matter content of soil resulting from long-term farmyard manure (FYM) applications. (a) *cv.* Julia, 1976–1979, (b) *cv.* Triumph, 1988–1991, (c) *cv.* Cooper, 1996–1999, (d) *cv.* Optic, 2004–2007. Reprinted from Johnston et al. (2009). With kind permission from Elsevier.

when more modern and N-responsive varieties were grown, this was no longer the case. For example, in the two latest periods shown (panels C and D) yields with FYM were about 2 t ha^{-1} greater than in 1976–1979 and could not be matched in the PK treatment even with the highest rate of N. The likely explanation is that the improved soil physical structure in the topsoil in the treatment receiving FYM permits more rapid and extensive growth of roots from the emerging seedlings. This leads to greater exploration of the soil, greater uptake of water and nutrients, and greater resilience to any stresses such as water shortage or disease occurring later in the growing season compared to plants in the more poorly structured soil of the PK treatment. Other factors may also contribute, including the larger and different microbial community in the FYM treatment compared to PK; for example, the composition of the rhizosphere population may give roots greater protection from soil-borne pathogens.

Of course, the same arguments can be applied to the situation of winter wheat plants in the Broadbalk Experiment – but here there is no unique benefit to crop yield from the higher SOM content in the FYM treatment. The most likely explanation of the difference between spring barley and winter wheat is that spring barley, having a growing season of only 5–6 months, is more sensitive to the impacts of soil structure than winter wheat, which grows for about 10 months.

4.9 Soil C and climate change

Soils contain a very large stock of carbon in organic forms, estimated by Batjes (1996) as 700 Pg to a depth of 30 cm and 1500 Pg to a depth of 100 cm. Thus SOC to a depth of 100 cm contains twice about the amount of C currently in atmospheric CO_2 (730 Pg) and nearly three times that in vegetation (500 Pg). Small changes in this large stock of SOC could have significant impacts upon future atmospheric CO_2 concentrations. If the SOC stock declines due to management practices or the impact of climate change, this means that additional CO_2 will be released to the atmosphere, adding to future climate change. Conversely, if management practices can be devised that cause SOC to increase over large areas, this could be a means of slowing the current rise in atmospheric CO_2 concentration and contribute to efforts to mitigate climate change.

Key factors to be considered in this context are:

1. What management practices or changes in land use tend to release additional C from SOC and should thus be avoided?
2. What management practices or changes in land use tend to increase SOC stock, either by slowing SOC decomposition or through increased atmospheric C input to soils via plants, and thus should be encouraged as climate change mitigation measures?
3. How is the current SOC stock likely to be influenced by climate change? This includes consideration of the impact of increased temperature, increased atmospheric CO_2 concentration and changes in soil moisture resulting from climate change.

4.9.1 Carbon losses from soil

Converting land of high SOC content, such as forest or grassland, to arable cropping will almost always lead to a loss of SOC and emission of additional CO_2 to the atmosphere. This is illustrated for grassland to arable conversions in Figures 4.4 and 4.6. Table 4.3 shows estimated changes in SOC stocks in four regions of the world over a 30 year period. Data are from a

Table 4.3 Projected changes in SOC stocks in four regions of the world due to expected changes in land use and management

	Current SOC stock	SOC change[a] predicted to 2030	
Region	**Tg**	**Tg**	**Comment**
Amazonia (Brazil)	30000	–4200	Continued deforestation
Indo-Gangetic Plain (India)	1300	Small loss	Some pasture to cropland conversion
Jordan	85	Small loss	Overgrazing of rangeland
Kenya	1500	–114	Savannah to subsistence agriculture conversion

[a]From expected changes in land use or management based on local expert knowledge.
Source: From Milne et al. (2007). Reproduced with kind permission from Elsevier.

project in which current SOC stocks in the selected regions were estimated as accurately as possible using available data and then local expert knowledge used to judge the most likely changes in land management over the next 30 years. SOC models (CENTURY or RothC) were then used to estimate the resulting changes in SOC (Milne et al., 2007). In the Brazilian Amazon, continued deforestation, albeit at a slower rate than in the past, was estimated to cause a 14% loss in SOC stock during 30 years representing a significant emission of CO_2 to the atmosphere in addition to that directly emitted from the destruction of trees. In Kenya an 8% loss in SOC was predicted due to ploughing of areas of savannah grassland for conversion to smallholder arable cropping. In the Indo-Gangetic Plains of India, only small changes in SOC stock were predicted because the area is already under intensive arable cropping, most native vegetation having been cleared long ago. In Jordan SOC stocks were low because of very limited plant growth and organic C inputs to soil under arid and semi-arid conditions. Only small changes were predicted though there was concern about overgrazing of semi-arid rangelands leading to decreased growth of pasture and hence decreased organic C inputs to soil.

Another land use change with serious consequences for SOC stocks, and hence for increased emission of CO_2 to the atmosphere and the exacerbation of climate change, is the drainage of peat soils. These soils hold vast stores of C, estimated at around 500 Pg C globally (Houghton et al., 1985; Gorham, 1991). Drainage of these soils leads to very large annual SOC losses that continue for hundreds of years; this has already occurred in the fens of eastern England. A current example is the drainage of tropical peatlands to grow additional oil palm for production of biofuels. After drainage and conversion to more aerobic conditions, peat oxidises at an increased rate leading to greatly increased emission of CO_2 for decades or longer. Fargione et al. (2008) estimated that the CO_2 emitted after clearance of peatland rainforest in southeast Asia to grow oil palm for biodiesel production would outweigh the greenhouse gas benefit from biodiesel replacing fossil fuel for over 400 years and termed this a 'carbon debt'.

4.9.2 Sequestration of carbon in soil to mitigate climate change

The converse of releasing C from soil to the atmosphere is the retention and stabilisation of additional C into SOC. The additional C retained is normally derived from CO_2 in the atmosphere through photosynthesis followed by transfer from plants to soil via roots or above-ground plant residues. Increasing SOC stock may be achieved by growing larger plants that deposit additional C into soil or changing land use from annual arable crops to perennials, grassland or forest. This process is commonly termed carbon sequestration and is promoted as a means of slowing climate change by decreasing atmospheric CO_2 concentration – or at least slowing the rate of increase. In some cases this is genuine, but there are nuances to the approach that are often overlooked. If an area of unproductive or degraded land can be revegetated with grass or trees, this undoubtedly leads to an increase in the stock of organic C, in both vegetation and soil, where this C would otherwise have been in the atmosphere as CO_2. However, if the same strategy were applied on land used for food production, wider implications have to be considered. It is true that SOC stock is increased at the site of land use change (arable to forest or grass), but, at least in some cases, some land is likely to be cleared elsewhere in the world to replace the lost production of food. The land use change elsewhere (forest or grassland to arable) will not be a direct response to a specific removal of agricultural land from production, but a general global trend. If the newly cleared land had native or semi-natural vegetation, and the soil a high SOC content, then the loss of C at such sites will tend to counteract the sequestration from afforestation at the initial site. This effect is termed indirect land use change (Searchinger et al., 2008); it has been discussed mainly in the context converting agricultural land to biofuel production but is equally valid when considering the global impacts of afforestation or other approaches involving removal of land from agriculture.

Soils used for agriculture, especially arable cropping, almost always have a lower SOC content than the same soil type under semi-natural vegetation. They are also subject to human management, so the *potential* for increasing SOC stock and mitigating climate change is large. Smith et al. (2008) summarise a range of management practices for increasing SOC stocks in arable and pasture systems. These authors point out that, in many cases, the *biophysical potential* for increasing SOC stock cannot be realised because the required practices are uneconomic, impractical or conflict with the demands of food security. But within arable agriculture, there are some management practices that have received attention as potentially practical measures to achieve C sequestration, one being reduced tillage. In both no-till (NT) and reduced till (non-inversion tillage) systems, organic C accumulates near the soil surface because it is not mixed into the plough layer. This effect can be seen in the near surface soil (0–10 cm) in Figure 4.18 showing data taken from a meta-analysis of 43 sites in North and South America comparing SOC in NT and ploughed soil (termed full inversion tillage, FIT, in this study). The meta-analysis (Angers and Eriksen-Hamel, 2008) showed a wide range of results. In many cases, the increased SOC content in surface soil under NT was offset by a slightly decreased amount compared to FIT in deeper layers. Many earlier studies have overlooked this offset, concentrating only on the additional SOC in surface soil, and this has led to an overemphasis on the potential of reduced tillage systems to sequester soil C and mitigate climate change. Nevertheless, it was concluded by Angers and Eriksen-Hamel (2008) that there was a small net accumulation of SOC under NT: on average, soil under NT contained 4.9 tC ha^{-1} more than that under FIT with the accumulation under NT tending to be more marked where the NT treatment had continued for at least 15 years. This is probably because continued disturbance tillage limits C stabilisation through the formation of macro-aggregates composed of micro-aggregates (Six et al., 2004 and discussed in Section 4.6.2). Thus, moving to reduced tillage in arable systems appears to have some role to play in sequestering C in soil, but less than was claimed in the past.

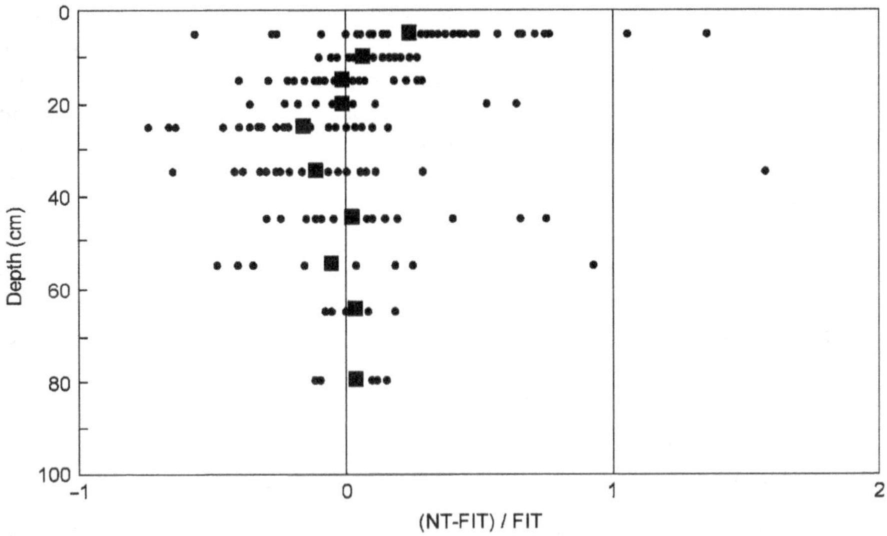

Figure 4.18 Relative change in soil C content under no-till (NT) compared with full inversion tillage (FIT) as a function of soil depth. Based on a meta-analysis of data from 43 sites where the two tillage treatments had been compared for at least 5 years (in many cases treatments had continued for >15 years). SOC data within a soil depth were not generally normally distributed. Large filled squares represent the geometric mean for each soil layer; small points show a range of results from individual sites. Results show an accumulation of SOC under the NT treatment in the near-surface layers (0–10 cm) but a greater amount of SOC in FIT treatment at the base of the plough layer (about 25 cm). At greater depths there was no significant difference between treatments. From Angers and Eriksen-Hamel (2008). with permission, copyright Soil Science Society of America.

Reduced tillage also reduces the energy (and hence CO_2 emissions from fuel) used for crop establishment and in many soils, though not all, delivers a range of other benefits for soil quality and functioning including decreased risk of erosion. However, in environments with relatively high rainfall and/or poorly drained soils moving to NT can increase nitrous oxide (N_2O) emissions compared to conventionally tilled soil (Rochette, 2008). Because of the high greenhouse warming potential of N_2O (almost 300 times that of CO_2 on a 100 year basis), a small increase in N_2O emission can easily outweigh the modest climate change benefit of NT through soil C sequestration. Wetter conditions (in terms of the categories defined by Rochette (2008)) are more prevalent in northwest Europe than in the central regions of North America; so whether or not NT or reduced till has an overall beneficial impact for climate change, mitigation will be region-specific. The discussion around this issue emphasises the need to take account of management impacts on the fluxes of all greenhouse gases (CO_2, N_2O and CH_4) when considering climate change mitigation through soil management – not to concentrate only on SOC stocks, as pointed out by Smith et al. (2008).

As described in Section 4.2, adding organic materials such as manures and crop residues to soil increases SOC content. But this does not necessarily imply sequestration of C and climate change mitigation – whether or not this is the case depends on the alternative fate of the organic materials. For example, manure addition increases SOC as shown in Figure. 4.3. But in practice virtually all animal manure is used in some way in agriculture; so if the manure applied to the field in Figure 4.3 had not been applied at that location, it would have been applied elsewhere and given an equivalent increase in SOC. So the increase shown for the field in Figure 4.3 is simply a movement of C from one soil location to another, not a net transfer of additional C from atmosphere to soil. The same is generally true of crop residue additions such as straw, unless the alternative practice is to burn the material. However, for organic wastes that would otherwise be disposed of by burning or placing in landfill sites, addition to soil where a fraction of the contained organic C is retained does represent a genuine increase in the stock of C in the soil, and this C would otherwise have been emitted to the atmosphere as CO_2 or CH_4. One example of a material that was often not previously used to increase SOC stock is green waste compost derived from municipal wastes such as plant material from tree pruning, grass cutting or household waste. Better utilisation of such materials through addition to soil can contribute to C sequestration as well as to improving soil properties (Powlson et al., 2012). The concept of soil C sequestration, of where it provides genuine climate change mitigation and where it does not, is discussed by Powlson et al. (2011c).

4.9.3 Climate change impacts on soil C stocks

Because of the large stock of organic C held in the world's soils, small losses of C from this large pool could have significant impacts upon future atmospheric carbon dioxide concentrations. Models that have coupled climate and carbon cycles show a large divergence in the size of the predicted biospheric feedback to the atmosphere. Central questions which still remain when attempting to reduce this uncertainty in the response of soils to climate change are (a) the temperature sensitivity of SOM, especially the more recalcitrant pools, (b) the balance between increased carbon inputs to the soil from increased plant production and increased losses due to increased rates of decomposition and (c) interactions between climate change and other aspects of global change including other climatic effects (e.g. changes in water balance), changes in atmospheric composition (e.g. increasing atmospheric carbon dioxide concentration) and land use change.

Climate change is likely to influence both the quantity of C entering soil from plant inputs and its rate of decomposition. An increased concentration of CO_2 in the atmosphere is likely to increase plant productivity and hence inputs of C from plant litter and roots, though plant growth will also be influenced by temperature and water availability. These influences could be either positive or negative in different regions, depending on whether climate change relieves a current constraint to growth or imposes a new constraint. Effects may well vary between seasons. It is generally thought that increased temperature will increase the rate of decomposition of SOC but there are numerous caveats to this. First, decomposition rate will be affected by other climate-related factors such as soil moisture which in turn is a result of rainfall amounts and patterns. Second, there is some debate about the actual impact of temperature on SOC decomposition, and it is possible that different SOC fractions (the more and less recalcitrant) are influenced differently. Given these uncertainties and the numerous possibilities for different factors to interact and cause opposing effects, it is not surprising that it is difficult to make unequivocal statements about the impact of climate change on SOC and the role of soil C in climate feedbacks.

In one study (Jones et al., 2005), the Hadley Centre General Circulation Model was linked with the RothC SOC dynamics model and run for a period of 240 years assuming certain climate changes but taking no account of land use change. The results showed large losses of soil C in some regions, especially the Amazon region of South America and the northwest of North America, and gains in some other regions. Overall it showed a loss of SOC of 54 Gt averaging 0.23 Gt year^{-1}. This is equivalent to an increase of about 3% on current anthropogenic CO_2 emissions (IPCC, 2007) and is far from negligible. For Europe, one modelling study indicates that increased plant productivity would likely counterbalance increased decomposition due to climate change up to 2080 (Smith et al., 2005).

Especially vulnerable soils

High-latitude regions are thought to be particularly vulnerable to SOC loss under climate change for a number of reasons. Firstly, high latitudes are projected to experience some of the greatest warming (see Mitchell et al., 2004). Secondly, at cooler temperatures, especially for those at around freezing point, a change in temperature has a much greater impact than an increase in temperate regions. This is particularly true of permafrost soils in the taiga and tundra, which hold around 500 Pg C and could lose this carbon rapidly under warming. The release of this huge stock of carbon to the atmosphere could significantly enhance further climate warming (Zimov et al., 2006).

Peatlands and wetlands are also expected to be particularly sensitive to climate warming. Peatlands also hold vast stores of carbon: estimates range from ~200 Pg C to 860 Pg C with most estimates in the region of around 500 Pg (e.g. Houghton et al., 1985; Gorham, 1991). Plant productivity in cool northern peatlands is relatively low compared to the large carbon losses that can occur when these soils are drained, with yearly losses of C of 0.8–8.3 t C ha^{-1} y^{-1} (Nykänen et al., 1995; Lohila et al., 2004; Maljanen et al., 2004) and losses occurring for over a hundred years (Lohila et al., 2004). If increasing temperature increases evapotranspiration (either coupled with or independently from changes in precipitation), peatlands would be expected to dry causing decomposition rate to increase further. Given the large stocks of carbon in peatland soils, losses of carbon will likely far exceed any increase in plant productivity arising from increases in temperature.

Other sensitive regions include currently hot arid areas. Though the soils in these areas tend to be low in carbon (Batjes, 1996), complete failure of plant growth in a warmer world would switch off all inputs of carbon to the soil, enhancing desertification and degradation. Even partial loss of vegetation integrity could make soils more vulnerable to degradation through other agents such as grazing and cultivation. Since soil C stocks are not as high in these regions, total potential C losses are lower, but these regions cover large areas of the arid tropics/subtropics.

Temperature impacts on SOC decomposition

The temperature sensitivity of litter and SOC decomposition is important, not only because of the rate of decomposition itself but also because it determines the magnitude of feedback from atmospheric CO_2. Additional CO_2 in the atmosphere would lead to increased global warming and still further additional SOC decomposition. In general, experimental results show that any soil moved to a higher temperature evolves more CO_2. However there is debate on this topic, largely due to the existence of confounding and feedback effects which may or may not be taken into account in different approaches (Davidson and Janssens, 2006; Kirschbaum, 2006; Pare et al., 2006; Briones et al., 2007; Czimczik and Trumbore, 2007). A specific issue is the sensitivities of the more recalcitrant fractions of SOC: in a warmer world, it is assumed that the more readily decomposable parts of total SOC will be selectively decomposed, leaving the more recalcitrant fractions as a greater proportion of the remaining C and thus having a larger impact on the overall sensitivity. Some studies suggest that recalcitrant C is not sensitive to temperature variation (Giardina and Ryan, 2000), others suggest that it is more sensitive to temperature than labile pools (Fierer et al., 2005, 2006; Knorr et al., 2005) or that recalcitrant and labile pools have a similar temperature sensitivity (Fang et al., 2005; Conen et al., 2006). This issue is a matter of ongoing research and debate.

References

Ahmad, R., Nelson, P.N., and Kookana, R.S. (2006) The molecular composition of soil organic matter as determined by ^{13}C NMR and elemental analyses and correlation with pesticide sorption. *European Journal of Soil Science*, **57**, 883–893.

Andrén, O. and Kätterer, T. (1997) ICBM: the introductory carbon balance model for exploration of soil carbon balances. *Ecological Application*, **7**, 1226–1236.

Angers, D.A. and Eriksen-Hamel, N.S. (2008) Full-inversion tillage and organic carbon distribution in soil profiles: a meta-analysis. *Soil Science Society of America Journal*, **72**, 1370–1374.

Angers, D.A., Recous, S., and Aita, C. (1997) Fate of C and nitrogen in water-stable aggregates during decomposition of ^{13}C and ^{15}N-labelled wheat straw in situ. *European Journal of Soil Science*, **48**, 295–300.

Baldock, J.A. and Oades, J.M. (1989) Effect of electrolyte concentration on glucose decomposition in soil. *Australian Journal of Soil Research*, **27**, 433–438.

Baldock, J.A., Oades, J.M., Nelson, P.N., Skene, T.M., Golchin, A., and Clarke, P. (1997) Assessing the extent of decomposition of natural materials using solid-state ^{13}C NMR spectroscopy. *Australian Journal of Soil Research*, **35**, 1061–1083.

Batjes, N.H. (1996) Total carbon and nitrogen in the soils of the world. *European Journal of Soil Science*, **47**, 151–163.

Bchmann, J., Guggenberger, G., Baumgartl, T., Ellerbrock, R.H., Urbanek, E., Goebel, M.O., Kaiser, K., Horn, R., and Fischer, W.R. (2008) Physical carbon-sequestration mechanisms under special conditions of soil wettability. *Journal of Plant Nutrition and Soil Science*, **171**, 14–26.

Beek, J. and Frissel, M.J. (1973) *Simulation of Nitrogen Behaviour in Soils*. Pudoc, Wageningen, the Netherlands.

Blair, G.J., Lefroy, R.D.B., and Lisle, L. (1995) Soil carbon fractions, based on their degree of oxidation, and the development of a carbon management index for agricultural systems. *Australian Journal of Agricultural Research*, **46**, 1459–1466.

Blair, N., Faulkner, R.D., Till, A.P., and Poulton, D.R. (2006) Long-term management impacts on soil C, N and physical fertility. Part I: broadbalk experiment. *Soil and Tillage Research*, **91**, 30–38.

Blaser, P. Kernebeek, P., Tebbens, L., Van Breemen, N., and Lister, J. (1997) Cryptopodzolic soils in Switzerland. *European Journal of Soil Science*, **48**, 411–423.

Bosatta, E. and Ågren, G.I. (1994) Theoretical analysis of microbial biomass dynamics in soils. *Soil Biology and Biochemistry*, **26**, 143–148.

Bosatta, E. and Ågren, G.I. (1995) Theoretical analyses of the interactions between inorganic nitrogen and soil organic matter. *European Journal of Soil Science*, **76**, 109–114.

Bossuyt, H., Six, J., and Hendrix, P.F. (2002) Aggregate protected carbon in no-tillage and conventional tillage agroecosystems. *Soil Science Society of America Journal*, **66**, 1965–1973.

Bossuyt, H., Six J., and Hendrix, P.F. (2005) Protection of soil carbon by microaggregates within earthworm casts. *Soil Biology and Biochemistry*, **37**, 251–258.

Briones, M.J.I., Ostle, N.J., and Garnett, M.H. (2007) Invertebrates increase the sensitivity of non-labile soil carbon to climate change. *Soil Biology and Biochemistry*, **39**, 816–818.

Brodowski, S., John, B., Flessa, H., and Amelung, W. (2006) Aggregate-occluded black carbon in soil. *European Journal of Soil Science*, **57**, 539–546.

Brookes, P.C., Powlson, D.S., and Jenkinson, D.S. (1984) Phosphorus in the soil microbial biomass. *Soil Biology and Biochemistry*, **16**, 169–175.

Bull, I.D., van Bergen, P.F., Nott, C.J., Poulton, P.R., and Evershed, R.P. (2000) Organic geochemical studies of soils from the Rothamsted classical experiment—V. The fate of lipids in different long-term experiments. *Organic Geochemistry*, **31**, 389–408.

Burke, I.C., Kaye, J.P., Bird, S.P., Hall, S.A., McCulley, R.L., and Sommerville, G.L. (2003) Evaluating and testing models of terrestrial biogeochemistry: the role of temperature in controlling decomposition. In: *Models in Ecosystem Science* (eds C.D. Canham, J.J. Cole, and W.K. Lauenroth), pp. 225–253. Princeton University Press, Princeton, New Jersey.

Cerri, C.E.P., Coleman, K.W., Jenkinson, D.S., Bernoux, M., Victoria, R., and Cerri, C.C. (2003) Modeling soil carbon from forest and pasture ecosystems of Amazon, Brazil. *Soil Science Society of America Journal*, **67**, 1879–1887.

Cerri, C.E.P., Easter, M., Paustian, K., Killian, K., Coleman, K., Bernoux, M., Falloon, P., Powlson, D.S., Batjes, N., Milne, E., and Cerri, C.C. (2007) Predicted soil organic carbon stocks and changes in the Brazilian Amazon between 2000 and 2030. *Agriculture, Ecosystems and Environment*, **122**, 58–72.

Chantigny, M.H., Angers, D.A., and Beauchamp, C.J. (2000) Decomposition of de-inking paper sludge in agricultural soils as characterized by carbohydrate analysis. *Soil Biology and Biochemistry*, **32**, 1561–1570.

Christensen, B.T. (1996) Carbon in primary and secondary organomineral complexes. In: *Advances in Soil Science*, Vol. 20 (eds M.R. Carter and B.A. Stewart), pp. 97–165. CRC Press, Boca Raton, Florida.

Coleman, K. and Jenkinson, D.S. (1996) RothC-26.3 – a model for the turnover of carbon in soil. In: *Evaluation of Soil Organic Matter Models Using Existing, Long-Term Datasets*, Vol. 38 (eds D.S. Powlson, P. Smith, and J.U. Smith), NATO ASI Series I, pp. 237–246. Springer-Verlag, Heidelberg, Germany.

Conen, F., Leifeld, J., Seth, B., and Alewell, C. (2006) Warming mineralises young and old soil carbon equally. *Biogeosciences*, **3**, 515–519.

Czimczik, C.I. and Trumbore, S.E. (2007) Short-term controls on the age of microbial carbon sources in boreal forest soils. *Journal of Geophysical Research – Biogeosciences*, **112**, G03001, doi:10.1029/2006JG000389.

Davidson, E.A. and Janssens, I.A. (2006) Temperature sensitivity of soil carbon decomposition and feedbacks to climate change. *Nature*, **440**, 165–173.

Denef, K., Zotarelli, L., Boddey, R.M., and Six, J. (2007) Microaggregate-associated carbon as a diagnostic function for management-induced changes in total organic carbon in two Oxisols. *Soil Biology and Biochemistry*, **39**, 1165–1172.

Fang, C., Smith, P., Moncrieff, J.B., and Smith, J.U. (2005) Similar response of labile and resistant soil organic matter pools to changes in temperature. *Nature*, **433**, 57–59.

Fargione, J., Hill, J., Tilman, D., Polasky, S., and Hawthorne, P. (2008) Land clearing and the biofuel carbon debt. *Science*, **319**, 1235–1238.

Fierer, N., Craine, J.M., McLauchlan, K., and Schimel, J.P. (2005) Litter quality and the temperature sensitivity of decomposition. *Ecology*, **86**, 320–326.

Fierer, N., Colman, B.P., Schimel, J.P., and Jackson, R.B. (2006) Predicting the temperature dependence of microbial respiration in soil: a continental-scale analysis. *Global Biogeochemical Cycles*, **20**, GB3026, doi:10.1029/2005GB002644.

Fontaine, S. and Barot, S. (2005) Size and functional diversity of microbe populations control plant persistence and long-term soil carbon accumulation. *Ecology Letters*, **8**, 1075–1087.

Fontaine, S., Barot, S., Barre, P., Bdioui, N., Mary, B., and Rumpel, C. (2007) Stability of organic carbon in deep soil layers controlled by fresh carbon supply. *Nature*, **450**, 277–280.

Friedlingstein, P., Cox, P., Betts, R., Bopp, L., Von Bloh, W., Brovkin, V., Cadule, P., Doney, S., Eby, M., Fung, I., Bala, G., John, J., Jones, C., Joos, F., Kato, T., Kawamiya, M., Knorr, W., Lindsay, K., Matthews, H.D., Raddatz, T., Rayner, P., Reick, C., Roeckner, E., Schnitzler, K.G., Schnur, R., Strassmann, K., Weaver, A.J., Yoshikawa, C., and Zeng, N. (2006) Climate-carbon cycle feedback analysis: results from the (CMIP)-M-4 model intercomparison. *Journal of Climate*, **19**, 3337–3353.

Giardina, C.P. and Ryan, M.G. (2000) Evidence that decomposition rates of organic carbon in mineral soil do not vary with temperature. *Nature*, **404**, 858–861.

Gillespie, A.W., Farrell, R.E., Walley, F.L., Ross, A.R.S., Leinweiber, P., Eckhardt, K.-U., Regier, T.Z., and Blyth, R.I.R. (2011) Glomalin-related soil protein contains non-mycorrhizal-related heat-stable proteins, lipids and humic materials. *Soil Biology and Biochemistry*, **43**, 766–777.

Gollany, H.T., Rickman, R.W., Liang, Y., Albrecht, S.L., Macahdo, S., and Kang, S. (2011) Predicting agricultural management influence on long-term soil organic carbon dynamics: implications for biofuel production. *Agronomy Journal*, **103**, 234–246.

Gorham, E. (1991) Northern peatlands: role in the carbon cycle and probable responses to climatic warming. *Ecological Applications*, **1**, 182–195.

Gregorich, E.G. and Ellert, B.U. (1993) Light fraction and macroorganic matter in mineral soils. In: *Soil Sampling and Methods of Analysis* (ed. M.R. Carter), pp. 397–407. CRC Press, Boca Raton, Florida.

Gregorich, E.G., Beare, M.H., Mckim, U.F., and Skjemstad, J.O. (2006) Chemical and biological characteristics of physically uncomplexed organic matter. *Soil Science Society of America Journal*, **70**, 975–985.

Houghton, R.A., Boone, R.D., Melillo, J.M., Palm, C.A., Woodwell, G.M., Myers, N., Moore, B., and Skole, D.L. (1985) Net flux of carbon-dioxide from tropical forests in 1980. *Nature*, **316**, 617–620.

Hunt, H.W., Coleman, D.C., and Ingham, R.E. (1987) The detrital food web in a shortgrass prairie. *Biology and Fertility of Soils*, **3**, 57–68.

IPCC (Intergovernmental Panel on Climate Change). (2007) Fourth Assessment Report, Synthesis Report. Summary for Policymakers. WMO, Geneva, Switzerland.

Jenkinson, D.S. (1973) Soil organic matter and its dynamics. In: *Russell's Soil Conditions and Plant Growth* (ed. A. Wild), 11th edn., pp. 564–607. Longman Scientific & Technical, Harlow, UK.

Jenkinson, D.S. and Coleman, K. (2008) The turnover of organic carbon in subsoils. Part 2. Modelling carbon turnover. *European Journal of Soil Science*, **59**, 400–413.

Jenkinson, D.S. and Powlson, D.S. (1976a) The effects of biocidal treatments on metabolism in soil. I. Fumigation with chloroform. *Soil Biology and Biochemistry*, **8**, 167–177.

Jenkinson, D.S. and Powlson, D.S. (1976b) The effects of biocidal treatments on metabolism in soil. V. A method for measuring soil biomass. *Soil Biology and Biochemistry*, **8**, 209–213.

Jenkinson, D.S., Adams, D.E., and Wild, A. (1991) Model estimates of CO2 emissions from soil in response to global warming. *Nature*, **351**, 304–306.

Jenkinson, D.S., Poulton, P.R., and Bryant, C. (2008) The turnover of organic carbon in subsoils. Part 1. Natural and bomb radiocarbon in soil profiles from the Rothamsted long-term field experiments. *European Journal of Soil Science*, **59**, 391–399.

Jenny, H. (1941) *Factors of Soil Formation. A System of Quantitative Pedology*. McGraw-Hill, New York.

Johnston, A.E. and Dawson, C.J. (2010) Physical, chemical and biological attributes of agricultural soils. In: *Proceedings of the International Fertilizer Society*, Vol. 675, York, UK, 40pp.

Johnston, A.E. and Poulton, P.R. (2005) Soil organic matter: its importance in sustainable agricultural systems. In: *Proceedings of the International Fertilizer Society*, Vol. 565, York, UK, 46pp.

Johnston, A.E., Poulton, P.R., and Coleman, K. (2009) Soil organic matter: its importance in sustainable agriculture and carbon dioxide fluxes. *Advances in Agronomy*, **101**, 1–57.

Jones, C., McConnell, C., Coleman, K., Cox, P., Falloon, P., Jenkinson, D., and Powlson, D. (2005) Global climate change and soil carbon stocks; predictions from two contrasting models for the turnover of organic carbon in soil. *Global Change Biology*, **11**, 154–166.

Kaiser, K. and Guggenberger, G. (2003) Mineral surfaces and soil organic matter. *European Journal of Soil Science*, **54**, 219–236.

Kalembasa, S.J. and Jenkinson, D.S. (1973) A comparative study of titrimetric and gravimetric methods for the determination of organic carbon in soil. *Journal of the Science of Food and Agriculture*, **24**, 1085–1090.

Kemmitt, S.J., Lanyon, C.V., Waite, I.S., Wen, Q., Addiscotte, T.M., Bird, N.R.A., O'Donnell, A.G., and Brookes, P.C. (2008) Mineralization of native soil organic matter is not regulated by the size, activity or composition of the soil microbial biomass – a new perspective. *Soil Biology and Biochemistry*, **40**, 61–73.

Killham, K., Amato, M., and Ladd, J.N. (1993) Effect of substrate location in soil and pore-water regime on carbon turnover. *Soil Biology and Biochemistry*, **25**, 57–62.

Kirschbaum, M.U.F. (2006) The temperature dependence of organic-matter decomposition – still a topic of debate. *Soil Biology and Biochemistry*, **38**, 2510–2518.

Knicker, H. (2009) Black carbon and thermally altered (pyrogenic) organic matter: chemical characteristics and their role in the environment. In: *Biophysico-Chemical Processes Involving Nonliving Organic Matter in Environmental Systems* (eds N. Senesi, B. Xing, and P.M. Huang), pp. 273–303. John Wiley & Sons, Inc., New York.

Knorr, W., Prentice, I.C., House, J.I., and Holland, E.A. (2005) Long-term sensitivity of soil carbon turnover to warming. *Nature*, **433**, 298–301.

Kogel-Knaber, J. (2002) The macromolecular organic composition of plant and microbial residues as inputs to soil organic matter. *Soil Biology and Biochemistry*, **34**, 139–162.

Kouno, K., Wu, J., and Brookes, P.C. (2002) Turnover of biomass C and P in soil following incorporation of glucose or ryegrass. *Soil Biology and Biochemistry*, **34**, 617–622.

Ladha, J.K., Reddy, C.K., Padre, A.T., and van Kessel, C. (2011) Role of nitrogen fertilizer in sustaining organic matter in cultivated soils. *Journal of Environmental Quality*, **40**, 1756–1766.

Lenton, T.M. and Huntingford, C. (2003) Global terrestrial carbon storage and uncertainties in its temperature sensitivity examined with a simple model. *Global Change Biology*, **9**, 1333–1352.

Liang, B., Lehmann, J., Sohi, S.P., Thies, J.E., O'Neill, B., Trujillo, L., Gaunt, J., Solomon, D., Grossman, J., Neves, E.G., and Luizao, F.J. (2010) Black carbon affects the cycling of non-black carbon in soil. *Organic Geochemistry*, **41**, 206–213.

Lohila, A., Aurela, M., Tuovinen, J.P., and Laurila, T. (2004) Annual CO_2 exchange of a peat field growing spring barley or perennial forage grass. *Journal of Geophysical Research*, **109**, D18116, doi:10.1029/2004JD004715.

Lynd, L.R., Weimer, P.J., van Zyl, W.H., and Pretorious, I.S. (2002) Microbial cellulose utilization: fundamentals and biotechnology. *Microbiology and Molecular Biology Reviews*, **66**, 506–577.

Mahieu, N., Randall, E.W., and Powlson, D.S. (1999) Statistical analysis of published 13C CPMAS NMR spectra of soil organic matter. *Soil Science Society of America Journal*, **63**, 307–319.

Major, J., Rondon, M., Molina, D., Riha, S.J., and Lehmann, J. (2010) Maize yield and nutrition during 4 years after biochar application to a Columbian savanna oxisol. *Plant and Soil*, **333**, 117–126.

Maljanen, M., Komulainen, V.M., Hytonen, J., and Silvola, J. (2004) Carbon dioxide, nitrous oxide and methane dynamics in boreal organic agricultural soils with different soil characteristics. *Soil Biology and Biochemistry*, **36**, 1801–1808.

McClougherty, C.A. and Linkins, E.A. (1988) Extractability of cellulases in forest litter and soil. *Biology and Fertility of Soils*, **6**, 322–327.

McGill, W.B. (1996) Review and classification of ten soil organic matter (SOM) models. In: *Evaluation of Soil Organic Matter Models Using Existing, Long-Term Datasets*, Vol. 38 (eds D.S. Powlson, P. Smith, and J.U. Smith), NATO ASI Series I, pp. 111–133. Springer-Verlag, Heidelberg, Germany.

McGill, W.B., Hunt, H.W., Woodmansee, R.G., and Reuss, J.O. (1981) PHOENIX, a model of the dynamics of carbon and nitrogen in grassland soil. In: *Terrestrial Nitrogen Cycles. Processes, Ecosystem Strategies and Management Impacts*, Vol. 33 (eds F.E. Clark and T. Roswall), Ecological Bulletins, pp. 49–115. Swedish Natural Science Research Council, Stockholm, Sweden.

Milne, E., Paustian, K., Easter, M., Sessay, M., Al-Adamat, R., Batjes, N.H., Bernoux, M., Bhattacharyya, T., Cerri, C.C., Cerri, C.E.P., Coleman, K., Falloon, P., Feller, C., Gicheru, P., Kamoni, P., Killian, K., Pal, D.K., Powlson, D.S., Williams, S., and Rawajfih, Z. (2007) An increased understanding of soil organic carbon stocks and changes in non-temperate areas: national and global implications. *Agriculture, Ecosystems and Environment*, **122**, 125–136.

Mitchell, T.D., Carter, T.R., Jones, P.D., Hulme, M., and New, M. (2004) A comprehensive set of high-resolution grids of monthly climate for Europe and the globe: the observed record (1901–2000) and 16 scenarios (2001–2100). Working Paper 55, Tyndall Centre for Climate Change Research, University of East Anglia, Norwich, UK.

Molina, J.A.E. and Smith, P. (1998) Modeling carbon and nitrogen processes in soil. *Advances in Agronomy*, **62**, 253 298.

Muller, C., Eickhout, B., Zaehle, S., Bondeau, A., Cramer, W., and Lucht, W. (2007) Effects of changes in CO_2, climate, and land use on the carbon balance of the land biosphere during the 21st century. *Journal of Geophysical Research – Biogeosciences*, **112**, Art. No. G02032.

Nannipieri, P. (2006) Role of stabilized enzymes in microbial ecology and enzyme extraction from soil with potential applications in soil proteomics. In: *Nucleic Acids and Proteins in Soil* (eds P. Nannipieri and K. Smalla), pp. 75–94. Springer, Heidelberg, Germany.

Neill, C. and Gignoux, J. (2006) Soil organic matter decomposition driven by microbial growth: a simple model for a complex network of interactions. *Soil Biology and Biochemistry*, **38**, 803–811.

Nelson, D.W. and Sommers, L.E. (1982) Total carbon, organic carbon, and organic matter. In: *Methods of Soil Analysis. Part 2. Chemical and Microbiological Properties* (eds A.L. Page, R.H. Miller, and D.R. Keeney), 2nd edn., pp. 539–579. American Society of Agronomy, Madison, Wisconsin.

Nykänen, H., Alm, J., Lang, K., Silvola, J., and Martikainen, P.J. (1995) Emissions of CH_4, N_2O and CO_2 from a virgin fen and a fen drained for grassland in Finland. *Journal of Biogeography*, **22**, 351–357.

Oades, J.M. and Waters, A.G. (1991) Aggregate hierarchy in soils. *Australian Journal of Soils Research*, **29**, 815–828.

Pare, D., Boutin, R., Larocque, G.R., and Raulier, F. (2006) Effect of temperature on soil organic matter decomposition in three forest biomes of eastern Canada. *Canadian Journal of Soil Science*, **86**, 247–256.

Paustian, K. (1994) Modelling soil biology and biochemical processes for sustainable agricultural research. In: *Soil Biota. Management in Sustainable Farming Systems* (eds C.E. Pankhurst, B.M. Doube, V.V.S.R. Gupta, and P.R. Grace), pp. 182–193. CSIRO Information Services, Melbourne, Victoria, Australia.

Poirier, N., Sohi, S.P., Gaunt, J.L., Mahieu, N., Randall, E.W., Powlson, D.S., and Evershed, R.P. (2005) The chemical composition of measurable soil organic matter pools. *Organic Geochemistry*, **36**, 1174–1189.

Powlson, D.S. (1994) The soil microbial biomass – before, beyond and back. In: *Beyond the Biomass: Compositional and Functional Analysis of Soil Microbial Communities* (eds K. Ritz, J. Dighton, and K.E. Giller), pp. 3–20. John Wiley & Sons, Ltd., Chichester, UK.

Powlson, D.S., Brookes, P.C., and Christensen, B.T. (1987) Measurement of soil microbial biomass provides an early indication of changes in total soil organic matter due to straw incorporation. *Soil Biology and Biochemistry*, **19**, 159–164.

Powlson, D.S., Glendining, M.J., Coleman, K., and Whitmore, A.P. (2011a) Implications for soil properties of removing cereal straw: results from long-term studies. *Agronomy Journal*, **103**, 279–287.

Powlson, D.S., Gregory, P.J., Whalley, W.R., Quinton, J.N., Hopkins, D.W., Whitmore, A.P., Hirsch, P.R., and Goulding, K.W.T. (2011b) Soil management in relation to sustainable agriculture and ecosystem services. *Food Policy*, **36**, S72–S87.

Powlson, D.S., Whitmore, A.P., and Goulding, K.W.T. (2011c) Soil carbon sequestration to mitigate climate change: a critical re-examination to identify the true and the false. *European Journal of Soil Science*, **62**, 42–55.

Powlson, D.S., Bhogal, A., Chambers, B.J., Coleman, K., Macdonald, A.J., Goulding, K.W.T., and Whitmore, A.P. (2012) The potential to increase carbon stocks through reduced tillage or organic material additions in England and Wales: a case study. *Agriculture, Ecosystems and Environment*, **146**, 23–33.

Pribyl, D.W. (2010) A critical review of the conventional SOC to SOM conversion factor. *Geoderma*, **156**, 75–83.

Rochette, P. (2008) No-till only increases N_2O emissions in poorly aerated soils. *Soil and Tillage Research*, **101**, 97–100.

de Ruiter, P.C., Van Veen, J.A., Moore, J.C., Brussard, L., and Hunt, H.W. (1993) Calculation of nitrogen mineralization in soil food webs. *Plant and Soil*, **157**, 263–273.

Schaumann, G.E. (2006) Soil organic matter beyond molecular structure. Part I: macromolecular and supramolecular characteristics. *Journal of Plant Nutrition and Soil Science*, **169**, 144–155.

Schimel, J.P. and Weintraub, M.N. (2003) The implications of exoenzyme activity on microbial carbon and nitrogen limitation in soil: a theoretical model. *Soil Biology and Biochemistry*, **35**, 549–563.

Schnitzer, M. and Monreal, C.M. (2011) *Quo vadis* soil organic matter research?: a biological link to the chemistry of humification. *Advances in Agronomy*, **113**, 139–213.

Schwezig, D., Kalbitz, K., and Matzner, E. (2003) Effects of aluminium on the mineralization of dissolved organic carbon derived from forest floors. *European Journal of Soil Science*, **54**, 311–322.

Searchinger, T., Heimlich, R., Houghton, R.A., Dong, F., Elobeid, A., Fabiosa, J., Tokgoz, S., Hayes, D., and Yu, T.H. (2008) Use of U.S. cropland for biofuels increase greenhouse gases through emissions from land-use change. *Science*, **319**, 1238–1240.

Simpson, A.J., Burdon, J., Graham, C.L., Hayes, M.B.H., Spencer, N., and Kingery, W.L. (2001) Interpretation of heteronuclear and multidimensional NMR spectroscopy of humic substances. *European Journal of Soil Science*, **52**, 495–509.

Six, J., Conant, R.T., Paul, E.A., and Paustian, K. (2002) Stabilization mechanism of soil organic matter: implications for C-saturation of soils. *Plant and Soil*, **241**, 155–176.

Six, J., Bossuyt, H., Degryze, S., and Denef, K. (2004) A history of research on the link between (micro) aggregates, soil biota, and soil organic matter dynamics. *Soil and Tillage Research*, **79**, 7–31.

Smith, P., Smith, J.U., Powlson, D.S., McGill, W.B., Arah, J.R.M., Chertov, O.G., Coleman, K., Franko, U., Frolking, S., Jenkinson, D.S., Jensen, L.S., Kelly, R.H., Klein-Gunnewiek, H., Komarov, A.S., Li, C., Molina, J.A.E., Mueller, T., Parton, W.J., Thornley, J.H.M., and Whitmore, A.P. (1997) A comparison of the performance of nine soil organic matter models using seven long-term experimental datasets. *Geoderma*, **81**, 153–225.

Smith, P., Falloon, P., Smith, J.U., and Powlson, D.S. (eds) (2001) *Soil Organic Matter Network (SOMNET): 2001 Model and Experimental Metadata*. GCTE Report 7 (2nd edn), GCTE Focus 3 Office, Wallingford, Oxon, UK, 224pp.

Smith, J.U., Smith, P., Monaghan, R., and Macdonald, A.R.J. (2002) When is a measured soil organic matter fraction equivalent to a model pool? *European Journal of Soil Science*, **53**, 405–416.

Smith, J.U., Smith, P., Wattenbach, M., Zaehle, S., Hiederer, R., Jones, R.J.A., Montanarella, L., Rounsevell, M.D.A., Reginster, I., and Ewert, F. (2005) Projected changes in mineral soil carbon of European croplands and grasslands, 1990–2080. *Global Change Biology*, **11**, 2141–2152.

Smith, P., Smith, J.U., Wattenbach, M., Meyer, J., Lindner, M., Zaehle, S., Hiederer, R., Jones, R., Montanarella, L., Rounsevell, M., Reginster, I., and Kankaanpää, S. (2006) Projected changes in mineral soil carbon of European forests, 1990–2100. *Canadian Journal of Soil Science*, **86**, 159–169.

Smith, P., Martino, D., Cai, Z., Gwary, D., Janzen, H.H., Kumar, P., McCarl, B., Ogle, S., O'Mara, F., Rice, C., Scholes, R.J., Sirotenko, O., Howden, M., McAllister, T., Pan, G., Romanenkov, V.,

Schneider, U., Towprayoon, S., Wattenbach, M., and Smith, J.U. (2008) Greenhouse gas mitigation in agriculture. *Philosophical Transactions of the Royal Society B*, **363**, 789–813.

Sohi, S.P., Mahieu, N., Arah, J.R.M., Powlson, D.S., Madari, B., and Gaunt, J.L. (2001) A procedure for isolating soil organic matter fractions suitable for modeling. *Soil Science Society of America Journal*, **65**, 1121–1128.

Sohi, S.P., Mahieu, N., Powlson, D.S., Madari, B., Smittenberg, R.H., and Gaunt, J.L. (2005) Investigating the chemical characteristics of soil organic matter fractions suitable for modeling. *Soil Science Society of America Journal*, **69**, 1248–1255.

Sohi, S.P., Krull, E., Lopez-Capel, E., and Bol, R. (2010) A review of biochar and its use and function in soil. *Advances in Agronomy*, **105**, 47–82.

Stockdale, E.A. and Brookes, P.C. (2006) Detection and quantification of the soil microbial biomass – impacts on the management of agricultural soil. *Journal of Agricultural Science*, **144**, 285–302.

Stevenson, F.J. (1994) *Humus Chemistry – Genesis, Composition, Reactions*, 2nd edn. John Wiley & Sons, Chichester, Ltd., UK.

Sus, O., Williams, M., Bernhofer, C., Béziat, P., Buchmann, N., Ceschia, E., Doherty, R., Eugster, W., Grünwald, T., Kutsch, W., Smith, P., and Watteenbach, M. (2010) A linked carbon cycle and crop development model: description and evaluation against measurements of carbon fluxes and carbon stocks at several European agricultural sites. *Agriculture, Ecosystems and Environment*, **139**, 402–418.

Swift, R.S. (1999) Macromolecular properties of soil humic substances: fact, fiction, and opinion. *Soil Science*, **164**, 790–802.

Vance, E.D., Brookes, P.C., and Jenkinson, D.S. (1987) An extraction method for measuring soil microbial biomass C. *Soil Biology and Biochemistry*, **19**, 703–707.

Waksman, S.A. (1936) *Humus*. Baillière, Tindall & Cox, London, U.K.

Watts, C.W., Clark, L.J., Poulton, P.R., Powlson, D.S., and Whitmore, A.P. (2006) The role of clay, organic carbon and long-term management on mouldboard plough draught measured on the Broadbalk wheat experiment at Rothamsted. *Soil Use and Management*, **22**, 334–341.

Webster, E.A., Hopkins, D.W., Chudek, J.A., Haslam, S.F.I., Šimek, M., and Pîcek, T. (2001) The relationship between microbial carbon and the resource quality of soil carbon. *Journal of Environmental Quality*, **30**, 147–150.

Wolters, V. (2000) Invertebrate control of soil organic matter stability. *Biology and Fertility of Soils*, **31**, 1–19.

Wu, J., Pommerening, B., Chaussod, R., and Brookes, P.C. (1990) Measurement of soil microbial biomass C by fumigation-extraction – an automated procedure. *Soil Biology and Biochemistry*, **22**, 1167–1169.

Wutzler, T. and Reichstein, M. (2007) Soils apart from equilibrium – consequences for soil carbon balance modelling. *Biogeosciences*, **4**, 125–136.

Zanella, A., Jabiol, B., Ponge, J.F., G., Sartori, G., De Waal, Van Delft, R.B., Graefe, U., Cools, N., Katzensteiner, K., Hager, H., Englisch, M., Brethes, A., Broll, G., Gobat, J.M., Brun, J.J., Milbert, G., Kolb, E., Wolf, U., Frizzera, L., P. Galvan, P., Kolli, R., Baritz, R., Kemmers, R., Vacca, A., Serra, G., Banas, D., Garlato, A., Chersich, S., E. Klimo, E., and Langohr, R. (2011) European humus forms reference base. http://hal.archivesouvertes.fr/docs/00/56/17/95/PDF/Humus_Forms_ERB_31_01_2011.pdf

Zimmermann, M., Leifeld, J., Schmidt, M.W.I., Smith, P., and Fuhrer, J. (2007) Measured soil organic matter fractions can be related to pools in the RothC model. *European Journal of Soil Science*, **58**, 658–667.

Zimov, S.A., Schuur, E.A.G., and Chapin, F.S. (2006) Permafrost and the global carbon budget. *Science*, **312**, 1612–1613.

5 Nitrogen and phosphorus cycles and their management

Phil M. Haygarth[1], Richard D. Bardgett[1], and Leo M. Condron[2]

[1] *Lancaster Environment Centre, Lancaster University, Lancaster, UK*
[2] *Agriculture and Life Sciences, Lincoln University, Canterbury, New Zealand*

5.1 Introduction

Nitrogen (N) and phosphorus (P) are the most important macronutrients that sustain soil, ecological and human life on Earth. All plants require adequate amounts of water, light, carbon dioxide and, critically, nutrients in order to allow them to grow to their maximum potential, and a shortage of nutrients can cause serious restrictions to crop growth (Brennan and Bolland, 2007; Cassman et al., 2002; Haygarth et al., 1998; Jarvis, 1993; Jarvis, 1998; Ladha et al., 2005). They give the basic nutrition for the production of crops, animals and fibre, providing the foundations upon which human existence is built. Soil provides the critical veneer through which the N and P cycling occurs, and plant growth is highly dependent on the availability of these nutrients (Haygarth and Ritz, 2009). Humans manage the N and P cycles in agricultural systems to enhance food production and to minimise outward consequences on the environment. The focus of this chapter is primarily, but not exclusively, on N and P cycles in managed agricultural soils.

Although the fundamental global need for N and P and their functions are similar, their properties are quite different. Nitrogen is group 15 in the periodic table. Critically from a soils perspective, N is a vital component of fertilisers and foods. Nitrogen gas and liquid is colourless and odourless, with the gas being generally inert and making up about 78% of the atmosphere by volume. When mixed with oxygen and subjected to electric sparks (hence its use in explosives), N forms nitric oxide (NO) and then nitric dioxide (NO_2). When heated under pressure with hydrogen in the presence of a suitable catalyst, ammonia forms (this is called the Haber process). Nitrogen can be 'fixed' from the atmosphere by free-living and symbiotic bacteria that associate with the roots of certain legume plants, such as clover, and non-legumes, such as alder, that have the actinomycete *Frankia* as their endophyte. Phosphorus is a colourless/red/silvery white element and has the atomic number 15 and an atomic weight of 30.97, with a standard state that is solid at 298 K (Emsley, 2000). It is group 15 in the periodic table (Parkes and Mellor, 1939) and is an essential component of living systems, playing a key role in the functioning of nervous tissue, bones and cell protoplasm. Nitrogen on the other hand has the atomic number 7 and an atomic weight of 14.01,

Soil Conditions and Plant Growth. Edited by Peter J. Gregory and Stephen Nortcliff.
© 2013 Blackwell Publishing Ltd. Published 2013 by Blackwell Publishing Ltd.

with a standard state that is gas at 298 K (Parkes and Mellor, 1939). As a consequence of these differing properties, the behaviours and cycling of N and P through soils and plants are quite different, with the N cycle having a larger and more significant atmospheric component than the P cycle, and N being biologically renewable via N fixation.

Although plants are known to access organic N forms, most N in agricultural systems is taken up by plants as ammonium (NH_4^+) or nitrate (NO_3^-) ions, but is then reduced in the plant and, when synthesised with carbohydrates, converted to amino acids mainly in the green leaf itself. Hence, as the extra level of N supply increases compared to other nutrients, the extra protein produced allows the plant leaves to grow larger and to have a larger surface area for photosynthesis. Thus, for many crops, the amount of leaf area available for photosynthesis is proportional to the amount of N supplied. In agricultural systems, N usually has a larger effect on crop growth, yield and crop quality than any other nutrient. Nitrogen is cycled from large atmospheric pools as dinitrogen gas into the soil ecosystem, either through biological fixation by N-fixing plants and free-living bacteria or by chemical fixation to produce N fertiliser (De Clercq et al., 2001). In the soil, the largest pool of N is in the organic matter as nitrogenous compounds arising from the decomposition of plant material and microorganisms and added organic matter (e.g. manure in agricultural systems). The soil food web, including microbes and fauna, plays an important role in making organic N available to plants through mineralisation, resulting in the release of NH_4^+, and via microbial immobilisation that reduces the availability of N to plants. Thus, the rates of mineralisation/immobilisation play an important role in controlling the soil N cycle and thus governing the transfer of N to the wider environment. The NH_4^+ produced during the mineralisation process is converted to NO_3^- by nitrification and involves oxidation of NH_4^+ to nitrite (NO_2^-) and then to NO_3^-, which, as a highly mobile ion, forms the basis upon which N transfer to the atmosphere or water can potentially occur. Figure 5.1 shows a simplified schematic of the N cycle through plants and soils.

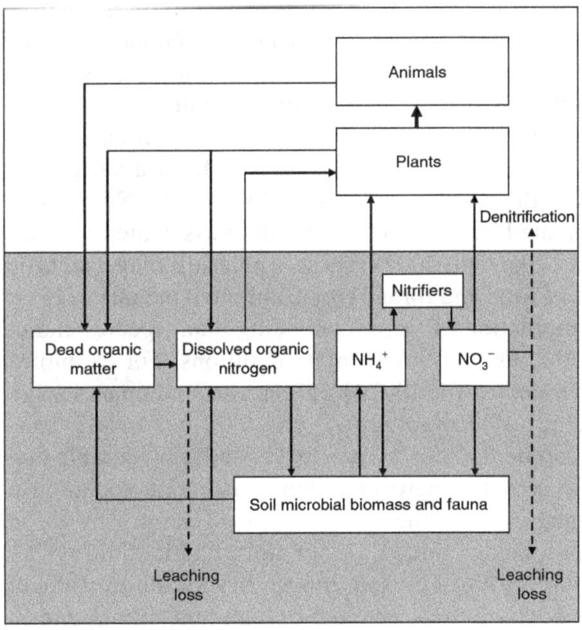

Figure 5.1 A simplified overview of the nitrogen cycle in managed agricultural systems. Reproduced from Bardgett (2005). By permission of Oxford University Press.

Phosphorus plays an essential role in agriculture and in all forms of life: respiration, photosynthesis in green leaves, microbial turnover and decomposing litter all require adequate levels of P in specialised forms (Cole et al., 1977). In agricultural cropping systems, adequate supplies of P are essential for seed and root formation, crop quality and strength of straw in cereals and for the accumulation and release of energy during cellular metabolism (Finkl jr and Simonson, 1979). Phosphorus in natural ecosystems is, like N, usually a scarce resource and is efficiently recycled: in contrast, in agricultural systems, P is removed in the crop or animal products. This means that P in fertilisers and animal fodders/concentrates is imported to the agricultural system in order to sustain productivity, but only 5–10% of fertiliser P that is added to soil is taken up by crops (Loehr 1974). The P cycle through managed soils and plants is principally controlled by inputs from mineral fertilisers and from animal dung and manures (which will also act as a conduit for 'new' farm P that is brought into the soil via P in the food concentrate supplement (common in dairy cows) – animal–soil pathway). In soil P processes are generally controlled by inorganic adsorption–desorption reactions, biologically controlled mobilisation and immobilisation by the microbial biomass, and the forms of P, principally, whether it is in an organic or inorganic form. Figure 5.2 shows a simplified schematic of the P cycle through plants and soils.

5.2 Sources of nitrogen and phosphorus to soil-plant systems

In natural ecosystems, N inputs are dominated by biological fixation. This process involves the conversion of atmospheric N_2 gas to ammonia by bacteria (predominantly *rhizobia*, *cyanobacter* spp.) that can freely exist in soil or are more commonly found in symbiotic association with plants such as legumes (*Fabaceae*) (Canfield et al., 2010). Biological N fixation inputs are not well quantified in these systems and are believed to range from 10 to 150 kg N ha^{-1} an^{-1}. In addition, smaller quantities of N (1–20 kg N ha^{-1} an^{-1}) are added in the form of direct atmospheric inputs of ammonium, ammonia, nitrate and particulate and organic N. These are derived from a variety of sources, including lightning, dust and anthropogenic pollution. Historically, N inputs to managed agroecosystems were dominated by biological N fixation associated with crop and pasture legumes (McNeill and Unkovich, 2007). These inputs still play a significant role in meeting the N requirements of food and forage crops, although the rate of N fixation varies widely depending on a combination of factors including plant species and cultivar, soil fertility (e.g. pH and P status) and environmental conditions. For example, pea crops such as chickpea (*Cicer arietinum*) fix 40–70 kg N ha^{-1} an^{-1} compared with 200–250 kg N ha^{-1} an^{-1} fixed by lucerne (*Medicago sativa*) and soya beans (*Glycine max*), while up to 600 kg N ha^{-1} an^{-1} can be fixed by some clovers under favourable conditions. The nature and quantities of N inputs to agroecosystems (and by association natural ecosystems) were changed dramatically in the early twentieth century by the development of the Haber process of producing ammonia from atmospheric dinitrogen gas (Canfield et al., 2010; Vitousek et al., 1997). The subsequent large-scale manufacture and use of synthetic N fertilisers such as urea, calcium ammonium nitrate and anhydrous ammonia contributed significantly to increased agricultural production across the globe. Inputs of

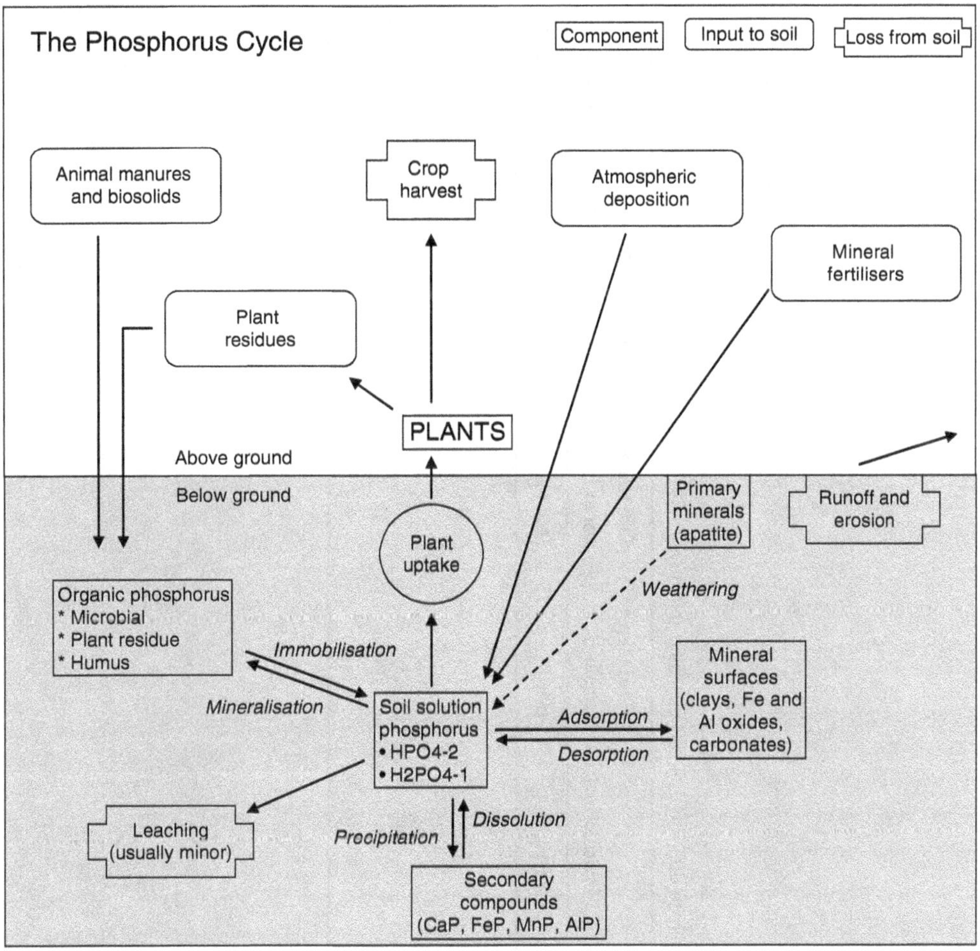

Figure 5.2 A simplified overview of the phosphorus cycle in managed agricultural systems.

fertiliser N increased steadily during the latter half of the twentieth century, in particular, to the current annual global total utilisation of 90 million tonnes, which will increase further in the future with expansion in demand linked to continued population growth. China currently consumes 25% of total N fertiliser production, followed by North America (16%), Europe (15%) and India (12%). To put these various N inputs into perspective, total annual global N inputs are estimated to be 170 million tonnes, which is sourced from a combination of industrial fixation (98), biological fixation (40), combustion (22) and lightning (10).

In contrast to N, atmospheric P inputs to natural and managed ecosystems in the form of dust are very limited at less than 1 kg P ha^{-1} an^{-1}, and P in natural ecosystems is almost exclusively derived from sparingly soluble primary minerals (apatite) present in the soil parent material (Vitousek et al., 2010). The manufacture of P fertilisers from phosphate rock (deposits of apatite P) was developed in the mid-nineteenth century and resulted in dramatic increases in the productivity of agroecosystems (Smit et al., 2009). Currently,

Table 5.1 Characteristics of some continents (and N. America) with respect to area of arable land (including permanent crops), yield and area of total cereals, P fertiliser consumption and rate (on arable land). P off-take with cereals is calculated assuming a content of 3.5 kg P/t (FAO statistics 2005/6).

	Area arable+ perm. crops	**Total cereals Area**	**Area**	**Production**	**P fertiliser applied**	**Off-take cereals**	**Cereal yield**	**Average P-rate**	**Cereal P off-take**
Continent	**(Mha)**	**(Mha)**	**(%)**	**(Mt)**	**(Mt P)**	**(Mt P)**	**($t\,ha^{-1}$)**	**($kg\,ha^{-1}$)**	**($kg\,ha^{-1}$)**
Africa	239	105	44	146	0.4	0.5	1.4	2	5
Asia	577	327	57	1141	10.0	4.0	3.5	17	12
Europe	296	121	41	395	1.8	1.4	3.3	6	11
N. America	229	78	34	463	2.5	1.6	5.9	11	21
Oceania	55	49	35	23	0.4	0.1	1.2	8	4
World	1562	700	45	2342	17.2	8.2	3.3	11	12

The P-rate is calculated for total arable area including permanent crops (not just cereals).
Source: Redrawn from Smit et al. (2009).

the majority of the 20 million tonnes of phosphate rock produced annually around the world are used in the manufacture of fertilisers. Most fertiliser P applied to soil is in the form of water-soluble P (e.g. monocalcium phosphate and diammonium phosphate) made by acidulation of phosphate rock. Some forms of phosphate rock can be used directly as fertiliser, although their effectiveness is limited in many situations compared with water-soluble fertiliser P. Table 5.1 shows the usage of P fertilisers around the continents of the world.

In grazed natural and managed ecosystems, a significant proportion of the N and P taken up by plants from soil are returned in the form of dung and urine. For example, in grazed pastoral systems, 70–85% of the ingested N and P are returned to the soil in excreta (Haynes and Williams, 1993). Moreover, given that the nature of grazing means that N and P in excreta are seldom if ever returned to the area of uptake, this leads to significant N and P inputs localised in small areas, which can have profound effects on nutrient dynamics and bioavailability. Dung deposits of grazing sheep represent N and P inputs of 130 and 35 kg ha^{-1}, respectively, while the corresponding N and P inputs for cattle dung deposits are 1000 and 280 kg ha^{-1}. Similarly, animal manure and slurry are important N and P inputs in intensive management systems where animals are housed indoors for part or all of the year. In some of these systems, most of the forage consumed by the housed animals is grown locally, and the N and P can therefore be effectively recycled by appropriate management and distribution of the manure and slurry resources. However, in high-intensity farming systems such as pig and poultry production in Europe and North America, almost all of the animal feed is imported from other parts of the respective continents or other countries. This large-scale dislocation of nutrients leads to significant net inputs of N and P when large quantities of manure are effectively disposed of by application to adjacent land. This results in the accumulation of large amounts of N and P in soil in excess of agronomic requirements, which in turn increases diffuse N and P transfer from the land to water bodies with consequent adverse environmental consequences linked to eutrophication.

5.3 Chemical and biological processes of nutrient cycling, transformations and bioavailability

5.3.1 Nitrogen mineralisation

Mineralisation is the process by which soil microbes break down soluble and insoluble organic matter and convert it into inorganic forms that are available for plant uptake (Figure 5.1). The majority of soil N (some 96–98%) is found in dead organic matter in the form of complex insoluble polymers such as proteins, nucleic acids and chitin. These polymers are too large to pass through microbial membranes, so microbes produce extracellular enzymes (e.g. proteinases, ribonucleases and chitinases) that break them down into smaller, water-soluble subunits that can be absorbed by microbial cells (e.g. amino acids). This material is called dissolved organic N (DON) and it makes up a large proportion of the total soluble N pool in natural ecosystems, especially those with highly organic soils (Schimel and Bennett, 2004). Even in fertilised agricultural soils, DON concentrations can be equal to, or even more than, inorganic N (Bardgett et al., 2003).

Most DON is absorbed by free-living soil microbes, which use C and N within it to support their growth. When microbes are C limited, they use the C from DON to support their energy needs and excrete plant-available ammonium (NH_4^+) as a waste product, i.e. N is mineralised. When DON is insufficient to meet microbial N demand, microbes absorb additional inorganic N (NH_4^+ and NO_3^-) from soil solution, i.e. N is immobilised by the microbial biomass which reduces inorganic N availability for plants. Both N immobilisation and mineralisation occur simultaneously, and a key factor regulating this balance is the relative demand by microbes for C and N, which is determined by the C:N ratio of the organic material used. Heterotrophic bacteria and fungi have C:N ratios ranging from 4:1 to 12:1, but as they break down organic matter they respire much of the C contained within it and use the remainder for biomass production (Kaye and Hart, 1997). There is a critical substrate C:N ratio of about 30:1, which is needed to meet microbial needs for N (Kaye and Hart, 1997); if substrate C:N ratios are greater, which is often the case for plant residues, microbes immobilise exogenous inorganic N and thereby reduce the availability of N for plants (Kaye and Hart, 1997).

Not all DON is utilised by microbes, and significant quantities may be leached from soil in drainage waters or taken up directly by plants as amino acids or peptides, thereby bypassing microbial mineralisation. This has been shown to be the case in many natural ecosystems, including arctic tundra (Chapin et al., 1993; Kielland, 1994) and boreal forest (Näsholm et al., 1998; Nordin et al., 2001). Agricultural plants also have the capacity to uptake amino acids (Näsholm et al., 1998; Weigelt et al., 2005) and peptides (Hill et al., 2011) directly, but this is thought to be of limited importance for plant nutrition due to rapid microbial turnover of DON in fertile agricultural soils (Bardgett et al., 2003; Owen and Jones, 2001).

5.3.2 Nitrification and denitrification

Nitrification is the process by which ammonium (NH_4^+) is oxidised to nitrite (NO_2^-) and then to nitrate (NO_3^-) (Figure 5.1). This process is carried out by autotrophic bacteria that are classified into two groups on the basis of whether they oxidise NH_4^+ to NO_2^- (*Nitrosomonas* and other *Nitros*-genera) or NO_2^- to NO_3^- (*Nitrobacter* and other *Nitro*-genera). Heterotrophic nitrification can also occur, especially by fungi in forest soils. The process of nitrification is influenced by several soil factors, but the availability of NH_4^+, which is the sole energy source for autotrophic nitrifiers, is of primary importance; as a result, rates of nitrification tend to be highest in heavily fertilised agricultural soils. Oxygen is also required, so factors that influence the diffusion of O_2 through soil, such as soil moisture status and structure, will also affect rates of nitrification. Nitrification is also affected by soil temperature, being optimal at 30–35°C, but it can also occur in cold, but unfrozen, soils (Brooks et al., 1996). Rates of nitrification are also affected by soil moisture, being very low in dry soils where thin water films restrict the diffusion of NH_4^+ to nitrifiers. Nitrification is an acidifying process because the oxidation of NH_4^+ to NO_3^- yields H^+ ions and is also sensitive to changes in soil pH being negligible below pH 4.5. An important recent development is the discovery that archaeal ammonium oxidisers – which perform the first step in the process of nitrification which was previously thought to be carried out exclusively by bacteria – are the most abundant ammonium oxidisers in soil and are likely to play a substantial role in the N cycle (Leininger et al., 2006).

Plants and microbes take up NO_3^- produced by nitrification, but large amounts of NO_3^- may also be lost from soil through leaching and denitrification. Compared to positively charged cations (e.g. NH_4^+) that are retained in soil by negatively charged soil colloids (e.g. clay minerals), NO_3^- is highly mobile in soil solution. As a result, when not taken up by microbes or plants, or when soils have a low anion exchange capacity, it is often lost from soil to drainage waters. This is especially common in coarse textured, sandy soils after rain events, in heavily fertilised soils and when soils are irrigated. Nitrate is also lost from soil by denitrification, which is the microbial reduction of NO_3^- to gaseous NO, N_2O and N_2. Denitrification is carried out by a wide range of microorganisms under conditions of high NO_3^- and C supply, and low oxygen, and is especially prevalent in wetlands that either receive inputs of NO_3^- or when they have a surface aerobic zone where nitrification occurs, such as in paddy fields. Denitrification is also common in fertilised grasslands in high rainfall regions, such as western Britain, where soils may be wet for much of the year. High rates of denitrification are of particular concern since it is a major source of N_2O, a potent greenhouse gas that contributes to the depletion of ozone in the stratosphere.

5.3.3 Nitrogen fixation

Some plants gain N via the process of N fixation, which is driven by microbes that have the capacity to reduce molecular N_2 to NH_3 and incorporate it into amino acids for protein synthesis. These microbes either live freely in soil or they live in symbiotic association with plants, forming nodules in the root where they receive carbohydrate from the plant to meet their energy needs and in turn supply the plant with amino acids formed from reduced N. Legumes are the most widely known N fixers and are associated with bacterium of the genus *Rhizobium*; however, non-legumes such as alder that have the actinomycete *Frankia* as their endophyte are also common. These symbiotic associations can supply large amounts of N into soil systems in both natural and agricultural systems, and their benefits for soil fertility have been recognised since pre-Roman times. In agricultural grasslands, legumes can contribute up to 150 kg N ha^{-1} an^{-1}, and the main route by which this N enters the soil is through the breakdown of plant litter inputs that are enriched in N but also via the exudation or leakage of N-rich exudates from roots. Another important pathway of N fixation is through free-living bacteria that can fix N while decomposing litter and soil organic matter (Vitousek and Hobbie, 2000). Non-symbiotic N fixers occur widely, but their contribution to the N cycle is of most importance in natural ecosystems where they fix relatively small, but significant, amounts of N (<3 kg N ha^{-1} an^{-1}) (Cleveland et al., 1999).

5.3.4 Phosphorus mineralisation and the role of organic forms

Organic forms of P account for 30–60% of total P in most soils and up to 90% in high organic matter soils. Soil organic P is derived from a combination of plant, animal and microbial residues, and the turnover of P through the organic and microbial biomass pools is an important component of P cycling in soil-plant systems (Condron et al., 2005). The quantities and dynamics of organic P in soil are influenced by a combination of soil and environmental conditions, together with land-use and land management practices. Continued inputs of P in mineral fertilisers and organic manures result in elevated levels of plant-available P in soil, especially in surface horizons. Phosphorus inputs accumulate in soil in inorganic and organic forms, and the latter can make a significant contribution to plant P nutrition via mineralisation, which is governed by the production of extracellular phosphatase enzymes by plant roots and soil microbes. One such example comes from a Mediterranean Antipodean study where the application of P fertiliser to pastures in ancient soils leads to the accumulation of organic P relative to inorganic P, with a particular accumulation of labile bioavailable forms (George et al., 2007) (Figure 5.3).

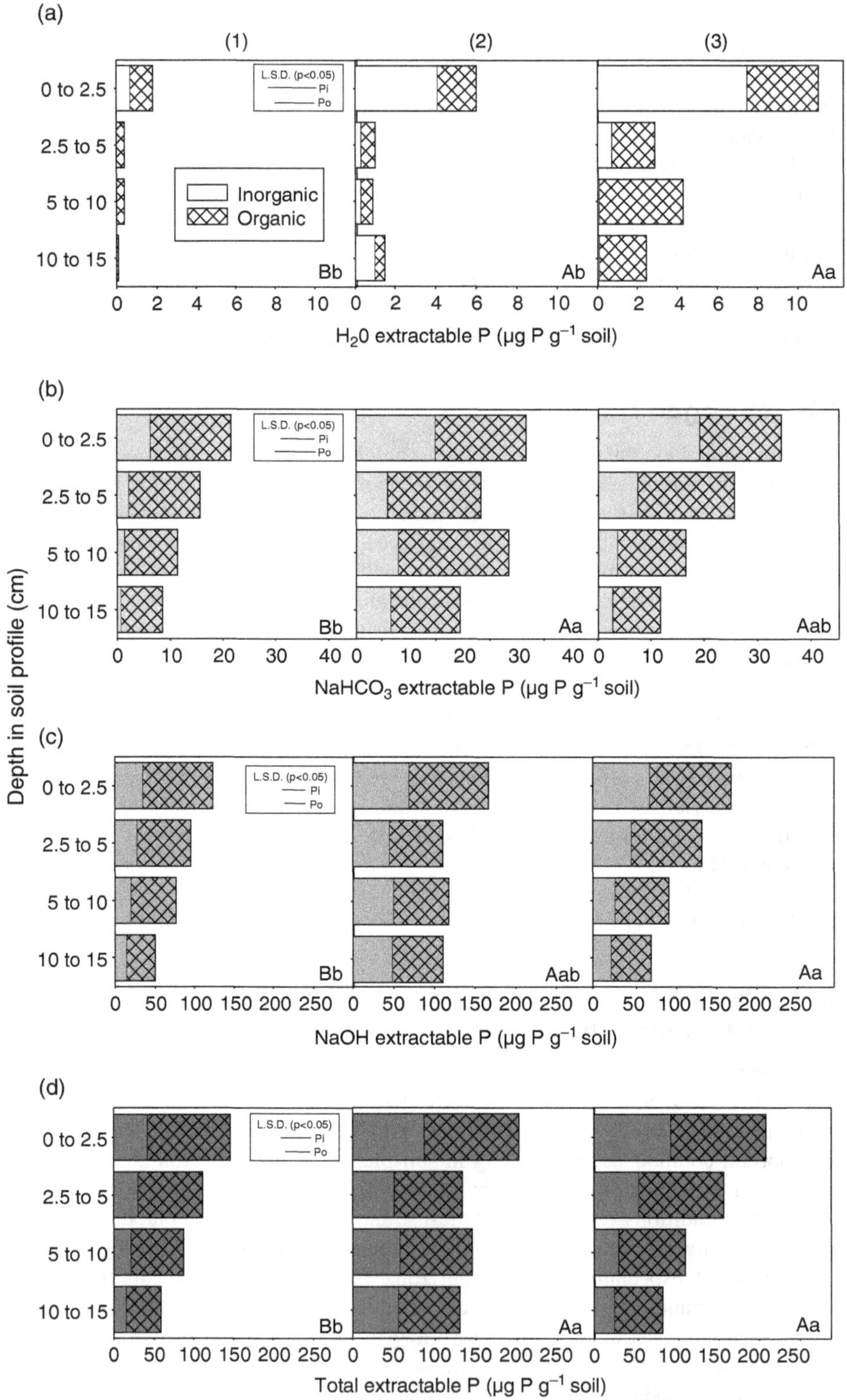

Figure 5.3 Phosphorus accumulation and forms under ancient Australian soils (Mediterranean climate), after fertiliser application (three treatments from a long-term pasture trial with either (1) unfertilised, (2) initially fertilised or (3) continuously fertilised). Data presented are the mean of 6 replicates with l.s.d. (P = 0.05) for both inorganic and organic fractions for each extract shown as bars. Panels with different letters (uppercase, inorganic P; lower case, organic P) within the same extract are significantly different between P treatments averaged over the entire profile. Reprinted from George et al. (2007). With kind permission from CSIRO.

As with N, the mineralisation of organic P is partly regulated by the C:P ratio of substrates. In general, when the C:P ratio is greater than 100, P is immobilised by microbes, which have a relatively high P requirement (1.5–2.5% P by dry weight compared to 0.05–0.5% for plants). As such, microbes compete strongly with plants for available P in soil. The importance of microbial immobilisation of P is illustrated by the fact that they often contain as much as 20–30% of the total soil organic P pool, which is considerably higher than the proportion of C (~ 1–2%) or N (~ 2–10%) contained in soil microbes. Microbial P is therefore an important source of available P in soil. Mycorrhizal fungi, which are associated with the roots of some 80% of terrestrial plants (Smith and Read, 1997), also play an important role in plant uptake of P (and N). Many studies demonstrate the ability of mycorrhizal fungi to enhance nutrient supply to plants, as well as to provide resistance to pathogens, insectivorous herbivores and drought (Smith and Read, 1997). For example, studies in grasslands show that arbuscular mycorrhizal (AM) fungi can enhance plant productivity by up to twofold (Van der Heijden et al., 1998), which is mostly due to increased plant uptake of P, although acquisition of N through AM fungi might also play a role (Hodge et al., 2001). Ectomycorrhizal fungi (EM) play a key role in plant nutrient uptake in boreal and temperate forests (Read and Perez-Moreno, 2003) where extensive ectomycorrhizal hyphal networks forage for nutrients in soil and excrete extracellular enzymes that degrade complex organic compounds. Mycorrhizal fungi increase P acquisition through two mechanisms: (1) they produce phosphatase enzymes that cleave ester bonds that bind P to C in organic matter, thereby releasing phosphate (PO_4^{3-}) that can be taken up by the fungus and passed on to the plant; and (2) they produce low molecular weight organic acids, such as oxalic acid, which enhance the availability of soil P by increasing weathering rates of sparingly soluble Ca, Fe and Al P minerals. For example, a study of P utilisation in the rhizosphere of AM ryegrass (*Lolium perenne*) compared with EM radiata pine (*Pinus radiata*) in New Zealand soils revealed that P depletion extended further from the root surface (rhizoplane) of pine (5 mm) than ryegrass (3 mm), while enhanced mineralisation of soil organic P in the pine rhizosphere was attributed to a combination of higher phosphatase enzyme activity, microbial activity and root exudate production (Chen et al., 2002) (Figure 5.4).

In an African system, George et al. (2002) demonstrated that in extremely P-deficient soils the growth of plants that exude phosphatase into the rhizosphere (i.e. plants that have a capacity to try to use organic P) leads to a depletion of organic P (NaOH-Po; Figure 5.5). This research confirmed the role of organic P in contributing to plant nutrient uptake when inorganic P supply is low (NaOH-Po was previously thought of as being recalcitrant or non-plant available) under field conditions.

Clearly, organic forms of P represent the key reservoirs for P mineralisation and recent research from tropical rainforests studied in Panama has confirmed the key role of organic forms in these highly productive systems (Turner and Engelbrecht, 2011). Figure 5.6 shows scatter plots of the relationship between concentrations of soil P fractions (determined by solution ^{31}P nuclear magnetic resonance (NMR) spectroscopy, a key means of separating and identifying organic P compounds) and soil chemical properties (total P, soil pH and total carbon).

5.3.5 Role of soil animals

The soil hosts a highly diverse community of invertebrates, including nematodes, collembolans, mites and enchytraeids, which gain most of their nutrition from roots and their exudates, organic matter and from feeding on soil microbes (Bardgett, 2005). Although most mineralisation of nutrients is governed by the activities of bacteria and fungi, their ability to do this is strongly affected by soil animals. Soil animals affect microbial communities either directly, through

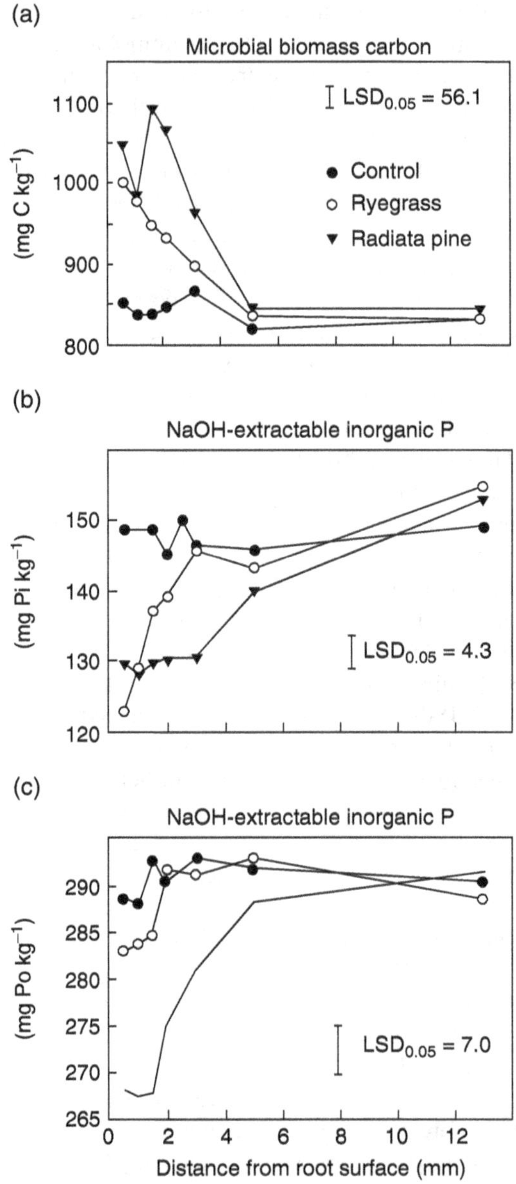

Figure 5.4 This shows the effect of distance from the root surface on microbial biomass carbon, inorganic and organic P forms determined in the rhizosphere of two New Zealand soils. There is clear depletion of both inorganic and organic forms of P close to the root, which is related to the presence of microbes. From Chen et al. (2002). With kind permission of Elsevier.

selectively feeding on fungi and bacteria, or indirectly by fragmenting of organic matter, disseminating microbial propagules and by altering nutrient availability (Bardgett, 2005). These interactions between microbes and animals drive processes of energy flow, decomposition and nutrient cycling, and therefore contribute to plant nutrient acquisition and plant growth.

As previously mentioned, microbes can immobilise significant quantities of N and P, making them unavailable for plant use. However, these nutrients can be liberated through

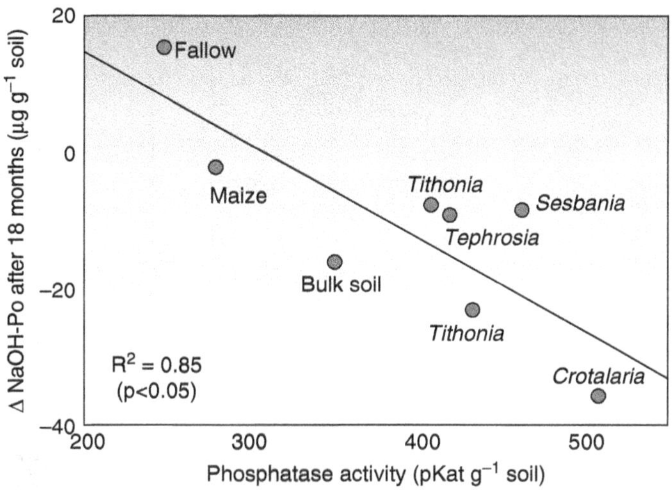

Figure 5.5 The relationship between phosphatase activity in African soils at 18 months plant growth and the decline in organic phosphorus (NaOH-Po) in the last 6 months of field growth (taken from George et al., 2002). With kind permission from Springer Science and Business Media.

the predation of microbes by protozoa, nematodes and microarthropods, which excrete nutrients that are in excess of their own requirements into the soil environment. This 're-mobilisation' of nutrients is termed the 'microbial loop' (Clarholm, 1985), and via this mechanism, soil animals can increase the availability and uptake of nutrients by plants, thereby stimulating plant growth (Bardgett, 2005; Ingham et al., 1985; Setälä and Huhta, 1991) (Figure 5.7). In addition, this grazing of microorganisms can lead to changes in the structure, size and activity of the soil microbial community, which in turn can indirectly affect processes of nutrient cycling and plant nutrient uptake.

Soil fauna also affect nutrient cycling by physical alteration of decomposing material and of the soil environment. For example, some animals consume plant detritus and egest this material into soil as faecal pellets, which provide a favourable environment for microbial growth, leading to enhanced rates of decomposition and nutrient release. Also, earthworms consume large amounts of plant litter along with soil mineral particles, and these two fractions are mixed together in the earthworm gut and then egested as surface or subsurface casts. These casts contain significantly greater numbers of microbes and have higher enzyme activities than the surrounding soil, leading to enhanced rates of N mineralisation. Phosphorus availability is also much greater in casts than in the surrounding soil because of the stimulation of phosphatase activity (Sharpley and Syers, 1976). While these positive effects of casting on microbial activity are often transitory (Lavelle and Martin, 1992), their net effect, along with other earthworm activities, is the stimulation of total soil nutrient availability and enhanced plant nutrient uptake (Edwards and Bohlen, 1996).

5.3.6 Influence of soil biodiversity

In recent years, a number of studies have examined whether the diversity of soil organisms affects processes in the soil, including nutrient cycling. There is some evidence that decomposition of organic matter can be positively influenced by the species diversity of saprophytic fungi and soil arthropods (Setälä and McLean, 2004). However, these effects mostly occur

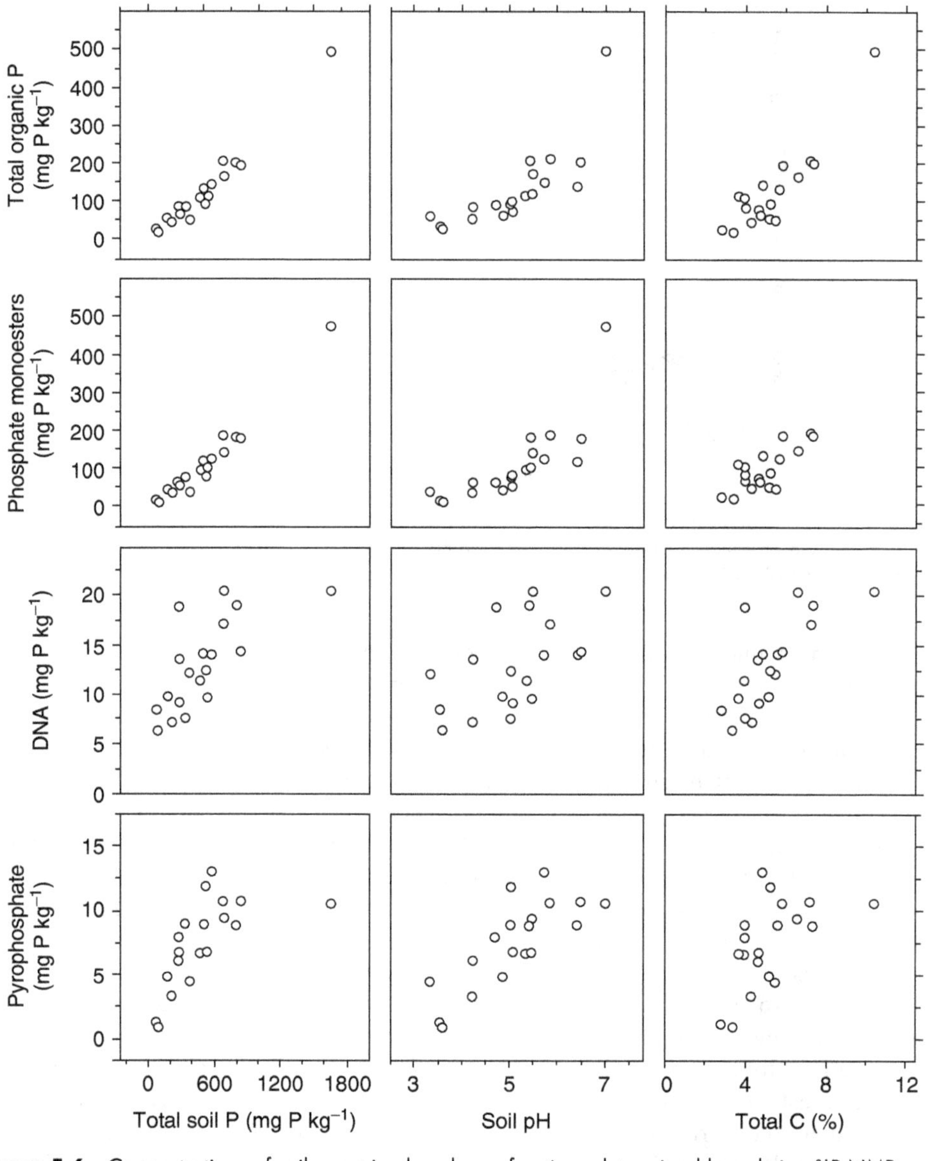

Figure 5.6 Concentrations of soil organic phosphorus fractions determined by solution ^{31}P NMR spectroscopy and chemical properties (total phosphorus, soil pH, total carbon) of soils under lowland tropical rain forest in central Panama. All correlations are statistically significant at $p = 0.05$. From Turner and Engelbrecht (2011). With kind permission from Springer Science and Business Media.

at the low diversity end of the diversity spectrum and at levels of diversity that are likely to be well below that which is found in real ecosystems (Wardle, 2002). There is also evidence that the diversity of ectomycorrhizal fungi (Jonsson et al. 2001) and arbuscular mycorrhizal fungi (Maherali and Klironomos, 2007; Van der Heijden et al., 1998; Vogelsang et al., 2006) can affect plant growth, with responses ranging from positive to negative. In general, however, the prevailing message that emerges from studies is that changes in the abundance of particular species and the nature of trophic interactions are more important drivers of

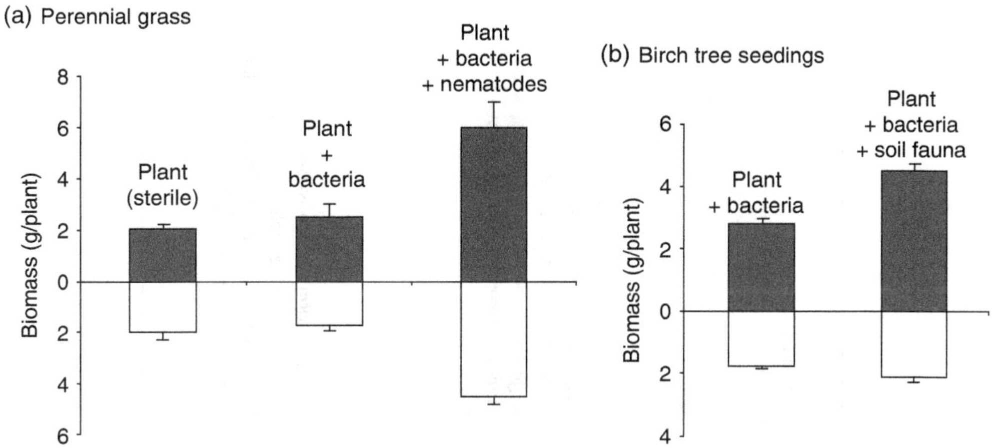

Figure 5.7 Influence of soil animals on plant growth. (a) Adding bacterial-feeding nematodes to soil increased shoot and root growth of the perennial grass *Bouteloua gracilis* (data from Ingham et al., 1985). (b) Addition of diverse fauna to soil increased root and shoot biomass of birch seedlings (*Betula pendula)*; data from Setälä and Huhta, 1991.

ecosystem function than is diversity *per se*. This is consistent with the view that there is a high degree of redundancy within soil communities, as evidenced by the prevalence of omnivory in soil food webs (Scheu and Setälä, 2002), although it is clear that some species are more redundant than others (Bardgett, 2005).

Very little is also known about the ecological effects of diversity of soil bacteria that perform specialised belowground functions, such as nitrogen fixation and nitrification. However, these types of functions are often performed by a physiologically and/or phylogenetically narrow range of taxa, when compared to 'broad' processes such as nitrogen mineralisation (Schimel et al., 2005). As a result, it is widely thought that processes such as N fixation and nitrification are more sensitive to changes in microbial community composition and diversity than is nitrogen mineralisation. For example, symbiotic nitrogen fixation in legumes is characteristically performed by a small number of strains of *Rhizobium*, and losses of a subset of these strains can greatly impair symbiotic nitrogen fixation and legume growth (Giller et al., 1998). It has been argued that 'broad' processes of mineralisation and immobilisation can be 'aggregated' into individual components, based on specific enzyme activities and spatial distribution of microbes in microsites that are sensitive to microbial community composition (Schimel et al., 2005). This emphasises the need for a better understanding of the spatial organisation of microbial communities within the soil and the need to simultaneously disentangle the specific functions of microbes involved in nutrient cycling.

5.4 Processes of nitrogen and phosphorus losses from soils

5.4.1 Nutrient losses to water and air

Nitrogen and P cause environmental problems when they are transferred from productive agricultural systems, both in water bodies (water quality) and the atmosphere (air quality). A summary of the problems and impacts associated is provided in Table 5.2 (modified from

Table 5.2 Impacts of phosphorus and nitrogen losses on the wider environment.

Substance of concern	Environmental and other impacts	Scale of agricultural contribution	On-farm sources	Scale of impact
Nitrate (NO_3^-)	Water quality • Eutrophication • Health Economic loss • Loss to farmers	Major source	Intensively managed land (inputs from fertiliser, manures, slurry, legumes and feeds)	Local: on-farm surface waters Regional: surface waters; catchment; aquifers National/international: maritime waters
Nitrite (NO_2^-)	Water quality • Fish stocks and health	Major source	Managed land	Local: on-farm surface waters Regional: surface waters and wells
Ammonia (NH_3)	'Acid rain' • Acidification of soils • Eutrophication of natural systems Direct toxicity	Major source (>85%)	Fertilisers (urea) Excreta Manures and slurry	Local: 'on-farm' deposition Regional: deposition into natural ecosystems National/international: cross-boundary transfer of NH_3 and deposition
Nitrous oxide (N_2O)	Greenhouse gas • Global warming Ozone interactions	Substantial (likely to increase in importance as other sources decrease)	N fertilisers Excreta	Global
Nitric oxide (NO)	Tropospheric ozone precursor	Probably minor	Combustion, fertilisers Manures and slurry	Global
Phosphorus (P)	Water quality • Eutrophication Health • Toxins from algal bloom Economic • Cost of removal	Substantial – increasing as industrial point sources decrease	Fertilisers Manures, slurry P-enriched soils	Local: on-farm surface waters Regional: surface waters, catchments National/international: maritime waters (cross-boundary transfer)

Source: Modified from Carton and Jarvis (2001).

Carton and Jarvis, 2001). When transferred into waterways, they play a role in the ecology of freshwater environments through acidification and eutrophication. Acidified waters are generally characterised by low pH, high SO_4^{2-} and NO_3^-, and both acidification and eutrophication result in major changes in the species composition of biotic communities, with reductions in species diversity and changes in productivity being common (Foy and Withers, 1995; Hornung, 1999; Vollenweider, 1968; Vollenweider and Krekes, 1982). Redfield (1934) reported that marine biomass tends to have a constant atomic ratio of 105:15:1 C:N:P and this is also applicable to freshwater systems (Uhlmann and Albrecht, 1968). The implication of this is that adding P to a water body will have a much greater effect on aquatic ecology than does adding either N or C (Hudson et al., 2000; Redfield, 1934; Uhlmann and Albrecht, 1968). Eutrophication describes the process by which lakes accumulate nutrients, with a resulting eutrophic state. This can lead to accelerated algal growth, but may have other effects such as changes in species composition, high oxygen demand and a loss of fish species and biodiversity (Foy and Withers, 1995; Hornung, 1999; Hudson et al., 2000; OSPAR, 1998; Uhlmann and Albrecht, 1968; Vollenweider, 1968; Vollenweider and Krekes, 1982). In addition to problems with eutrophication, NO_3^- has been suggested as the cause of methaemoglobinaemia in young infants and also with stomach cancer, but there is some controversy over the extent of this (Addiscott, 1999), while NO_2^- is known to contribute to stomach cancer and is toxic to fish, especially salmonids.

Ammonia (NH_3) is a reactive gas that has effects both on atmospheric chemistry and on terrestrial and aquatic ecosystems (Pain and Jarvis, 1999). After reaction with acidic compounds in the atmosphere, deposition of NH_3 on land as a gas or as ammonium salts can add N to nutrient-poor soils (e.g. heathlands and moorlands) and change the types of plants and thus the biodiversity. Ammonium salts can also be transported over long distances in the atmosphere and contribute to acid rain effects. It is estimated that the worldwide N_2O emitted *directly* from agricultural fields is 20–30% of the total N_2O emitted annually (Brown and Jarvis, 2001; Mosier, 1994). Greenhouse gases trap heat in the lower atmosphere, reflecting infrared energy back to the surface, partly responsible for global warming. Many soils of the world that are naturally low in nutrients are therefore particularly sensitive to enrichment by, for example, atmospheric inputs of N or inputs of fertiliser. Losses of P from soil to the atmosphere do occur as phosphine (PH_3) gas (Roels and Verstraete, 2001). Phosphine emissions are unlikely to have a major impact because transport is likely to be short range and localised, but there remains as yet a lack of substantial research evidence either way here.

5.4.2 Processes of nitrogen emissions to air and water

Under anaerobic soil conditions, denitrification occurs when microorganisms are capable of using the oxygen from NO_2^-, NO_3^- and N_2O in the place of elemental oxygen with the emerging gaseous end products of N_2O, dinitrogen (N_2) and NO_x (Jarvis et al., 2001). In some soils, chemical denitrification can also occur (see also Section 5.3.2). Most denitrification takes place in the upper layer of soils and the gases produced are released into the atmosphere. The rate of denitrification depends on soil characteristics, soil moisture and aeration, the supply of carbon, temperature and available substrate (NO_2^- and NO_3^-). It is most likely to occur in poorly drained, fine-textured soil and in situations with high water tables where anaerobic conditions are most likely to be present (De Clercq et al., 2001). A separate process by which N is released to the atmosphere is volatilisation of NH_3 from soils, added fertiliser, excreta and manure (Pain and Jarvis, 1999). In excreta and manure, the main source of NH_3 is urea, a major nitrogenous constituent of animal urine and some inorganic fertilisers, especially prevalent in the aqueous phase. Therefore, the application of liquid-based animal manures or inorganic fertilisers can increase the potential for volatilisation depending on the land application method. High soil pH, dry conditions and high temperatures also increase the potential for the NH_3 volatilisation.

Nitrogen leaching can be a major means of N loss from the soil ecosystem (De Clercq et al., 2001; Jarvis, 2000; Wilson and Ball, 1999). Because it is negatively charged, NO_3^- is not readily adsorbed to soil colloids and thus it is commonly present in soil solutions and easily leached away. Some NH_4^+ and organic N compounds are also easily leached, but not to the same extent as NO_3^-. Leaching potential is governed by soil type and structure (higher potential in easily drained soils), rainfall patterns and the supply of readily available N. Higher levels of mineralisation that occur in autumn after dry summer periods can result in substantial quantities of N for leaching, as the assimilative capacities of crops decline (Scholefield et al., 1993).

5.4.3 Processes of phosphorus emissions to air and water

The atmospheric component of the P cycle is small, with only a small amount of phosphine (PH_3) gas being generated (Stein et al., 1996). Possible sources of PH_3 include decaying organic matter and breakdown of metal compounds (Roels and Verstraete, 2001). In terms of mass balance, there is little available information to make any difference, but in the absence of further information it is thought that the soil–atmospheric loss pathway is generally insignificant. Conversely, processes of P transfer to water are more substantial, with three mechanisms being solubilisation, detachment and incidental – described by Haygarth et al. (2005) as a transfer continuum (Figure 5.8) (Haygarth et al., 2005; Withers and Haygarth,

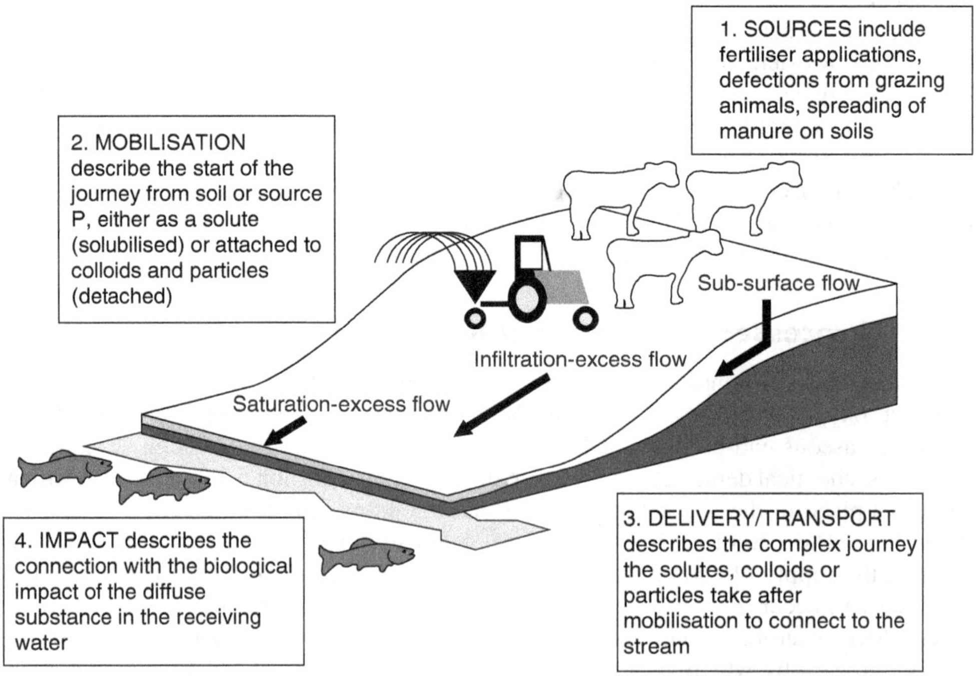

Figure 5.8 The source–mobilisation–delivery 'transfer continuum' as described by Haygarth et al. (2005). Figure modified from Withers and Haygarth (2007). With kind permission from John Wiley & Sons.

2007). The continuum concept was developed to describe P flows from agricultural land to water and helps promote the notion of connectivity from fine-scale soil processes and the wider river catchment. Although it was devised for P, it has generic value for N. 'Solubilisation' refers to the release of molecules or macromolecules of P from soil surfaces and soil biota into soil water. It reflects long-term management history, and there is strong evidence that solubilisation potential increases with increased soil P status. Heckrath et al. (1995) used Olsen-P measurements to define a 'change point' value of 60-mg kg^{-1} soil in an arable soil, above which there is a much-enhanced tendency to release P to land drains. Detachment is the movement of soil particles with P attached, often associated with soil erosion, and is a physical mechanism for mobilising P from soil to waters (Kronvang, 1990; Sharpley and Smith, 1990). Soil erosion (Burnham and Pitman, 1986; Elliot et al., 1991; Evans, 1990; Heathwaite and Burt, 1992; MAFF, 1997; Morgan, 1980; Quinton, 1997) plays the role of particle transfer in P loss (Catt et al., 1994; Kronvang, 1990; Sharpley and Smith, 1990; Zobisch et al., 1994). The role of small colloids in the detachment and transport of P has also been described (Haygarth et al., 1997), reflecting the point that P strongly adheres to surfaces. Incidental transfers are the circumstances by which applications of manure or fertiliser coincide with conditions favouring fast discharges of excess water in overland or preferential flow (Haygarth and Jarvis, 1999).

5.5 Nitrogen and phosphorus use in agricultural systems

The amounts and forms of N and P inputs in managed agricultural systems have changed dramatically over the past 150 years as a consequence of the development and use of N and P fertilisers (Vitousek et al., 2009). Global N fertiliser use is currently 90 million tonnes which represents 50% of total N inputs, while most of the 20 million tonnes of phosphate rock mined annually is used in fertilisers. However, it is important to note that benefits of these N and P fertiliser inputs have not been evenly distributed around the world. Increased fertiliser use has been concentrated in Europe and North America, with recent increases in emerging economies such as China, India and Brazil. In intensively managed arable systems, continued large inputs of fertilisers and manures have resulted in the accumulation of significant amounts of N and P in soil, often to levels that exceed agronomic optima. This N and P loading has in turn led to increases in N and P losses from soil in drainage and runoff, with consequent adverse impacts on water quality related to accelerated eutrophication. In stark contrast, the availability and use of N and P fertilisers in large areas of Africa, Asia and Central and South America have been and remain limited, to the extent that crop productivity is commonly constrained by the lack of adequate N and P inputs on often degraded soils.

As a consequence of the potential environmental impacts of losses from the soil and, in some systems, the limited availability, there is an increasing emphasis in developed intensive arable systems on improving the utilisation efficiency of N and P inputs with the objective of reducing environmentally damaging N and P losses (Christensen, 2004; Frossard et al., 2009). This desire to improve efficiency is also being driven by recent and predicted increases in the cost of N and P fertilisers. Somewhat ironically, improving the efficiency of N and P inputs in intensive systems will also potentially benefit low-input arable systems where the focus is to maximise the agronomic benefit of limited N and P fertiliser resources.

There are fundamental differences in the pathways and dynamics of N and P cycling in the soil-plant environment which will determine the nature and efficacy of any management regime designed to improve nutrient-use efficiency and reduce losses. Biological processes primarily control the transformations and bioavailability of N in soil, and there are significant potential nitrate-N losses from soil during drainage as well as gaseous losses as dinitrogen, ammonia and nitrous oxide. Nitrogen losses from soil are important from both agronomic (nitrate) and environmental (nitrate, nitrous oxide) perspectives (Jenkinson, 2001; Liu et al., 2010). On the other hand, P dynamics are controlled by a combination of biological and chemical processes, and while drainage losses are limited there are important issues regarding restricted availability and utilisation associated primarily with the chemical reactivity of phosphate in the soil (Frossard et al., 2000).

The nature of N dynamics in the soil-plant system means that enhanced spatial and temporal coincidence between availability and plant uptake is necessary to optimise the utilisation of soil and applied N. As P is less mobile than N, optimisation efforts have focused on enhancing the solubility and potential bioavailability of soil and fertiliser P. Based on extensive research, a variety of best management practices have been developed to tighten N and P cycles in different agroecosystems and thereby increase nutrient-use efficiency, lower fertiliser inputs and reduce the risk of adverse environmental impacts. Best management practices for N include the selection of appropriate crop rotations and varieties to optimise productivity, maintenance of plant cover to minimise soil erosion and nutrient loss in drainage, use of soil analysis in combination with nutrient budget models to calculate appropriate levels of fertiliser and manure inputs (including timing and methods of application) and the use of appropriate pest and disease control strategies (Goulding et al., 2008). For P, the focus has been on improving the management of P inputs in intensive, integrated livestock-cropping systems to reduce diffuse P losses (Condron, 2004). This includes identifying critical P source areas within a farm or catchment based on a combination of soil type, P status and hydrology, and then reducing the risk of P loss by restricting P inputs to these areas and by installing reactive buffers such as riparian strips where appropriate. In most cases, the implementation of these practices can result in improved management of N and P resources and inputs in relation to crop requirements, which in turn reduces loss and enhances long-term sustainability.

The use of short-rotation cover crops and green manures also has the potential to significantly enhance the efficiency of N and P use in a wide range of temperate and tropical cropping systems (Cherr et al., 2006; Fageria, 2007; Thorup-Kristensen et al., 2003). Table 5.3 shows the impacts of catch crops and green manures in different input cropping systems. In temperate regions, cover crops such as grasses and brassicas are grown during autumn and winter between the main crops, and their primary function is to capture and retain mineral N released from soil organic matter and crop residues, which is susceptible to loss by leaching. Green manures are most commonly legume cover crops (e.g. clover and medic) that capture N but also add N to the system via biological fixation. Cover crops and green manures are not harvested and are incorporated prior to main crop establishment to provide N and other nutrients that are released by mineralisation. Green manures are used extensively in tropical cropping systems to increase N inputs, reduce soil erosion and increase soil organic matter. Above-ground accumulation of N in cover crops and green manures ranges from 20 to over 600 kg N ha^{-1}, not including the N accumulated in the root biomass. The amount of N in cover crops and green manures depends on a combination of factors including species (legume and non-legume), climate (temperate and tropical), growth period and soil fertility. For example, temperate legumes such as vetch (*Vicia*), clover (*Trifolium*) and medic (*Medicago*) can accumulate 150–250 kg N ha^{-1}. The release of N during cover crop and green manure decomposition following incorporation can

Table 5.3 Impacts of catch crops and green manures in different input cropping systems.

	Potential Positive Impacts	Potential Negative Impacts
Catch crops in high input systems	Reduced N leaching loss Improved soil fertility Erosion control Improved soil physical conditions	Increased N fertiliser requirement due to cover crop uptake and immobilisation during residue decomposition
Green manures in low-input systems	Increased N supply stability Consistent crop yields Increased soil organic matter Increased soil fertility, including N and P mineralisation Erosion control Improved soil physical conditions Improved crop rooting depth	Increased N loss due to poor temporal coincidence with crop demand Loss of cash crop as a result of undersowing green manure

Source: Adapted from Thorup-Christensen et al. (2003).

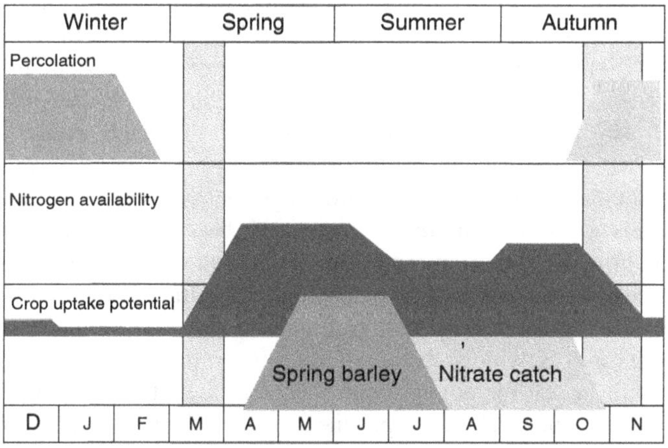

Figure 5.9 The seasonal dynamics of potential nitrate leaching (percolation), mineralisation (availability) and crop uptake under northwest European conditions for a spring barley undersown with ryegrass as a nitrate catch crop. The vertical grey zones indicate periods susceptible to elevated nitrate leaching. Taken from Chrisensen (2004). With permission from CAB International.

significantly reduce or even eliminate the requirement for N fertiliser. The agronomic effectiveness of cover crop and green manure N depends on the spatial and temporal coincidence between N release and crop demand, which is determined mainly by a combination of plant maturity (C:N), the timing of incorporation and prevalent soil and environmental conditions (temperature and moisture) during residue decomposition (Figure 5.9). In this example, only a proportion of the total N accumulated in cover crops and green manures will be utilised by the subsequent crop. For example, it has been found that 15–20% of cover crop N is taken up by the succeeding crop when incorporated in spring, compared with only 5–15% for autumn incorporation. Most of the N taken up or fixed by a single cover or green manure crop is retained in the soil and thereby makes a continuing long-term contribution to crop N requirements. The inclusion of a green manure crop can also enhance soil P availability and subsequent crop P acquisition. A number of legumes have been shown to increase the dissolution and bioavailability of sparingly soluble P minerals including phosphate rocks, which are poorly utilised by most crop species. This has been mainly attributed to the enhanced exudation of organic acids (e.g. citric acid). Furthermore, recycling of labile green manure biomass containing significant amounts of N and P may increase the rate of organic N and P mineralisation in soil, which may contribute to reducing fertiliser requirements. The use of green manures may be particularly useful in organic cropping systems, where inputs of soluble N and P fertilisers are not permitted and N and P requirements are met by biological fixation and by the application of sparingly soluble phosphate rock (Condron, 2004).

Various soil treatments may also be used to improve the efficiency of N and P cycling in arable cropping systems. In this context, soil treatment includes the addition of chemical compounds or microbial inoculants that are designed to improve nutrient retention and acquisition and thereby reduce fertiliser inputs. For example, the application of the nitrification inhibitor dicyandiamide to intensively managed grassland soils has been shown to significantly reduce nitrate leaching and nitrous oxide emission, which in turn reduced N fertiliser requirement (Moir et al., 2007). It is possible that nitrification inhibitors may be used in some cropping systems to tighten N cycling. Inoculation of the plant rhizosphere with microorganisms has the potential to enhance the bioavailability of mineral and organic forms of P in soil (Richardson, 2001). For example, under laboratory conditions, it has been shown that a range of soil bacteria and fungi (*Bacillus, Pseudomonas, Penicillium, Aspergillus* spp.) can increase dissolution of sparingly soluble P minerals found in soil, which was attributed to the enhanced production of low molecular weight organic acids such as oxalic and citric acids. However, while there is some evidence to show that plants responded positively to soil inoculation with some of these microorganisms, this may have been due to direct growth promotion rather than improved soil P acquisition.

Certain soil amendments may also play a role in tightening N cycling in soil. Biomass-derived black carbon (charcoal) in soil originates from a combination of natural and anthropogenic fires and is widely acknowledged to be the most stable form of organic carbon in soil, with an estimated turnover time and a half-life of thousands of years. Biochar is a specific form of charcoal produced by pyrolysis of organic material, which involves low temperature (700 °C) combustion in limited oxygen. Biochar can be made from a wide range of crop and wood residue materials at a variety of scales with up to 50% of the feedstock carbon being converted to biochar. Biochar amendment of soil at high rates of application can potentially be a means of increasing total soil carbon. There is an emerging debate that says that biochar production has the potential to be used to sequester atmospheric carbon dioxide and convert it to stable soil carbon and thereby contribute to the mitigation of ongoing and predicted global climate change (Woolf et al. 2010). It has been estimated that widespread adoption of biochar production from agricultural and forest waste biomass could reduce global greenhouse gas emissions by 1.8 gigatonnes per annum, which is equivalent to 12% of current anthropogenic emissions. In addition to the potential for carbon sequestration, there is the possibility that biochar amendment of soil could improve soil fertility and plant productivity enhancing nutrient retention and bioavailability, including biological N fixation. Until recently, most of the research on the benefits of charcoal amendment on soil fertility was based on observations made on Amazonian Dark Earths (Terra Preta), which were formed by continued additions of large quantities of charcoal to soil several thousand years ago. The beneficial effects of contemporary charcoal amendment of agricultural soils, especially under temperate conditions using biochar made by pyrolysis, remain to be established. However, with regard to the effect of biochar amendment on soil N dynamics, there is increasing evidence that it can significantly reduce the rate of nitrification and enhance the retention and plant availability of ammonia and ammonium added to soil (Clough and Condron, 2010).

5.6 Future soil nutrient cycles and environmental change

5.6.1 Future of phosphorus fertilisers

Unlike N, P is a non-renewable resource and the productivity of managed ecosystems is heavily dependent on continued inputs of P in the form of mineral fertilisers manufactured from phosphate rock (Condron, 2004; Frossard et al., 2000). This fertiliser P is required to maintain the high levels of production in developed agroecosystems and to increase the productivity of underdeveloped and degraded soils. The future nature and use of P fertiliser will be largely determined by the availability and supply of phosphate rock. It is widely acknowledged that there are finite reserves of readily accessible phosphate rock and that these are being exhausted at an increasing rate to meet increased demand linked to a combination of population growth and accelerated economic development (Cordell et al., 2009; Smit et al., 2009). Figure 5.10 shows the P flows through the global food system. The current annual global production of phosphate rock is 20 million tonnes, and known phosphate rock reserves are estimated to be 2,400 million tonnes. The latter represents 120 years of supply at current production rates, but it has been predicted that continued increases in demand could mean that global production of phosphate rock may begin to decline from as early as 2030. While phosphate rock deposits are found in many locations around the world, the largest reserves are located in Morocco and China (60% of total), with significant quantities present in the USA and South Africa (20%). It is likely that increased demand and declining reserves will lead to pressure on supply and price, including increased volatility. For example, in the USA, the cost of superphosphate fertiliser was relatively stable over the 30 year period from 1975 at \$140–300 t^{-1}(average \$215 t^{-1}), although by 2008, the cost had increased to \$800 t^{-1} in response to increased demand and shortage of supply. The potential impact of declining supply and increasing the cost of P fertilisers in the future will be especially acute in areas with high population growth such as much of Africa, where the availability of P fertiliser is already limited by a combination of economic and logistic constraints.

At least part of the increased pressure on P resources could be alleviated by improved utilisation of fertiliser P and recycling of P contained in various agricultural and urban waste streams (Frossard et al., 2009). There is an urgent need to further investigate and quantify practical means of improving the effectiveness and utilisation of existing and alternative P fertilisers (e.g. phosphate rock and waste P). For example, the inclusion of selected legume green manure crops could be used to increase the dissolution and utilisation of sparingly soluble forms of soil and fertiliser P. In intensive animal production systems, there is considerable potential to develop new and innovative waste management regimen designed to enhance P reuse and recycling and thereby reduce excessive P loading onto soils from organic waste disposal. This may require waste treatment and transport to areas where the agronomic benefit of P addition can be realised. In the long-term, it would be advantageous if the feed production system and related animal production facilities could be located in the same area that would enable efficient P utilisation and recycling. The ongoing expansion in

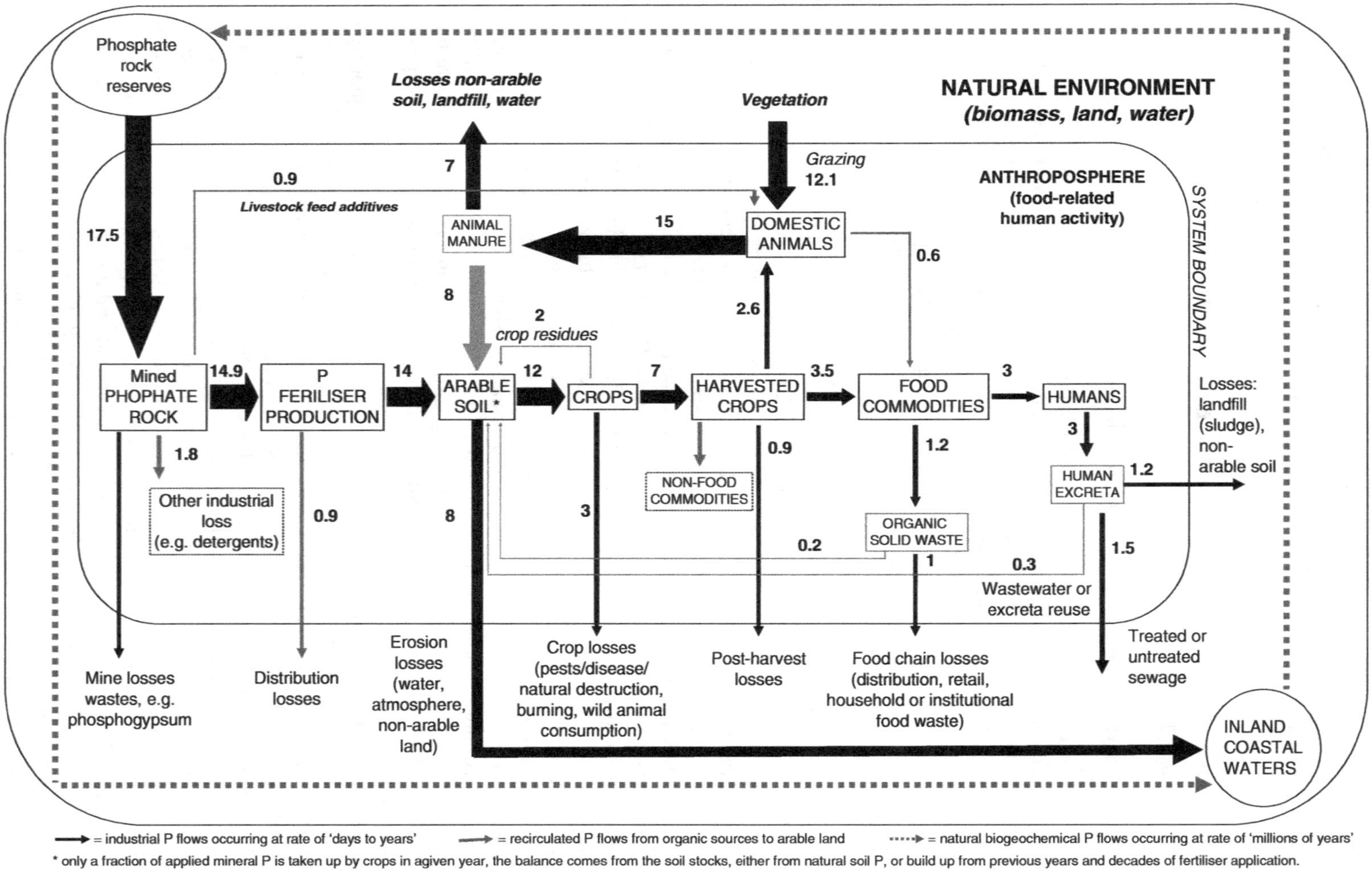

Figure 5.10 Phosphorus flows through the global food system (redrawn from Cordell et al. (2009)). With kind permission from Elsevier.

urbanisation across the globe will further increase net transfer of P from farmland to cities on regional and transnational scales. Given the predicted increasing value of P, it should be possible to recover a significant proportion of the P present in liquid and solid municipal and industrial waste streams and return these to rural areas as fertiliser. However, this will require considerable investigation and investment in new waste management technologies related to the collection, processing and transport of waste P resources. This includes increasing the agronomic effectiveness and fertiliser value of sparingly soluble forms of P commonly found in many waste products such as biosolids.

5.6.2 Future of phosphorus and nitrogen cycles and land-use change

Soils perform the key function of harbouring and regulating the cycling of N and P and critically support the increasing global demand for food and fibre (crops, milk, meat, wool and hide production) from an expanding human population. Phosphorus and N cycles through soils and plants have always underpinned civilisation, as their role is fundamental to the provision of food. But there are future changes on the horizon. On the one hand, the global N cycle is increasing as a result of anthropogenic interference (Gruber and Galloway, 2008), but, on the other hand, there are projections for reductions in P supply to soil (Elser and Bennett, 2011). Confusingly, at the same time there are predictions for increasing demands on soil, not as agricultural land, but as land for providing other services to society, especially in densely populated countries where multiple demands by the society are made. There is an imperative for soil managers to protect productive land and the ability of the soil and the landscape to undertake their nutrient cycling functions (Haygarth and Ritz, 2009).

References

Addiscott, T.M. (1999) Nitrate and health: introductory comments. In: *Managing Risks of Nitrates to Humans and the Environment* (eds W.S. Wilson et al.), pp. 247–249. The Royal Society of Chemistry, Cambridge, UK.

Bardgett, R.D. (2005) *The Biology of Soil: A Community and Ecosystem Approach*. Oxford University Press, Oxford, UK.

Bardgett, R.D., Streeter, T.C., and Bol, R. (2003) Soil microbes compete effectively with plants for organic-nitrogen inputs to temperate grasslands. *Ecology*, **84**, 1277–1287.

Brennan, R.F. and Bolland, M.D.A. (2007) Soil and tissue tests to predict the phosphorus requirements of canola in southwestern Australia. *Journal of Plant Nutrition*, **30**, 1767–1777.

Brooks, P.D., Williams, M.W., and Schmidt, S.K. (1996) Microbial activity under alpine snowpacks, Niwot Ridge, Colorado. *Biogeochemistry*, **32**, 93–113.

Brown, L. and Jarvis, S.C. (2001) Estimation of nitrous oxide emissions from UK agriculture. In: *IGER Innovations 2001* (ed. T. Gordon), pp. 60–63. Institute of Grassland of Environmental Research, Aberystwyth, UK.

Burnham, C.P. and Pitman, J.I. (1986) *Soil Erosion*. The Ashford Press, Ashford, Kent, UK.

Canfield, D.E., Glazer, A.N., and Falkowski, P.G. (2010) The evolution and future of Earth's nitrogen cycle. *Science*, **330**, 192–196.

Carton, O.T. and Jarvis, S.C. (2001) Nitrogen and phosphorus cycles in agriculture. In: *Nutrient Management Legislation in European Countries* (eds P. De Clercq et al.), pp. 3–13. Wageningen Press, Wageningen, the Netherlands.

Cassman, K.G., Dobermann, A., and Walters, D.T. (2002) Agroecosystems, nitrogen-use efficiency, and nitrogen management. *Ambio*, **31**, 132–140.

Catt, J.A., Quinton, J.N., Rickson, R.J., and Styles, P. (1994) Nutrient losses and crop yields in the Woburn erosion reference experiment. In: *Conserving Soil Resources: European Perspectives* (ed. R.J. Rickson), pp. 94–104. CAB International, Wallingford, UK.

Chapin, F.S., Moilanen, L., and Kielland, K. (1993) Preferential use of organic nitrogen for growth by a non-mycorrhizal arctic sedge. *Nature*, **361**, 150–153.

Chen, C.R., Condron, L.M., Davis, M.R., and Sherlock, R.R. (2002) Phosphorus dynamics in the rhizosphere of perenial ryegrass (*Lolium perenne*) and radiata pine (*Pinus radiata* D. Don). *Soil Biology and Biochemistry*, **34**, 487–489.

Cherr, C.M., Scholberg, J.M.S., and McSorely, R. (2006) Green manure approaches to crop production: a synthesis. *Agronomy Journal*, **98**, 302–319.

Christensen, B.T. (2004) Tightening the nitrogen cycle. In: *Managing Soil Quality: Challenges in Modern Agriculture* (eds P. Schønning et al.), pp. 47–67. CAB International, Wallingford, UK.

Clarholm, M. (1985) Interactions of bacteria, protozoa and plants leading to mineralization of soil nitrogen. *Soil Biology and Biochemistry*, **17**, 181–187.

Cleveland, C.C., Townsend, A.R., Schimel, D.S., Fisher, H., Howarth, R.W., Hedin, L.O., Perakis, S.S., Latty, E.F., Von Fischer, J.C., Elseroad, A., and Wasson M.F. (1999) Global patterns of terrestrial nitrogen (N2) fixation in natural ecosystems. *Global Biogeochemical Cycles*, **13**, 623–645.

Clough, T.J. and Condron, L.M. (2010) Biochar and the nitrogen cycle: introduction. *Journal of Environmental Quality*, **39**, 1218–1223.

Cole, C.V., Innis, G.S., and Stewart, J.W.B. (1977) Simulation of phosphorus cycling in semi-arid grassland. *Ecology*, **58**, 1–15.

Condron, L.M. (2004) Phosphorus – surplus and deficiency. In: *Managing Soil Quality – Challenges in Modern Agriculture* (eds P. Schjønning et al.), pp. 69–84. CAB International, Wallingford, UK.

Condron, L.M., Turner, B.L., and Cade-Menun, B.J. (2005) Chemistry and dynamics of soil organic phosphorus. In: *Phosphorus: Agriculture and the Environment* (eds J.T. Sims and A.N. Sharpley), pp. 87–121. ASA/CSSA/SSSA, Madison, Wisconsin.

Cordell, D., Dranget, J.-O., and White, S. (2009) The story of phosphorus: global food security and food for thought. *Global Environmental Change*, **19**, 292–305.

De Clercq, P., Gertis, A.C., Hofman, G., Jarvis, S.C., Neetson, J.J., and Sinabell, F. (2001) *Nutrient Management Legislation in European Countries*. Wageningen Press, Wageningen, the Netherlands.

Edwards, C.A. and Bohlen, P.J. (1996) *Biology and Ecology of Earthworms*, 3rd edn. Chapman and Hall, London.

Elliot, W.J., Foster, G.R., and Elliot, A.V. (1991) Soil erosion: processes, impacts and prediction. In: *Soil Management for Sustainability* (eds R. Lal and F.J. Pierce), pp. 25–34. Soil and Water Conservation Society, Edmonton, Alberta, Canada.

Elser, J. and Bennett, E. (2011) Phosphorus cycle: a broken biogeochemical cycle. *Nature*, **478**, 29–31.

Emsley, J. (2000) *The Shocking History of Phosphorus. A biography of the Devil's Element*. Macmillan, London.

Evans, R. (1990) Soils at risk of accelerated erosion in England and Wales. *Soil Use and Management*, **6**, 125–131.

Fageria, N.K. (2007) Green manuring in crop production. *Journal of Plant Nutrition*, **30**, 691–719.

Finkl, Jr, C.W. and Simonson, R.W. (1979) Phosphorus cycle. In: *The Encyclopedia of Soil Science* (eds R.W. Fairbridge and C.W. Finkl, Jr). Dowden, Hutchinson and Ross, Inc., Stroudsburg, Pennsylvania.

Foy, R.H. and Withers, P.J.A. (1995) The contribution of agricultural phosphorus to eutrophication. *Proceedings of the Fertiliser Society*, **365**, 1–32.

Frossard, E., Condron, L.M., Oberson, A., Sinaj, S., and Fardeau, J.C. (2000) Processes governing phosphorus availability in temperate soils. *Journal of Environmental Quality*, **29**, 15–23.

Frossard, E., Bünemann, E., Jansa, A., Oberson, A., and Feller, C. (2009) Concepts and practices of nutrient management in agro-ecosystems: can we draw lessons from history to design future sustainable agricultural production systems? *Austrian Journal for Food and Environment (Die Bodenkultur BOKU) Land Management*, **60**, 43–60.

George, T.S., Gregory, P.J., Robinson, J.S., Buresh, R.J., and Jama, B. (2002) Utilisation of soil organic P by agroforestry and crop species in the field, western Kenya. *Plant and Soil*, **246**, 53–63.

George, T.S., Simpson, R.J., Hadobas, P.A., Marshall, D.J., and Richardson, A.E. (2007) Accumulation and phosphatase-lability of organic phosphorus in fertilised pasture soil. *Australian Journal of Agricultural Research*, **58**, 47–55.

Giller, K.E., Witter, E., and McGrath, S.P. (1998) Toxicity of heavy metals to microorganisms and microbial processes in agricultural soils – a review. *Soil Biology and Biochemistry*, **30**, 1389–1414.

Goulding, K., Jarvis, S., and Whitmore, A. (2008) Optimizing nutrient management for farm systems. *Philosophical Transactions of the Royal Society B*, **363**, 667–680.

Gruber, N. and Galloway, J.N. (2008) An Earth-system perspective of the global nitrogen cycle. *Nature*, **451**, 293–296.

Haygarth, P.M. and Jarvis, S.C. (1999) Transfer of phosphorus from agricultural soils. *Advances in Agronomy*, **66**, 195–249.

Haygarth, P.M. and Ritz, K. (2009) The future of soils and land use in the UK: soil systems for the provision of land-based ecosystem services. *Land Use Policy*, **26**, S187–S197. DOI: 10.1016/j.landusepol.2009.09.016.

Haygarth, P.M., Warwick, M.S., and House, W.A. (1997) Size distribution of colloidal molybdate reactive phosphorus in river waters and soil solution. *Water Research*, **31**, 439–448.

Haygarth, P.M., Chapman, P.J., Jarvis, S.C., and Smith, R.V. (1998) Phosphorus budgets for two contrasting grassland farming systems in the UK. *Soil Use and Management*, **14**, 160–167.

Haygarth, P.M., Condron, L.M., Heathwaite, A.L., Turner, B.L., and Harris, G.P. (2005) The phosphorus transfer continuum: linking source to impact with an interdisciplinary and multi-scaled approach. *Science of the Total Environment*, **344**, 5–14.

Haynes, R.J. and Williams, P.H. (1993) Nutrient cycling and soil fertility in the grazed pasture ecosystem. In: *Advances in Agronomy* (ed. L.S. Donald), pp. 119–199. Academic Press, San Diego, California.

Heathwaite, A.L. and Burt, T. (1992) The evidence for past and present erosion in the Slapton catchment. In: *Soil Erosion, Past and Present: Archaeological and Geographical Perspectives* (eds M. Bell and J. Boardman), pp. 89–100. Oxbow Monograph, Oxford, UK.

Heckrath, G., Brookes, P.C., Poulton, P.R., and Goulding, K.W.T. (1995) Phosphorus leaching from soils containing different phosphorus concentrations in the broadbalk experiment. *Journal of Environmental Quality*, **24**, 904–910.

Hill, P.W., Quilliam, R.S., DeLuca, T.H., Farrar, J., Farrell, M., Roberts, P., Newsham, K.K., Hopkins, D.W., Bardgett, R.D., and Jones, D.L. (2011) Acquisition and assimilation of nitrogen as peptide-bound and D-enantiomers of amino acids by wheat. *PLoS ONE*, **6**, e19220.

Hodge, A., Campbell, C.D., and Fitter, A.H. (2001) An arbuscular mycorrhizal fungus accelerates decomposition and acquires nitrogen directly from organic material. *Nature*, **413**, 297–299.

Hornung, M. (1999) The role of nitrates in the eutrophication and acidification of surface waters. In: *Managing Risks of Nitrates to Humans and the Environment* (eds W.S. Wilson et al.), pp. 155–174. Royal Society of Chemistry, Cambridge, UK.

Hudson, J.J., Taylor, W.D., and Schindler, D.W. (2000) Phosphate concentrations in lakes. *Nature*, **406**, 54–56.

Ingham, R.E., Trofymow, J.A., Ingham, E.R., and Coleman, D.C. (1985) Interactions of bacteria, fungi, and their nematode grazers: effects on nutrient cycling and plant growth. *Ecological Monographs*, **55**, 119–140.

Jarvis, S.C. (1993) Nitrogen cycling and losses from dairy farms. *Soil Use and Management*, **9**, 99–105.

Jarvis, S.C. (1998) Nitrogen management for sustainability. In: *Grass for Dairy Cattle* (eds J.H. Cherney and D.J.R. Cherney), pp. 161–192. CAB International, Wallingford, UK.

Jarvis, S.C. (2000) Progress in studies of nitrate leaching from grassland soils. *Soil Use and Management*, **16**, 152–156.

Jarvis, S.C., Yamulki, S., and Brown, L. (2001) Source of nitrous oxide emissions in intensive grassland managements. *Phyton (Horn, Austria)*, **41**, 107–118.

Jenkinson, D.S. (2001) The impact of humans on the nitrogen cycle, with focus on temperate arable agriculture. *Plant and Soil*, **228**, 3–15.

Jonsson, L.M., Nilsson, M.-C., Wardle, D.A., and Zackrisson, O. (2001) Context dependent effects of ectomycorrhizal species richness on tree seedling productivity. *Oikos*, **93**, 353–364.

Kaye, J.P. and Hart, S.C. (1997) Competition for nitrogen between plants and soil microorganisms. *Trends in Ecology & Evolution*, **12**, 139–143.

Kielland, K. (1994) Amino acid absorption by arctic plants: implications for plant nutrition and nitrogen cycling. *Ecology*, **75**, 2373–2383.

Kronvang, B. (1990) Sediment associated phosphorus transport from two intensively farmed catchment areas. In: *Soil Erosion on Agricultural Land* (eds J. Boardman et al.), pp. 131–330. John Wiley & Sons, Ltd., Chichester, UK.

Ladha, J.K., Pathak, H., Krupnik, T.J., Six, J., and van Kessel, C. (2005) Efficiency of fertilizer nitrogen in cereal production: retrospects and prospects. *Advances in Agronomy*, **87**, 85–156.

Lavelle, P. and Martin, A. (1992) Small-scale and large-scale effects of endogeneic earthworms on soil organic matter dynamics in soil and the humid tropics. *Soil Biology and Biochemistry*, **24**, 1491–1498.

Leininger, S., Urich, T., Schloter, M., Schwark, L., Qi, J., Nicol, G.W., Prosser, J.I., Schuster, S.C., and Schleper, C. (2006) Archaea predominate among ammonia-oxidizing prokaryotes in soils. *Nature*, **442**, 806–809.

Liu, J., You, L., Aminin, M., Obersteiner, M., Herrero, M., Zehnder, A.J.B., and Yang, H. (2010) A high resolution assessment of global nitrogen flows in cropland. *Proceedings of the National Academy of Sciences*, **107**, 8035–8040.

Loehr, R.C. (1974) Characteristics and comparative magnitude of non point sources. *Journal WPFC*, **46**, 1849–1872.

MAFF. (1997) *Controlling Soil Erosion*. Her Majesty's Stationary Office, London.

Maherali, H. and Klironomos, J.N. (2007) Influence of phylogeny on fungal community assembly and ecosystem functioning. *Science*, **316**, 1746–1748.

McNeill, A. and Unkovich, M. (2007) The nitrogen cycle in terrestrial ecosystems. In: *Nutrient Cycling in Terrestrial Ecosystems* (eds P. Marschner and Z. Rengel), pp. 37–64. Springer, Heidelberg, Germany.

Moir, J.L., Cameron, K.C., and Di, H.J. (2007) Effects of the nitrification inhibitor dicyandiamide on soil mineral N, pasture yield, nutrient uptake and pasture quality in a grazed pasture system. *Soil Use and Management*, **23**, 111–120.

Morgan, R.P.C. (1980) Soil erosion and conservation in Britain. *Progress in Physical Geography*, **4**, 24–47.

Mosier, A.R. (1994) Nitrous oxide emissions from agricultural soils. *Nutrient Cycling in Agroecosystems*, **37**, 191–200.

Näsholm, T., Ekblad, A., Nordin, A., Giesler, R., Högberg, M., and Högberg, P. (1998) Boreal forest plants take up organic nitrogen. *Nature*, **392**, 914–916.

Nordin, A., Högberg, P., and Näsholm, T. (2001) Soil nitrogen form and plant nitrogen uptake along a boreal forest productivity gradient. *Oecologia*, **129**, 125–132.

OSPAR. (1998) *OSPAR Strategy to Combat Eutrophication*. Oslo and Paris Commissions, Oslo and Paris, France.

Owen, A.G. and Jones, D.L. (2001) Competition for amino acids between wheat roots and rhizosphere microorganisms and the role of amino acids in plant N acquisition. *Soil Biology and Biochemistry*, **33**, 651–657.

Pain, B. and Jarvis, S. (1999) Ammonia emissions from agriculture. In: *IGER Innovations* (ed. T. Gordon). Cambrian Printers, Aberystwyth, UK.

Parkes, G.D. and Mellor, J.W. (1939) *Mellor's Modern Inorganic Chemistry*. London, UK.

Quinton, J.N. (1997) Reducing predictive uncertainty in model simulations: a comparison of two methods using the European Soil Erosion Model (EUROSEM). *Catena*, **30**, 101–117.

Read, D.J. and Perez-Moreno, J. (2003) Mycorrhizas and nutrient cycling in ecosystems – a journey towards relevance? *New Phytologist*, **157**, 475–492.

Redfield, A.C. (1934) On the proportions of organic derivatives in sea water and their relationship to the composition of plankton. In: *James Johnston Memorial Volume* (ed. L.U. Press), pp. 176–192. Liverpool University Press, Liverpool, UK.

Richardson, A.E. (2001) Prospects for using soil microorganisms to improve the acquisition of phosphorus by plants. *Australian Journal of Plant Physiology*, **28**, 897–906.

Roels, J. and Verstraete, W. (2001) Biological formation of volatile phosphorus compounds, a review paper. *Bioresource Technology*, **79**, 243–250.

Scheu, S. and Setälä, H. (2002) Multitrophic interactions in decomposer food webs. In: *Multitrophic Level Interactions* (eds T. Tscharntke and B. Hawkins), pp. 223–264. Cambridge University Press, Cambridge, UK.

Schimel, J.P. and Bennett, J. (2004) *Nitrogen Mineralization: Challenges of a Changing Paradigm*. Ecological Society of America, Washington, DC, ETATS-UNIS.

Schimel, J.P., Bennett, J., and Frierer, N. (2005) Microbial community composition and soil nitrogen cycling: is there really a connection? In: *Biological Diversity and Function in Soil* (eds R.D. Bardgett et al.). Cambridge University Press, Cambridge, UK.

Scholefield, D., Tyson, K.T., Garwood, E.A., Armstrong, A.C., Hawkins, J., and Stone, A.C. (1993) Nitrate leaching from grazed grassland lysimeters: effects of fertilizer input, field drainage and patterns of weather. *European Journal of Soil Science*, **44**, 601–614.

Setälä, H. and Huhta, V. (1991) Soil fauna increase Betula pendula growth: laboratory experiments with coniferous forest floor. *Ecology*, **72**, 665–671.

Setälä, H. and McLean, M.A. (2004) Decomposition rate of organic substrates in relation to the species diversity of soil saprophytic fungi. *Oecologia*, **139**, 98–107.

Sharpley, A.N. and Smith, S.J. (1990) Phosphorus transport in agricultural runoff: the role of soil erosion. In: *Soil Erosion on Agricultural Land* (eds J. Boardman et al.), pp. 351–366. John Wiley & Sons, Ltd., Chichester, UK.

Sharpley, A.N. and Syers, J.K. (1976) Potential role of earthworm casts for the phosphorus enrichment of runoff in waters. *Soil Biology and Biochemistry*, **8**, 341–346.

Smit, A.L., Bindraban, P.S., Schröder, J.J., Conjiin, J.G., and van der Meer, H.G. (2009) *Phosphorus in Agriculture: Global Resources, Trends and Developments*, p. 36. Plant Research International BV, Wageningen, the Netherlands.

Smith, S.E. and Read, D.J. (1997) *Mycorrhizal Symbiosis*, 2nd edn. Academic Press, London.

Stein, M.S., Bergman, A., and Glidemann, D. (1996) Microbiological phosphine production. *Microbiology Europe*, **4**, 14–18.

Thorup-Kristensen, K., Magid, J., and Jensen, L.S. (2003) Catch crops and green manures as biological tools in nitrogen management in temperate zones. *Advances in Agronomy*, **79**, 227–302.

Turner, B.L. and Engelbrecht, B.M.J. (2011) Soil organic phosphorus in lowland tropical rain forests. *Biogeochemistry*, **103**, 297–315.

Uhlmann, D. and Albrecht, E. (1968) Biogeochemische Faktoren der Eutrophierung von Trinkwassen-Talsperren. *Limnologica (Berlin)*, **6**, 225–245.

Van der Heijden, M.G.A., Klironomos, J.N., Ursic, M., Moutoglis, P., Streitwolf-Engel, R., Boller, T., Wiemken, A., and Sanders, I.R. (1998) Mycorrhizal fungal diversity determines plant biodiversity, ecosystem variability and productivity. *Nature*, **396**, 72–75.

Vitousek, P.M. and Hobbie, S. (2000) Heterotrophic nitrogen fixation in decomposing litter: patterns and regulation. *Ecology*, **81**, 2366–2376.

Vitousek, P.M., Aber, J.D., Howarth, R.W., Likens, G.E., Matson, P.A., Schindler, D.W., Schlesinger, W.H., and Tilman, D.G. (1997) Human alteration of the global nitrogen cycle: sources and consequences. *Ecological Applications*, **7**, 737–750.

Vitousek, P.M., Nayler, R., Crews, T., David, M.B., Brinkwater, L.E., Holland, E., Johnes, P.J., Katzenberger, J., Martinelli, L.A., Matson, P.A., Nziguheba, G., Ojima, D., Palm, C.A., Robertson, G.P., Sanchez, P.A., Townsend, A.R., and Zhang, F.S. (2009) Nutrient imbalances in agricultural development. *Science*, **324**, 1519–1520.

Vitousek, P.M., Porder, S., Houlton, B.Z., and Chadwick, O.A. (2010) Terrestrial phosphorus limitation: mechanisms, implications, and nitrogen–phosphorus interactions. *Ecological Applications*, **20**, 5–15.

Vogelsang, K.M., Reynolds, H.L., and Bever, J.D. (2006) Mycorrhizal fungal identity and richness determine the diversity and productivity of a tallgrass prairie system. *New Phytologist*, **172**, 554–562.

Vollenweider, R.A. (1968) *Scientific Fundamentals of the Eutrophication of Lakes and Flowing Waters, with Particular Reference to Nitrogen and Phosphorus as Factors of Eutrophication*. Organisation for Economic Co-operation and Development, Paris, France.

Vollenweider, R.A. and Krekes, J.J. (1982) *Eutrophication of Waters, Monitoring, Assessment and Control*. Organisation for Economic Co-operation and Development, Paris, France.

Wardle, D.A. (2002) *Communities and Ecosystems: Linking the Aboveground and Belowground Components*. Princeton University Press, Princeton, New Jersey.

Weigelt, A., Bol, R., and Bardgett, R.D. (2005) Preferential uptake of soil nitrogen forms by grassland plant species. *Oecologia*, **142**, 627–635.

Wilson, W.S. and Ball, A.S. (eds). (1999) *Managing the Risks of Nitrates to Humans and the Environment*. Royal Society of Chemistry, Cambridge, UK.

Withers, P.J.A. and Haygarth, P.M. (2007) Agriculture, phosphorus and eutrophication: a European perspective. *Soil Use and Management*, **23**, 1–4.

Woolf, D., Amonette, J.E., Street-Perrott, A., Lehmann, J., and Joseph, S. (2010) Sustainable biochar to mitigate global climate change. *Nature Communications*, **1**, 1–9.

Zobisch, M.A., Richter, C., Heiligtag, B., and Schlott, R. (1994) Nutrient losses from cropland in the central highlands of Kenya due to surface runoff and soil erosion. *Soil and Tillage Research*, **33**, 109–116.

6 Properties and management of cationic elements for crop growth

Philip J. White[1] and Duncan J. Greenwood[2]

[1]The James Hutton Institute, Dundee, UK
[2]University of Warwick, Warwick, UK

This chapter describes the properties and management of both the 'macronutrient' cationic elements (potassium, calcium and magnesium) and the 'micronutrient' cationic elements (zinc, iron, copper, manganese and nickel) for crop nutrition. It begins by noting that transpiration-driven mass flow of the soil solution is required to deliver many of these elements to the root surface for adequate plant nutrition. The amounts of each element required by plants and their essential functions are then reviewed, and the molecular mechanisms by which root cells take up essential cationic elements from the rhizosphere solution are described. Aspects of soil geochemistry that determine the phytoavailability of essential mineral cations in the soil solution are presented, including the interactions between cations in the soil and the effects of soil pH. Management practices for optimising the delivery of essential mineral cations for crop production and quality are discussed, and the chapter concludes by considering the prospects for harnessing crop genetics to improve the use of cationic elements in crop production.

6.1 The movement of essential cationic elements to the root surface

Plants acquire many of the mineral elements required for their growth and fecundity from the soil solution in their cationic forms (Table 6.1; White, 2003; Maathuis, 2009). These essential cationic elements include potassium (K), calcium (Ca), magnesium (Mg), iron (Fe), manganese (Mn), zinc (Zn), copper (Cu) and nickel (Ni). The phytoavailability of these elements in the soil solution is often determined by pH (Figure 6.1; Taiz and Zeigler, 2006).

The movement of cationic elements in the soil solution is determined primarily by the processes of mass flow and diffusion (Barber, 1995; Fageria et al., 2011). Mass flow, which is also known as mass transfer or bulk flow, is the term given to the movement of solutes that is proportional to the flow of the solution in which they are dissolved. The mass flow of cations to the root surface is driven by transpiration from the shoot and the consequent uptake of water by the plant. The delivery of a particular cation to the root surface is calculated as the product of the concentration of the cation in the soil solution and the rate of transpiration. Mass flow of the soil solution will deliver ample quantities of cations that are present at high

Soil Conditions and Plant Growth. Edited by Peter J. Gregory and Stephen Nortcliff.
© 2013 Blackwell Publishing Ltd. Published 2013 by Blackwell Publishing Ltd.

Table 6.1 The functions of mineral elements acquired as cations from the soil solution and the critical leaf concentrations for their sufficiency and toxicity in non-tolerant crop plants.

Element	Form acquired	Physiological functions	Critical leaf concentrations ($mg\,g^{-1}$ DM)	
			Sufficiency	Toxicity
Potassium (K)	K^+	Enzyme activation; cellular osmoticum; counter-cation for anion accumulation and electrogenic transport	5–40	>50
Calcium (Ca)	Ca^{2+}	Membrane and cell-wall structure; counter-cation for anion accumulation in vacuoles; cytosolic signalling	0.5–10	>100
Magnesium (Mg)	Mg^{2+}	Constituent of chlorophyll; photosynthetic charge separation; enzyme cofactor; nucleic acid stabilisation	1.5–3.5	>15
Iron (Fe)	Fe^{2+}, Fe^{3+} chelates	Photosynthesis; mitochondrial respiration; C and N metabolism; production and scavenging of reactive oxygen species; regulation of transcription and translation; hormone biosynthesis	$50–150 \times 10^{-3}$	>0.5
Manganese (Mn)	Mn^{2+} Mn chelates	Photosystem II; enzyme activation in photosynthesis, C and N metabolism, RNA polymerase	$10–20 \times 10^{-3}$	0.2–5.3
Copper (Cu)	Cu^+, Cu^{2+}, Cu chelates	Photosynthesis; mitochondrial respiration; C and N metabolism; protection against oxidative stress	$1–5 \times 10^{-3}$	$15–30 \times 10^{-3}$
Zinc (Zn)	Zn^{2+}, Zn chelates	Structural stability of proteins; regulation of transcription and translation; oxidoreductases and hydrolytic enzymes	$15–30 \times 10^{-3}$	$100–300 \times 10^{-3}$
Nickel (Ni)	Ni^{2+}, Ni chelates	Constituent of urease	0.1×10^{-3}	$20–30 \times 10^{-3}$
Sodium (Na)	Na^+	Osmotic replacement of K^+; counter-cation for anion accumulation and electrogenic transport; C4 and CAM metabolism	Beneficial	2–5
Aluminium (Al)		Herbivore defence; prevention of Fe–toxicity	Beneficial	$40–200 \times 10^{-3}$
Cobalt (Co)	Co^{2+}	Nitrogen fixation	Beneficial	$10–20 \times 10^{-3}$
Lead (Pb)	Pb^{2+}	–	–	$10–20 \times 10^{-3}$
Cadmium (Cd)	Cd^{2+}, Cd chelates	–	–	$5–10 \times 10^{-3}$

The critical concentration for sufficiency is defined as the concentration in a diagnostic tissue that allows a crop to achieve 90% of its maximum yield. The critical concentration for toxicity is defined as the concentration in a diagnostic tissue above which yield is decreased by more than 10%. It should be recognised that critical tissue concentrations depend upon the exact solute composition of the soil solution and can differ greatly both between and within plant species. The latter differences reflect both ancestral habitats and ecological strategies.

Source: Data compiled from Maathuis (2009), Hänsch and Mendel (2009), Pilon-Smits et al. (2009) and White and Brown (2010).

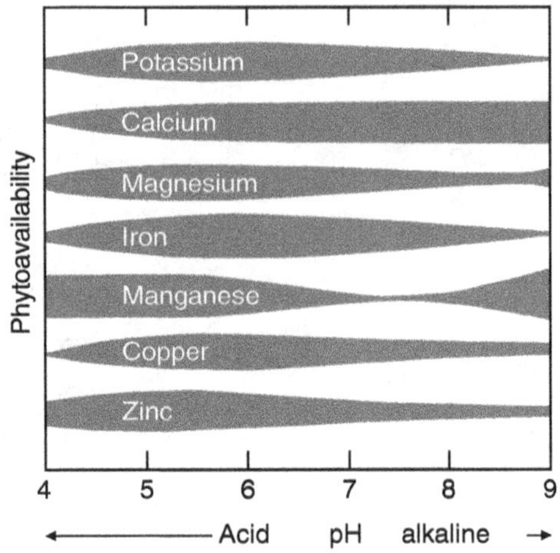

Figure 6.1 Effects of pH on the phytoavailability of seven essential cationic elements in the soil solution (Taiz and Zeigler, 2006).

Table 6.2 Quantities of cations delivered to the root surface of a maize plant growing in a fertile alfisol by mass flow, diffusion and interception (Fageria et al., 2011).

Cation	Mass flow	Diffusion	Interception
Potassium	18	80	2
Calcium	375	0	150
Magnesium	222	0	33
Iron	66	21	13
Manganese	22	35	43
Copper	219	0	6
Zinc	230	0	43

Data are expressed as percentages of the total quantities acquired by the plant.

concentrations in the soil solution relative to their concentrations in plant tissues, but often insufficient quantities of cations that are present at low concentrations in the soil solution relative to their concentrations in plant tissues, for the mineral nutrition of crops (Table 6.2). Diffusion is defined as the movement of a solute from a region of high concentration to a region of low concentration. When the uptake of a particular cation by a plant exceeds the amounts delivered by mass flow, a concentration gradient develops for that cation in the rhizosphere. In principle, the diffusion of a particular cation is calculated as the product of the concentration gradient and the diffusion coefficient of the cation, as described by Fick's First Law (White, 2003). However, since soil is neither homogeneous nor a solution, to calculate the diffusion of a solute to the root surface, this equation must be modified to include a tortuosity factor and a term for the soil volume occupied by water (Barber, 1995; Fageria et al., 2011). Diffusive movement of cations in the soil solution is generally slow and is effective in delivering cations to the root only from distances of 0.1–15 mm from the root surface (Barber, 1995). Nevertheless, the delivery of K, Fe and Mn to the root surface is often determined by local diffusion in the vicinity of the root (Table 6.2). In addition to mass flow and diffusion, cations can be intercepted during the growth of a root system. This can contribute significantly to the mineral nutrition of the plant for nutrients whose concentrations in the soil solution approach or exceed those found in plant tissues, such as Ca, Mg, Mn and Zn (Table 6.2).

6.2 Cationic elements in plants

6.2.1 Potassium

Potassium (K^+) is the most abundant inorganic cation in plants, and shoot concentrations up to 100 mg g^{-1} dry matter (DM) have been reported (Broadley et al., 2004; Watanabe et al., 2007). Potassium is present at high concentrations in most plant tissues but is particularly concentrated in growing tissues and reproductive organs. It is required by plants for various biochemical and biophysical functions (Table 6.1; Maathuis, 2009; White and Karley, 2010). In particular, K^+ is required for the activation of many enzymes, as a cellular osmoticum for rapidly expanding cells and as a counter-cation for anion accumulation and electrogenic transport processes (Leigh and Wyn Jones, 1984; Maathuis, 2009; White and Karley, 2010). The K^+ concentrations in metabolically active cellular compartments are maintained at approximately 100 mM, whilst vacuolar K^+ concentrations vary considerably from several hundred millimolar to a minimum of about 20 mM (White and Karley, 2010). Apoplastic K^+ concentrations fluctuate according to K^+ availability and transpirational water flows (White and Karley, 2010). Crop demand for K is directly related to its N demand, and the K requirements of plants are linearly related to those of N as they grow. Potassium deficiency is often observed in rapidly growing plants and K-fertilisers are, therefore, frequently applied in crop production. Potassium is delivered to the roots through a combination of transpiration-driven mass flow of the soil solution and, since K concentrations in the soil solution are generally high, local diffusion in the rhizosphere (Barber, 1995).

6.2.2 Calcium

Shoot Ca concentrations generally range from 1 to 50 mg g^{-1} DM, reflecting both Ca availability in the soil solution and the Ca requirements of different plant species (Broadley et al., 2003; Watanabe et al., 2007). Tissue Ca concentrations are lower in commelinoid monocots (cereals and grasses) than in non-commelinoid monocots or other angiosperm species growing in the same environment, which appears to be directly correlated with the cation exchange capacity of cell walls and the free carboxyl groups of cell-wall pectins (White and Broadley, 2003). Calcium is required as a divalent cation (Ca^{2+}) for various structural roles in the cell wall and cellular membranes, as a counter-cation for organic and inorganic anions in the vacuole and for coordinating cellular responses to developmental cues and environmental challenges through changes in cytosolic Ca^{2+} concentration (White and Broadley, 2003). Since Ca^{2+} is cytotoxic, submicromolar concentrations of Ca^{2+} must be maintained in the cytosol, whereas supramillimolar Ca^{2+} concentrations can be present in the vacuole and apoplast (White and Broadley, 2003). In the cytosol, Ca^{2+} is buffered by chelation with soluble organic compounds, whereas in the vacuole Ca is often present as soluble or insoluble oxalate or phytate salts (White and Broadley, 2003). Calcium deficiency is rare in nature, but can occur on soils with low cation exchange capacity (CEC). By contrast, Ca toxicity can occur on calcareous soils, although the flora of these soils is most often determined by insensitivity to P and Fe deficiencies (Lee, 1999). Calcium concentrations in the soil solution are generally in the millimolar range, and sufficient Ca for plant nutrition can be provided by transpiration-driven mass flow of the soil solution (Barber, 1995).

6.2.3 Magnesium

Shoot Mg concentrations range from 1 to 10 mg g^{-1} DM in plants with an adequate Mg supply (Wilkinson et al., 1990; Broadley et al., 2004). When growing in the same environment, most angiosperm species have a similar shoot Mg/Ca quotient, with the exception of members of the Caryophyllales order, which often have a higher shoot Mg/Ca quotient than other angiosperm species (Broadley et al., 2004, 2008; Watanabe et al., 2007). Magnesium is required for the structure and conformation of nucleic acids, as a cofactor for numerous enzymes, as a counter-cation during photosynthetic charge separation and as a constituent of chlorophyll. About 75% of leaf Mg appears to be associated with protein synthesis, through its roles in ribosomal structure and function, and between 15% and 20% is associated with chlorophyll (Wilkinson et al., 1990; White and Broadley, 2009). Most cellular Mg is associated with organic acids and proteins (Mengel et al., 2001), and a cytosolic Mg^{2+} concentration of about 0.4 mM is generally maintained (White et al., 1990). Magnesium deficiencies can occur in plants growing on shallow or course-textured soils, where Mg^{2+} leaches through the soil profile; on strongly acidic soils, where Al^{3+} and Mn^{2+} can inhibit Mg^{2+} uptake; on sodic or saline soils, where Na^{+} can inhibit Mg^{2+} uptake; and on alkaline soils, where carbonate formation and excess Ca^{2+}, K^{+} and Na^{+} can reduce Mg^{2+} availability and inhibit Mg^{2+} uptake (Wilkinson et al., 1990; Römheld and Kirkby, 2007; White and Broadley, 2009). The concentration of Mg^{2+} in the soil solution generally lies between 125 μM and 8.5 mM, which is sufficient for transpiration-driven mass flow to supply sufficient Mg to roots for adequate plant nutrition (Wilkinson et al., 1990; Barber, 1995).

6.2.4 Iron

Shoot Fe concentrations above 50–150 μg g^{-1} DM are required for adequate plant nutrition (Mengel et al., 2001; White and Brown, 2010). Iron is a redox-active metal that functions in photosynthesis, mitochondrial respiration, nitrogen assimilation, production and scavenging of reactive oxygen species, regulation of transcription and translation, and in the biosynthesis of plant signals such as ethylene, gibberellic acid and jasmonic acid (Hänsch and Mendel, 2009; Guerinot, 2010). It is an integral component of many proteins and enzymes either as a haem group, as occurs in cytochromes, peroxidases and catalases; as non-haem iron bound to sulphur atoms, as occurs in ferredoxin, nitrite reductase, sulphite reductase, glutamate synthase and iron-storage proteins; or as non-haem non-sulphur bound iron, as occurs in the plastidic Fe-storage protein ferritin (Hänsch and Mendel, 2009; Guerinot, 2010). Up to 90% of the Fe in leaf cells is contained in chloroplasts and is distributed equally between stroma and thylacoids (Hänsch and Mendel, 2009; Guerinot, 2010). Iron deficiency occurs on a variety of soils and is generally caused by low Fe phytoavailability rather than low abundance of Fe in the soil (Follett et al., 1981; Shuman, 1998; Schmidt, 1999; Frossard et al., 2000; Fageria, 2009; White and Broadley, 2009). The amount of Fe in the soil solution is decreased by an increase in redox potential and/or pH and is determined by the presence of $Fe(OH)_3$ in well-oxidised soils, by $Fe_3(OH)_8$ in moderately oxidised soils and by $FeCO_3$ in highly reduced soils (Lindsay and Schwab, 1982). The Fe concentrations in solutions from most agricultural soils range from 10^{-8} to 10^{-6} M, but Fe concentrations in solutions from alkaline or calcareous soils rarely exceed 10^{-10} M (Frossard et al., 2000; Fageria, 2009). The low Fe concentration in the soil solution, together with its small diffusion coefficient, makes plants reliant on transpiration-driven mass flow of the soil solution for the delivery of sufficient Fe to their roots for nutrition (Barber, 1995). Iron deficiency associated with soil alkalinisation is referred to as lime-induced chlorosis. It is estimated that up to one-third of the world's agricultural soils are calcareous and susceptible to lime-induced Fe chlorosis (Follett et al., 1981; FAO 2010). At the other extreme, Fe-toxicities can occur in waterlogged or flooded soils, in which Fe^{3+} is reduced to Fe^{2+}, which increases Fe^{2+} availability and Fe uptake by plants (Schmidt, 1999; Fageria, 2009).

6.2.5 Manganese

Shoot concentrations above 10–20 μg Mn g^{-1} DM are required for plant growth and fecundity (Mengel et al., 2001; White and Brown, 2010). Manganese serves as a catalytically active metal in various enzymes, including the water-splitting reaction of photosystem II, superoxide dismutase and oxalate oxidase, and also regulates the activity of several enzymes catalysing reactions in intermediary metabolism (malic enzyme, isocitrate dehydrogenase and PEP carboxykinase), fatty acid biosynthesis, nitrogen metabolism (glutamine synthetase and arginase), the shikimic acid pathway and subsequent pathways leading to the formation of aromatic amino acids, lignins, flavonoids and the phytohormone indole acetic acid, gibberellic acid biosynthesis and RNA polymerase (Hänsch and Mendel, 2009; Williams and Pittman, 2010). Manganese deficiency can occur in plants growing on organic soils, in which Mn^{2+} is chelated by organic material; alkaline and calcareous soils, in which Mn is precipitated as manganese oxides (MnO_2, Mn_2O_3, Mn_3O_4); and in coarse-textured sandy soils, in which Mn^{2+} is subject to leaching (He et al., 2005; Fageria, 2009). Since most Mn in soils is associated with organic matter, it is often concentrated in the topsoil (Shuman, 1985; Havlin et al., 2005; Fageria, 2009). Global crop production is restricted by Mn toxicity to a greater extent than Mn deficiency (Mengel et al., 2001; He et al., 2005). Tissue concentrations above 0.2 mg Mn g^{-1} DM can be toxic to some plants, although the ability of an individual plant to tolerate Mn toxicity is influenced by plant genotype, plant growth rate and the concentration of silicic acid in the soil solution (Table 6.1; Horst, 1988; White and Brown, 2010). Several plants that hyperaccumulate Mn to concentrations greater than 10 mg Mn g^{-1} DM are also known (Baker and Brooks, 1989; Whiting et al., 2005). Intriguingly, plants lacking sufficient Zn often accumulate excess Mn, suggesting a physiological interaction between these two elements (Brady and Weil, 2002). Manganese toxicity is prevalent in plants grown on acid mineral soils and can also occur in plants growing on waterlogged or flooded soils as redox potential becomes more negative (Mengel et al., 2001; Kochian et al., 2004; Fageria, 2009). On acid soils, Mn toxicity can be corrected by increasing soil pH through liming.

6.2.6 Zinc

Shoot Zn concentrations vary greatly, from 50 to 200 μg Zn g^{-1} DM in agricultural plants supplied with adequate Zn to over 10 mg Zn g^{-1} DM in Zn-hyperaccumulator plants growing in soils with high Zn concentrations (Broadley et al., 2001, 2007). Zinc occurs in plants as Zn^{2+}, which is associated with a multitude of proteins. For example, in *Arabidopsis thaliana*, Zn^{2+} is associated with over 2,300 proteins in 181 gene families (Broadley *et al.*, 2007). These proteins include a large number of zinc-finger-containing proteins and transcription factors, oxidoreductases and hydrolytic enzymes such as protein kinases, metalloproteases and phosphatases (Broadley et al., 2007; Hänsch and Mendel, 2009; Clemens, 2010). Most agricultural soils contain 10–300 μg Zn g^{-1}, which would be sufficient for adequate crop nutrition if it were phytoavailable, and Zn^{2+} concentrations in the soil solution generally range from 10^{-8} to 10^{-6} M (Alloway, 1995; Barber, 1995; Welch, 1995; Frossard et al., 2000; Broadley et al., 2007). Zinc has a small diffusion coefficient and sufficient Zn for plant nutrition must be provided to the roots by transpiration-driven mass flow of the soil solution (Barber 1995). Zinc deficiency is the most common micronutrient deficiency in crops (Cakmak, 2002, 2004; Alloway, 2004; White and Broadley, 2009). High pH often limits Zn phytoavailability, and Zn deficiencies occur in plants growing on eroded calcareous or alkaline soils, especially in arid and semiarid environments, where Zn^{2+} concentrations in the soil solution can be as low as 10^{-11}–10^{-9} M (Hacisalihoglu and Kochian, 2003; Broadley et al., 2007; Cakmak, 2008; Clemens, 2010). Zinc toxicity is less common than Zn deficiency, but can occur in crops growing on acidic soils enriched by anthropogenic Zn inputs, such as agricultural soils treated with sewage sludge, soils of urban and peri-urban areas and land contaminated by mining or smelting activities (Chaney, 1993; Broadley et al., 2007).

6.2.7 Copper

Plants require concentrations above 1–5 µg Cu g^{-1} DM for their growth and fecundity (Mengel et al., 2001; White and Brown, 2010). Copper is a redox-active metal and, consequently, has roles in many aspects of plant biochemistry including photosynthesis and mitochondrial respiration, carbon and nitrogen metabolism, protection against oxidative stress and cell-wall remodelling (Burkhead et al., 2009; Hänsch and Mendel, 2009). Copper is the catalytic metal in many oxidases because of its high affinity for dioxygen molecules. In *Arabidopsis thaliana*, Cu is associated with over 100 different proteins, including Cu-binding proteins/chaperones and various enzymes (Kramer and Clemens, 2005). Several of these enzymes are important for mounting plant defence responses to pests and pathogens (Fageria et al., 2011). About 50% of leaf Cu is associated with chloroplasts, where it participates in photosynthetic reactions (Hänsch and Mendel, 2009). For some essential functions, either Cu or Fe can be utilised and, depending upon their Cu and Fe nutrition, plants can switch their metabolism between enzymes catalysing the same biochemical reaction containing either Cu or Fe (Hänsch and Mendel, 2009). Examples include Cu-nitrite versus heme-nitrite reductase, Cu/Zn-superoxide dismutase versus Fe-superoxide dismutase and cytochrome oxidase versus di-iron oxidase. In legumes, Cu deficiency reduces nodulation and N_2 fixation. High pH is often the major factor restricting Cu phytoavailability, and Cu deficiency is often observed on calcareous or alkaline soils (Lindsay, 1979; He et al., 2005). Copper deficiency is also observed in plants growing on highly organic soils, since organic matter chelates Cu tightly (He et al., 2005). The affinity of organic matter for Cu is also the reason why, in many soils, Cu is concentrated in the topsoil (Fageria, 2009). Copper has a small diffusion coefficient and transpiration-driven mass flow of the soil solution must provide Cu for plant nutrition (Barber, 1995). Typical Cu^{2+} concentrations in the soil solution range from 10^{-9} to 10^{-6} M (Barber, 1995; Welch, 1995). Excessive Cu is detrimental to plant growth and tissue concentrations above 15–30 µg g^{-1} DM are toxic to most plants, with the exception of Cu-hyperaccumulator plants in which leaf Cu concentrations over 1 mg Cu g^{-1} DM have been reported (Table 6.1; Baker and Brooks, 1989; Broadley et al., 2001; Mengel et al., 2001; He et al., 2005; Whiting et al., 2005; White and Brown, 2010). Copper rarely reaches toxic concentrations in natural soils, but anthropogenic activities, such as the mining and refining metal ores, the repeated use of Cu-containing fungicides, pesticides or herbicides, and the application of Cu-rich manures from commercial pig or poultry farms, have contaminated both natural and agricultural land (Welch et al., 1991; He et al., 2005).

6.2.8 Nickel

Trace amounts of Ni are sufficient for adequate plant nutrition and tissue concentrations above 2–4 ng g^{-1} DM suffice for plant growth and fecundity (Brown et al., 1987; Tejada-Jiménez et al., 2009; White and Brown, 2010). However, crop plants generally contain tissue concentrations of 0.1–5 µg Ni g^{-1} DM (Mengel et al., 2001). Nickel is required by plants for the activation of the enzyme urease, which catalyses the hydrolysis of urea to carbon dioxide and ammonia. Tissue concentrations exceeding 10–50 µg Ni g^{-1} DM are toxic to most plants, with the exception of Ni-hyperaccumulator plants, which can tolerate in excess of 1 mg Ni g^{-1} DM (Broadley et al., 2001; Mengel et al., 2001; He et al., 2005; Whiting et al., 2005; Reeves, 2006; Fageria, 2009; Tejada-Jiménez et al., 2009; White and Brown, 2010). Nickel deficiency is rarely seen in nature and Ni concentrations in the soil solution from most agricultural soils are less than 10 µM Ni, which does not produce Ni toxicity (Welch, 1995; Mengel et al., 2001). However, Ni toxicity can occur worldwide on soils derived from Ni-rich serpentine or ultrabasic rocks or as a consequence of anthropogenic activities, such as the mining and refining of metal ores or the application of Ni-rich biosolids or municipal composts to agricultural land (He et al., 2005; Fageria, 2009). Like Fe, Zn, Mn and Cu, Ni has a small diffusion coefficient and transpiration-driven mass flow of the soil solution must provide Ni for plant nutrition (Barber, 1995).

6.2.9 Beneficial and phytotoxic elements

In addition to these essential mineral elements, several mineral elements that are considered beneficial to plants, which promote growth for various taxa under certain environmental conditions, are also acquired from the soil solution as cations (Table 6.1). These include Na, which can replace K in its biophysical roles and is required by C4 and CAM plants for the regeneration of PEP from pyruvate; Co, which is essential for nitrogen fixation in legumes and may enhance drought resistance and herbivore defence; and Al, which has also been implicated in herbivore defence and the prevention of Fe-toxicity (Mengel et al., 2001; Pilon-Smits et al., 2009). Plant roots are exposed to these and various other non-essential elements that are present as cations in the soil solution and can reach toxic concentrations in plants. Excessive concentrations of Na, Al, Co, cadmium (Cd) and lead (Pb) commonly reach toxic concentrations in the soil solution either as a consequence of natural or anthropogenic activities (see Chapter 7).

Sodium is present in high concentrations in saline and sodic soils, which occur worldwide but mostly in the arid subtropics where rainfall is low (Frossard et al., 2000; FAO, 2010). In these soils, excess soluble salts accumulate near the surface of the soil. They occur both naturally and anthropogenically when water containing high salt concentrations is used for irrigation. Saline soils are defined as having an electrical conductivity greater than 4 dS m^{-1}, whereas sodic soils are defined simply as having a high ratio of Na to Ca and Mg in the soil solution (see Chapter 7). Sodic soils can be acid or alkaline, saline or non-saline, although most are alkaline soils with the dominant anion being bicarbonate (Qadir and Schubert, 2002). Sodium toxicity is thought to limit crop production on between 5% and 15% of potential agricultural land (Munns and Tester, 2008). Plants vary appreciably in their ability to tolerate high Na concentrations in the environment (see Chapter 7). At one extreme are halophytes and at the other extreme are sensitive plants, whose growth is impaired at leaf concentrations below 2–5 mg Na g^{-1} DM (Munns and Tester, 2008). Symptoms of Na toxicity in plants are caused both directly, through effects of Na^+ on plant metabolism, and indirectly, through the osmotic effects of Na^+ or Na^+ inhibition of transport of other cations, especially Ca^{2+} (Munns and Tester, 2008). Crop yields on saline soils can be improved by management practices that reduce Na^+ concentrations in the soil solution or reduce the ratio of Na^+ to other cations in the soil solution and/or the cultivation of Na-tolerant crops. Traditionally, saline soils are remediated by leaching soluble salts from the soil profile by irrigation with freshwater, whereas sodic soils are remediated through the application of Ca^{2+}, often as gypsum, followed by flushing the soil with freshwater (see Chapter 7).

Aluminium limits crop production in acidic soils, which constitute over 40% of the world's arable land (Von Uexküll and Mutert, 1995; Sumner and Noble, 2003). It is present in the soil solution as Al^{3+}, $Al(OH)^{2+}$ and $Al(OH)_2^+$ (Kinraide, 1991; Mengel et al., 2001; Kochian et al., 2004). Aluminium toxicity is manifest primarily by the inhibition of root elongation and altered root architectures, which are both associated with the effects of toxic concentrations of Al^{3+} at the root tip (Kinraide, 1991; Mengel et al., 2001). Leaf concentrations greater than 40–200 µg Al g^{-1} DM are toxic to most plants, although some plants can tolerate leaf concentrations in excess of 1 mg Al g^{-1} DM (Table 6.1; Barceló and Poschenrieder, 2002; Watanabe et al., 2007). These Al-hyperaccumulator plants occur in about 45 families, belonging mostly to the basal branches of the rosid and asterid clades (Barceló and Poschenrieder, 2002; Jansen et al., 2002; Watanabe et al., 2007). Many native plants growing on acidic soils have developed Al-tolerance through extracellular or symplastic detoxification mechanisms (Barceló and Poschenrieder, 2002; Kochian et al., 2004; Pilon-Smits et al., 2009). Plant roots reduce Al phytoavailability by secreting organic acids or mucilage to chelate Al, raising rhizosphere pH and binding Al to cell-wall components (Ma et al., 2001; Kochian et al., 2004; Delhaize et al., 2007). Aluminium entering plant cells is rendered non-toxic by sequestration in the vacuole complexed with organic acids (Ma et al., 2001; Kochian et al., 2004). Liming, especially with dolomitic lime ($CaMg(CO_3)_2$), is an effective way to raise soil pH and avoid Al and Mn toxicities and Ca and Mg deficiencies in acid soils (Fageria, 2009). However, the cultivation of Al-excluding or Al-tolerant crops, or plants generally tolerant of acid soils, can also facilitate agricultural production on these soils. Such crops are generally native to acid soils and are constitutively adapted to overcome this constraint (Fageria, 2009).

Cobalt is not very abundant in most soils. Its concentration generally ranges from 1 to 40 mg kg^{-1} soil and its concentration in the soil solution approximates 0.2–0.7 µM (Mengel et al., 2001). Leaf concentrations are typically in the range of 0.1–10 µg Co g^{-1} DM, and concentrations greater than 10–20 µg Co g^{-1} DM are toxic to most plants (Table 6.1; Palit et al., 1994; Mengel et al., 2001; Watanabe et al., 2007). Cobalt toxicity can be observed in plants growing in natural soils formed from serpentine or other ultrabasic rocks. Nevertheless, several plants native to these areas can tolerate leaf concentrations in excess of 1 mg Co g^{-1} DM (Baker et al., 2000). The majority of these Co-hyperaccumulator plants belong to the Lamiaceae, Scrophulariaceae, Asteraceae and Fabaceae (Baker et al., 2000).

High concentrations of Cd and Pb occur both naturally and through anthropogenic activities. Natural mineral outcrops can be enriched in Cd through the weathering of Cd-rich rocks and Cd pollution of the environment has occurred through the mining and refining of metal ores or through the application of Cd-containing phosphate fertilisers, sewage sludge and municipal composts to agricultural soils (He et al. 2005; Kirkby and Johnson, 2008). Leaf concentrations greater than 5–10 µg Cd g^{-1} DM are toxic to most plants, although some plants have adapted to grow on sites with high soil Cd concentrations and can tolerate leaf concentrations in excess of 100 µg Cd g^{-1} DM (Table 6.1; Baker and Brooks, 1989; Broadley et al., 2001). The majority of the plants that hyperaccumulate Cd occur in the order

Brassicales (Broadley et al., 2001). Soils with high Pb concentrations can occur naturally, but the mining and refining of metal ores, incineration of coal and waste, and the presence of Pb in automobile fuel have made a significant contribution of Pb accumulation in agricultural soils (Colburn and Thornton, 1978; Bacon et al., 1996). Leaf concentrations greater than 10–20 μg Pb g^{-1} DM are toxic to most plants, with the exception of several Pb-hyperaccumulator plants (Table 6.1; Broadley et al., 2001).

6.3 The uptake of essential cationic elements from the rhizosphere

The concentration of a cation in the soil solution affects both the rate at which it can be delivered to the root system and the transport mechanism required for its uptake into a root cell. Some cations, such as K^+, Ca^{2+} and Mg^{2+}, are sufficiently available and mobile in well-fertilised agricultural soils to supply plant demand, but the root system must forage for less readily available cations, such as Fe^{2+}, Fe^{3+}, Mn^{2+}, Zn^{2+}, Cu^+, Cu^{2+} and Ni^{2+}, which have low concentrations in the soil solution and small diffusion coefficients (Barber, 1995). With the exception of K^+, the delivery of essential mineral cations to the root surface is largely determined by transpiration-driven mass flow of the soil solution, rather than by local diffusion (Table 6.2; Barber, 1995; Fageria et al., 2011). Hence, the production of an extensive root system to provide a large surface area for cation uptake in unexplored volumes of soil and the modification of rhizosphere chemistry and biology by root exudates to increase the concentrations of mineral cations in the soil solution improve the acquisition of essential mineral cations (see Section 6.6).

Mineral cations entering roots from the soil solution can reach the xylem, for delivery to the shoot, via intracellular (symplastic) or extracellular (apoplastic) pathways. Most essential mineral cations enter the root symplast through transport proteins located in the plasma membrane of epidermal and/or cortical cells and are transported across the root through the cytoplasm of root cells connected by plasmodesmata (Karley and White, 2009; Puig and Peñarrubia, 2009; Tejada-Jiménez et al., 2009; White and Broadley, 2009). They are then loaded into the xylem by transport proteins located in the plasma membrane of stellar parenchyma cells. The formation of a Casparian band in the root endodermis at an early stage of development restricts the apoplastic movement of solutes to the xylem. This enforces entry of most essential mineral cations to the symplast and, therefore, the selectivity of cation transport to the shoot. Among the essential cationic mineral elements, only Ca is thought to reach the xylem via a predominantly apoplastic pathway, although a significant proportion of the Fe and Zn acquired by plant roots can reach the xylem by this route in particular plant species and/or when these elements occur at high concentrations in the soil solution (White, 2001; White et al., 2002; Taiz and Zeigler, 2006; Broadley *et al.*, 2007). In addition, since the apoplastic pathway is relatively non-selective, toxic cations such as Na^+ and Cd^{2+} can reach the xylem by this route (Lux et al., 2004; Plett and Møller, 2010). The acquisition of cationic elements reaching the xylem via an apoplastic pathway is generally restricted to the extreme root tip and to regions in which lateral roots are being initiated (White, 2001).

The uptake of cationic mineral elements that are delivered to the xylem via a symplastic pathway generally exhibits both saturatable and non-saturatable components (Figure 6.2; White, 2003). The concentration-dependence of cation uptake and relative magnitude of the saturatable and non-saturatable components reflect the properties, activity and abundance of transport proteins in the plasma membrane of root cells. A multitude of transport

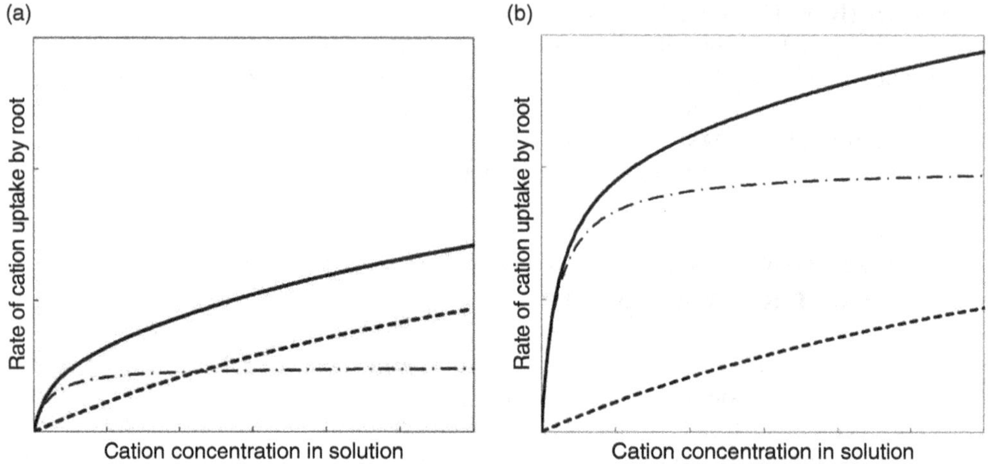

Figure 6.2 General relationship between the uptake of a cation and its concentration in solution in (a) plants with sufficient cation and (b) plants lacking the cation. Total uptake (solid line) is represented as the sum of a 'high-affinity' saturatable (-· -) and a non-saturatable (- - -) component. The capacity of the saturatable component is increased in plants lacking the cation.

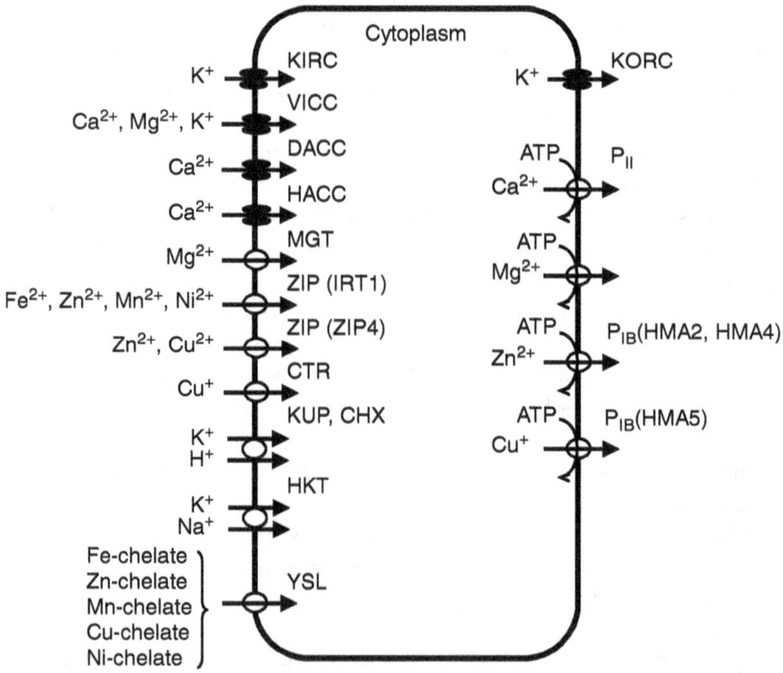

Figure 6.3 Transport proteins in the plasma membrane of root cells. See text for details.

proteins in the plasma membrane of root cells catalysing uptake from the soil solution and xylem loading of essential cationic minerals have been identified (Figure 6.3). Potassium can enter root cells through voltage-gated inward-rectifying K^+-selective channels (KIRC), such as those encoded by members of the "shaker" gene family, non-selective cation channels (VICC), such as those encoded by members of the cyclic nucleotide-gated channel (CNGC) and glutamate receptor (GLR) gene families; or 'high-affinity' K^+ transporters, such as K^+/H^+ symporters (HAK/KUP), K^+/Na^+ co-transporters (HKT/Trk) and cation-H^+ exchangers (CHX) in the plasma membrane (Karley and White, 2009; White and Karley, 2010). Potassium is loaded into the xylem through voltage-gated, outward-rectifying K^+-selective channels (KORC) present in the plasma membrane of root pericycle and stelar parenchyma cells. Calcium can enter root cells through Ca^{2+}-permeable cation channels including hyperpolarisation-activated Ca^{2+} channels (HACC), voltage-independent non-selective cation channels (VICC) and depolarisation-activated Ca^{2+} channels (DACC; White and Broadley, 2003; Karley and White, 2009; McAinsh and Pittman, 2009; Dodd et al., 2010). Efflux of Ca^{2+} from root cells to the apoplast is catalysed by Ca^{2+}-ATPases in the plasma membrane. Magnesium enters root cells either through non-selective cation channels or through Mg^{2+}-transporters (MGT) in their plasma membrane (Karley and White, 2009; White and Broadley, 2009). It is speculated that ATPases catalyse Mg^{2+} efflux from root cells into the xylem. Since many proteins catalysing K^+ and, especially, Ca^{2+} transport are non-selective cation channels, Na^+ can also enter root cells through these transporters (White, 1999; Munns and Tester, 2008).

Plants exhibit two strategies for acquiring Fe from the soil (Figure 6.4; Schmidt, 1999; Mengel et al., 2001; Guerinot, 2010). Roots of non-graminaceous species (Strategy I Plants) acidify the rhizosphere and release organic acids and phenolic compounds to increase Fe^{3+} concentrations in the soil solution. These compounds chelate Fe^{3+}, which is subsequently reduced to Fe^{2+} by ferric reductases in the plasma membrane of root epidermal cells and Fe^{2+} is taken up by Fe^{2+} transporters encoded by members of the ZIP gene family (Puig and Peñarrubia, 2009; White and Broadley, 2009; Guerinot, 2010). Cereals and grasses (Strategy II Plants) can take up Fe^{2+} through Fe^{2+} transporters, but they also release phytosiderophores (structural derivatives of mugineic acid) into the rhizosphere to chelate Fe^{3+} and take up Fe^{3+}-phytosiderophore complexes through homologues of the maize yellow stripe protein (YSL) in the plasma membrane of root cells (Puig and Peñarrubia, 2009; White and Broadley, 2009; Guerinot, 2010). Zinc can be taken up across the plasma membrane of root cells as Zn^{2+}, catalysed by Zn^{2+} transporters encoded by members of the ZIP gene family, or, in cereals and grasses, as a Zn-phytosiderophore complex, catalysed by YSL proteins (Grotz and Guerinot, 2006; Broadley et al., 2007; Puig and Peñarrubia, 2009; White and Broadley, 2009; Clemens, 2010). Members of the heavy metal P_{1B}-ATPase family load Zn^{2+} into the xylem. The uptake of Cu by root cells occurs through high-affinity Cu^+ transporters of the copper transporter (CTR) family and/or by Cu^{2+} transporters encoded by members of the ZIP gene family, and Cu^{2+} is loaded into the xylem by heavy metal P_{1B}-ATPases (Grotz and Guerinot, 2006; Burkhead et al., 2009; Puig and Peñarrubia, 2009; White and Broadley, 2009; Clemens, 2010). In cereals and grasses, Cu can also be taken up as a Cu-phytosiderophore complex, catalysed by YSL proteins. Manganese uptake by root cells is catalysed by Mn^{2+} transporters encoded by members of the ZIP gene family (Puig and Peñarrubia, 2009; Williams and Pittman, 2010) and it has been speculated that Ni^{2+} also enters root cells via these transporters (Pilon-Smits et al., 2009; Tejada-Jiménez et al., 2009). Furthermore, since some ZIPs, such as IRT1, are relatively non-selective between cations, they can transport Co^{2+} and Cd^{2+} into root cells. It is also conceivable that Mn and Ni can be taken up by root cells as phytosiderophore complexes catalysed by YSL proteins (Puig and Peñarrubia, 2009; Tejada-Jiménez et al., 2009; Guerinot, 2010).

Both the activity of transport proteins and the expression of genes that encode them are regulated dynamically to ensure appropriate tissue concentrations of essential mineral

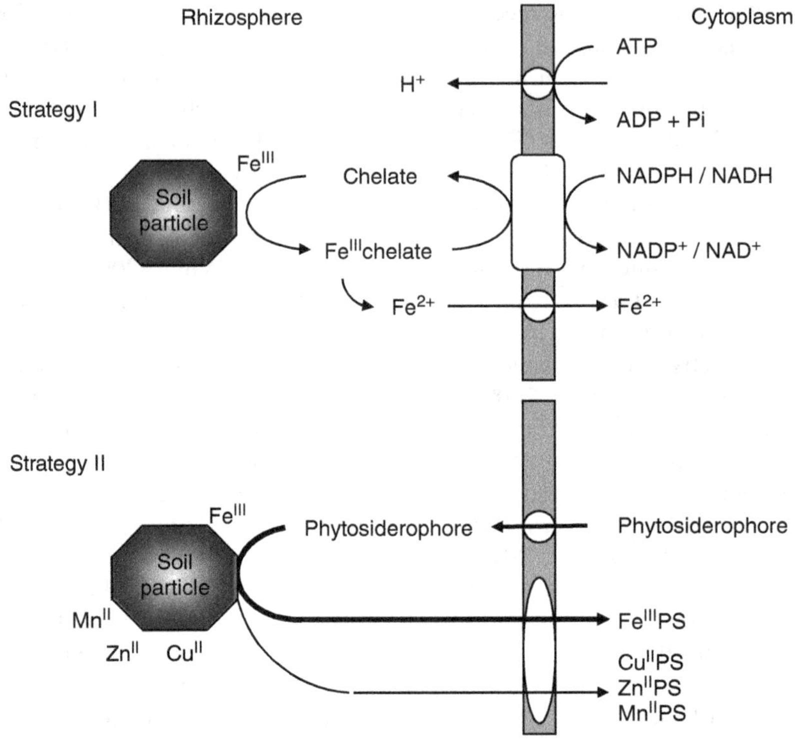

Figure 6.4 Root responses to iron deficiency in non-graminaceous (Strategy I) and graminaceous (Strategy II) plant species. In Strategy I plants, increased acidification of the rhizosphere by H^+-ATPases and the induction of ferric reductase activity to reduce Fe(III) chelates in the rhizosphere releases Fe^{2+} for uptake across the plasma membrane of root cells by Fe-deficiency-inducible, high-affinity Fe^{2+} transporters. In Strategy II plants, enhanced synthesis and release of phytosiderophores (PS) into the rhizosphere chelate Fe^{3+}, Zn^{2+}, Cu^{2+} and Mn^{2+}, and these chelates are transported across the plasma membrane by Fe-deficiency-inducible transport proteins. Adapted from White (2012). With kind permisison of Academic Press London.

elements. The abundance and/or activity of specific high-affinity transport proteins are increased in plants lacking K, Fe, Zn, Cu or Mn and enable the acquisition of these elements from soil solutions containing low concentrations of their cations (Figure 6.2). In addition, when Fe, Zn, Cu or Mn are in short supply, roots efflux more protons, organic acids, phytosiderophores and enzymes capable of degrading organic compounds to increase the concentrations of these elements in the soil solution (Broadley et al., 2007; Pilon et al., 2009; Puig and Peñarrubia, 2009; White and Broadley, 2009; Clemens, 2010; Guerinot, 2010; White and Karley, 2010). These root traits are often complimented by associations with mycorrhizal fungi and by fostering beneficial microbial communities in the rhizosphere (Rengel et al., 1999; Harrier and Watson, 2003; Barea *et al.*, 2005; Morgan et al., 2005; Broadley et al., 2007; Smith and Read, 2007; Cavagnaro, 2008; White and Broadley, 2009). Since the delivery of cationic elements to the root surface is generally determined by their concentrations in the soil solution, their mobilisation from recalcitrant soil constituents can have a major effect on their acquisition by plants, whereas increased root uptake capacity alone often has little effect on the acquisition of these elements by field crops (Barber, 1995).

6.4 Cationic elements in the soil solution

Differences in the complement and concentrations of cationic elements in soil solutions reflect the enormous diversity of soils across the world. Base cations are found in rock fragments, in minerals, in aqueous solution and on the negatively charged surfaces of soil particles. The quantities of base cations on negatively charged surfaces vary from about 1 to 200 cmol kg^{-1} soil and are often used for estimating fertiliser requirements (Havlin et al., 2005; Fageria, 2009). Some of this variation results from differences in agronomic practice, but for the most part it results from processes within the soil and their dependence on the parent material and the extent of weathering. Weathering breaks down rocks and rock fragments by water- and temperature-dependent physical processes, such as freezing and thawing. Chemical changes in soil take place by hydrolysis and oxidation and by the effects of atmospheric deposition (e.g. of S and Na), carbon dioxide and organic acids and other compounds produced by the activities of animals, plants and microorganisms. Initially, weathering produces small pieces of rocks and mineral grains, which have similar composition as the parent rocks, but, as weathering continues, clays and amorphous materials are produced, which contain less mineral elements than the parent rocks. Even these products are further transformed by subsequent weathering. In addition, organic matter is accumulated in soils as a result of the formation and degradation of plant debris. However, despite the complexity of the soil-forming processes, widely applicable principles are emerging that govern the formation of the different components of soil and their cationic properties. In this section, these principles are introduced by first discussing the cationic properties of the main components of soil and the key stages of their formation. The quantitative relationships governing the dependence of base cation concentrations in solution on each other, on the CEC of the soil (the sum of all cation equivalents held on exchange sites) and on pH are then discussed.

6.4.1 The properties and formation of soils

The characteristics of mineral fragments vary enormously depending on the rocks from which they are derived. For example, fragments derived from some forms of magma that have cooled slowly contain few base-rich cations, whereas fragments derived from magma that has cooled rapidly are generally rich in base cations (Wild, 1993). There is similar diversity in mineral fragments derived from the two other major rock types, namely sedimentary rocks (such as limestone) which, as the name implies, are formed by sedimentation, and metamorphic rocks, which are formed by recrystallisation under high temperatures and pressures. Generally, fragments derived directly from parent rocks are large and occur within the sand and silt particle size fractions. Consequently, their surface area per unit mass is small and their CEC is less than those of clays of similar composition (Table 6.3). Nevertheless, if the parent material is base-rich, the CEC of the sand and silt particles can be substantial. Average values of 19 cmol kg^{-1} for the sand fraction and 33 cmol kg^{-1} for the silt fraction in the upper 20–30 cm layers of soil have been reported (McAleese and McConaghy, 1957). These values are significant even when compared with smectite clays that are derived from similar base-rich rocks.

Mineral particles less than 2 μm in diameter are classified as clays. First, it must be noted that, although the examples given here illustrate the main processes of clay formation, clays can vary widely in both their physical and chemical properties. In general, clays

Table 6.3 Typical values for the cation exchange capacity (CEC) of major components of soils.

Soil colloid	CEC (cmol kg^{-1})[a]
Soil organic matter	110–290
Smectite	60–150
Kaolinite	3–15
Sesquioxides	2–4

[a]CEC varies with pH.
Source: Data compiled from Parfitt et al. (1995), Tan (1998), Corrado (2004) and Pusch and Yong (2006).

consist of two distinct units: silica–oxygen and aluminium hydroxide. The silica–oxygen unit consists of Si^{4+} at the centre of a tetrahedron of three O^{2-} ions (Mott, 1988a). These tetrahedra link with one another in different ways, but in all cases there is a Si–O–Si bond. In clays, these bonds form two-dimensional paper-like sheets that are only one Si layer thick. The aluminium hydroxide unit is arranged geometrically so that Al^{3+} is at the centre of an octahedral arrangement of six OH^- that are polymerised into sheets one Al atom thick. The formation of these sheets and the way they are combined affect the cationic properties of clays, as is illustrated by kaolinite and smectite. Kaolinite is formed from one Si sheet adjacent to one Al sheet. The bonds between the two sheets are strong, shrinking and swelling is small, and the clay is resistant to weathering. It is formed in acid environments (Wilson, 1999). It has a low CEC (Table 6.3), contains few base cations and has almost zero permanent charge. However, since it has exposed OH^- groups, its surface charge is pH dependent (Tan, 1998). By contrast, smectite consists of one Al sheet sandwiched between two Si sheets held together by Si–O–Al links (Mott, 1988a). The bonding between sheets is weak, shrinking and swelling is considerable, and it is readily degraded by weathering into kaolinite (Wilson, 1999). It is formed in base-rich environments and cations such as Mg^{2+} and Fe^{2+} become entrapped within 'holes' of the Al sheet and also replace some of the Si^{4+} and Al^{3+} ions during its formation (Mott, 1988a). The mix of cations involved in this amorphous replacement varies in different forms of smectite, but, since these cations have lower valencies than Si^{4+} or Al^{3+}, all these forms have a high negative charge and a high CEC of, typically, 100 cmol kg^{-1} (Table 6.3). It is noteworthy that Ca^{2+} never takes part in amorphous replacement because it is too large to fit within the available spaces. The weak bonding between layers also means that K^+ and NH_4^+, uniquely among cations, can become entrapped between layers and are then said to be fixed. Fixed K^+ and NH_4^+ are only slowly available to plants, and the significance of this is discussed later in Section 6.5. Another important group of clays is hydrated Fe and Al oxides. These clays have weak electronegative charges in acid conditions and an electropositive charge in alkaline conditions, which allows them to adsorb anions such as phosphate. They also form covalent bonds with heavy metal cations such as Cu^{2+}, Zn^{2+} and Pb^{2+}. The adsorption capacity of these clays ranges from 3 to 30 cmol kg^{-1} (Tan, 1998).

The elucidation of the effects of weathering, and the sequence of soil formation, is complicated by the difficulty in finding soil pairs with the same parent materials that have experienced different intensities of weathering. One difficulty is the huge disparity in age and weathering between soils developed after an ice cap and those where none has ever been present. The processes of soil formation have also been confounded by sufficient deposition of elements from the atmosphere to sustain vegetation on some highly weathered soils

(Jackson et al., 1971; Chadwick et al., 1999). Nevertheless, studies have been made on enough suitable pairs of soils to suggest that, as the duration or intensity of weathering increases, the order of events is as follows: (1) Hydrolysis of the primary rocks with a loss of base cations, which is initially rapid but then decreases sharply with time (Syers and Walker, 1969; Walker and Syers, 1976; Bain et al., 1993). (2) Formation of smectite- or kaolinite-type clays, depending on the pH of the surrounding media (Fisher and Ryan, 2006). (3) Breakdown of smectite clays into kaolinite and then into gibbsite, $Al(OH)_3$ (Wilson, 1999; Fisher and Ryan, 2006). (4) Leaching of silica, but not of coarse-grained quartz, decomposition of kaolinite and an increase in the proportions of Fe and Al oxides in soil (Aniku and Singer, 1990; Troeh and Thompson, 2005; Uehara, 2006). (5) Leaching of Fe and Al oxides, with the consequence that the remainder consists of nutrient-free coarse-grained quartz (Jackson et al., 1971; Troeh and Thompson, 2005). Thus, weathering results in a loss of base cations and the ability of soils to retain added cations. Weathering is accelerated by high temperatures and rainfall.

In addition to its inorganic constituents, soil also contains organic matter, which, because of its low bulk density, can constitute a substantial proportion of the total soil volume. Soil organic matter exists in many different forms (see Chapter 4). In the context of this chapter, and as a general rule, the CEC of soil organic matter varies with pH and is typically 200 cmol kg^{-1} (Table 6.3). Some components of soil organic matter form strong complexes with cations such as Cu^{2+}, Fe^{2+} and Mn^{2+}. These are generally insoluble, but they can be solubilised under some environmental conditions (Tan, 1998).

6.4.2 Relationships between cations on exchange sites and in the soil solution

The solid matrix of soils has both negatively and positively charged components. Negative charges adsorb base cations and dominate temperate soils. Two key parameters defining cation exchange between the solid matrix and the soil solution are the CEC of the soil and its percent base saturation. The CEC of a soil defines its capacity to adsorb and exchange all types of inorganic cations including protons and aluminium. The percent base saturation is the percentage of the total CEC that is saturated with base cations (i.e. cations other than protons and aluminium). Both are important as they influence the extent to which soils can meet plant demands for cations such as K^+ and Mg^{2+} without frequent applications of these macronutrient elements. The percent base saturation is strongly correlated with soil pH (Figure 6.5) and, together with soil CEC, these parameters have a decisive influence on lime requirement. Provided the soil CEC is high, when the soil pH is over 7 it is buffered strongly by the reaction of H^+ on base carbonates, such as $CaCO_3$, and when soil pH is between 5 and 6 it is buffered by the replacement of Ca^{2+}, Mg^{2+}, K^+ and Na^+ on the exchange sites with H^+. Below pH 4, the soil is often strongly buffered by hydroxyl–aluminium complexes that have formed during the breakdown of clay minerals (Goulding and Annis, 1998). Subsoil acidity is a serious problem and very difficult to correct. Much emphasis is often placed on maintaining the pH of the ploughed layer by liming which, among other things, should prevent the spread of acidity to the subsoil.

In this section, for the sake of clarity, ideal relationships between cations on exchange sites and those in dilute soil solutions are presented initially to avoid the complications associated with the effects of solution ionic strength on the chemical activity of each cation. Adjustments that need to be made for the effects of solution ionic strength, and the consequences of the formation of uncharged ion pairs, are considered subsequently.

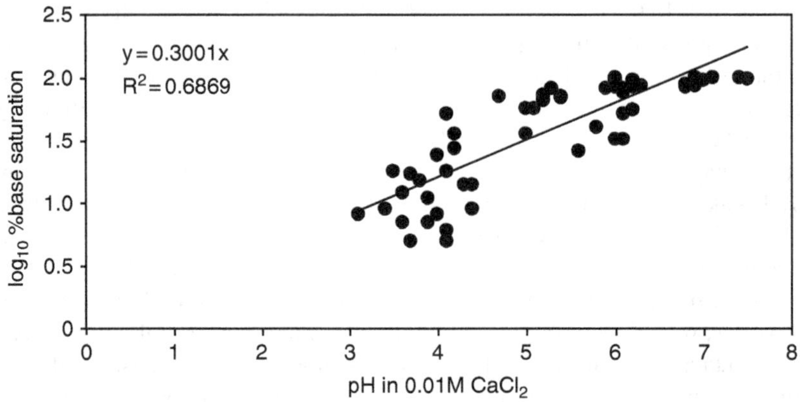

Figure 6.5 Relationship between percent base saturation of soil and the pH of the soil solution. Data are taken from Avery (1990) for each horizon of the profiles that he identified as representative of brown and gley soil. Data from three soils for which pH <4 are omitted.

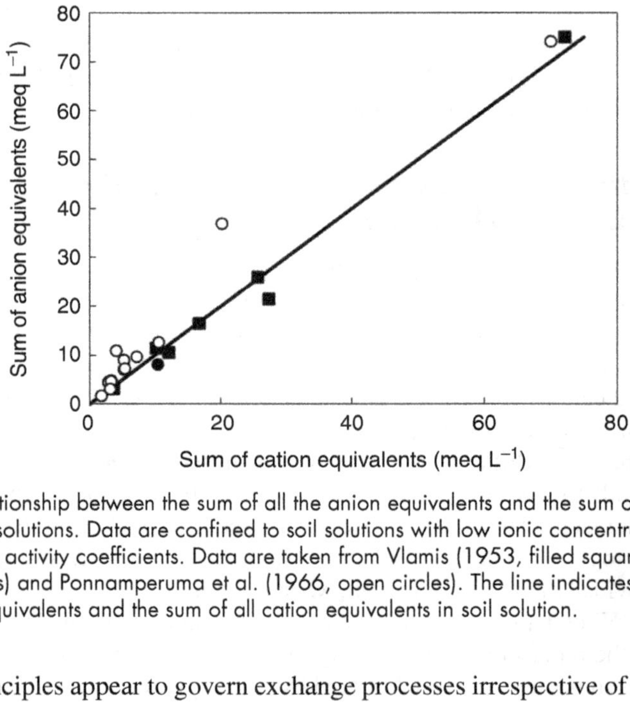

Figure 6.6 Relationship between the sum of all the anion equivalents and the sum of all the cation equivalents in soil solutions. Data are confined to soil solutions with low ionic concentrations to avoid complications from activity coefficients. Data are taken from Vlamis (1953, filled squares), Eaton et al. (1960, filled circles) and Ponnamperuma et al. (1966, open circles). The line indicates equality in the sum of all anion equivalents and the sum of all cation equivalents in soil solution.

The same principles appear to govern exchange processes irrespective of the nature of the exchange sites. Within the micro-volume of solution adjacent to the negative charges on the solid matrix, there is a diffuse double layer in which the cation concentrations decrease and the anion concentrations increase with distance. Beyond this minute distance from the surface, the sum of the positive charges of the cations must always equal the sum of the negative charges of the anions in the soil solution to maintain electrical neutrality (Figure 6.6). Strict rules govern the interchange between cations on the exchange sites (denoted by the subscript 'exchange') and those in the soil solution (denoted by subscript 'solution'). According to the law of mass action, when equilibrium is established for the monovalent cations A^+ and B^+:

$$\frac{A^{+}_{\text{solution}}}{B^{+}_{\text{solution}}} = K_{ex}\left(\frac{A_{\text{exchange}}}{B_{\text{exchange}}}\right)$$

where K_{ex} is a selectivity index. The larger the K_{ex}, the greater the relative adsorption of B^+ compared with A^+.

A number of more empirical relationships have been advanced for the exchange of monovalent, divalent and trivalent cations with one another. Divalent cations, such as Ca^{2+} and Mg^{2+}, are absorbed less strongly than monovalent cations, such as H^+ and K^+. Schofield developed a fundamental law governing the ratio of the concentrations of two cations M1 and M2 of valency Z1 and Z2 in soil solution (Schofield, 1947; Schofield and Taylor, 1955a, b). It states that the ratio $[M1]^{1/Z1}/[M2]^{1/Z2}$ remains constant irrespective of soil/water ratio provided that the cation contents on the exchange complex remain approximately constant, which is usually the case as the amounts of cations in the soil solution are usually small compared with those on the exchange sites. If the ionic concentrations are other than dilute, they must be replaced with ionic activities as described in Section 6.4.3 (Schofield and Taylor, 1955a, b). The predictions of this relationship have been confirmed experimentally for the following ratios: K:Ca, Ca:Al, K:Al and Na:Al (Bower, 1959; Salmon, 1964; Bache, 1970). Thus, the quotient $[K^+]/([Ca^{2+}] + [Mg^{2+}])^{0.5}$ has a fundamental basis as a measure of the K-status of a soil and has proved useful in elucidating the K dynamics of soils (Evangelou *et al.*, 1994). The total cation concentrations in solutions from many soils are dominated by the concentrations of Ca^{2+} and Mg^{2+}, and the sum of their charge equivalents is, therefore, approximately proportional to the sum of the charge equivalents of anions. The approximate constancy of $[K^+]/([Ca^{2+}] + [Mg^{2+}])^{0.5}$ over a wide range of soil water contents, together with the quantitative charge-equivalent relationship between Ca^{2+}+Mg^{2+} and anions, has proved useful in developing simulation models for the responses of crops to K-fertilisers (Greenwood and Karpinets, 1997a, b) and the way in which these are affected by the application of N-fertilisers (Zhang et al., 2007).

An important aspect of the K economy of soils concerns K^+ fixed between some clay sheets and its release into the soil solution (Karpinets and Greenwood, 2003). This process buffers the available K^+ in the soil solution. Typically, the soil solution contains about 0.1–0.2% of the total K in the soil, a further 1–2% is present as exchangeable K^+ held in cation exchange sites and about 1–10% is present as fixed K^+ associated with clay lattices (Mengel et al., 2001; Fageria, 2009). The remainder of the soil K is effectively unavailable over the growth period of a typical arable crop. Interchange between K^+ in solution and exchangeable K^+ is very rapid, but the exchange between K^+ in solution and fixed K^+ is slow. The amounts of each of these components, and the rates of interchange between them, are important for the self-organisation of soil K pools, in which the solution and exchangeable forms of K^+ tend toward critical values characteristic of a soil (Karpinets and Greenwood, 2003). If a crop acquires K^+, the concentration of K^+ in solution will fall which, if the concentration falls below the critical value, prompts the slow release of fixed K^+ into solution. This can replenish the exchangeable K^+ pool. If K-fertiliser is added to the soil, it will rapidly increase K^+ in solution and replenish the exchangeable K^+ pool and, if these rise above the critical values, some K^+ will enter the fixed K^+ pool. Changes in the net balance between crop K^+ acquisition and the addition of K-fertilisers will seldom equal the sum of the changes in solution and exchangeable K^+ because of the buffering effect of fixed K^+.

An important group of soils derived from sedimentary rocks consist largely of $CaCO_3$. Calcium carbonate has no cation exchange properties, but it reacts with carbonic and other acids to form soluble Ca salts. Carbon dioxide dissolved in water forms carbonic acid which, though a weak acid, reacts with $CaCO_3$ to release Ca^{2+} ions into the solution. It has long been recognised how the various processes can be defined in terms of classical chemical equilibria (Knox, 1912). The key equations and values of the corresponding coefficients are as follows (Hodgman, 1953):

Equilibria	Coefficients at 25 °C	
$\left[Ca^{2+}\right]\left[CO_3^{2-}\right] = S$	4.95×10^{-9}	(6.1)
$\dfrac{\left[H_2CO_3\right]}{\left[P\right]} = K1$	3.29×10^{-2}	(6.2)
$\dfrac{\left[H^+\right]\left[HCO_3^-\right]}{\left[H_2CO_3\right]} = K2$	4.3×10^{-7}	(6.3)
$\dfrac{\left[H^+\right]\left[CO_3^{2-}\right]}{\left[HCO_3^-\right]} = K3$	5.61×10^{-11}	(6.4)

where P is the partial pressure of CO_2 expressed in atmospheres and concentrations are expressed in molarities. Division of Equation 6.3 by Equation 6.4 gives:

$$\frac{[HCO_3^-]^2}{[H_2CO_3][CO_3^{2-}]} = K2/K3 \qquad (6.5)$$

Multiplication of Equation 6.5 by Equation 6.1 gives:

$$\frac{[HCO_3^-]^2[Ca^{2+}]}{[H_2CO_3]} = K2S/K3 \qquad (6.6)$$

Electrical neutrality must be maintained and thus:

$$2\left[Ca^{2+}\right]+\left[H^+\right]=\left[HCO_3^-\right]+2\left[CO_3^{2-}\right]$$

As the concentrations of $[H^+]$ and $\left[CO_3^{2-}\right]$ are small compared with those of $[Ca^{2+}]$ and $\left[HCO_3^-\right]$, they may be neglected so that $2\left[Ca^{2+}\right]\approx\left[HCO_3^-\right]$.

Substitution of these values in Equation 6.6 gives:

$$\frac{4[Ca^{2+}]^3}{[H_2CO_3]} = K2S/K3 \qquad (6.7)$$

Elimination of $[H_2CO_3]$ by Equation 6.2 gives:

$$\left[Ca^{2+}\right]\approx\sqrt[3]{<S.K1.P.K2/(4*K3)>} \qquad (6.8)$$

which after substitution of the above coefficients gives:

$$\left[Ca^{2+}\right]\approx 0.00713.\sqrt[3]{P} \qquad (6.9)$$

Multiplication of Equations (6.2), (6.3) and (6.4) and substitution of CO_3 by Equation 6.1 gives:

$$P = \frac{[H^+]^2.S}{[Ca^{2+}]K1K2K3}$$

and substitution of Ca^{2+} in Equation 6.9 gives:

$$P^{1.33} = \frac{[H^+]^2 S}{7.13\times10^{-3}\,K1K2K3} \quad (6.10)$$

Equations (6.8) and (6.10) are of crucial importance for understanding the cationic relationships in these soils. Equation (6.8) indicates that Ca^{2+} concentration increases approximately in proportion to the cube root of the partial pressure of CO_2. So, in the surface layers of soil where there is high metabolic production of CO_2, there will be higher concentrations of both CO_2 and Ca^{2+} than in the subsoil, where metabolic activity is lower. Leaching will result in the movement of Ca^{2+} from the surface soil to the subsoil, where it will be precipitated. Another consequence of low CO_2 concentrations and, thus, CO_2 partial pressures in the subsurface soil is that the H^+ concentration will fall and the pH will rise as defined by Equation 6.10.

Soils often receive strong acids, such as nitric acid from the nitrification of ammonia in soil and sulphuric acid from rainfall that dissolve $CaCO_3$ to form soluble salts. Such acids are much stronger than carbonic acid and, in consequence, can give rise to higher concentrations of Ca^{2+} and greater Ca^{2+} leaching down the soil profile.

6.4.3 Caveats concerning high ionic strength and uncharged complexes

The concentration of an ion gives a value to the upper limit of its reactivity. If the ionic strength of a solution rises, then the reactivity of an ion is reduced. To correct for this, concentrations are multiplied by an activity coefficient for an appropriate ionic strength of solution. As an approximation to the Debye–Huckel equation, the activity coefficient can be calculated from:

$$-\log_{10} f_i = \left(AZ_i^2 I^{½} / \left(1 + a_i B I^{½}\right)\right)$$

where f_i is the activity coefficient of ion i, of valency Z_i, A and B are functions of temperature and a_i is the effective diameter of the ion. I is the ionic strength given by:

$$I = ½\sum C_i Z_i^2$$

where C_i is the molar concentration of ion i. The activity coefficients of trivalent ions are depressed more than those of divalent ions which are depressed more than those of monovalent ions by increase in ionic strength (Tan, 1998).

Another complication in estimating the phytoavailability of cations in a soil solution arises because some pairs of cations and anions combine to form uncharged complexes or ion pairs. An example of this is calcium sulphate. In solutions containing a concentration less than that at which precipitation occurs, there exists in solution the following species:

$$CaSO_4^{o} \leftrightarrow Ca^{2+} + SO_4^{2-}$$

where $CaSO_4^o$ is uncharged. The equilibrium constant, Keq, is given by:

$$Keq = (Ca^{2+})(SO_4^{2-}) / (CaSO_4^{o})$$

where the terms in parenthesis are activities.

The complicated calculations required to determine the concentrations and activities of cations in soil solutions have led to the development of several sophisticated computer programs. Prominent ones include GEOCHEM (Sposito and Mattigod, 1980; Parker et al., 1995; Shaff et al., 2009) and PHREEQC (Parkhurst and Appelo, 1999), both of which include many inorganic and organic species in widely different environments including those with solid cation exchange sites. Although, recent models and associated experimental work reveal some discrepancies between model predictions and actual measurements (e.g. Huber and Montserrat, 2002; Unsworth et al., 2006), these programmes have been fundamental to developing our understanding of the chemistry of soil solutions. Further development of these computer programmes, including improvements in the treatment of activity coefficients and ion pairs, will be especially valuable for studying the effects of aluminium and cationic micronutrients in the soil solution as they are prone to speciation (Mott, 1988b; Parker et al., 1988; Kinraide, 1991; Percival et al., 1999; Ge et al., 2000; Cancès et al., 2003; Almås et al., 2007).

6.5 Management practices increasing the acquisition of essential mineral cations

6.5.1 Potassium

Most soils contain K concentrations between about 0.3 and 25 $g\,kg^{-1}$ soil, and soil solutions often contain K^+ concentrations between 0.1 and 1 mM (Mengel et al., 2001; Fageria, 2009). The K^+ available for plant growth – which is present as K^+ in the soil solution, as exchangeable K^+ associated with cation exchange sites and as fixed K^+ associated with clay lattices – amounts to less than 2–10% of the total K in the soil, a large proportion of which resides in the topsoil (Section 6.4.2). The requirements for K-fertiliser are determined primarily by the ability of the soil to provide sufficient K^+ to satisfy the demand of a particular crop. Although crops differ in their K requirements (Section 6.6), and in some plant species Na^+ can substitute for K^+ in its biophysical roles (Section 6.2), their basic K requirements are ultimately established by their rate of growth, which is generally determined by the availability of N. Soils vary greatly in their ability to supply K^+ to a crop. Some soils can release more K^+ than is required by crops such that no additional K-fertiliser is necessary, whilst other soils have no capacity to retain or supply K^+ and must receive sufficient K-fertiliser each year to meet the entire crop requirement. Some soils are so well buffered that the optimum strategy is to attempt to maintain a given level of exchangeable K^+ through regular applications of K-fertilisers. To provide a more quantitative basis for K-fertiliser practice, the interrelationships between soil K^+ pools and K^+ concentration in the soil solution, the movement of K^+ to the root and its acquisition by the crop and the growth-driven K demand of a crop have been incorporated into predictive models. These models have been tested against data from field experiments and are now embedded in decision support tools for K-fertiliser practice (Zhang et al., 2007).

The main additional sources of K for plant nutrition are crop residues and composts (0.30–0.45% DM; Defra, 2010), animal manures (0.65–1.04% DM; Defra, 2010),

Table 6.4 Commonly applied inorganic fertilisers containing essential cationic elements (Fageria 2009; White and Broadley 2009).

(A) Macronutrients			
Potassium	**Calcium**	**Magnesium**	
KCl, KNO_3	Calcitic lime [$CaCO_3$]	Epsom salts, kieserite [$MgSO_4$]	
K_2SO_4; $K_2SO_4{\cdot}MgSO_4$	Dolomitic lime [$CaMg(CO_3)_2$]	Dolomite lime [$MgCO_3$]	
KPO_3	Burned lime [CaO]	Dolomitic lime [$CaMg(CO_3)_2$]	
Kainit [$MgSO_4$ + KCl + NaCl]	Slaked lime [$Ca(OH)_2$]	MgO	
	Basic slag [$CaSiO_3$]	Struvite [$(NH_4)Mg(PO_4){\cdot}6(H_2O)$]	
	Wood ash		
(B) Micronutrients			
Iron	**Manganese**	**Copper**	**Zinc**
$FeSO_4.7H_2O$; $Fe_2(SO_4)_2.4H_2O$	$MnCl_2$	$CuCl_2$	$ZnSO_4.H_2O$; $ZnSO_4.7H_2O$
$(NH_4)_2SO_4.FeSO_4.6H_2O$	$MnSO_4.H_2O$	$CuSO_4.H_2O$; $CuSO_4.5H_2O$	$ZnCO_3$
$Fe(NH_4)PO_4.H_2O$; $Fe(NH_4)HP_2O_7$	$MnCO_3$	CuO; Cu_2O	$Zn_3(PO_4)_2$
FeO; Fe_2O_3	MnO	Cu-silicates	ZnO
Fe-silicates	MnEDTA	$Na_2CuEDTA$	ZnS
NaFeEDTA	Mn-silicates	NaCuHEDTA	Zn-silicates
NaFeHEDTA	Mn-polyflavenoid		Zn-EDTA chelates
NaFeEDDHA	Mn-methoxyphenylpropane		Zn-HEDTA chelates
			Raplex zinc [ZnPF]

K-bearing minerals and K-fertilisers. Most K-fertilisers are applied to field crops as KCl, KNO_3, K_2SO_4, K_2SO_4, $MgSO_4$ or KPO_3 (Table 6.4; Lægrid et al., 1999; Fageria, 2009). Strategies for optimising the use of K-fertilisers include: (1) Liming acid soils, which not only increases the retention and availability of applied K^+ by replacing Al^{3+} on cation exchange sites with Ca^{2+} and K^+ but also improves root growth and, thereby, the uptake of water and mineral nutrients. (2) Maintaining adequate phytoavailable K^+ in the soil solution by applying K-fertilisers at an appropriate rate and time for a particular crop (Eckert, 1987; Kopittke and Menzies, 2007). (3) In wide-row crops, using a banded placement of K-fertiliser close to the developing root system in a K-deficient soil, rather than broadcasting K-fertiliser. (4) Supplying adequate moisture to allow K^+ acquisition by roots in regions of the soil where K^+ is available. (5) Incorporating crop residues after harvest, which can contain up to 80% of the K acquired by a cereal crop and up to 50% of the K acquired by a legume crop (Fageria, 2009). (6) Improving the organic matter content of soil, using crop residues, composts or animal manures, which, in addition to supplying K and other essential mineral elements, improve soil structure, water relations and mitigate against erosion (see Chapter 8). (7) Using crop species and genotypes within species that acquire and utilise K^+ most efficiently to produce a commercial yield (see Section 6.6).

6.5.2 Calcium and magnesium

Most soils contain sufficient Ca and Mg concentrations for adequate crop nutrition (White and Broadley, 2003, 2009). Soil solutions often contain Ca^{2+} concentrations in excess of 1 mM and Mg^{2+} concentrations between 125 μM and 8.5 mM (Mengel et al., 2001; White and Broadley, 2003, 2009). A lack of Ca seldom restricts the growth of arable crops, but Mg deficiencies can occur on sandy soils in the early stages of crop growth (Wilkinson et al., 1990; White and Broadley, 2009). In a similar manner to K^+, the Mg^{2+} immediately available for plant growth is present as Mg^{2+} in the soil solution and as exchangeable Mg^{2+} associated with cation exchange sites, which can constitute up to 5% of the Mg in the soil (Mengel et al., 2001). The most effective agronomic practice to increase Ca^{2+} and Mg^{2+} concentrations in the soil solution is liming. Calcite lime ($CaCO_3$), dolomite lime ($MgCO_3$) and dolomitic lime ($CaMg(CO_3)_2$), which increase soil pH, are the most appropriate Ca-fertilisers and Mg-fertilisers for acid soils, although burned lime (CaO), slaked lime ($Ca(OH)_2$), basic slag ($CaSiO_3$) and wood ash are also commonly used Ca-fertilisers (Table 6.4). Other Ca-fertilisers include gypsum ($CaSO_4$), $Ca(NO_3)_2$, superphosphates (14–20% Ca), rock phosphates and farmyard manures (0.5–2.3% Ca). Other common Mg-fertilisers include composts (~0.24% Mg), farmyard manures (0.2–0.3% Mg), $MgSO_4$ (Epsom salts or kieserite) and MgO (Draycott and Allison, 1998; Mengel et al., 2001; Fageria, 2009; White and Broadley, 2009; Defra, 2010). In addition, magnesium ammonium phosphate (struvite) has recently received attention as an alternative Mg-fertiliser, since it has potential as a sustainable P source for agriculture (Parsons and Smith, 2008). The application of lime increases Ca and/or Mg in the topsoil, whereas water-soluble fertilisers distribute these elements throughout the soil profile. Prophylactic foliar applications of soluble Ca-fertilisers are often made to horticultural crops to prevent Ca-deficiency disorders (Shear, 1975; Ho and White, 2005) and foliar applications of $MgSO_4$ are also common on some crops (Draycott and Allison, 1998). Applying Ca-fertilisers and Mg-fertilisers at an appropriate time and rate, as well as maintaining appropriate Ca and Mg concentrations and Ca/Mg quotients in the soil solution, are important factors for crop production (Eckert, 1987; Kopittke and Menzies, 2007). Supplying adequate moisture and improving the organic matter content of soil also benefit the acquisition of Ca^{2+} and Mg^{2+}.

6.5.3 Micronutrients

Although the total amounts of Fe, Zn, Cu and Mn in most soils are sufficient for crop nutrition, the phytoavailability of these mineral elements restricts production in many regions of the world (Follett et al., 1981; Alloway, 1995; Barber, 1995; Welch, 1995; Loneragan, 1997; Shuman, 1998; Schmidt, 1999; Graham et al., 1999; Frossard et al., 2000; Rengel, 2001; Broadley et al., 2007; Fageria, 2009; White and Broadley, 2009). Iron rapidly becomes unavailable in alkaline and calcareous soils. For this reason, the application of many inorganic Fe-fertilisers (Table 6.4), including $FeSO_4$, $Fe_2(SO_4)_3$, FeO, Fe_2O_3, $(NH_4)_2SO_4$, $FeSO_4$, $Fe(NH_4)HP_2O_7$, $Fe(NH_4)PO_4$ and Fe-frits, to soils is usually ineffective in combating Fe chlorosis. However, the application of Fe chelates, such as NaFeEDTA, NaFeHEDTA and NaFeEDDHA, to soils is effective (Mortvedt, 1991; Shuman, 1998; Rengel *et al.*, 1999). In addition, the availability of Fe in the rhizosphere can be increased by soil acidification with elemental S (Shuman, 1998). By contrast, the application of soluble fertilisers to the soil is often an effective way to correct Zn, Cu and Mn deficiencies (Shuman, 1998; Harris et al., 2007; Cakmak, 2008; Fageria, 2009; White and Broadley, 2009). Common Zn-fertilisers include $ZnSO_4$, ZnO, $ZnCO_3$, ZnS, $Zn_3(PO_4)_2$, zinc silicate frits, $Na_2ZnEDTA$, NaZnHEDTA and raplex zinc, ZnPF (Table 6.4). Copper sulphate ($CuSO_4$), CuO, Cu_2O, $CuCl_2$, $Na_2CuEDTA$, NaCuHEDTA and Cu-frits are common Cu-fertilisers (Table 6.4). Copper can also be supplied to crops in sewage sludges and animal manures. Manganese sulphate ($MnSO_4$), MnO, MnO_2, MnEDTA, $MnCO_3$, $MnCl_2$ and Mn-frits are used as Mn-fertilisers and are often applied in combination with N-fertilisers and/or P-fertilisers that induce a local acidification of the rhizosphere soil (Fageria, 2009). Foliar applications of soluble mineral fertilisers, usually $FeSO_4$, $ZnSO_4$, $CuSO_4$ or $MnSO_4$, are also used as a temporary or emergency measure to correct micronutrient deficiencies in crops (Gupta, 1979; Loneragan, 1997; Rengel et al., 1999; Cakmak, 2002, 2004, 2008; Fang et al., 2008; Fageria et al., 2009). The trace amounts of Ni required by plants can be supplied as $NiCl^2$, $Ni(NO_3)_2$ or NiO. Additional sources of Fe, Zn, Cu and Mn for crop nutrition include crop residues and organic manures, which not only provide phytoavailable sources of these elements but also improve soil structure and water relations.

As an additional or alternative strategy to the application of mineral fertilisers, the yields of crops sensitive to deficiencies in Fe, Zn, Cu or Mn can be increased either by intercropping with plants that are better able to access and mobilise these elements or by including such plants in crop rotations (Rengel et al., 1999; Jolley et al., 2004; Naeve, 2006; Graham et al., 2007; Inal et al., 2007). Soil microorganisms can also be exploited to increase the phytoavailability of these elements for crop production (see Chapter 3; Rengel et al., 1999; Barea et al., 2005; Morgan et al., 2005; Lynch, 2007; Smith and Read, 2007; White and Broadley, 2009). Several studies have found that both mycorrhizal associations (Kothari et al., 1991; Caris et al., 1998; Rengel et al., 1999; Harrier and Watson, 2003; Cavagnaro, 2008) and inoculants of growth-promoting bacteria (Rengel, 2001; Whiting et al., 2001; Barea et al., 2005) can increase tissue Fe, Zn and Cu concentrations and yields of crop plants. Manganese occurs in more than one oxidation state (Mn^{4+}, Mn^{3+}, Mn^{2+}) and the redox potential of the soil solution affects the solubility and availability of Mn markedly (Mengel et al., 2001). Thus, microbes that oxidise Mn can have a great influence on its phytoavailability (Nogueira et al., 2007).

6.6 Genetic strategies to improve the efficiency of acquisition and physiological use of essential cationic elements

6.6.1 Plant traits improving the acquisition of essential cationic elements

Plant growth can be restricted by the lack of any essential mineral element. Although a lack of Ca or Ni rarely limits crop production, deficiencies of K, Mg, Fe, Zn, Cu and Mn can occur on agricultural soils worldwide (Pretty and Stangel, 1985; Fageria, 2009; White and Broadley, 2009). Where there is an absolute lack of a mineral element in the soil, it must be supplied directly to crops to increase production and the objective of plant breeding is to maximise the uptake and physiological utilisation of the mineral element(s) supplied as fertiliser. In situations where sufficient mineral element is present in the soil but crop production is constrained by its phytoavailability, plant breeding seeks, in addition, to produce genotypes with a greater ability to mobilise the mineral element from its recalcitrant forms in the soil.

The delivery of most cationic mineral elements to the root surface is determined largely by transpiration-driven mass flow of the soil solution (see Sections 6.1 and 6.2). Nevertheless, their acquisition is enhanced by (1) the production of an extensive root system in unexplored volumes of soil, which reduces the path of water movement through the soil and increases the surface area available for cation uptake, and (2) the modification of soil chemistry and biology by root exudates, which increases the concentrations of cationic mineral elements in the soil solution. For cationic elements that are concentrated at the soil surface, such as Mn, Cu and Ni in some soils, a topsoil foraging phenotype can be beneficial. However, for cationic elements that are present at high concentrations in the soil solution, such as K, Ca and Mg, or elements that are distributed evenly within the soil profile, such as Fe and Zn, proliferation of roots throughout the soil volume is most appropriate. For elements that are taken up predominantly at the tip of roots, such as Ca and in some plant species Fe, the number of root apices in moist soil is of critical importance. For the latter elements, extensive root branching and the generation and maintenance of root tips in moist soil are advantageous.

6.6.2 Differences between plant species in their nutritional requirements

Plant species differ in their demand for essential mineral elements (see Section 6.2). For example, the nutritional demand for Ca is less in grasses and cereals than in other plant species and is generally higher in plants native to calcareouos soils (Broadley et al., 2003; White and Broadley, 2003). Similarly, the nutritional demand for Mg is less in grasses and cereals than in other plant species, and members of the Caryophyllales appear to have a greater demand for Mg than other eudicot species (Broadley et al., 2004, 2008). The demand for K is high in all plant species, but leaf K concentrations expressed on a fresh weight basis are often highest in members of the Poales, Ranunculales and Rosales (Broadley et al., 2004; Watanabe et al., 2007). By contrast, the demand for essential micronutrients is low and, although some higher level phylogenetic trends have been reported in leaf concentrations of Ni, Zn and Cu (Broadley et al., 2001, 2004, 2007; Watanabe et al., 2007), and with the notable exception of the rare hyperaccumulator plants, these may have little practical consequence for plant mineral nutrition. There is also considerable variation both between and within plant species in their ability to acquire and utilise essential cationic elements for growth and seed production and, consequently, their ability to grow on soils with low phytoavailability of essential cationic elements (Broadley et al., 2007; Fageria, 2009; White and Broadley, 2009). There is also variation in the ability of plants to tolerate mineral toxicities arising in acidic, alkaline, saline, sodic and calcareous soils (Fageria, 2009). Variation in these characteristics can be used to develop novel crops for infertile and/or hostile soils.

6.6.3 Differences between and within plant species in the acquisition and utilisation of essential cationic elements

Several definitions are available in the literature relating to the efficiency of use of mineral elements in agriculture (White et al., 2005). Here, the efficiency of agronomic use of an essential mineral element (MUE) is defined as the dry matter yield of a crop per unit of a mineral element available to it (g DM g^{-1} M_a). This is numerically equal to the product of crop content (M) per unit mineral element available (g M g^{-1} M_a), which is referred to as a crop's uptake efficiency (MUpE), and crop yield per unit content (g DM g^{-1}M), which is referred to as a plant's utilisation efficiency (MUtE).

The ability to acquire and utilise K^+ is fundamental to reducing the amount of K-fertiliser used in agriculture. In general, grasses and cereals often achieve their growth potential with a lower K^+ supply than legumes, vegetable or tuber crops and also compete best in K-limited ecosystems (Pretty and Stangel, 1985; Mengel et al., 2001; Fageria, 2009). Some arable crops show marked responses to K-fertiliser. These include potatoes, sugar beet and some vegetables. Differences in response to K-fertiliser and agronomic KUE have also been demonstrated among genotypes of several crop plants (Trehan, 2005; Fageria, 2009). Variation in K uptake efficiency (KUpE) has been observed among genotypes of many plant species (Trehan, 2005; Rengel and Damon, 2008; Fageria, 2009; White et al., 2010). Greater KUpE has been attributed by various authors to (1) increased K^+ uptake capacity of root cells, which reduces the K^+ concentration in the rhizosphere solution and increases diffusional K^+ fluxes to the root surface; (2) release of more non-exchangeable K^+ by root exudates, which increases the K^+ concentration in the soil solution; (3) proliferation of roots into the soil volume, which not only increases the area for K^+ uptake but also reduces the distance required for K^+ diffusion and water flow; and (4) higher transpiration rates, which accelerate the mass flow of the soil solution to the root surface (Jungk and Claassen, 1997; Baligar et al., 2001; Høgh-Jensen and Pedersen, 2003; Trehan, 2005; Rengel and Damon, 2008; Fageria, 2009; White et al., 2010). Within-species genetic variation in tissue K concentration and KUtE has also been reported (Baligar et al., 2001; Rengel and Damon, 2008; Fageria, 2009; White et al., 2010). Greater KUtE has been attributed to better K^+ redistribution within the plant, from older tissues to younger tissues with immediate K requirements, and/or a better ability to maintain appropriate cytoplasmic K^+ concentrations, either by anatomical adaptations or by the substitution of K^+ for other solutes in the vacuole (Rengel and Damon, 2008; White et al., 2010).

In most environments, the concentrations of Ca^{2+} and Mg^{2+} in the soil solution are high enough for sufficient Ca and Mg to be delivered to the root system by transpiration-driven mass flow of the soil solution for adequate plant nutrition (Wilkinson et al., 1990; Barber, 1995; White and Broadley, 2003 and 2009). However, since human dietary deficiencies of both Ca and Mg are prevalent in both developing and developed countries (Thacher et al., 2006; Broadley and White, 2010), efforts to increase Ca and Mg concentrations in edible portions through plant breeding are currently being pursued (White and Broadley, 2009). Considerable variation has been observed among genotypes of several species in their growth responses to Ca^{2+} and Mg^{2+} in the soil solution and also in their tissue Ca and Mg concentrations, which often reflect differences in the acquisition and accumulation of these elements (White and Broadley 2003, 2009; Havlin et al., 2005; Broadley et al., 2008). For example, calcifuge ecotypes achieve their growth potential at lower Ca^{2+} concentrations in the soil solution than calcicole ecotypes and are also generally more tolerant of acid soils (White and Broadley, 2003). Similarly, crops that are adapted to acid soils generally have lower Ca requirements (Fageria, 2009).

The total concentrations of Fe, Zn, Cu and Mn in most soils would be sufficient for crop nutrition if these elements were available in the soil solution (see Section 6.5). Hence, there is considerable interest in breeding crops that produce higher yields on previously infertile soils through greater acquisition and/or better physiological utilisation of these elements (Graham et al., 2007; Lynch, 2007; White and Broadley, 2009). The acquisition of mineral elements with restricted mobility in the soil, such as Fe, Zn, Cu and Mn can be improved by (1) investing more biomass in the root system; (2) producing a greater number and more even spread of roots; (3) developing a more extensive root system, with longer, thinner roots with more root hairs; and (4) proliferating lateral roots in mineral-rich patches (White et al., 2005; Lynch, 2007; White and Broadley, 2009). In addition, the exudation of protons, phytosiderophores and organic acids by roots, which displace cations from their binding sites in the soil, and the secretion of enzymes capable of degrading organic compounds that chelate mineral cations, such as phytases, phosphatases and proteases, can also improve their phytoavailability and acquisition by plants (Welch, 1995; Rengel, 2001; Abadía et al., 2002; Alloway, 2004; Ismail et al., 2007; Broadley et al., 2007; Lynch, 2007; Uren, 2007; Degryse et al., 2008; Fageria, 2009; White and Broadley, 2009). Increasing Fe(III) reductase activity in tandem with Fe^{2+} transport capacity can increase Fe acquisition by Strategy I Plants, whilst greater production and exudation of phytosiderophores in tandem with an increased capacity for Fe-phytosiderophore uptake can increase Fe acquisition by Strategy II Plants (White and Broadley, 2009). There is considerable within-species genetic variation in root architecture, root exudation and root uptake capacity that might be harnessed to improve the acquisition of Fe, Zn, Cu and Mn from unfertilised soils (White et al., 2005; Lynch, 2007; Fageria, 2009). Genotypic differences have been observed both between and within plant species in their growth responses to Fe, Zn, Cu and Mn availability in the soil solution. Crops that are highly susceptible to Fe deficiency and respond well to the addition of Fe-fertilisers, include sorghum, soybean, common bean, proso millet and rice, whilst Fe-efficient crops include wheat, sunflower and amaranth (Fageria, 2009). Among cereals, maize responds more to the application of Zn-fertiliser than does millet, rice or wheat (Rashid and Fox, 1992). Oats, spinach, wheat, alfalfa and barley generally respond well to Cu-fertilisers, whilst beans, peas, brassica, rye, grass, potatoes and soybeans respond less well (Mengel et al., 2001). Oats are also extremely susceptible to Mn deficiency and respond well to Mn-fertiliser applications, whilst rye, which is efficient in Mn uptake and utilisation, responds less dramatically to the application of Mn-fertilisers (Welch et al., 1991; Havlin et al., 2005). There is also considerable within-species variation in the uptake of Fe, Zn, Cu and Mn, in tissue concentrations of these elements and in the physiological utilisation of these elements to produce yield (Graham, 1988; Chaney et al., 1989; Cakmak et al., 1998; Rengel, 1999; Hacisalihoglu and Kochian, 2003; Grusak and Cakmak, 2005; Jolley et al., 2006; Evans et al., 2007; Pfeiffer and McClafferty, 2007; Wu et al., 2007; Fageria, 2009; White and Broadley, 2009; Broadley et al., 2010).

6.7 Concluding remarks

Plants acquire eight essential mineral elements from the soil solution in their cationic form (Table 6.1). Three macronutrients, K, Ca and Mg, are required at tissue concentrations of several $mg\,g^{-1}$ DM and five micronutrients, Zn, Fe, Cu, Mn and Ni, are required at lower tissue concentrations. In addition, several other mineral elements present as cations in the soil solution are considered beneficial for the growth of various taxa under certain environmental conditions. These include Na, Co and Al. Under most agricultural conditions, cationic elements are taken up by all regions of the root through transport proteins located in the plasma membrane of epidermal or cortical cells and are delivered to the xylem via a symplastic pathway. Notable exceptions to this generalisation are Ca and, in some plant species, Fe, whose uptake is restricted to the root apex and whose movement to the xylem is predominantly through the apoplast. In addition, when a high concentration of a particular cation is present in the soil solution, a substantial proportion reaching the shoot might reach the xylem though an apoplastic pathway. Many genes encoding transport proteins catalysing the influx of essential cationic elements to the root symplast and their delivery to the xylem have been identified (Figure 6.3).

Most cationic mineral elements are delivered to the root surface by transpiration-driven mass flow of the soil solution (Sections 6.1 and 6.2). Thus, their absolute concentrations in the soil solution are critical for adequate plant nutrition. The concentrations of cations in the soil solution are determined by the chemistry of the parent rocks, the processes of weathering, agronomic practices and vegetation processes (Sections 6.4, 6.5 and 6.6). Appropriate management strategies, such as the manipulation of soil pH or the application of mineral fertilisers, crop residues or animal manures, can increase the acquisition of cationic mineral elements by plants. In acid soils, which restrict crop production in many areas of the world, the application of dolomitic lime is recommended to raise soil pH, alleviate Al and Mn toxicities and prevent Ca and Mg deficiencies. The pH of alkaline calcareous soils, where crop production is limited by high pH, Ca^{2+} and bicarbonate concentrations, can be lowered by the addition of S, Fe sulphates or aluminium sulphate and fertilisers containing urea, ammonium or phosphate, although the effects of these amendments are often short lived. Saline soils can be remediated by leaching soluble salts from the soil profile by flushing with freshwater, and sodic soils can be remediated through the application of Ca^{2+}, generally as gypsum, followed by flushing with freshwater (see Chapter 7).

Where crop production is restricted by low phytoavailability of Fe, Zn, Cu or Mn, these elements can be supplied as soil or foliar fertilisers (Section 6.5). Alternatively, genotypes that produce higher yields on infertile soils, whether through greater acquisition and/or better physiological utilisation of these elements, can be grown (Section 6.6). Genotypes with greater acquisition of Fe, Zn, Cu and Mn often possess an extensive root system that exploits the spatial and temporal variations in mineral phytoavailability, and their roots release protons, phytosiderophores, organic acids and diverse enzymes into the rhizosphere to increase the concentrations of Fe, Zn, Cu and Mn in the soil solution. The cultivation of companion plants or soil microorganisms that release protons, phytosiderophores, organic acids and hydrolytic enzymes into the rhizosphere can also be exploited to increase the phytoavailability of these elements for crop production.

References

Abadía, J., López-Millán, A.-F., Rombolà, A., and Abadía, A. (2002) Organic acids and Fe deficiency: a review. *Plant and Soil* **241**, 75–86.

Alloway, B.J. (1995) *Heavy Metals in Soils*, 2nd edn. Blackie Academic & Professional, London.

Alloway, B.J. (2004) *Zinc in Soils and Crop Nutrition*. International Zinc Association, Brussels, Belgium.

Almås, Å.R., Lofts, S., Mulder, J., and Tipping, E. (2007) Solubility of major cations and Cu, Zn and Cd in soil extracts of some contaminated agricultural soils near a zinc smelter in Norway: modelling with a multisurface extension of WHAM. *European Journal of Soil Science* **58**, 1074–1086.

Aniku, J.R.F. and Singer, M. (1990) Pedogenic iron oxide trends in a marine terrace chronosequence. *Soil Science Society of America Journal* **54**, 147–152.

Avery, B.W. (1990) *Soils of the British Isles*. C.A.B. International, Wallingford, UK.

Bache, B.W. (1970) Determination of pH, lime potential and aluminium hydroxide potential in acid soils. *Journal of Soil Science* **21**, 28–37.

Bacon, J.R., Jones, K.C., McGrath, S.P., and Johnston, A.E. (1996) Isotopic character of lead deposited from the atmosphere at a grassland site in the United Kingdom since 1860. *Environmental Science and Technology* **30**, 2511–2518.

Bain, D.C., Mellor, A., Robertson-Rintoul, M.S.E., and Buckland, S.T. (1993) Variations in weathering processes and rates with time in a chronosequence of soils from Glen Feshie, Scotland. *Geoderma* **57**, 275–293.

Baker, A.J.M. and Brooks, R.R. (1989) Terrestrial higher plants which hyperaccumulate metallic elements: a review of their distribution, ecology and phytochemistry. *Biorecovery* **1**, 81–126.

Baker, A.J.M., McGrath, S.P., Reeves, R.D., and Smith, J.A.C. (2000) Metal hyperaccumulator plants: a review of the ecology and physiology of a biological resource for phytoremediation of metal polluted soils. In: *Phytoremediation of Contaminated Soil and Water* (eds N. Terry and G.S. Banuelos), pp. 85–107. CRC Press, Boca Raton, Florida.

Baligar, V.C., Fageria, N.K., and He, Z.L. (2001) Nutrient use efficiency in plants. *Communications in Soil Science and Plant Analysis* **32**, 921–950.

Barber, S.A. (1995) *Soil Nutrient Bioavailability*, 2nd edn. John Wiley & Sons, Inc., New York.

Barceló, J. and Poschenrieder, C. (2002) Fast root growth responses, root exudates, and internal detoxification as clues to the mechanisms of aluminium toxicity and resistance: a review. *Environmental and Experimental Botany* **48**, 75–92.

Barea, J.M., Pozo, M.J., Azcón, R., and Azcón-Aguilar, C. (2005) Microbial co-operation in the rhizosphere. *Journal of Experimental Botany* **56**, 1761–1778.

Bower, C.A. (1959) Cation-exchange equilibria in soils affected by sodium salts. *Soil Science* **88**, 32–35.

Brady, N.C. and Weill, R.R. (2002) *The Nature and Properties of Soil*, 13th edn. Longman, Prentice-Hall, London.

Broadley, M.R. and White, P.J. (2010) Eats roots and leaves. Can edible horticultural produce address dietary calcium, magnesium and potassium deficiencies? *Proceedings of the Nutrition Society* **69**, 601–612.

Broadley, M.R., Willey, N.J., Wilkins, J.C., Baker, A.J.M., Mead, A., and White, P.J. (2001) Phylogenetic variation in heavy metal accumulation in angiosperms. *New Phytologist* **152**, 9–27.

Broadley, M.R., Bowen, H.C., Cotterill, H.L., Hammond, J.P., Meacham, M.C., Mead, A., and White, P.J. (2003) Variation in the shoot calcium content of angiosperms. *Journal of Experimental Botany* **54**, 1431–1446.

Broadley, M.R., Bowen, H.C., Cotterill, H.L., Hammond, J.P., Meacham, M.C., Mead, A., and White, P.J. (2004) Phylogenetic variation in the shoot mineral concentration of angiosperms. *Journal of Experimental Botany* **55**, 321–336.

Broadley, M.R., White, P.J., Hammond, J.P., Zelko, I., and Lux A. (2007) Zinc in plants. *New Phytologist* **173**, 677–702.

Broadley, M.R., Hammond, J.P., King, G.J., Astley, D., Bowen, H.C., Meacham, M.C., Mead, A, Pink, D.A.C., Teakle, G.R., Hayden, R.M., Spracklen, W.P., and White, P.J. (2008) Shoot calcium and magnesium concentrations differ between subtaxa, are highly heritable, and associate with potentially pleiotropic loci in *Brassica oleracea*. *Plant Physiology* **146**, 1707–1720.

Broadley, M.R., Ó Lochlainn, S., Hammond, J.P., Bowen, H.C., Çakmak, I., Eker, S., Erdem, H., King, G.J., and White, P.J. (2010) Shoot zinc (Zn) concentration varies widely within *Brassica oleracea* L. and is affected by soil Zn and phosphorus (P). *Journal of Horticultural Science and Biotechnology* **85**, 375–380.

Brown, P.H., Welch, R.M., and Cary, E.E. (1987) Nickel: A micronutrient essential for higher plants. *Plant Physiology* **85**, 801–803.

Burkhead, J.L, Reynolds, K.A.G, Abdel-Ghany, S.E., Cohu, C.M., and Pilon, M. (2009) Copper homeostasis. *New Phytologist* **182**, 799–816.

Cakmak, I. (2002) Plant nutrition research: priorities to meet human needs for food in sustainable ways. *Plant and Soil* **247**, 3–24.

Cakmak, I. (2004) *Proceedings of the International Fertiliser Society 552. Identification and Correction of Widespread Zinc Deficiency in Turkey - a Success Story (a NATO-Science for Stability Project).* International Fertiliser Society, York, UK.

Cakmak, I. (2008) Enrichment of cereal grains with zinc: agronomic or genetic biofortification? *Plant and Soil* **302**, 1–17.

Cakmak, I., Torun, B., Erenolgu, L., Ozturk, H., Marschner, H., Kalayci, M., Ekiz, H., and Yilmaz, A. (1998) Morphological and physiological differences in the response of cereals to zinc deficiency. *Euphytica* **100**, 349–357.

Cancès, B., Ponthieu, M., Castrec-Rouelle M., Aubry, E., and Benedetti, M.F. (2003) Metal ions speciation in a soil and its solution: experimental data and model results. *Geoderma* **113**, 341–355.

Caris, C., Hordt, W., Hawkins, H.J., Romheld, V., and George E. (1998) Studies of iron transport by arbuscular mycorrhizal hyphae from soil to peanut and sorghum plants. *Mycorrhiza* **8**, 35–39.

Carrado, K.A. (2004) Introduction: clay structure, surface acidity and catalysis. In: *Handbook of Layered Materials* (eds S.M. Auerbach, K.A. Carrado and P.K. Dutta), pp. 1–38. Marcell Dekker Inc., New York.

Cavagnaro, T.R. (2008) The role of arbuscular mycorrhizas in improving plant zinc nutrition under low soil zinc concentrations: a review. *Plant and Soil* **304**, 315–325.

Chadwick, O.A., Derry, L.A., Vitousek, P.M., Huebert, B.J., and Hedin L.O. (1999) Changing sources of nutrients during four million years of ecosystem development. *Nature* **397**, 491–497.

Chaney, R.L. (1993) Zinc phytotoxicity. In: *Zinc in Soil and Plants* (ed A.D. Robson), pp. 135–150. Kluwer Academic Publishers, Dordrecht, the Netherlands.

Chaney, R.L., Bell, P.F., and Coulombe, B.A. (1989) Screening strategies for improved nutrient uptake and use by plants. *HortScience* **24**, 565–572.

Clemens, S. (2010) Zn - a versatile player in plant cell biology. In: *Plant Cell Monographs 17, Cell Biology of Metals and Nutrients* (eds R. Hell and R.-R. Mendel), pp. 281–298. Springer, Dordrecht, the Netherlands.

Colburn, P. and Thornton, I. (1978) Lead pollution in agricultural soils. *European Journal of Soil Science* **29**, 513–526.

Degryse, F., Verma, V.K., and Smolders, E. (2008) Mobilization of Cu and Zn by root exudates of dicotyledonous plants in resin-buffered solutions and in soil. *Plant and Soil* **306**, 69–84.

Delhaize, E., Gruber, B.D., and Ryan, P.R. (2007) The roles of organic anion permeases in aluminium resistance and mineral nutrition. *FEBS Letters* **581**, 2255–2262.

Department for Environment, Food and Rural Affairs, UK [Defra] (2010) *Fertiliser Manual (RB209)*, 8th edn. The Stationery Office, Norwich, UK.

Dodd, A.N., Kudla, J., and Sanders, D. (2010) The language of calcium signalling. *Annual Review of Plant Biology* **61**, 593–620.

Draycott P., and Allison M. (1998) *Proceedings of the International Fertiliser Society 412. Magnesium Fertilisers in Soil and Plants: Comparisons and Usage*. International Fertiliser Society, York, UK.

Eaton, F.M., Harding, R.B., and Ganje, J.J. (1960) Soil solution extractions at tenth bar moisture percentages. *Soil Science* **90**, 253–258.

Eckert, D.J. (1987) Soil test interpretations: basic cation saturation ratios and sufficiency levels. In: *Soil Testing: Sampling, Correlation, Calibration, and Interpretation* (ed J.R. Brown), pp. 53–64. SSSA, Madison, Wisconsin.

Evangelou, V.P., Wang, J., and Phillips, R.E. (1994) New developments and perspectives on soil potassium quantity intensity relationship. *Advances in Agronomy* **52**, 173–227.

Evans, I., Solberg, E., and Huber, D.M. (2007) Copper and plant disease. In: *Mineral Nutrition and Plant Disease* (eds L.E. Datnoff, W.H. Elmer and D.M. Huber), pp.177–188. The American Phytopathological Society, St Paul, Minnesota. .

Fageria, N.K. (2009) *The Use of Nutrients in Crop Plants*. CRC Press, Boca Raton, Florida.

Fageria, N.K., Barbosa, M.P., Moreira, A., and Guimaraes, C.M. (2009) Foliar fertilization of crop plants. *Journal of Plant Nutrition* **32**, 1044–1064.

Fageria, N.K., Baligar, V.C., and Jones, C.A. (2011) *Growth and Mineral Nutrition of Field Crops*. CRC Press, Boca Raton, Florida.

Fang, Y., Wang, L., Xin, Z., Zhao, L.Y., An, X.X., and Hu, Q.H. (2008) Effect of foliar application of zinc, selenium, and iron fertilizers on nutrients concentration and yield of rice grain in China. *Journal of Agricultural and Food Chemistry* **56**, 2079–2084.

Fisher, G.B. and Ryan, P.C. (2006) The smectite-to-disordered kaolinite transition in a tropical soil chronosequence, Pacific coast, Costa Rica. *Clays and Clay Minerals* **54**, 571–586.

Follett, R.H., Murphy, L.S., and Donahue, R.L. (1981) *Fertilizers and Soil Amendments*. Prentice-Hall, Englewood Cliffs, NJ.

Food and Agriculture Organization of the United Nations [FAO] 2010. *Land Resources*. http://www.fao.org/nr/land/databasesinformation-systems/en/ [last accessed 31st March 2010]

Frossard, E., Bucher, M., Mächler, F., Mozafar, A., and Hurrell, R. (2000) Potential for increasing the content and bioavailability of Fe, Zn and Ca in plants for human nutrition. *Journal of the Science of Food and Agriculture* **80**, 861–879.

Ge, Y., Murray, P., and Hendershot, W.H. (2000) Trace metal speciation and bioavailability in urban soils. *Environmental Pollution* **107**, 137–144.

Goulding, K.W.T. and Annis, B. (1998) *Proceedings of the International Fertiliser Society 410. Lime, Liming and the Management of Soil Acidity*. The International Fertiliser Society, York, UK.

Graham, R.D. (1988) Genotype differences in tolerance to manganese deficiency. In: *Manganese in Soils and Plants* (eds R.D. Graham, R.J. Hannam and N.C. Uren), pp. 261–276. Kluwer Academic, Dordrecht, the Netherlands.

Graham, R., Senadhira, D., Beebe, S., Iglesias, C., and Monasterio, I. (1999) Breeding for micronutrient density in edible portions of staple food crops: conventional approaches. *Field Crops Research* **60**, 57–80.

Graham, R.D., Welch, R.M., Saunders, D.A., Ortiz-Monasterio, I., Bouis, H.E., Bonierbale, M., de Haan, S., Burgos, G., Thiele, G., Liria, R., Meisner, C.A., Beebe, S.E., Potts, M.J., Kadian, M., Hobbs, P.R., Gupta, R.K., and Twomlow, S. (2007) Nutritious subsistence food systems. *Advances in Agronomy* **92**, 1–74.

Greenwood, D.J. and Karpinets, T.V. (1997a) Dynamic model for the effects of the K-fertiliser on crop growth, K-uptake, and soil-K in arable cropping. 1. Description of the model. *Soil Use and Management* **13**, 178–183.

Greenwood, D.J. and Karpinets, T.V. (1997b) Dynamic model for the effects of the K-fertiliser on crop growth, K-uptake, and soil-K in arable cropping. 2. Field test of the model. *Soil Use and Management* **13**, 184–189.

Grotz, N. and Guerinot, M.L. (2006) Molecular aspects of Cu, Fe and Zn homeostasis in plants. *Biochimica et Biophysica Acta* **1763**, 595–608.

Grusak, M.A. and Cakmak, I. (2005) Methods to improve the crop-delivery of minerals to humans and livestock. In: *Plant Nutritional Genomics* (eds M.R. Broadley and P.J. White), pp. 265–286. Blackwell Publishing Ltd., Oxford, UK.

Guerinot, M.L. (2010) Iron. In: *Plant Cell Monographs 17, Cell Biology of Metals and Nutrients* (eds R. Hell and R.-R. Mendel), pp. 75–94. Springer, Dordrecht, the Netherlands.

Gupta, U.C. (1979) Copper in agricultural crops. In: *Copper in the Environment* (ed J.O. Nriagu), pp. 255–287. John Wiley & Sons, Inc., New York.

Hacisalihoglu, G. and Kochian, L.V. (2003) How do some plants tolerate low levels of soil zinc? Mechanisms of zinc efficiency in crop plants. *New Phytologist* **159**, 341–350.

Hänsch, R. and Mendel, R.R. (2009) Physiological functions of mineral micronutrients (Cu, Zn, Mn, Fe, Ni, Mo, B, Cl). *Current Opinion in Plant Biology* **12**, 259–266.

Harrier, L.A. and Watson, C.A. (2003) The role of arbuscular mycorrhizal fungi in sustainable cropping systems. *Advances in Agronomy* **20**, 185–225.

Harris, D., Rashid, A., Miraj, G., Arif, M., and Shah H. (2007) 'On-farm' seed priming with zinc sulphate solution - A cost-effective way to increase the maize yields of resource-poor farmers. *Field Crops Research* **102**, 119–127.

Havlin, J.L., Tisdale, S.L., and Nelson, W.L. (2005) *Soil Fertility and Fertilizers. An Introduction to Nutrient Management*, 7th edn. Pearson Prentice Hall, New York.

He, Z.L., Yang, X.E., and Stoffella, P.J. (2005) Trace elements in agroecosystems and impacts on the environment. *Journal of Trace Elements in Medicine and Biology* **19**, 125–140.

Ho, L. and White, P.J. (2005) A cellular hypothesis for the induction of blossom end rot in tomato fruit. *Annals of Botany* **95**, 571–581.

Hodgman, C.D. (1953) Editor. *Handbook of Chemistry and Physics*, 35th ed. Chemical Rubber Publishing Company, Cleveland, Ohio.

Høgh-Jensen, H., and Pedersen, M.B. (2003) Morphological plasticity by crop plants and their potassium use efficiency. *Journal of Plant Nutrition* **26**, 969–984.

Horst, W.J. (1988) The physiology of manganese toxicity. In: *Manganese in Soils and Plants* (eds R.D. Graham, R.J. Hannam and N.C. Uren), pp. 75–88. Kluwer Academic Publishers, Dordrecht, the Netherlands.

Huber, C. and Montserrat, F. (2002) Computer modelling of trace metal ion speciation: practical implementation of a linear continuous function for complexation by natural organic matter. *Computers and Geosciences* **28**, 587–596.

Inal, A., Gunes, A., Zhang, F., and Cakmak, I. (2007) Peanut/maize intercropping induced changes in rhizosphere and nutrient concentrations in shoots. *Plant Physiology and Biochemistry* **45**, 350–356.

Ismail, A.M., Heuer, S., Thomson, M.J., and Wissuwa M. (2007) Genetic and genomic approaches to develop rice germplasm for problem soils. *Plant and Soil* **65**, 547–570.

Jackson, M.L., Levelt, T.W.M., Syers, J.K., Rex, R.W., Clayton, R.N., Sherman, G.D., and Uehara, G.R. (1971) Geomorphological relationships of tropospherically derived quartz in the soils of the Hawaiian Islands. *Soil Science Society of America Proceedings* **35**, 515–525.

Jansen, S., Broadley, M.R., Robbrecht. E., and Smets E. (2002) Aluminum hyperaccumulation in angiosperms: A review of its phylogenetic significance. *Botanical Review* **68**, 235–269.

Jolley, V.D., Hansen, N.C., and Shiffler, A.K. (2004) Nutritional and management related interactions with iron-deficiency stress response mechanisms. *Soil Science and Plant Nutrition* **50**, 973–981.

Jolley, V.D., Cook, K.A., Hansen, N.C., and Stevens, W.B. (2006) Plant physiological responses for genetic evaluation of iron efficiency in Strategy I and Strategy II plants: A review. *Journal of Plant Nutrition* **19**, 1241–1255.

Jungk, A. and Claassen, N. (1997) Ion diffusion in the soil-root system. *Advances in Agronomy* **61**, 53–110.

Karley, A.J. and White, P.J. (2009) Moving cationic minerals to edible tissues: potassium, magnesium, calcium. *Current Opinion in Plant Biology* **12**, 291–298.

Karpinets, T.V. and Greenwood, D.J. (2003) Potassium dynamics. In: *Handbook of Processing and Modeling of Soil Plant System*, (eds D.K. Benbi and R. Nieder), pp. 525–559. Howarth Press, New York.

Kinraide, T.B. (1991) Identity of the rhizotoxic aluminium species. *Plant and Soil* **134**, 167–178.

Kirkby, E.A. and Johnson, A.E. (2008) Soil and fertilizer phosphorus in relation to crop nutrition. In: *The Ecophysiology of Plant-Phosphorus Interactions* (eds P.J. White and J.P. Hammond), pp. 177–223. Springer, Dordrecht, the Netherlands.

Knox, J. (1912) *Physico-Chemical Calculations*. Methuen, London.

Kochian, L., Hoekenga, O.A., and Piñeros, M.A. (2004) How do crop plants tolerate acid soils? Mechanisms of aluminum tolerance and phosphorous efficiency. *Annual Review of Plant Biology* **55**, 459–493.

Kopittke, P.M. and Menzies, N.W. (2007) A review of the use of the basic cation saturation ratio and the "ideal" soil. *Soil Science Society of America Journal* **71**, 259–265.

Kothari, S.K., Marschner, H., and Romheld, V. (1991) Contribution of the VA mycorrhizal hyphae in acquisition of phosphorus and zinc by maize grown in a calcareous soil. *Plant and Soil* **131**, 177–185.

Kramer, U. and Clemens, S. (2005) Function and homeostasis of zinc, copper, and nickel in plants. *Topics in Current Genetics* **14**, 215–271.

Lægreid, M., Bøckman, O.C., and Kaarstad, O. (1999) *Agriculture, Fertilizers and the Environment*. CABI Publishing, Wallingford, UK.

Lee, J.A. (1999) The calcicole-calcifuge problem revisited. *Advances in Botanical Research* **29**, 1–30.

Lee, J.A. and J. H. Tallis JH (1973) Regional and historical aspects of lead pollution in Britain. *Nature* **245**, 216–218.

Leigh, R.A. and Wyn Jones, R.G. (1984) A hypothesis relating critical potassium concentrations for growth to the distribution and functions of this ion in the plant cell. *New Phytologist* **97**, 1–13.

Lindsay, W.L. (1979) *Chemical Equilibria in Soils*. John Wiley & Sons, Inc., New York.

Lindsay, W.L. and Schwab, A.P. (1982) The chemistry of iron in soils and its availability to plants. *Journal of Plant Nutrition* **5**, 321–340.

Loneragan, J.F. (1997) Plant nutrition in the 20th and perspectives for the 21st century. *Plant and Soil* **196**, 163–174.

Lux, A., Sottníková, A., Opatrná, J., and Greger M (2004) Differences in structure of adventitious roots in Salix clones with contrasting characteristics of cadmium accumulation and sensitivity. *Physiologia Plantarum* **120**, 537–545.

Lynch, J.P. (2007) Roots of the second green revolution. *Australian Journal of Botany* **55**, 493–512.

Ma, J.F., Ryan, P.R., and Delhaize, E. (2001) Aluminium tolerance in plants and the complexing role of organic acids. *Trends in Plant Science* **6**, 273–278.

Maathuis, F.J.M. (2009) Physiological functions of mineral macronutrients. *Current Opinion in Plant Biology* **12**, 250–258.

McAinsh, M.R. and Pittman, J.K. (2009) Shaping the calcium signature. *New Phytologist* **181**, 275–294.

McAleese, D.M. and McConaghy, S. (1957) Studies on the basaltic soils of Northern Ireland. II. Contributions from the sand, silt and clay separates to cation-exchange properties. *Journal of Soil Science* **8**, 135–140.

Mengel, K., Kirkby, E.A., Kosegarten, H., and Appel T. (2001) *Principles of Plant Nutrition*. Kluwer Academic, Dordrecht, the Netherlands.

Morgan, J.A.W., Bending, G.D., and White, P.J. (2005) Biological costs and benefits to plant-microbe interactions in the rhizosphere. *Journal of Experimental Botany* **56**, 1729–1739.

Mortvedt, J.J. (1991) Correcting iron deficiencies in annual and perennial plants. *Present technologies and future prospects. Plant and Soil* **130**, 273–279.

Mott, C.J.B. (1988a) The inorganic components of soil In: *Russell's Soil Conditions and Plant Growth*, 11th edn (ed A. Wild), pp. 213–238. New York Longman Scientific and Technical, New York.

Mott, C.J.B. (1988b) Surface chemistry of soil particles. In: *Russell's Soil Conditions and Plant Growth* (ed A. Wild), 11th edn, pp. 239–281. New York Longman Scientific and Technical, New York.

Munns, R. and Tester, M. (2008) Mechanisms of salinity tolerance. *Annual Review of Plant Biology* **59**, 651–81.

Naeve, S.L. (2006) Iron deficiency chlorosis in soybean: Soybean seeding rate and companion crop effects. *Agronomy Journal* **98**, 1575–1581.

Nogueira, M.A., Nehls, U., Hampp, R., Poralla, K., and Cardoso, E.J.B.N. (2007) Mycorrhiza and soil bacteria influence extractable iron and manganese in soil and uptake by soybean. *Plant and Soil* **298**, 273–284.

Palit, S., Sharma, A., and Talukder, G. (1994) Effect of cobalt on plants. *Botanical Review* **60**, 149–181.

Parfit, R.L., Giltrap, D.J., and Whitton J.S. (1995) Contribution of organic matter and clay minerals to the cation exchange capacity of soils. *Communications in Soil Science and Plant Analysis* **26**, 1343–1355.

Parker, D.R., Kinraide, T.B., and Zelazny, L.W. (1988) Aluminium speciation and phytotoxicity in dilute hydroxyl-aluminium solutions. *Soil Science Society of America Journal* **52**, 438–444.

Parker, D.R., Norvell, W.A., and Chaney, R.L. (1995) GEOCHEM-PC a chemical speciation program for IBM and compatible personal computers. In: *Soil Science Society of America Journal, Special Publication 42, Chemical Equilibrium Reaction Models* (eds R.H. Loeppert *et al.*), pp. 253–269. SSSA, Madison, Wisconsin.

Parkhurst, D.L. and Appelo, C.A.J. (1999) *User Guide to PHREEQC (Version 2) - A Program for Speciation, Batch Reaction, One Dimensional Transport and Inverse Geochemical Calculations*. US Geological Survey, Denver, Colorado.

Parsons, S.A. and Smith, J.A. (2008) Phosphorus removal and recovery from municipal wastewaters. *Elements* **4**, 109–112.

Percival, H.J., Speir, T.W., and Parshotam, A. (1999) Soil solution chemistry of contrasting soils amended with heavy metals. *Australian Journal of Soil Research* **37**, 993–1004.

Pfeiffer, W.H. and McClafferty, B. (2007) HarvestPlus: breeding crops for better nutrition. *Crop Science* **47**, S88-S105.

Pilon, M., Cohu, C.M., Ravet, K., Abdel-Ghany, S.E., and Gaymard, F. (2009) Essential transition metal homeostasis in plants. *Current Opinion in Plant Biology* **12**, 347–357.

Pilon-Smits, E.A.H., Quinn, C.F., Tapken, W., Malagoli, M., and Schiavon, M. (2009) Physiological functions of beneficial elements. *Current Opinion in Plant Biology* **12**, 267–274.

Plett, D.C and Møller, I.S. (2010) Na^+ transport in glycophytic plants: what we know and would like to know. *Plant, Cell and Environment* **33**, 612–626.

Ponnamperuma, F.N., Tianco, E.M., and Loy, T.A.A. (1966) Ionic strengths of the solutions of flooded soils and other natural aqueous solutions from specific conductance. *Soil Science* **102**, 408–413.

Pretty, K.M. and Stangel, P.J. (1985) Current and future use of world potassium. In: *Potassium in Agriculture* (ed R.D. Munson), pp. 99–128. ASA, CSSA and SSSA, Madison, Wisconsin.

Puig, S. and Peñarrubia, L. (2009) Placing metal micronutrients in context: transport and distribution in plants. *Current Opinion in Plant Biology* **12**, 299–306.

Pusch, R. and Yong, R.N. (2005) *Microstructure of Smectite Clays and Engineering Performance*. Taylor and Francis, Abingdon, UK.

Qadir, M. and Schubert, S. (2002) Degradation processes and nutrient constraints in sodic soils. *Land Degradation and Development* **13**, 275–294.

Rashid, A. and Fox, R.L. (1992) Evaluating internal zinc requirements of grain crops by seed analysis. *Agronomy Journal* **84**, 469–474.

Reeves, R. (2006) Hyperaccumulation of trace elements by plants. In: *Phytoremediation of Metal-Contaminated Soils* (eds J.L. Morel, G. Echevarria and N. Goncharova), pp. 25–52. Springer, Dordrecht, the Netherlands.

Rengel, Z. (1999) Physiological mechanisms underlying differential nutrient efficiency of crop genotypes. In: *Mineral Nutrition of Crops: Fundamental Mechanisms and Implications* (ed Z. Rengel), pp. 227–265. Howarth Press, New York.

Rengel, Z. (2001) Genotypic differences in micronutrient use efficiency in crops. *Communications in Soil Science and Plant Analysis* **32**, 1163–1186.

Rengel, Z. and Damon, P.M. (2008) Crops and genotypes differ in efficiency of potassium uptake and use. *Physiologia Plantarum* **133**, 624–636.

Rengel, Z., Batten, G.D., and Crowley, D.E. (1999) Agronomic approaches for improving the micronutrient density in edible portions of field crops. *Field Crops Research* **60**, 27–40.

Römheld, V. and Kirkby, E.A. (2007) *Proceedings of the International Fertiliser Society 616. Magnesium functions in crop nutrition and yield*. International Fertiliser Society, York, UK.

Salmon, R.C. (1964) Cation exchange reactions. *Journal of Soil Science* **15**, 271–283.

Shaff, J., Schultz, B., Craft, E., Clark, R.T., and Kochian, L.V. (2009). Geochem-EZ: a chemical speciation program with greater power and flexibility. *Plant and Soil* **330**, 207–214.

Schmidt, W. (1999) Mechanisms and regulation of reduction-based iron uptake in plants. *New Phytologist* **141**, 1–26.

Schofield, R.K. (1947) A ratio law governing the equilibrium of cations in soil solution. *Proceeding 11th International Congress of Applied Chemistry* **3**, 257–261.

Schofield, R.K. and Taylor, A.W. (1955a) Measurements of the bases in soils. *Journal of Soil Science* **6**, 137–146.

Schofield, R.K. and Taylor, A.W. (1955b) The measurement of soil pH. *Soil Science America Proceedings* **19**, 164–167.

Shear, C.B. (1975) Calcium-related disorders of fruits and vegetables. *HortScience* **10**, 361–365.

Shuman, L.M. (1985) Fraction method for soil microelements. *Soil Science* **140**, 11–22.

Shuman, L.M. (1998) Micronutrient fertilizers. *Journal of Crop Production* **1**, 165–195.

Smith, S.E. and Read, D.J. (2007) *Mycorrhizal symbiosis*, 3rd edn. Elsevier, London.

Sposito, G. and Mattigod, S.V. (1980) *GEOCHEM: A Computer Program for the Calculation of Chemical Equilibria in Soil Solution and Other Natural Water Systems*. Kearney Foundation of Soil Science, University of California, Riverside, California.

Sumner, M.E. and Noble, A.D. (2003) Soil acidification: The world story. In: *Handbook of Soil Acidity* (ed Z. Rengel), pp. 1–28. Marcel Dekker, New York.

Syers, J.K. and Walker, T.W. (1969) Phosphorus transformations in a chronosequence of soils developed in wind-blown sand in New Zealand. *Journal of Soil Science* **20**, 57–64.

Taiz, L. and Zeigler, E. (2006) *Plant Physiology*, 4th edn. Sinauer Associates, Sunderland, Massachusetts.

Tan, K.H. (1998) *Principles of Soil Chemistry*, 3rd edn. Marcell Dekker, New York.

Tejada-Jiménez, M., Galvan, A., Fernández, E., and Llamas, A. (2009) Homeostasis of the micronutrients Ni, Mo and Cl with specific biochemical functions. *Current Opinion in Plant Biology* **12**, 358–363.

Thacher, T.D., Fischer, P.R., and Strand, M.A., and Pettifor, J.M. (2006) Nutritional rickets around the world: causes and future directions. *Annals of Tropical Paediatrics* **26**: 1–16.

Trehan, S.P. (2005) Nutrient management by exploiting genetic diversity of potato - a review. *Potato Journal* **32**, 1–15.

Troeh, F.R. and Thompson L.M. (2005) *Soil and Soil Fertility*, 6th revised edn. Blackwell Publishing Ltd., London.

Uehara G (2006) Chemical composition. In: *Encyclopedia of Soil Science*(ed R. Lal), 2nd edn, pp. 213–217. CRC Press, Boca Raton, Florida.

Unsworth, E.R., Warnken, K.W., and Zhang, H. *et al.* (2006) Model predictions of metal speciation in freshwaters compared to measurements by *in situ* techniques. *Environmental Science and Technology* **40**, 1942–1949.

Uren, N.C. (2007) Types, amounts, and possible functions of compounds released into the rhizosphere by soil-grown plants. *The Rhizosphere: Biochemistry and Organic Substances at the Soil-Plant Interface*(eds R. Pinton, Z. Varanini and P. Nannipieri), 2nd edn, pp. 1–22. CRC Press, Boca Raton, Florida.

Vlamis, J. (1953) Acid soil infertility as related to soil solution and solid-phase effects. *Soil Science* **75**, 383–394.

Von Uexküll, H.R. and Mutert, E. (1995) Global extent, development and economic impact of acid soils. *Plant and Soil* **171**, 1–15.

Walker, T.W. and Syers, J.K. (1976) The fate of phosphorus during pedogenesis. *Geoderma* **15**, 1–9.

Watanabe, T., Broadley, M.R., Jansen, S., White, P.J., Takada, J., Satake, K., Takamatsu, T., Tuah, S.J., and Osaki, M. (2007) Evolutionary control of leaf element composition in plants. *New Phytologist* **174**, 516–523.

Welch, R.M. (1995) Micronutrient nutrition of plants. *Critical Reviews in Plant Science* **14**, 49–82.

Welch, R.M., Allaway, W.H., House, W.A., and Kubota, J. (1991) Geographic distribution of trace element problems. In: *Micronutrients in Agriculture*(eds J.J. Mortvedt, F.R. Cox, L.M. Shuman and R.M. Welch), 2nd edn, pp. 31–57. Soil Science Society of America, Madison, Wisconsin.

White, P.J. (1999) The molecular mechanism of sodium influx to root cells. *Trends in Plant Science* **4**, 245–246.

White, P.J. (2001) The pathways of calcium movement to the xylem. *Journal of Experimental Botany* **52**, 891–899.

White, P.J. (2003) Ion transport. In: *Encyclopaedia of Applied Plant Sciences* (eds B. Thomas, D.J. Murphy and B.G. Murray), pp. 625–634. Academic Press, London.

White, P.J. (2012) Chapter 2. Ion uptake mechanisms of individual cells and roots: Short-distance transport. In: *Marschner's Mineral Nutrition of Higher Plants*(ed P. Marschner), 3rd edn, pp. 7–47. Academic Press, London.

White, P.J. and Broadley, M.R. (2003) Calcium in plants. *Annals of Botany* **92**, 487–511.

White, P.J. and Broadley, M.R. (2009) Biofortification of crops with seven mineral elements often lacking in human diets – iron, zinc, copper, calcium, magnesium, selenium and iodine. *New Phytologist* **182**, 49–84.

White, P.J. and Brown, P.H. (2010) Plant nutrition for sustainable development and global health. *Annals of Botany* **105**, 1073–1080.

White, P.J. and Karley, A.J. (2010) Potassium. In: *Plant Cell Monographs 17, Cell Biology of Metals and Nutrients* (eds R. Hell and R.-R. Mendel), pp. 199–224. Springer, Dordrecht, the Netherlands.

White, P.J., Marshall, J., and Smith, J.A.C. (1990) Substrate kinetics of the tonoplast H^+-translocating inorganic pyrophosphatase and its activation by free Mg^{2+}. *Plant Physiology* **93**, 1063–1070.

White, P.J., Whiting, S.N., Baker, A.J.M., and Broadley, M.R. (2002) Does zinc move apoplastically to the xylem in roots of *Thlaspi caerulescens*? *New Phytologist* **153**, 201–207.

White, P.J., Broadley, M.R., Greenwood, D.J., and Hammond, J.P. (2005) *Proceedings of the International Fertiliser Society 568. Genetic modifications to improve phosphorus acquisition by roots*. International Fertiliser Society, York, UK.

White, P.J., Hammond, J.P., King, G.J., Bowen, H.C., Hayden, R.M., Meacham, M.C., Spracklen, W.P., and Broadley, M.R. (2010) Genetic analysis of potassium use efficiency in *Brassica oleracea*. *Annals of Botany* **105**, 1199–1210.

Whiting, S.N., de Souza, M.P., and Terry, N. (2001) Rhizosphere bacteria mobilize Zn for hyperaccumulation by *Thlaspi caerulescens*. *Environmental Science and Technology* **35**, 3144–3150.

Whiting, S.N., Reeves, R.D., Richards, D.G., Johnson, M.S., Cooke, J.A., Malaisse, F., Paton, A., Smith, J.A.C., Angle, J.S., Chaney, R.L., Ginocchio, R., Jaffré, T., Johns, B., McIntyre, T., Purvis, O.W., Salt, D.E., Schat, H., Zhao, F.J., and Baker, A.J.M. (2005) Use of plants to manage sites contaminated with metals. In: *Plant Nutritional Genomics* (eds M.R. Broadley and P.J. White), pp. 287–315. Blackwell Publishing Ltd., Oxford, UK.

Wild, A. (1993) *Soils and the Environment. An Introduction*. Cambridge University Press, Cambridge, UK.

Wilkinson, S.R., Welch, R.M., Mayland, H.F. and Grunes, D.L. (1990) Magnesium in plants: Uptake, distribution, function, and utilization by man and animals. *Metal Ions in Biological Systems* **26**, 33–56.

Williams, L.E. and Pittman, J.K. (2010) Dissecting pathways involved in manganese homeostasis and stress in higher plant cells. In: *Plant Cell Monographs 17, Cell Biology of Metals and Nutrients* (eds R. Hell and R.-R. Mendel), pp. 95–117. Springer, Dordrecht, the Netherlands.

Wilson, M.J. (1999). The origin and formation of clay minerals in soils: past present and future perspectives. *Clay Minerals* **34**, 7–25.

Wu, J., Schat, H., Sun, R., Koornneef, M., Wang, X. and Aarts, M.G.M. (2007) Characterization of natural variation for zinc, iron and manganese accumulation and zinc exposure response in *Brassica rapa* L. *Plant and Soil* **291**, 167–180.

Zhang, K., Greenwood, D.J., White, P.J. and Burns, I.G. (2007) A dynamic model for the combined effects of N, P and K fertilizers on yield and mineral composition: description and experimental test. *Plant and Soil* **298**, 81–98.

7 Managing adverse soil chemical environments

Mark E. Hodson[1] and Erica Donner[2]

[1]Environment Department, University of York, York, UK
[2]Centre for Environmental Risk Assessment and Remediation, University of South Australia, Mawson Lakes, South Australia, Australia

7.1 Potentially toxic metals (PTMs)

All metals are potentially toxic. As the fifteenth to sixteenth century physician Paracelsus said 'the dose determines the response'. However, within studies of contaminated soils/ adverse environments, there is a clear perception of typical potentially toxic metals. These are often erroneously referred to as heavy metals, an ill-defined term (Duffus, 2002; Hodson, 2004) that is used as a synonym for nasty metals that have a toxic effect. In many respects, the phrase 'potentially toxic metal' is as unsatisfactory as 'heavy metal' since, as Paracelsus tells us, all metals, indeed all substances, can be toxic. However, the phrase does at least make the point that the metals in question though present need not be having a toxic effect. The majority of studies concerned with potentially toxic metals are concerned with Al, Cd, Cr, Cu, Hg, Ni, Pb and Zn, and examples of case studies in this section will focus on these elements.

7.1.1 Sources of PTMs

PTMs are naturally occurring elements and are therefore ubiquitous in the soil and wider environment. They cause a problem when present at concentrations which are elevated above critical levels. These critical levels are the subject of much research and indeed legislation and depend on a number of factors such as soil properties but also proposed land use.

Natural sources

Metals are present in rocks and, therefore, through natural pedogenic processes end up in soils. Concentrations of PTMs in 'typical' uncontaminated soils are listed in Table 7.1. Fuller listings are given in, for example, Kabata-Pendias (2000), Bowen (1979) and Alloway (2013a).

Soil Conditions and Plant Growth. Edited by Peter J. Gregory and Stephen Nortcliff.
© 2013 Blackwell Publishing Ltd. Published 2013 by Blackwell Publishing Ltd.

Table 7.1 Ranges of PTMs in 'typical' uncontaminated soils.

Element	Normal range in soil/mg kg^{-1}
Aluminium (Al)	10 000–300 000
Arsenic (As)	0.1–40
Cadmium (Cd)	0.01–20
Cobalt (Co)	0.5–65
Chromium (Cr)	5–1500
Copper (Cu)	2–250
Mercury (Hg)	0.01–0.5
Manganese (Mn)	20–10 000
Molybdenum (Mo)	0.1–40
Nickel (Ni)	2–750
Lead (Pb)	2–300
Thallium (Tl)	0.1–0.8
Zinc (Zn)	1–900

Source: From Bowen (1979).

Airborne dust and aerosols are also increasingly recognised as inputs to soil systems (e.g. Derry and Chadwick, 2007). Soils have naturally elevated concentrations of PTMs when developed on metal-enriched rocks, for example soils developed on ophiolite complexes and mineral deposits (Bowen, 1979; Alloway, 2013a; Kabata-Pendias, 2000) and when weathering processes have led to the concentration of specific elements. In the latter case, this is normally classically immobile elements such as Al and Fe in highly weathered tropical soils leading to the formation of laterites (e.g. Brimhall et al., 1991; Samouelian and Cornu, 2008; Fandeur et al., 2009).

Anthropogenic sources

Many textbooks have been written on the anthropogenic enrichment of metal concentrations in soils (e.g. Alloway, 2013a; Kabata-Pendias, 2000). In brief, metal enrichment can occur due to elevated concentrations of metals in soil amendments and/or accidental addition of metals to soils due to industrial activities and spillage. Classic examples of increased metal concentrations due to soil amendments are the occurrence of Cd in phosphate fertilisers (Smolders and Mertens, 2013) and elevated concentrations of PTMs in sewage sludge applied to land (Alloway, 2013b). This latter example, at least in Europe, is historic. Present-day sewage sludges that are licensed for land disposal have very low concentrations of metals.

7.1.2 Controls on mobility/availability of PTMs

A wide variety of factors control the mobility and availability of PTMs to plants in soils. Ultimately, mobility depends upon the PTMs being present in the soil solution and movement of that solution through the soil, whilst availability depends upon the PTMs being in solution and to a certain extent the speciation, that is the chemical form, of the PTM in the solution. In this section, we will briefly review the impacts of pH, adsorption, Eh, organic matter and speciation on mobility and availability.

pH is perhaps the most important variable governing mobility and availability. Metals, with the exceptions of those that form oxyanions such as Cr (e.g. CrO_4^{3-}), are more soluble under acidic than alkaline conditions. In a pure solution comprising dissolved metal M, water, H^+ and OH^- as pH increases from acid to alkali, metals typically change their speciation from cations to neutrally charged hydroxide molecules to hydroxyanions, i.e.:

$$M^{n+}_{(aq)} + nOH^-_{(aq)} \Rightarrow M(OH)_n{}^0{}_{(aq)}; M(OH)_n{}^0{}_{(aq)} + mOH^-_{(aq)} \Rightarrow M(OH)_{n+m}{}^{m-}{}_{(aq)} \quad (7.1)$$

If the concentration of the neutrally charged hydroxide ion is sufficiently great, precipitation will occur. In reality, of course the situation is more complex. Firstly, the presence of other cations and anions in solution affects what complexes form and secondly ions interact with surfaces. However, the generalisation holds that PTMs are more mobile and available at lower pH. Speciation impacts availability and will be discussed at the end of this section. The impact of surfaces on metal mobility will now be briefly touched upon.

Surfaces provide sites where metals (or indeed any molecule in solution) may precipitate out of solution, provided concentrations are high enough. An energy barrier (activation energy) has to be overcome for precipitation to occur. This barrier is lower when precipitation occurs on a surface. However, the main way in which surfaces impact on PTM mobility and availability is through adsorption. Adsorption can be defined as the 'accumulation of substance/material at an interface between the solid surface and the bathing solution' (Sparks, 1995). There are two main varieties of adsorption: specific and non-specific (Figure 7.1).

Specific adsorption is considered to be the more permanent of the two. It involves covalent-type bonds between specific (hence the name) surfaces and molecules in solution and renders PTMs immobile and nonavailable. Non-specific sorption involves ionic-type bonds between exchange sites on surfaces and charged molecules in solution. Non-specific sorption is reversible. It therefore immobilises PTMs, but only temporarily, and ions held on exchange sites are often thought to be available to plants (and other soil organisms). Protons compete with cationic PTMs for negatively charged exchange sites and typically at lower pHs such sites are dominated by H^+. Although published about 30 years ago, the textbook of Lindsay (1979) is still an excellent reference that contains partition coefficients, phase diagrams and solubility data for a range of PTMs, which allow prediction of the impact of pH both on solubility and exchange site occupancy.

As well as competing for exchange sites, pH is also important in generating such sites. Soil particles often have either a negative or positive charge on their surface. This charge may be pH independent or pH dependent. pH independent charge arises from charge imbalances in the solid due to the isomorphous substitution of ions of differing charge, for example Al replacing Si or Mg replacing Al in mineral structures, resulting in a deficit of positive charge and therefore generating a negatively charged surface. Such pH independent charge is typical of 2:1 clays such as smectite. pH dependent charge is more typical of 1:1 clays such as kaolinite and oxyhydroxides of Al, Fe and Si. At low pH, the hydroxyl groups on the surface of such solids protonate, resulting in a positively charged surface, whilst at higher pH the hydroxyls deprotonate thereby generating negative surface charge. The pH at which such surfaces have a neutral charge is known as the point of zero charge (PZC) and this varies from solid to solid. Typically, the PZC of Fe-oxyhydroxides is 7–9, of Al-oxyhydroxides is 8–9.2, of Mn-oxyhydroxides is 1.5–4.6 and of clay minerals is 5–6. Thus, in a 'typical' slightly acidic soil Fe- and Al-oxyhydroxides have a positive charge and Mn-oxyhydroxides have a negative charge.

It is difficult to make generalisations about the role that organic matter plays in PTM mobility and availability. The dissociation of carboxylic and phenolic acid groups can provide exchange sites for PTMs. However, PTMs can also bond specifically to organic matter, often forming ring structures in which the PTM bonds to two or more hydroxyl groups (Figure 7.2).

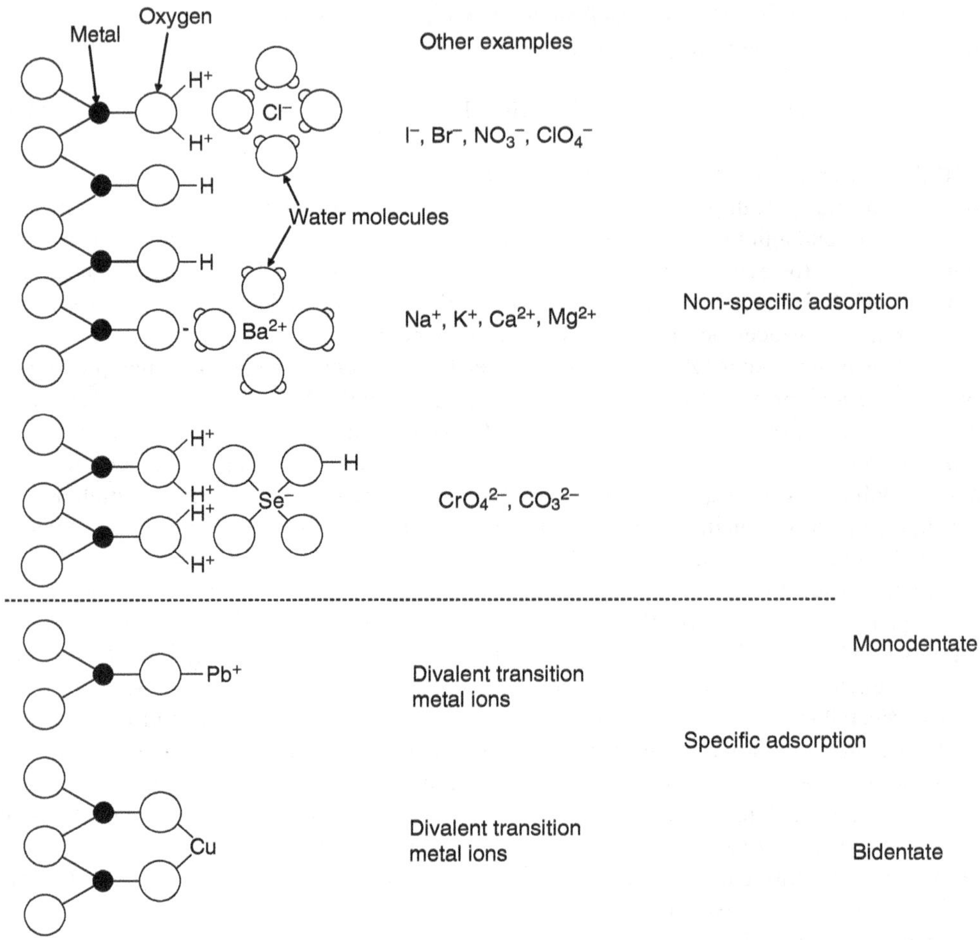

Figure 7.1 Cartoon of non-specific and specific adsorption to hydroxyl groups on mineral surfaces. Modified from Hayes (1987).

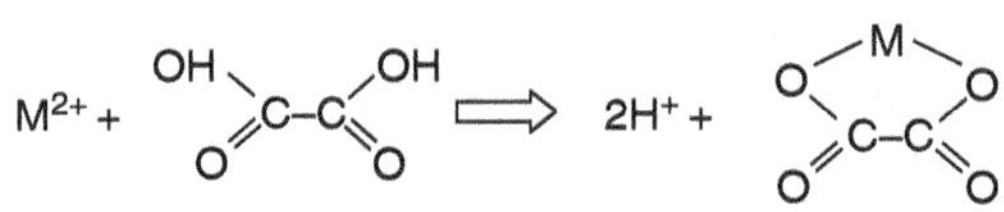

Figure 7.2 The chelation of a metal ion (M^{2+}) by oxalic acid, an organic compound that occurs naturally in soil solution through the action of biological activity.

These so-called chelate complexes are very stable and, when associated with solid organic matter, lead to the immobilisation of PTMs. When organic matter is present in a dissolved state, complexed PTMs can be held in solution at concentrations far higher than predicted in inorganic systems. Although the mobility of PTMs held in solution like this may be high, their availability is usually low (see later).

The redox state of soils (how reducing or oxidising they are) impacts on metal mobility and availability in a number of ways. Under reducing conditions, highly sorptive oxyhydroxides such as Fe- and Mn-oxyhydroxides will dissolve and any adsorbed PTMs will be released into solution. In addition, generally the lower valance state of multivalent cations is more soluble. These increases in solubility can be offset, to a certain extent by the precipitation of reduced species such as sulphides, though clearly this depends on S concentrations in the soil. Typically, reducing conditions develop when soils are poorly drained. In such cases the mobility of the PTMs will be relatively low although they will be available for uptake by plants and animals that can tolerate the reducing conditions.

Finally, in this section speciation must be mentioned. There is increasingly good evidence that, at least in the short term, the speciation of metals governs their availability for uptake (Di Toro et al., 2001; Van Gestel and Koolhaas, 2004; Steenbergen et al., 2005; Thakali et al., 2006a,b; Arnold et al., 2007; Lock et al., 2007; Allen et al., 2008). It is now widely accepted that metals are available for uptake when present as free metal ions and also as simple inorganic ions such as chlorides whilst metals present as larger organo-metal complexes are not available because the metals are held too tightly to the organic ion to sorb onto, and then pass through, biological interfaces. Thus, generally as ionic strength and dissolved organic carbon content of soil solution increases, a decrease in bioavailability would be expected. However, metal complexes will dissociate to release free metal ions and maintain steady-state conditions as free metal ions are taken up by organisms. Thus, although metals in organo-metal complexes may be unavailable in the short term, they do represent a reservoir of potentially available metals held in the soil.

7.1.3 Impacts of PTMs on plant growth: toxic effects including case studies

Potentially toxic metals can be subdivided into those which are essential for plant growth and those which are not. The impacts of deficiencies of those elements which are essential have been dealt with in Chapter 6. This section will therefore be concerned solely with the impacts of excesses of PTMs on plants. In broad terms, an excess of any PTM will have a toxic effect with severity of that effect increasing with the availability and concentration of the PTM. The actual value of an 'excess' of a PTM which results in harm will depend on soil properties (see Section 7.1.2) and also the plant. Plants that have evolved to tolerate unusually high concentrations of PTMs are discussed in Section 7.2.

Kabata-Pendias (2000) provides a useful table summarising the toxic effects of different PTMs on plants, which is reproduced in Table 7.2.

The causes of these toxic effects are summarised as:

- changes in permeability of the cell membrane (Ag, Au, Br, Cd, Cu, F, Hg, I, Pb, UO_2);
- reactions of thiol groups with cations (Ag, Hg, Pb);
- competition for sites with essential metabolites (As, Sb, Se, Te, W, F);
- affinity for reacting with phosphate groups and active groups of ADP or ATP (Al, Be, Sc, Y, Zr, lanthanides and possibly all other PTMs);
- replacement of essential ions (mainly major cations) (Cs, Li, Rb, Se, Sr);
- occupation of sites for essential groups such as phosphate and nitrate (arsenate, fluorate, borate, bromate, selenate, tellurate and tungstate).

Although toxicity varies from plant to plant and is dependent on soil conditions, in broad terms, the most toxic PTMs to plants are Hg, Cu, Ni, Pb, Co, Cd and possibly Ag, Be and Se (Kabat-Pendias, 2000). Due to variability in soil conditions and plant tolerance, it is not possible to state with any degree of accuracy the concentration of these elements which causes

Table 7.2 General effects of trace elements on common cultivars.

Element	Symptoms	Sensitive crop
Al	Overall stunting, dark green leaves, purpling of stems, death of leaf tips and coralloid and damaged root system	Cereals
As	Red-brown necrotic spots on old leaves, yellowing or browning of roots, depressed and wilting of new leaves	Legumes, onions, spinach, cucumbers, bromgrass, apricots, peaches
B	Margin or leaf tip chlorosis, browning of leaf points, decaying growing points and wilting and dying-off of older leaves. In severely affected pine trees, necrosis occurs on needles near the ends of shoots and in the upper half of the tree	Cereals, potatoes, tomatoes, cucumbers, sunflowers, mustard, apple, apricots, citrus, walnut
Be	Inhibition of seed germination and reduced growth, degradation of protein enzymes	–
Cd	Brown margin of leaves, chlorosis, reddish veins and petioles, curled leaves and brown stunted roots. Severe reduction in growth of roots, tops and number of tillers (in rice). Reduced conductivity of stem, caused by deterioration of xylem tissues. Reduction of chlorophyll and carotenoids	Legumes, spinach, radish, carrots, oats
Co	Interveinal chlorosis in new leaves followed by induced Fe chlorosis and white leaf margins and tips and damaged root tips	–
Cr	Chlorosis of new leaves, necrotic spots and purpling tissues, injured root growth	–
Cu	Dark green leaves followed by induced Fe chlorosis, thick, short or barbed-wire roots, depressed tillering. Changes in lipid content and losses of polypeptides involved in photochemical activities	Cereals, legumes, spinach, citrus seedlings, gladiolus
F	Margin and leaf tip necrosis and chlorotic and red-brown points of leaves	Gladiolus, grapes, fruit trees, pine trees
Fe	Dark green foliage, stunted growth of tops and roots, dark brown to purple leaves of some plants (e.g. Bronzing disease of rice)	Rice, tobacco
Hg	Severe stunting of seedlings and roots, leaf chlorosis and browning of leaf points	Sugar beets, maize, roses
Li	Chlorotic and necrotic spots on leaves and injured root growth	Citrus
Mn	Chlorosis and necrotic lesions on old leaves, blackish-brown or red necrotic spots, accumulation of MnO_2 particles in epidermal cells, drying tips of leaves, and stunted roots and plant growth	Cereals, legumes, potatoes, cabbage
Mo	Yellowing or browning of leaves, depressed root growth, depressed tillering	Cereals
Ni	Interveinal chlorosis (caused by Fe-induced deficiency) in new leaves, grey-green leaves and brown and stunted roots and plant growth	Cereals
Pb	Dark green leaves, wilting of older leaves, stunted foliage and brown short roots	–
Rb	Dark green leaves, stunted foliage and increasing amount of shoots	–
Se	Interveinal chlorosis or black spots at Se concentration of about 4 ppm and complete bleaching or yellowing of younger leaves at higher Se content, pinkish spots on roots	–
Ti	Chlorosis and necrosis of leaves, stunted growth	Beans
Tl	Impairment of chlorophyll synthesis, mild chlorosis and slight cupping of leaves, reduced germination of seeds and growth of plants	Tobacco, cereals
Zn	Chlorotic and necrotic leaf tips, interveinal chlorosis and leaves, retarded growth of entire plant and injured roots resemble barbed wire	Cereals, spinach

Source: Reproduced from Kabata-Pendias (2000). Reproduced with permission from Taylor & Francis Group LLC.

toxicity in plants, but concentrations in the order of single to a few hundred ppm in plant leaf tissue have a toxic effect. Plants have evolved a variety of methods for tolerating excessive concentrations of PTMs (Figure 7.3) and these mechanisms include (Kabata-Pendias, 2000; Memon and Schroder, 2009):

- selective uptake of ions;
- reducing the permeability of cell walls or modifying membranes either structurally or functionally;
- isolating ions in an immobile form via the precipitation of minerals or fixation by ligands;
- modification of the impacted metabolic pathway, for example by increasing the enzyme system that is affected or bypassing the affected metabolic pathway;
- alteration of enzymes so that the PTM is incorporated without enzyme functionality being affected;
- enhanced excretion of the PTM via leaf shedding, guttation, root exudates, volatilisation, etc.

Thus, one strategy when confronted with soil containing elevated concentrations of PTMs is simply to grow plants that can tolerate these concentrations. However, there is then the potential of moving the PTMs through the food chain, particularly if the plants grown are used as a food source. Thus, the most usual route is to remediate the soil. There are many possible remedial treatments for contaminated land, which are dealt with in specialist texts (e.g. La Grega et al., 2001; Nathanail and Bardos, 2004). However, Section 7.1.4 will consider phytoremediation, the use of plants to remediate contamination. Phytoremediation may be applied to both PTMs and organic contaminants. The latter are outwith the scope of this chapter. Application of phytoremediation to organic contaminants is well described in, for example, Gerhardt et al. (2009), Cunningham et al. (1996) and Schnoor et al. (1995).

7.1.4 Phytoremediation

Phytoremediation is the use of plants to remediate contaminated soils. The concept has its origins in the measurement of high concentrations of metals (primarily Ni) in the tissues of plants growing in soils, which in turn contained elevated concentrations of PTMs (e.g. Minguzzi and Vergano, 1948; Doksopulo, 1961; Severne and Brooks, 1972; Cole,1973; Jaffré and Schmidt, 1974; Jaffré et al., 1976; Brooks et al., 1977). It was later realised that such plants could be transferred from their native habitats to contaminated environments where, by accumulating similar concentrations of metals in their tissues, the plants would gradually remove the contaminants from the soil (Yamada et al., 1975; Chaney, 1983; Baker and Brooks, 1989). Since the initial exploration of phytoremediation, several distinct concepts have evolved which are explored later. Useful recent reviews of phytoremediation include Zhao and McGrath (2009), Dickinson et al. (2009) and Robinson et al. (2009).

Phytoextraction

Phytoextraction is the term used for the basic method of phytoremediation. Plants which extract metals into their tissues, thereby removing the contaminants from the soil, are grown. The method relies on the identification of hyperaccumulators, the name given to plants that can accumulate high concentrations of metals in their tissues. Hyperaccumulator

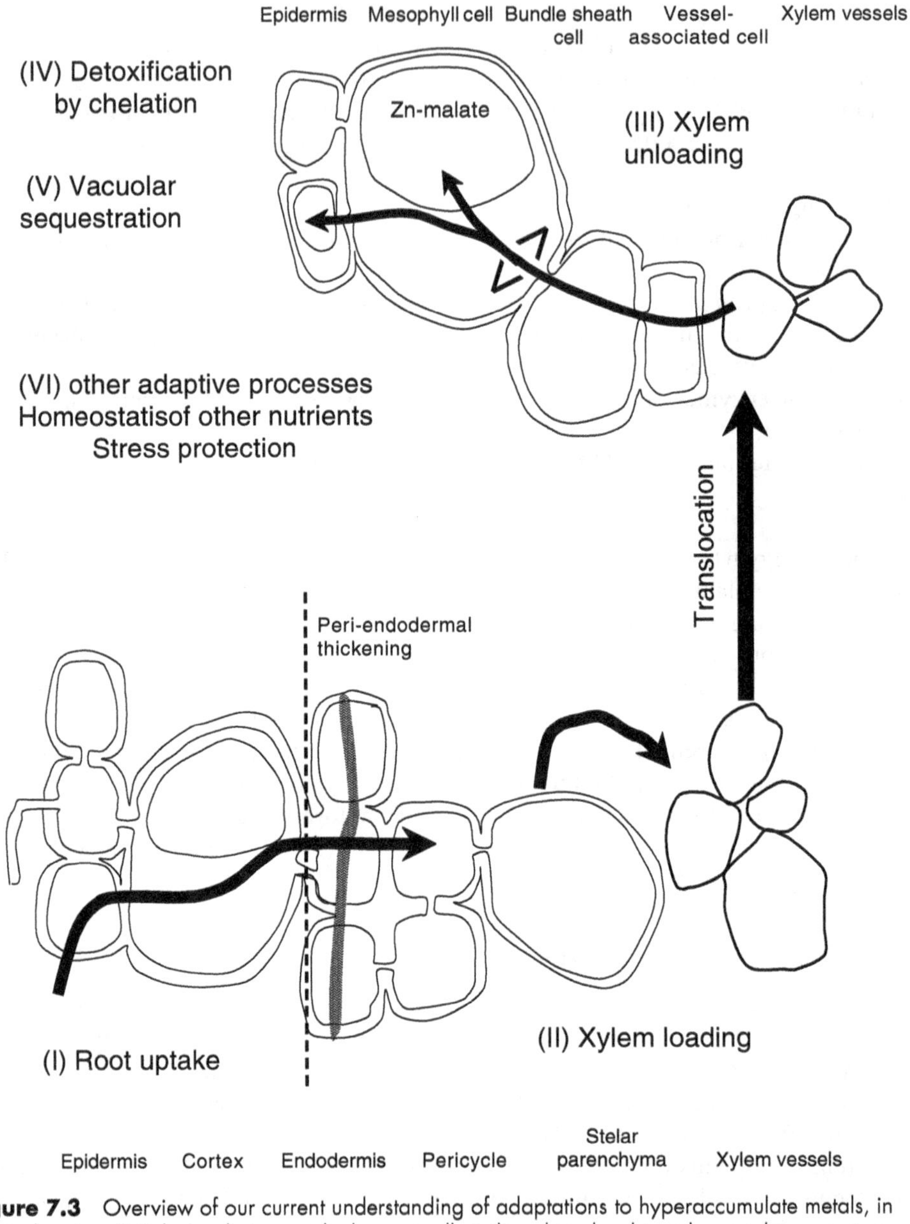

Figure 7.3 Overview of our current understanding of adaptations to hyperaccumulate metals, in particular zinc. (I) Enhanced zinc uptake by root cells is thought to be driven by metal transport genes such as ZIP4. In some hyperaccumulators such as *Thalspi caerulescens* a peri-endodermal thickening (apoplastic barrier along the root axis) is present. (II) Active xylem loading depends on reduced vacuolar root sequestration (main zinc storage is thought to be in the cortex) and enhanced activity of a P-type metal ATPase gene, HMA4 in stellar parenchyma, leading to efficient efflux of zinc from the symplasm. Efflux of citrate seems to play a role in zinc translocation. Loading of zinc in the xylem can occur as Zn-nicotianamine complexes. (III) In leaves Zn is unloaded in vessel-associated cells. (IV) Detoxification is assumed to be operated by chelation of metals. < > refers to chelation. Possible ligands of zinc in the cytoplasm are histidine and nicotianamine. (V) Vacuolar sequestration in the leaves is thought to be the main pathway of detoxification of metals. Zn is mainly stored in vacuoles or mesophyll and/or epidermal cells. In the vacuoles, a large pool of malate favours the formation of Zn-malate complexes. (VI) Other adaptive processes include homeostasis of other nutrients, in particular iron and phosphorus. Adapted from Verbruggen et al. (2009). Copyright John Wiley & Sons.

Table 7.3 Numbers of species identified as hyperaccumulators in Reeves and Baker (2000) and the PTMs which they hyperaccumulate.

PTM	Number of hyperaccumulating species
Zn	10
Pb	14
Cd	2
Ni	303
Co	26
Cu	35
Mn	9
Se	20

is a loosely defined term. Brooks et al. (1977) defined hyperaccumulators, with reference to Ni, as those plants in which tissue concentrations exceed 1000 mg kg^{-1}. Subsequent authors have suggested that the tissue concentrations required to classify a plant as a hyperaccumulator depend on the element with concentrations of 100 (e.g. Cd, Baker et al., 1994), 1000 (e.g. Ni, Brooks et al., 1977) and 10 000 mg kg^{-1} (e.g. Zn, McGrath, 1998) being proposed. Kabata-Pendias (2000) describes hyperaccumulators as plants that accumulate metals to concentrations greater than background levels and more than other species grown in the same soil. A range of hyperaccumulators for a range of metals have been identified. Metals that can be hyperaccumulated include As, Cd, Cr, Cu, Mn, Ni, Pb, Se and Zn (e.g. Baker et al., 1994; Kumar et al., 1995; Reeves and Baker, 2000; LaCoste et al., 2001; Ma et al., 2001; Galeas et al., 2007; Casado et al., 2008; Fernando et al. 2008). Table 7.3 summarises numbers of hyperaccumulators identified by Reeves and Baker (2000). The high number of Ni hyperaccumulators reflects the large amount of research carried out on plants growing in soils derived from serpentinite and other ultramafic rocks. Table 7.4 gives selected examples of more recent studies that have identified hyperaccumulators. Additional plant genera that include hyperaccumulators include Arabidopsis, Astragalus, Brassica, Garcinia, Helianthus, Maytenus, Pelargonium, Pteris, Stanleya and Thlaspi (Cunningham and Ow, 1996; McGrath, 1998; Galeas et al., 2007; Fernando et al., 2008; Gonzaga et al., 2009; Liang et al., 2009).

Many laboratory studies have been published in the peer-reviewed literature in which plants have been shown to accumulate high concentrations of metals from aqueous solution or from soils amended with metal salts. In both cases, the metals have a high degree of availability. There are comparatively fewer studies in which plants have been shown to accumulate metals from contaminated sites (as opposed to naturally metal-enriched soils) in the field. In the authors' opinion, this demonstrates that phytoremediation is not yet a fully developed technology (e.g. Van Nevel et al., 2007) though the high costs associated with field trials and potentially the commercial sensitivity of field trial results may also explain the relative paucity of published results from field trials compared to pot and greenhouse trials. A few examples of field studies are given in Table 7.5.

Phytoextraction has two clear advantages over many remedial methods – it is of low cost and low impact. Amongst its disadvantages are that it can take a long time and that most hyperaccumulators are element specific and do not address the range of contaminants found at a site. The issue of time is largely a result of the fact that the mass of metal removed from a contaminated site is a function of both the concentration of metal in the plant tissues and the biomass of the plant. The majority of hyperaccumulators have a low biomass and a low growth rate, a

Table 7.4 A selection of plant species and PTMs that they hyperaccumulate.

Plant species	PTMs	Reference
Pteris vittata L.	As	Bui et al. (2011)
Pityrogramma calomelanos L.	As	Bui et al. (2011)
Thelypteris palustris	As	Anderson et al. (2011)
Pteris vittata	As, Cr	Kalve et al. (2011)
Viola principis H. de Boiss	As, Pb	Lei et al. (2008)
Baccharis sarothroides Gray	As, Cr, Cu, Ni, Pb, Zn	Haque et al. (2008)
Solanum photeinocarpum	Cd	Zhang et al. (2011)
Cynodon dactylon L.	Pb, Zn	Bui et al. (2011)
Lonicera japonica Thunb.	Cd	Liu et al. (2011a)
Potentilla griffithii Hook	Cd, Zn	Qiu et al (2011)
Sedum alfredii Hance	Cd, Zn	Wu et al. (2007)
Leersia hexandra Swartz	Cr	Liu et al (2011b)
Spartina argentinensis	Cr	Redondo-Gomez et al. (2011)
Phytolacca acinosa	Mn	Xu et al. (2007)
Ricinus communis L.	Pb	Liu et al. (2008)
Tephrosia candida	Pb	Liu et al. (2008)
Debregeasia orientalis	Pb	Liu et al. (2008)
Eleusine indica L.	Pb, Zn	Bui et al. (2011)
Cyperus rotundus L.	Pb, Zn	Bui et al. (2011)
Equisetum ramosissimum (Vauch)	Pb, Zn	Bui et al. (2011)

result of the energy cost of accumulating metals. One phytoremediation method that attempts to get around the high concentration–low biomass problem is assisted phytoextraction.

Assisted phytoextraction

In this method, soil chemistry is manipulated to render the contaminants more available to plants so that normally non-accumulating, high biomass plants can be grown and used to remove metals from soil. This technique has its origins in attempts to render Pb, which is generally immobile in soils, available to plants so that it might be extracted rather than being left immobile in the soil. In this authors' view, there is a certain irony in the amount of research effort that has been expended towards mobilising otherwise relatively inert contaminant lead in soils. In the majority of studies to date, soil chemistry is manipulated by the addition of a chelation agent, most usually EDTA (e.g. Huang and Cunningham, 1996; Lombi et al., 2001; Lopez et al., 2007; Shibata et al., 2007; Dourmett et al., 2008; Saifullah et al., 2009; Sun et al., 2009) but also (in recognition that EDTA is a significant pollutant and relatively recalcitrant) ammonium thiocyanate (Anderson et al., 1999), citric acid, NH_4-citrate, oxalic acid, SS-ethylenediamine discussinic acid (EDDS), nitrilotriacetic acid (NTA) (Duquene et al., 2009), EDTA, tartrate and glutamate (Doumett et al., 2008), the phytohormones indole-3-acetic acid, gibberellic acid and kinetin (Lopez et al., 2007), citric acid and NTA (Irtelli and Navari-Izzo, 2006). The purpose of the chelating agent is to move more of the contaminant into solution from which it may be taken into plant tissues. Whilst assisted phytoremediation can result in high metal concentrations in high biomass plants thereby reducing the time required for remediation, the fate of the soil amendment and the mobilised contaminants requires careful monitoring. There is the possibility that both components could leak from the system into groundwater (Nowack et al., 2006).

Table 7.5 Examples of phytoextraction field trials.

Soil type	Metals present	Plant type	Location of trial	Results	Reference
Paddy field polluted by Zn-refining plant dust	Cd	Indica rice cultivars MORETSU and IR-8	Southwest Japan	18% reduction in total Cd over 2 years	Ibaraki et al. (2009)
Soil mixed with pyrite ore roasting wastes	As, Co, Cu, Pb, Zn	*Populus alba* L., *Populus nigra* L., *Populus tremula* L., *Salix alba* L.	Torviscosa, Udine, Italy	No significant phytoextraction but soil was stabilised and metals immobilised	Vamerali et al. (2009)
Historically sewage sludge amended soils	Cd, Zn	*Thlaspi caerulescens*, *Arabidopsis halleri*	Woburn, Bedfordshire, UK	1.0–8.7% Cd and 0.2–2.2 Zn removed by *T. caerulescens*, 0.01% Cd and 0.04% Zn by *A. halleri*	McGrath et al. (2006)
Thermal power plant fly ash mixed with soil and lime	Cr, Cu, Pb, Zn, Mn, Fe, Ni, Cd	*Pongamia pinnata*, *Tectona grandis*, *Delbergia sissoo*, *Cassia siamea*, *Eucalyptus hybrida*, *Dendrocalamus strictus*	Khaperkheda, Maharashtra, India	Limited metal uptake	Jambhulkar and Juwarkar (2009)
Contaminated soil	Cd, Cu, Pb, Zn	*Salix viminalis*	Copenhagen, Denmark	Extraction of 0.13% Cd and 0.29% Zn per year from moderately contaminated soil	Jensen et al. (2009)
Sewage sludge disposal cite	Cd, Zn	Nine varieties of *Salix*	Nottingham, UK	0.06% extraction of Cd predicted over 25 years	Maxted et al. (2007)

Phytovolatilisation

Phytovolatilisation has been used for the remediation of both inorganic (e.g. Se and Hg) and organic (e.g. BTEX, TCE and MTBE) contaminated soils. It depends on contaminants being taken up into the plant, remaining in solution and being carried to the leaves from where the contaminants volatilise (Chaney et al., 1997; Baeder-Bederski-Anteda, 2003). In the case of Hg phytovolatilisation, genetic modification of *Arabidopsis* and *Nicotinia tabacum* is carried out to produce plants which express merA and merB genes and which take up methylmercury from soil solution, convert it to Hg(II) which is then volatilised as elemental Hg (e.g. Raskin, 1996; Rugh et al., 1998; Ruiz and Daniell, 2009). The volatilised Hg is significantly less toxic than methylmercury. For Se, it is the inorganic form of the element that is taken up and then converted to a less toxic organic form (dimethyl selenite) which is volatilised. The Se uptake path is related to that of S and a wide variety of genetically engineered pathways have been exploited to enhance the uptake and volatilisation of Se (e.g. Kubachka et al, 2007; Pilon-Smits and Leduc, 2009).

7.2 Salt-affected soils (saline, sodic and saline-sodic soils)

There are more than 930 million hectares of salt-affected land worldwide, equating to more than 7% of the world's total land area (FAO, 1998). The management of these soils has been challenging humans since the advent of irrigated agriculture many millennia ago (Umali, 1993). They have even been implicated in the decline of historic civilisations and societies, such as Babylon and Sumeria in Asia, Carthage in North Africa and the Hohokam Indians in present-day Arizona (Umali, 1993; Schofield et al., 2001). An indication of the global distribution of salt-affected soils can be obtained from Figure 7.4.

7.2.1 Definitions of saline and sodic soils

The US salinity laboratory has defined saline soils as those with a saturated paste extract electrical conductivity (ECe) of>4 dS m^{-1} (Richards, 1954). This value corresponds to a total cation concentration (TCC) of 40 $mmol_c L^{-1}$ or an osmotic potential of −145 kPa (Sumner et al., 1998) and approximates the level at which many plant species are negatively affected by salt accumulation (Figure 7.5).

Although this value is widely agreed upon in the literature, the measurement of ECe is relatively laborious, hence the EC of a 1:5 soil:water extract ($EC_{1:5}$) is often used instead of the ECe. In such cases, the saturation water content can be estimated on the basis of soil texture and a conversion factor applied to predict the ECe. The accuracy of this conversion varies according to soil type. Typically, the ions responsible for the high electrical conductivity are chlorides and sulphates of calcium, magnesium, sodium and potassium, i.e. soluble salts that can reach high concentrations without precipitation occurring.

Sodic soils are generally defined by the parameter of exchangeable sodium percentage (ESP), which expresses the concentration of Na^+ held on soil exchange sites as a percentage of the cation exchange capacity of the soils as shown in Equation 7.2 (concentrations in $cmol_c\ kg^{-1}$):

$$ESP = 100 \times \frac{\text{Exchangeable Na}}{\text{Cation Exchange Capacity}} \tag{7.2}$$

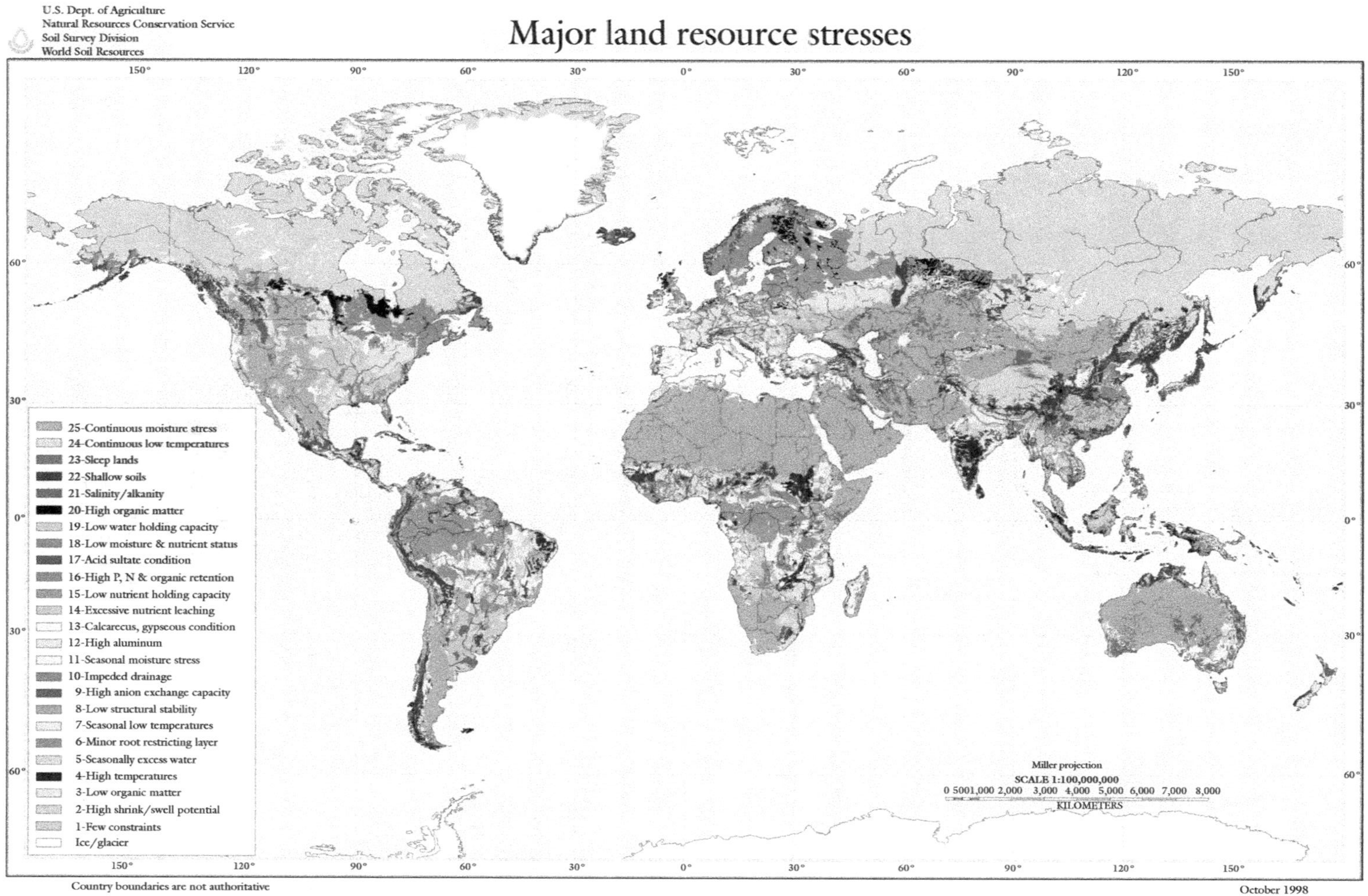

Figure 7.4 Global distribution of saline/sodic soils (these soils would fall within the category of salinity/alkalinity according to this map key). http://soils.usda.gov/use/worldsoils/mapindex/stresses.html. With permission from USDA-NRCS, Soil Survey Division, World Soil Resources. For a colour version of this figure, please see Plate 7.1.

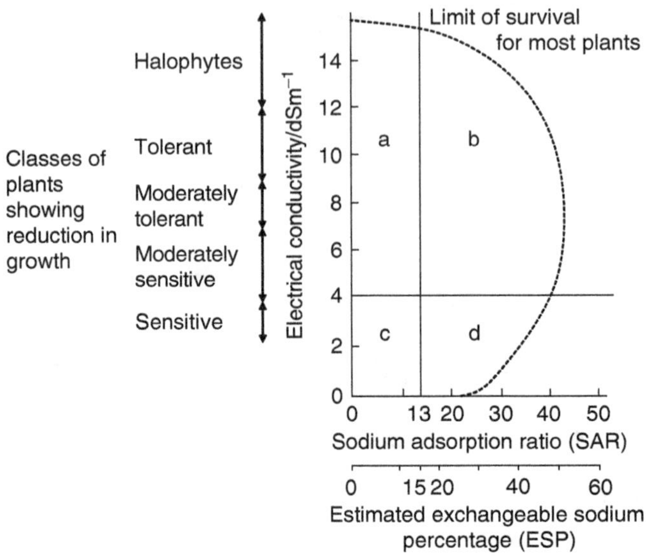

Figure 7.5 Diagram illustrating the classification of normal, saline, saline-sodic and sodic soils in relation to soil pH, electrical conductivity (EC), sodium adsorption ratio (SAR) and exchangeable sodium percentage (ESP). Also shown are the ranges for different degrees of sensitivity of plants to salinity. a, sodic soil, pH < 8.5; b, saline-sodic soil, soil pH generally < 8.5; c, normal soils, ph < 8.5; d, sodic soils, pH > 8.5. Reprinted from Brady and Weil (2010) by permission of Pearson Education, Inc., Upper Saddle River, NJ.

It is worth noting that the term 'sodic soil' has now effectively replaced the old term 'alkali soil' as examples of acid sodic soils were increasingly being reported in the literature (e.g. Ford et al., 1993; McKenzie et al., 1993). Other expressions that have been previously used in reference to sodic soils include the terms black alkali, solonetz, slick-spot, non-saline sodic, saline sodic and saline alkali (Letey, 1984; Sumner et al., 1998; Qadir and Schubert, 2002).

The sodicity of irrigation water and soil solutions is defined using the parameter of sodium adsorption ratio (SAR) which is a measure of the relative concentrations of Na^+, Ca^{2+} and Mg^{2+} in solution and can be calculated as shown in Equation 7.3 (concentrations in $mmol_c\ L^{-1}$):

$$\mathrm{SAR} = \frac{\left[\mathrm{Na}^+\right]}{\left[\mathrm{Ca}^{2+} + \mathrm{Mg}^{2+}\right]^{1/2}} \tag{7.3}$$

Rengasamy and Marchuk (2011) recently proposed the introduction of a 'cation ratio of soil structural stability' (CROSS). This new concept is analogous to the SAR but incorporates both the differential dispersive effects of Na^+ and K^+ and the differential flocculating powers of Ca^{2+} and Mg^{2+}. The validity of this concept as an index of structural stability across a wide range of soils requires further testing. It is quite likely however that CROSS may be a more effective predictor of soil structural behaviour where levels of K and Mg are high (e.g. in soils irrigated with recycled water).

Unlike saline soils, total concentrations of ions in sodic soils can be relatively low with an EC of <4 dS m^{-1}. However, the ratio of Na^+ to Ca^{2+} and Mg^{2+} ions is high. The original US salinity laboratory definition of sodic soils as those with ESP>15 (Richards, 1954) has proved controversial over the years, partly due to the large-scale variability in salinity and sodicity levels across continents and partly due to a lack of consideration of the effects of electrolyte concentration on sodic soil behaviour, for example the onset of soil dispersion (Szabolcs, 1979, Halliwell et al., 2001). As a result, there is currently no universally accepted ESP threshold value for sodic soils. In the USA, the value of 15 is still generally accepted, whereas in Australia, an ESP of 6 is most commonly used to denote sodic soils (Northcote and Skene, 1972). Sumner (1993) emphasised that sodic behaviour (e.g. dispersion) can even be apparent in soils with ESP values <1, demonstrating that sodicity is not confined to the traditional definitions of ESP presented earlier. The problems associated with ESP- and SAR-based sodicity definitions have prompted several authors (e.g. Sumner 1993; Shaw et al. 1994; Sumner et al. 1998) to recommend that new classification systems based on soil behaviour (e.g. dispersibility) and its limitations for productivity to be devised.

Regardless of issues relating to classification, broad definitions of sodic soils indicate that their key defining factor is the presence of a high proportion of exchangeable and/or soluble sodium ions relative to other cations (Shaw et al., 1994). Furthermore, these excess sodium ions are often at levels that can adversely affect soil structure and crop growth and are characteristically present in the form of salts capable of causing alkaline hydrolysis on cation exchange complexes (Schofield et al., 2001). Sodic soils can be acid or alkaline, saline or non-saline (Qadir and Schubert, 2002); however, most are alkaline with the dominant anion being the bicarbonate ion.

Some soils demonstrate characteristics of both saline and sodic soils and are hence termed 'saline sodic'. These soils not only have elevated levels of neutral soluble salts but also demonstrate an excess of sodium salts capable of alkaline hydrolysis. In such cases, crop growth may be adversely affected not only by osmotic stress and the effects of excess salts, but also by the high Na^+ levels and associated structural degradation of the soil (Burrow et al., 2002). It depends on a combination of factors such as the SAR, overall salt content and mechanical energy inputs (e.g. tillage practices) as to whether the soil will be flocculated or dispersive and whether or not other poor physical properties such as crusting will be exhibited (Sumner et al., 1998).

7.2.2 Origins and properties of saline and sodic soils

The formation of saline and sodic soils is intimately related to the water balance and salt distribution of the soil system. Both the salinisation and sodification of soils are caused by the accumulation of excess soluble salts at or near the soil surface. The source of the salts has frequently been used to distinguish between primary and secondary salinity/sodicity (Schofield et al. 2001; Qadir and Schubert, 2002), where primary salinity/sodicity relates to natural sources of salts, and secondary salinity/sodicity relates to anthropogenically sourced salts (e.g. pumping saline groundwater or effluent onto land as irrigation water). Rengasamy (2006) provides an alternative classification, with three major types of salinity defined on the basis of soil and groundwater processes; these being groundwater associated salinity, non-groundwater associated salinity and irrigation associated salinity.

Saline soils

Naturally occurring saline soils are most common in arid and semi-arid regions of the world, where low rainfall and leaching rates can lead to the accumulation of excess salts. The source of these salts can be varied, including Quaternary weathering of geological parent materials (Ford et al., 1993), aeolian sources (Shaw et al., 1994), windborne deposition from sea water and saltwater ingress through underground aquifers (Abrol et al., 1988).

The term *dryland salinity* is used to denote the occurrence of saline soils in a non-irrigated environment. This does not mean that dryland salinity is not anthropogenically induced, however, as although saline soils do occur naturally in some areas, agricultural practices in arid and semi-arid regions have been linked to a substantial increase in dryland salinity (Rengasamy, 2002, 2006). One of the major causes of increasing salinisation has been large-scale land clearance and deforestation for agricultural production, with the replacement of deep-rooted perennial species by shallow-rooted annual species allowing increased infiltration of water which then leads to rising watertables (Abrol et al., 1988; Fitzpatrick et al., 1994). In such cases, it is common for the soluble salts present in the soil to also rise up through the profile, concentrating in the soil surface layers and affecting plant growth. The amount of additional deep drainage or recharge necessary to cause salinisation can vary from as little as 5 mm to over 130 mm depending on the original salt/water balance and the system dynamics (Fitzpatrick et al., 1994). Where an impermeable layer is present in the soil profile, saline seeps may also occur, with infiltrating water being intercepted and conducted laterally to discharge on low lying land or in landscape depressions (Abrol et al., 1988).

The introduction of new irrigation systems and the expansion of existing systems is an ongoing worldwide trend (FAO, 2002) and *irrigation-induced salinity* is increasing hand-in-hand with the global increase in irrigated land area. Abrol et al. (1988) reported that almost 50% of the irrigated land in arid and semi-arid regions of the world was affected to some degree by salinisation and/or sodification. One of the major causes of increasing irrigation-induced salinity has been inadequate attention to drainage systems design and maintenance (Umali, 1993) as inadequate drainage may result in rising groundwater levels (Endo et al., 2002). In such a case, the soil may not only become waterlogged following rainfall, but may also become salinised through the movement and accumulation of salts from naturally saline soil strata, from saline irrigation water and/or from the capillary rise of groundwater and soluble salts (Abrol et al., 1988; Umali, 1993). Salt accumulation and translocation is also determined partly by irrigation method, water quality and quantity (Pessarakli, 1991; Schofield et al., 2001).

Even in areas where drainage is adequate, irrigation-induced salinity may occur as a result of inadequate leaching. If salt build-up has occurred and insufficient volumes of good quality irrigation waters are applied, leaching, which is dependent on soil properties and permeabilities (Endo et al., 2002) may be inadequate to remove the salts from the root zone. It is worth noting that the re-use of wastewaters and effluents for irrigation is increasing in many countries around the world (Balks et al., 1998; Bond, 1998). Effluents typically contain relatively high salt loadings, which often include high sodium concentrations (Balks et al., 1998), suggesting there is a risk of the soils becoming degraded by effluent irrigation (Jayawardane et al., 2001; Tillman and Surapaneni, 2002).

Sodic soils

In some ways, the formation of sodic soils closely parallels that of saline soils, with the major difference being that the accumulating salts are high in Na^+. Sodic soils present particularly complex problems for management however, in that not only the chemical properties but also the physical properties of the soils may be seriously affected. Some of the environmental conditions that promote the genesis of sodic soils include the presence of shallow saline groundwater, textural discontinuities throughout the profile (i.e. duplex soils), the occurrence of perched watertables close to the surface, low slope gradients and impeded drainage (Fitzpatrick et al., 1994; Bui et al., 1998). Increasing soil sodification can lead to serious degradation of soil structural and physical properties, with concomitant impacts on plant growth and productivity (Rengasamy and Olsson, 1991). This topic is dealt with in detail in Chapter 9, but to summarise, as the Na^+ concentration in a soil increases, repulsion forces between clay particles also increase, and at a unique level of sodicity and salinity for each soil (depending on the interactions of factors such as particle size, clay mineralogy, hydration, pH, ESP, EC and organic matter content (Qadir and Schubert, 2002)), clay swelling and dispersion will occur (Oster and Shainberg, 2001). Slaking may also result, with soil macro-aggregates breaking down to micro-aggregates upon wetting (Levy et al., 1998). These processes can lead to reduced porosity, as clay swelling and slaking causes a reduction in the number and size of pores at the soil surface and thereby limits the infiltration of rain or irrigation waters (Qadir and Schubert, 2002). The dispersion of clay particles can further reduce hydraulic conductivity and soil permeability as clay platelets are released from soil aggregates, blocking pores and plugging soil-water flow paths (Endo et al., 2002). Dispersion can result in a very unstable soil structure, and hence conditions that may be highly undesirable for agricultural land uses. Other soil physical changes and effects that are known to occur in association with sodic soils include increased susceptibility to surface crusting and sealing, hardsetting, runoff, erosion, tunnelling and poor aeration (Rengasamy and Olsson, 1991; Ford et al., 1993; Shaw et al., 1994; Qadir and Schubert, 2002).

7.2.3 Plant growth in saline and sodic soils

Plants differ substantially in their tolerance of salt stress (e.g. Figure 7.6), ranging from sensitive species such as *Oryza sativa* (rice) to highly tolerant (halophytic) species such as *Atriplex amnicola* (saltbush) which can even maintain good growth at salinities greater than sea water (Munns and Tester, 2008).

Halophytes can be defined as plant species which can reproduce (i.e. complete the life-cycle) under natural conditions of salt concentration of at least 200 mM NaCl and as such are distinguished from those species which are relatively salt tolerant but do not usually live

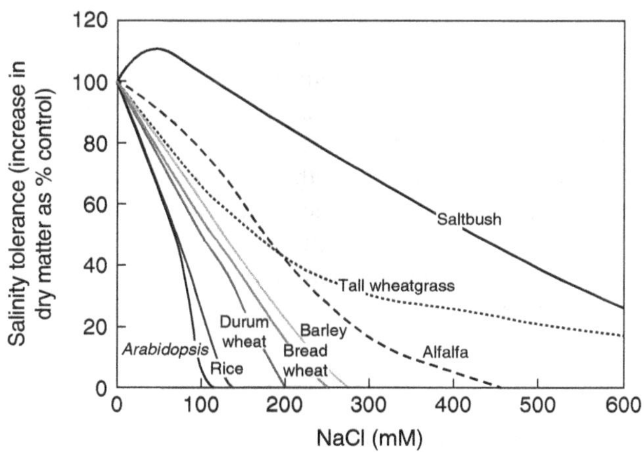

Figure 7.6 Diversity in the salt tolerance of different plant species, shown as increases in shoot dry matter after growth in solution or sand culture containing NaCl for at least 3 weeks, relative to plant growth in the absence of NaCl. For a colour version of this figure, please see Plate 7.2.

under saline conditions (Flowers and Colmer, 2008). More information on salinity tolerance in natural halophytes, constituting about 1% of the world's flora, can be found in Flowers and Colmer, (2008), whilst Maas and Grattan (1999), Shannon and Grieve (1999) and Steppuhn et al. (2005) present data showing the salinity tolerance of a very extensive range of agricultural crop species. With some exceptions (e.g. sugar beet), most crop plants are relatively salt tolerant at germination even though the germination process may be delayed (e.g. Maas and Poss, 1989). Sensitivity of plants typically increases during emergence and vegetative development, although research investigating this to date has frequently been difficult to interpret due to differences between controlled experimental conditions and realistic field conditions (Läuchli and Grattan, 2007; Tavakkoli et al., 2010). As soil salinity and sodicity are commonly accompanied by additional abiotic and biotic stresses, numerous interactive effects may be relevant, yet many of these (e.g. salinity and boron toxicity interactions) are still only poorly understood (Läuchli and Grattan, 2007). Figure 7.7 summarises key properties of the major types of salt-affected soils and some possible mechanisms of impact on plant growth occurring in these different types of soil.

Typically, saline and sodic soils impact on plant growth via one (or more) of three effects, resulting in reduced yields:

1. physiological drought due to a reduction in water availability caused by lowering of the osmotic potential (see Chapter 9 for details);
2. ionic imbalances in the plant cells; and
3. toxicity due to specific ions, most usually Na^+, Cl^- or B (as a variety of oxyanions, predominantly $B(OH)_4^-$ and at pH>7 $H_2BO_3^-$ and $B_4O_7^{2-}$).

In many cases, high pH may also be a significant cause of plant stress in these environments, impacting on nutrient availability and plant metabolism. However, this is independent of the direct salt effects on plant growth, and as mentioned previously, not all salt-affected soils are alkaline.

These major effects on plant growth in salt-affected soils are summarised below, but the interested reader may also refer to several comprehensive reviews published in the literature, including those by Hasegawa et al. (2000), Zhu (2001), Sairam and Tyagi (2004), Bartels and Sunkar, (2005), Chinnusamy et al. (2005) and Munns and Tester (2008). Plant growth response to salt stress has been described as a two-phase process, with the initial response shortly after exposure attributed predominantly to osmotic stress and the second, much slower effect over the following days, weeks or months, attributed to salt accumulation in the plant (Munns, 2002, 2005). This salt-specific effect is particularly noticeable in the older leaves of the plant and may result in leaf death or premature leaf senescence with corresponding reductions in total photosynthetic leaf area and overall plant growth (Munns, 2002, 2005). It should be noted that this two-phase model is heavily based on results from glasshouse trials and hydroponic systems and therefore may be less representative for plants growing under field conditions.

Physiological drought

The initial phase of growth reduction following sudden exposure of a growing plant to salt stress (e.g. under experimental conditions) is primarily due to osmotic stress as the increased ionic strength (and hence osmotic potential) of the solution outside the root causes changes in cell–water relations. The initial response to this osmotic effect is cell dehydration and

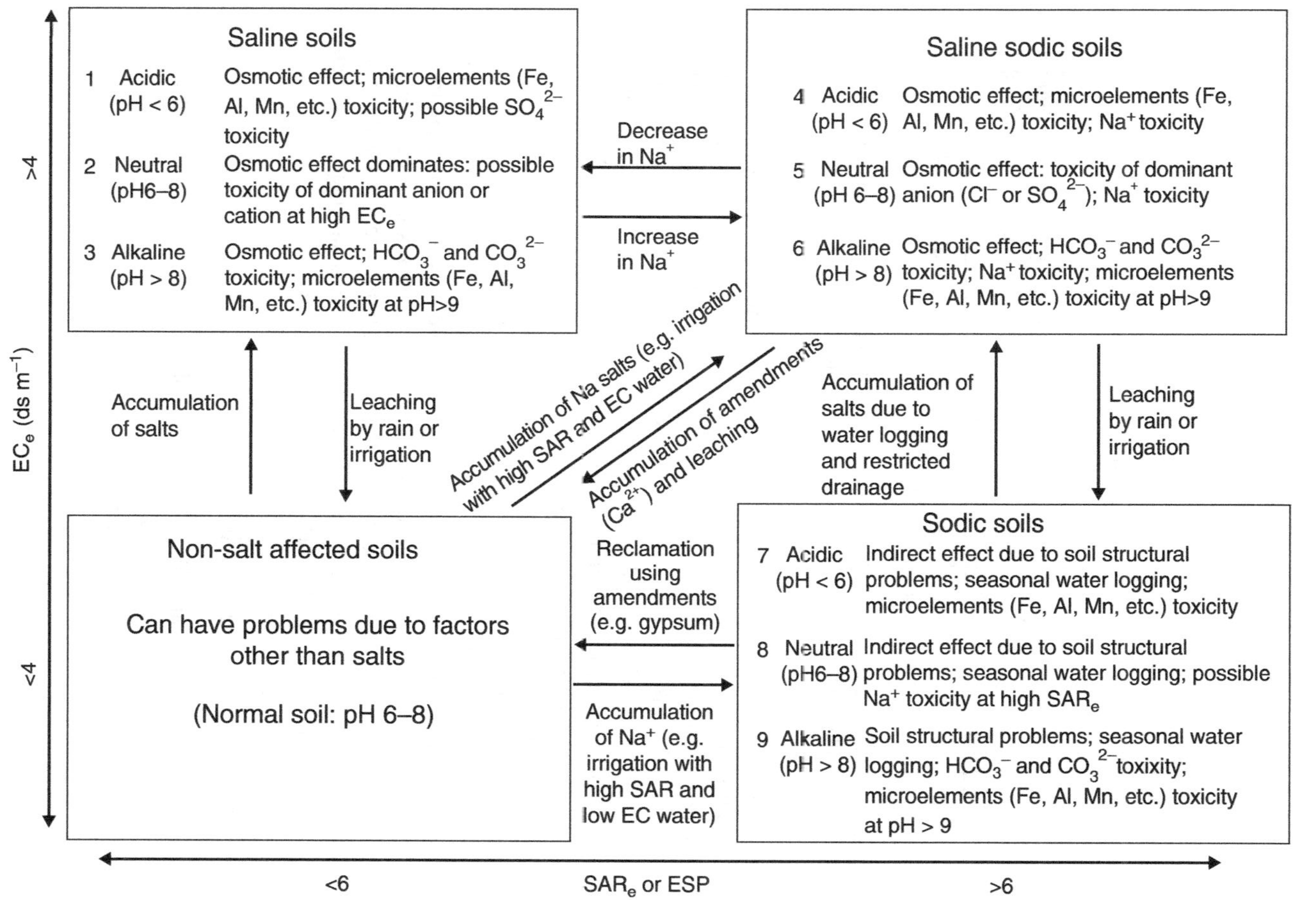

Figure 7.7 Key properties of the major types of salt-affected soils and possible mechanisms of impact on plant growth. Soils are categorised based on sodium absorption ration (SAR_e) and electrical conductivity (ECe) measured in soil saturation extracts and pH1:5 measured in soil:water suspension. The value of SARe (SARe ≈ ESP) of 6 and above to classify a soil as sodic is based on the Australian soil classification criteria (Isbell, 1996); however, the general categories shown are of relevance worldwide. Toxicity, deficiency or ion-imbalance due to other elements (e.g. B, K, N, P) will depend on the ionic composition of the soil solution and the particular species of plant.

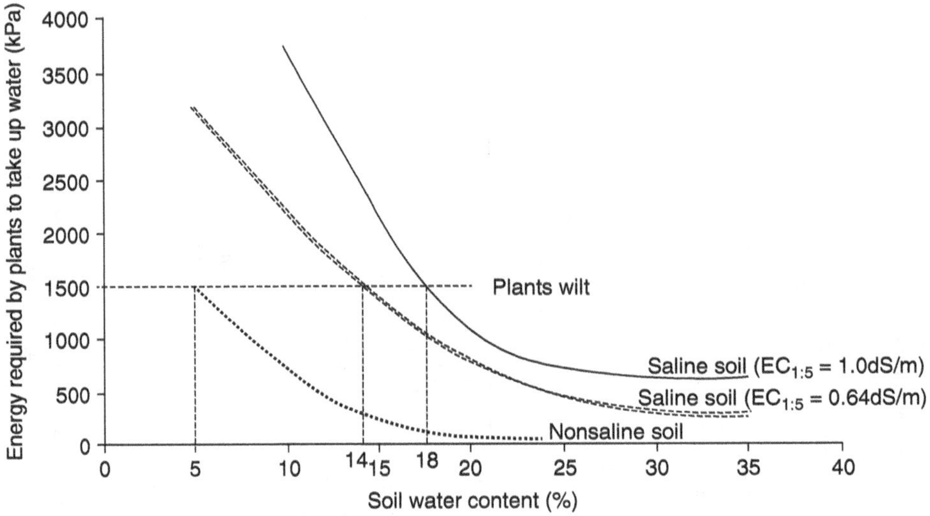

Figure 7.8 Energy (equivalent to soil matric plus osmotic potential) required by plants to take up water from a loamy soil as influenced by EC1:5 and % soil-water content.

shrinkage as the ability of the plant to absorb water is compromised. This occurs within moments of salinisation, is very similar to the effects of water stress and differs both within and between genotypes (Munns, 2002, 2005; James et al., 2008; Rajendran et al., 2009; Tavakkoli et al., 2010). The effect is transient and recovery is rapid. By using a pressurisation technique to maintain experimental plants at maximum water status whilst the soil was progressively salinised, Passioura and Munns (2000) were able to show that the transient growth reduction following initial application of salt to growing plants was due solely to changes in cell–water relations. Under conditions of increasing salinity, plants must expend more and more energy to maintain turgor. For example, Figure 7.8 illustrates the energy input required for plants to access soil water in a sandy loam soil under different conditions of soil moisture and salinity. In a non-saline soil, plants were able to access sufficient moisture to maintain turgor as long as the soil contained as little as 5% soil-water content, whereas with increasing soil salinity the energy required by the plant to extract water from the soil increases. At an $EC_{1:5}$ of 1 dS m^{-1} in the same sandy loam soil, plants required at least 18% soil-water content to take up sufficient moisture to prevent wilting.

Impacts of high pH on nutrient availability and plant metabolism

As discussed in Chapter 6, nutrient availability is linked to soil pH. The high pH of most sodic soils (>8.5) and many saline soils means that plant availability of essential cationic nutrients are often limited, thus resulting in nutrient deficiencies in plants growing in these soils.

Ion toxicity and imbalance

Given that the dominant salt in salt-affected soils is NaCl, high concentrations of both Na^+ and Cl^- ions are likely to be present. For most plants to tolerate salinity, Na^+ and Cl^- uptake must be restricted and the uptake of macronutrients such as K^+, NO_3^- and Ca^{2+} must be maintained. Chloride toxicity occurs in sensitive plants (many citrus and woody plant species) when solution concentrations exceed 20 mM, whereas tolerant species (halophytes) are unaffected (e.g. *Atriplex hastate*) or even stimulated (e.g. *Sueda maritime*) by concentrations in excess of 200 mM (Greenway and Munns1980; Marschner, 1995). Detailed information about Cl^- uptake and movement within plants and the mechanisms of Cl^- transport contributing to salt tolerance in plants can be found in White and Broadley (2001) and Teakle and Tyerman (2010), respectively, whilst Na^+ tolerance and transport in higher plants has been reviewed by Tester and Davenport (2003). Considerable variation in plant tolerance is also exhibited for Na^+, with plants ranging from Na^+ tolerant or natrophilic to Na^+ intolerant or natrophobic (Marschner, 1995). It is difficult to provide a value at which Na soil concentrations become toxic. This is because high Na^+ concentrations in salt-affected soils typically coincide with high Cl^- and possibly also high HCO_3^- concentrations (i.e. high pH), and the toxic effect of Na is also strongly linked to the ratio of Na to Ca and its impact on Ca uptake (Marschner, 1995). Chloride and Na^+ toxicity can be expressed as a decrease in yield, chlorosis and decreased photosynthesis (Sykes, 1992; Maas, 1993; Marschner, 1995; Xu et al., 2000). High NaCl concentrations can also inhibit Ca, Mn and NO_3^- uptake and therefore cause deficiency/imbalance in these nutrients. It can also lead to enhanced uptake of P and subsequent P toxicity (Marschner, 1995).

Boron (B) is also often found in high concentrations in saline/sodic soils, particularly in the sub-surface layers of these soils, where saline-sodic conditions prevail. Toxicity has been reported in sensitive species at soil solution concentrations of 1 mg L^{-1}, whilst concentrations of 10–15 mg L^{-1} are toxic even to tolerant plants (Ryan et al., 1977; Chaudan and Powar, 1978; Bonilla et al., 1980; Johnson and Fixen, 1990). Visual symptoms of boron toxicity are initially dark green wilted leaves (Kabata-Pendias, 2000). However, as concentrations of B increase, chlorosis of leaf tips and/or margins and necrosis occurs. Toxicity appears to be due to metabolic impairment due to formation of complexes with NAD^+ or the ribose of RNA as well as more specific metabolic pathway inhibitions (Loomis and Durst, 1992; Lukaszewski et al., 1992). Tolerance appears to depend partly on the ability to maintain low B shoot accumulation, and this has also been linked to mechanisms of active B efflux from roots (Sutton et al., 2007)

7.2.4 Amelioration and management of saline and sodic soils

Two major strategies are available for the amelioration of saline and sodic soils. The first strategy is based on the cultivation of salt-resistant crops. As already mentioned, the salinity tolerance of crop species varies significantly (Maas and Grattan, 1999, Shannon and Grieve, 1999; Steppuhn et al., 2005) and genetic manipulation may also lead to the development of tolerant cultivars (e.g. Zhang and Blumwald, 2001; Munns, 2009). The use of these salt tolerant species can significantly boost the productivity of agriculture in salt-affected soils. However, in areas where salt loads are continuing to rise (e.g. due to ongoing use of low quality irrigation waters) this must really be viewed as a short-term management option. Unless it is allied to other management actions, the soil will continue to become progressively more saline or sodic and will eventually be rendered infertile. A historic example of this can be seen in the Sumarian civilisation of Mesopotamia (Jacobsen, 1982; Wild, 2003). As their soils became increasingly saline/sodic, there was a change in agriculture from wheat to more tolerant barley but with no associated change in irrigation methods. As a result, the soil continued to degrade and ultimately became infertile to barley as well. Starved of cereal crops the civilisation collapsed.

The more successful long-term strategy for ameliorating salt-affected soils requires the salt concentration to be lowered by flushing out the problem salts. For saline soils, this is theoretically quite straightforward, simply requiring a sufficient volume of good quality irrigation water to wash the salts out of the soil. In practice, however, it is rarely so simple, as good quality irrigation water in salt-affected areas is often in very short supply. Moreover, for sodic soils, matters are frequently complicated by the degradation of soil structure (Chapter 9) and a source of Ca^{2+} ions to replace the dominant Na^+ ions on the soil exchange sites is also generally required. The costs of sodic soil remediation may therefore be prohibitively high. Further information about salt-affected soil amelioration can be found in Qadir et al. (2000).

Salt tolerant crops

Crops are able to tolerate saline/sodic conditions through a variety of mechanisms (Flowers and Colmer, 2008; Munns and Tester, 2008; Türkan and Demiral, 2009; Munns, 2009), including:

1. selective exclusion of ions;
2. selective inclusion of ions:
   (a) to maintain cell turgor
   (b) to maintain K^+ uptake
   (c) to partition Na^+ and Cl^- into specific tissues or organelles (e.g. vacuoles);
3. induction and up-regulating of antioxidative enzymes;
4. accumulation of 'compatible organic solutes' (e.g. betaines, polyols, sugars and amino acids) to increase hyperosmotic tolerance; and
5. rapid and efficient salt excretion via salt glands or tissue shedding.

Generally, salt tolerant plants are known to have lower rates of Na^+ and Cl^- transport to leaves compared with salt-sensitive plants, as well as greater ability to tolerate these ions and compartmentalise them into vacuoles, thereby preventing their build-up in the cytoplasm or cell walls (Munns, 2002; Munns and Tester, 2008).

Traditional methods of improving the salt tolerance of plants have focused on screening cultivars and using cross-breeding to produce cultivars with improved levels of both tolerance and yield. More recently, however, the genomics revolution has made genetic engineering of salt tolerant plants possible (e.g. Zhang and Blumwald, 2001) and has also facilitated more efficient molecular screening of cultivars for use in cross-breeding programmes (e.g. Munns and James, 2003; Lindsay et al., 2004; James et al., 2006). Nevertheless, genetic engineering for increased salt tolerance is not a simple procedure and agriculturally relevant progress in this direction has so far been quite limited. For instance, single gene modifications can be used to confer salt tolerance in transgenic plants, but they are also likely to incur significant fitness costs in terms of plant development and physiology (Denby and Gehring, 2005). Salt tolerance is a genetically complex trait, governed by multiple genes with additive, dominance and reciprocal effects (Genc et al., 2007), and, as noted earlier, numerous different mechanisms may potentially contribute to salt tolerance. It is also a developmentally regulated process, with tolerance of plants at one stage of development not always correlated with tolerance at other stages (Yamaguchi and Blumwald, 2005). Despite the challenges, efforts to confer favourable traits to improve crop salt tolerance are continuing (e.g. Choi et al., 2011; Gao et al., 2011) and further success in this direction is to be anticipated (Mittler and Blumwald, 2010; Roy et al., 2011). By studying the naturally occurring variation of salt stress tolerance in varieties, landraces and wild relatives of crops, and studying the traits that contribute to tolerance, the genetic loci determining these traits can be determined. Once the molecular basis of traits contributing to tolerance have been identified, marker-assisted breeding and genetic modification technologies can be used to introduce these genes into current, high yielding elite cultivars.

Reducing saline/sodic conditions

For saline soils, good quality irrigation water (see 'Ion toxicity and imbalance') can be used to flush out salts from the soil profile. To achieve this, the volume of water applied needs to be sufficient to meet the water needs of the crops and to leach out the salts. The fraction of water required to pass through the root zone to leach salts to an acceptable level has traditionally been known as the *leaching requirement*, whilst the fraction of infiltrated water that actually percolates below the root zone is called the *leaching fraction*. This can be calculated from the depths of the irrigation and drainage waters and is also (under certain steady-state conditions) approximately equal to the ratio of the electrical conductivity of the irrigation water to the drainage water (Qadir et al., 2000). Details of the necessary calculations to determine the amount of irrigation water to apply are given by Ayers and Westcott (1985) and Rowell (1994). It is important to ensure that the flushed drainage water does not cause deterioration of the quality of drainage water downstream. To alleviate sodic and saline-sodic conditions, the key objective is to reduce the concentration of Na^+ in the soil; this is most usually done by additions of Ca^{2+}, usually in the form of gypsum followed by flushing of the soil (Richards, 1954). Good drainage has to be maintained for this strategy to work and sometimes this requires addition of non-sodic soils or manures together with specialised ploughing regimes (e.g. deep ripping) to create a permeable upper soil layer.

Preventing salinity/sodicity problems

In order to avoid or ameliorate soil salinity/sodicity problems in the first place, good quality irrigation water should be used and the water table kept low. Irrigation water quality guidelines (e.g. Table 7.6) can be used to indicate water composition ranges suitable for restricted and/or unrestricted use in agriculture.

Maintenance of a low water table requires good drainage. For example, open or till drains or drainage wells may be used to maintain the water table at a depth of >2 m in arid and semi-arid regions to prevent groundwater rising by capillarity and evaporating, thereby increasing ion concentrations in the rooting zone of plants. As with flushing of soils, the drainage water must be disposed of with care in order not to create salinity/sodicity problems elsewhere (Ayers and Westcot, 1985).

Practices such as 'partial root zone drying' may also be practised in order to minimise the amount of salt added to the soil via irrigation (Kaman et al., 2006). Other relevant irrigation-related management approaches that have the potential to aid the management of sodicity/salinity problems such as optimal timing and duration of irrigation are also addressed in Chapter 9. In rain-fed agriculture, suitable crop rotation practices (e.g. rotation of shallow-rooted annual crops with deep-rooted perennial species) may help restore the balance between rainfall and water use, thus preventing/slowing the rise of salts to the surface (Munns, 2002).

Table 7.6 Guidelines for interpretation of water quality for irrigation.

Potential irrigation problem	Units	Degree of restriction on use None	Slight to moderate	Severe
Salinity				
Electrical conductivity	dS m^{-1}	<0.7	0.7–3.0	>3.0
Or				
Total dissolved solids	mg L^{-1}	<450	450–2000	>2000
Infiltration and electrical conductivity				
SAR 0–3		>0.7	0.7–2.0	<0.2
3–6		>1.2	1.2–0.3	<0.3
6–12		>1.9	1.9–0.5	<0.5
12–20		>2.9	2.9–1.3	<1.3
20–40		>5.0	5.0–2.9	<2.9
Specific ion toxicity (sensitive crops)				
Sodium				
Surface irrigation	SAR	<3	3–9	>9
Sprinkler irrigation	meq L^{-1}	<3	>3	
Chloride				
Surface irrigation	meq L^{-1}	<4	4–10	>10
Sprinkler irrigation	meq L^{-1}	<3	>3	
Boron	mg L^{-1}	<0.7	0.7–3.0	>3.0
Miscellaneous effects (susceptible crops)				
Nitrogen (NO3 as N)	mg L^{-1}	<5	5–30	>30
Bicarbonate (overhead sprinkling)	meq L^{-1}	<1.5	1.5–8.5	>8.5
pH		Normal range 6.5–8.4		

Source: Retabulated from Ayers and Westcot (1985).

7.3 Acid and alkaline soils

7.3.1 Designation of acid and alkaline soils

Soils may be designated as either acid or alkaline on the basis of soil pH, a parameter well recognised as a master variable affecting both soil chemistry and plant nutrition. Although pH 7.0 can technically be used to differentiate an acid soil from an alkaline soil, from a practical perspective soils with pH close to 7.0 (i.e. pH 6.5–7.5) are often considered as neutral. By contrast, soils with pH 5.5–6.5 (measured in 0.01 M $CaCl_2$) can be classed as slightly acidic, those with pH 4.5–5.5 as moderately acidic and those with pH<4.5 as strongly acidic. Similarly, pH 7.5–8.5 can be considered slightly alkaline, pH 8.5–9.5 moderately alkaline and pH 9.5 and above strongly alkaline (Charman and Murphy, 2000). Delineating between these different degrees of acidity and alkalinity is of extensive practical significance in terms of land management, crop growth and related issues. For instance, key aspects of soil chemistry regulated by pH and relevant to plant growth include element availability and toxicity (Barber, 1995; Impellitteri et al., 2001) microbial activity (Entry et al., 2002) and root growth (Pavlovkin et al., 2009).

Soils tend to vary in terms of acidity/alkalinity both spatially and temporally due to differences in local soil forming processes (e.g. precipitation, temperature), parent materials (i.e. basic vs. siliceous rock), landscape characteristics (e.g. slope) and land management regimes (e.g. use of soil amendments, tillage practices, etc.), but most naturally occurring soils have a pH somewhere between pH 4 and pH 10. The actual soil pH in a given area is essentially a function of the soil composition (i.e. the nature and relative proportions of mineral and organic constituents) and related ion exchange and hydrolysis reactions (Ulrich and Sumner, 1991). Soil pH values <4 are relatively uncommon because the pH is buffered in this range by the dissolution of aluminosilicate and oxide minerals (Ulrich, 1991). Nevertheless, in special cases such as acid sulphate soils, soil pH values well below this level have been recorded (Melville and White, 2000; Fitzpatrick et al., 2009).

The major source of alkalinity in soils is the dissolution of carbonates and subsequent production of OH^- ions. Alkaline soils are 100% saturated with exchangeable base cations (e.g. Ca^{2+}, Mg^{2+} and Na^+) and may have a very large acid-neutralising capacity depending on their carbonate content.

Soils containing free calcium carbonate ($CaCO_3$) are referred to as calcareous soils and are typically slightly to moderately alkaline. Under the FAO World Reference Base soil classification system, most calcareous soils fall within the group of Calcisols, comprising soils containing a calcic horizon with substantial secondary accumulation of carbonates (FAO, 1998). Calcareous soils may also contain free magnesium carbonate ($MgCO_3$). They frequently occur in arid/semi-arid regions (FAO 1998; Candy and Black, 2009) where low precipitation and biological activity equate to relatively low acid input and leaching, hence resulting in the net accumulation of alkalinity. However, alkaline soils may also occur in higher rainfall areas, where young soils are forming on alkaline parent materials (e.g. limestone). Calcic horizons may form via the leaching and precipitation of carbonates under natural precipitation regimes (e.g. seasonal leaching and precipitation cycles) and/or under irrigation scenarios whereby salts move upwards from the water table and precipitate near the top of the capillary fringe. They exhibit a wide range of morphologies and pedogenic origins (Candy and Black, 2009) and are sometimes densely cemented with poor drainage characteristics. These hardened calcic horizons are often referred to as calcretes and may present problems for plant root development and irrigation management.

Highly alkaline soils (pH=8.5–10) are generally associated with the dissolution of sodium carbonate ($NaCO_3$) which is more soluble than $CaCO_3$ and thus typically results in greater production of OH^- ions. These soils are also often sodic (see Section 7.2).

7.3.2 Distribution of acid soils

Acid soils occur naturally in many areas of the world (Figure 7.9) but acidification may also be anthropogenically induced.

Naturally occurring acid soils tend to be found in relatively humid areas where high rainfall leads to extensive leaching and mineral weathering such that base cations are depleted and the soil neutralising capacity exceeded. Acid soils with pH<5.5 in their upper horizons reportedly occupy around 30% (~3950 million ha) of the world's ice-free land area (see Table 7.7) and are located predominantly in two main geographical belts, one in the humid northern temperate zone and the other in the humid tropics (von Uexküll and Mutert, 1995). Approximately 75% of these areas are also affected by subsoil acidity. The predominant natural vegetation communities occurring on these soils are coniferous forest in the northern belt and savannah and tropical rainforest in the southern belt (von Uexküll and Mutert, 1995).

Acid soils occur in all major regions of the world (see Table 7.8; FAO, 1991); however, the majority (~60%) are located in the humid tropics. From a global perspective, the largest

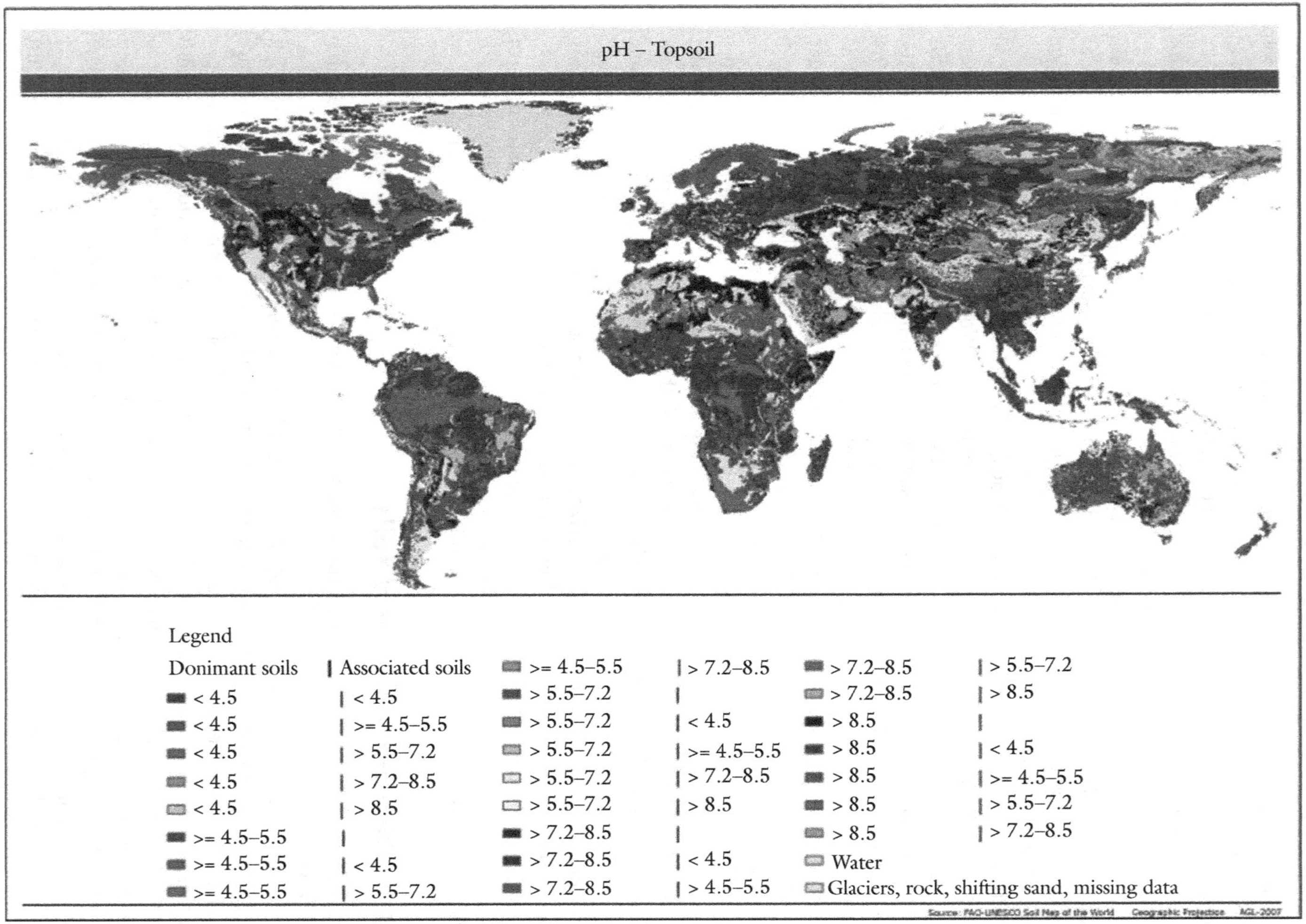

Figure 7.9 Global distribution of topsoil pH. Used with permission. FAO-UNESCO Soil Map of the World 2007. (http://www.fao.org/fileadmin/templates/nr/images/resources/images/Maps/geonetwork/ph_t.png). For a colour version of this figure, please see Plate 7.3.

Table 7.7 Global extent of acid soils classified according to major soil groups.

FAO soil group	Soil taxonomy (approximate match)	Area (million ha)
Fluvisols	Fluvents	50
Gleysols	Aquents, Aquepts	402
Regosols	Psamments	293
Arenosols	Psamments	280
Rankers	Haplumbrepts	61
Andosols	Andisols	34
Cambisols	Ochrepts, Tropepts	299
Podzoluvisols	Boralfs, Aqualfs	255
Podzols	Spodosols	415
Planasols	Aquults	15
Acrisols	Ultisols	731
Nitosols	Ultisols	118
Ferralsols	Oxisols	727
Histosols	Histosols	270
	Total	3950 (≅30% of global land area)

Source: From von Uexküll and Mutert (1995). Reproduced with kind permission from Springer Science and Business Media.

Table 7.8 Global distribution of acid soils.

Region	Area (million ha)	Area (%)
Africa	659	16.7
Australia and New Zealand	239	6.1
Europe	391	9.9
Asia	1044	26.4
America	1616	40.9
Total	3950	100

Source: From von Uexküll and Mutert (1995). Reproduced with kind permission from Springer Science and Business Media.

areas of acid soils are located in North and South America (40.9% of the world's acid soils by area) and Asia (26.4% of the world's acid soils by area). From a regional perspective, the highest proportion of acid soils is found in south-east Asia where 63% of the region's soils are classed as acid soils. By contrast, 35% of North American soils and 57% of South American soils are acidic (von Uexküll and Mutert, 1995).

Comparison of global land use statistics and acid soil distribution data indicate that a major portion of the world's acid soils still remain under natural vegetation (von Uexküll and Mutert, 1995; Sumner and Noble, 2003). Nevertheless, with modern agricultural techniques for the amelioration and management of soil acidity (e.g. lime amendment and phosphate fertilisation), large areas of uncultivated acid soils are now considered potentially suitable for arable production. Indeed, these areas of undeveloped acid soils have even been referred to as agriculture's last frontier (Borlaug and Dowswell, 1993), with recent land clearing and agricultural development of areas such as the *Cerrado* (savanna) region of South America used as an example of the latent potential for highly productive agricultural production on acid soils (e.g. Lopes, 1996; Yamada, 2005). Yet, although it is true that with the appropriate inputs of lime and fertiliser high agricultural yields may potentially be obtained on such soils, it should also be noted that these remaining areas of uncultivated acid soils support important habitat for numerous flora and fauna species of notable conservation significance, whilst also delivering a variety of other ecosystem services (Klink and Machado, 2005). Moreover, large areas of acid soils currently under cultivation are still delivering suboptimal returns due to insufficient liming and soil management regimes, whilst potentially slipping further towards the development of subsoil acidity (AACM, 1995). Given these conditions, it is clear that the agricultural sector can still benefit from further efforts to improve the nutrient supply capacity of acid soils in existing agricultural areas in order to sustain and improve agricultural productivity (Fageria and Baligar, 2003). Further clearing of marginal lands requiring extensive inputs to facilitate efficient agriculture should thus be very carefully considered, as potential side effects include pollution of water resources and increased greenhouse gas emissions.

7.3.3 Anthropogenic drivers of soil acidification

Precipitation is naturally acidic when in equilibrium with atmospheric CO_2 and is thus a continuous source of acidity to soils. In addition, areas affected by industrial pollution (e.g. heightened atmospheric levels of oxides of sulphur and nitrogen) may also be subject to the precipitation of 'acid rain' which can contribute substantially to progressive soil acidification (Bini and Bresolin, 1998). Models and analyses of the acid–base status of global precipitation indicate that acid precipitation is of particular concern in North America, Europe and China (Greenfelt and Galloway, 1996; Galloway, 2001; Rodhe et al., 2002), where detrimental effects on soil quality have been observed (Tomlinson, 2003; Hicks et al., 2008; Lehmann et al., 2008). Much less is known about the extent of acid deposition in other regions of the world (WMO, 1996), although the results of Rodhe et al. (2002) indicate that potential problem areas include several regions with sensitive soils in southern, south-eastern and eastern Asia, and in central South America. These sensitive soils include those which have low base saturation and cation exchange capacity and hence limited buffering capacity to resist the effects of acid deposition (Kuylenstierna et al., 2001).

In practice, the study of acid rain induced soil acidification is complicated by the ongoing parallel process of natural acidification and the relatively long time scales involved in the acidification process (Stuanes et al., 1995; Matzner and Davis, 1996). Yet, although it is difficult to clearly discriminate between pollution-induced acidification and natural acidification on the basis of fundamental parameters such as low base saturation, low pH and low alkalinity, there are certain characteristics such as high soil solution concentrations of SO_4^{2-} and NO_3^-, which specifically indicate the contribution of acid rain. Whereas these are amongst the key acidifying ions derived from acid rain, organic anions and HCO_3^- are the primary drivers of natural acidification (Alewell, 2003). Agricultural practices can also accelerate the process of soil acidification. For example, the application of nitrogen-containing inorganic fertilisers is well recognised as a contributing factor to the development of acid soils (Stumpe and Vlek, 1991; Jensen and Hauggaard-Nielsen, 2003). This is because the nitrification of fertiliser ammonia effectively releases H^+ to the soil system as indicated in Equation 7.4:

$$2NH_4^+ + 4O_2 = 2NO_3^- + 4H^+ + 2H_2O \tag{7.4}$$

The continual loss of alkaline grain, pasture and animal products due to the removal of material during harvest is a further driver in the acidification of agricultural soils (von Uexküll and Mutert, 1995) as is the increasing prevalence of legume-based pasture systems

(Tang et al., 1997). Together these causes of increasing acidity on extensively managed agricultural lands can lead to ongoing land degradation and significant reductions in farm productivity. For example, acidification due to the prevalence of legume-based pasture systems is a growing problem in large parts of Australia whilst acidification of intensively cropped land is a serious problem in large areas of the USA and former USSR (von Uexküll and Mutert, 1995; Garvin and Carver, 2003).

7.3.4 Acid sulphate soils

One key group of acid soils which are particularly challenging to manage are the acid sulphate soils (ASS). These are soils which either contain sulphide minerals or are affected by the transformation of sulphide minerals (Fanning, 2002). As such, ASS either contain sulphuric acid or have the potential to form sulphuric acid (most typically through the oxidation of iron pyrite) in quantities greatly exceeding the acid-neutralising capacity of the soil (Dent and Pons, 1995).

Sulphidic soil materials are naturally occurring in many areas of the world (see Ljung et al., 2009), especially in low lying coastal areas (e.g. mangrove swamps), and over 17 million ha of ASS have been identified worldwide (Andriesse and van Mensvoort, 2006; Ljung et al., 2009). The oxidation of sulphidic soil materials can occur naturally as a result of soil drying during times of drought (Dent and Pons, 1995) but is also frequently instigated by human activities, such as the draining of floodplains and backswamps for agriculture, aquaculture or housing developments (Dent and Pons, 1995; Österholm and Åström, 2004; Ljung et al., 2009).

The course of pyrite oxidation is complex, involving both chemical and microbiological reactions. Complete oxidation of pyrite to form sulphuric acid involves up to 22 electron transfer reactions and proceeds via a number of steps (Nordstrom, 1982). However, the overall oxidation process can be represented by the following equation):

$$FeS_2 + 15/4O_2 + 7/2H_2O \rightarrow Fe(OH)_3 + 2SO_4^{2-} + 4\,H^+ \qquad (7.5)$$

The pH of oxidised ASS can be as low as pH 2 (Dent and Pons, 1995) and acid leaching from these soils can have very serious consequences for soil nutrient availability (Golez and Kyuma, 1997), surface water quality (Sammut et al., 1996; Corfield, 2000), toxic metal availability (Åström, 2001), human health (Ljung et al., 2009), vegetation and aquatic life (Sammut et al., 1996; Hyne and Wilson, 1997; Lin et al., 2001) and even civil infrastructure (e.g. the corrosion of bridges, roads and pipelines) (Fitzpatrick et al., 2001). Careful management of these materials based on appropriate risk assessment when considering and/or undertaking their development is paramount to the prevention of serious environmental degradation, health effects and financial implications associated with unmitigated acid sulphate soil drainage.

7.3.5 Managing plant growth in acid soils

Although many plant species are well suited to growth in slightly acid soils, acid soil infertility and growth inhibition is a very common management issue in soils with pH <5.6 and is particularly problematic when pH is <5. Plants that are highly sensitive to acidity (e.g. lucerne and barley) tend to be adversely affected below pH 5, whilst more tolerant plants can grow normally until around pH 4.5. Below this, only very acid tolerant plants such as oats will grow without a significant reduction in production. Problems typically associated with acid soils include Al and Mn toxicities (Scott et al. 2000; Barcelo and Poschenrieder, 2002), nutrient deficiencies (e.g. Mo, Ca, Mg), legume nodulation failures (Bordeleau and Prevost, 1994) and inhibition of root growth (Marschner, 1991).

A key aspect of managing acid soils for plant growth is the pH amendment of the soil (Edmeades and Ridley, 2003). This is commonly practised in both agricultural and garden settings, generally via the application of acid-neutralising materials such as lime and dolomite and can be a simple and successful management strategy (e.g. Table 7.9).

Whilst some acid soils, particularly those occurring in tropical regions, are acid throughout the whole profile, others which have been exposed to more moderate weathering and leaching comprise acid upper layers underlain by alkaline layers with substantial acid-neutralising capacity at depth (Farina et al., 2000a, b). As liming is most effective in situations where only the surface soil layer is acidic, it is now recognised that liming to prevent acidification (particularly the development of subsoil acidification) is as important as liming for the purpose of acid soil amelioration, because this ensures that the land remains profitable in order to offset the cost of liming. By contrast, once acidification develops, the initial capital input required to return the soil to a suitably productive state may be preventative (Edmeades and Ridley, 2003).

Individual soils respond differently to liming and there is no optimal soil pH that will suit all crops or plant species (Edmeades and Ridley, 2003). Acid amelioration practices should thus be planned in accordance with the nature of the soil and required land use. Research in this area has seen a move away from the concept of soil lime requirement (laboratory-based determinations of the lime required to increase soil pH by a specific amount) towards the concept of biological lime requirement, meaning the amount of lime required to eliminate restrictions to plant growth (e.g. Slattery and Coventry, 1993). However, there are many factors at play in determining this level and crop responses to liming have repeatedly proved difficult to predict. Progress in the development of predictive relationships between relevant soil parameters and crop yields has been made (Edmeades et al., 1984; Dierolf et al., 1999), but further research efforts are required. In some cases, the expense of lime amendment can be managed by determining which of the factors that potentially limit plant growth on acid soils (e.g. Al and Mn toxicity, Ca deficiency, P deficiency, etc.) are of greatest relevance in a given setting and specifically addressing those issues. For example, Edmeades and Ridley (2003) cite cases of legume-based pastures on acid soils responding equally well to relatively small amounts of Mo fertilisation as to large dressings of lime. In such cases, an initial campaign of Mo fertilisation to improve productivity could help support the cost of ongoing pH management. Fageria and Baligar (2003) show increases in rice and bean yield in Cerrado and Varzea acid soils from Brazil in response to N fertiliser applications, N levels having previously been reduced in the soils due to net N removal due to harvesting and soil organic matter reductions (Table 7.10).

In situations where Al toxicity is identified as the primary concern, relatively small applications of lime to bring the pH to around 5.5 may effectively return plant growth to near-maximum productivity (Kamprath, 1970). It is clear that site-specific diagnosis and management of acid soil infertility can minimise the expense of amelioration.

Efforts should also be made to slow the rate of acidification through more efficient fertiliser use (to minimise nitrate leaching) and to deliberately choose less acidifying fertilisers (i.e. replacing fertilisers such as sulphate of ammonium and mono-ammonium phosphate with urea or legume fixed nitrogen). It may also be necessary to choose acid tolerant species and cultivars to grow, particularly in agricultural situations where cash

Table 7.9 Range and typical responses of crops and pastures to surface application of topsoil incorporated lime.

Country	Crop	Soil acidity	Absolute range and typical range[a] of response to lime (% relative to control)	Reference
New Zealand	White clover-based pasture (mineral soils)	pHW 5.0–6.0	0–10 (about 5)	Edmeade et al. (1984)
	Clover-based pasture organic soils	pHW 4.0–5.0	10–140 (50–100)	
Australia, NSW, Vic	Subterranean clover-based pasture	pHCa 4.2–4.7	−15 to 170 (20–60)	Scott et al. (2000)
	Phalaris pastures	pHCa 4.1–5.0	0–900 (30–100)	
	Cocksfoot pastures	pHCa 4.1–5.0	0–70 (about 20)	
Australia, NSW, Vic	Wheat	pHW 5.0–6.0	0–38	Coventry et al. (1989)
USA, southern	Cotton	pHW 5.0–5.4	0–400 (10–30)	Adams (1984), Adams and Pearson (1967)
	Corn	pHW 4.7–5.5	11–56 (10–30)	
	Soybean	pHW 4.7–5.8	0–56 (30–50)	
	Alfalfa	pHW 4.7–5.8	150 to > 1000 (100–200)	
	Other forage crops[b]	pHW 4.3–5.5	0–250 (10–30)	
USA, Midwest	Corn	pHW 5.0	75–96c	McLean and Brown (1984)
	Soybean	pHW 5.0	70–91[c]	
	Alfalfa	pHW 5.0	20–84[c]	
Various[d]	Soybean	A1 base saturation 75%[e]	25–95[f]	Dierolf et al. (1999)
	Corn	A1 base saturation 75%[e]	66–95[f]	
	Various[g]	A1 base saturation 75%[e]	80–95[f]	

[a]Typical value by visual inspection of the data.
[b]White clover, crimson clover, Bermuda grass, millet, sorghum, ryegrass.
[c]Response at pH 5.0 relative to maximum yield.
[d]All tropical soils (Oxisols and Ultiols) from ten countries.
[e]Al base saturation % of ECEC.
[f]Response at Al saturation of 75%, relative to maximum yield.
[g]Groundnut, rice, cowpea, mungbean, pigeon pea, cassava, tanier, yams, sugarcane.
Source: Modified and used with permission from Edmeades and Ridley (2003).

Table 7.10 Responses to fertilisation of rice and common bean grown in rotation in Cerrado and Varzea acid soils.

Fertility level	Rice grain yield/t ha^{-1}	Common bean grain yield/t ha^{-1}
	Cerrado soil	
Low	1.7^{b}	1.2^{c}
Medium	2.1^{a}	1.8^{b}
High	2.1^{a}	2.2^{a}
Medium + green manure	2.4^{a}	1.5^{a}
F test	*	**
	Varzea soil	
Low	4.3^{b}	2.9^{b}
Medium	5.5^{a}	6.6^{a}
High	5.5^{a}	8.5^{a}
Medium + green manure	6.3^{a}	8.2^{a}
F test	**	**

*and ** indicate significance at the 0.05 and 0.01 probability levels. Within the same column, means followed by the same letter do not differ significantly at the 0.05 probability level by Tukey's test.
Source: Modified and used with permission from Fageria and Baligar (2003).

flow is needed to facilitate further soil liming. In so doing, it should not be forgotten that specific toxicities and deficiencies are likely to be involved and the selected species/cultivars will need to be suitable from this perspective also. It will not do to select a cultivar that is relatively acid tolerant on some soils but which is highly sensitive to Mo deficiency if that is a key cause of the acid soil infertility for the field in question. In any case, given the increasing use of gene technology it is likely that additional advances in the production of acid tolerant crop species will continue to occur. Furthermore, isolation of key genes such as those underlying plant root Al tolerance are not only useful for genetic engineering applications but can also be used as markers to help select for tolerance traits in standard (i.e. traditional) plant breeding trials. Research has shown there to be a wide range of genetic diversity for acid soil and Al toxicity tolerance, which is not only conducive to crop improvement via hybridization and phenotypic selection but also provides a rich basis for the molecular manipulation of plant genotypes (Garvin and Carver, 2003).

In the special case of highly acidic ASS, management for plant and crop growth must be carefully balanced around the hydrological regime (Wilson et al., 1999). For example, shallow drains for agricultural land management are considered preferable to deep drains as they impart reduced risk of additional subsoil pyrite oxidation and acid production. In all cases, the further disturbance and exposure of unoxidised materials is to be avoided. Acid tolerant crops (e.g. sugar cane) can potentially be successfully farmed on ASS, with regular liming and good drainage management key to the success of such ventures. Where major acidification has resulted in scalded areas with visually distinctive sparse patches of 'burnt' or 'scalded' vegetation, rehabilitation is particularly difficult (Lin et al., 2001). Flushing with freshwater may be of some benefit although research in this area is still ongoing as there is considerable potential for the standing water after flooding to become highly acidic prior to discharge into local receiving waters (Melville and White, 2000).

7.3.6 Managing plant growth in alkaline soils

Moderately, alkaline soils frequently provide the basis for highly productive agricultural land. However, at higher levels of alkalinity problems are increasingly encountered, and soils with pH 9.5 and above are generally considered unsuitable for agriculture (Charman and Murphy, 2000). Typical problems encountered in alkaline soils include plant nutrient deficiencies (e.g. Fe deficiency chlorosis), root growth inhibition and stunting (Gruber and Kosegarten, 2002). In addition, as highly alkaline soils are also often sodic, some of the problems associated with alkaline soils are in fact derived from their salt-affected nature and associated management issues (e.g. poor soil structure, poor drainage) as covered in Section 7.2, rather than from the alkalinity per se.

Respiring roots and microorganisms release CO_2 which hydrolyses in soil water to produce carbonic acid, thereby forming localised areas of acidity in alkaline soils which can aid in the plant uptake of key nutrients. Nevertheless, these and other biologically driven processes to solubilise key nutrients (such as the excretion of siderophores, protons and enzymes from roots) (Hedley et al., 1982; Grinsted et al., 1982; Dakora and Phillips, 2002; Hinsinger et al., 2003) are not always sufficient to overcome alkaline soil infertility, and active management is frequently required to mitigate the associated retardation of plant growth. The most common nutrient deficiency in plants growing on alkaline soils is iron (Fe) deficiency; however, numerous other plant nutrients such as P, Mn, Zn and Cu also have limited bioavailability in the alkaline pH range (Chapter 6).

Some arid and semi-arid zone plants are relatively well adapted to growing in alkaline soils (e.g. Abd El-Ghani, 1998; Buxbaum and Vanderbilt, 2007) because, as indicated earlier, the low precipitation and restricted leaching conditions common to arid areas are often a contributing factor to the formation of alkaline soils. By contrast, many important agricultural plant species have a low tolerance to this type of soil environment (e.g. Kerley et al., 2002). Despite this, it is far less common to practise wide-scale pH amendment of alkaline soils than it is to undertake liming and pH amendment of acid soils. Acidifying reactions can, however, be brought about through the application of elemental sulphur (S), which is subsequently oxidised to sulphuric acid by both chemical and microbially mediated reactions (Lindemann et al., 1991; Cifuentes and Lindemann, 1993; Neilsen et al., 1993). For the purposes of gardening, aluminium sulphate ($Al_2(SO_4)_3$) and iron sulphate ($FeSO_4$) are commonly used for alkaline soil amendment.

Given the limited broad-scale pH amendment of alkaline soils, it is often important for crops to be specially selected as suitable for alkaline soil cropping. Indeed, even within species some varieties are likely to be better adapted to alkaline soils than others (e.g. Brand et al., 1999). In addition to the careful selection of existing species and cultivars with useful adaptive characteristics, genetic engineering technology can also facilitate the identification, isolation and transfer of key genes from native species that are particularly well adapted to alkaline soil conditions. Micronutrient fertilisers may also be applied as a strategy for improving crop performance, either as a soil amendment or in foliar sprays or trunk implants (Wallace, 1991).

As calcareous soils are particularly common in regions of low rainfall, irrigation is frequently required in order to facilitate plant growth for agricultural production. This in itself may be linked to management problems if drainage through the calcic horizon is impeded and inorganic salts from the irrigation water accumulate within the soil profile, potentially leading to the development of salinity and/or sodicity.

References

AACM International Pty Ltd. (1995) *Social and Economic Feasibility of Ameliorating Soil Acidification – A National Review*. Land and Water Resources Research and Development Corporation, Australia.

Abd El-Ghani, M.M. (1998) Environmental correlates of species distribution in arid desert ecosystems of eastern Egypt. *Journal of Arid Environments*, **38**, 297–313.

Abrol, I.P., Yadav, J.S.P., and Massoud, F.I. (1988) *Salt-Affected Soils and Their Management*. FAO Soils Bulletin No. 39. Food and Agriculture Organisation of the United Nations, Rome.

Adams, F. (1984) Crop responses to lime in the southern United States. In: *Soil Acidity and Liming* (ed. F. Adams), 2nd edn., pp. 211–265. Agronomy monograph No. 12, American Society of Agronomy, Madison, Wisconsin.

Adams, F. and Pearson, R.W. (1967) Crop responses to lime in the southern United States and Puerto Rico. In: *Soil Acidity and Liming* (eds R.W. Pearson and F. Adams), pp. 161–206. Agronomy monograph No. 12, American Society of Agronomy, Madison, Wisconsin.

Alewell, C. (2003) Acid inputs into the soils from acid rain. In: *Handbook of Soil Acidity* (ed. R. Zdenko), pp. 83–116. Marcel Dekker Inc., New York.

Allen, H.E., Yanqing, L., and Di Toro, D.M. (2008) Ecotoxicity of Ni in soil. *Mineralogical Magazine*, **72**, 367–371.

Alloway, B.J. (2013a) Heavy Metals in Soil. Springer, Berlin.

Alloway, B.J. (2013b) Sources of heavy metals and metalloids in soil. In: Heavy Metals in Soils (ed. B.J. Alloway), pp. 11–50. Springer, Berlin.

Anderson, C.W.N., Brooks, R.R., Chiarucci, A., LaCoste, C.J., Leblanc, M., Robinson, B.H., Simcock, R., and Stewart, R.B. (1999) Phytomining for nickel, thallium and gold. *Journal of Geochemical Exploration*, **67**, 407–415.

Anderson, L.L., Walsh, M., Roy, A., Bianchetti, C.M., and Merchan, G. (2011) The potential of *Thelypteris palustris* and Asparagus sprengeri in phytoremediation of arsenic contamination. *International Journal of Phytoremediation*, **13**, 177–184.

Andriesse, W. and van Mensvoort, M.E.F. (2006) Acid sulphate soils: distribution and extent. In: *Encyclopedia of Soil Science*, pp. 14–19. Taylor and Francis, New York.

Arnold, R.E., Hodson, M.E., and Comber, S. (2007) Effect of organic complexation on the toxicity of Cu to the earthworm *Eisenia fetida*. *Applied Geochemistry*, **22**, 2397–2405.

Åström, M. (2001) The effect of acid soil leaching on trace element abundance in a medium sized stream W. Finland. *Applied Geochemistry*, **16**, 387–396.

Ayers, R.S. and Westcot, D.W. (1985) *Water Quality for Agriculture*. Irrigation and drainage paper 29. Rev. 1. FAO, Rome.

Baeder-Bederski-Anteda, O. (2003) Phytovolatilisation of organic chemicals. *Journal of Soils and Sediments*, **3**, 65–71.

Baker, A.J.M. and Brooks, R.R (1989) Terrestrial higher plants which hyperaccumulate metal elements–a review of their distribution, ecology and phytochemistry. *Biorecovery*, **1**, 81–126.

Baker, A.J.M., McGrath, S.P., Sidoli, C.M.D., and Reeves, R.D. (1994) The possibility of in situ heavy metal decontamination of polluted soils using crops of metal-accumulating plants. *Resources, Conservation and Recycling*, **11**, 41–49.

Balks, M.R., Bond, W.J., and Smith, C.J. (1998) Effects of sodium accumulation on soil physical properties under an effluent-irrigated plantation. *Australian Journal of Soil Research*, **36**, 821–830.

Barber, S. (1995) *Soil Nutrient Bioavailability: a Mechanistic Approach*, 2nd edn. John Wiley & Sons, Inc., New York.

Barcelo, J. and Poschenrieder, C. (2002). Fast root growth responses, root exudates, and internal detoxification as clues to the mechanisms of aluminium toxicity and resistance: a review. *Environmental and Experimental Botany*, **48**, 75–92.

Bartels, D. and Sunkar, R. (2005) Drought and salt tolerance in plants. *Critical Reviews in Plant Sciences*, **24**, 23–58.

Bini, C. and Bresolin, F. (1998) Soil acidification by acid rain in forest ecosystem: a case study in northern Italy. *Science of the Total Environment*, **222**,1–15.

Bond, W.J. (1998) Effluent irrigation – an environmental challenge for soil science. *Australian Journal of Soil Research*, **36**, 543–555.

Bonilla, I., Cadania, C., and Carpena, O. (1980) Effects of boron on nitrogen metabolism and sugar levels of sugar beet. *Plant and Soil* **57**, 3–9.

Bordeleau, L.M. and Prevost, D. (1994) Nodulation and nitrogen fixation in extreme environments. *Plant and Soil*, **161**, 115–125.

Borlaug, N.E. and Dowswell, C.R. (1993) Fertilizer: to nourish infertile soil that feeds a fertile population the crowds a fragile world. Keynote address at the 61st Annual Conference, International Fertilizer Industry Association (IFA), May 24–27, 1993, New Orleans, cited in: Lopes, A.S. (1996). Soils under Cerrado: A success story in soil management. *Better Crops International*, **10**, 9–15.

Bowen, H. J. M. (1979). *Environmental Chemistry of the Elements*. Academic Press, London.

Brady, N.C. and Weil, R.R. (2010) *Elements of the Nature and Properties of Soil*, 3rd Edition, p. 306. Pearson Education, Inc., Upper Saddle River, NJ.

Brand, J.D., Tang, C.. and Rathgen, A.J. (1999) Adaptation of Lupinus augustifolius L. And L-Pilosus Murr to calcareous soil. *Australian Journal of Agricultural Research*, **50**, 1027–1033.

Brimhall, G.H., Lewis, C.J., Ford, C., Bratt, J., Taylor, G., and Warin, O. (1991) Quantitative geochemical approach to pedogenesis – importance of parent material reduction, volumetric expansion, and eolian influx in lateritization. *Geoderma*, **51**, 51–91.

Brooks, R.R., Le, J., Reeves, R.D., and Jaffre, T. (1977) Detection of nickeliferous rocks by analysis of herbarium specimens of indicator plants. *Journal of Geochemical Exploration*, **7** 49–57.

Bui, E.N., Krogh, L., Lavado, R.S., Nachtergaele, F.O., Toth, T., and Fitzpatrick, R.W. (1998) Distribution of sodic soils: the world scene. In: *Sodic Soils: Distribution, Properties, Management and Environmental Consequences* (eds M.E. Sumner, and R. Naidu), pp.19–34. Oxford University Press, New York.

Bui, T.K.A., Kim, D.D., Tua, I.V., Kien, N.T., and Anh, D.T. (2011) Phytoremediation potential of indigenous plants from Thia Nguyen province, Vietnam. *Journal of Environmental Biology*, **32**, 257–262.

Burrow, D.P., Surapaneni, A., Rogers, M.E., and Olsson, K.A. (2002) Groundwater use in forage production: the effect of saline-sodic irrigation and subsequent leaching on soil sodicity. *Australian Journal of Experimental Agriculture*, **42**, 237–247

Buxbaum, C.A.Z. and Vanderbilt, K. (2007) Soil heterogeneity and the distribution of desert and steppe plant species across a desert-grassland ecotone. *Journal of Arid Environments*, **69**, 617–632.

Candy, I. and Black, S. (2009) The timing of Quaternary calcrete development in semi-arid southeast Spain: investigating the role of climate on calcrete genesis. *Sedimentary Geology*, **218**, 6–15.

Casado, M., Anawar, H.M., Garcia-Sanchez, A., and Regina, I.S. (2008) Cadmium and zinc in polluted mining soils and uptake by plants (El Losar mine, Spain). *International Journal of Environment and Pollution*, **33**,146–159.

Chaney, R.L. (1983) Plant uptake of inorganic waste constitutes. In: *Land Treatment of Hazardous Wastes* (eds J.F. Parr, P.B. Marsh, and J.M. Kla), pp. 50–76. Noyes Data Corp, Park Ridge, Illinois.

Chaney, R.L., Malik, M., Li, Y.M., Brown, S.L., Brewer, E.P., Angle, J.S., and Baker, A.J.M. (1997) Phytoremediation of soil metals. *Current Opinion in Biotechnology*, **8**, 279–284.

Charman, P.E.V. and Murphy, B.W. (2000) *Soils: Their Properties and Management*, 2nd edn. Oxford University Press, Oxford.

Chaudan, R.P.S. and Powar, S.L. (1978) Tolerance of wheat and pea to boron in irrigated water. *Plant and Soil*, **50**, 145–149.

Chinnusamy, V., Jagendorf, A., and Zhu, J.K. (2005) Understanding and improving salt tolerance in plants. *Crop Science*, **45**, 437–448.

Choi, J.Y., Seo, Y.S., Kim, S.J., Kim, W.T., and Shin, J.S. (2011) Constitutive expression of CaXTH3, a hot pepper xyloglucan endotransglucosylase/hydrolase, enhanced tolerance to salt and drought stresses without phenotypic defects in tomato plants (Solanum lycopersicum cv. Dotaerang). *Plant Cell Reports*, **30**, 867–877.

Cifuentes, F.R. and Lindemann, W.C. (1993) Organic matter stimulation of elemental sulphur oxidation in a calcareous soil. *Soil Science Society of America Journal*, **57**, 727–731.

Cole, M.M. (1973) Geobotanical and biogeochemical investigations in the schlerophyllous woodland and shrub associations of the Eastern Goldfields area of Western Australia with particular reference to the role of *Hybanthus floribundus* (Lindl.) F. Muell. As a nickel indicator and accumulator plant. *Applied Ecology*, **10**, 269–320.

Corfield, J. (2000) The Effects of Acid Sulphate Run-off on a Sub-Tidal Estuarine Macro Benthic Community in the Richmond River, NSW, Australia. *ICES Journal of Marine Science*, **57**, 1517–1523.

Coventry, D.R., Walker, B.R., Morrison, G.R., Hyland, M.T., Avery, J.C., and Maden, J.J.L. (1989) Yield responses to lime of wheat and barley on acid soils in north-eastern Victoria. *Australian Journal of Experimental Agriculture*, **29**, 209–214.

Cunningham, S.D. and Ow, D.W. (1996) Promises and prospects of phytoremediation. *Plant Physiology*, **110**, 715–719.

Cunningham, S.D., Anderson, T.A., Schwab, A.P., and Hsu, F.C. (1996) Phytoremediation of soils contaminated with organic pollutants. *Advances in Agronomy*, **56**, 55–114.

Dakora, F.D. and Phillips D.A. (2002) Root exudates as mediators of mineral acquisition in low-nutrient environments. *Plant and Soil*, **245**, 35–47.

Denby, K. and Gehring, C. (2005) Engineering drought and salinity tolerance in plants: lessons from genome-wide expression profiling in Arabidopsis. *Trends in Biotechnology*, **23**, 547–552.

Dent, D.L. and Pons, L.J. (1995) A world perspective on acid sulphate soils. *Geoderma*, **67**, 263–276.

Derry, L.A. and Chadwick, O.A. (2007) Contributions from Earth's atmosphere to soil. *Elements*, **3**, 333–338.

Di Toro, D.M., Allen, H.E., Bergman, H.L., Meyer, J.S., Paquin, P.R., and Santore, R.C. (2001) Biotic ligand model of the acute toxicity of metals.1. Technical basis. *Environmental Toxicology and Chemistry*, **20**, 2383–2396.

Dickinson, N.M., Baker, A.J.M., Doronila, A., Laidlaw, S., and Reeves, R. (2009) Phytoremediation of inorganics: realism and synergies. *International Journal of Phytoremediation*, **11**, 97–114.

Dierolf, T.S., Amien, L.I., and Yost, R.S. (1999) Testing and validation of an expert system for making lime decisions. *Nutrient Cycling in Agroecosystems*, **54**, 9–20.

Doksopulo, EP (1961) *Nickel in Rocks, Soils, Waters and Plants Adjacent to the Chorchanskaya Group* (in Russian). University of Tbilisi, Tbilisi, Georgia.

Dourmett, S., Lamperi, L., Checchini, L., Azzarello, E., Mugnai, S., Mancuso, S., Petruzzelli, G., and Bubba, M. (2008) Heavy metal distribution between contaminated soil and Paulownia tomentosa, in a pilot-scale assisted phytoremediation study: influence of different complexing agents. *Chemosphere*, **72**, 1481–1490.

Duffus, J.H. (2002) Heavy metals – a meaningless term? *Pure and Applied Chemistry*, **74**, 793–807.

Duquene, L., Van den Hove, H., Tack, F., Meers, E., Baeten, J., and Wannijn, J. (2009) Enhanced phytoextraction of uranium and selected heavy metals by Indian mustard and ryegrass using biodegradable soil amendments. *Science of the Total Environment*, **407**, 1496–1505.

Edmeades, D.C., and Ridley, A.M. (2003) Using lime to ameliorate topsoil and subsoil acidity. In: *Handbook of Soil Acidity* (ed. R. Zdenko), pp. 297–336. Marcel Dekker Inc., New York.

Edmeades, D.C., Pringle, R.M., Shannon, P.W., and Mansell, G.P. (1984) Effects of lime on pasture production on soils in the North Island of New Zealand. 4. Predicting lime responses. *New Zealand Journal of Agricultural Research*, **27**, 371–382.

Endo, T., Yamamoto, S., Honna, T., and Eneji, A.E. (2002) Sodium-calcium exchange selectivity as influenced by clay minerals and composition. *Soil Science*, **167**,117–125.

Entry, J.A., Rygiewicz, P.T., Watrud, L.S., and Donnelly, P.K. (2002) Influence of adverse soils conditions on the formation and function of Arbuscular mycorrhizas. *Advances in Environmental Research*, **7**, 123–138.

Fageria, N.K. and Baligar, V.C. (2003) Fertility management of tropical acid soils for sustainable crop production. In: *Handbook of Soil Acidity* (ed. Z. Rengel), pp. 359–385. Marcel Dekker, New York.

Fandeur, D., Juillot, F., Morin, G., Olivi, L., Cognigni, A., Webb, S.M., Ambrosi, J.P., Fritsch, E., Guyot, F., and Brown, G.E. (2009) XANES evidence for oxidation of Cr(III) to Cr(VI) by Mn-Oxides in a lateritic regolith developed on serpentinized ultramafic rocks of New Caledonia. *Environmental Science and Technology*, **43**, 7384–7390.

Fanning, D.S. (2002) Acid sulphate soils. In: *Encyclopedia of Soil Science* (ed. R. Lal), pp. 11–13. Marcel Dekker, New York.

FAO (1991) *World Soil Resources Report 66*. Rome, Italy.

FAO (1998) *World Reference Base for Soil Resources*. Food and Agriculture Organization of the United Nations, Rome, Italy.

FAO (2002) *World Agriculture: Towards 2015/2030*. Summary Report, Food and Agriculture Organization of the United Nations, Rome, Italy.

Farina, M.P.W., Channon, P., and Thibaud, G.R. (2000a) A comparison of strategies for ameliorating subsoil acidity: I. Long-term growth effects. *Soil Science Society of America Journal*, **64**, 646–651.

Farina, M.P.W., Channon, P., and Thibaud, G.R. (2000b) A comparison of strategies for ameliorating subsoil acidity: I. Long-term soil effects. *Soil Science Society of America Journal*, **64**, 652–658.

Fernando, D.R., Marshall, A.T., Gouget, B., Carriere, M., Collins, R.N., Woodrow, I.E., and Baker, A.J. (2008) Novel pattern of foliar metal distribution in a manganese hyperaccumulator. *Functional Plant Biology*, **35**, 193–200.

Fitzpatrick, R.W., Boucher, S.C., Naidu, R., and Fritsch, E. (1994) Environmental consequences of soil sodicity. *Australian Journal of Soil Research*, **32**, 1069–1093.

Fitzpatrick, R.W., Merry, R., Williams, J., White, I., Bowman, G., and Taylor, G. (2001) *Acid Sulphate Soils Assessment, Coastal, Inland and Minesite Conditions*. CSIRO Land and Water, Australia.

Fitzpatrick, R.W., Shand, P., and Merry, R.H. (2009) Acid sulphate soils. In: *Natural History of the Riverland and Murraylands* (ed. J.T. Jennings), pp. 66–110. Royal Society of South Australia (Inc.), Adelaide, South Australia.

Flowers, T.J. and Colmer, T.D. (2008) Tansley review: salinity tolerance in halophytes. *New Phytologist*, **179**, 945–963.

Ford, G.W., Martin, J.J., Rengasamy, P., Boucher, S.C., and Ellington, A. (1993) Soil sodicity in Victoria. *Australian Journal of Soil Research*, **31**, 869–909

Galeas, M.L., Zhang, L.H., Freeman, J.L., Wegner, M., and Pilon-Smits, E.A.H. (2007) Seasonal fluctuations of selenium and sulphur accumulation in selenium hyperaccumulators and related nonaccumulators. *New Phytologist*, **173**, 517–525.

Galloway, J.N. (2001) Acidification of the world: natural and anthropogenic. *Water Air and Soil Pollution*, **130**, 17–24.

Gao, S.Q., Chen, M., Xu, Z.S., Zhao, C.P., Li, L.C., Xu, H.J., Tang, Y.M., Zhao, X., and Ma, Y.Z. (2011) The soybean GmbZIP1 transcription factor enhances multiple abiotic stress tolerances in transgenic plants. *Plant Molecular Biology*, **75**, 537–553.

Garvin, D.F. and Carver, B.F. (2003) Role of the genotype in tolerance to acidity and aluminium toxicity. In: *Handbook of Soil Acidity* (ed. R. Zdenko), pp. 387–406. Marcel Dekker Inc., New York.

Genc, Y., McDonald, G.K., and Tester, M. (2007) Reassessment of tissue Na+concentration as a criterion for salinity tolerance in bread wheat. *Plant, Cell and Environment*, **30**, 1486–1498.

Gerhardt, K.E,, Huang, X.D., Glick, B.R., and Greenberg, B.M. (2009) Phytoremediation and rhizoremediation of organic soil contaminants: potential and challenges. *Plant Science*, **176**, 20–30.

Golez, N.V. and Kyuma, K. (1997) Influence of pyrite oxidation and soil acidification on some essential nutrient elements. *Aquacultural Engineering*, **16**, 107–124.

Gonzaga, M.I.S., Ma, L.Q., Santos, J.A.G., and Matias, M.I.S. (2009) Rhizosphere characteristics of two arsenic hyperaccumulating Pteris ferns. *Science of the Total Environment*, **407**, 4711–4716.

Greenfelt, P. and Galloway, J.N. (1996) In: *Global Acid Deposition Assessment* (eds D.M. Whelpdale and M.S. Kaiser). WMO/GAW Report No. 106.

Greenway, H. and Munns, R. (1980) Mechanisms of salt tolerance in nonhalophytes. *Annual Review of Plant Physiology*, **31**, 149–190.

Grinsted, M.J., Hedley, M.J., White, R.E., and Nye, P.H. (1982) Plant induced changes in the rhizosphere of Rape (Brassica napus var. Emerald) seedlings. I. pH change and the increase in P concentration in the soil solution. *New Phytologist*, **91**, 19–29.

Gruber, B. and Kosegarten, H. (2002) Depressed growth of non-chlorotic vine grown in calcareous soil is an iron deficiency symptom prior to leaf chlorosis. *Journal of Plant Nutrition and Soil Science*, **165**, 111–117.

Halliwell, D.J., Barlow, K.M., and Nash, D.M. (2001) A review of the effects of wastewater sodium on soil physical properties and their implications for irrigation systems. *Australian Journal of Soil Research*, **39**, 1259–1267

Haque, N., Peralta-Videa, J.R., Jones, G.L., Gill, T.E., and Gardea-Torresdey, J.L. (2008) screening the phytoremediation potential of desert broom (Baccharis sarothroides Gray) growing on mine tailings in Arizona, USA. *Environmental Pollution*, **153**, 362–368.

Hasegawa, P.M., Bressan, R.A., Zhu, J.K., and Bohnert, H.J. (2000) Plant cellular and molecular responses to high salinity. *Annual Review of Plant Physiology and Plant Molecular Biology*, **51**, 463–499.

Hayes, K.F. (1987) Equilibrium, spectroscopic and kinetic studies of ion adsorption at the oxide/aqueous interface. PhD Dissertation, Stanford University, Stanford, California.

Hedley, M.J., Nye, P.H., and White, R.E. (1982) Plant induced changes in the rhizosphere of rape (*Brassica napus* var. emerald) seedlings. II. Origin of the pH change. *New Phytologist*, **91**, 31–44.

Hicks, W.K., Kuylenstierna, J.C.I., Owen, A., Dentener, F., Seip, H.M., and Rodhe, H. (2008) Soil sensitivity to acidification in Asia: status and prospects. *AMBIO*, **37**, 295–303.

Hinsinger, P., Plassard, C., Tang, C., and Jaillard, B. (2003) Origins of root-induced pH changes in the rhizosphere and their responses to environmental constraints: a review. *Plant and Soil*, **248**, 43–59.

Hodson, M.E. (2004) Heavy metals – geochemical bogey men? *Environmental Pollution*, **129**, 341–343.

Huang, J.W. and Cunningham, S.D. (1996) Lead phytoextraction: species variation in lead uptake and translocation. *New Phytologist*, **134**, 75–84.

Hyne, R. and Wilson, S.P. (1997) Toxicity of acid sulphate soil leachate and aluminium to the embryos and larvae of Australian Bass (Maquaria novemaculeata) in estuarine water. *Environmental Pollution*, **97**, 221–227.

Ibaraki, T., Kuroyanagi, N., and Murakami, M. (2009) Practical phytoextraction in cadmium-polluted paddy fields using a high cadmium accumulating rice plant cultured by early drainage of irrigation water. *Soil Science and Plant Nutrition*, **55**, 421–427.

Impellitteri, C.A., Allen, H.E., Yin, Y., You, S-J., and Saxe, J.K. (2001) Soil properties controlling metal partitioning. In: *Heavy Metals Release in Soils* (eds H.M. Selim and D. Sparks), pp. 149–165. Lewis Publishers, New York.

Irtelli, B. and Navari-Izzo, F. (2006) Influence of sodium nitrolotriacetate (NTA) and citric acid on phenolic and organic acids in *Brassica juncea* grown in excess cadmium. *Chemosphere*, **65**, 1348–1354.

Isbell R.F. (1996). *The Australian Soil Classification*. CSIRO Publishing, Melbourne.

Jacobsen, T. (1982) *Salinity and Irrigation Agriculture in Antiquity. Diyala Basin Archaeological Projects: Report on Essential Results 1957–8*. Bibliotheca Mesopotamia No. 14. Udena Publications, Malibu, California.

Jaffré, T. and Schmidt, M. (1974) Comptes Rendus de l'Académie des Sciences. *Série D: Sciences Naturelles*, **278**, 1727–1730

Jaffré, T., Brooks, R.R., Lee, J., and Reeves, R.D. (1976) *Serbia acuminate* a hyperaccumulator of nickel from New Caledonia. *Science*, **193**, 579–580.

Jambhulkar, H.P. and Juwarkar, A.A. (2009) Assessment of bioaccumulation of heavy metals by different plant species grown on fly ash dump. *Ecotoxicology and Environmental Safety*, **72**, 1122–1128.

James, R.A., Davenport, R., and Munns, R. (2006) Physiological characterisation of two genes for Na^+ exclusion in wheat: *Nax1* and *Nax2*. *Plant Physiology* **142**, 1537–1547.

James, R.A., von Caemmerer, S., Condon, A.G., Zwart, A.B., and Munns, R. (2008) Genetic variation in tolerance to the osmotic stress component of salinity stress in durum wheat. *Functional Plant Biology*, **35**, 111–123.

Jayawardane, N.S., Biswas, T.K., Blackwell, J., and Cook, F.J. (2001) Management of salinity and sodicity in a land FILTER system, for treating saline wastewater on a saline-sodic soil. *Australian Journal of Soil Research*, **39**, 1247–1258

Jensen, E.S. and Hauggaard-Nielsen, H. (2003) How can increased use of biological N-2 fixation in agriculture benefit the environment? *Plant and Soil*, **252**, 177–186.

Jensen, J.K., Holm, P.E., Mejrup, J., Larsen, M.B., and Borggaard, O.K. (2009) The potential of willow for remediation of heavy metal polluted calcareous urban soils. *Environmental Pollution*, **157**, 931–937.

Johnson, G.V. and Fixen, P.E. (1990) Tests for sulphur, boron, molybdenum and chlorine. In: *Soil Testing and Plant Analysis* (ed. R.L. Westerman), pp. 265–273. Soil Science Society of America, Wisconsin.

Kabata-Pendias, A. (2000) *Trace Elements in Soils and Plants, pp 423*. CRC Press, Boca-Raton, Florida.

Kalve, S., Sarangi, B.K., Pandey, R.A., and Chakrabarti, T. (2011) Arsenic and chromium hyperaccumulation by an ecotype of Pteris vittata I prospective for phytoextraction from contaminated water and soil. *Current Science*, **100**, 888–894.

Kaman, H., Kirda, C., Cetin, M., and Topcu, S. (2006). Salt accumulation in the root zones of tomato and cotton irrigated with partial root-drying technique. *Irrigation and Drainage*, **55**, 533–544.

Kamprath, E.J. (1970) Exchangeable aluminium as a criterion for liming leached mineral soils. *Soil Science Society of America Proceedings*, **34**, 252–254.

Kerley, S.J., Norgaard, C., Leach, J.E., Christiansen, J.L., Huyghe, C., and Romer, P. (2002) The development of potential screens based on shoot calcium and iron concentrations for the evaluation of tolerance in Egyptian genotypes of white lupin (Lupinus albus L.) to limed soils. *Annals of Botany*, **89**, 341–349.

Klink, C.A. and Machado, R.B. (2005) Conservation of the Brazilian Cerrado. *Conservation Biology*, **19**, 707–713.

Kubachka, K.M., Meija, J., Leduc, D.L., Terry, N., and Caruso, J.A. (2007) Selenium volatiles as proxy to the metabolic pathways of selenium in genetically modified Brassica juncea. *Environmental Science and Technology*, **41**, 1863–1869.

Kumar, P.B.A.N., Dushenkov, V., Motto H., and Raskin I. (1995) Phytoextraction- the use of plants to remove heavy-metals from soils. *Environmental Science and Technology*, **29**, 1232–1238.

Kuylenstierna, J.C.I., Rodhe, H., Cinderby, S., and Hicks, K. (2001) Acidification in developing countries: Ecosystem sensitivity and the critical load approach on a global scale. *Ambio*, **30**, 20–28.

LaCoste, C., Robinson, B., and Brooks, R. (2001) Uptake of thallium by vegetables: Its significance for human health, phytoremediation, and phytomining. *Journal of Plant Nutrition*, **24**, 1205–1215

LaGrega, M.D., Buckingham, P.L., Evans, J.C., and Environmental Resources Management (2001) *Hazardous Waste Management*, p. 1202. McGraw Hill, New York.

Läuchli, A. and Grattan, S.R. (2007) Chapter 1. Plant growth and development under salinity stress. In: *Advances in Molecular Breeding Toward Drought and Salt Tolerant Crops* (eds M.A. Jenks et al.), pp. 1–32. Springer..

Lehmann, J., Solomon, D., Zhao, F.J., and McGrath, S.P. (2008) Atmospheric SO2 emissions since the late 1800s change organic sulfur forms in humic substance extracts of soils. *Environmental Science and Technology*, **42**, 3550–3555.

Lei, M. Chen, T.B., Huang, Z.C., Wang, Y.D., and Huang, Y.Y. (2008) simultaneous compartmentalisation of lead and arsenic in co-hyperaccumulator *Viola principis* H. de Boiss: an application of SRXRF microprobe. *Chemo*, **72** 1491–1496.

Letey, J (1984) Impact of salinity on the development of soil science. In: *Soil Salinity under Irrigation: Processes and Management* (eds I. Shainberg and J. Shalhevet), pp. 1–11. Springer-Verlag, Berlin, Germany.

Levy, G.J., Shainberg, I., and Miller, W.P. (1998) Physical properties of sodic soils. In: *Sodic Soils: Distribution, Properties, Management and Environmental Consequences* (eds M.E. Sumner and R. Naidu), pp. 77–94. Oxford University Press, New York.

Liang, H.M., Lin, T.H., Chiou, J.M., and Yeh, K.C. (2009) Model evaluation of the phytoextraction potential of heavy metal hyperaccumulators and non-accumulators. *Environmental Pollution*, **157**, 1945–1952.

Lin, C., Rosicky, M., McConchie, D., Sullivan, L.A., and Lancaster, G. (2001) Coastal land scalding in NSW, Australia: soil chemical characteristics and their implications for remediation of the scalded lands. *Land Degradation and Development*, **12**, 293–303.

Lindemann, W.C., Aburto, J.J., Haffner, W.M., and Bono, A.A. (1991) Effect of sulphur source on sulphur oxidation. *Soil Science Society of America Journal*, **55**, 85–90.

Lindsay, W.L. (1979) *Chemical Equilibira in Soils*. John Wiley & Sons, Ltd., Chichester, UK.

Lindsay, M.P., Lagudah, E.S., Hare, R.A., and Munns, R. (2004) A locus for sodium exclusion (Nax1), a trait for salt tolerance, mapped in durum wheat. *Functional Plant Biology*, **31**, 1105–1114.

Liu, X.H., Gao, Y.T., Khan, S., Duan, G., Chen, A.K. Ling, L., Zhao, L., Liu, Z., and Wu, X.C. (2008) Accumulation of Pb, Cu and Zn in native plants growing on contaminated sites and their potential accumulation capacity in Heqing, Yunnan. *Journal of Environmental Science- China*, **20**, 1469–1474.

Liu, Z., He, X., and Chen, W. (2011a) effects of cadmium hyperaccumulation on the concentrations of four trace elements in *Lonicera japonica* Thunb. *Ecotoxicology*, **20**, 698–705.

Liu, J., Duan, C.Q., Zhang, X.H., Zhu, Y.N.A., and Lu, X.Y. (2011b) Potential of Leersia hexandra Swartz for phytoextraction of Cr from soil. *Journal of Hazardous Materials*, **188**, 85–91.

Ljung, K., Maley, F., Cook, A., and Weinstein, P. (2009). Acid sulfate soils and human health – a millenium ecosystem assessment. *Environment International*, **35**, 1234–1242.

Lock, K., Van Eeckhout, H., De Schamphelaere, K.A.C., Criel, P., and Janssen, C.R. (2007) Development of a biotic ligand model (BLM) predicting nickel toxicity to barley (Hordeum vulgare). *Chemosphere*, **66**, 1346–1352.

Lombi, E., Zhao, F.J., Dunham, S.J., and McGrath, S.P. (2001) Phytoremediation of heavy metal-contaminated soils: natural hyperaccumulation versus chemically enhanced phytoextraction. *Journal of Environmental Quality*, **30**, 1919–1926.

Loomis, W.D. and Durst, R.W. (1992) Boron and cell walls. *Current topics in Plant Biochemistry and Physiology*, **10**, 149–178.

Lopes, A.S. (1996) Soils under Cerrado: a success story in soil management. *Better Crops International*, **10**, 9–15.

Lopez, M.L., Peralta-Videa, J.R., Parsons, J.G., Benitez, T., and Gardea-Torresdey, J.L. (2007) Gibberellic acid, kinetin and the mixture indole-3-acetic acid assisted with EDTA-induced lead hyperaccumulation in alfalfa plants. *Environmental Science and Technology*, **41**, 8165–8170.

Lukaszewski, K.M., Blevins, D.G., and Randall, D.D. (1992) asparagine and boric acid cause allantoate accumujlation in soybean leaves by inhibiting manganese dependent allantoaet amidohydrolase. *Plant Physiology*, **99**, 1670–1676.

Ma, L.Q., Komar, K.M., Tu, C., Zhang, W., Cai., Y., and Kennelley, ED (2001) A hardy, versatile, fast-growing plant helps remove arsenic from contaminated soils. *Nature* **409**, 579.

Maas, E.V. (1993) Salinity and citriculture. *Tree physiology*, **12**, 195–216.

Maas, E.V. and Grattan, S.R. (1999) Crop yields as affected by salinity. In: *Agricultural Drainage, Agronomy Monograph No. 38*, pp. 55–108. American Society of Agronomy, Crop Science Society of America, Soil Science Society of America, Madison, Wisconsin.

Maas, E.V. and Poss, J.A. (1989) Salt sensitivity of wheat at different growth stages. *Irrigation Science*, **10**, 29–40.

Marschner, H. (1991) Mechanisms of adaptation of plants to acid soils. *Plant and Soil*, **134**, 1–20.

Marschner, H. (1995) *Mineral Nutrition of Higher Plants*, pp. 889. Academic Press, London.

Matzner, E. and Davis, M. (1996) Chemical soil conditions in pristine Nothofagus forests of New Zealand as compared to German forest. *Plant and Soil*, **186**, 285–291.

Maxted, A.P., Black, C.R., West, H.M., Crout, N.M.J., McGrath, S.P., and Young, S.D. (2007) Phytoextraction of cadmium and zinc by Salix from soil historically amended with sewage sludge. *Plant and Soil*, **290**, 157–172.

McGrath, S.P. (1998) Phytoextraction for soil remediation. In: *Plants that Hyperaccumulate Heavy Metals* (ed. R.R. Brooks), pp. 261–287. CAB International, Wallingford, UK.

McGrath, S.P., Lombi, E., Gray, C.W., Caille, N., Dunham, S.J., and Zhao, F.J. (2006) Field evaluation of Cd and Zn phytoextraction potential by the hyperaccumulators Thlaspi caerulescens and *Arabidopsis halleri*. *Environmental Pollution*, **141**, 115–125.

McKenzie, D.C., Abbott, T.S., Chan, K.Y., Slavich, P.G., and Hall, D.J.M. (1993) The nature, distribution and management of sodic soils in New South Wales. *Australian Journal of Soil Research*, **31**, 839–868.

McLean, E.O. and Brown, J.R. (1984) Crop responses to lime in the Midwestern United States. In: *Soil Acidity and Liming* (ed. F. Adams), 2nd edn., pp. 267–303. Agronomy monograph No. 12. American Society of Agronomy, Madison, Wisconsin.

Melville, M.D. and White, I. (2000) Acid sulphate soils. In: *Soils: Their Properties and Management* (eds P.E.V. Charman and B.W. Murphy), 2nd edn., pp. 249–252. Oxford University Press, Oxford, UK.

Memon, A.R. and Schroder, P. (2009) Implications of metal accumulation mechanisms to phytoremediation. *Environmental Science and Pollution Research International*, **16**, 162–175.

Minguzzi, C. and Vergnano, O. (1948) Il contenuto di nichel nelle ceneri di *Alyssum bertolonii*. *Atti della Società Toscana di Scienze Naturale*, **55**, 49–74.

Mittler, R. and Blumwald, E. (2010) Genetic engineering for modern agriculture: challenges and perspectives. *Annual Review of Plant Biology*, **61**, 443–462.

Munns, R. (2002) Comparative physiology of salt and water stress. *Plant, Cell and Environment*, **25**, 239–250.

Munns, R. (2005) Genes and salt tolerance: bringing them together. *New Phytologist*, **167**, 645–663.

Munns, R. (2009) Strategies for crop improvement in saline soils. In: *Salinity and Water Stress: Improving Crop Efficiency* (eds M. Ashraf, M. Ozturk, and H.R. Athar), pp. 99–110. Tasks for Vegetation Science 41. Springer. DOI: 10.1007/978-1-4020-9065-3.

Munns, R. and James, R.A. (2003) Screening methods for salinity tolerance: a case study with tetraploid wheat. *Plant and Soil*, **253**, 201–218.

Munns, R., and Tester, M. (2008) Mechanisms of salinity tolerance. *Annual Review of Plant Biology*, **59**, 651–681.

Nathanail, P.C. and Bardos, R.P. (2004) *Reclamation of Contaminated Land*, p. 238. John Wiley & Sons, Ltd., Chichester, UK.

Neilsen, D., Hogue, E.J., Hoyt, P.B., and Drought, B.G. (1993) Oxidation of elemental sulphur and acidulation of calcareous orchard soils in southern British Columbia. *Canadian Journal of Soil Science*, **73**, 103–114.

Nordstrom, D.K. (1982). Aqueous pyrite oxidation and the consequent formation of secondary iron minerals. In *Acid sulphate weathering* (eds. J.A. Kittrick, D.S. Fanning and L.R Hossner), vol 10, special publication, pp 37–56. Soil Science Society of America, Madison, Wisconsin.

Northcote, K.H., and Skene, J.K.M. (1972) Australian soils with saline and sodic properties. Soil publication no. 27. Commonwealth Scientific and Industrial research orgnaisation (CSIRO), Melbourne, Australia.

Nowack, B., Schulin, R., and Robinson, B. H. (2006). Critical assessment of chelant-enhanced metal phytoextraction. *Environmental Science and Technology*, **40**, 5225–5232.

Oster, J.D. and Shainberg, I. (2001) Soil responses to sodicity and salinity: challenges and opportunities. *Australian Journal of Soil Research*, **39**, 1219–1224.

Österholm, P. and Åström, M. (2004). Quantification of current and future leaching of sulphur and metals from Boreal acid sulphate soils, western Finland. *Australian Journal of Soil Research*, **42**, 547–551.

Passioura, J.B. and Munns, R. (2000). Rapid environmental changes that affect leaf water status induce transient surges or pauses in leaf expansion rate. *Australian Journal of Plant Physiology*, **27**, 941–948.

Pavlovkin, J., Pal'ove-Balang, P., Kolarovic, L., and Zelinova, V. (2009) Growth and functional responses of different cultivars of Lotus corniculatus to aluminum and low pH stress. *Journal of Plant Physiology*, **166**, 1479–1487.

Pessarakli, M. (1991) Water utilisation and soil salinity control in arid-zone agriculture. *Communications in Soil Science and Plant Analysis*, **22**, 1787–1796.

Pilon-Smits, E.A.H. and Leduc, D.L. (2009) Phytoremediation of selenium using transgenic plants. *Current Opinions in Biotechnology*, **20**, 207–212.

Qadir, M., and Schubert, S. (2002) Degradation processes and nutrient constraints in sodic soils. *Land Degradation and Development*, **13**, 275–294.

Qadir, M., Ghafoor, A., and Murtaza, G. (2000) Amelioration strategies for saline soils: a review. *Land Degradation Development*, **11**, 501–521.

Qiu, R.L., Thangavel, P., Hu, P.J., Senthilkumar, P., Ying, R.R., and Tang, Y.T. (2011) Interaction of cadmium and zinc on accumulation and sub-cellular distribution in leaves of hyperaccumulator *Potentilla griffithii*. *Journal of Hazardous Materials*, **186**, 1425–1430.

Rajendran, K., Tester, M., and Roy, S.J. (2009) Quantifying the three main components of salinity tolerance in cereals. *Plant, Cell and Environment*, **32**, 237–249.

Raskin, I. (1996) Plant genetic engineering may help with environmental cleanup – commentary. *Proceedings of the National Academy of Science, USA*, **93**, 3164–3166.

Redondo-Gomez, S., MAteos-Naranjos, E., Vecino-Bueno, I., and Feldman, S.R. (2011) Accumulation and tolerance characteristics of chromium in a cordgrass Cr-hyperaccumulator, *Spartina argentinensis*. *Journal of Hazardous Materials*, **185**, 862–869.

Reeves, R.D. and Baker, A.J.M. (2000) Metal-accumulating plants. In: (*Phytoremediation of Toxic Metals: Using Plants to Clean Up the Environment* (eds I. Raskin and B.D. Ensley), pp.193–229. John Wiley & Sons, Inc., New York.

Rengasamy, P. (2002) Transient salinity and subsoil constraints to dryland farming in Australian sodic soils: an overview. *Australian Journal of Experimental Agriculture*, **42**, 351–361.

Rengasamy, P. (2006) World salinization with emphasis on Australia. *Journal of Experimental Botany*, **57**, 1017–1023.

Rengasamy, P. and Marchuk, A. (2011) Cation ratio of soil structural stability (CROSS). *Soil Research*, **49**, 280–285.

Rengasamy, P. and Olsson, K.A. (1991) Sodicity and soil structure. *Australian Journal of Soil Research*, **29**, 935–952.

Richards, L.A. (ed.) (1954) *Diagnosis and Improvement of Saline and Sodic Soils*. USDA Agric. Handbook No. 60, Washington, District of Colombia.

Robinson, B.H., Bañuelos, G., Conesa, H.M., Evangelous, M.W.H., and Rainer, S. (2009) The phytomanagement of trace elements in soil. *Critical Reviews in Plant Sciences*, **28**, 240–266.

Rodhe, H., Dentener, F., and Schulz, M. (2002) The global distribution of acidifying wet deposition. *Environmental Science and Technology*, **36**, 4382–4388.

Rowell, D.L. (1994) *Soil Science: Methods and Applications*, p. 350. Prentice Hall, Harlow, Essex.

Roy, S.J., Tucker, E.J., and Tester, M. (2011) Genetic analysis of abiotic stress tolerance in crops. *Current Opinion in Plant Biology*, **14**, 232–239.

Rugh, C.L., Senecoff, J.F., Meagher, R.B., and Merkle, S.A. (1998) Development of transgenic yellow poplar for mercury phytoremediation. *Nature Biotechnology*, **16**, 925–928.

Ruiz, O.N. and Daniell, H. (2009) Genetic engineering to enhance mercury phytoremediation. *Current Opinion in Biotechnology*, **20**, 213–219.

Ryan, J., Miyamoto, S., and Stroehlein, J.L. (1977) Relation of solute and orbed boron to the boron hazard in irrigation water. *Plant and Soil*, **47**, 253–256.

Saifullah, Meers, E., Qadir, M., de Caritat, P., Tack, F.M.G., Du Laing, G., and Zia, M.H. (2009) EDTA-assisted Pb phytoextraction. *Chemosphere*, **74**, 1279–1291.

Sairam, R.K. and Tyagi, A. (2004) Physiology and molecular biology of salinity stress tolerance in plants. *Current Science*, **86**, 407–421.

Sammut, J., White, I., and Melville, M.D. (1996) Acidification of an estuarine tributary in eastern Australia due to drainage of acid sulphate soils. *Marine Freshwater Research*, **47**, 669–684.

Samouelian, A. and Cornu, S. (2008) Modelling the formation and evolution of soils, towards an initial synthesis. *Geoderma*, **62**, 401–409.

Schnoor, J.L., Licht, L.A. , McCutcheon, S.C., Wolfe, N.L. and Carreira, L.H. (1995). Phytoremediation of Organic and nutrient contaminants. *Environmental Science and Technology*, **29**, 318A–323A.

Schofield, R., Thomas, D.S.G., and Kirkby, M.J. (2001) Causal processes of soil salinisation in Tunisia, Spain and Hungary. *Land Degradation and Development*, **12**, 163–181.

Scott, B.J., Ridley, A.M., and Conyers, M.K. (2000) Management of soil acidity in long-term pastures of south-eastern Australia: a review. *Australian Journal of Experimental Agriculture*, **40**, 1173–1198.

Severne, B.C. and Brooks, R.R. (1972) A nickel-accumulating plant from Western Australia. *Planta*, **103**, 91–94.

Shannon, M.C. and Grieve, C.M. (1999) Tolerance of vegetable crops to salinity. *Scientia Horticulurae*, **78**, 5–38.

Shaw, R., Brebber, L., Ahern, C., and Weinand, M. (1994) A review of sodicity and sodic soil behaviour in Queensland. *Australian Journal of Soil Research*, **32**, 143–172.

Shibata, M., Konno, T., Akaike, R., Xu, Y., Shen, R.F., and Ma, J.F. (2007) Phytoremediation of Pb contaminated soil with polymer-coated EDTA. *Plant and Soil*, **290**, 201–208.

Slattery, W.J. and Coventry, D.R. (1993) Response of wheat, triticale, barley and canola to lime on four soil types in north-eastern Victoria. *Australian Journal of Experimental Agriculture*, **33**, 609–618.

Smolders, E. and Mertens, J. (2013) Cadmium. In: Heavy Metals in Soils (ed. B.J. Alloway), pp. 283–312. Springer, Berlin.

Sparks, D.L. (1995) *Environmental Soil Chemistry*, p. 267. Academic Press, London.

Steenbergen, N.T.T.M., Iaccino, F., de Winkel, M., Reijnders, L., and Peijnenburg, W.J.G.M. (2005) Development of a biotic ligan model and a regression model predicting acute copper toxicity to the earthworm Aporrectodea caliginosa. *Environmental Science and Technology*, **39**, 5694–5703.

Steppuhn, H., van Genuchten, M. Th., and Grieve, C.M. (2005) Root-zone salinity: II. Indices for tolerance in agricultural crops. *Crop Science*, **45**, 221–232.

Stuanes, A.O., Abrahamsen, G., and Rosberg, I. (1995) Acidification of soils in five catchments in Norway. *Water Air and Soil Pollution*, **85**, 635–640.

Stumpe, J.M. and Vlek, P.L.G. (1991) Acidification induced by different nitrogen-sources in columns of selected tropical soils. *Soil Science Society of America Journal*, **55**, 145–151.

Sumner, M.E. (1993) Sodic soils: new perspectives. *Australian Journal of Soil Research*, **31**, 683–750.

Sumner, M.E. and Noble, A.D. (2003) Soil acidification: the world story. In: *Handbook of Soil Acidity* (ed. R. Zdenko), pp. 1–28. Marcel Dekker Inc., New York.

Sumner, M.E., Rengasamy, P., and Naidu, R. (1998) Sodic soils: a reappraisal. In: *Sodic Soils: Distribution, Properties, Management and Environmental Consequences* (eds M.E. Sumner and R. Naidu), pp. 3–18. Oxford University Press, New York.

Sun, Y.B., Zhou, Q.X., An, J., Liu, W.T., and Liu, R. (2009) Chelator-enhanced phytoextraction of heavy metals from contaminated soil irrigated by industrial wastewater with the hyperaccumulator plant (Sedum alfredii Hance). *Geoderma*, **150**, 106–112.

Sutton, T., Baumann, U., Hayes, J., Collins, N.C., Shi, B-J., Schnurbusch, T., Hay, A., Mayo, G., Pallotta, M., Tester, M., and Langridge, P. (2007) Boron-toxicity tolerance in barley arising from efflux transporter amplification. *Science*, **318**, 1446–1449.

Sykes, S.R. (1992) The inheritance of salt exclusion in woody perennial fruit species. *Plant and Soil*, **146**, 123–129.

Szabolcs, I. (1979) *Review of Research on Salt-Affected Soils*. United Nations Educational, Scientific and Cultural Organisation (UNESCO), Paris, France.

Tang, C., McLay, C.D.A., and Barton, L. (1997) A comparison of proton excretion of twelve pasture legumes grown in nutrient solution. *Australian Journal of Experimental Agriculture*, **37**, 563–570.

Tavakkoli, E., Rengasamy, P., and Mcdonald, G.K. (2010) The response of barley to salinity stress differs between hydroponics and soil systems. *Functional Plant Biology*, **37**, 621–633.

Teakle, N.L. and Tyerman, S.D. (2010) Mechanisms of Cl– transport contributing to salt tolerance. *Plant, Cell and Environment*, **33**, 566–589.

Tester, M. and Davenport, R. (2003) Na+tolerance and Na+transport in higher plants. *Annals of Botany*, **91**, 503–527.

Thakali, S., Allen, H.E. Di Toro, D.M., Ponizovsky, A.A., Rooney, C.P., Zhao, F.-J., and McGrath, S.P. (2006a) A terrestrial biotic ligand model. 1. Development and application to Cu and Ni toxicities to barley root elongation in soils. *Environmental Science and Technology*, **40**, 7085–7093.

Thakali, S., Allen, H.E. Di Toro, D.M., Ponizovsky, A.A., Rooney, C.P., Zhao, F.-J., McGrath, S.P., Criel, P., van Eeckhout, H., Janssen, C., Oorts, K., and Smolders, E. (2006b) A terrestrial biotic ligand model. 2. Application of Ni and Cu toxicities to plants, invertebrates and microbes in soil. *Environmental Science and Technology*, **40**, 7094–7100.

Tillman, R.W. and Surapaneni, A. (2002) Some soil-related issues in the disposal of effluent on land. *Australian Journal of Experimental Agriculture*, **42**, 225–235.

Tomlinson, G.H. (2003) Acidic deposition, nutrient leaching and forest growth. *Biogeochemistry*, **65**, 51–81.

Turkan, I. and Demiral, T. (2009). Recent developments in understanding salinity tolerance. *Environmental and Experimental Botany*, **67**, 2–9. DOI: 10.1016/j.envexpbot.2009.05.008.

Ulrich, B. (1991) An ecosystem approach to soil acidification. In: *Soil Acidity* (eds B. Ulrich and M.E. Sumner). Springer-Verlag, Berlin, Germany.

Ulrich, B. and Sumner, M.E. (1991) *Soil Acidity*. Springer-Verlag, Berlin, Germany.

Umali, D.L. (1993) *Irrigation-induced salinity. A growing problem for development and the environment.* World Bank Technical Paper No. 215. The World Bank, Washington, D.C.

Vamerali, T., Bandiera, M., Coletto, L., Zanetti, F., Dickinson, M., and Mosca, G. (2009) Phytoremediation trials on metal- and arsenic-contaminated pyrite wastes (Torviscosa, Italy). *Environmental Pollution*, **157**, 887–894.

Van Gestel, C.A.M., and Koolhaas, J.E. (2004) Water-extractability, free ion activity, and pH explain cadmium sorption and toxicity to Folsomia candida (Collembola) in seven soil-pH combinations. *Environmental Toxicology and Chemistry*, **23**, 1822–1833.

Van Nevel, L., Mertens, J., Oorts, K., and Verheyen, K. (2007) Phytoextraction of metals from soils: How far from practice? *Environmental Pollution*, **150**, 34–40.

Verbruggen, N., Hermans, C., and Schat, H. (2009) Molecular mechanisms of metal hyperaccumulation in plants. *New Phytologist*, **181**, 759–776.

Von Uexküll, H.R. and Mutert, E. (1995) Global extent, development and economic impact of acid soils. *Plant and Soil*, **171**, 1–15.

Wallace, A. (1991) Rational approaches to control of iron-deficiency other than plant-breeding and choice of resistant cultivars. *Plant and Soil*, **130**, 281–288.

White P.J. and Broadley M.R. (2001) Chloride in soils and its uptake and movement within the plant: a review. *Annals of Botany*, **88**, 967–988.

Wild, A. (2003) *Soils, Land and Food*. Cambridge University Press, p. 246. Cambridge, UK.

Wilson, B.P., White, I., and Melville, M.D. (1999). Floodplain hydrology, acid discharge and change in water quality associated with a drained acid sulfate soil. *Marine and Freshwater Research*, **50**, 149–157.

WMO (1996) *World Meteorological Organisation/Global Atmospheric Watch* (eds D.M. Whelpdale and M.S. Kaiser). Report No. 106.

Wu, F.Y., Ye, Z.H., Wu, S.C., and Wong, M.H. (2007) Metal accumulation and arbuscular mycorrhizal status in metallicolous and nonmetallicolous populations of *Pteris vittata* L. and *Sedum alfredii* Hance. *Planta*, **226**, 1363–1378.

Xu, G., Magen, H., Tarchitzky, J., and Kafkafi U. (2000) Advances in chloride nutrition of plants. *Advances in Agronomy*, **68**, 97–150.

Xu, X.H., Chen, X.C., Shi, J.Y., Chen, Y.X., Wu, W.X., and Perara, A. (2007) Effects of manganese on uptake and translocation of nutrients in a hyperaccumulator. *Journal of Plant Nutrition*, **30**, 1737–1751.

Yamada, T. (2005). The Cerrado of Brazil: a success story of production on acid soils. *Soil Science and Plant Nutrition*, **51**, 617–620.

Yamada, K., Miyahara, K., and Kotoyori, T. (1975) Studies on soil pollution caused by heavy metals. Part IV: soil purification by plants that absorb heavy metals. *Gamm Ken Nogyo Shienjo Hokuku*, **15**, 39–54.

Yamaguchi, T. and Blumwald, E. (2005) Developing salt-tolerant crop plants: challenges and opportunities. *Trends in Plant Science*, **10**, 615–620.

Zhang, H.X. and Blumwald, E. (2001) Transgenic salt-tolerant tomato plants accumulate salt in foliage but not in fruit. *Nature Biotechnology*, **19**, 765–768.

Zhang, X., Xia, H., Li, Zhi'an, Zhuang, P., and Gao, B. (2011) Identification of a new potential Cd-hyperaccumulator *Solanum photeinocarpum* by soil seed-bank metal concentration gradient method. *Journal of Hazardous Materials*, **189**, 414–419.

Zhao, F.J. and McGrath, S.P. (2009) Biofortification and phytoremediation. *Current Opinion in Plant Biology*, **12**, 373–380

Zhu, J.K. (2001) Plant salt tolerance. *Trends in Plant Science*, **6**, 66–71.

8 Managing the soil physical environment for plants

Paul D. Hallett[1] and A. Glyn Bengough[1,2]

[1] *The James Hutton Institute, Dundee, UK*
[2] *Division of Civil Engineering, University of Dundee, Dundee, UK*

8.1 Introduction

The performance of plants is affected considerably by the soil physical environment. High-quality soil is characterised by a porous structure that drains well but stores water to support plants, breaks up easily and allows roots to elongate so that they can capture nutrients and water. If soil is damaged through compaction or drains poorly, the yield of crops, forest growth or quality of sports fields can decrease. The physical environment of soil drives competition between species and therefore vegetation ecology. People invest considerable time and money manipulating the physical conditions of soil to improve plant productivity. One of the greatest expenditures in agriculture is soil cultivation to improve seedbed conditions for seed germination, shoot emergence and root growth. Foresters also adopt soil cultivation if soil is damaged by harvesting machinery or has inherent physical properties that may hinder tree establishment and productivity. Land for amenity purposes, such as golf courses, sport fields and parkland, is sometimes modified with imported soil where the texture, packing and drainage are engineered to produce improved conditions for plant growth and recovery after mechanical stress.

This chapter describes the physical properties of soil that influence plants and how these properties can be managed. It also describes natural processes, induced by weathering or biological processes, that may improve soil physical conditions. Large areas of land have been disturbed by tillage, intensive grazing, vehicle damage and engineering projects with the long-term use of soils dependent on the natural physical resilience of soil. Careful soil management, through natural processes and sometimes mechanical intervention, makes it possible to enhance the soil physical environment for plants.

8.2 Basic physical properties of soil

This chapter begins with basic definitions of soil physical properties. Much of the information presented also requires an understanding of soil water physics (Chapters 2 and 9).

Soil Conditions and Plant Growth. Edited by Peter J. Gregory and Stephen Nortcliff.
© 2013 Blackwell Publishing Ltd. Published 2013 by Blackwell Publishing Ltd.

Information on processes underlying the formation of soil physical structure can be found in Chapters 3 (Concepts of soil fertility), 4 (Soil organic matter) and 10 (Plant-induced changes to soil processes).

8.2.1 Porosity, density and void ratio

Plant roots grow in the pores between soil particles. The abundance, interconnectivity, size distribution and tortuosity of the soil pore system therefore affect the capacity of soil to support plants. The most widely used soil property related to the abundance of pores is bulk density, ρ_d, defined from the mass of soil, M_s, in a given volume V_t:

$$\rho_d = \frac{M_s}{V_t} \tag{8.1}$$

Bulk density is simple to measure from an extracted core of soil and is often used to describe compaction. As in equation 8.1, it is usually expressed in terms of dry mass of soil, but can also be expressed in terms of wet mass, which is more common for some potting mix manufacturers and geotechnical engineers. Mineral soils usually have bulk densities ranging from 1.0 to 1.9 $Mg\,m^{-3}$ depending on texture and soil compaction (Keller and Hakansson, 2010). Organic soils have much smaller bulk densities because of the small density of organic compounds. Particle density, ρ_s, is defined as $\rho_s = M_s / V_s$, where V_s is the volume of soil particles alone (i.e. air space removed). For mineral soil, ρ_s varies from 2.6 to 2.7 $Mg\,m^{-3}$ (Blanco-Canqui et al., 2006), but appreciable amounts of organic carbon can reduce this to about 1.5 $Mg\,m^{-3}$ (Ruhlmann et al., 2006). Potting media used in horticulture have a much smaller density because of the greater proportion of organic compounds present. Wiberg et al. (2005) found a 76% coefficient of variation in the bulk density of 24 brands of potting media, mainly due to poor quality control. Moreover, the compressibility of potting media and temporal changes due to decomposition complicate the maintenance of prescribed bulk densities (Blok and Wever, 2008).

A simpler to interpret physical property is soil porosity, φ,

$$\varphi = \frac{V_p}{V_t} \tag{8.2}$$

where V_p is the volume of pores. Unlike bulk density, particle density does not affect the measurement, so it is a better description of pore space in soil. Porosity can be determined from ρ^s and ρ_d^b by

$$\varphi = \frac{\rho_s - \rho_d}{\rho_s} \tag{8.3}$$

Geotechnical engineers often describe the abundance of pores as void ratio, e, defined as

$$e = \frac{V_p}{V_s} \tag{8.4}$$

As the denominator, V_s, is fixed and thus remains constant regardless of soil density or porosity, it offers mathematical simplicity for soil-management studies (Dexter, 1997). ρ_d^b and φ, in contrast, are derived from V_t in the denominator, which changes with fluctuations in porosity. The conversion from φ to e is

$$e = \frac{\varphi}{1 - \varphi} \tag{8.5}$$

8.2.2 Soil water

Water resides within soil pores, with bulk abundance defined by either gravimetric, w, or volumetric, θ, water content,

$$w = \frac{M_w}{M_s} \text{ and} \tag{8.6}$$

$$\theta = \frac{V_w}{V_t} \tag{8.7}$$

where M_w and V_w are the mass and volume of water, respectively. Conversions between gravimetric and volumetric water content are a potential source of operator error, particularly if the property is poorly defined as 'water content' only and expressed as a percentage. Typically, manual measurements in the laboratory report gravimetric water content, whereas most soil water sensors report volumetric water content. If ρ_d is known, θ can be derived from w by

$$\theta = w\rho_d / \rho_w, \tag{8.8}$$

where ρ_w is the density of water.

Chapter 9 (Section 9.2.2) describes water retention in soil. Capillary forces between air, liquid and solid interfaces result in water held at negative matric potentials in unsaturated soils. There is an inverse relationship between the magnitude of matric potential (or suction) and pore radius, resulting in the gradual drainage of pores from largest to smallest as suction increases and the soil dries. The proportion of pores filled with water influences both aeration and the amount of stored water available for plants.

8.2.3 Soil strength

The rate of root growth is influenced considerably by soil strength, which is imparted by capillary forces and particle bonds in soil (Pärtel and Wilson, 2001; Bengough, 2003). Poor soil management that results in compaction (Glab, 2008; Alameda and Villar, 2009) or losses in organic matter (Emerson et al., 1994; Hati et al., 2006) generally increases soil strength, resulting in decreased root growth. A common field method for measuring soil strength is the use of a field penetrometer, which measures the resistance experienced by a metal cone being pushed into soil over a range of depths (Fritton, 2008; Sinnett et al., 2008). Another approach is the use of a shear vane, which has been advocated by some as a better measurement to describe mechanical resistance experienced by roots in the field (McKenzie and McBratney, 2001), but it is sometimes difficult to use for shallow measurements. Field measurements with either approach are influenced considerably by soil water content, because of changes in matric potential and degree of saturation (McKenzie and McBratney, 2001; Whalley et al., 2007). An alternative approach is to test soil cores equilibrated to prescribed matric potentials with a miniature penetrometer in the laboratory (Bengough and Mullins, 1990). This allows for testing at multiple matric potentials and removes the impact of water content on the variability of field results. Another limitation to penetrometer readings is the frictional resistance of the metal cone, so Bengough et al. (1991) developed a rotating-tip penetrometer to reduce its influence. In subsequent work, they found that using a rotating-tip penetrometer to reduce the frictional resistance reduced penetration resistance by more than 50% (Bengough et al., 1997).

Physical limitations to root growth are described in Section 8.5.

8.3 Soil structure

Soil structure is the physical framework in which all soil processes occur. The arrangement of solids, gases, liquids, voids, organic matter and organisms defines soil structure. It thus has a large influence on the capacity of plant roots to grow through soil, extract water and nutrients, exchange gases and interact with other plants, herbivores and pathogens (Angers and Caron, 1998). The range of sizes of different components of soil structure is enormous, varying by a factor of $>10^8$ between the diameter of small clay particles (typically 10^{-7} m) and soil hydrological features at field scale. Dexter (1988) described soil structure as the 'spatial heterogeneity of the different components or properties of soil', which embraces the complex environment created by the myriad of mineral particles, organic substances, organisms, weathering processes and anthropogenic impacts underlying soil formation. This definition also embraces the complexity of the resulting pore structure and the potential influence on the growth and functioning of plant roots. Young and Crawford (2004) described soil as 'the most complicated biomaterial on the planet'.

A surge in research beginning in the 1990s showed conclusively that omitting soil structure from examinations of chemical, physical and biological processes in the natural soil environment often produces results that differ greatly from reality (Horn et al., 1994; Connolly, 1998). In the same decade, the rapid increase in available computer power combined with new mathematical approaches allowed soil structure to be defined far more accurately than was previously possible (Rieu and Sposito, 1991; Crawford and Matsui, 1996; Bird et al., 2000). Models of water and solute transport (Kosugi, 1999; Vogel et al., 2006; Zhang and Lv, 2009), biochemical transformations (Totsche et al., 2010) and plant root development (Pierret et al., 2007) among others have been developed where a major component is a detailed description of the soil pore structure.

Figure 8.1 illustrates the complex physical structure of soil, ranging in scale from the agglomeration of clay into domains to large-scale cracks. Textural porosity is derived from spaces left from the imperfect packing of soil minerals. Loam soils tend to have larger densities than clay soils because the greater array of particle sizes results in the in-filling of textural pores (Table 8.1). Structural pores form at small-scale (μm) from the aggregation of clay and organic matter complexes (Chapter 4). At larger scales, cracking and spaces left from decomposing roots or earthworm burrows form macropores. In general, textural pores provide water storage for plant roots, whereas structural pores provide rapid transport pathways.

The importance of structural pores is often underestimated. They are the large volume singularities in soil and consequently usually air filled. This spacious, aerobic environment is required by many micro-organisms (Emerson and Greenland, 1990). Structural pores are the weak point in soil that fractures under the action of tillage implements (Hallett et al., 1995), so their abundance and arrangement is important to the energy expended in creating seedbeds. Table 8.2 lists the size classification of soil pores and their relevance in terms of soil properties.

8.3.1 Soil structure formation

Soil structure formation is driven by various hydrological, mechanical and biological processes (Horn et al., 1994; Angers and Caron, 1998; Cosentino et al., 2006). The hydrological processes are the frequency and rate of wetting and drying of soil (Emerson

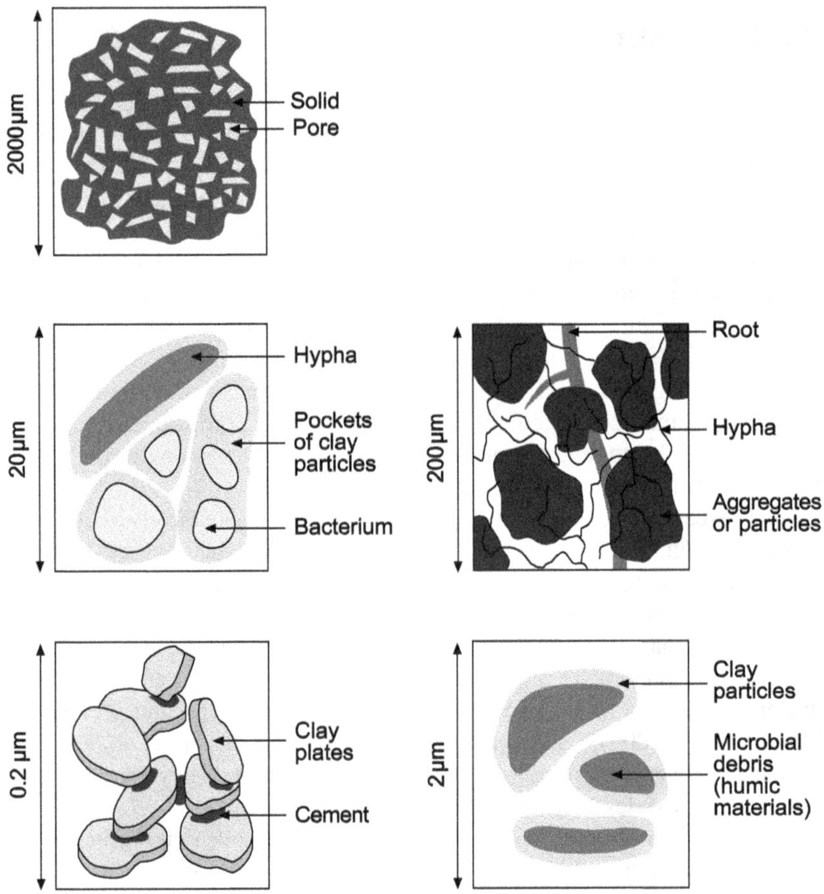

Figure 8.1 Soil physical structure across a wide range of scales. Redrawn from Tisdall and Oades (1982). With kind permission from John Wiley & Sons. For a colour version of this figure, please see Plate 8.1.

Table 8.1 Typical densities and porosities of soils with different textures and their parent material.

Description	Bulk density (Mg m^{-3})	Porosity
Clay texture	1.12	0.58
Loam texture	1.28	0.52
Sandy texture	1.61	0.39
Sandy loam (compacted)	1.90	0.29
Sandstone	2.12	0.20
Soil minerals	2.65	0.00

and Grundy, 1954). Mechanical processes include the bonding of soil particles and external stresses from overburden soil, compaction or tillage. Hydrological and mechanical processes interact in the capillary cohesion of soil particles by water menisci in pores, shrinking and swelling dynamics, and the weight of overburden soil (Or and Ghezzehei, 2002). Biological processes include the mechanical deformation of soil by roots or fauna, exudates that alter

Table 8.2 Soil pore sizes and their relevance to soil properties.

Size, m	Example
10^{-2}	Macropores: cracks, clods
10^{-3} (mm)	Tilled seedbed
10^{-4}	Smallest diameter for root penetration
10^{-5}	Largest water filled pores at field capacity
10^{-6} (μm)	Smallest size accessible to bacteria
10^{-7}	Plants can access water held in these pores
10^{-8}	Space between clay plates (can be 50% of pore space)
10^{-9} (nm)	Thickness of three layers of water molecules on clay surface

bond energy, water surface tension or wetting rates, and organic matter inputs that serve as nuclei for aggregation and may alter soil physical behaviour (Cosentino et al., 2006).

Cycles of drying and wetting induce shrinking and swelling in non-rigid soils and can cause soils to slake or crack (Yoshida and Hallett, 2008). Cracks form macropores in soil that serve as rapid transmission pathways between upper and lower soil layers (Morris, 1992). They form if sufficient mechanical energy is transmitted to soil to break bonds between soil particles (Lima and Grismer, 1994). Plant roots exploit macropores as conduits through plough pans to allow access to subsoil water (White and Kirkegaard, 2010). In row-planted crops, greater water uptake close to plants can cause a hydromechanical gradient that drives the formation of deep inter-row cracks, particularly in heavy clays such as in paddy fields (Yoshida and Adachi, 2001). This can be important to field drainage in wet regions, but in dry regions, it can result in hydrologically isolated volumes of soil and the rapid transmission of surface water to groundwater (Lima and Grismer, 1994). Large surface cracks are a very visible soil structure feature, but cracks also occur at much smaller scales; they are the boundaries defining incipient soil aggregates that coalesce under the mechanical action of tillage to produce tilth (Hallett et al., 1995).

Tillage is the most abrupt form of mechanical disturbance that impacts upon soil structure. It has a desired outcome of fragmenting soil into an array of aggregate sizes that provide conditions to support plant growth from seed germination through to the longer term uptake of water, nutrients and oxygen by roots (Warkentin, 2008). Stresses imparted from wheels during tillage or other operations, however, may cause reductions in soil porosity through compaction (Horn et al., 1994) and decrease continuity of soil pores through shear damage (O'Sullivan et al., 1999). If pore continuity decreases, transmission pathways from upper to lower soil layers decrease, with implications for gas and water transport.

Just as the rate of water transport depends greatly on the soil pore structure, the ingress of water into soil pores can result in air pressure build-up that causes soils to slake (Figure 8.2). A well-structured soil will be able to dissipate air pressure build-up through connected air-filled pores (Henin and Santamari, 1975). Bonding between particles also controls the resistance of soil to slaking, similar to the resistance to cracking described earlier. Organic compounds in soil can increase the bond energy and may also decrease the wetting rate through hydrophobicity (Czarnes et al., 2000). Consequently, a depletion of organic matter can result in soils with a much greater susceptibility to slaking.

Slaking, cracking and mechanical deformation are influenced considerably by the type and amount of clay minerals in soils. The negative surface charge of many clays forms electrostatic forces of attraction that influence bond energy (Barzegar et al., 1995). Interactions with biological compounds, such as polysaccharides, can increase bond energy

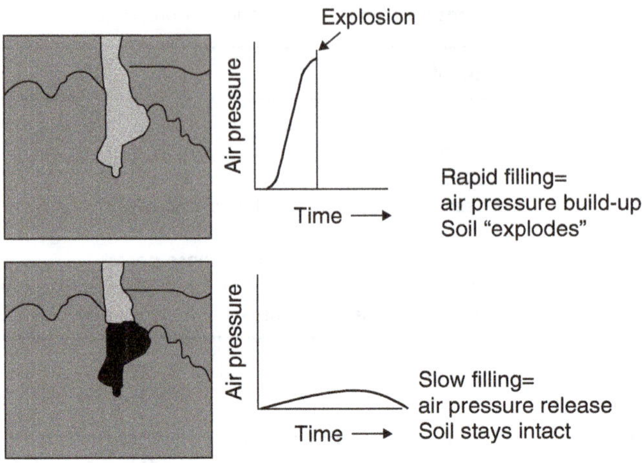

Figure 8.2 Build-up and dissipation of air pressure of soil is a major process leading to slaking of soil aggregates. For a colour version of this figure, please see Plate 8.2.

significantly (Chenu, 1989; Chenu and Guérif, 1991). Clays with a 2:1 mineral structure (e.g. smectites) have a greater propensity to shrink and swell under cycles of drying and wetting than 1:1 minerals (e.g. kaolinite). The differential swelling of clays during wetting is another process that can cause slaking.

8.3.2 Aggregation

The aggregation of soil particles is a primary driver of soil structure formation (Emerson and Greenland, 1990). Chapter 4 described mineral–organic matter complexes that produce aggregated structures in soil. For plant roots, aggregation produces inter-aggregate pore spaces that ease penetration and improve aeration and intra-aggregate pore spaces that retain water and nutrients (Angers and Caron, 1998). These pore classes were defined in Section 8.3 as structural and textural, respectively. Aggregated soils generally have greater structural stability, resulting in less temporal fluctuation in pore structure over time (Le Bissonnais, 1996; Or et al., 2000). This is extremely important for the stability of cultivated seedbeds (Atkinson et al., 2007) as structural degradation can lead to decreased water infiltration and consequently greater soil erosion (Canton et al., 2009; Vermang et al., 2009). Moreover, physically degraded seedbeds are more prone to drought and waterlogging (Or and Ghezzehei, 2002).

The concept of 'aggregates within clods' was proposed by Currie (1966) and later refined to a proposed hierarchical order of soil aggregation in the highly cited paper of Tisdall and Oades (1979). It is now widely accepted that soils possess an aggregate hierarchy that underlies soil structure (Six et al., 2004). Figure 8.1 illustrates the hierarchical arrangement of soil particles. The lowest hierarchical order can be classified as individual mineral particles such as clay plates. This order combines to form floccules or domains of clay that represent the next hierarchical order. Subsequent combinations of orders form larger units in the aggregate hierarchy. Aggregates, as a topic, receive considerable attention in this chapter because the term has become analogous with soil structure and the concept underlies a large body of research related to root growth and rhizosphere development.

Aggregates are often demarcated into size classes, with agreement among many researchers that macro-aggregates are >250 μm, meso-aggregates are 53–250 μm and micro-aggregates are <53 μm (Chan et al., 1994; Six et al., 2000). Some researchers only use 250 μm as the demarcation between macro-aggregate and micro-aggregate, with meso-aggregates not defined (Six et al., 2004; De Gryze et al., 2006; Fristensky and Grismer, 2008; Wick et al., 2009). Stable aggregates are separated from bulk soil by imparting stresses either by mechanical agitation (Ashman et al., 2009), wet sieving (Le Bissonnais, 1996) or ultrasonic dispersion (Fristensky and Grismer, 2008). The concept is that unstable soil will fragment, whereas stable soil will remain aggregated and not pass through a sieve. Immersion in water and agitation simulates the action of rapid wetting (i.e. slaking) or raindrop impact in the field. As the amount of energy imparted to soil and its spatial distribution will influence the extent of fragmentation, aggregates collected through these approaches are thought by some to be biased considerably by the fractionation procedure (Ashman et al., 2003). This debate has been going on for some time (Beare and Bruce, 1993), but the fractionation approach described by Le Bissonnais (1996) is accepted by many to be the most robust sieving approach. There is a resurgence of interest in using ultrasonic dispersion because energy inputs are more controlled (Fristensky and Grismer, 2008; Zhu et al., 2009), although North (1979) found the approach to be insensitive for identifying changes in aggregation resulting from long-term soil-management practices.

Soil aggregate stability

The structure of soil can be a highly dynamic property, governed by a range of soil properties and external stresses. Freshly tilled seedbeds, for instance, can change over time as the tilth coalesces due to raindrop impact, fluctuations in soil water content (Dexter, 1997) and compaction from machinery (O'Sullivan and Simota, 1995). Dispersion and slaking of surface soil can lead to sealing with considerable effects on sediment loss through erosion (Skidmore and Powers, 1982). In a given soil, the quantity, quality and distribution of carbon are the major variables controlling stability (Zhang and Horn, 2001; Gerke and Kohne, 2002; Goebel et al., 2005; Park and Smucker, 2005), in addition to the amount of clay minerals and their mineralogy (Denef and Six, 2005). With the massive losses in carbon experienced in many managed soils, particularly those under arable production (Bellamy et al., 2005), declines in soil stability are posing a major threat to the environment and crop productivity (Diacono and Montemurro, 2010). The Great Dust Bowl that plagued the Midwest of the USA in the 1930s and dust storms that affect Beijing presently are examples where degradation of soil structural stability can have devastating effects.

There are numerous approaches for measuring soil stability, with considerable overlap with the techniques referred to in the Section 8.3.2 to fractionate soil into different sized 'aggregates'. Many approaches have been developed since Yoder (1936) developed a standardized method of immersing soil in water on an oscillating nest of six sieves with decreasing aperture from top to bottom. He defined the lowering and raising of the sieves by 3.18 cm, at a rate of 30 cycles/min for a period of 30 min. This method forms the standard used by a large body of research on the stability of soil to wetting stresses (Kemper and Rosenau, 1986). After wet sieving, the size distribution of aggregates is assessed from the measured mass retained on each sieve size above 0.1 mm and uses Stokes Law to estimate smaller aggregate sizes. The size distribution of soil aggregates can be expressed as mean weight diameter (MWD) by

$$MWD = \sum_{i=1}^{n} \overline{x_i} w_i, \tag{8.9}$$

where $\overline{x_i}$ is the mean diameter of each sieve fraction, w_i the mass of soil retained and n the size fraction. An increase in MWD indicates greater soil stability. Other methods to characterise aggregate size distribution have been proposed, such as the use of fractals (Perfect et al., 1992).

The wet sieving approach to assess aggregate stability has been modified by altering the capillary stress during wetting (Amezketa et al., 1996) or the use of different wetting liquids (e.g. ethanol) to isolate fragmentation mechanisms (Henin, 1948; Emerson, 1954; Le Bissonnais, 1996). It is often argued that the energy input from wetting soil is difficult to quantify, so mechanical agitation through shaking (Watts and Dexter, 1997) or ultrasonic dispersion (North, 1979; Fristensky and Grismer, 2008; Zhu et al., 2009) have also been used. Another test of wet aggregate stability developed by Emerson (1954) uses a visual examination of the dispersion of clay and slaking of small air-dry soil aggregates immersed in distilled water.

Dry soil stability is also measured, generally to evaluate the percentage of soil aggregates that are not erodible by wind (Chepil, 1952; Fryrear, 1985). The tests agitate samples of dry soil over one or more sieves with no exposure to water. Aggregates <0.42 mm are characterised as wind erodible, while aggregates >0.84 mm are non-erodible (Chepil, 1951). More recent research has examined the influence of the size distribution of aggregates on wind erodibility in greater detail, using measurements such as geometric mean diameter or fractals to characterise the distribution of aggregate size (Chepil, 1951; Popham et al., 2003). Dry soil stability is not only important for assessing the potential environmental impacts of poor soil management but also the abrasive impact of wind blown soil particles on plants (Michels et al., 1993; Baker, 2007). In extreme instances, damage of crops by blown soil during establishment can require complete replanting. Climate change is anticipated to increase periods of drought, prompting concern about the potential implications of wind erosion (Merrill et al., 1999). Wind breaks offer some protection, although this common feature of fields in the early twentieth century has declined as mechanised agriculture allowed for larger areas of land to be managed.

A plethora of research has been published on soil aggregate stability, with various critiques over the years that identify problems such as the lack of a unified fractionation technique and physical meaning of collected soil aggregates (Beare and Bruce, 1993; Amezketa, 1999; Ashman et al., 2003; Warkentin, 2008). However, aggregate stability tests do provide a useful measure of the propensity of soil to structurally collapse or erode, thereby allowing for comparison between different land-management practices. They also provide some insight into the structural arrangement of soil, such as the clay–organic matter complexes that form the building blocks of soil aggregates.

8.3.3 Approaches to measuring soil structure

Measures of soil aggregates and their stability are often used as a surrogate measurement of soil structure. This provides useful insight into mechanisms underlying the formation of soil structure, but the approach is destructive, so it is difficult to reconstruct the original structure of the soil. Using aggregates to describe soil structure has been compared to studying architecture by levelling a building with a wrecking ball. Other approaches include state-of-the-art non-invasive three-dimensional (3-D) imaging and practical field-based assessments of soil structure.

Non-invasive imaging

Recent advances in non-invasive imaging using X-ray computer-aided tomography (CT) and nuclear magnetic resonance (NMR) allows for the 3-D physical structure of soil to be visualised on intact specimens (Taina et al., 2008). X-ray CT was first used with soil to estimate bulk density (Petrovic et al., 1982) and subsequently water content (Hainsworth and Aylmore, 1983). The fuzzy two-dimensional slices with poor resolution found in this early research have been replaced by 3-D images with a resolution as small as 3 μm in micro-focus scanners (Tippkotter et al., 2009) and 0.84 μm in nano-focus scanners (Sleutel et al., 2008). Dual energy source X-ray CT scanners are now available from various manufacturers as bench-top instruments that allow for the discrimination of soil, air and water (Rogasik et al., 1999). They are also beginning to allow for the discrimination of organic matter (Sleutel et al., 2008) and plant roots (Perret et al., 2007), although there is still considerable scope to improve the technique. A major limitation to the analysis of roots is the size of sample that is required and the time required for reconstruction of images assembled from huge data sets. This can limit sample scanning to <10/day and diminishes the resolution so that only macropores and large roots can be detected, although the technology is improving very rapidly (Gregory et al., 2009).

Synchrotron-source X-ray CT decreases scanning time while retaining resolution (Peth et al., 2008). Access time for users is limited as Synchrotrons are large-scale national facilities, but they have become a major research tool of some soil scientists (Feeney et al., 2006). Capture rates on small specimens can be rapid enough to monitor dynamic processes such as a water flow (Prunty et al., 2003).

NMR can discriminate the 3-D distribution of elements within soil but it is hindered by interference from ferromagnetic inclusions. The approach was used in the 1970s to characterise soil water content (Prebble and Currie, 1970) and later used to investigate pore size distribution, connectivity (Bird et al., 2005) and water retention characteristics (Jaeger et al., 2009). NMR is also used extensively to characterise organic matter in soil, including in studies of rhizosphere development (Chen et al., 2006) and root tissue decomposition (Mathers et al., 2007).

Soil thin sections

The 3D imaging tools discussed thus far are beginning to replace the older approach of thin sections (Kubiëna, 1938). In this approach, water is replaced with a solvent in intact soil specimens, the sample is then impregnated with a resin and finally cut and polished into a thin slice of soil set on a glass slide. Polishing of one surface can decrease the sample thickness to <20 μm. There are two major advantages to using thin sections over non-invasive techniques. First, microscopic techniques allow for the discrimination of the spatial distribution of clays, sand and organic matter; and second, the thin sections can also be sampled at the micro-scale to investigate the distribution of elements, either by removing material or laser ablation (Bruneau et al., 2002). Another advancement in thin sections is an ability to identify micro-organisms either through indiscriminate staining with calcofluor (Nunan et al., 2001) or the detection of specific organisms using molecular probes (Eickhorst and Tippkotter, 2008).

Intrusion and extrusion of liquids

Soil structure can also be inferred from the pore size distribution estimated from mercury porosimetry or water retention characteristics. Mercury porosimetry injects mercury into soil under pressure and allows for pores as small as 0.003 μm to be detected (Bartoli et al., 1999). The measurement of water retention characteristics is described in Chapter 9. Briefly, soils are desaturated on tension tables for smaller (less negative; i.e. 0 to −20 kPa) matric potentials or in pressure plate apparatus for larger (more negative; i.e. −20 kPa to −1500 kPa) matric potentials. Shrinkage of soil during drying influences the results, but methods have been developed to correct for this effect (Braudeau et al., 1999). Both water retention and mercury porosimetry are limited by the impacts of pore connectivity and shape so they do not provide a direct measurement of pore structure. However, equipment to measure water retention is commonplace in most soil physics laboratories and the data collected is also relevant to the accessibility of water by plants. Dexter and Bird (2001) proposed the use of water retention characteristics to provide a qualitative description of soil structure known as the S factor. This is obtained from the slope at the inflection point of a water retention curve and reflects the spread of pore sizes.

Visual indicators of soil structure

More direct, qualitative measurements of soil structure examine distinct visual features (Mueller et al., 2009). In many national soil surveys, the shapes of aggregates prised from a soil profile were characterised. These data are simple to collect and provide valuable information on the physical condition of soil, thereby identifying soil types, regions or management practices where the soil physical conditions for plant growth may be limited. There are five major classes of soil structure: platy, prismatic, columnar, granular and blocky. Subclasses, such as subangular blocky, can also be defined. Soil with a granular structure, characteristic of grasslands high in organic matter, can change to a columnar or platy structure if carbon mineralisation and compaction from machinery damages aggregation. Platy structures found in plough pans limit penetration by plant roots and water. Readers should refer to general soil science textbooks such as Rowell (1994) and Marshall et al. (1996) for further information.

Soil physical quality indicators have been developed that characterise various features of a block of soil extracted with a spade. In the Visual Soil Structure Quality Assessment, a score is assigned based on the difficulty encountered when pushing the spade into the soil and extracting the block, the shape, strength and size of the largest aggregates, and the presence of gleyed anaerobic soil (Ball and Douglas, 2003). The block of soil is broken apart by hand to assess its friability and examine the porosity of larger aggregates. Scores from 1–5 are assigned, with 1 indicating a highly friable soil that provides an excellent environment for roots and 5 indicating a very compact soil, containing at best a few cracks. More detailed visual scoring of soil structure can be done with the Peerlkamp test or the French method 'Le profil cultural' (Mueller et al., 2010). The different approaches have been compared to each other, examined for operator variability and used on many soil types across the globe (Mueller et al., 2009). There is some need to refine the approach for particular regions or land uses. Water content and the time since tillage operations will also influence the results.

8.4 Soil tilth and seedbeds

In a review of research on soil structure, Warkentin (2008) quotes Cato in the ancient textbook 'Roman Farm Management':

> *What is the first principle of good agriculture? To plough well.*
> *What is the second? To plough again; and then to manure.*

Agriculturists like Cato, and his successors up to the nineteenth century, viewed soil structure as the tilth produced by soil cultivation. Although knowledge of soil structure produced by different cultivation implements or under different environmental conditions is plentiful, it involves mainly observational studies with minimal quantification of *in situ* properties and their variability with time (Atkinson et al., 2009). There has been abundant research investigating the distribution of clod and aggregate sizes following tillage (Braunack and Dexter, 1989a; Slowinskajurkiewicz, 1994; Guerif et al., 2001). Earlier research sieved the tilth to separate size fractions, while other studies impregnated soil with resin and then studied macrostructure in large blocks or micro-structure to 20 μm pore sizes in thin sections (Braunack and Dexter, 1989a). Much of this research was largely observational, although there are some examples of predicting physical properties, such as saturated hydraulic conductivity, from soil thin sections (Zhang et al., 2005). The importance of the size and geometry of soil structural units in relation to transport to roots was estimated by Passioura (1991). He predicted that the time taken for roots to extract water increased by up to two orders of magnitude for roots clumped within subsoil macropores as compared with uniformly distributed roots in a seedbed.

8.4.1 Soil tillage

Since Sir Edward John Russell published the first edition of this book in 1912, the practices used to cultivate soil have been revolutionised. On many soils around the globe, the plough has been replaced by reduced tillage systems; for example, over 20% of USA and almost half of Argentinian agricultural land is under zero tillage (Farage et al., 2007). Several drivers have prompted this change, including the ability to control weeds with herbicides, recognition of the deleterious impact of mechanical disturbance from cultivation on soil carbon mineralisation and a desire to save costs on fuel and labour (Triplett and Dick, 2008). The FAO and others have promoted reduced tillage for its capacity to decrease soil degradation, particularly in regions with less fertile, fragile or impoverished soils (Knowler and Bradshaw, 2007).

Figure 8.3 illustrates the various operations in cultivating soil using conventional, reduced and no-till tillage systems. Under conventional tillage, the soil is first inverted to a depth of about 20 cm with a mouldboard plough. This buries residue and causes the initial fragmentation of the soil. A second pass with a disc breaks the clods produced from ploughing into smaller aggregates, which can be broken down further by a pass with tines or harrowing. Planting is done by drilling seeds into the cultivated seedbed. Section 8.4.2 describes the optimal seedbed conditions that farmers strive for with conventional tillage. Cultivation may be done after crop establishment for the mechanical control of weeds.

Conservation tillage disrupts soil to a shallower depth and is a generic term applied to any cultivation practice that is not conventional tillage. Essentially, it removes the use of a plough. The variety of practices defined as conservation tillage is enormous, varying from the direct drilling of seeds into uncultivated seedbeds in zero till to elaborate minimum tillage implements that subsoil, cultivate, plant and fertilise in one pass.

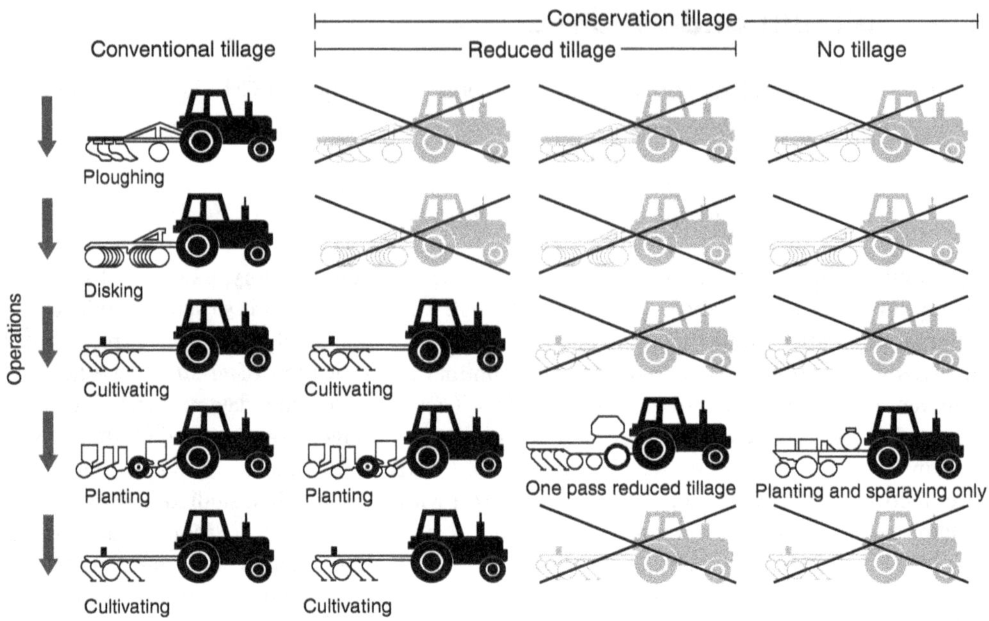

Figure 8.3 Field operations involved under different forms of soil cultivation (redrawn from FAO, 2003).

Table 8.3 provides an overview of the advantages and disadvantages of soil tillage, particularly in relation to ploughing.

8.4.2 Desired physical properties of seedbeds

Considerable energy can be invested in creating seedbeds that optimise germination, plant growth and yield. Over 50% of the root mass of a developed plant are found in the tilled layer forming the seedbed, so its physical properties are very important to the production of food and biomass (Finney and Knight, 1973). Seedbed properties vary considerably depending on soil management, residue incorporation, tillage approaches, soil type and drainage (Guérif et al., 2001). In a good seedbed, almost all viable seeds will germinate and emerge (Nasr and Selles, 1995). Soil temperature is one of the most important physical properties of a seedbed that regulates germination (Addae and Pearson, 1992). Many models that simulate germination therefore use thermal time based on cumulative degree days. As water becomes scarcer, however, the significance of drought has been incorporated in hydrothermal time models (Rinaldi et al., 2005).

Loosening soil through cultivation decreases mechanical impedance to seedling emergence and root growth (Braunack and Dexter, 1989a). This has to be balanced against good seed–soil contact, which can be improved through rolling of the soil surface after sowing using pressures great enough to pack the soil but small enough to not impede plants (Passioura and Leeper, 1963; Adem et al., 1984). In general, an 'optimal' seedbed in cultivated soils consists of aggregates 1–5 mm in size (Russell, 1961). The packing of aggregates should form sufficient pores >75 μm in diameter that freely drain so that oxygen exchange maintains an aerobic seedbed (Adem et al., 1984). In addition, pores 0.2–30 μm in diameter are needed as a reservoir to store water that can be used by plants.

Table 8.3 Advantages of disadvantages of soil tillage as summarised by Triplett and Dick (2008).

Tillage advantages

1. Reduces competition from weeds
2. Creates a seedbed that promotes uniform and efficient crop establishment
3. Buries residues and reduces disease inoculum
4. Reduces extreme surface roughness and thus facilitates more efficient equipment use
5. Promotes water infiltration in loosened soil – at least initially
6. Stimulates organic matter mineralisation and nutrient release
7. Leads to more rapid warming in the spring if tillage is applied before cold dormant period
8. Mixes fertiliser nutrients throughout the rooting depth of the crop
9. Disrupts habitat and life cycle of harmful pests
10. Breaks up surface soil crusts
11. Reduces surface and subsurface compaction

Tillage disadvantages

1. Disrupts aggregates and reduces soil structure
2. Supports greater extremes in soil temperatures
3. Exposes the soil to raindrop impact
4. Promotes soil erosion
5. Destroys macropores
6. Destroys cracking patterns in soils with vertic characteristics
7. Promotes increased soil drying and reduces water use efficiency
8. Causes compaction and tillage pan formation
9. Disrupts life cycles of beneficial organisms such as earthworms
10. Decreases the store of organic nutrients in the soil
11. Fragments fungal hyphal networks
12. Accelerates oxidation of labile organic matter and interferes with soil carbon sequestration
13. Causes increased fertiliser immobilisation/fixation
14. Requires greater inputs of fuel and energy
15. Requires greater investment in equipment

A great deal of research has investigated the optimal size distribution of aggregates for seedbeds (Braunack and Dexter, 1989b). It has been suggested that aggregates should be about one-fifth to one-tenth the size of the seed (Hadas and Russo, 1974), but for smaller seeds, this may not be practical and result in erosion. Whereas cereals perform well close to the 1 mm threshold described by Russell (1961), maize and sugar beet perform better at the larger 5 mm threshold. However, in the various studies reviewed by Braunack and Dexter (1989b), they noted that soil type and climate had a very large impact on the results, as shown in studies done in multiple years when conditions were either wet or dry. In general, a finer tilth results in better performance in drier conditions because the risk of anaerobic conditions is insignificant and seed–soil contact is greater.

Soil water potential regulates the rate that seeds imbibe water and this rate varies markedly between plants and crop cultivars. Al-Karaki (1998) demonstrated this impact for lentil seeds of different sizes (Figure 8.4). Smaller seeds (30 mg) imbibed water more rapidly than

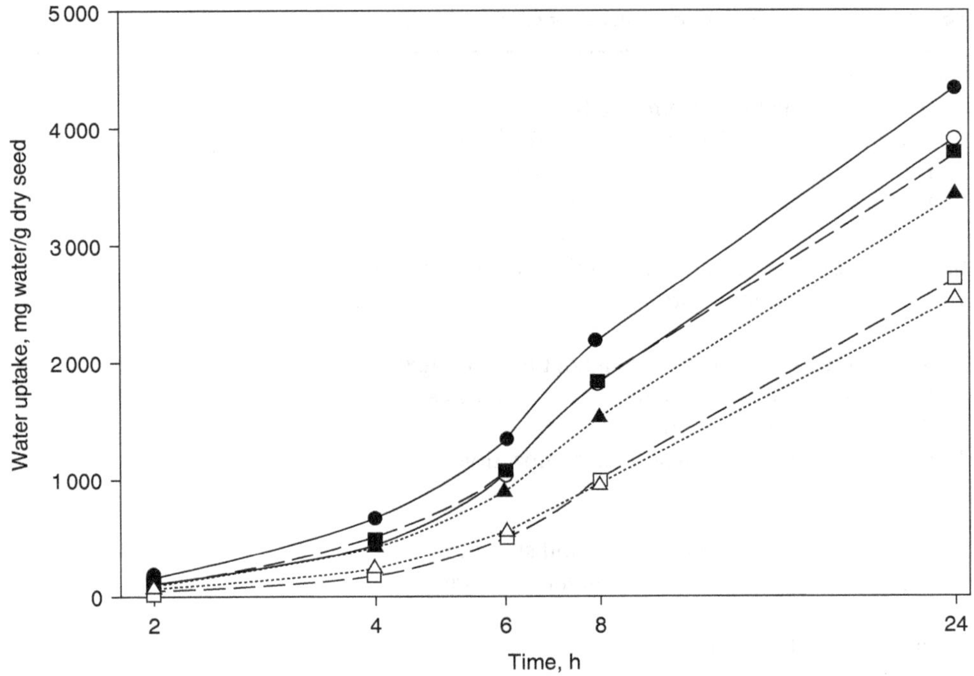

Figure 8.4 Water uptake by small (filled symbols, 30 mg) and large (empty symbols, 58 mg) lentil seeds over time for different water potentials of PEG (circle –.03 MPa, square –0.15 MPa and triangle –0.30 MPa). Data taken from Al-Karaki (1998). With kind permission from John Wiley & Sons.

larger seeds (58 mg), and a change in water potential from –0.03 MPa to –0.30 MPa decreased water uptake over 24 h by 40%. Research by Wuest (2007) suggests that vapour transport from soil to seed plays a substantial role in seed imbibition in unsaturated soils. Once germinated, root elongation is also affected by soil water potential, either by the potential gradient for water uptake or hypoxia if the soil is anaerobic or through mechanical impedance (Whitmore and Whalley, 2009).

8.4.3 Factors regulating soil fragmentation

The fragmentation of soil in the formation of seedbeds is controlled by a multitude of processes but the major drivers are: (1) internal pore structure, (2) water content, (3) mechanical and hydrological properties of the soil and (4) the mechanical action of the tillage implement. These drivers interact and generally result in decreased tillage efficiency in terms of energy input and fragmentation if soil is damaged through compaction or organic matter loss (Adem et al., 1984; Keller et al., 2007). In the soil thin sections shown in Figure 8.5, these impacts are visually apparent. Years of ploughing that depleted carbon and physically damaged the soil shown on the right has resulted in a narrower size distribution of pores and units of denser soil. A change in soil management, as shown for the soil on the left, can produce a broader range of pore sizes and improve aggregation of the soil. Fragmentation occurs through the coalescence of pre-existing pores and cracks in the soil, so the soil shown on the left should fragment into smaller aggregates with less energy input than the soil on the right (Grant et al., 1990).

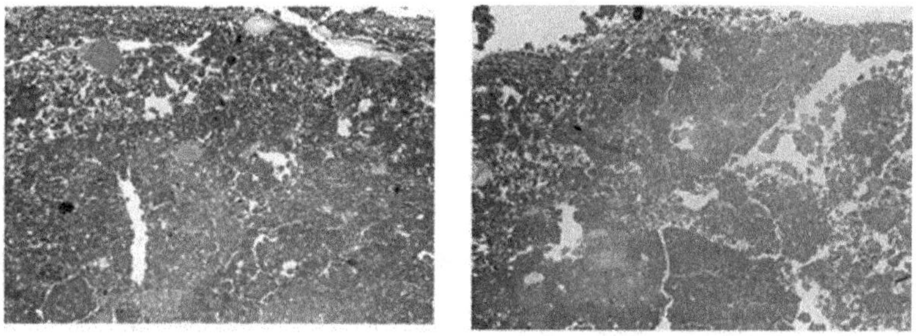

Figure 8.5 Thin sections of topsoil from plots after 11 years of zero tillage (left) and conventional tillage by ploughing and discing (right). From VandenBygaart et al. (1999). With permission from Elsevier. For a colour version of this figure, please see Plate 8.3.

The pore structure of the soil also affects water drainage and the presence of air-filled pores (Keller et al., 2007). Stresses imposed by tillage concentrate at the edges of air-filled pores, whereas the stresses will transmit through water-filled pores (Hallett et al., 1995). With increasing water content, fewer air-filled pores exist in the soil so fragmentation can decrease (Grant et al., 1990).

However, there is a trade-off between the influence of soil water content on the strength of the soil and the presence of air-filled pores (Dexter and Birkas, 2004; Keller et al., 2007). If soil is too wet, tillage deforms the soil plastically at the expense of the fragmentation of clods into smaller aggregates. When soil is dry, much greater energy needs to be expended in tillage and the resulting seedbed may require irrigation for germination. Considerable research has attempted to define the optimal water content for tillage. One approach is based on the transition between brittle, plastic and liquid behaviour, defined by Atterberg (1911):

Plastic limit, PL (or lower plastic limit) – transition from brittle to plastic behaviour.
Liquid limit, LL (or upper plastic limit) – transition from plastic to liquid behaviour.
Plasticity index = liquid limit – plastic limit

Specific standards exists for measuring both LL and PL. For both tests the soil is initially sieved to either 425 μm (geotechnical standard) or sometimes 2 mm, remoulded and then tested at a range of water contents. PL is defined by the water content at which a thread of soil rolled to 3 mm begins to crumble. It is a difficult test to perform that requires considerable practice to avoid operator bias. LL can be determined using either a drop cone penetrometer or Casagrande apparatus. With a drop cone penetrometer, the sharp tip of a cone is placed in contact with the surface of the soil and then released. A penetration depth of 20 mm defines the LL. The Casagrande apparatus consists of a round brass cup that moves up and then drops 1 cm through the rotation of a cam. A paste of soil is placed in the cup and a groove is scratched down the centre. LL occurs at the water content when 25 drops closes the groove in the soil paste. The drop cone penetrometer is generally thought to provide more reliable results as it is simpler and induces less dynamic stress during testing.

When soil water content is at about 0.8 PL, the combined effects of limited plastic deformation, soil adhesion and the presence of air-filled pores can result in optimal tillage efficiency (Keller et al., 2007). The wettest water content for tillage is roughly the field capacity of the soil (water content following free drainage; see Chapter 9), whereas the

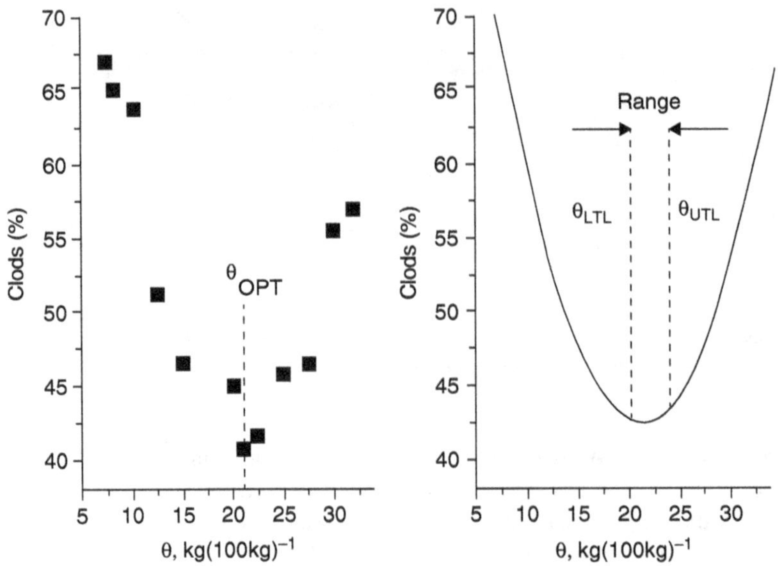

Figure 8.6 Presence of soil clods following tillage at a range of water contents. Raw data are shown on the left, including the minimum value of clods corresponding to the optimum water content for tillage, θ_{OPT}. The right-hand graph shows the fitted quadratic equation with the upper and lower tillage limits, θ_{UTL} and θ_{LTL} and the range of water contents for tillage. Reprinted from Dexter et al. (2005). With permission from Elsevier.

driest cut-off is more arbitrary and has been taken as the water content where the soil strength is double its value for the optimum water content (Dexter and Bird, 2001).

Another approach to identify the optimum water content for tillage uses water retention characteristics to characterise air-filled pore structure, soil composition and bulk density (Dexter et al., 2005). Figure 8.6 shows raw data and a fitted quadratic equation for the breakdown of soil clods over a range of water contents. Using pedotransfer functions, the optimum, lower limit and upper limit water content for tillage, θ_{OPT}, θ_{LTL} and θ_{UTL} respectively, can be estimated. Due to the interdependency of soil properties (e.g. clay content and bulk density) and reliability of pedotransfer functions, caution is needed when adopting this type of approach (Dexter et al., 2005).

Good soil management strives to maintain or increase the range of water contents available for tillage. The major variables influencing the size of the range of water contents are organic matter (see Section 8.2.4 on soil aggregates) and pore structure. Both of these properties can be degraded through intensive soil cultivation that enhances carbon mineralisation, disrupts aggregates and decreases their stability (Six et al., 2004; Diacono and Montemurro, 2010) and mechanically deforms the soil through wheel damage or the action of an implement (Horn et al., 1994). Odell et al. (1960) found a positive correlation between PL and soil organic carbon, which suggests that soils depleted of carbon need to be tilled when drier. From a practical standpoint, land managers can be faced with a small window of opportunity to till soils efficiently, particularly in the case of hard-setting soils (Mullins et al., 1988).

8.4.4 Soil friability

Friability describes how soil breaks up into tilth, whether it is from the action of cultivation implements or simply crumbling soil in your hand when assessing physical structure in the field. Christensen (1930) developed the first index to describe soil friability based on the variability of the strength of soil aggregates. This was refined by Braunack et al. (1979) who applied extreme statistics to describe the variability and scaling of strength of different-sized soil aggregates. This produced a highly flexible approach that produced fitting parameters that are related to the spread of strength between and within levels of the soil aggregate hierarchy.

The strength of soil aggregates can be assessed with a crushing test, where soil is compressed between flat parallel plates (Hallett et al., 2000). Failure force, F, relates to tensile failure stress, σ by

$$\sigma = c\frac{F}{l^2} \tag{8.10}$$

where l is the geometric mean of the three principal axis lengths of the aggregate and c is a constant equal to 0.576 (Dexter and Kroesbergen, 1985). The distribution of failure stress is evaluated as,

$$\log\sigma = -k\log V + A \tag{8.11}$$

where

$$A = \log[\sigma_0 V_0^k \Gamma(1+k)] \tag{8.12}$$

The parameters σ and V are the strength and volume of the aggregate being tested, while σ_0 and V_0 are the strength and volume of the basic elements of soil that bind to form aggregates. The Gamma function, Γ can be obtained from tables and k is a fitting parameter that is proportional to the scatter of the failure stress. k was used by Utomo and Dexter (1981) as an index of soil friability. More complex models have been proposed that use energy rather than stress of fracture (Perfect et al., 1998) or relate more directly to the scaling of pore structure (Hallett et al., 2000). A simpler, more practical approach based on the coefficient of variation of aggregate strength has also been developed (Watts and Dexter, 1998). Field methods based on drop shatter tests (Schjonning et al., 2007) and simpler laboratory methods for assessing aggregate breakdown (Ding et al., 2009) have also been developed to provide a measure of soil friability.

Utomo and Dexter (1981) proposed the following classification of soil friability based on observations from a range of soils:

$k<0.05$	Not friable
$k=0.050$–0.10	Slightly friable
$k=0.10$–0.25	Friable
$k=0.25$–0.40	Very friable
$k>0.40$	Mechanically unstable

They found that cycles of wetting and drying or freezing and thawing increased k because of the production of micro-cracks that decreased aggregate strength. Other studies have shown soil to be most friable as the water content approaches the plastic limit (Watts and Dexter, 1998), with more recent models of friability incorporating the impact of water content directly (Munkholm et al., 2007). Watts and Dexter (1998) made many observations of the impact of soil management on friability with the following major findings: (1) there was a positive correlation between friability and organic matter; (2) deep soil was less friable

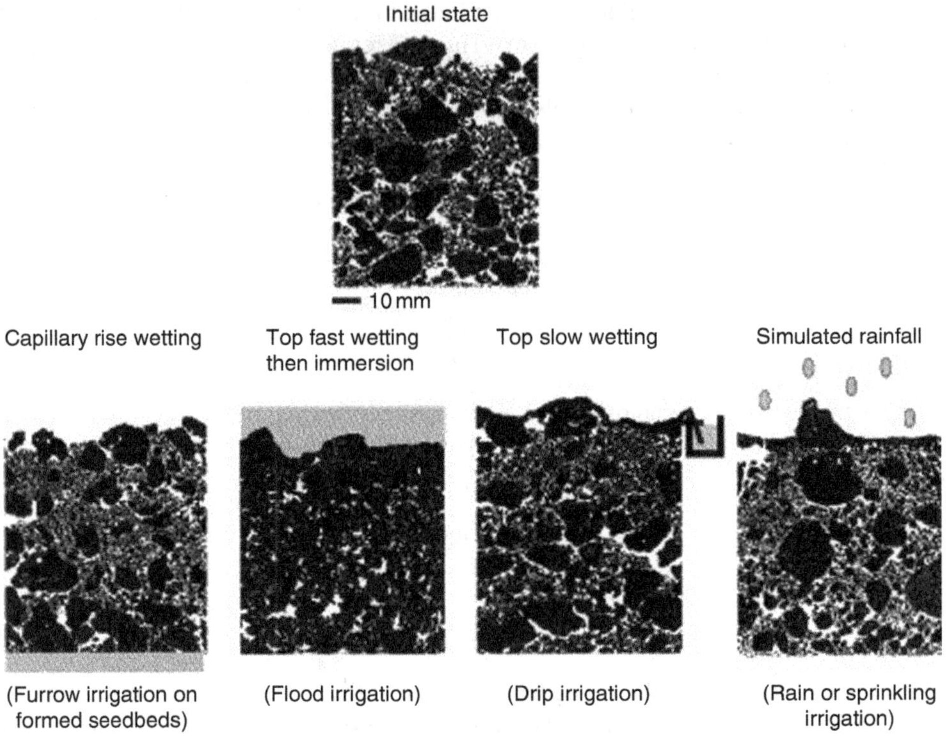

Figure 8.7 Binary images of vertical slices (pixel size 107 μm, solid in black): initial seedbed, bottom slow wetting (BS), top fast wetting then immersion (TFI), top slow wetting (TS) and rainfall simulation (R). From Bresson and Moran (2004). With permission from Elsevier.

than shallow soil; (3) tillage of wet soil decreased the friability of soils following subsequent drying and (4) soil friability changed with seasons.

8.4.5 Post-tillage seedbed dynamics

At water contents generally much greater than θ_{UTL}, soil can begin to deform under its own weight and the action of capillarity forces from pore water (Ghezzehei and Or, 2003). Weathering processes can also disrupt unstable soil aggregates through slaking and raindrop impact (Le Bissonnais, 1996). This produces a seedbed that varies over the season and can degrade rapidly if the soil is poorly managed or inherently unstable due to its composition. Consequently, the tilth produced by tillage can be short lived, giving roots a limited window of opportunity to benefit from the energy exerted in soil cultivation. Hard setting of the soil surface can occur rapidly following cultivation, resulting in a severe restriction to seedling emergence (Rengasamy et al., 2003). This process is characterised by an initial dispersion of the surface soil from wetting followed by the formation of continuous, strong crust on drying. Processes described previously that improve soil aggregation also decrease hard-setting in soils.

Changes in tilth over time for different wetting conditions of soil are shown in Figure 8.7 (Bresson and Moran, 2004). Under slow wetting, aggregates coalesce through their own weight, overburden and the action of capillary bonding with adjacent aggregates (Or, 1996).

Under more rapid wetting, aggregates disperse and break down, with rainfall inducing crusting. Arguably one of the greatest challenges to soil physics is the development of predictive models that describe seedbed dynamics from cultivation to harvest. Although commendable models have been developed to predict surface crusting and its influence on water infiltration (Assouline and Mualem, 2000), and to some extent the temporal change in pore structure over time (Or et al., 2000), research investment in this area has been minimal.

8.5 Critical soil physical conditions for plant growth

Soil physical conditions can limit both root and shoot growth if the soil is too hard, too dry, does not contain sufficient oxygen or is far from the optimal temperature (Kaspar and Bland, 1992; daSilva et al., 1994; Bengough et al., 2006a). There is much literature describing both root and shoot growth responses to these stresses, particularly in annual crops, and the main effects are summarised in Table 8.4. The main aspects of these responses will be considered here, and readers are referred to recent reviews on individual topics for more detailed study.

Table 8.4 Effects of soil physical limitations on plant growth and diagnostic soil measurements.

Stress	Effects on plant growth	Soil physical measurement
Mechanical impedance (Bengough et al., 2011)	Slows root elongation (Figure 8.8) Increase in root diameter (up to two-fold) Slows leaf elongation/expansion at larger values of penetrometer resistance (Figure 8.9)	Penetrometer resistance Direct measurement of root penetration resistance
Water stress (Davies et al., 2005)	Slows root elongation (Figure 8.8) Small decrease in root diameter Slows leaf elongation/expansion (Figure 8.9)	Soil matric potential (tensiometer or psychrometer)
Waterlogging (Colmer and Voesenek, 2009): Hypoxia Anoxia	Slows or stops root elongation (Figure 8.8) Slows or stops leaf elongation/expansion in susceptible species, though shoot elongation may be substantially enhanced in prolonged submergence of tolerant species	Oxygen diffusion rate Oxygen concentration
Non-optimal temperature: Too hot Too cold (Kaspar and Bland, 1992)	Slows root elongation (Figure 8.8) Slows leaf elongation/expansion Changes to root orientation	Direct measurement of soil temperature (thermometer, thermocouples and thermistor)
Multiple stresses: Water stress and mechanical impedance Waterlogging and mechanical impedance Waterlogging and soil temperature	Interactions due to: Increase in soil strength as soil dries Decreased turgor in water-stressed roots restricts the maximum growth pressure that roots can exert Decrease in oxygen diffusion rate with increase in soil compaction Increased oxygen demand from roots waterlogged in warm soil	As described in this column in the cells above

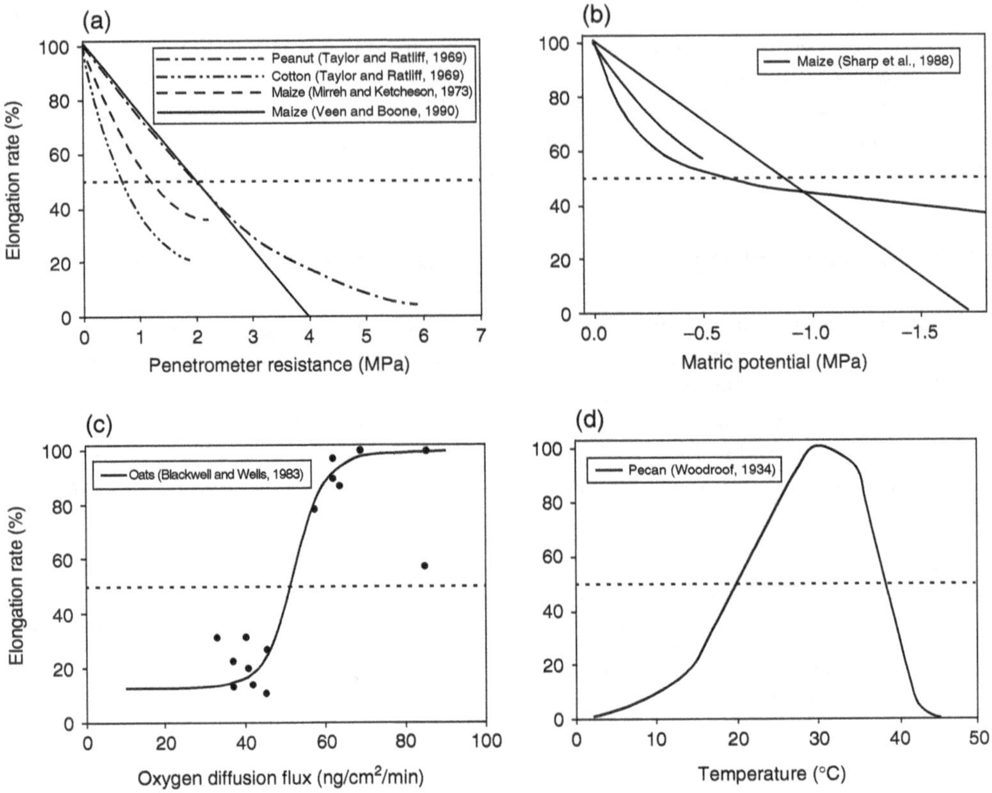

Figure 8.8 Root elongation rate expressed as a percentage of the unimpeded elongation rate in relation to (a) penetrometer resistance, (b) matric potential, (c) oxygen diffusion flux and (d) temperature.

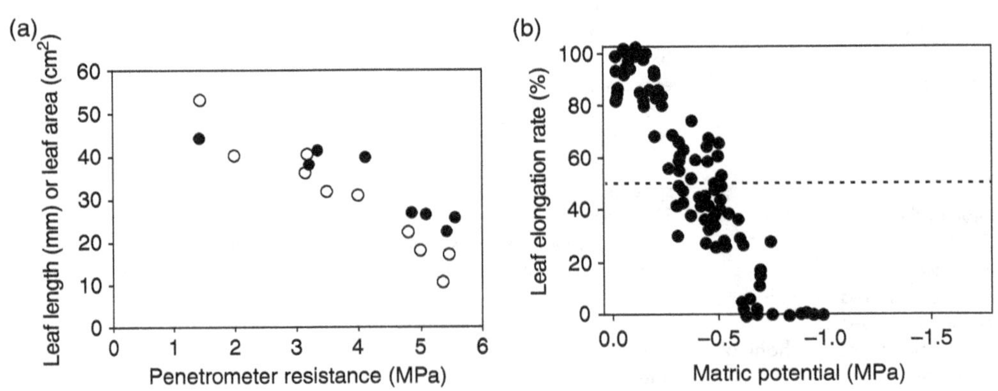

Figure 8.9 (a) Leaf length and leaf area (hollow) of wheat seedlings as a function of penetrometer resistance at two time points (data from Masle and Passioura, 1987); (b) leaf elongation rate (% of unstressed) as a function of matric potential (data from Parent et al., 2010).

8.5.1 Local root response to soil conditions

Roots respond to the local soil environment due to the influence of the soil on cell division and expansion rates in the root meristem and elongation zones. Seedling root elongation is shown as a function of individual stresses in Figure 8.8. Here it can be seen that root elongation is slowed to approximately half of its unimpeded rate by soils with a penetrometer resistance of between 1 and 2 MPa; a matric potential of between −0.5 and −1 MPa for maize roots; a oxygen diffusion rate of between 45 and 60 ng/cm^2/min for oat roots and a temperature diverging more than 10 °C from the optimum for pecan roots (this will be very species dependent). These represent substantial physical limitations to root elongation, where a seedling root system might only achieve half of the vertical penetration into the soil attained by an unimpeded root (though it should be noted that the angle of root growth may also be affected by stress [e.g. Sheppard and Miller, 1977]). In the field, plants often experience more than one physical stress. Soil strength often increases rapidly as the soil dries so that roots will be confronted with both water stress and mechanical impedance (Whitmore and Whalley, 2009; Bengough et al., 2011). Recent surveys of physical properties from a range of 19 soils indicated that mechanical impedance often poses a greater restriction to root elongation in drying soil than water stress (Bengough et al., 2011): the increase in soil strength on drying was such that more than half the penetrometer resistances measured were in excess of 2 MPa at a matric potential of −200 kPa.

Root elongation rate is essentially the product of the rate of adding new cells onto a cell file and the final cell length. The rate of cell wall yielding influences both the rate of cell partitioning in the meristem and also the final cell length and is thought to be primarily determined by the stiffness of the cell walls, as long as there is sufficient turgor to drive growth. The rate of root elongation (*dl/dt*) is then given by (Greacen, 1986; Bengough et al., 2006a)

$$\left(\frac{dl}{dt}\right) = l.m(\sigma,\psi)\left(P - Y(\sigma,\psi) - \sigma(\psi)\right) \tag{8.13}$$

where, l is the length of the zone of elongating root tissue, m is cell wall extensibility, P is turgor pressure, Y is cell wall yield threshold, σ is the root penetration resistance of the soil and ψ is the soil matric potential. P, Y, σ and ψ all have dimensions of pressure, while m is strain rate per unit pressure. Yield threshold and extensibility are both aspects of cell wall stiffness that are related to the rates of making and breaking tethers between cell wall microfibrils and are therefore probably interrelated (Passioura and Fry, 1992). By considering this equation in more detail, root responses can start to be understood at the cellular scale and the growth trajectories of individual cells can be deduced as they move from the meristem through the elongation zone to finally form mature root tissue (Chavarria-Krauser and Schurr, 2004).

Cell wall stiffness is strongly influenced by both soil strength and matric potential, as indicated in Equation 8.13. In the case of mechanical impedance, cell walls stiffen axially in hard soils, causing cells to expand radially and the root to thicken. Such stiffening is evident when roots grow through a hard soil layer into looser soil or are transferred into a non-impeding medium such as hydroponics (Bengough and Young, 1993; Croser et al., 1999). Root elongation rates may take a period of 2–5 days to recover from the stress, allowing time for cells formed in the absence of the stress to enter the elongation zone. For water-stressed roots, rapid local growth (strain) rates are maintained close to the root apex, despite a decrease in turgor caused by the negative soil matric potential (Sharp et al., 2004). This arises because of local cell wall loosening near the root apex, in contrast with stiffening of the cell wall that occurs in the basal region of the elongation zone.

In the case of hypoxia, slower root growth is mainly due to decreased cell length, with cells becoming shorter and fatter (Eavis, 1972). As oxygen concentrations decrease further to approach anoxia, cell production in the root meristem slows too (e.g. <0.3 atm oxygen partial pressure for pea roots). In relatively waterlogging tolerant species, the formation of aerenchyma (continuous gas spaces formed by cell lysis or by cells separating) within the root cortex allows transport of oxygen to the root tip to be maintained, decreasing the likelihood of root tip death (Colmer and Voesenek, 2009). The formation of new, especially stem-borne, roots is a very common response to flooding and the new roots formed will generally contain a greater proportion of aerenchyma tissue.

In the field, roots may experience multiple stresses due to spatial and temporal variation in soil physical conditions, particularly soil water and soil density. Soil surface layers may become sufficiently dry that mechanical impedance and water stress may limit root growth and crop yield, while water is available deeper in the profile (Whalley et al., 2006). The least limiting water range is a method for determining the range of water contents for particular soil conditions where root growth is free from major physical limitation (daSilva et al., 1994). The water content at which penetrometer resistance exceeds 2 MPa is often taken as the soil strength limit; if the soil is any drier than the limit, the roots are likely to be slowed by substantial mechanical impedance to growth. The water content corresponding to −1.5 MPa has been used for the water stress limit, while the water content corresponding to 10 cm^3cm^{-3} is taken as the aeration limit. Field capacity water content is often taken as an additional wet end limit, where root growth will be restricted if it is exceeded, although this is less obviously linked to a physical limitation to root growth. By combining this threshold approach with soil water data, it is then possible to characterise for how much of a growing season a root system has been exposed to particular physical stresses (da Silva and Kay, 2004; Bengough et al., 2006b), although it is clear that these threshold values may benefit from further refinement (Mohammadi et al., 2010). This type of approach is promising in relation to starting to scale up responses from the individual root tip in the laboratory to whole crop performance.

8.5.2 Root–shoot signalling and feedback responses

Soil physical conditions influence shoot growth directly via root–shoot signalling and may decrease growth of the whole plant if plant access to water and nutrients becomes limited (e.g. if the root system is too stunted to explore sufficient soil volume).

There has been substantial work on how soil water status influences shoot growth directly via root–shoot signalling. In loose growth media, seedling shoot growth is relatively more sensitive to drying matric potential than is root elongation (Sharp et al., 1988). Maize shoot elongation virtually ceased in vermiculite at −0.81 MPa, whereas root elongation was slowed by only 50%. Leaf elongation rate decreases approximately linearly with decreasing matric potential (Figure 8.9; Parent et al., 2010). In most drying soils, it is likely that a combination of mechanical impedance and water stress are involved in regulating leaf expansion rates, although this is a matter of ongoing debate, particularly because of the difficulty in separating the effects of soil strength and matric potential (Passioura, 2002). Classic experiments by Masle and Passioura (1987) showed that leaf area decreased linearly with increasing penetrometer resistance for wheat plants grown in compacted soil, while there was no evidence that water or nutrients were limiting leaf growth (Figure 8.9). In barley, leaf elongation rate decreased rapidly when the strength of the growing medium (a sand) was increased by applying an external compressive pressure, suggesting a direct role of soil strength (Young et al., 1997). The nature of root-sourced signals and their control of leaf expansion is likely to involve multiple interactions between ABA and ethylene and to be modulated by factors such as xylem sap pH (Davies et al., 2005).

References

Addae, P.C. and Pearson, C.J. (1992) Thermal requirements for germination and seedling growth of wheat. *Australian Journal of Agricultural Research*, **43**, 585–594.

Adem, H.H., Tisdall, J.M., and Willoughby, P. (1984) Tillage management changes size-distribution of aggregates and macro-structure of soils used for irrigated row-crops. *Soil and Tillage Research*, **4**, 561–576.

Alameda, D. and Villar, R. (2009) Moderate soil compaction: implications on growth and architecture in seedlings of 17 woody plant species. *Soil and Tillage Research*, **103**, 325–331.

Al-Karaki, G.N. (1998) Seed size and water potential effects on water uptake, germination and growth of lentil. *Journal of Agronomy and Crop Science-Zeitschrift fur Acker und Pflanzenbau*, **181**, 237–242.

Amezketa, E. (1999) Soil aggregate stability: a review. *Journal of Sustainable Agriculture*, **14**, 83–151.

Amezketa, E., Singer, M.J., and LeBissonnais, Y. (1996) Testing a new procedure for measuring water-stable aggregation. *Soil Science Society of America Journal*, **60**, 888–894.

Angers, D.A. and Caron, J. (1998) Plant-induced changes in soil structure: processes and feedbacks. *Biogeochemistry*, **42**, 55–72.

Ashman, M.R., Hallett, P.D., and Brookes, P.C. (2003) Are the links between soil aggregate size class, soil organic matter and respiration rate artefacts of the fractionation procedure? *Soil Biology and Biochemistry*, **35**, 435–444.

Ashman, M.R., Hallett, P.D., Brookes, P.C., and Allen, J. (2009) Evaluating soil stabilisation by biological processes using step-wise aggregate fractionation. *Soil and Tillage Research*, **102**, 209–215.

Assouline, S. and Mualem, Y. (2000) Modeling the dynamics of soil seal formation: analysis of the effect of soil and rainfall properties. *Water Resources Research*, **36**, 2341–2349.

Atkinson, B.S., Sparkes, D.L., and Mooney, S.J. (2007) Using selected soil physical properties of seedbeds to predict crop establishment. *Soil and Tillage Research*, **97**, 218–228.

Atkinson, B.S., Sparkes, D.L., and Mooney, S.J. (2009) Effect of seedbed cultivation and soil macrostructure on the establishment of winter wheat (Triticum aestivum). *Soil and Tillage Research*, **103**, 291–301.

Atterberg, A. (1911) Die plastizität der tone. *Internationale Mitteilungen fuer Bodenkunde*, **1**, 10–43.

Baker, J.T. (2007) Cotton seedling abrasion and recovery from wind blown sand. *Agronomy Journal*, **99**, 556–561.

Ball, B.C. and Douglas, J.T. (2003) A simple procedure for assessing soil structural, rooting and surface conditions. *Soil Use and Management*, **19**, 50–56.

Bartoli, F., Bird, N.R.A., Gomendy, V., Vivier, H., and Niquet, S. (1999) The relation between silty soil structures and their mercury porosimetry curve counterparts: fractals and percolation. *European Journal of Soil Science*, **50**, 9–22.

Barzegar, A.R., Oades, J.M., Rengasamy, P., and Murray, R.S. (1995) Tensile strength of dry, remoulded soils as affected by properties of the clay fraction. *Geoderma*, **65**, 93–108.

Beare, M.H. and Bruce, R.R. (1993) A comparison of methods for measuring water-stable aggregates – implications for determining environmental-effects on soil structure. *Geoderma*, **56**, 87–104.

Bellamy, P.H., Loveland, P.J., Bradley, R.I., Lark, R.M., and Kirk, G.J.D. (2005) Carbon losses from all soils across England and Wales 1978–2003. *Nature*, **437**, 245–248.

Bengough, A.G. (2003) Root growth and function in relation to soil structure, composition, and strength. In: *Root Ecology* (eds H. de Kroon and E.J.W. Visser), pp. 151–171. Springer-Verlag, Berlin.

Bengough, A.G. and Mullins, C.E. (1990) Mechanical impedance to root-growth – a review of experimental-techniques and root-growth responses. *Journal of Soil Science*, **41**, 341–358.

Bengough, A.G. and Young, I.M. (1993) Root elongation of seedling peas through layered soil of different penetration resistances. *Plant and Soil*, **149**, 129–139.

Bengough, A.G., Mullins, C.E., Wilson, G., and Wallace, J. (1991) The design, construction and use of a rotating-tip penetrometer. *Journal of Agricultural Engineering Research*, **48**, 223–227.

Bengough, A.G., Mullins, C.E., and Wilson, G. (1997) Estimating soil frictional resistance to metal probes and its relevance to the penetration of soil by roots. *European Journal of Soil Science*, **48**, 603–612.

Bengough, A.G., Bransby, M.F., Hans, J., McKenna, S.J., Roberts, T.J., and Valentine, T.A. (2006a) Root responses to soil physical conditions; growth dynamics from field to cell. *Journal of Experimental Botany*, **57**, 437–447.

Bengough, A.G., Bransby, M.F., Hans, J., McKenna, S.J., Roberts, T.J., and Valentine, T.A. (2006b) Root responses to soil physical conditions; growth dynamics from field to cell. *Journal of Experimental Botany*, **57**, 437–447.

Bengough, A.G., McKenzie, B.M., Hallett, P.D., and Valentine, T.A. (2011) Root elongation, water stress, and mechanical impedance: a review of limiting stresses and beneficial root tip traits. *Journal of Experimental Botany*, **62**, 59–68.

Bird, N.R.A., Perrier, E., and Rieu, M. (2000) The water retention function for a model of soil structure with pore and solid fractal distributions. *European Journal of Soil Science*, **51**, 55–63.

Bird, N.R.A., Preston, A.R., Randall, E.W., Whalley, W.R., and Whitmore, A.P. (2005) Measurement of the size distribution of water-filled pores at different matric potentials by stray field nuclear magnetic resonance. *European Journal of Soil Science*, **56**, 135–143.

Blackwell, P.S. and Wells, E.A. (1983) Limiting oxygen flux densities for oat root extension. *Plant and Soil*, **73**, 129–139.

Blanco-Canqui, H., Lal, R., Post, W.M., Izaurralde, R.C., and Shipitalo, M.J. (2006) Organic carbon influences on soil particle density and rheological properties. *Soil Science Society of America Journal*, **70**, 1407–1414.

Blok, C. and Wever, G. (2008) Experience with selected physical methods to characterize the suitability of growing media for plant growth. *Acta Horticulture*, **779**, 239–250.

Braudeau, E., Costantini, J.M., Bellier, G., and Colleuille, H. (1999) New device and method for soil shrinkage curve measurement and characterization. *Soil Science Society of America Journal*, **63**, 525–535.

Braunack, M.V. and Dexter, A.R. (1989a) Soil Aggregation in the seedbed – a review. 1. Properties of aggregates and beds of aggregates. *Soil and Tillage Research*, **14**, 259–279.

Braunack, M.V. and Dexter, A.R. (1989b) Soil aggregation in the seedbed – a review. 2. Effect of aggregate sizes on plant-growth. *Soil and Tillage Research*, **14**, 281–298.

Braunack, M.V., Hewitt, J.S., and Dexter, A.R. (1979) Brittle-fracture of soil aggregates and the compaction of aggregate beds. *Journal of Soil Science*, **30**, 653–667.

Bresson, L.M. and Moran, C.J. (2004) Micromorphological study of slumping in a hardsetting seedbed under various wetting conditions. *Geoderma*, **118**, 277–288.

Bruneau, P.M.C., Ostle, N., Davidson, D.A., Grieve, I.C., and Fallick, A.E. (2002) Determination of rhizosphere C-13 pulse signals in soil thin sections by laser ablation isotope ratio mass spectrometry. *Rapid Communications in Mass Spectrometry*, **16**, 2190–2194.

Canton, Y., Sole-Benet, A., Asensio, C., Chamizo, S., and Puigdefabregas, J. (2009) Aggregate stability in range sandy loam soils. Relationships with runoff and erosion. *Catena*, **77**, 192–199.

Chan, K.Y., Heenan, D.P., and Ashley, R. (1994) Seasonal-changes in surface aggregate stability under different tillage and crops. *Soil and Tillage Research*, **28**, 301–314.

Chavarria-Krauser, A. and Schurr, U. (2004) A cellular growth model for root tips. *Journal of Theoretical Biology*, **230**, 21–32.

Chen, Y.M., Wang, M.K., Zhuang, S.Y., and Chiang, P.N. (2006) Chemical and physical properties of rhizosphere and bulk soils of three tea plants cultivated in Ultisols. *Geoderma*, **136**, 378–387.

Chenu, C. (1989) Influence of a fungal polysaccharide, scleroglucan, on clay microstructures. *Soil Biology and Biochemistry*, **21**, 299–305.

Chenu, C. and Guérif, J. (1991) Mechanical strength of clay minerals as influenced by an adsorbed polysaccharide. *Soil Science Society of America Journal*, **55**, 1076–1080.

Chepil, W.S. (1951) Properties of soil which influence wind erosion. 5. Mechanical stability of structure. *Soil Science*, **72**, 465–478.

Chepil, W.S. (1952) improved rotary sieve for measuring state and stability of dry soil structure. *Soil Science Society of America Proceedings*, **16**, 113–117.

Christensen, O.S. (1930) An index of friability of soils. *Soil Science*, **29**, 119–136.

Colmer, T.D. and Voesenek, L.A. (2009) Flooding tolerance: suites of plant traits in variable environments. *Functional Plant Biology*, **36**, 665–681.

Connolly, R.D. (1998) Modelling effects of soil structure on the water balance of soil-crop systems: a review. *Soil and Tillage Research*, **48**, 1–19.

Cosentino, D., Chenu, C., and Le Bissonnais, Y. (2006) Aggregate stability and microbial community dynamics under drying-wetting cycles in a silt loam soil. *Soil Biology and Biochemistry*, **38**, 2053–2062.

Crawford, J.W. and Matsui, N. (1996) Heterogeneity of the pore and solid volume of soil: distinguishing a fractal space from its non-fractal complement. *Geoderma*, **73**, 183–195.

Croser, C., Bengough, A.G., and Pritchard, J. (1999) The effect of mechanical impedance on root growth in pea (Pisum sativum). I. Rates of cell flux, mitosis, and strain during recovery. *Physiologia Plantarum*, **107**, 277–286.

Currie, J.A. (1966) The volume and porosity of soil crumbs. *Journal of Soil Science*, **17**, 24–35.

Czarnes, S., Hallett, P.D., Bengough, A.G., and Young, I.M. (2000) Root- and microbial-derived mucilages affect soil structure and water transport. *European Journal of Soil Science*, **51**, 435–443.

Davies, W.J., Kudoyarova, G., and Hartung, W. (2005) Long-distance ABA signaling and its relation to other signaling pathways in the detection of soil drying and the mediation of the plant's response to drought. *Journal of Plant Growth Regulation*, **24**, 285–295.

De Gryze, S., Six, J., and Merckx, R. (2006) Quantifying water-stable soil aggregate turnover and its implication for soil organic matter dynamics in a model study. *European Journal of Soil Science*, **57**, 693–707.

Denef, K. and Six, J. (2005) Clay mineralogy determines the importance of biological versus abiotic processes for macroaggregate formation and stabilization. *European Journal of Soil Science*, **56**, 469–479.

Dexter, A.R. (1988) Advances in characterization of soil structure. *Soil and Tillage Research*, **11**, 199–238.

Dexter, A.R. (1997) Physical properties of tilled soils. *Soil and Tillage Research*, **43**, 41–63.

Dexter, A.R. and Kroesbergen, B. (1985) Methodology for determination of tensile-strength of soil aggregates. *Journal of Agricultural Engineering Research*, **31**, 139–147.

Dexter, A.R. and Bird, N.R.A. (2001) Methods for predicting the optimum and the range of soil water contents for tillage based on the water retention curve. *Soil and Tillage Research*, **57**, 203–212.

Dexter, A.R. and Birkas, M. (2004) Prediction of the soil structures produced by tillage. *Soil and Tillage Research*, **79**, 233–238.

Dexter, A.R., Czyz, E.A., Birkas, M., az-Pereira, E., Dumitru, E., Enache, R., Fleige, H., Horn, R., Rajkaj, K., de la Rosa, D., and Simota, C. (2005) SIDASS project – Part 3. The optimum and the range of water content for tillage – further developments. *Soil and Tillage Research*, **82**, 29–37.

Diacono, M. and Montemurro, F. (2010) Long-term effects of organic amendments on soil fertility. A review. *Agronomy for Sustainable Development*, **30**, 401–422.

Ding, Q.S., Pan, G.X., Ding, W.M., and Ji, C.Y. (2009) A laboratory rig and a scale- and energy-controlled procedure for tender soil fragmentation test. *Soil Science Society of America Journal*, **73**, 1286–1290.

Eavis, B.W. (1972) Soil physical conditions affecting seedling root growth. 3. Comparisons between root growth in poorly aerated soil and at different oxygen partial pressures. *Plant and Soil*, **37**, 151–158.

Eickhorst, T. and Tippkotter, R. (2008) Improved detection of soil microorganisms using fluorescence in situ hybridization (FISH) and catalyzed reporter deposition (CARD-FISH). *Soil Biology and Biochemistry*, **40**, 1883–1891.

Emerson, W.W. (1954) The determination of the stability of soil crumbs. *Journal of Soil Science*, **5**, 233–250.

Emerson, W.W. and Grundy, G.M.F. (1954) The effect of rate of wetting on water uptake and cohesion of soil crumbs. *Journal of Agricultural Science*, **44**, 249–253.

Emerson, W.W. and Greenland, D.J. (1990) Soil aggregates-formation and stability. In: *Soil Colloids and Theri Associations in Aggregates* (eds M.F. De Boodt, M.H.B. Hayes, and A. Herbillon), pp. 485–511. Plenum Press, New York.

Emerson, W.W., Foster, R.C., Tisdall, J.M., and Weissmann, D. (1994) Carbon content and bulk-density of an irrigated natrixeralf in relation to tree root-growth and orchard management. *Australian Journal of Soil Research*, **32**, 939–951.

FAO (2003) *Explore On-Farm: On-Farm Trials for Adapting and Adopting Good Agricultural Practices*. Food and Agriculture Organization of the United Nations, Crop and Grassland Service , FAO, Rome.

Farage, P.K., Ardo, J., Olsson, L., Rienzi, E.A., Ball, A.S., and Pretty, J.N. (2007) The potential for soil carbon sequestration in three tropical dryland farming systems of Africa and Latin America: a modelling approach. *Soil and Tillage Research*, **94**, 457–472.

Feeney, D.S., Crawford, J.W., Daniell, T., Hallett, P.D., Nunan, N., Ritz, K., Rivers, M., and Young, I.M. (2006) Three-dimensional microorganization of the soil-root-microbe system. *Microbial Ecology*, **52**, 151–158.

Finney, J.R. and Knight, B.A.G. (1973) Effect of soil physical conditions produced by various cultivation systems on root development of winter-wheat. *Journal of Agricultural Science*, **80**, 435–442.

Fristensky, A. and Grismer, M.E. (2008) A simultaneous model for ultrasonic aggregate stability assessment. *Catena*, **74**, 153–164.

Fritton, D.D. (2008) Evaluation of pedotransfer and measurement approaches to avoid soil compaction. *Soil and Tillage Research*, **99**, 268–278.

Fryrear, D.W. (1985) Determining Soil aggregate stability with a rapid rotary sieve. *Journal of Soil and Water Conservation*, **40**, 231–233.

Gerke, H.H. and Kohne, J.M. (2002) Estimating hydraulic properties of soil aggregate skins from sorptivity and water retention. *Soil Science Society of America Journal*, **66**, 26–36.

Ghezzehei, T.A. and Or, D. (2003) Pore-space dynamics in soil aggregate bed under a static external load. *Soil Science Society of America Journal*, **67**, 12–19.

Glab, T. (2008) Effects of tractor wheeling on root morphology and yield of lucerne (*Medicago sativa* L.). *Grass and Forage Science*, **63**, 398–406.

Goebel, M.O., Bachmann, J., Woche, S.K., and Fischer, W.R. (2005) Soil wettability, aggregate stability, and the decomposition of soil organic matter. *Geoderma*, **128**, 80–93.

Grant, C.D., Dexter, A.R., and Huang, C. (1990) Roughness of soil fracture surfaces as a measure of soil microstructure. *Journal of Soil Science*, **41**, 95–110.

Greacen, E.L. (1986) Root response to soil mechanical properties. *Transactions of the 13th Congress of the International Society of Soil Science*, **5**, 20–47.

Gregory, P.J., Bengough, A.G., Grinev, D., Schmidt, S., Thomas, W.T.B., Wojciechowski, T., and Young, I.M. (2009) Root phenomics of crops: opportunities and challenges. *Functional Plant Biology*, **36**, 922–929.

Guérif, J., Richard, G., Durr, C., Machet, J.M., Recous, S., and Roger-Estrade, J. (2001) A review of tillage effects on crop residue management, seedbed conditions and seedling establishment. *Soil and Tillage Research*, **61**, 13–32.

Hadas, A. and Russo, D. (1974) Water uptake by seeds as affected by water stress, capillary conductivity, and seed-soil water contact. 1. Experimental study. *Agronomy Journal*, **66**, 643–647.

Hainsworth, J.M. and Aylmore, L.A.G. (1983) The use of computer-assisted tomography to determine spatial-distribution of soil-water content. *Australian Journal of Soil Research*, **21**, 435–443.

Hallett, P.D., Dexter, A.R., and Seville, J.P.K. (1995) identification of preexisting cracks on soil fracture surfaces using dye. *Soil and Tillage Research*, **33**, 163–184.

Hallett, P.D., Dexter, A.R., Bird, N.R.A., and Seville, J.P.K. (2000) Scaling of the structure and strength of soil aggregates. *Subsoil Compaction: Distribution, Processes and Consequences*, **32**, 22–31.

Hati, K.M., Mandal, K.G., Misra, A.K., Ghosh, P.K., and Bandyopadhyay, K.K. (2006) Effect of inorganic fertilizer and farmyard manure on soil physical properties, root distribution, and water-use efficiency of soybean in Vertisols of central India. *Bioresource Technology*, **97**, 2182–2188.

Henin, S. (1948) Essai d'une theorie de la stabilite de la structure des sols. *Plant and Soil*, **1**, 167–178.

Henin, S. and Santamari, R. (1975) Observation comparée sur le comportement des fragments de terre soumis à l'action de l'eau de différentes solutions et de l'alcool. *Science du Sol*, **3**, 171–182.

Horn, R., Taubner, H., Wuttke, M., and Baumgartl, T. (1994) Soil physical-properties related to soil-structure. *Soil and Tillage Research*, **30**, 187–216.

Jaeger, F., Bowe, S., Van As, H., and Schaumann, G.E. (2009) Evaluation of 1 H NMR relaxometry for the assessment of pore-size distribution in soil samples. *European Journal of Soil Science*, **60**, 1052–1064.

Kaspar, T.C. and Bland, W.L. (1992) Soil-temperature and root-growth. *Soil Science*, **154**, 290–299.

Keller, T. and Håkansson, I. (2010) Estimation of reference bulk density from soil particle size distribution and soil organic matter content. *Geoderma*, **154**, 398–406.

Keller, T., Arvidsson, J., and Dexter, A. R. (2007) Soil structures produced by tillage as affected by soil water content and the physical quality of soil. *Soil and Tillage Research*, **92**, 45–52.

Kemper, W.D. and Rosenau, R.C. (1986) Aggregate stability and size distribution. In: *Methods of Soil Analysis: Part 1. Physical and Mineralogical Methods* (ed A. Klute), pp. 425–442. ASA and SSSA, Madison, WI.

Knowler, D. and Bradshaw, B. (2007) Farmers' adoption of conservation agriculture: A review and synthesis of recent research. *Food Policy*, **32**, 25–48.

Kosugi, K. (1999) General model for unsaturated hydraulic conductivity for soils with lognormal pore-size distribution. *Soil Science Society of America Journal*, **63**, 270–277.

Kubiëna, W.L. (1938) *Micropedology*. Collegiate Press Inc. Ames, IA, 243 pp.

Le Bissonnais, Y. (1996) Aggregate stability and assessment of soil crustability and erodibility: I. Theory and methodology. *European Journal of Soil Science*, **47**, 425–437.

Lima, L.A. and Grismer, M.E. (1994) Application of fracture mechanics to cracking of saline soils. *Soil Science*, **158**, 86–96.

Marshall, T.J., Holmes, J.W. and Rose, C.W. (1996) *Soil Physics*. Cambridge University Press, Cambridge.

Masle, J. and Passioura, J.B. (1987) The effect of soil strength on the growth of young wheat plants. *Australian Journal of Plant Physiology*, **14**, 643–656.

Mathers, N.J., Jalota, R.K., Dalal, R.C., and Boyd, S.E. (2007) C-13-NMR analysis of decomposing litter and fine roots in the semi-arid Mulga Lands of southern Queensland. *Soil Biology and Biochemistry*, **39**, 993–1006.

McKenzie, D.C. and McBratney, A.B. (2001) Cotton root growth in a compacted Vertisol (Grey Vertosol). I. Prediction using strength measurements and 'limiting water ranges'. *Australian Journal of Soil Research*, **39**, 1157–1168.

Merrill, S.D., Black, A.L., Fryrear, D.W., Saleh, A., Zobeck, T.M., Halvorson, A.D., and Tanaka, D.L. (1999) Soil wind erosion hazard of spring wheat-fallow as affected by long-term climate and tillage. *Soil Science Society of America Journal*, **63**, 1768–1777.

Michels, K., Sivakumar, M.V.K., and Allison, B.E. (1993) Wind erosion in the southern sahelian zone and induced constraints to pearl-millet production. *Agricultural and Forest Meteorology*, **67**, 65–77.

Mirreh, H.F. and Ketcheson, J.W. (1973) Influence of soil water matric potential and resistance to penetration on corn root elongation. *Canadian Journal of Soil Science*, **53**, 383–388.

Mohammadi, M.H., Asadzadeh, F., and Vanclooster, M. (2010) Refining and unifying the upper limits of the least limiting water range using soil and plant properties. *Plant and Soil*, **334**, 221–234.

Morris, P.H. (1992) Cracking in drying soils. *Canadian Geotechnical Journal*, **29**, 263–277.

Mueller, L., Kay, B.D., Hu, C.S., Li, Y., Schindler, U., Behrendt, A., Shepherd, T.G., and Ball, B.C. (2009) Visual assessment of soil structure: evaluation of methodologies on sites in Canada, China and Germany Part I: comparing visual methods and linking them with soil physical data and grain yield of cereals. *Soil and Tillage Research*, **103**, 178–187.

Mueller, L., Schindler, U., Mirschel, W., GrahamShepherd, T., Ball, B.C., Helming, K., Rogasik, J., Eulenstein, F., and Wiggering, H. (2010) Assessing the productivity function of soils. A review. *Agronomy for Sustainable Development*, **30**, 601–614.

Mullins, C.E., Ley, G.J., Lal, R., Young, I.M., Costigan, P.A., and Bengough, A.G. (1988) Hard-setting soils. *Journal of the Science of Food and Agriculture*, **44**, 121.

Munkholm, L.J., Perfect, E., and Grove, J. (2007) Incorporation of water content in the Weibull model for soil aggregate strength. *Soil Science Society of America Journal*, **71**, 682–691.

Nasr, H.M. and Selles, F. (1995) Seedling emergence as influenced by aggregate size, bulk-density, and penetration resistance of the seedbed. *Soil and Tillage Research*, **34**, 61–76.

North, P.F. (1979) Rothamsted studies of soil structure. 6. Assessment of the ultrasonic method of determining soil structural stability in relation to soil-management properties. *Journal of Soil Science*, **30**, 463–472.

Nunan, N., Ritz, K., Crabb, D., Harris, K., Wu, K.J., Crawford, J.W., and Young, I.M. (2001) Quantification of the in situ distribution of soil bacteria by large-scale imaging of thin sections of undisturbed soil. *Fems Microbiology Ecology*, **37**, 67–77.

Odell, R., Thornburn, T.H., and McKenzie, L.J. (1960) Relationships of Atterberg limits to some other properties of Illinois soils. *Soil Science Society of America Journal*, **24**, 297–300.

O'Sullivan, M.F. and Simota, C. (1995) Modeling the environmental impacts of soil compaction – a review. *Soil and Tillage Research*, **35**, 69–84.

O'Sullivan, M.F., Robertson, E.A.G. and Henshall, J.K. (1999) Shear effects on gas transport in soil. *Soil and Tillage Research*, **50**, 73–83.

Or, D. (1996) Wetting-induced soil structural changes: the theory of liquid phase sintering. *Water Resources Research*, **32**, 3041–3049.

Or, D. and Ghezzehei, T.A. (2002) Modeling post-tillage soil structural dynamics: A review. *Soil and Tillage Research*, **64**, 41–59.

Or, D., Leij, F.J., Snyder, V.A., and Ghessehei, T.A. (2000) Stochastic model for posttillage soil pore space evolution. *Water Resources Research*, **36**, 1641–1652.

Parent, B., Suard, B., Serraj, R., and Tardieu, F. (2010) Rice leaf growth and water potential are resilient to evaporative demand and soil water deficit once the effects of root system are neutralized. *Plant Cell and Environment*, **33**, 1256–1267.

Park, E.J. and Smucker, A.J.M. (2005) Erosive strengths of concentric regions within soil macroaggregates. *Soil Science Society of America Journal*, **69**, 1912–1921.

Pärtel, M. and Wilson, S.D. (2001) Root and leaf production, mortality and longevity in response to soil heterogeneity. *Functional Ecology*, **15**, 748–753.

Passioura, J.B. (1991) Soil structure and plant-growth. *Australian Journal of Soil Research*, **29**, 717–728.
Passioura, J.B. (2002) Soil conditions and plant growth. *Plant, Cell and Environment*, **25**, 311–318.
Passioura, J.B. and Leeper, G.W. (1963) Soil compaction and manganese deficiency. *Nature*, **200**, 29–30.
Passioura, J.B. and Fry, S.C. (1992) Turgor and cell expansion: beyond the Lockhart equation. *Australian Journal of Plant Physiology*, **19**, 565–576.
Perfect, E., Rasiah, V., and Kay, B.D. (1992) Fractal dimensions of soil aggregate-size distributions calculated by number and mass. *Soil Science Society of America Journal*, **56**, 1407–1409.
Perfect, E., Zhai, Q., and Blevins, R. L. (1998) Estimation of Weibull brittle fracture parameters for heterogenous materials. *Soil Science Society of America Journal*, **62**, 1212–1219.
Perret, J.S., Al-Belushi, M.E., and Deadman, M. (2007) Non-destructive visualization and quantification of roots using computed tomography. *Soil Biology and Biochemistry*, **39**, 391–399.
Peth, S., Horn, R., Beckmann, F., Donath, T., Fischer, J., and Smucker, A.J.M. (2008) Three-dimensional quantification of intra-aggregate pore-space features using synchrotron-radiation-based microtomography. *Soil Science Society of America Journal*, **72**, 897–907.
Petrovic, A.M., Siebert, J.E., and Rieke, P.E. (1982) Soil bulk-density analysis in 3 dimensions by computed tomographic scanning. *Soil Science Society of America Journal*, **46**, 445–450.
Pierret, A., Doussan, C., Capowiez, Y., Bastardie, F., and Pages, L. (2007) Root functional architecture: a framework for modeling the interplay between roots and soil. *Vadose Zone Journal*, **6**, 269–281.
Popham, T.W., Skidmore, E.L., Lamb, J.A., Merrill, S.D., Lindstrom, M.J., Mokma, D.L., and Yoder, R.E. (2003) Aggregate-mean diameter and wind-erodible soil predictions using dry aggregate-size distributions. *Soil Science Society of America Journal*, **67**, 425–436.
Prebble, R.E. and Currie, J.A. (1970) Soil water measurement by a low-resolution nuclear magnetic resonance technique. *Journal of Soil Science*, **21**, 273–288.
Prunty, L., Bell, J., and Rivers, M. (2003) Synchrotron radiation for detecting movement of a volatile soil liquid. *Journal of Environmental Engineering-ASCE*, **129**, 479–482.
Rengasamy, P., Chittleborough, D., and Helyar, K. (2003) Root-zone constraints and plant-based solutions for dryland salinity. *Plant and Soil*, **257**, 249–260.
Rieu, M. and Sposito, G. (1991) Fractal fragmentation, soil porosity, and soil-water properties. 1. Theory. *Soil Science Society of America Journal*, **55**, 1231–1238.
Rinaldi, M., Di Paolo, E., Richter, G.M., and Payne, R.W. (2005) Modelling the effect of soil moisture on germination and emergence of wheat and sugar beet with the minimum number of parameters. *Annals of Applied Biology*, **147**, 69–80.
Rogasik, H., Crawford, J.W., Wendroth, O., Young, I.M., Joschko, M., and Ritz, K. (1999) Discrimination of soil phases by dual energy X-ray tomography. *Soil Science Society of America Journal*, **63**, 741–751.
Rowell, D.L. (1994) *Soil Science: Methods and Applicaitons*. Prentice Hall, Essex
Ruhlmann, J., Korschens, M., and Graefe, J. (2006) A new approach to calculate the particle density of soils considering properties of the soil organic matter and the mineral matrix. *Geoderma*, **130**, 272–283.
Russells, E.J. (1961) *Soil Conditions and Plant Growth*. Longmans, London.
Schjonning, P., Munkholm, L.J., Elmholt, S., and Olesen, J. E. (2007) Organic matter and soil tilth in arable farming: management makes a difference within 5–6 years. *Agriculture Ecosystems and Environment*, **122**, 157–172.
Sharp, R.E., Poroyko, V., Hejlek, L.G., Spollen, W.G., Springer, G.K., Bohnert, H.J., and Nguyen, H. T. (2004) Root growth maintenance during water deficits: physiology to functional genomics. *Journal of Experimental Botany*, **55**, 2343–2351.
Sharp, R.E., Silk, W.K., and Hsiao, T.C. (1988) Growth of the primary maize root at low water potentials. I. Spatial distribution of expansive growth. *Plant Physiology*, **87**, 50–57.
Sheppard, S.C. and Miller, M.H. (1977) Temperature changes and the geotropic reaction of the radicle of *Zea mays* L. *Plant and Soil*, **47**, 631–644.
da Silva, A.P. and Kay, B.D. (2004) Linking process capability analysis and least limiting water range for assessing soil physical quality. *Soil and Tillage Research*, **79**, 167–174.
daSilva, A.P., Kay, B.D., and Perfect, E. (1994) Characterization of the least limiting water range of soils. *Soil Science Society of America Journal*, **58**, 1775–1781.
Sinnett, D., Morgan, G., Williams, M., and Hutchings, T.R. (2008) Soil penetration resistance and tree root development. *Soil Use and Management*, **24**, 273–280.
Six, J., Paustian, K., Elliott, E.T., and Combrink, C. (2000) Soil structure and organic matter: I. Distribution of aggregate-size classes and aggregate-associated carbon. *Soil Science Society of America Journal*, **64**, 681–689.

Six, J., Bossuyt, H., Degryze, S., and Denef, K. (2004) A history of research on the link between (micro) aggregates, soil biota, and soil organic matter dynamics. *Soil and Tillage Research*, **79**, 7–31.

Skidmore, E.L. and Powers, D.H. (1982) Dry soil-aggregate stability – energy-based index. *Soil Science Society of America Journal*, **46**, 1274–1279.

Sleutel, S., Cnudde, V., Masschaele, B., Vlassenbroek, J., Dierick, M., Van Hoorebeke, L., Jacobs, P., and De Neve, S. (2008) Comparison of different nano- and micro-focus X-ray computed tomography set-ups for the visualization of the soil microstructure and soil organic matter. *Computers and Geosciences*, **34**, 931–938.

Slowinskajurkiewicz, A. (1994) Changes in the structure and physical-properties of soil during spring tillage operations. *Soil and Tillage Research*, **29**, 397–407.

Taina, I.A., Heck, R.J., and Elliot, T.R. (2008) Application of X-ray computed tomography to soil science: a literature review. *Canadian Journal of Soil Science*, **88**, 1–20.

Taylor, H.M. and Ratliff, L.H. (1969) Root elongation rates of cotton and peanuts as a function of soil strength and water content. *Soil Science*, **108**, 113–119.

Tippkotter, R., Eickhorst, T., Taubner, H., Gredner, B., and Rademaker, G. (2009) Detection of soil water in macropores of undisturbed soil using microfocus X-ray tube computerized tomography (μ CT). *Soil and Tillage Research*, **105**, 12–20.

Tisdall, J.M. and Oades, J.M. (1979) Stabilization of soil aggregates by the root systems of ryegrass. *Australian Journal of Soil Research*, **17**, 429–441.

Tisdall, J.M. and Oades, J.M. (1982) Organic matter and water-stable aggregates in soils. *Journal of Soil Science*, **33**,141–163.

Totsche, K.U., Rennert, T., Gerzabek, M.H., Kogel-Knabner, I., Smalla, K., Spiteller, M., and Vogel, H.J. (2010) Biogeochemical interfaces in soil: the interdisciplinary challenge for soil science. *Journal of Plant Nutrition and Soil Science-Zeitschrift fur Pflanzenernahrung und Bodenkunde*, **173**, 88–99.

Triplett, G.B. and Dick, W.A. (2008) No-tillage crop production: a revolution in agriculture! *Agronomy Journal*, **100**, S153–S165.

Utomo, W.H. and Dexter, A.R. (1981) Soil friability. *Journal of Soil Science*, **32**, 203–213.

VandenBygaart, A.J., Protz, R., Tomlin, A.D., and Miller, J.J. (1999) Tillage system effects on near-surface soil morphology: observations from the landscape to micro-scale in silt loam soils of southwestern Ontario. *Soil and Tillage Research*, **51**, 139–149.

Veen, B.W. and Boone, F.R. (1990) The influence of mechanical resistance and soil-water on the growth of seminal roots of maize. *Soil and Tillage Research*, **16**, 219–226.

Vermang, J., Demeyer, V., Cornelis, W.M., and Gabriels, D. (2009) Aggregate stability and erosion response to antecedent water content of a Loess soil. *Soil Science Society of America Journal*, **73**, 718–726.

Vogel, H.J., Cousin, I., Ippisch, O., and Bastian, P. (2006) The dominant role of structure for solute transport in soil: experimental evidence and modelling of structure and transport in a field experiment. *Hydrology and Earth System Sciences*, **10**, 495–506.

Warkentin, B.R. (2008) Soil structure: a history from tilth to habitat. *Advances in Agronomy*, **97**, 239–272.

Watts, C.W. and Dexter, A.R. (1997) The influence of organic matter in reducing the destabilization of soil by simulated tillage. *Soil and Tillage Research*, **42**, 253–275.

Watts, C.W. and Dexter, A.R. (1998) Soil friability: theory, measurement and the effects of management and organic matter content. *European Journal of Soil Science*, **49**, 73–84.

Whalley, W.R., Clark, L.J., Gowing, D.J.G., Cope, R.E., Lodge, R.J., and Leeds-Harrison, P.B. (2006) Does soil strength play a role in wheat yield losses caused by soil drying? *Plant and Soil*, **280**, 279–290.

Whalley, W.R., To, J., Kay, B.D., and Whitmore, A.P. (2007) Prediction of the penetrometer resistance of soils with models with few parameters. *Geoderma*, **137**, 370–377.

White, R.G. and Kirkegaard, J.A. (2010) The distribution and abundance of wheat roots in a dense, structured subsoil – implications for water uptake. *Plant Cell and Environment*, **33**, 133–148.

Whitmore, A.P. and Whalley, W.R. (2009) Physical effects of soil drying on roots and crop growth. *Journal of Experimental Botany*, **60**, 2845–2857.

Wiberg, A., Koenig, R., and Cerny-Koenig, T. (2005) Variability in the physical and chemical properties of retail potting media. *Horttechnology*, **15**, 752–757.

Wick, A.F., Stahl, P.D., Ingram, L.J., and Vicklund, L. (2009) Soil aggregation and organic carbon in short-term stockpiles. *Soil Use and Management*, **25**, 311–319.

Woodroof, J.G. (1934) Pecan root growth and development. *Journal of Agricultural Research*, **49**, 511–530.

Wuest, S. (2007) Vapour is the principal source of water imbibed by seeds in unsaturated soils. *Seed Science Research*, **17**, 3–9.

Yoder, R.E. (1936) A direct method of aggregate analysis of soils and a study of the physical nature of erosion losses. *Agronomy Journal*, **28**, 337–351.

Yoshida, S. and Adachi, K. (2001) Effects of cropping and puddling practices on the cracking patterns in paddy fields. *Soil Science and Plant Nutrition*, **47**, 519–532.

Yoshida, S. and Hallett, P. D. (2008) Impact of hydraulic suction history on crack growth mechanics in soil. *Water Resources Research*, **44**, WC001

Young, I.M. and Crawford, J.W. (2004) Interactions and self-organization in the soil-microbe complex. *Science*, **304**, 1634–1637.

Young, I.M., Montagu, K., Conroy, J., and Bengough, A.G. (1997) Mechanical impedance of root growth *directly* reduces leaf elongation rates of cereals. *New Phytologist*, **135**, 613–619.

Zhang, B. and Horn, R. (2001) Mechanisms of aggregate stabilization in Ultisols from subtropical China. *Geoderma*, **99**, 123–145.

Zhang, X.X. and Lv, M. (2009) The nonlinear adsorptive kinetics of solute transport in soil does not change with pore-water velocity: demonstration with pore-scale simulations. *Journal of Hydrology*, **371**, 42–52.

Zhang, X.X., Deeks, L.K., Bengough, A.G., Crawford, J.W., and Young, L.M. (2005) Determination of soil hydraulic conductivity with the lattice Boltzmann method and soil thin-section technique. *Journal of Hydrology*, **306**, 59–70.

Zhu, Z.L., Minasny, B., and Field, D.J. (2009) Measurement of aggregate bond energy using ultrasonic dispersion. *European Journal of Soil Science*, **60**, 695–705.

9 Soil water and its management

Anne Verhoef[1] and Gregorio Egea[2]

[1]*Department of Geography and Environmental Science, Soil Research Centre, University of Reading, Reading, UK*
[2]*Area of Agro-forestry Engineering, School of Agricultural Engineering, University of Seville, Seville, Spain*

9.1 Soil water management – a general overview

A plant requires water for photosynthesis, for tissue rigidity (turgidity) and to produce carbohydrates (Jones, 1992; Roberts, 2005). The leaves draw up water from the soil, via the roots, through the plant's xylem. The driving force behind this process is transpiration: the loss of water from microscopic openings in the leaves (stomata). The main purpose of stomatal opening is to exchange CO_2 and O_2 with the atmosphere; CO_2 is needed to build carbohydrates, allowing plants to grow, while O_2 is expelled during photosynthesis and taken up during plant respiration (during which CO_2 is disposed of), a process that serves to maintain existing tissue. The water obtained from the soil also contains vital mineral nutrients which plants require for biochemical processes.

Transpiration is, in many ways, an undesirable trade-off of canopy gas exchange; it entails a significant loss of water, much more than is needed to maintain turgidity and for photosynthesis. A plant therefore needs a continuous supply of water, with adequate nutrients dissolved within it, to maintain healthy functioning. In areas where the vegetation's water demand is frequently larger than the amount of water available in the soil, plants will suffer. This will result in reduced growth or even plant mortality. For natural vegetation, these periods of drought define the distinctive plant communities that have adapted to and can survive under these dry conditions.

When considering agricultural and horticultural field crops, supplementing water stored in soil via irrigation may be required to meet plant water requirements (see Sections 9.2.4 and 9.2.5), or water harvesting (Zhao et al., 2009) may need to be practised. In some cases (generally in regions that have an arid climate), virtually all water is supplied via irrigation, often from deep subsurface stores, because surface water stores will be rapidly depleted or are not available.

The irrigating farmer or horticulturalist ideally provides their crop with enough water to grow successfully, with the aim to optimise water use efficiency, WUE (see Section 9.3.2), while at the same time avoiding waste or creation of excess water around the plant roots. The latter will hamper gas exchange between roots and soil and between soil and atmosphere. Furthermore, it can cause toxic effects such as nitric oxide being formed during hypoxia of root tissues. Highly water-saturated soils cause oxygen starvation of plant roots (and soil microorganisms) due to the

Soil Conditions and Plant Growth. Edited by Peter J. Gregory and Stephen Nortcliff.
© 2013 Blackwell Publishing Ltd. Published 2013 by Blackwell Publishing Ltd.

fact that gases diffuse much more slowly through water than through air. Hence, removal of CO_2 and replenishment of O_2 are slowed down or cease entirely (see Visser et al., 2003).

When excess water interferes with respiration for prolonged periods, thereby hampering crop growth, removal of water is required. This can be through active removal of soil water, by installing artificial drainage systems or by enhancing the natural drainage capability of the land, using land-management practices.

Often, irrigation and drainage systems are installed together, in particular, when flushing of the soils is needed to remove excess salts in (semi-) arid regions (see Chapter 7). Some crops are purposely grown under ponded conditions, for example rice, so here removal of 'excess' water would not be desirable, apart from during the harvesting period.

Excess water can be the result of over-irrigation (whether accidentally or on purpose, i.e. to remove salts); it can also occur naturally (although is often exacerbated through adverse human intervention). It is caused by a combination of climate, soil, vegetation, hydrological and topographical factors (see Section 9.6.1).

The practices of irrigation and the improvement of natural drainage, as well as the conservation of soil water, through suppressing evaporative and drainage losses or through practices such as runoff control, are collectively referred to as soil water management. Water conservation can also be practised by introducing drought-resistance species or varieties that have a high WUE or via tillage or soil amendments.

Agricultural soil water management is crucial in the context of meeting food demand and improving food security (see Rosegrant et al., 2009 and Chapter 1). Experiences in the past have shown that soil water management needs to take place in an integrated and sustainable way. Soil water management forms an important part of 'integrated water resources management' (IWRM), which can be defined as a systematic process for the sustainable development, allocation and monitoring of water resource use in the context of social, economic and environmental objectives.

This chapter will focus on the biophysical aspects of soil water management only.

This chapter will address the issues related to soil water management by starting with a detailed description of the soil water balance. It continues with the definition of key soil physical concepts, properties and variables related to water in soil. The chapter also addresses how these variables can be measured or estimated and the ways in which they play a role in soil water management. This naturally requires a better understanding about the interaction between soil water and plants, so some important physiological concepts are briefly discussed as well as the physiological background to water shortage/excess water and indicators of water stress. Finally, there is a discussion of the two major branches of soil water management: irrigation and drainage, only briefly addressing soil water conservation techniques. This chapter should not be viewed as a summary of key irrigation and drainage handbooks. Rather it discusses the overarching concepts and refers to major reference works in the literature, as well as providing a brief summary of tools (e.g. software packages) available to aid the soil water management process.

9.2 Key concepts related to water in soil

9.2.1 Soil water balance

A key equation when discussing shortage or excess of water in soils is the water balance for the root zone, which is given by (see Hillel, 1998):

$$\Delta\Theta = P + I + C - E - T - D - R \tag{9.1}$$

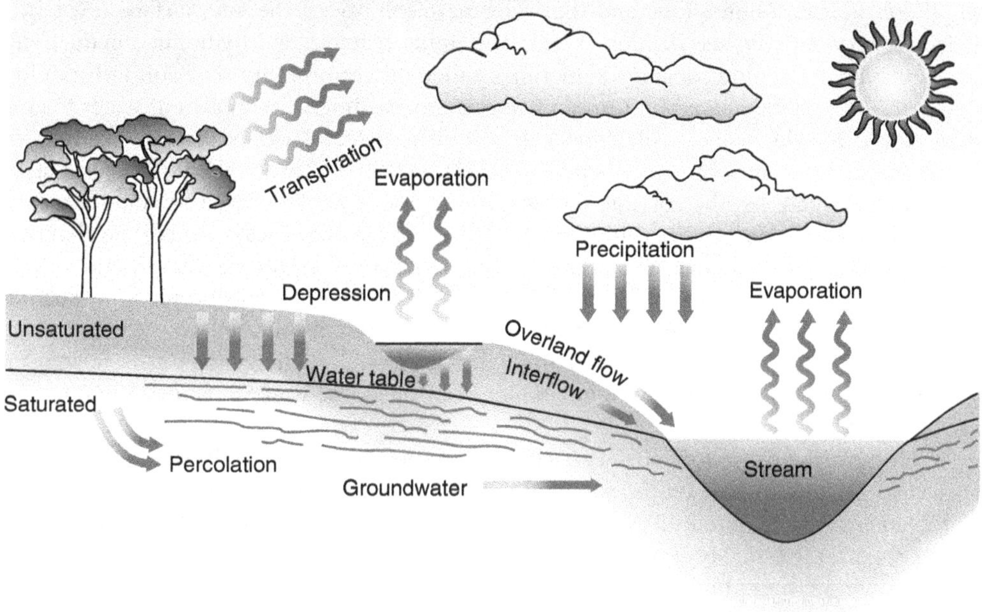

Figure 9.1 The hydrological cycle; focus is on key fluxes related to the soil water balance. From http://soer.justice.tas.gov.au/2003/image/267/index.php; image owner: Government of Tasmania. For a colour version of this figure, please see Plate 9.1.

where $\Delta\Theta$ is the change in soil moisture content during a given period, P is the precipitation, I is the amount of irrigation, C is the capillary rise, E is the soil evaporation, T is the transpiration, D is the deep drainage and R is the (surface and subsurface) runoff or runon (Figure 9.1). P, I and C are gain terms and E, T and D are loss terms, whereas R can be either a gain (negative R, i.e. runon) or a loss (i.e. runoff). In Equation 9.1 the terms generally have units of depth of water per unit of time, typically mm H_2O day^{-1}. A soil profile will gain water when the combined gain terms are larger than the loss terms and *vice versa*. Hence, $\Delta\Theta$ will generally exhibit strong seasonal and interannual variations.

Equation 9.1 describes the soil water store, one of the stores of the terrestrial part of the hydrological cycle; other land stores are ice, groundwater and surface water (rivers, lakes and swamps). The soil water store is relatively small compared to, for example, the ice and groundwater stores, but it is this water that is used by the plants (including our crops) to sustain their growth, so it is of vital importance to mankind. Water is transferred between the land and the atmosphere via transpiration and evaporation. It also flows from the land to the oceans via rivers (Figure 9.1; Farr and Henderson, 1986; Price, 1996).

P (and/or I) that is intercepted by the vegetation and does not drip off to reach the ground (throughfall), or reach the ground via the stems or tree-trunks (stemflow), can be evaporated back into the atmosphere; this is called interception loss, which can be significant for forests but is generally less so for crops (Herbst et al., 2008; Siles et al., 2010). The remainder of the water falling from the sky will directly reach the soil surface, where it can then infiltrate, soil conditions permitting.

The amount of water per unit area per unit time that enters the soil is called the infiltration rate and its value is determined by the precipitation/irrigation rate (the supply rate), the initial soil moisture conditions and the water permeability of the soil surface layer (the hydraulic conductivity, see Section 9.2.5). The initial wetness and hydraulic conductivity determine the infiltration capacity, sometimes called the infiltrability. The soil infiltrability varies; this means that at certain times a soil can cope with larger supplies of water than at other times. Starting with a relatively dry soil profile, there will be a large gradient in wetness (and hence in soil water suction, see Section 9.2.2) between the thin surface layer where water is entering the soil and the underlying drier soil. This large driving force (see Section 9.2.5) will cause infiltrability to be relatively large initially. As the soil profile wets up, the suction gradient, and hence the flow rate, will decrease, although at the same time the hydraulic conductivity will increase, which will counteract the decrease in suction gradient to some extent. When the topsoil reaches saturation, infiltration rate will equal the saturated hydraulic conductivity, K_s. Although the vertical suction gradient is now zero, gravity will still be acting on this water, which will 'push' the water in at a steady rate (see Section 9.2.5).

At this time, the soil has reached its final infiltration capacity or steady-state infiltrability. This means that from now on the soil can only accept precipitation or irrigation rates $\leq K_s$. Water being supplied at rates higher than K_s will pond immediately when the surface is flat or will run off if the soil surface is sloping. Organic matter, slope, soil texture and antecedent soil moisture are important factors influencing infiltration capacity (see, e.g. Wilcock and Essery, 1984).

However, before runoff occurs, excess water will first collect in surface depressions, whose total volume is called the surface storage capacity. The rate of runoff will depend on the excess of rain/irrigation intensity over the infiltration rate, as well as on the slope and the roughness of the soil (Hillel, 1971). Runoff is undesirable from a water management point of view as it negatively affects the amount of water available in the rooting zone for plant growth. Furthermore, runoff can cause erosion, which can result in loss of fertile topsoil (Lal, 2003). Runoff is often called surface water flow, to distinguish it from below-ground lateral flow. Below-ground lateral flow is often split into groundflow, which occurs below the water table, and interflow, where water that has accumulated within the topsoil layers starts to move laterally (this will only happen when the surface is sloping). When this interflow reaches the surface, it may become ponded or surface water (Figure 9.1; Farr and Henderson, 1986).

The infiltration process results in downward water movement in the soil, causing vertical and temporal changes of soil water content in the root zone, which will influence the water available to plants (see Section 9.2.4). The surface layers, initially having gained water, will now lose this water to deeper layers. In the absence of groundwater, or when dealing with a deep groundwater table, this process is called redistribution. With a groundwater table present, this water movement is often called internal drainage (Hillel, 1971). The duration of this post-infiltration movement (whether redistribution or internal drainage) strongly depends on the soil type.

Water eventually flowing out of the root zone, and therefore out of reach of the roots, is frequently referred to as deep drainage, *D*, or seepage. This water passes on to deeper substrate layers and eventually aquifers (e.g. sediments or rock formations), generally well beyond the reach of the plant roots. Drainage continues until the water reaches a vertical level below which all voids or pores are water-filled. The upper surface of this completely saturated zone is called the groundwater table (GWT; Farr and Henderson, 1986). When no, or little, lateral drainage is present (whether natural or through artificial drainage systems), continuing vertical inflow from above will fill the available pore spaces, causing the groundwater table to rise progressively through the subsoil, then the topsoil, until it reaches the surface, when ponding occurs. The depth of the GWT will vary with time due to its interaction with the water balance terms (Equation 9.1 and Figure 9.1).

If an impermeable layer (e.g. a hardpan) is present within the subsoil, this layer will prevent internal drainage resulting in saturated soil conditions and eventual ponding of water on the surface, even though the horizons below the impermeable layer are not saturated. This is referred to as a perched water table.

Finally, water can also travel upwards from the fully saturated zone to the root zone, via a process called capillary rise, C in Equation 9.1. This is the result of the fact that soil pores act as capillaries that can transport water upwards, as will be explained in Section 9.2.2. Capillary rise can contribute significantly to root zone water content in fine-textured soils and will allow water lost via evaporation during the day to be replenished during the night.

Having discussed how water enters and drains through the profile and how the root profile interacts with groundwater or deep soil water stores (losing to or gaining water from it), the focus switches to the evaporative process, which returns water to the atmosphere. The process of evaporation involves the change of liquid water into water vapour, which requires energy (resulting from the net radiation reaching the land surface, see Section 9.4.3) to break the bonds of the water molecules (so-called latent heat of vaporisation, λ: ~2.45×10^6 J kg^{-1} water at 20°C); this water vapour is then transferred to the atmosphere, either via soil pores (soil evaporation, E), stomata (transpiration, T) or via interception loss. The combined processes of soil evaporation and transpiration are often grouped together and termed evapotranspiration, although an increasing number of authors prefer the general term evaporation instead, because the physical processes behind both evaporation and transpiration are identical. During the evaporation process, water vapour will also have to negotiate a thin boundary layer close to the leaf and soil surfaces where air-flow is close to being laminar, and the main transfer process is diffusion, after which it can be transported, via turbulent transfer, to higher atmospheric layers.

The continuous mixing of the air, a result of the turbulent nature of air flow near the land surface, ensures a gradient of water vapour pressure between the evaporating surface and the near-surface atmosphere, where the atmospheric vapour pressure is lower than that at the surface. The combined effects of supply of energy and continual removal of water vapour (to maintain the water vapour pressure gradient) together determine the evaporative demand of the atmosphere. The evaporative demand, sometimes called evaporativity, therefore depends on the available energy, the relative humidity, wind speed and temperature. Together these variables determine the maximum possible rate of evapotranspiration from the evaporating surface. This maximum rate is often called the potential evaporation, E_p (this generally includes E and T). Evaporation can only occur at its potential rate if water is continually supplied from below to the site of evaporation, that being the soil pores or the stomata. If this supply falls below the evaporative demand, evaporation will occur at a rate below the potential rate: the actual evaporation, E_a. Hence, $E_a \leq E_p$. There are cases when $E_a > E_p$, this occurs when additional energy is supplied by advection, for example, when well-irrigated cropland is situated next to fallow dryland.

Soil evaporation is an important soil water loss term when the soil surface is bare. Under annual field crops, this may occur during fallow periods and throughout periods when soil tillage, planting, sowing and germination take place. Soil evaporation also plays an important role during the initial crop growth stages, when the canopy only covers a small percentage of the field surface area. It can also be considerable for sparse canopies, e.g. orchards, where a lot of exposed bare soil exists below and between the individual trees. Covering the soil with a mulch layer, or tilling the soil in such a way as to decrease evaporation (see Section 9.5.5), will avoid unnecessary water loss and improve crop water use efficiency (see Section 9.3.2).

When considering soil evaporation two cases can be distinguished: steady-state evaporation in the presence of a relatively shallow GWT and evaporation in the absence of a GWT. The latter will be a transient-state process, resulting in a net loss of soil water, thereby drying the soil profile. The first process may cause cumulative salinisation near the surface, a problem often encountered in (semi-) arid regions (see Section 9.5.4 and Chapter 7). The steady rate evaporation depends on depth of the water table and the dryness of the soil at the surface as well as on the hydraulic properties of the soil (Hillel, 1971).

Soil drying (soil evaporation in the absence of a GWT) generally occurs in two stages. During the first stage, the energy-limited stage, soil evaporation is entirely determined by the evaporative demand; it assumes a saturated soil surface, $E_a = E_p$. During the second stage, the water limited stage, $E_a < E_p$ and E_a continues to decrease, until eventually $E_a \sim 0$. This point will be reached relatively rapidly in coarse-textured soils, whereas soil evaporation in medium- to fine-textured soil will persist much longer. The reasons for this will be elaborated upon in Sections 9.2.2 and 9.2.5.

Transpiration is intimately linked to the process of photosynthesis and the extent of stomatal opening, which in turn partly depends on how much water is available for plants in the root zone; this process is discussed in Sections 9.2.4 and 9.3.2.

9.2.2 Water retention in soils

Not all of the soil water that has collected in the soil pores of the root zone is available for transpiration. To explain the reasons behind this, the ways in which soil water is retained in pores need to be understood. For plant roots to take up water from soil they need to overcome the force with which the soil matrix holds on to the water. These so-called matric forces do not just hold water against root water uptake, but also against evaporation and drainage; this is called water retention.

There are three mechanisms for binding of water to the solid matrix (Koorevaar et al., 1983): direct adhesion of water molecules to solid surfaces by London-van der Waals forces; capillary binding of water, due to a combination of adhesive and cohesive forces; and osmotic binding of water in diffuse electric double layers. Here, adhesion is the attraction between water molecules and the solid soil particles. Water molecules also experience strong intermolecular attractive forces with neighbouring water molecules; this is called cohesion. At the water surface there is an enhancement of the intermolecular attractive forces, which is called the surface tension, σ (which has a value 0.0728 Nm^{-1} at 20°C). The combination of these adhesive forces, and cohesive forces between water molecules, cause capillary binding in soils; the most important mechanism of water retention in coarse to medium-textured soils. Only in fine-textured soils may osmotic binding exceed capillary binding (Koorevaar et al., 1983).

The concept of capillary binding, caused by capillary action, deserves further explanation, because pores in soils may be viewed as irregularly shaped non-cylindrical capillaries, with radius r (m). In soil, pores exist in different sizes, hence we are dealing with a 'bunch' of capillaries with a range of radii. The pore-size distribution (see Chapter 8), largely determined by texture and soil structure, therefore plays an important role in soil water retention. Capillary action occurs when the adhesion of water to the walls of a capillary is stronger than the cohesive forces between the water molecules. The height, h_w (in m), to which capillary action will take water (capillary rise) in a uniform circular tube with radius r is given by

$$h_w = \frac{2\sigma}{\rho_w r g} \qquad (9.2)$$

(see, e.g. Koorevaar et al., 1983, for a derivation of this formula) where g is the acceleration of gravity (N kg^{-1}) and ρ_w is the density of water (kg m^{-3}).

This equation determines the extent of the capillary fringe, i.e. the thickness of the soil layer above the groundwater table that remains largely saturated (very shallow for sandy soils and up to about 0.4 m thick for clay soils). This capillary fringe may lead to a root zone that experiences excess water, despite the fact that the water table is below the root zone. However, the occurrence of capillary rise also means that the soil will be able to transport water from deeper saturated layers towards the root zone, resulting in replenishment (C in Equation 9.1). Soil water is also transported as a result of gradients in soil matric potential; this will be discussed in Section 9.2.5.

The theory of capillary binding also determines when a soil pore of radius r (m) will empty when a root exerts a certain suction to overcome the soil water suction, S (in *Pa*, i.e. Nm^{-2}). This relationship is given by:

$$S = \frac{2\sigma}{r} \qquad (9.3)$$

Hence, the narrower the pore, the larger the root suction required to empty it, i.e. to enable air to enter the pore. Soil water suction, S, is the amount by which the pressure in the soil water, P_w, is less than atmospheric pressure, P_a (atmospheric pressure has been given a reference value of $P_a = S = 0$); it is related to how tightly the water is held in the soil, as discussed in more detail in Section 9.2.3.

9.2.3 Key soil moisture contents and water retention curves

In a saturated soil, i.e. a soil where all pores are filled with water and moisture content, θ, equals θ_s, the saturated water content. This is the maximum soil moisture content possible in a soil, although in practice some trapped air will be present causing θ_s to be slightly less than the total porosity. The suction required at which the first (i.e. the biggest) pores start to release their water, and hence air enters into these pores, is the so-called air-entry value. Once the largest pores have emptied, soil water will remain in progressively smaller pores and it will become harder for roots to extract water from these ever smaller water-filled pores, as they need to overcome larger and larger soil water suctions (see Equation 9.3). Water will also be held in the necks between larger air-filled pores and as thin films held on to the particle surfaces enclosing air-filled pores.

Uptake of soil water by plant roots becomes more difficult as the soil water suction increases (i.e. as the matric potential, ψ_m, decreases, see Section 9.2.5). Most plant roots are generally unable to take up any more water when S has increased to about 1500 kPa or ~150 m, equivalent to a pore diameter of ~2 μm. Hence, this particular suction value, and its related soil moisture content, is called permanent wilting point, PWP, often defined as the point at which plants wilt but do not recover overnight. There are several factors that could affect PWP, including; crop species, soil type and crop stage of growth. Nevertheless, PWP is generally assumed to occur at $S = 1500$ kPa.

Another salient soil water suction-moisture content point in relation to root water uptake is that of field capacity, FC, although its usefulness, physical basis and definition are often disputed (see Hillel, 1971; Nachabe, 1998). Generally, this is defined as the water content in a soil profile at which internal drainage or redistribution has nearly stopped (is 'very slow'); often 2–3 days after the soil was thoroughly wetted by heavy rain or irrigation. At this stage, all transmission pores (pores with diameter > 50–100 μm) have been emptied. The suction at field capacity is often defined as $S = |\psi m| = 3.3$ m (33 kPa), although 10 kPa is another widely used value to denote FC. Also, some scientists prefer to define it by using a hydraulic conductivity (e.g. the soil water suction when hydraulic conductivity $K = 0.05$ mm day^{-1}).

The relationship between soil water content (θ) and soil water suction (or matric potential, ψ_m) is known as the water release curve, water release characteristic or water retention curve. The shape of the water retention curve is largely determined by a soil's textural type, because of the predominance of different mechanisms causing binding of water to the solid matrix. Capillary binding of water can only take place up to suctions of 10 m. Suctions larger than 10 m are caused exclusively by adhesive and osmotic binding. Hence, the pore size distribution, as affected by soil texture and structure, will determine the unique shape of the water retention curve for each soil (see Figure 9.2). The different distribution of pore sizes in soils causes sandy soils to release most of their water at relatively low suctions, while clay soils lose very little water until field capacity and still have a lot of water left at PWP.

The determination of water retention curves is briefly discussed in Section 9.4.2. While the curve is a unique property of each soil, it demonstrates hysteresis (Koorevaar et al., 1983). This means that the water retention curve derived from progressive extraction of soil water starting from saturation (desorption) is different from that obtained by gradual addition of water to air-dry soil (absorption). Typically, soil water characteristics are obtained through desorption. Hysteresis can be caused by the irregularity of soil pores' sizes and shapes, the entrapment of air during wetting, the lower contact angle of the meniscus during wetting than drying and different arrangements of clay particles during wetting and drying (Koorevaar et al., 1983).

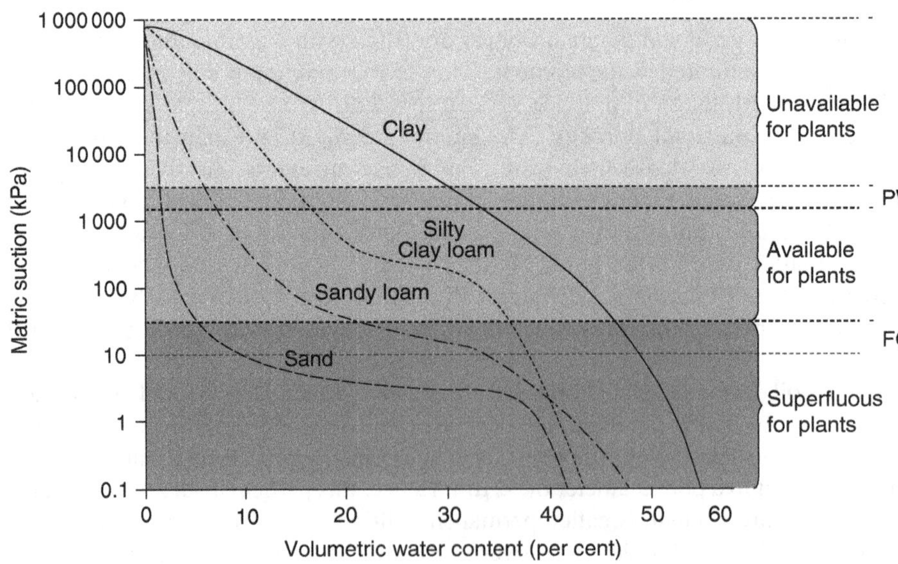

Figure 9.2 Typical water retention curves for a number of soils. Reprinted from Hugget et al. (2004). With permission from Hodder Arnold.

9.2.4 Available soil water

The amount of soil water available for plant growth, often called (plant) available soil water or available water capacity, is generally expressed as the difference between θ at FC and θ at PWP: $\theta_{FC} - \theta_{PWP}$; this concerns water in the storage pores, with approximate diameters between 2–50 μm.

Table 9.1, derived from Rowell (1994), shows typical values of θ_{FC} and θ_{PWP}, as well as available soil water, for ten different soil textural classes. Available soil water ranges between 0.10 for the sandy soil and 0.22 for loamy soils, although these minimum and maximum values are indicative only. Soils with the largest available water capacities are medium-textured. Although clay soils contain a lot of remaining pore water after initial drainage (large θ_{FC}), the considerable number of pores of small diameter (<2 μm) results in a large amount of soil moisture remaining once permanent wilting point is reached; this is not available to plant roots. These soils can only be dried out further through the process of evaporation.

Sandy soils mainly have pores in the transmission category (pore size > 50 μm) and hence will have lost most of their water already between saturation and field capacity. At PWP very little water will be left in the soil.

When used in crop water management (for irrigation purposes), the quantity $\theta_{FC} - \theta_{PWP}$ is often multiplied by the thickness of the root zone and then referred to as total available water, *TAW* (see Allen et al., 1998). Next, to indicate that soil water availability actually decreases with decreasing soil moisture content (e.g. due to a reduced rate of water transfer from the bulk soil to the root zone, largely as a result of decreased hydraulic conductivity values, see Section 9.2.5), it is sometimes split into readily available water (*RAW*) and a decreasingly available range, the latter starting when θ has reached a threshold value, θ_t. In crop models these concepts are often incorporated via a water stress coefficient, β, (called K_s in Allen et al., 1998) a dimensionless transpiration reduction factor (ranging from 0 to 1):

$$\beta = \frac{TAW - D_r}{TAW - RAW} = \frac{TAW - D_r}{(1-p)TAW} \tag{9.4}$$

where D_r is the root zone depletion (total water content in the root zone relative to field capacity; $D_r = 0$ at $\theta = \theta_{FC}$) and p is the fraction of *TAW* that a crop can extract from the root zone, without suffering water stress, i.e. before crop evapotranspiration gets reduced. The factor p differs between crops. Equation 9.4 indicates a linear decrease in β starting at θ_t, although some authors assume a curvi-linear decline. Similar to β, Sinclair (2005) and Metselaar and Sinclair (2007), introduced the concept of a relative daily transpiration rate, *RT*, given by:

$$RT = 1 - \psi_{m,s} \left(\theta / \theta_s\right)^{-b} / \psi_l \tag{9.5}$$

Here, θ and θ_s are as defined before and $\psi_{m,s}$ is the soil matric potential at saturation (or rather, at air-entry point). Parameters b and $\psi_{m,s}$ are texture-dependent parameters for which standard values can be found (based on Clapp and Hornberger, 1978). Parameter ψ_l is the hydrostatic pressure in leaves during periods of active transpiration. *RT* is defined as the water flux of a plant growing on drying soil relative to a plant growing on well-watered soil, and it is calculated between the upper, θ_u, and lower, θ_l, limits of soil water available to support plant transpiration. *RT* is a function of the fraction of transpirable soil water, FTSW:$(\theta - \theta_l)/(\theta_u - \theta_l)$. Sinclair (2005) defined this lower limit as θ were $RT < 0.1$; it was

Table 9.1 Soil moisture contents (θ) at field capacity (FC) and permanent wilting point (PWP) for ten typical soils ranging in texture from clay to sand.

		θ(cm³ cm⁻³)	
Texture	**PWP**	**FC**	**FC-PWP**
Clay	0.28	0.44	0.16
Silty clay	0.28	0.44	0.16
Clay loam	0.23	0.44	0.21
Silty clay loam	0.20	0.42	0.22
Sandy clay loam	0.16	0.36	0.20
Loam	0.14	0.36	0.22
Silt loam	0.14	0.36	0.22
Sandy loam	0.08	0.22	0.14
Loamy sand	0.06	0.18	0.12
Sand	0.05	0.15	0.10

The last column shows the plant available soil water.
Source: Data taken from Rowell (1994).

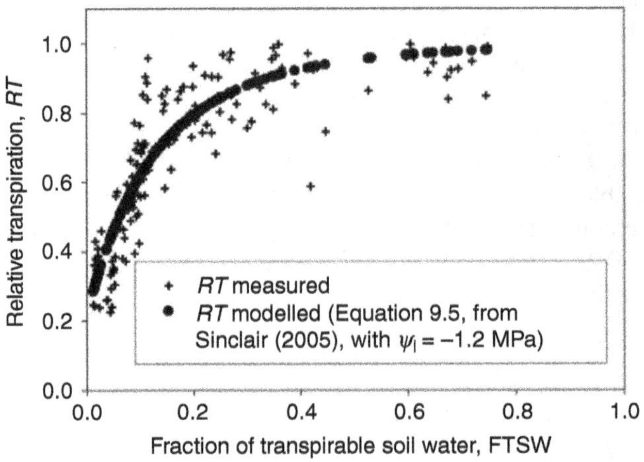

Figure 9.3 The relative transpiration rate as measured for well-watered and water-stressed dwarf sunflowers, using load-cell lysimeters (see Section 9.4.3). Also indicated is the result of Equation 9.5 with $\psi_l = -1.2$ MPa. The good fit between Equation 9.5 and the measurements illustrate the validity of Sinclair's (2005) theory, at least for this particular experiment.

found to range between 0.070 and 0.289, for sandy soil and clay soil, respectively, with the soil types as defined by Clapp and Hornberger (1978); see also Section 9.2.5. The upper limit was found to be 0.13 greater than the lower limit, for most soils, but these values were 0.08 and 0.15, respectively, for sand and silt soils. In this context, FC and PWP are not defined, although the lower and upper limits calculated in this way are very similar to those given in Table 9.1.

Equation 9.5 is an approximation of Sinclair's original formula, which involved depth of root water extraction, a variable for the geometry of soil water extraction around roots, the hydraulic conductivity and the daily transpiration rate on well-watered soil. It appeared (Sinclair, 2005), from sensitivity analyses, that all of these factors played a relatively minor role and that *RT* was largely determined by soil texture and ψ_l, as summarised in Equation 9.5.

Figure 9.3 gives an example of the relative transpiration rate as measured for well-watered and water-stressed dwarf sunflowers. The experimental set-up involved mini-lysimeters (see Section 9.4.3 and Diaz-Espejo et al. (2005)).

9.2.5 Water flow in soils

As previously mentioned, apart from water retention, the rate of water flow in soils is also important in the context of crop water uptake. When roots extract soil water for plant growth and maintenance, it needs to be replenished again, to sustain this continual process and to avoid (permanent) wilting. Furthermore, water flow in soils is important when discussing infiltration and drainage rates. In considering horizontal water movement, soil water will move in the direction of increasing suction, i.e. from places where it is less tightly held by the soil matrix to where it is more strongly retained. However, during vertical redistribution of water, both gravity and suction difference will affect movement and it will depend on the vertical suction profile and vertical location within the soil profile whether water will move and in what direction it will move. The effects of gravity and soil water suction are often discussed in terms of potential energy; water will flow from where its potential energy is higher to where its potential energy is lower. It is assumed that high elevation and low suction equate to high potential energy.

In this framework of potential energy, the major factors influencing movement are the gravitational potential, ψ_g, and the suction potential, more often called the matric potential, ψ_m. The sum of these two is called the hydraulic potential:

$$\psi_h = \psi_m + \psi_g \tag{9.6}$$

Here ψ_h is a measure of the overall potential energy status of the water. Soil water moves to decreasing values of ψ_h (more negative values). The potentials in Equation 9.6 can be expressed in different units: J kg^{-1}, J m^{-3} (= Pa) or in m, i.e. expressed as an equivalent head of water (see e.g. Koorevaar et al., 1983; Rowell, 1994).

The potential calculations require a reference point, for both potentials. For the gravitational potential, this is usually the soil surface ($\psi_g=0$ at $z=0$). The rule is then, when working with potential expressed as an equivalent head of water, which is often the easiest, that $\psi_g=z$ (with z, the depth in the soil, taken as negative). The reference for ψ_m is free water (an open dish of water under atmospheric pressure). This means that $\psi_m=0$ at the water table or when the soil is saturated. As soon as $\theta<\theta_s$, ψ_m becomes negative. *In situ* values of ψ_m can be determined using tensiometers (see Section 9.4.2) or derived from soil moisture measurement, assuming the water retention curve is known. Note that below the groundwater table, soil water pressure becomes positive and hence ψ_m (now referred to as ψ_p, the pressure potential, because the matric forces play no role anymore) increases with depth below the GWT, as a result of the increasing weight of the water column above the point under consideration.

Plotting hydraulic potential, ψ_h, against soil depth is helpful to determine which parts of the soil profile experience upward flow (as a result of evaporation and transpiration) or downward flow (drainage); this is called a potential diagram. It is the gradient of ψ_h with depth, ($\Delta\psi_h/\Delta z$, in mm^{-1}), that will determine the rate of soil water flow, F_w (ms^{-1}) together with the hydraulic conductivity, $K(\theta)$ (ms^{-1}), as a function of θ; this is given by Darcy's law:

$$F_w = -K(\theta)\frac{\Delta\psi_h}{\Delta z} \tag{9.7}$$

When the soil is saturated, hydraulic conductivity, K, is at its largest: the saturated hydraulic conductivity, K_s. As soon as θ decreases, $K(\theta)$ will decrease, for the following reasons. Firstly, as a soil dries there is much less cross-sectional area of water-filled pores through which water can flow. Secondly, water now has to take a more convoluted path; this drier soil is said to have a higher tortuosity, τ. Thirdly, the size of the water-filled pores plays an important role because the volume flow rate of water (the discharge rate, Q, in $m^3\ s^{-1}$) is more rapid through a large pore; Poiseuille's law dictates that Q is proportional to r^4. Finally, the internal friction in the water, as characterised by the viscosity, η, increases considerably with decreasing temperature, although this effect is often ignored in soil physics. Taking all these factors into account, $K(\theta)$ can be approximated by (Koorevaar et al., 1983):

$$K(\theta) = \frac{1}{8\eta\tau} \Sigma(\Delta\theta)_i r_i^2 \tag{9.8}$$

where i represents the ith pore size class. Hence, the hydraulic conductivity of a porous medium can be estimated from the size-distribution of the water-filled pores. The pore size distribution can be estimated from the soil water characteristic, if we consider the volumes of water (i.e. θ) released for increments of suction are measured (the latter can be directly related to r, see Rowell, 1994). This method works best in sandy soils, where soil water is largely present in capillary pores and the water retention curve exhibits a clear 'chair-shape' (see Figure 9.2). For medium- to-fine-textured soils this approach works less well. Generally, K is either measured (see Section 9.4.4) or calculated using mathematical expressions. Equation 9.8 explains why $K(\theta)$ decreases so sharply with drying and swelling/shrinking. It also explains why K_s and $K(\theta)$ vary so much between soil types.

One phenomenon worth mentioning here is preferential flow. Gerke (2006) states that 'preferential flow comprises all phenomena where water and solutes move along certain pathways, while bypassing other volume fractions of the porous soil matrix'. Coppola et al. (2009) distinguish four types of preferential flow: (1) preferential flow in non-capillary pores (macropores); (2) preferential flow in interaggregate pores; (3) 'fingering' at the interface of a less permeable layer above a more permeable one and (4) preferential flow due to spatial irregularities or temporal dynamics in soil water repellency.

Preferential flow may cause rain or irrigation water to bypass large regions of the root profile, so that water is lost from it via deep drainage, and not used to decrease the soil water deficit. However, some types of preferential flow (e.g. of type 4) may cause wet bulbs to occur in the root-zone, which in fact improve water availability. It is thought that some trees (e.g. needle-leaf trees or shrubs in dry climates) release water-repellent resin below their crown to induce this wet bulb (see Robinson et al., 2010).

The soil water characteristic and $K(\theta)$ functions define a soil's hydraulic properties. When water retention data are not available to calculate $\theta_{FC} - \theta_{PWP}$, the equations proposed by Clapp and Hornberger (1978) and Cosby et al. (1984) can be used to calculate $\psi_m(\theta)$ and $K(\theta)$. Parameters required are $\psi_{m,s}$, the matric potential at saturation (m), parameter b (unit-less) denoting the slope of the water retention curve, K_s and θ_s. By setting ψ_m to FC (generally taken to be 3.3 m) and PWP (150 m) values, the corresponding θ-values (and hence available water) can be calculated if b and θ_s are known. Standard parameter values can be taken from the literature (Clapp and Hornberger, 1978; Cosby et al., 1984) or values can be calculated from pedotransfer functions (e.g. Cosby et al., 1984). In this case the parameters are calculated as a function of percentage sand, silt and clay content, using equations based on a large numbers of soil samples. Although the Clapp and Hornberger equations are widely used, an alternative parameterisation, the Van Genuchten equations (Van Genuchten, 1980), generally provides a better fit to the water retention and hydraulic conductivity data as a result of their mathematical form and the fact that they have three extra tunable parameters (residual soil moisture content, θ_r, and fitting parameters l and m). Typical parameter sets for the Van Genuchten approach, and related pedotransfer functions, can be found in Schaap and Leij (1998) and Wösten et al. (1999), for example.

9.3 Soil water and plants

9.3.1 Introduction to soil–plant water relations

Water and air both reside in the soil pores and their relative proportion is crucially important for healthy plant growth. The plant root system provides the aerial parts (i.e. stem and leaves) with water and dissolved minerals. Under optimum soil moisture conditions, plant roots will extract soil water at the potential rate to meet the evaporative demand of the atmosphere. Evaporation of water from vegetation takes place through the leaves' stomata, in a process known as transpiration, T, which can be considered as a necessary 'cost' associated with the inward diffusion of CO_2 gas from the air and the process of photosynthesis or CO_2 assimilation. The extent of opening of the stomata is called stomatal conductance, g_s.

A limited water supply in the soil is sensed by roots and communicated to the shoot to mediate (partial) stomatal closure (causing a decrease in g_s) and permit timely adjustment of the plant's water losses by transpiration. The inter-dependence between photosynthesis and T means that as g_s decreases there is a decrease in biomass production, although this adaptation to water stress greatly depends on how efficiently plants use water under limited soil water availabilities. This plant behavioural trait is termed water use efficiency (WUE) and a detailed description of this concept is provided in Section 9.3.2. Due to the close link between photosynthesis and T, a drop in T will cause a decrease in growth. Consequently, there is the need for irrigation to safeguard crop productivity and farmer's profitability. Irrigation represents a key feature in the optimization process of on-farm water resources in water-limited areas.

Although there are still some gaps in our knowledge regarding stomatal regulation (Lawson, 2009), it is now widely accepted that stomatal aperture is regulated by both internal physiological factors (e.g. leaf water status, plant hormones such as abscisic acid, ABA) and external environmental factors, such as light (in the photosynthetically active waveband), air CO_2 concentration and air humidity (Farquhar and Sharkey, 1982; Buckley, 2005). The physiological mechanisms controlling the response of stomata to environmental stresses, such as reduced soil water availability, are complex and still not fully understood which led to g_s initially being calculated from multiplicative factors depending on environmental variables (Jarvis, 1976; Stewart, 1988). However, more recently, various authors have attempted to model stomatal behaviour in a more physiologically explicit fashion (Dewar, 2002; Gao et al., 2002; Buckley et al., 2003).

Many plant physiological processes (e.g. cell expansion and leaf photosynthesis) are significantly affected by leaf water status. As with soils, leaf water status is commonly expressed as leaf water potential, the thermodynamic expression of water status of leaves. Leaf water potential at a given time of the day is, therefore, not only the result of soil water status but is also directly linked to transpiration rate. The classical equation of water flux (Van den Honert, 1948) synthesizes the role that both factors play in leaf water potential (ψ_l):

$$\psi_l = \psi_s - \left(R_s + R_p\right)J_w \tag{9.9}$$

where ψ_s is the soil water potential (negative) given by the sum of ψ_h (Equation 9.6) and the osmotic potential (ψ_o), R_s and R_p are the hydraulic resistances from bulk soil to roots, a function of $K(\theta)$, and from roots to shoot, respectively, and J_w is the water flux through the plant (i.e. transpiration rate, later on in this chapter referred to as T), which is greatly determined by stomatal regulation. Water transport through the soil–plant–atmosphere continuum requires the osmotic component to be considered. In soils there are no semi-permeable membranes (such as those found in plant tissues) so the osmotic potential does not play a role in water transport and hence hydraulic potential is the driving force. However, in the case of water flow from soils to plants, water needs to pass through semi-permeable membranes and hence the osmotic effect plays a role (this also explains why the transpirational stream is reduced in saline soils).

The major factors influencing the leaf water potential are ion concentration, pressure and gravity. Consequently, ψ_l may be expressed in terms of its individual components, usually written as the following sum:

$$\psi_l = \psi_o + \psi_p + \psi_g \tag{9.10}$$

where ψ_o (negative), ψ_p (positive) and ψ_g (positive) denote the effects of solutes (osmotic potential), hydrostatic pressure (turgor) and gravity (gravitational potential), respectively, on the free energy of water (Taiz and Zeiger, 2002). The gravitational component is generally omitted because it is negligible compared to the osmotic potential and the hydrostatic pressure of plant tissues. These different components of leaf potential can be used when monitoring plant water stress (see Section 9.3.3).

Soil salinity also plays a major role in canopy gas exchange and plant water relations in general. Dissolved solutes in the rooting zone generate a low (more negative) osmotic potential (ψ_o) that lowers the soil water potential. Therefore, the effect of dissolved solutes is similar (as a first approximation) to that of a soil water deficit, and most plants respond to excessive levels of soil salinity in the same way as for water stress. A major difference between low water potential environments caused by salinity and water shortage is that the total amount of soil water may be much higher in a saline soil, albeit it is not available for the plant. Further details on soil salinity and its management are provided in Chapter 7 and Section 9.5.3.

Soil drying may affect plant growth through physiological effects on plants, as described earlier, but also by direct effects on the soil physical environment. As reviewed by Whitmore and Whalley (2009), soil water shortage may affect vegetation in a number of interacting ways, highlighting that 'soil drought' is not a single and simple stress (see Chapter 8 for an account of how physical properties affect plant growth).

Excess soil water also adversely affects plant functioning, in this case through anaerobic stress. For detailed background information of the processes involved and plant mechanisms that underlie short- and long-term tolerance to external anaerobic conditions refer to the reviews by Vartapetian and Jackson (1997) and Visser et al. (2003).

9.3.2 Concepts of water use efficiency

The term 'Water Use Efficiency (WUE)' has traditionally been used as an indicator of plant performance. WUE can be defined in two different ways: from an eco-physiological or agronomical point of view (Gregory, 2004; Katerji et al., 2008).

The eco-physiological approach is based on the analysis, at a given instant, of the relationship between net CO_2 assimilation and transpiration (T), which can be used at the field-, plant- or leaf-scales (Steduto et al., 1997; Chen and Coughenour, 2004; Katerji et al., 2008; Morison et al., 2008).

For agronomists, WUE or 'crop water productivity' (Katerji et al., 2008; Geerts and Raes, 2009) is usually a value defined by yield per unit area, Y, divided by water used to produce yield. The denominator may refer to water lost by evapotranspiration (ET) or just T by excluding direct evaporation (E) from the soil and plant surfaces (via evaporation of intercepted rainfall). When this is the case, the expression is then normally termed 'transpiration efficiency'.

The amount of water used can also be written in terms of all of the loss components in the soil water balance equation (see also Equation 9.1 in Section 9.2.1):

$$WUE = \left(\frac{Y}{T}\right)\left(\frac{1}{\left[1+(E+D+R)/T\right]}\right) \tag{9.11}$$

This equation has been normalised with regards to transpiration (Gregory, 2004). Interpretation of Equation 9.11 makes it clear that any enhancement in WUE must come about by exploiting three key processes: (1) agronomic management practices that maximize T by reducing the other losses; (2) acquiring more carbon (biomass) in exchange for the water transpired by the crop (i.e. improving crop transpiration efficiency) and (3) partitioning more of the achieved biomass into the harvested product (Gregory et al., 1997; Condon et al., 2004; Gregory, 2004).

The pressing need to improve WUE of agricultural systems is not only focused on irrigated cropping systems, but also on rainfed systems. The sharp increase in wheat yields and rainfall-use efficiency observed in Australia during the last two decades is considered to be one third attributable to new (more efficient) cultivars and two thirds attributable to the adoption of improved agronomic practices (Angus et al., 2001; Stephens, 2002; Turner, 2004)

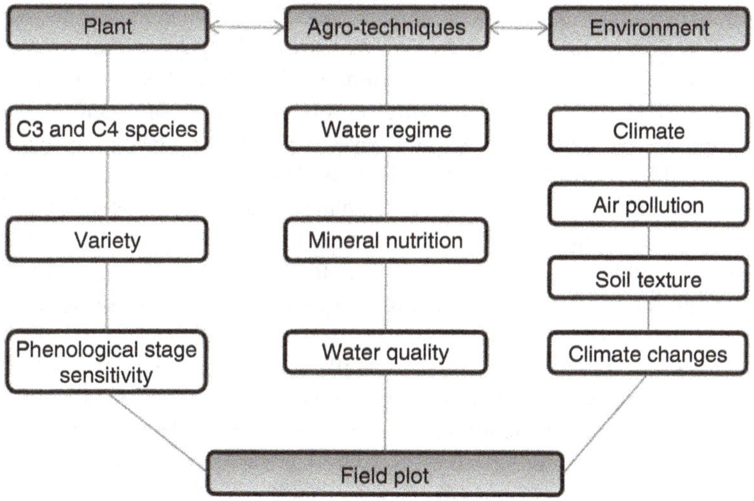

Figure 9.4 Schematic representation of the main determinants of WUE. Adapted from Katerji et al. (2008). With permission from Elsevier.

(see Section 9.5.4). One agronomic practice that has proven effective to reduce evaporation from the soil surface is 'mulching' with crop residues, both irrigated and rainfed systems (Gregory, 2004; Deng et al., 2006; see also Section 9.5.4). As Hsiao et al. (2007) noted, there is still much room for improving WUE in irrigated cropping systems, in part owing to the low water storage and conveyance (≈ 70%) and irrigation (≈ 37%) efficiencies globally (Wallace and Gregory, 2002; Gregory, 2004). As reviewed by Katerji et al. (2008), the sources of variability of WUE are numerous and sometimes of a very distinct nature and may be grouped into three categories related to (1) plant, (2) agro-techniques and and (3) environment (see Figure 9.4).

Among the latter, soil texture has proven to play a significant role in determining WUE. Katerji and Mastrorilli (2009) in a long-term experiment, reported that WUE of potato, corn, sunflower and sugar beet was reduced by 22–25% when crops were grown on clay soil as compared to those grown on loam soils. This was largely related to the available water, as explained in Section 9.2.4.

9.3.3 Plant indicators of water stress

Irrigation techniques for which the decision to irrigate is based on measured levels of soil moisture content will be presented in Sections 9.5.1 and 9.5.2. As outlined by Jones (2004), a potential problem with all soil-water-based approaches is that many plant physiological attributes respond directly to changes in plant tissue water status rather than to changes in the bulk soil water content (or water potential). This is because leaf water potential at a given time of the day is not only the result of soil water status but also is directly linked to evaporative demand through its influence on transpiration rate (see Equation 9.8).

Another reported drawback of soil-based irrigation management systems relates to the high soil heterogeneity combined with the limited volume of soil that some soil moisture devices sample (Jones, 2004; Verhoef and Campbell, 2005), causing inaccurate estimates of soil water stress. This fact causes many researchers to consider 'plant stress sensing' as the most reliable approach for precise irrigation scheduling.

As reviewed by Jones (2004), plant-based water stress indicators can be grouped into two categories: first, direct measurements of some aspect of plant water status; and second, other plant processes that respond sensitively to water stress. The first group comprises all the available techniques to measure tissue (leaf, stem, etc.) water status (e.g. pressure chamber, or psychrometer). The principles underlying the quantification and measurement of plant water status have been described elsewhere (Boyer, 1995; Nobel, 1999; Kirkham, 2004) and these sources should be consulted for more detail and background theory. The second group comprise gas exchange indicators such as stomatal conductance, which can be directly measured with specialized equipment (e.g. porometers) or indirectly derived through thermal sensing, sap flow measurements (to estimate transpiration rate), micromorphometric variations of plant organs (i.e. leaf, branch, trunk or fruit) or other physiologically based but remotely sensed indicators.

Direct estimation of plant water status is usually done via measurements of tissue water potential by means of psychrometric methods (Jones, 1992) or by means of the pressure chamber (Figure 9.5, top left) (Scholander et al., 1965; Hsiao, 1990). Psychrometer-based measurements of water potential, although valuable and with the possibility of automation, require sophisticated equipment and high level of technical skill, limiting their usefulness for irrigation scheduling purposes. Leaf water potential, ψ_l, has been recommended for many years as a sensitive index for irrigation control in both field crops and fruit trees

Figure 9.5 (top left) Scholander-type pressure bomb; (top right) LVDT sensor installed on a fruit tree trunk; (bottom left) Sap flow probe installed on a fruit tree trunk; (bottom right) infrared thermometer installed above a fruit tree canopy. Pictures taken at the experimental station of the University of Cartagena (Spain).

(Grimes and Yamada, 1982; Peretz et al., 1984). However, the interactions between ψ_l and the temporal fluctuations of the surrounding environment (Fereres and Goldhamer, 1990; Jones, 1990) make it difficult to identify a threshold ψ_l for use in irrigation scheduling.

As outlined by Jones (2004), a further limitation of plant water status (and ψ_l in particular) as an indicator of water stress results from the fact that it is subject to some physiological control. In extreme cases, plants termed as 'isohydric' (e.g. maize or poplar plants; Tardieu and Simonneau, 1998) present good endogenous control systems and may maintain a relatively stable leaf water status over a wide range of evaporative demand or soil water supplies. Under these circumstances, and especially in strongly isohydric species, the usefulness of plant water status as an indicator of water stress is limited. This is in contrast with those species such as sunflower or barley, which appear to have less effective control of leaf water status and have been termed 'anisohydric' (Tardieu and Simmoneau, 1998).

None of the aforementionned plant-based methods, although suitable as plant stress indicators in research studies, are well adapted for automation of irrigation scheduling or control and all are labour-intensive (therefore expensive, especially for predawn measurements). Several indirect methods for continuously measuring or monitoring water status are currently available as alternatives to direct measurement. Jones (2004) has made an extensive and comparative review of such methods.

Among these plant-based indicators, a substantial amount of effort and research has been devoted to develop precise irrigation scheduling based on changes in the stem diameter, although variations in diameter may also occur in other parts of the plant (Ueda and Shibata, 2001; Cermak et al., 2007; Carminati et al., 2009). The fundamentals and usefulness of this technique for irrigation scheduling have been extensively reviewed by Fernández and Cuevas (2010) and Ortuño et al. (2010). The trunk of woody plants (e.g. fruit trees or vines) represents a major plant water store that may contribute substantially to its transpiration stream even under well-watered conditions (Zweifel et al., 2007). Diurnal and seasonal changes in the water content of the extensible tissues of the stem which are driven by changes in xylem water potential cause diurnal variations in the trunk diameter (stem diameter variations: SDV) that may be used to derive some stress indices, for example in a number of fruit

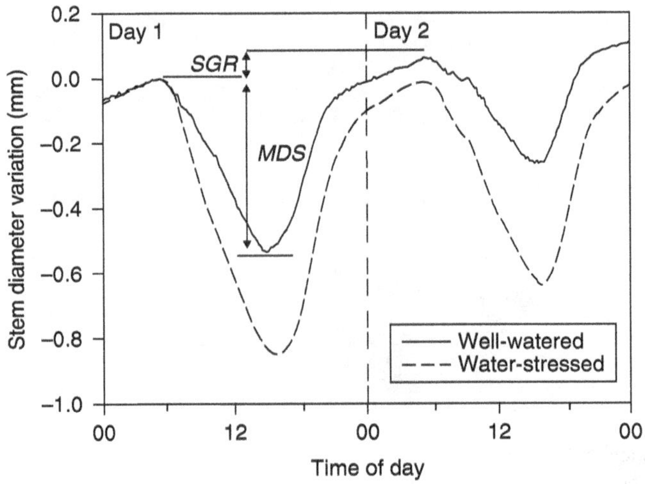

Figure 9.6 Stem diameter variations recorded during two consecutive days in well-watered and water stressed almond trees in SE Spain (Egea, 2008). MDS, maximum daily shrinkage; SGR, stem growth rate.

tree species (Huguet et al., 1992; Goldhamer and Fereres, 2001; Intrigliolo and Castel, 2004; Nortes et al. 2005; see Figure 9.6).

Most SDV-records for irrigation scheduling during the last two decades have used linear variable differential transformers (LVDT) (Figure 9.5, top right), because of their robustness and high precision (Fernández and Cuevas, 2010), although other devices like strain-gauges have also proven successful for continuous measurement of stem, branch or even fruit diameters (Ueda et al., 1996; Link et al., 1998). One of the limitations of the method lies in the requirement of expert knowledge to interpret SDV records. This limitation results from the range of factors, besides water stress, that affect SDV, such as variations in growth pattern due to the tree's age (Goldhamer and Fereres, 2001), crop load (Intrigliolo and Castel, 2007), tree size (Intrigliolo and Castel, 2006) or crop management (Fernández and Cuevas, 2010). Nonetheless, when the method is properly used, it presents great potential for automatically scheduling irrigation in commercial fruit tree orchards (Vélez et al., 2007; Egea et al., 2009).

Another plant-based indicator with great potential for irrigation scheduling is sap flow (SF). Various methods are available to calculate mass flow of water in the transpiration stream by using heat as a tracer [Figure 9.5, bottom left; for reviews (see Swanson, 1994; Smith and Allen, 1996; Kostner et al., 1998)]. Sap flow systems are easily automated, robust and reliable enough for operation in the field over extended periods of time. Fernández et al. (2008) demonstrated that the 'transpiration ratio', as defined by the ratio between the actual daily water use (measured with SF sensors) divided by the potential daily water use of well-watered plants, appeared to be a reliable tool to schedule irrigation in commercial fruit tree orchards and grapevines. However, keeping plants 'well-watered' can present problems related to over-irrigation (e.g. hypoxia), which can make the use of these plants as representatives of the orchard unrealistic.

There has also been an increasing interest in using infrared measurements of canopy temperature as an indicator of stomatal closure, transpiration and 'crop stress', since the closure of stomata leads to increases in leaf temperature through the well-reported transpirational cooling effect (Ehrler, 1973; Jones et al., 2002). Thermal methods have the advantage of being more rapid than conventional gas-exchange measurements and of not interfering with the stomatal responses through physical contact with the leaves (Guilioni et al., 2008). With the development and availability of infrared thermometers (Monteith and Szeicz, 1962), there have been many attempts to relate canopy temperature to plant water stress conditions to provide guidance for irrigation scheduling. Most of these methods use stress index approaches, such as the crop water stress index (CWSI, Idso et al., 1981; Jones, 2004), that are indirectly or empirically related to stomatal conductance.

CWSI is based on measured canopy-air temperature difference ($\Delta T = T_c - T_a$). After obtaining the two limits of ΔT for potential (non-water stressed, ΔT_{nws}) and null transpiration of a crop (ΔT_{dry}), or similarly, for maximum and minimum canopy conductance, under given environmental conditions, the CWSI can be calculated as follows (Idso et al., 1981):

$$\mathrm{CWSI} = \frac{\Delta T - \Delta T_{nws}}{\Delta T_{dry} - \Delta T_{nws}} \tag{9.12}$$

The ΔT_{nws} and ΔT_{dry} limits used in Equation 9.12 can be obtained using different methods. A widely used one is the method by Idso et al. (1981), where two baselines are obtained empirically by relating ($T_c - T_a$) to air vapour pressure deficit values (see Figure 9.7).

An alternative approach makes use of direct infrared measurements over wet or dry reference surfaces, either natural or artificial (Jones, 1999a,b; Jones et al., 2002; Wang et al., 2005; Moller et al., 2007). Although this method has great potential, as long as caution is taken in the design and placement of the reference surfaces, presently it is not suitable to be used directly and continuously for irrigation scheduling. Testi et al. (2008), by using continuously recorded Infra-Red Thermometers (IRT) installed above the tree canopy (Figure 9.5, bottom right), demonstrated the potential of CWSI as a plant-based water stress indicator in irrigation scheduling of pistachio trees. Besides CWSI, other thermal indices, such as the one proposed by Jones (1999a) and successfully tested, among others, by Leinonen and Jones (2004) and Grant et al. (2007), have also demonstrated its accuracy as stress indicators.

The viewing angle and the sample of foliage targeted by the IRT devices that Testi et al. (2008) used can also be obtained with high-resolution satellite or air-borne remote sensing equipment. This highlights the possibility of remote thermal detection of water stress for irrigation applications, as demonstrated by Sepulcre-Cantó et al. (2006) with high-resolution airborne thermal images flown over non-homogenous orchard canopies. Traditionally, critical issues such as the optimum spatial and spectral resolutions, the turnaround time and repeat cycle have been major limiting factors of the usefulness of remote sensing products (e.g. infrared thermal images) for precision crop management. Airborne sensors can deliver higher spatial and spectral resolutions and are more flexible in terms of revisit time than satellite-based products (Suárez et al., 2008; Sobrino et al., 2009), but the high operating costs and the long turnaround times due to high volume of data processing acquired in large airborne campaigns have limited their use. Nonetheless, developments conducted on unmanned aerial vehicles (Berni et al., 2009) have opened the gate to provide useful remote sensing products in agriculture. In addition to canopy temperature, these systems may also provide other physiological and structural indicators for remote sensing detection of water stress, such as the physiological reflectance index (PRI) which is sensitive to xanthophyll pigment detection under water stress conditions (Suárez et al., 2008) or solar-induced chlorophyll fluorescence emission (Zarco-Tejada et al., 2009), due to the strong correlation demonstrated between steady-state chlorophyll fluorescence and stomatal conductance (Flexas et al., 2002).

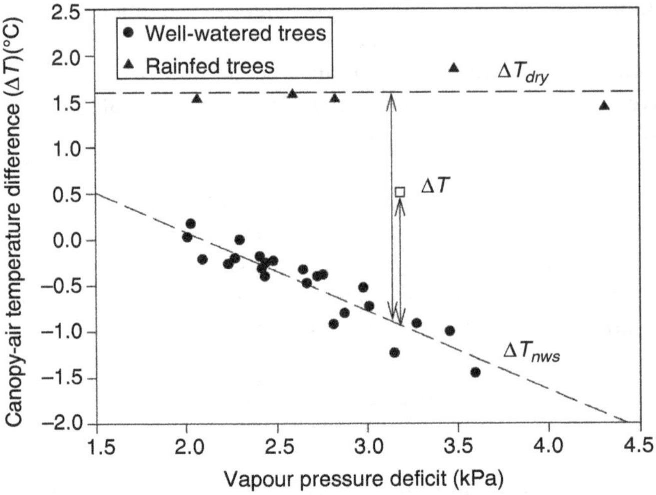

Figure 9.7 Illustration of calculation of Idso's Crop Water Stress Index (CWSI, Equation 9.12) showing the dependence of ΔT_{dry} and ΔT_{nws} on air vapour pressure deficit (VPD). ΔT_{dry} and ΔT_{nws} baselines were determined from leaf-scale measurements of temperature performed on rainfed and well-watered almond trees in SE Spain, respectively. Adapted from Egea (2008).

9.4 Determination of soil water balance and related variables

A wide range of soil-based, as opposed to plant-based, irrigation scheduling methods is currently available. Despite the fact that plant-based water-stress indicators promise great potential, soil-based techniques to inform irrigation scheduling are currently still much more common. The soil-based irrigation-scheduling approaches broadly fall under two categories: either direct measurement techniques or water balance methods. Direct measurement techniques measure soil water content or matric potential *in situ*. Irrigation is then scheduled according to a target soil water content (or deficit) or matric potential. Water balance methods keep a running balance of the water in the soil profile from measurements of rainfall and irrigation amounts and estimates of evapotranspiration (Hess and Knox, 2002). Hence, this section will describe measurement techniques to determine θ and ψ_m, as well as key components of the water balance.

Monitoring of soil moisture content (and groundwater levels, using piezometers, see Section 9.4.2) is also essential once drainage systems have been installed, to ensure the artificially lowered water levels and lower moisture contents in the root zone are within the intended range and not adversely affecting crop growth.

9.4.1 Determination of soil water content

Techniques for measuring soil water content are often classified as either direct or indirect. The direct methods refer to those methods that involve removal of water from the soil matrix to directly measure the amount of water present in the sample. This has traditionally been achieved by recording the change in mass of a soil sample after it has been dried out to a constant mass at 105°C. This method requires the use of convective oven drying and is commonly referred to as the gravimetric method (see also Chapter 8). Water content can be expressed either on a mass basis (gravimetric soil water content, θ_g, in kg water kg^{-1} dry soil) or on a volumetric basis (volumetric soil water content, θ, water volume to soil bulk volume, m^3 water m^{-3} soil). They are related through

$$\theta = \left(\frac{\rho_d}{\rho_w}\right)\theta_g \tag{9.13}$$

where ρ_d is the dry bulk density of the soil (kg m^{-3}) and ρ_w the water density (kg m^{-3}).

The gravimetric method has the advantage of being accurate and inexpensive, but it is a destructive, time-consuming method, that does not allow for repetitive measurements in the exact same location. Hence, it is less suitable when (near-) continuous measurements are required, e.g. for scientific purposes of for precision irrigation.

Indirect methods are based on the measurement of some θ-dependent physical or chemical properties. These properties include the dielectric constant (relative permittivity), electrical conductivity, heat capacity, hydrogen content or magnetic susceptibility (see Durner and Lipsius, 2005 for a detailed overview). Some of the most relevant available methods for indirectly estimating θ are summarized in the rest of the section.

Neutron moderation or neutron thermalization

The neutron moderation technique is a long-standing method which represented, for many years, the benchmark field method to determine θ. The neutron probe has a radioactive source of high-energy (fast neutrons) and a detector for slow, thermalized neutrons and is lowered into the soil via a preinstalled access tube. The neutrons that are emitted from the radioactive source lose energy as they collide with soil particles; these neutrons slow down dramatically when they collide with hydrogen nuclei, which are part of H_2O molecules and become thermalized. Hence, a measure of the quantity of thermalized neutrons returning to the detector within a given time period gives a good measure of θ. The method requires a linear calibration between the count rate of thermalized neutrons and θ obtained from soil samples taken at the site. Unlike other devices, the neutron probe is not affected by salinity or air gaps. This method offers the opportunity of frequent and undisturbed revisiting of an experimental field, thereby offering the opportunity of long-term monitoring over a (number of) growing season(s). Unfortunately, the measurements cannot be automated. The soil sensing volume is large as compared to other methods, with a sphere of influence of variable radius: about 0.15 m in wet soil and up to 0.5 m in dry soil, thus limiting measurements to > 0.3 m depth below the soil surface without special calibration for surface layers (Topp and Ferre, 2005).

Due to regulations and hazards concerning the use of radioactive sources, the method requires certified personnel and in recent years regulations and controls have become more and more stringent. This has led to many contractors withdrawing technical support, often causing owners to decommission their neutron probes. This issue, together with the fact that the instrument is expensive and somewhat cumbersome, has stimulated the development and use of other methods, largely electromagnetic.

Electromagnetic methods

These techniques are becoming widely adopted because they provide instantaneous measurements, do not require maintenance, and can provide continuous readings through automation (Muñoz-Carpena, 2004). The fundamental principle of the electromagnetic (EM) methods relies on the strong relationship between soil bulk dielectric permittivity (ε) and θ found for many mineral soils (Topp et al., 1980; Malicki et al., 1996). This strong dependence stems from the high permittivity of free water ($\varepsilon \approx 80$) as compared to that of mineral soil solids ($\varepsilon \approx 2$–9) and air ($\varepsilon \approx 1$). These methods estimate θ from a property (e.g. wave travel time, impedance, capacitor charge time, or frequency shift) related to the response of an EM signal (travelling through the soil either on a transmission line or in the soil medium directly) to changes in the apparent permittivity of the soil, ε_a.

Electromagnetic methods, whether time domain reflectometry (TDR), frequency domain reflectrometry (FDR), or capacitance methods, all estimate water content based on the permittivity of the soil medium. The accuracy of these estimates depends on a two-stage calibration. The first being the measurement of ε_a and the second a calibration between ε_a and estimated θ (Blonquist et al., 2005). A common approach to establish the relationship between ε_a and θ has been the empirical equation derived by Topp et al. (1980), among others (Topp and Reynolds, 1998), although most sensors have their own calibration curve provided by the manufacturer. This equation has proven to be accurate for many soils, but does not account for factors such as a temperature effect on determined water contents in some soils (see Verhoef et al., 2006; Fernandez-Galvez et al., 2007) or the loss of accuracy in soils containing high ion-exchange-capacity clays (Topp et al., 2000; Ferré and Topp, 2002). At low frequencies, EM sensors are salinity-dependent or more correctly dependent on the temperature- and water content-dependent bulk electrical conductivity that results from salinity (Robinson et al., 1998; Kelleners et al., 2004). In those cases, a soil-specific $\varepsilon_a - \theta$ relationship is required (Muñoz-Carpena, 2004).

9.4.2 Soil water pressure sensors

Positive pressures (i.e. below the GWT) can be determined using piezometers, tubes of a few cm diameter (to avoid capillary rise), of known length, open at both ends, installed in a soil profile where a water table is present. The water level in the piezometer is used to calculate the positive pressure potential at the bottom of the tube, z_b ($\psi_p = 0$ at the water level, z_w, and hence $\psi_p = |z_b - z_w|$ (Koorevaar et al., 1983). As discussed in Section 9.2.1, water in unsaturated soils is retained at suction, $S = -\psi_m$, which may be determined either by direct measurement of soil water pressure (tensiometers), by equilibrium water vapour pressure in soil air (thermocouple psychrometers), or by inference from θ of a reference porous medium in equilibrium with the surrounding soil (heat dissipation sensors, gypsum blocks, granular matrix sensors, filter paper).

Tensiometry is the most widely used technique to measure the soil matric potential under relatively moist soil conditions. A tensiometer is a device that consists of a porous cup connected to a pressure-sensing element (pressure gauge or transducer) through a water-filled rigid body tube. A Bourdon-type vacuum gauge is commonly employed for water potential measurements, although mercury manometers, if a greater accuracy is required, or pressure transducers, if automatic and continuous readings need to be recorded, can also be used. The porous cup is normally made of ceramic material because of its structural strength as well as permeability to water flow.

Tensiometer length may be variable and the use of a specific length will depend on the user's needs according to factors such as crop type (i.e. soil depth explored by the root system) or the irrigation criteria to be adopted by the irrigator. It is important for tensiometers to be installed properly, with the ceramic cup firmly in contact with the surrounding soil. Then, the soil solution can flow into or out of the tensiometer through the small pores in the ceramic cup until the liquid in the cup comes into equilibrium with the matric potential of the soil water around the cup. As the soil dries, soil matric potential decreases (suction/tension increases) and the tensiometer vacuum gauge reading increases. The opposite happens when an increase in soil water content (e.g. from irrigation or rainfall events) takes place.

The suction generated within the cup is transmitted through the liquid column to the gauge. The vacuum gauge does not provide direct measurements of the tension in the cup when there is a difference in height between the two, since the liquid between the cup and the gauge is in static equilibrium and, thus, the pressure in the liquid column increases linearly with depth (Koorevaar et al., 1983). Therefore, ψ_m in the tensiometer cup (here expressed in pressure units) will be given by the expression (Koorevaar et al., 1983):

$$\psi_m = P_{gauge} + \rho_w g \Delta z \tag{9.14}$$

where P_{gauge} is the pressure, relative to atmospheric pressure, of the liquid in the vacuum gauge and Δz is the difference in height between the cup and the gauge.

Tensiometers are instruments that have been extensively used for irrigation scheduling purposes because they provide direct measurements of soil moisture status, are easily managed and relatively inexpensive. However, they also present some drawbacks that might limit their use. In this regard, the typical measurement range (ψ_m ~ down to −80 kPa or −8 m) remains a severe limitation for soil studies and for some irrigation schemes such as deficit irrigation scheduling (see Section 9.5.1). Furthermore, the intimate soil contact they require around the ceramic cup for consistent readings is not always guaranteed in swelling or coarse soils.

An alternative to tensiometers is thermocouple psychrometers (TCPs), reliable and accurate devices (provided that appropriate precautions are taken) for determination of water potential (sum of matric and osmotic potential). Separation of matric and osmotic water potential components can be achieved with one of the methods described in Andraski and Scanlon (2002). TCPs infer the water potential of the liquid phase of a soil sample from measurements within the vapour phase that is in equilibrium with the liquid soil water. As soil dries, the relative humidity of the soil air decreases. Water potential in the air phase in equilibrium with the soil sample can thus be determined from the existing theoretical relationship between relative humidity and water potential (see, e.g. Andraski and Scanlon, 2002; Durner and Or, 2005, for a detailed description of the technique). As pointed out by Andraski and Scanlon (2002), the main difficulties of this technique are first, the range of variation of relative humidity in the soil gas phase is very narrow as compared to the typical range of interest of soil water potential (≈−0.01 to −1.5 MPa); and second, temperature differences in the sensor-sample system may lead to large errors in the determination of water potential.

For accurate determinations, the method requires calibration of TCPs, consistency in measurement technique and careful cleaning and handling. A wide range of instruments has been developed and are commercially available for use in laboratory, glasshouse and field experiments (Andraski and Scanlon, 2002). However, notwithstanding the reliability and accuracy of these devices, the degree of complexity of the technique as well as the number of precautions that need to be taken has limited its broad application in field studies.

The last group of methodologies indirectly infers soil matric potential from water content of a reference porous medium in equilibrium with the surrounding soil. Techniques used are the heat dissipation method, electrical resistance method and filter paper method. In general, heat dissipation sensors consist of a heating element and an accurate thermocouple embedded within a porous ceramic matrix (Flint et al., 2002). The working principle is based on the fact that thermal conductivity of a porous material is proportional to its water content. These sensors are relatively inexpensive and easy to read, but require individual calibration and temperature correction (Flint et al., 2002).

The electrical resistance method consists of electrodes embedded in a porous medium (block). Variations in electrical resistance of the porous media are due to variations in θ and hence to variations in the matric potential of the surrounding soil. These sensors have low resolution, are temperature dependent and require calibration. However, they are simple, inexpensive and are suited to schedule deficit irrigation (Muñoz-Carpena, 2004). Gypsum blocks are an example of electrical resistance sensors, which use gypsum as porous material. Gypsum serves as a buffer (to a specific level) against soil salinity changes, since variations in soil electrical conductivity would affect electrical resistance measurements. Major limitations of gypsum blocks are the degradation they undergo with time and the fact that they are not suitable for measurements under high soil moisture conditions (matric potential > −30 kPa). Some of the disadvantages of gypsum blocks have been overcome by Granular Matrix Sensors (Irmak and Haman, 2001).

9.4.3 Estimation of the soil water balance

The soil water balance, SWB, was given in Equation 9.1. The way in which this equation is generally employed in irrigation scheduling is to assume a root zone soil moisture depletion, D_r (see Equation 9.4), of zero when the soil is at field capacity. Thereafter, evaporative losses and deep drainage increase D_r and rainfall and irrigation decrease it. Irrigation will take place when D_r has reached a certain level; this level mainly depends on crop type and crop stage, rooting depth, plant spacing and irrigation water availability.

Consequently, this kind of irrigation scheduling method (see Section 9.5.1) requires measurements/estimates of gain terms P and I (through the use of rain gauges or knowledge of sprinkler/trickle/dripper rates) and of loss terms D and ET (ideally of E and T separately). For a relatively flat surface, R can be ignored. Often, capillary rise, C, is assumed to be negligible, which is a reasonable assumption if the soil is coarse-textured and groundwater levels are relatively deep. Drainage (D) below the root zone is usually ignored for irrigation scheduling purposes because $\theta < \theta_{FC}$; water flow will have diminished considerably, see Section 9.2.3 (but not stopped entirely, especially in fine-textured soils in which deep drainage and redistribution will continue for a considerable length of time).

Hence, the key water balance component to estimate is the evapotranspiration, ET. The most direct way to obtain some measure of evaporation is by using a pan evaporimeter or open-pan: a container, filled with water to a few centimetres below the rim. Some pans are traditionally mounted above the ground surface, others are sunk in the ground. Their shape (generally cylindrical), depth, diameter, material (usually metal) and colour vary as well. A fixed-point gauge in a stilling well serves to indicate the level of water in the pan. The new water level is found (e.g. daily) and the change from the previous reading is the pan evaporation, after which the water level will be brought to the fixed point again. The reservoir is often covered with wire-mesh netting to protect the pan from wild-life and wind-blown debris. The results from identical pans vary according to their position and exposure. A pan evaporimeter determines potential evaporation, not the actual evapotranspiration, and can, depending on location, overestimate potential evaporation because of advection effects. Hence, it is not ideal for irrigation or drainage scheduling.

Another method is weighing lysimetry, which involves monitoring weight changes of an extracted, undisturbed soil column (although in some cases repacked soil columns have

Figure 9.8 Lysimeter set-up based on loadcells in an olive plantation near Seville; changes in weight (with a resolution of 0.01 mm) were measured using three 50 kg capacity load cells per lysimeter from which evaporation loss was derived. (Top left) Perspex lysimeter box placed on loadcells bolted on to aluminium baseplate, (top right) holes to place inner (near tree trunk) and outer lysimeters, (bottom left), placement of lysimeters, now reinforced around edges and equipped with tubes to feed through instrument cables, (bottom right) a photo of the finished installation. Also visible are some meteorological sensors. Photos courtesy of Dr Antonio Diaz-Espejo, CSIC, Seville, Spain. For a colour version of this figure, please see Plate 9.2.

been used). Evaporation is calculated from differences between regularly measured masses. Weighing can be automated and continuous, using balance beam, hydraulic or mechanical load-cell based lysimeters and can be corrected for precipitation, irrigation and often drainage. Alternatively manually weighed microlysimeters can be used (Boast and Robertson, 1982). These smaller lysimeters, consisting of cores, which weigh typically 1–3 kg, with diameters ranging between 5 and 20 cm and depths of between 5 and 30 cm, are used for measurements of soil evaporation below canopies.

Lysimeters vary in size and depth, mainly depending on the type of vegetation and economic considerations (see Klocke et al., 1985; Dugas and Bland, 1989; Allen and Fisher, 1990; Daamen et al., 1993). Large lysimeters, the ones in theory most useful for irrigation and drainage planning purposes, are relatively expensive and difficult to install and maintain. They are therefore mainly used for research, although their numbers have dwindled in recent years, to make way for micrometeorological methods (see Verhoef and Campbell 2005, for more information on lysimeters and meteorological methods for estimation of *ET*). Figure 9.8 shows the installation procedure and arrangement of eight lysimeters (0.55 m × 0.55 m × 0.25 m, uniformly packed with the sandy loam soil obtained *in situ*)

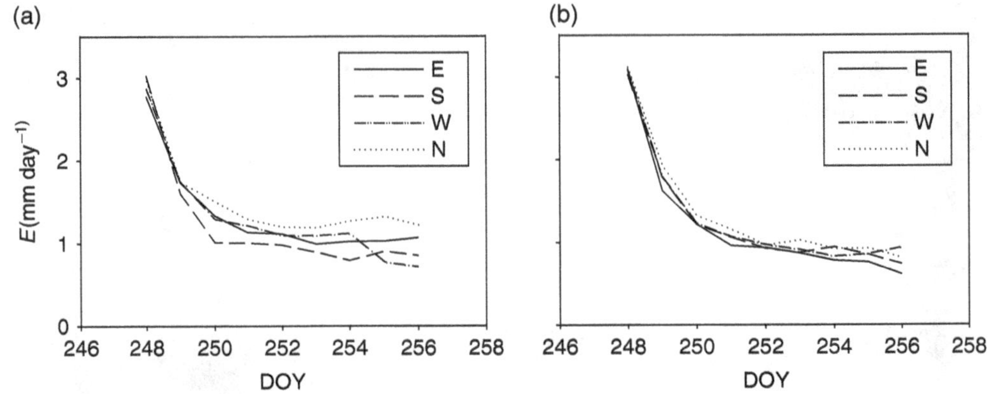

Figure 9.9 Soil evaporation as measured using the load cell-based lysimeter set-up illustrated in Figure 9.8 for (a) the outer (OLS) and (b) the inner (ILS) lysimeters (referring to those more than 1 m away from the tree trunk and those right next to the tree, respectively, see Figure 9.8), placed at different cardinal directions (East, E; South, S; West, W; and North, N).

installed in an orchard planted with 35-year-old olive trees (ground cover ~50%), located at the CSIC experimental farm La Hampa (Coria del Rio, Sevilla, Spain). These lysimeters were installed at different orientations (N, E, S, W) around and distances from (1 m: the inner set of lysimeters, (ILS) and 2 m: the outer set of lysimeters (OLS)) the trunk of a single olive tree. Each lysimeter was made of perspex, a material with a very low thermal conductivity to avoid lateral heat conduction through the lysimeter casing.

Figure 9.9 shows the soil evaporation results for a 9-day drying period after an irrigation event on DOY 248 (5 September) in 2005, for the outer (OLS) and inner (ILS) lysimeters. For both sets of lysimeters, evaporation decreased sharply after irrigation, reaching a near-constant level of about 1 mm day^{-1}. Because this was a sandy soil located in a hot and dry climate, first stage drying (see Section 9.2.1) would only have lasted for a few hours after irrigation; the daily sum of E was below E_p for all days, so that Figure 9.9 illustrates the second stage of drying.

The inner lysimeters were more or less continuously shaded and hence very little difference between their E-values was observed. For the OLS, radiation load was much more variable and depended strongly on orientation of the lysimeter around the tree. This is reflected in the larger variability in E for the OLS. Figure 9.9 indicates that it is worth keeping irrigated areas as small as possible because large water losses by E can occur, even for shaded soil; this can be achieved by micro-irrigation (Section 9.5.1).

In the soil physical approaches described, $\Delta\Theta$ is measured using the type of soil moisture sensors described in Section 9.4.1 to indirectly estimate ET, from Equation 9.1, assuming P, I, D and C are known or can be neglected (see Fernández-Galvez et al., 2007). This is not the same as scheduling according to a target soil water content, where the absolute values of θ are used to guide irrigation scheduling, rather than the changes in the entire water balance.

A final way to obtain estimates of evapotranspiration is by employing meteorological theory and equipment. Two main classes of micrometeorological methods can be defined: the first focuses on obtaining experimentally *in situ* evaporative flux measurements (and those can be split into direct and indirect methods), whereas the second class concerns calculation of ET from well-known equations, for example the Penman-Monteith equation, driven by easily available weather variables. These meteorological methods rely on measurement of the surface energy balance (SEB), which is given by:

$$R_n - G = H + \lambda ET \tag{9.15}$$

where λ is the latent heat of vaporisation (~ 2.45 MJ kg^{-1}). Equation 9.15 is an important boundary condition for energy exchange processes at the land surface, determining how available energy (net radiation flux density, R_n, minus soil heat flux density, G) is partitioned between sensible heat flux density, H (the flux of heat from the earth's surface to the atmosphere that is not associated with phase changes of water) and latent heat flux density, i.e. the evapotranspiration, λET. The *SEB* is directly related to the *SWB* through the evaporation term, λET (see Brutsaert, 1982; Hillel, 1998). In the *SEB* energy flux is expressed in units of Wm^{-2}. *ET* in mm day^{-1} can be converted to λET in Wm^{-2}, with knowledge of λ and given that 1 kg H_2O m^{-2} = 1 mm H_2O.

The first group of meteorological methods to determine *ET*, or rather λET, experimentally, is the direct method using Eddy Covariance (EC). This method is based on turbulence theory. The EC method relies on the fact that turbulence is random in nature and hence an atmospheric variable A may be separated into its turbulent part, A', and its non-turbulent, average part, $\bar{A}$, ($A = \bar{A} + A'$). This method calculates λET from the covariance between the vertical wind speed, w (ms^{-1}) and the specific humidity, q (kg$_{water}$ kg$^{-1}{}_{air}$); $\overline{w'q'}$. By obtaining fast-response measurements of state variables w (using a three-axis sonic anemometer) and q (from an adjacent Infra Red Gas Analyser, *IRGA*, see Figure 9.10) time series can be generated from which the perturbation values of the data points can be calculated (i.e. the fluctuations around the mean, as indicated by the primes).

Figure 9.10 Detail of eddy covariance (EC) equipment, comprising a three-axis sonic anemometer and the adjacent Infra Red Gas Analyser (IRGA). Picture taken at the experimental station of the University of Cartagena (Spain). For a colour version of this figure, please see Plate 9.3.

IRGAs are available in either a closed path (e.g. LI-COR's LI-7000) or an open path (e.g. LI-7500) set up and have associated advantages and disadvantages (e.g. Leuning and Judd, 1996). Acquisition and post-processing software are required to obtain and record the raw data (about 20 MB a day), to generate the covariances and to apply the corrections (see, for example, www.geos.ed.ac.uk/abs/research/micromet/edisol/edidownl.html).

EC instruments are expensive, require post-processing software and some insight into meteorological theory. Consequently, the EC method is generally only used for research purposes. It is sometimes used to derive or test crop coefficients (e.g. Paco et al., 2006), which will inform crop models.

An indirect method is the Bowen ratio energy balance (BREB) method, which infers the evaporation from combining measurements of air temperature, T_a, and actual vapour pressure (e in mbar), measured at two levels, with the *SEB*. This yields the Bowen Ratio, β', which is defined by $H/\lambda ET$:

$$\beta' = \gamma \frac{\Delta \overline{T}}{\Delta \overline{e}} \tag{9.16}$$

where γ is the psychrometric constant (~ 0.66 mbar K^{-1}). λET is then given by

$$\lambda ET = \frac{R_n - G}{1 + \beta'} \tag{9.17}$$

The *BREB* technique requires two high-precision atmospheric humidity and temperature sensors (installed at different heights), a net radiometer and an estimate of soil heat flux (see Sauer and Horton, 2005, for example). Its robustness and simplicity result in relatively reliable estimates of evaporation, and the comparatively low cost of the sensors make it a useful method for estimates of crop water use (e.g. Todd et al., 2000).

An alternative, indirect method of estimating evaporation is by modelling. Models range from those calculating actual evaporation by multiplication of the potential evaporation by a crop coefficient; a factor depending on crop type, development stage and soil water status (Allen et al., 1998), to detailed mechanistic Soil Vegetation Atmosphere Transfer models (Shuttleworth and Wallace, 1988; Verhoef and Allen, 2000).

A reliable equation used widely for estimation of evapotranspiration for the purpose of irrigation and drainage scheduling (see Allen et al., 1998) is the Big-Leaf model or Penman-Monteith (Monteith, 1965) equation given by:

$$\lambda ET = \frac{\Delta(R_n - G) + \rho_a c_p VPD / r_a}{\Delta + \gamma(1 + r_s / r_a)} \tag{9.18}$$

where r_a is the aerodynamic resistance (sm^{-1}) and r_s (sm^{-1}) the canopy surface resistance. *VPD* (mbar) is the vapour pressure deficit (e_s-e; with e_s being the saturation vapour pressure in mbar) at reference level, z_r (m). Furthermore, Δ (mbar K^{-1}) is the slope of the vapour pressure temperature curve, γ is the psychrometric constant (mbar K^{-1}), ρ_a (kg m^{-3}) is the density of air and c_p (J kg^{-1} K^{-1}) is the specific heat of air at constant pressure.

The meteorological variables: R_n, G, T_a (to get Δ and γ), u (to calculate r_a) and *VPD* are relatively easy to obtain using standard meteorological instrumentation. However, the aerodynamic and surface resistance is less straightforward to determine. The aerodynamic resistance, r_a, is a function of the rate of turbulent mixing and can be calculated from measurements of wind speed, estimates of atmospheric stability and knowledge of surface roughness parameters (roughness length and displacement height; see Brutsaert (1982), Raupach (1994) or Verhoef et al. (1997) on how to derive these variables from standard estimates of canopy structure). Standard meteorological textbooks can be referred to for calculation of r_a (e.g. Garratt, 1992).

If evaporation from the soil is assumed to be negligible, then r_s can be equated to the bulk stomatal resistance of the leaves in the canopy: often calculated with $r_s = 1/(g_s LAI)$ where g_s is the average leaf stomatal conductance (see Section 9.3.1) and LAI is the leaf area index. This formula assumes that the air within the canopy is well-mixed, with all the leaves of the canopy being exposed to exactly the same atmospheric properties. Refer to Jones (1992) and Jones (2004) for overviews of methods to estimate stomatal conductance, g_s (required to obtain r_s).

The exchange between the atmosphere and a dense (i.e. where the canopy covers the entire ground surface) sole crop or canopy is relatively well understood, and Equation 9.18 can yield very precise estimates of evapotranspiration. However, when the surface is sparsely vegetated, for example, before crop canopy closure has taken place, a significant percentage of the surface consists of exposed undergrowth or bare soil, so ideally a two-component or sparse canopy model is used. Such models introduce more explicit descriptions of aerodynamic transfer through the leaf boundary layer, and they have proved useful in describing a variety of sparse canopies, including crops and natural vegetation (Wallace et al., 1990; Van den Hurk et al., 1995; Allen et al., 1998; Verhoef and Allen, 2000). In some of these two-or multi-component models, there is no interaction between the component latent or sensible fluxes. However, there are many situations where fluxes from the substrate can significantly affect the fluxes from the upperstorey canopy. For example, Wallace et al. (1990) showed that, in a sparse millet canopy, transpiration could vary by up to 40%, depending on the fluxes of heat and water vapour from the soil below.

9.4.4 Estimation of hydraulic conductivity

Hydraulic conductivity is an important variable when describing water flow within soils (Equation 9.7). Knowledge of this variable is also essential in drainage studies. The design and functioning of subsurface drainage systems depends to a great extent on the soil's saturated hydraulic conductivity (K_s); all drain-spacing equations make use of this parameter. To design or evaluate a drainage project, K_s has to be determined as accurately as possible (Oosterbaan and Nijland, 1994).

Methods to determine (un)saturated hydraulic conductivity generally employ formulae based on the Darcy equation and the boundary conditions of the flow; these flow conditions are generally artificially imposed. The K-value is calculated from the formula using the values of hydraulic head ($\Delta\psi_h/\Delta_z$) and discharge observed (F_w) under the imposed conditions (Equation 9.7). These methods can be grouped into *in situ* field methods and laboratory methods. In the latter case an undisturbed soil sample is extracted from the field and analysed under lab conditions. A variety of methods can be found in Klute (1972), Oosterbaan and Nijland (1994), Rowell (1994) and Bagarello et al. (2004).

In the last two decades, the Multistep Outflow method (see Van Dam et al., 1994; Vereecken et al., 1997), based on inverse parameter optimization for soil hydraulic functions [e.g. those presented in Van Genuchten (1980)], combined with transient outflow measurements, has been developed to make the measurement of laboratory hydraulic properties faster, more reliable and less laborious.

9.5 Water management to alleviate soil water shortage

Agricultural activity dominates the use of freshwater and accounts for some 70% of withdrawals from water resources globally. As most of the water withdrawn by agriculture is lost by evapotranspiration, in marked contrast to domestic and industrial withdrawals, agriculture accounts for 80–90% of freshwater used by humans (FAO, 2002; Morison et al., 2008). This highlights the importance of agriculture in the challenge of making the Earth's available water serve the needs of its growing number of users. Furthermore, there is widespread agreement that climate change will exacerbate the present shortages of water in certain areas and may lead to increased drought (IPCC, 2007). Thus, reducing agricultural water use and making water resources more sustainable are increasingly urgent matters. This challenge requires combined efforts from multiple disciplines, addressing agronomic, physiological, biotechnological/genetic and engineering issues (Morison et al., 2008; Farooq et al., 2009). The utilization of monitoring techniques (such as those described in Section 9.4.3) to assess the level of soil water deficit, and the use of such measurements to control irrigation systems, will assist us to move towards precision agriculture. In this context, the aim is to satisfy plant water (and nutritional) requirements or alternatively achieve a desired and controlled level of water stress – so called deficit irrigation (see Section 9.5.1), while minimizing environmental impacts caused by excess applied water and subsequent leaching, and/or avoiding plant diseases related to excess water. Moreover, improving irrigation efficiency can contribute greatly to reducing production costs of vegetables and crops, making the industry more competitive and sustainable.

It is notable that irrigated agriculture is not sustainable in many (semi-)arid regions unless drainage is used simultaneously to provide salinity control (see Chapter 7 and Section 9.5.3) as well as for optimal root respiration.

9.5.1 Types of irrigation and irrigation scheduling

The irrigation system employed is a major determinant of on-farm irrigation performances and may be characterised by one of three categories: (1) surface irrigation, (2) sprinkler irrigation and (3) micro or localized irrigation. Surface irrigation systems are the most widespread worldwide, representing about 84.5% of the total irrigated area globally, followed by sprinkler irrigation (≈ 13.5%) and localized irrigation (≈ 2%) (AQUASTAT, 2010). The main indicator of farm irrigation efficiency is the application efficiency, which can be defined as the depth or volume of water added to the root zone store expressed as a ratio of the depth or volume of water applied to the field. In general, surface irrigation systems (e.g. furrow-, border- or basin-irrigation) apply large irrigation depths, and as these systems have traditionally been operated manually, over-irrigation is often practised (Smith et al., 2005). Consequently, on a global basis, on-farm application efficiency is estimated to be ~37%, which means that 63% of the water delivered to fields is lost as runoff and/or drainage (Wallace, 2000). Nonetheless, if the rate of water application is correctly matched to the infiltration rate of the soil and slope of the land, surface irrigation systems may significantly improve the crop's performance.

The other two types of irrigation systems (i.e. sprinkler and localized) belong to the group of pressurized systems. Sprinkler systems, which include set, travelling rain-guns and continuously moving lateral systems (Pereira et al., 2002), are able to achieve better irrigation performances than surface systems.

Localized or micro-irrigation systems (i.e. micro-sprinkler, drip or subsurface irrigation systems) are designed to apply small and frequent irrigations. The working pressure of these systems is lower than in sprinkler systems. A common characteristic of these 'high-tech' schemes is that there is less wet soil surface exposed to the atmosphere and hence less soil evaporation (as well as less interception loss, which would occur after sprinkler irrigation). Localised irrigation systems also allow maintenance of a required/desired moisture level in the root zone, enable targeted fertiliser application ('fertigation'), reduce the risk of disease by keeping the foliage dry and facilitate farm operations due to the fact that the soil surface remains dry. As with sprinkler irrigation, the application efficiency depends mainly on a number of system and management variables, the latter related to the duration of the irrigation and irrigation frequency (Pereira et al., 2002).

Subsurface drip irrigation systems are particularly advantageous as they can increase crop WUE because less water is lost as evaporation from the soil surface (Ayars et al., 1999). These systems involve permanent drip tapes below the soil surface. Depending on factors such as soil texture, crop type, the permanence of the system and whether or not the irrigation system is going to be used for plant germination, the installation depth may normally vary between 0.1 and 0.5 m (Ayars et al., 1999), although deeper installation depths have also been reported (Hutmacher et al., 1996). These systems present some additional advantages, such as a higher uniformity of application, less operation cost if well maintained, less weeds and unrestricted field access (e.g. for spraying or harvesting). In contrast, the main limitation of these systems is the susceptibility of the emitters to clogging by root penetration, rust, microorganisms or chemical precipitates, although with appropriate design and preventive practices this limitation might be overcome. The joint use of deficit irrigation programs (see later) and subsurface irrigation systems has proven to be a reliable strategy to significantly increase WUE in dry areas (Romero et al., 2004).

The advent of these localised irrigation systems has helped to reduce the water required in agri-horticultural crops but, at the same time, has highlighted the need for new methods of accurate irrigation scheduling. The choice of irrigation scheduling method depends to a large extent on the irrigator's objectives and the irrigation system available. When water supplies are limited, as is the case in arid and semi-arid regions, farmers should place their emphasis on maximising WUE (see Section 9.3.2); these scheduling principles clearly deviate from those aimed at maximising yield by fully meeting crop water requirements.

In this regard, deficit irrigation (DI) strategies have demonstrated that, under water scarcity scenarios and with sufficient knowledge to manage DI optimally, the objective of maintaining or even increasing farmers' profit may be attainable while reducing irrigation water use (Fereres and Soriano, 2007). The reasons why DI can improve crop water productivity are diverse and have been explained elsewhere (English and Raja, 1996; Fereres and Soriano, 2007). DI can be understood as any irrigation strategy applying water below the crop ET requirements (English, 1990). Until the development of precise irrigation methods (e.g. drip irrigation), there had been little room to manage irrigation with the degree of precision needed in optimal DI programmes, where controlling the timing of application is essential for avoiding the detrimental effects of stress.

Depending on factors such as the pattern of stress imposed, the intensity of water stress or the way irrigation water is delivered to the root-zone, various types of 'controlled' DI strategies can be practised to increase WUE: sustained deficit irrigation (SDI), regulated deficit irrigation (RDI), partial root zone drying (PRD) and supplemental irrigation (SI). The term SDI refers to those DI strategies that impose a uniform degree of water shortage throughout the growing season to deliberately seek a slow development of water stress

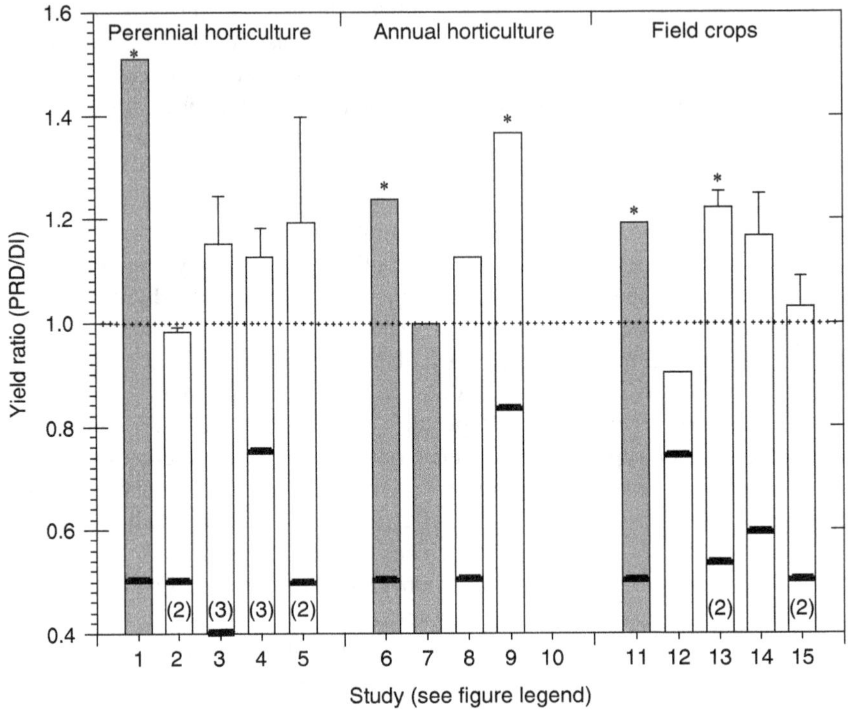

Figure 9.11 Crop yield ratio of PRD to DI at similar irrigation volumes (a ratio of 1 indicates that yield with both techniques is equivalent). The fraction of full irrigation is given as a thick line in each column. Columns are means ± SE of the number of experiments/seasons given in parentheses at the base of each column. Shaded columns denote pot experiments where the root system was confined. Significant differences between PRD and DI ($P < 0.05$) are indicated with an asterisk (*). From Dodd (2009). By permission of Oxford University Press.

(as soil water reserves deplete) and a progressive plant adaptation to water deficits (Fereres and Soriano, 2007).

RDI strategies, as defined by Chalmers et al. (1981), are based on applying a certain degree of water deficit during specific periods of the growth cycle when the crops are less sensitive to water stress (non-critical periods), while full water needs are met during the remaining periods (critical periods). Experiments with RDI have been successful in many fruit and nut tree species, including almond (Goldhamer and Viveros, 2000; Egea et al., 2010), apricot (Pérez-Pastor et al., 2009) and citrus (Domingo et al., 1996) trees.

PRD is an irrigation technique that has been extensively studied in recent years (Dry et al., 1996). This approach is based on the generation of wet-dry cycles in the root-zone by alternating irrigation between different parts of the root system. This irrigation strategy seeks to promote the production of chemical signals (e.g. the plant hormone abscisic acid, ABA) from roots in dry soil and their transmission to the shoots to restrict water use (through triggering partial stomatal closure) while maintaining crop water supply from roots in the wet soil fraction (Stoll et al., 2000; Dodd, 2005). Considering the cost and management complexity of implementing PRD, adoption of this technique (in contrast to conventional DI, where the whole root system is irrigated) requires evidence that it stimulates beneficial physiological and agronomic responses, as compared to DI strategies receiving the same volumes of water. Dodd (2009) recently performed a comparative study between PRD and conventional DI based upon 15 experiments comprising 10 different crop species. His results indicated that in no case did PRD significantly decrease yield and that, in six studies, PRD significantly increased yield by more than 15% (Figure 9.11). Sadras (2009) showed that for a broad range of horticultural crops and environments, yield per unit of applied water was, on average, significantly, but modestly (5%) higher in PRD than with conventional deficit irrigation.

Supplemental irrigation (SI), is used in some humid to temperate areas as a tactical measure to complement reasonably ample rainfall and stabilise production (Hsiao et al., 2007). In dry zones, where water supply is very limited, it is used as a form of DI by applying irrigation water in certain critical periods of the growing season to significantly improve crop performance relative to the rainfed situation. Examples of how SI may benefit yield and water use efficiency of crops grown in water-limited environments are given in Oweis et al. (2000) and Turner (2004).

To reduce uncertainty and risk associated with the employment of irrigation strategies, computer models that simulate crop performance can assist water managers in optimising a limited supply of irrigation water; Equation 9.18, the Penman-Monteith equation, is invariably at the heart of this type of models. The new FAO crop model AquaCrop represents an example of these useful tools (Raes et al., 2009; Steduto et al., 2009). AquaCrop simulates attainable yields of major herbaceous crops as a function of water consumption under rainfed, supplemental, deficit and full irrigation conditions. Another existing FAO tool is CROPWAT 8.0, a computer program for the calculation of crop water and irrigation requirements (including schedules for different management conditions and varying crop patterns) based on soil, climate and crop data. CROPWAT 8.0 can also be used to evaluate farmers' irrigation practices and to estimate crop performance under both rainfed and irrigated conditions (see http://www.fao.org/nr/water/infores_databases_cropwat.html).

Two other well-established models are CropSyst (Stockle et al., 1997) and WOFOST (Diepen et al., 1989). Todorovic et al. (2009) compared the performance of AquaCrop with that of CropSyst and WOFOST in simulating sunflower (*Helianthus annuus* L.) growth under different water regimes in a Mediterranean environment. Their results indicated that despite the fact that AquaCrop requires less input information than CropSyst and WOFOST, it performed similarly in simulating both biomass and yield at harvesting.

9.5.2 Automated irrigation scheduling

The degree of automation of the irrigation systems described in Section 9.5.1, as well as their related advantages, is variable. An automated irrigation system saves time and effort for the irrigation manager and also allows higher flexibility in scheduling farm work. The irrigation control systems of these modern systems also offer the possibility to store information on control variables, which represent a valuable feedback to identify problems (e.g. water leakages, undesirable pH or electrical conductivity levels in the irrigation water, etc.).

There are three different ways to perform automated irrigation: time-based, volume-based or based on other (soil, plant or climate related) variables. In the former two groups the irrigation scheduler plays a fundamental role by setting either the time and duration of irrigation (time-based) or the volume of water to be supplied to crops (volume-based). In the latter, however, the system is designed to work more autonomously.

In time-based automation, the irrigation interval is determined from the amount of water to be supplied, application rate of the emitters and number of emitters. Volume-based automation represents another approach for automating irrigation and consists of measuring the water supplied during each irrigation event to automatically interrupt the water flow once the applied water volume has met the target volume. Both time-based and volume-based irrigation systems require an early estimate of crop evapotranspiration, ideally from one of the methodologies described in Section 9.4.3. An effective procedure to adjust estimated to real crop water needs is to monitor one or more of the available soil and plant water status indicators (reviewed in Section 9.3.3).

Irrigation automation based on measurements of one or more variables of the soil–plant–atmosphere continuum emerged from the need to increase precision in irrigation control to achieve specific objectives in crop management and to maximize irrigation efficiency. Effective operation of such systems requires a sensing system that determines irrigation need in real time; this prerequisite rules out large-scale manual monitoring programmes for such purposes and indicates a need for automated monitoring systems (Jones, 2004). These systems set the irrigation calendar by assessing one or more control variables that are continuously recorded. Edaphic factors, such as soil moisture content, meteorological factors, such as atmospheric evaporative demand, relative humidity, air temperature or solar radiation, or plant physiology-based factors, such as leaf temperature, micromorphometric variations of some plant organs or sap flow have been proposed as indicators to develop self-controlled automatic irrigation systems (Section 9.3.3).

As indicated by Fernández and Cuevas (2010), the characteristics of a good indicator for irrigation scheduling can be summarised as follows. The variable sensed by the indicator must be closely related to production parameters of economic importance (e.g. crop yield and fruit quality). The indicator response to water stress must be quick (earliness) and marked (intensity). It must also be highly sensitive, provide high signal intensity (i.e. very responsive to water stress) and a relatively low coefficient of variation (reduced 'noise' between sampling points). The sensors must allow automated and continuous data collection and transmission and must be reliable, robust, inexpensive, easy to install, operate and maintain. Finally, the sensor outputs must be easy to interpret and suitable for use in automatic irrigation control.

Irrigation scheduling based on soil water determination has been widely reported. A soil water-based irrigation control system uses real-time information on the soil water status to bypass a time-based preprogrammed schedule or to maintain soil water content in the root zone within a specified moisture range that is optimal for plant growth. These two approaches are *bypass* and *on-demand*, respectively (Muñoz-Carpena and Dukes, 2005). Effectiveness of these soil sensor-based irrigation technologies has been successfully assessed on field crops (Meron et al., 2001; Wang et al., 2007; Muñoz-Carpena et al., 2008), on residential green areas (McCready et al., 2009) or on ornamental potted plants (Nemali and van Iersel, 2006; Burnett and van Iersel, 2008).

Occasionally, automated irrigation is not devoted only to schedule irrigation under standard conditions related to crop water requirements, but also to deal with sporadically occurring situations, such as irrigation to prevent frost damage, or cutting off sprinkler irrigation under severe wind.

As pointed out by Greenwood et al. (2010), the use of sensor-based procedures for soil water management varies with the type of crop and environment. Accordingly, shallow-rooted crops (e.g. potted plants, lawns or urban landscapes) could be managed with soil water sensors installed at one depth only whereas deep-rooted agricultural crops would require a vertical profile of sensors to allow irrigation according to the distribution of soil water down the profile. This ensures optimal use of the stored water in the soil profile and hence substantial saving of irrigation water. Figure 9.12 summarizes a sequential methodology described

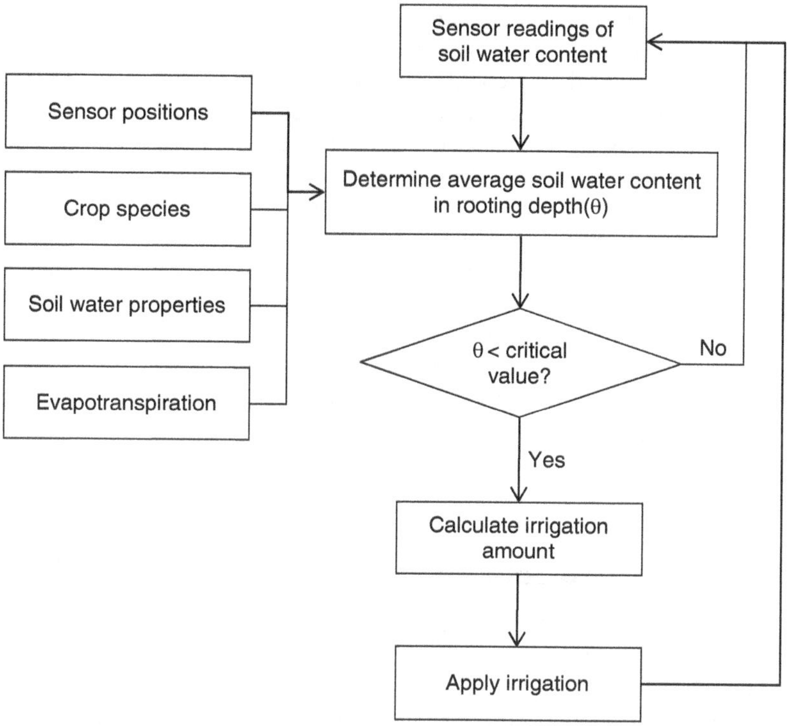

Figure 9.12 Flow chart of irrigation scheme based on sensor readings and model predictions. Adapted from Greenwood et al. (2010). Reproduced with permission, copyright Cambridge Journals.

by Greenwood et al. (2010) to estimate timing and irrigation doses in sensor-based irrigation scheduling.

Nemali and van Iersel (2006) devised an irrigation system that used calibrated, dielectric moisture sensors, interfaced with a datalogger and solenoid valves, to irrigate a soil substrate to a set-point (volumetric water content, θ) and maintain it close to that target for prolonged periods. The irrigation system was able to maintain θ for a long period within an acceptable range of the set-point, despite large variations in environmental conditions and plant size, which resulted in little or no wastage of water. Likewise, Blonquist et al. (2006) used a dielectric moisture content sensor to schedule irrigation for turfgrass. To accomplish this, they defined a preset threshold θ value (θ_t, see also Section 9.2.4 and Equation 9.4) determined from both soil physical properties and crops to be grown:

$$\theta_t = \theta_{FC} - MAD(\theta_{FC} - \theta_{PWP}) \tag{9.19}$$

where *MAD* (equivalent to *p* in Section 9.2.4) is the management allowed depletion value (Cuenca, 1989). The *MAD* is the fraction of *TAW* that is allowed to be removed from the plant root zone before irrigation is required. Cuenca (1989) reported typical *MAD* values as 0.33 for shallow-rooted, high value crops; 0.5 for medium-rooted, moderate value crops and 0.67 for deep-rooted, low value crops.

The irrigation controller was set to irrigate each day for a specific time and duration, but with the only possibility to open the solenoid valve when θ measured by the sensor was below θ_t. Relative to irrigation recommendations based on evaporation estimates, the θ

sensor-based system applied approximately 16% less water, while relative to a fixed irrigation rate of 50 mm week^{-1} the system applied approximately 53% less water. Many research papers have also documented the enhanced irrigation efficiency that can be attained in many field crops by means of soil water sensors without any noticeable detrimental effects on crop productivity. A few examples are those published by Zotarelli et al. (2009) and Muñoz-Carpena et al. (2008), who found that soil moisture sensor-based irrigation systems significantly reduced crop water requirements, water percolation and nitrate leaching from tomato plants.

When deficit irrigation strategies (Section 9.5.1) are used, monitoring the soil (or the plant) water status becomes even more critical for minimisation of risks. These irrigation strategies normally exploit some plant physiological traits or sensitivity differences to water stress of specific phenological stages to enhance water use efficiency (Fereres and Soriano, 2007). However, the fulfilment of the desired crop response under these strategies usually requires very precise irrigation water management, since the crop response may dramatically deviate from the expected response in a narrow range of soil moisture availabilities.

9.5.3 Irrigation and salinity

A major consequence of irrigated agriculture is the risk of increased soil salinity and its impact on the sustainability of the irrigation. Although salinization may be driven by different factors (Rengasamy, 2006), irrigation-associated salinity represents a major limitation to agriculture in all semi-arid regions. All irrigation waters contain salts and, as water evaporates, salts concentrate in the soil profile and must be displaced below the root zone before they reach a concentration that limits crop production. The salinization process on irrigated land may be accelerated under conditions of poor quality irrigation water, low hydraulic conductivity of soil layers, as found in heavy clay and sodic soils, and high evaporative conditions (Rengasamy, 2006). Under these circumstances, some of the irrigation water losses are unavoidable and are needed to maintain the salt balance, although they can be minimized with efficient irrigation methods and by appropriate management (see Chapter 7).

Soil salinization may be a major drawback in deficit irrigation (DI) programs (Section 9.5.1), where irrigation water use is reduced and the hazard of increased salinity due to decreased leaching is amplified. Therefore, the question as to whether DI can be used as a tactical measure to reduce irrigation water use over long time periods when supplies are limited by droughts or other factors is currently unresolved (Fereres and Soriano, 2007). More research in this field is required to investigate the sustainability of DI (and especially when low quality irrigation water, such as reclaimed water, is used) and to what extent it can contribute to the permanent reduction of water use in irrigated systems.

The mechanisms by which saline soil water affect plant growth can be divided into two phases (Munns, 2002). The first phase of the plant growth response is due to the osmotic effect of the salt in the soil solution, which reduces the ability of the plant to take up water and generates in the plant a similar response to that under water stress. The second phase takes place if excessive amounts of salt enter the plant tissues, since they will eventually rise to toxic levels in the older transpiring leaves, causing premature senescence and hence a reduction in the assimilate synthesis that may further limit growth (Munns, 2002). The ability of plants to tolerate salt stress and to maintain (or to improve) WUE when irrigated with saline water or grown in salinized soils is highly variable. Katerji et al. (2003) studied the

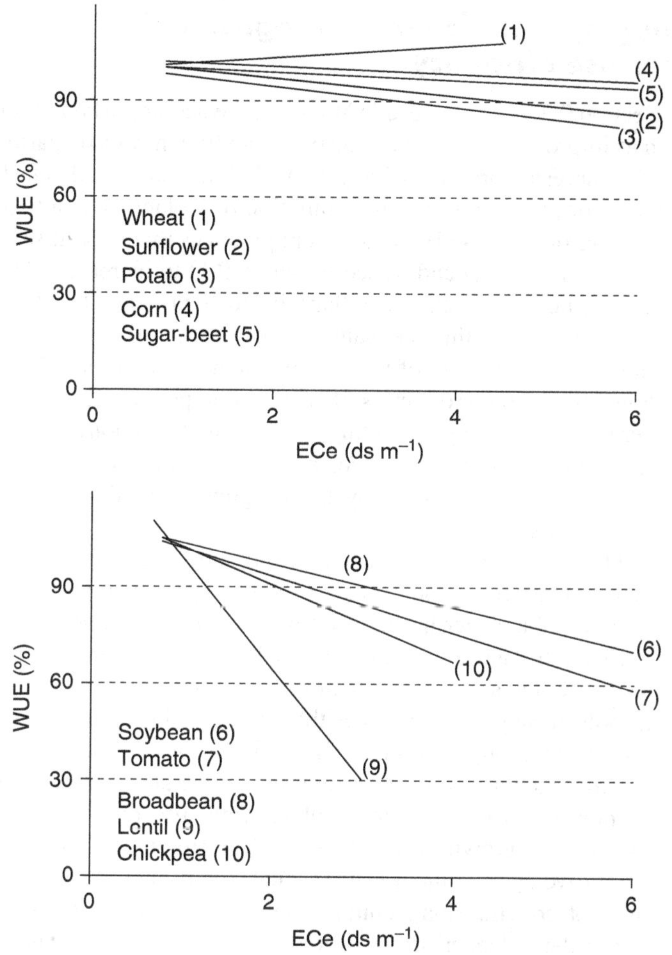

Figure 9.13 Relative (with respect to the non-saline treatment) water use efficiency (WUE) of different crops as a function of soil salinity during the crop season (ECe: electrical conductivity of the saturated paste). Redrawn from Katerji et al. (2008) who derived the relationships from data obtained by Katerji et al. (2003). With kind permission from Elsevier.

plant response to increasing soil salinity in ten species. They identified two groups of species; those which are salt-tolerant and that maintain, or slightly improve, their WUE with an increase in salinity (wheat, sunflower, potato, maize and sugar beet) and a second group of species (tomato, lentil, broad bean and chickpea) which are salt-sensitive and reduce their WUE with the deterioration of water quality (Figure 9.13). These results emphasise the importance of carefully choosing the best species to be grown in saline soils or with saline irrigation water. Indeed, breeding programs for salt tolerance are deemed to be an important way to cope with salinity in dry-land agriculture (Munns et al., 2006). Another potential risk of using poor quality irrigation water is the 'sodium hazard', leading to sodic soils which cause a decline in the structure of fine-textured soils, leading to poor infiltration of water and thus to salinization (through evaporation of irrigation water at the soil surface) and waterlogging (Munns et al., 2006). This underlines the close link existing between the processes of soil salinization, waterlogging and evaporation.

9.5.4 Managing soils to improve agronomic water use efficiency

As explained in Section 9.3.1, an increase in WUE can be accomplished by managing plant-related traits such as improvement of crop transpiration efficiency or by partitioning more of the biomass into the harvested product. Changes in WUE achieved in this fashion are invariably related to choice of plant variety or manipulation of a plant's phenological stage. This section will focus on agronomic (soil) management practices that maximise transpiration, T, by reducing the other losses (E, D and R, see Equation 9.1) to increase WUE, as illustrated by Equation 9.11, together with techniques that aim to improve soil water availability for crop transpiration (either by affecting the water regime or the soil's water holding capacity).

Evapotranspiration is a component of the energy balance (Equation 9.15), and a separate balance can be drawn up for the crop and soil surface component. Hence, any alteration in the available energy (e.g. through shading) will affect evapotranspiration (see also Equation 9.18). The aerodynamic resistance, r_a, also determines evaporation (again, see Equation 9.18) and its value is affected by the roughness length: the smoother the soil surface the lower the evaporation.

Hatfield et al. (2001) review how soil-management practices can improve WUE. They conclude that soil management affects the process of soil evapotranspiration by modifying the available energy, the available water in the soil profile, or the exchange rate between the soil and the atmosphere. Their literature survey revealed a large variation in measured WUE across a range of climates, crops and soil management practices. It is possible to increase WUE by 25 to 40% through soil-management practices that involve tillage. Overall, precipitation use efficiency can be enhanced through adoption of more intensive cropping systems in semi-arid environments and increased plant populations in more temperate and humid environments.

Soil tillage is a powerful tool to increase plant water use by alleviating soil physical constraints that hamper root growth. For example, 'deep ripping' can increase the depth of rooting and hence improve the amount of water available to the crop, when poor soil structure physically impedes root growth. This technique will only be useful if water infiltrating the soil can actually reach these deeper layers. Rotations, alternating deep-and shallow-rooting crops, can lead to 'biopores' for a subsequent crop (Turner, 2004). The use of minimum tillage or conservation tillage involves leaving crop residues on the surface while controlling weeds by herbicides rather than tillage. This approach also allows earlier planting as delays resulting from using tillage to remove weeds are reduced. Next, the seed for the follow-on crop is sown with minimum disturbance of the soil surface. This practice has led to reduced losses of water by soil evaporation and increased yields (Unger, 1978; Cornish and Pratley, 1991; Stewart and Robinson, 1997; Turner, 2004). However, the greater retention of incoming rainfall through minimum tillage may also increase water losses through deep drainage that are detrimental in a landscape in which secondary salinity can develop (Turner, 2004).

Total available soil water ($\theta_{FC} - \theta_{PWP}$) in the shallow subsoil can be improved by adding soil organic matter (SOM). Adding SOM enhances water retention and hence will raise θ_{FC} because of its hydrophilic nature and its beneficial effects on soil structure through increased soil aggregate formation and aggregate stability (Oades, 1984; Huntington, 2006). This leads to a larger amount of storage pores compared to transmission pores (> 50 μm) and residual pores (<2 μm). These changes in the physical properties also result in higher infiltration rates (Franzluebbers, 2002) and lower susceptibility to erosion (Wilson et al., 2009).

As already discussed in Section 9.3.2, increased water use by the crop itself via transpiration at the expense of water loss by weeds or from loss terms in the water balance equation (soil evaporation, deep drainage, surface runoff, or lateral subsurface flow) will improve water use efficiency. Ways to achieve this include earlier planting to more closely match rainfall and the use of fertilizers to maximise yield and hence decrease soil evaporation losses. Similarly, high plant density and more even planting increases crop-water use and reduces soil evaporation, but only where ample soil water is available from rainfall or supplement irrigation. When crops rely on water stored in the root zone, low planting densities are preferred to ensure a larger availability of water per plant (Turner, 2004).

Fallowing land to conserve moisture is another widely adopted practice to improve yields of subsequent crops in water-limited environments. The benefits of this practice vary with season, soil type and management of the fallow land (O'Leary and Connor, 1997; Stewart and Robinson, 1997).

Another way to optimise soil water availability is by practising intercropping (the agricultural practice of concurrently cultivating two or more crops in the same field) or agroforestry (the combination of trees or shrubs with crops). Agroforestry has been promoted as a possible means of increasing the productive use of rainfall in water-limited environments, by using the water that is usually inaccessible to conventional cropping systems (i.e. soil water reserves that are below the crop rooting zone and/or rainfall occurring outside the normal cropping seasons). However, any potential increase in rainfall utilisation by agroforestry systems, when compared to either woodlots or open crop fields, must be offset against the evaporation of rainfall intercepted by the likely greater canopy size (tree and crop) and not therefore available for crop growth (Jackson, 2000). On the other hand, increased shading of the soil by the trees will decrease soil evaporative losses (Wallace et al., 1999).

Another practice aimed at reducing the amount of soil water lost is the covering of the soil surface by different types of materials; this is called mulching. The main aim is to prevent soil evaporation, either by blocking the soil pores entirely (e.g. by using plastic) or by creating a hydraulic discontinuity through using a porous medium with large pores. Many materials have been used as mulch, inorganic ones such as plastic film, sand and gravel, and organic ones such as crop residue, straw and compost. Mulch at the soil surface affects hydrological processes involving rainfall interception, infiltration, evaporation and dew deposition (Li, 2003). A mulch layer therefore changes the soil's water balance. Mulching also affects the surface radiation budget (Lei et al., 2004) an effect largely related to the colour and nature of the mulch, which affects albedo (Kemper et al., 1994; Jonathan et al., 1998; Jalota et al., 2001; Anikwe et al., 2007). Hence, a mulch affects the energy available for evaporation (Novaka et al., 2000; Lei et al., 2004), thereby indirectly affecting the soil's water balance. Through these combined effects a mulch layer will also affect a soil's temperature regime. The influence of mulching on evaporation depends on the type of mulch, its texture (Xie et al., 2006) and layer thickness (Kemper et al., 1994). Its effectivity also depends on the stage of evaporation (Li, 2003).

Figure 9.14 shows that the evaporation from bare soil was higher compared to that from mulched soil during the first 2 days after irrigation (first stage of drying), after each irrigation event. This was caused by the mulch layer creating a hydraulic discontinuity, forcing the water to move in the vapour phase only. However, from day 3 (after irrigation) onwards, the evaporation from mulched soil was equal to or greater than that for bare soil. The reason for this is that the mulch-covered soil lost less water by evaporation during the first stage so that the moisture content in the mulched soil was higher than that in bare soil, thereby increasing hydraulic conductivity. Also, the bare soil forms a thin dry layer (a 'self-mulching' layer) hindering evaporation through its poor hydraulic properties. Similar observations were reported by other researchers when they used various mulch materials to reduce the

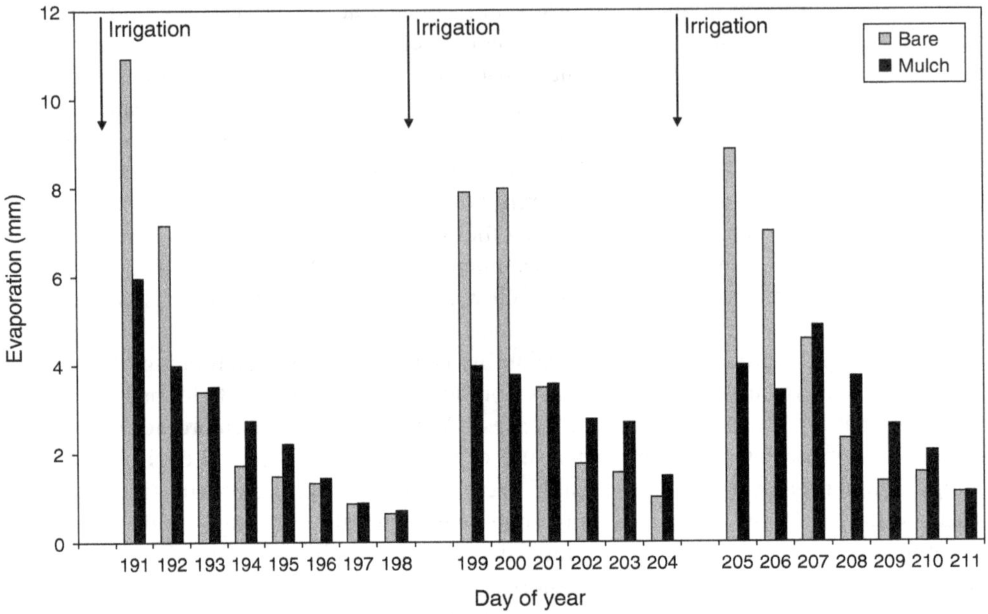

Figure 9.14 Evaporation as measured by microlysimeters for bare soil and soil covered by a mulch consisting of dried, shredded date palm leaves, in a date palm plantation at Qassim University farm, Saudi Arabia (data obtained during summer of 2009). Figure courtesy of Dr Abdulaziz Al Harbi. Reprinted from Al Harbi, 2010.

evaporation from bare soil (e.g. Modaihsh et al., 1985; Groenevelt et al., 1989; Van Wesemael et al., 1996; Mellouli et al., 2000).

Mulching can also change aerodynamic resistance and cause an enhanced turbulent transport of water vapour (and sensible heat) due to the relatively large roughness length of mulch layers such as gravel and plant residues (Nachtergaele et al., 1998; Xie et al., 2006). Additionally, mulching also avoids structural soil damage through water drop impacts, which can cause slaking and crusting of soil (and hence affect infiltration rate), as well as erosion (see Section 9.6.1).

A final point needs to be made here. Most of this section has focussed on how to increase and/or conserve the soil moisture available in the pore space of the root zone, for situations when the GWT is situated so far below the root zone that it does not influence the available soil water. In many cases, however, shallow water tables can be a valuable resource to provide crops with water (Kahlown et al., 2005; Qureshi et al., 2010). This will result in reductions in irrigation and hence improvements in (irrigation) water use efficiency. Ideally, this practice should be used only in areas where shallow water tables are a low salinity risk or where the local hydrology results in net recharge (e.g. Hurst et al., 2004).

9.6 Water management to alleviate excess water

9.6.1 Causes of excess water and remediation solutions

Root zone saturation can be caused by a variety of natural and anthropogenic factors, often occurring in combination. Natural factors comprise the climate, soil factors, plant factors, topography, geomorphological and landscape factors (Section 9.2.1). Examples of anthropogenic influences that can cause suboptimal values of root-zone aeration, which often adversely affects crop growth, are over-irrigation and drainage of uplands resulting in flooding of lower-situated lands (D'haeze et al., 2003). Also, clearing of native vegetation for annual crops and pastures is recognised as a major cause of waterlogging (and secondary salinity), for example in southern Australia (Turner and Ward, 2002).

Focussing on soil factors, excess water in the root environment can be the result of a number of causes that are often attributable to poor physical soil quality. Firstly, below-optimal final infiltration capacity, combined with high-intensity rainstorms or high irrigation rates, may result in temporary ponding. Infiltration capacity largely depends on the soil texture type and initial water distributions (Giraldez and Sposito, 1985; Chu and Mariño, 2005). Low infiltration capacities can be the result of destruction of the surface tilth, generally caused by the mechanical impact of rain-drops on structurally unstable soils. The resulting finer particles fill up the larger pores or cause a crust to form (see, for example, Stroosnijder and Hoogmoed, 1984). Alternatively, runoff or runon leads to water erosion, thereby clogging up larger pores due to the deposition of finer particles. Yimer et al. (2008) found that changes in soil structure caused by surface soil compaction because of tillage and animal trampling, coupled with a smaller soil organic carbon content, caused a decline in infiltration capacity and soil moisture content after conversion of forest to cultivation and grazing in the highlands of Ethiopia.

Even if the infiltration rate is adequate, low hydraulic conductivity in the upper soil layers may cause poor internal drainage or slow redistribution, causing waterlogging within the soil profile, resulting in poor aeration around roots. Water can move through cracks or discontinuities in the soil structure, but there is reduced movement through the pores within the soil matrix. Drainage can only take place around the structural elements, not through them. In extreme cases, impermeable layers such as a plough pan, may lead to a perched water table, as discussed in Section 9.2.1.

Excess water can be alleviated by preventing or correcting low infiltration capacity by, for example, adding organic matter to the soil or covering the soil with crop residue to prevent damage to the soil structure. If insufficient coarse pores are available in the surface layers (i.e. the soil has a low hydraulic conductivity), cultivation can remedy this. A compacted subsoil can be ameliorated by deep tillage (using deep ploughs, deep cultivators or subsoilers), thereby destroying compacted layers or subsurface pans, although results with respect to crop yield are variable (e.g. Carter et al., 1996).

Alternatively, a rising groundwater table may give rise to anaerobic conditions. The extreme case is when the actual groundwater or the capillary fringe can reach the surface. A related problem here is secondary salinity; in southern Australia, for example, the replacement of deep-rooted perennial native vegetation with shallow-rooted annual crops has resulted in a rising water table and the development of a major secondary salinity problem (Barrett-Lennard, 2002; Turner and Ward, 2002). Establishment of tree belts, drains and perennial pasture, or a combination of these, will positively affect the water balance and ground water level and thereby alleviate water logging and secondary salinity in the rejuvenated landscape of southwestern Australia, while maintaining crop production at near-current levels (Turner and Ward, 2002). Rundle and Rundle (2002) presented a case study of farm-based solutions to water logging and secondary salinity in southwestern Australia. The plan included the grading of interceptor drains and the locating of dams along the drains to hold water for the dry summer months. Belts of three or four species of trees were planted on the down-slope side of the drains after leveling the spoil from the drain and deep-ripping the soil.

Noorduijn et al. (2010) assessed the positive effects of alley-farming agroforestry on groundwater levels through a 13-year alley farming trial incorporating different combinations of belt width, alley width and revegetation density. However, despite the reintegration of perennial plants back into farming systems in many regions the salinisation processes continues because the available perennial options are generally less profitable than annual crops. A possible solution is the partial restoration of productivity of saltland through revegetation with salt tolerant halophytes (Barrett-Lennard, 2002).

The highest permissible levels of the GWT below the surface depend on the crop and the soil type (as the latter will determine capillary rise). For agricultural crops, the depth of the GWT is best below 90 cm, although pastures and meadows can cope with higher watertables. The depth of the GWT becomes particularly important in the case of saline groundwater: the capillary fringe should then be well below bottom of the crop's root system. For certain ecologically important plant species, like those growing on SSSIs (Sites of Special Scientific Interest), high water tables are essential, and drainage is undesirable (Kalusová et al., 2009; Araya et al., 2010).

Overbank flooding may also give rise to anaerobic conditions and hence adversely affect survival and growth of riparian species in floodplain ecosystems, for example during floodplain restoration. Natural resource practitioners should therefore take into account flood frequency and duration in their planting plans to increase the likelihood of restoration success (McCurry et al., 2010).

If the deeper subsoil is too impermeable to allow water to percolate below the crop root zone quickly enough, the field must be drained artificially. The efficiency of the drainage system can often be improved by deep tillage, as this increases the number of vertical waterways and the water can pass through from the surface to the drains. The success of the drainage system depends on the stability of the soil structure reflected in the temporal stability of the cracks and fissures.

9.6.2 Agricultural drainage systems

Excess water that cannot be alleviated by tillage or other soil-management practices can be remediated by artificially improving the natural drainage of the land. This is achieved by installing "Agricultural drainage systems (ADS)'. These are artificial systems that enhance water flow from the land, so that agriculture can benefit from the subsequently reduced water levels, via increased crop yields, improved diversification and increased intensification (where required), effective use of inputs, such as fertilisers, and improved mechanisation (Oosterbaan, 1994). ADS have the purpose of either reducing the flow of water over the soil surface (surface flow) or through the underground (inter- and groundflow), which leads to a distinction between 'surface drainage systems' and 'subsurface drainage systems'. A surface drainage system is applied when ponding frequently occurs on the soil surface, whereas a subsurface drainage system is applied when the water logging occurs in the soil (i.e. shallow GWTs or perched water tables; Oosterbaan, 1994). When both surface and subsurface water logging occur, a combined surface/subsurface drainage system is required.

This section briefly considers the configuration of sub-surface drainage systems, as surface systems are outside of the scope of this chapter. Subsurface field drainage systems consist of horizontal or slightly sloping channels made in the soil. They can be open ditches, buried pipe drains, or mole drains into and within which groundwater flows because of hydraulic gradients existing in the soil. The flow rate is affected by a number of environmental and technical factors (Tanji and Kielen, 2002).

From a soil physical/hydrological point of view, the hydraulic conductivity of the soil as well as the configuration of the GWT and the hydraulic pressure of the groundwater are important. Furthermore, the rate of groundwater recharge by internal drainage or by horizontal groundflow from an external source needs to be considered. Physical characteristics of the drainage system that play an important role are the location of the drains (in relation to the groundwater level and the soil surface), the diameter of the drains, the slope of the drains and the spacing between the drains (Hillel, 1971; Van Schilfgaarde, 1974; Farr and Henderson, 1986; Oosterbaan, 1994).

Direct and indirect effects of agricultural land drainage

ADS will have an effect on the physical, chemical, biological and hydrological soil properties and variables, as well as on the soil water and energy balance components. These will, in turn, affect soil temperature and moisture content.

The direct effects of installing a drainage system in waterlogged land are: a reduction in the average amount of water stored on or in the soil (inducing drier soil conditions and reducing water logging) and an increased flow (discharge) of water through the system (Oosterbaan, 1994). These direct effects trigger a series of indirect effects, which can be either positive or negative. Some examples are given in the rest of this section (largely based on Oosterbaan, 1994).

Positive effects include increased aeration of the soil so that crops can develop a deeper root system. This can give winter-sown crops an advantage in spring as it will allow them to access a larger available soil volume for nutrient and water uptake. On heavy soils, the increased transpiration, due to increased soil water, can dry out the soil more and increase the amount of cracking, so that better drainage occurs during following rainstorms. In this way, vigorous crops on heavy land can help create and stabilise the natural drainage system. It will also improve anchorage of trees and crops (see Chapter 8).

Having drier top soils will also reduce the risk of frost heaving (which causes crop damage) because soils need to have a considerable amount of soil water to heave. Furthermore, drier top soils will generally result in a more stable soil structure, which will improve infiltrability and hence decrease surface runoff.

Agronomic advantages are higher availability of nitrogen in the soil due to increased nitrification (Feddes and van Wijk, 1977), higher and more diversified crop production, better workability of the land (Wind and Buitendijk, 1979), earlier planting dates due to increased spring temperatures, reduction of peak discharges by an increased temporary storage of water in the soil and removal of salts or other harmful substances from the soil due to an increased discharge (Section 9.5.3). This drainage water can be used for various purposes including irrigation (Sharma et al., 1990).

Negative effects comprise: (1) increased decomposition of organic matter, because of increased temperatures and reduced soil moisture contents affecting the root and microbial respiration; (2) soil subsidence (Todd, 1980); (3) acidification of potential acid sulphate soils (AARD and LAWOO, 1992); (4) increased risk of drought through reduced capillary rise from the ground water table (Section 9.6.1); (5) ecosystem damage if lowering of the water table adversely affects plant species and (6) problems caused downstream by the discharge of salty or otherwise polluted drainage water.

The indirect effects of drier soil conditions on weeds, pests and plant disease can be both positive and negative; the net result depends on the environmental conditions.

9.6.3 Design and implementation of drainage systems

Drainage should only be undertaken with clear agricultural aims (so-called *agricultural object factors*) in mind. In the context of this chapter: increasing crop yield (e.g. by extending the growth season or by being able to introduce new crops) or enabling soil tillage.

To achieve the agricultural targets requires *a priori* knowledge of *criterion factors* such as the extent of the saturation conditions and the soil salinity. Criterion factors are factors that are affected by the drainage system and at the same time influence the object factors. A more specific description of a criterion factor is called a *criterion index*, such as (1) the average depth of the watertable during the cropping season; (2) crop-season average salinity levels of the root zone and (3) the minimum number of days that the soil is workable during a critical period.

Decisions on the design and implementation of ADS are also influenced by environmental factors such as irrigation, precipitation, soil hydraulic properties, topography and aquifer conditions. These factors must be specified as *environmental indices*, such as the average seasonal rainfall, the maximum daily rainfall, the mean and vertical distribution of hydraulic conductivity and the variation in hydraulic conductivity with depth in the soil.

For further reading on land drainage for agriculture see Hillel (1971), Van Schilfaarde (1974), Wither and Vipond (1974), Farr and Henderson (1986) and Oosterbaan (1994). Reliable software, such as EnDrain to perform calculations on horizontal subsurface drainage systems in agriculture can be downloaded from http://www.waterlog.info/software.htm. SaltMod is a mathematical, numerical simulation model that describes the relations between crop rotation, water balance components, groundwater levels, soil salinity and subsurface drainage and SahysMod combines Saltmod with a model of ground water flow and hydraulics to account for large spatial variation through a network of polygons.

References

AARD and LAWOO (Agency for Agricultural Research and Development and Land and Water Research Group). (1992) *Acid Sulphate Soils in the Humid Tropics: Water Management and Soil Fertility*. AARD, Jakarta, Indonesia, 283pp.

Al Harbi, A.B. (2010) The effect of mulching on the surface energy balance and key soil physical variables, under sub-humid and semi-arid conditions. PhD thesis, University of Reading, Reading, UK, 225pp.

Allen, R.G. and Fisher, D.K. (1990) Low-cost electronic weighing lysimeters. *American Society of Agricultural Engineers* **33**, 1823–1833.

Allen, R.G., Pereira, L.S., Raes, D., and Smith, M. (1998) Crop evapotranspiration-Guidelines for computing crop water requirements. *FAO Irrigation and Drainage Paper 56*. FAO, Rome, Italy.

Andraski, B.J. and Scanlon, B.R. (2002) Thermocouple psychrometry. In: *Methods of Soil Analysis. Part 4. Physical Methods* (eds J.H. Dane and G.C. Topp), pp. 609–642. Soil Science Society of America, Book series No. 5. Madison, Wisconsin.

Angus, J.F., Kirkegaard, J.A., and Peoples, M.B. (2001) Rotation sequence and phase: research on crop and pasture systems. In: *Proceedings of the 10th Australian Agronomy Conference*, Hobart, Australia, http://www.regional.org.au/au/asa/2001. Accessed on January 2011.

Anikwe, M.A.N., Mbah, C.N., Ezeaku, P.I., and Onyia, V.N. (2007) Tillage and plastic mulch effects on soil properties and growth and yield of cocoyam (*Colocasia esculenta*) on an ultisol in southeastern Nigeria. *Soil and Tillage Research* **93**, 264–272.

AQUASTAT (2010) FAO's information system on water and agriculture. Last accessed in January 2010. http://www.fao.org/nr/water/aquastat/main/index.stm.

Araya, Y.N., Gowing, D.J.G., and Dise, N.B. (2010) A controlled water-table depth system to study the influence of fine-scale differences in water regime for plant growth. *Aquatic Botany* **92**, 70–74.

Ayars, J.E., Phene, C.J., Hutmacher, R.B., Davis, K.R., Schoneman, R.A., Vail, S.S., and Mead, R.M. (1999) Subsurface drip irrigation of row crops: a review of 15 years of research at the Water Management Research Laboratory. *Agricultural Water Management* **42**, 1–27.

Bagarello, V., Iovino, M., and Elrick, D. (2004) A simplified falling-head technique for rapid determination of field-saturated hydraulic conductivity. *Soil Science Society of America Journal* **68**, 66–73.

Barrett-Lennard, E.G. (2002) Restoration of saline land through revegetation. *Agricultural Water Management* **53**, 213–226.

Berni, J.A.J., Zarco-Tejada, P.J., Suárez, L., and Fereres, E. (2009) Thermal and narrowband multispectral remote sensing for vegetation monitoring from an unmanned aerial vehicle. *IEEE Transactions on Geoscience and Remote Sensing* **47**, 722–738.

Blonquist, J.M., Jones, S.B., and Robinson, D.A. (2005) Standardizing characterization of electromagnetic water content sensors: part 2. Evaluation of seven sensing systems. *Vadose Zone Journal* **4**, 1059–1069.

Blonquist, J.M., Jones, S.B., and Robinson, D.A. (2006) Precise irrigation scheduling for turfgrass using a subsurface electromagnetic soil moisture sensor. *Agricultural Water Management* **84**, 153–165.

Boast, C.W. and Robertson, T.M. (1982) A 'micro-lysimeter' method for determining evaporation from bare soil: description and laboratory evaluation. *Soil Science Society of America Journal* **46**, 689–696.

Boyer, J.S. (1995) *Measuring the Water Status of Plants and Soils*. Academic Press, London.

Brutsaert, W.H. (1982) *Evaporation into the Atmosphere. D.* Reidel Publishing company, Dordrecht, Holland, the Netherlands.

Buckley, T.N. (2005) The control of stomata by water balance. *New Phytologist* **168**, 275–291.

Buckley, T.N., Mott, K.A., and Farquhar, G.D. (2003) A hydromechanical and biochemical model of stomatal conductance. *Plant Cell and Environment* **26**, 1767–1785.

Burnett, S.E. and van Iersel, M.W. (2008) Morphology and irrigation efficiency of Gaura lindheimeri grown with capacitance sensor-controlled irrigation. *HortScience* **43**, 1555–1560.

Carminati, A., Vetterlein, D., Weller, U., Vogel, H.J., and Oswald, S.E. (2009) When roots lose contact. *Vadose Zone Journal* **8**, 805–809.

Carter, M.R., Holmstrom, D.A., Cochrane, L.M., Brenton, P.C., van Roestel, J.A., Langille, D.R., and Thomas, W.G. (1996) Persistence of deep loosening of naturally compacted subsoils in Nova Scotia. *Canadian Journal of Soil Science* **76**, 541–547.

Cermak, J., Kucera, J., Bauerle, W.L., Phillips, N., and Hinckley, T.M. (2007) Tree water storage and its diurnal dynamics related to sap flow and changes in stem volume in old-growth Douglas-fir trees. *Tree Physiology* **27**, 181–198.

Chalmers, D.J., Mitchell, P.D., and Van Heek, L.A.G. (1981) Control of peach tree growth and productivity by regulated water supply, tree density, and summer prunning. *Journal of the American Society for Horticultural Science* **106**, 307–312.

Chen, D.X. and Coughenour, M.B. (2004) Photosynthesis, transpiration, and primary productivity: scaling up from leaves to canopies and regions using process models and remotely sensed data. *Global Biogeochemical Cycles* **18 (GB4033)**, 1–15.

Chu, X. and Mariño, M.A. (2005) Determination of ponding condition and infiltration into layered soils under unsteady rainfall. *Journal of Hydrology* **313**, 195–207.

Clapp, R.B. and Hornberger, G.M. (1978) Empirical equations for some soil hydraulic properties. *Water Resources Research* **14**, 601–604.

Condon, A.G., Richards, R.A., Rebetzke, G.J., and Farquhar, G.D. (2004) Breeding for high water-use efficiency. *Journal of Experimental Botany* **55**, 2447–2460.

Coppola, A., Kutilek, M., and Frind, E.O. (2009) Transport in preferential flow domains of the soil porous system: measurement, interpretation, modelling, and upscaling. *Journal of Contaminant Hydrology* **104**, 1–3.

Cornish, P.S. and Pratley, J.E. (1991) Tillage practices in sustainable farming systems. In: *Dryland Farming: A Systems Approach* (eds V. Squires and P. Tow), pp. 76–101. Sydney University Press, South Melbourne, Australia.

Cosby, B.J., Hornberger, G.M., Clapp, R.B., and Ginn, T.R. (1984) A statistical exploration of the relationships of soil moisture characteristics to the physical characteristics of soils. *Water Resources Research* **20**, 682–690.

Cuenca, R.H. (1989) *Irrigation System Design: An Engineering Approach*. Prentice Hall, Inc., Englewood Cliffs, New Jersey. p. 552.

Daamen, C.C., Simmonds, L.P., Wallace, J.S., Laryea, K.B., and Sivakumar, M.V.K. (1993) Use of microlysimeters to measure evaporation from sandy soils. *Agricultural and Forest Meteorology* **65**, 159–173.

Deng, X.P., Shan, L., Zhang, H., and Turner, N.C. (2006) Improving agricultural water use efficiency in arid and semiarid areas of China. *Agricultural Water Management* **80**, 23–40.

Dewar, R.C. (2002) The Ball-Berry-Leuning and Tardieu-Davies stomatal models: synthesis and extension within a spatially aggregated picture of guard cell function. *Plant Cell and Environment* **25**, 1383–1398.

D'haeze, D., Deckers, J., Raes, D., Phong, T.A., and Chanh, N.D.M. (2003) Over-irrigation of Coffea canephora in the Central Highlands of Vietnam revisited: simulation of soil moisture dynamics in Rhodic Ferralsols. *Agricultural Water Management* **63**, 185–202.

Diaz-Espejo, A., Verhoef A., and Knight, R. (2005) Illustration of micro-scale advection using grid-pattern mini-lysimeters. *Agricultural and Forest Meteorology* **129**, 39–52.

Diepen, C.A., Wolf, J., Keulen, H., and Rappoldt, C. (1989) WOFOST: a simulation model of crop production. *Soil Use and Management* **5**, 16–24.

Dodd, I.C. (2005) Root-to-shoot signalling: assessing the roles of 'up' in the up and down world of long-distance signalling in planta. *Plant and Soil* **274**, 251–270.

Dodd, I.C. (2009) Rhizosphere manipulations to maximize 'crop per drop' during deficit irrigation. *Journal of Experimental Botany* **60**, 2454–2459.

Domingo, R., Ruiz-Sánchez, M.C., Sánchez-Blanco, M.J., and Torrecillas, A. (1996) Water relations, growth and yield of Fino lemon trees under regulated deficit irrigation. *Irrigation Science* **16**, 115–123.

Dry, P.R., Loveys, B.R., Düring, H., and Botting, D.G. (1996) Effects of partial rootzone drying on grapevine vigour, yield, composition of fruit and use of water. In: *Proceedings 9th Australian Wine Industry Technology Conference* (eds C.S. Stockley, A.N. Sas, R.S. Johnstone and T.H. Lee), pp. 128–131. Winetitles, Adelaide, Australia.

Dugas, W.A. and Bland, W.L. (1989) The accuracy of evaporation measurements from small lysimeters. *Agricultural and Forest Meteoroogy* **46**, 119–129.

Durner, W. and Lipsius, K. (2005) Determining soil hydraulic properties. In: *Encyclopedia of Hydrological Sciences* (eds M.G. Anderson and J.J. McDonnell) John Wiley & Sons, Ltd., Chichester, UK.

Durner, W. and Or, D. (2005) Soil water potential measurement. In: *Encyclopedia of Hydrological Sciences* (eds M.G. Anderson and J.J. McDonnell). John Wiley & Sons, Ltd., Chichester, UK.

Egea, G. (2008) Characterization and modelling of the agronomical and physiological responses of almond trees to deficit irrigation. PhD Thesis, Polytechnic University of Cartagena, Cartagena, Spain.

Egea, G., Pagan, E., Baille, A., Domingo, R., Nortes, P.A., and Perez-Pastor, A. (2009) Usefulness of establishing trunk diameter based reference lines for irrigation scheduling in almond trees. *Irrigation Science* **27**, 431–441.

Egea, G., Nortes, P.A., Gonzalez-Real, M.M., Baille, A., and Domingo, R. (2010) Agronomic response and water productivity of almond trees under contrasted deficit irrigation regimes. *Agricultural Water Management* **97**, 171–181.

Ehrler, W.L. (1973) Cotton leaf temperatures as related to soil-water depletion and meteorological factors. *Agronomy Journal* **65**, 404–409.

English, M. (1990) Deficit irrigation. I. Analytical framework. *Journal of Irrigation and Drainage Engineering* **116**, 399–412.

English, M. and Raja, S.N. (1996) Perspectives on deficit irrigation. *Agricultural Water Management* **32**, 1–14.

FAO (2002) Crops and drops: making the best use of water for agriculture. *FAO. Information Brochure*. FAO, Rome, Italy. p. 28.

Farooq, M., Kobayashi, N., Wahid, A., Ito, O., and Basra, S.M.A. (2009) Strategies for producing more rice with less water. *Advances in Agronomy* **101**, 351–388.

Farquhar, G.D. and Sharkey, T.D. (1982) Stomatal conductance and photosynthesis. *Annual Review of Plant Physiology and Plant Molecular Biology* **33**, 317–345.

Farr, E. and Henderson, W.C. (1986) *Land Drainage*. Longman, London, p. 251.

Feddes, R.A. and van Wijk, A.L.M. (1977) An integrated model approach to the effect of water management on crop yields. *Agricultural Water Management* **1**, 3–20.

Fereres, E. and Goldhamer, D.A. (1990) Deciduous fruit and nut trees. In: *Irrigation of Agricultural Crops* (eds B.A. Stewart and D.R. Nielsen), pp. 987–1017. American Society of Agronomy, Madison, Wisconsin.

Fereres, E. and Soriano, M.A. (2007) Deficit irrigation for reducing agricultural water use. *Journal of Experimental Botany* **58**, 147–159.

Fernández, J.E. and Cuevas, M.V. (2010) Irrigation scheduling from stem diameter variations: a review. *Agricultural and Forest Meteorology* **150**, 135–151.

Fernández, J.E., Green, S.R., Caspari, H.W., Diaz-Espejo, A., and Cuevas, M.V. (2008) The use of sap flow measurements for scheduling irrigation in olive, apple and Asian pear trees and in grapevines. *Plant and Soil* **305**, 91–104.

Fernández-Gálvez, J., Verhoef, A., and Barahona, E. (2007) Estimating soil water fluxes from soil water records obtained using dielectric sensors. *Hydrological Processes* **21**, 2785–2793.

Ferré, P.A. and Topp, G.C. (2002) Time domain reflectometry. In: *Methods of Soil Analysis. Part 4. Physical Methods* (eds J.H. Dane and G.C. Topp), pp. 434–446. Soil Science Society of America, SSSA Book Series No. 5, Madison, Wisconsin.

Flexas, J., Escalona, J.M., Evain, S., Gulias, J., Moya, I., and Osmond, C.B. et al. (2002) Steady-state chlorophyll fluorescence (Fs) measurements as a tool to follow variations of net CO_2 assimilation and stomatal conductance during water-stress in C-3 plants. *Physiologia Plantarum* **114**, 231–240.

Flint, A.L., Campbell, G.S., Ellett, K.M., and Calissendorff, C. (2002) Calibration and temperature correction of heat dissipation matric potential sensors. *Soil Science Society of America Journal* **66**, 1439–1445.

Franzluebbers, A.J. (2002) Water infiltration and soil structure related to organic matter and its stratification with depth, *Soil and Tillage Research* **66**, 197–205.

Gao, Q., Zhao, P., Zeng, X., Cai, X., and Shen, W. (2002) A model of stomatal conductance to quantify the relationship between leaf transpiration, microclimate and soil water stress. *Plant Cell and Environment* **25**, 1373–1381.

Garratt, J.R. (1992) *The Atmospheric Boundary Layer*. Cambridge atmospheric and space science series. Cambridge University Press, Cambridge, UK.

Geerts, S. and Raes, D. (2009) Deficit irrigation as an on-farm strategy to maximize crop water productivity in dry areas. *Agricultural Water Management* **96**, 1275–1284.

Gerke, H.H. (2006) Preferential flow descriptions for structured soils. *Journal of Plant Nutrition and Soil Science* **169**, 382–400.

Giraldez, J.V. and Sposito, G. (1985) Infiltration in swelling soils. *Water Resources Research* **21**, 33–44.

Goldhamer, D.A. and Viveros, M. (2000) Effects of preharvest irrigation cutoff durations and postharvest water deprivation on almond tree performance. *Irrigation Science* **19**, 125–131.

Goldhamer, D.A. and Fereres, E. (2001) Irrigation scheduling protocols using continuously recorded trunk diameter measurements. *Irrigation Science* **20**, 115–125.

Grant, O.M., Tronina, L., Jones, H.G., and Chaves, M.M. (2007) Exploring thermal imaging variables for the detection of stress responses in grapevine under different irrigation regimes. *Journal of Experimental Botany* **58**, 815–825.

Greenwood, D.J., Zhang, K., Hilton, H.W., and Thompson, A.J. (2010) Opportunities for improving irrigation efficiency with quantitative models, soil water sensors and wireless technology. *Journal of Agricultural Science* **148**, 1–16.

Gregory, P.J. (2004) Agronomic approaches to increasing water use efficiency. In: *Water Use Efficiency in Plant Biology* (ed. M.A. Bacon), pp. 142–170. Blackwell Publishing Ltd., Oxford, UK.

Gregory, P.J., Simmonds, L.P., and Warren, G.P. (1997) Interactions between plant nutrients, water and carbon dioxide as factors limiting crop yields. *Philosophical Transactions of the Royal Society London B* **352**, 987–996.

Grimes, D.W. and Yamada, H. (1982) Relation of cotton growth and yield to minimum leaf water potential. *Crop Science* **22**, 134–139.

Groenevelt, P.H., Van Straaten, P., Rasiah, V., and Simpson, J. (1989) Modifications in evaporation parameters by rock mulches. *Soil Technology* **2**, 279–285.

Guilioni, L., Jones, H.G., Leinonen, I., and Lhomme, J.P. (2008) On the relationship between stomatal resistance and leaf temperatures in thermography. *Agricultural and Forest Meteorology* **148**, 1908–1912.

Hatfield, J.L., Sauer, T.J., and Prueger, J.H. (2001) Managing soils to achieve greater water use efficiency: a review. *Agronomy Journal* **93**, 271–280.

Herbst, M., Rosier, P.T.W., McNeil, D.D., Harding, R.J., and Gowing, D.J. (2008) Seasonal variability of interception evaporation from the canopy of a mixed deciduous forest. *Agricultural and Forest Meteorology* **148**, 1655–1667.

Hess, T.M. and Knox, J.W. (2002) Irrigation advisory services: experiences in UK. *Workshop on Irrigation Advisory Services and Participatory Extension in Irrigation Management*, FAO-ICID. Proceedings on line: http://www.fao.org. 24th July, Montreal, Quebec, Canada.

Hillel, D. (1971) *Soil and Water: Physical Principles and Processes*. Academic Press, New York.

Hillel, D. (1998) *Environmental Soil Physics*. Academic Press, San Diego, California.

Hsiao, T.C. (1990) Measurements of plant water status. In: *Irrigation of Agricultural Crops* (eds B.A. Steward and D.R. Nielsen), pp. 243–279. American Society of Agronomy, Agronomy Monograph. No. 30, Madison, Wisconsin.

Hsiao, T.C., Steduto, P., and Fereres, E. (2007) A systematic and quantitative approach to improve water use efficiency in agriculture. *Irrigation Science* **25**, 209–231.

Huggett, R., Lindley, S., and Gavin, H. (2004) *Physical Geography: A Human Perspective*, Hodder Arnold, London, p. 520.

Huguet, J.G., Li, S.H., Lorendeau, J.Y., and Pelloux, G. (1992) Specific micromorphometric reactions of fruit-trees to water-stress and irrigation scheduling automation. *Journal of Horticultural Science* **67**, 631–640.

Huntington, T.G. (2006) Available water capacity and soil organic matter. In: *Encyclopedia of Soil Science* (ed. R. Lal), 2nd edn. Oxon, UK.

Hurst, C.A., Thorburn, P.J., Lockington, D., and Bristow, K.L. (2004) Sugarcane water use from shallow water tables: implications for improving irrigation water use efficiency. *Agricultural Water Management* **65**, 1–19.

Hutmacher, R.B., Mead, R.M., and Shouse, P. (1996) Subsurface drip: improving alfalfa irrigation in the west. *Irrigation Journal* **45**, 48–52.

Idso, S.B., Jackson, R.D., Pinter, P.J., Reginato, R.J., and Hatfield, J.L.(1981) Normalizing the stress-degree-day parameter for environmental variability. *Agricultural Meteorology* **24**, 45–55.

Intrigliolo, D.S. and Castel, J.R. (2004) Continuous measurement of plant and soil water status for irrigation scheduling in plum. *Irrigation Science* **23**, 93–102.

Intrigliolo, D.S. and Castel, J.R. (2006) Usefulness of diurnal trunk shrinkage as a water stress indicator in plum trees. *Tree Physiology* **26**, 303–311.

Intrigliolo, D.S. and Castel, J.R. (2007) Crop load affects maximum daily trunk shrinkage of plum trees. *Tree Physiology* **27**, 89–96.

IPCC (2007) Climate change 2007: the physical science basis. In: *Contribution of Working Group I to the Fourth Assessment Report of the Intergovernmental Panel on Climate Change* (eds S.D. Solomon, M. Qin, M. Manning, Z. Chen, M. Marquis, K.B. Averyt, M. Tignor and H.L. Miller). Cambridge University Press, Cambridge, UK and New York.

Irmak, S. and Haman, D.Z. (2001) Performance of the Watermark (R) granular matrix sensor in sandy soils. *Applied Engineering in Agriculture* **17**, 787–795.

Jackson, N.A. (2000) Measured and modelled rainfall interception loss from an agroforestry system in Kenya. *Agricultural and Forest Meteorology* **100**, 323–336.

Jalota, S. K., Khera, R., and Chahal, S.S. (2001) Straw management and tillage effects on soil water storage under field conditions. *Soil Use and Management* **17**, 282–287.

Jarvis, P.G. (1976) Interpretation of variations in leaf water potential and stomatal conductance found in canopies in field. *Philosophical Transactions of the Royal Society of London B* **273**, 593–610.

Jonathan, P., Rochefort, L., and Quinty, F. (1998) Energy and moisture considerations on cutover peatland: surface microtopography, mulch cover and Sphagnum regeneration. *Ecological Engineering* **10**, 293–312.

Jones, H.G. (1990) Physiological aspects of the control of water status in horticultural crops. *HortScience* **25**, 19–26.

Jones, H.G. (1992) *Plants and Microclimate: A Quantitative Approach to Environmental Plant Physiology* (ed. H.G. Jones), p. 411. Cambridge University Press, Cambridge, UK.

Jones, H.G. (1999a) Use of infrared thermometry for estimation of stomatal conductance as a possible aid to irrigation scheduling. *Agricultural and Forest Meteorology* **95**, 139–149.

Jones, H.G. (1999b) Use of thermography for quantitative studies of spatial and temporal variation of stomatal conductance over leaf surfaces. *Plant Cell and Environment* **22**, 1043–1055.

Jones, H.G. (2004) Irrigation scheduling: advantages and pitfalls of plant-based methods. *Journal of Experimental Botany* **55**, 2427–2436.

Jones, H.G., Stoll, M., Santos, T., de Sousa, C., Chaves, M.M., and Grant, O.M. (2002) Use of infrared thermography for monitoring stomatal closure in the field: application to grapevine. *Journal of Experimental Botany* **53**, 2249–2260.

Kahlown, M.A., Ashraf, M., and Zia-ul-Haq (2005) Effect of shallow groundwater table on crop water requirements and crop yields. *Agricultural Water Management* **76**, 24–35.

Kalusová, V., Le Duc, M.G., Gilbert, J.C., Lawson, C.S., Gowing, D.J.G., and Marrs, R.H. (2009) Determining the important environmental variables controlling plant species community composition in mesotrophic grasslands in Great Britain. *Applied Vegetation Science* **12**, 459–471.

Katerji, N. and Mastrorilli, M. (2009) The effect of soil texture on the water use efficiency of irrigated crops: results of a multi-year experiment carried out in the Mediterranean region. *European Journal of Agronomy* **30**, 95–100.

Katerji, N., van Hoorn, J.W., Hamdy, A., and Mastrorilli, M. (2003) Salinity effect on crop development and yield, analysis of salt tolerance according to several classification methods. *Agricultural Water Management* **62**, 37–66.

Katerji, N., Mastrorilli, M., and Rana, G. (2008) Water use efficiency of crops cultivated in the Mediterranean region: review and analysis. *European Journal of Agronomy* **28**, 493–507.

Kelleners, T.J., Soppe, R.W.G., Ayars, J.E., and Skaggs, T.H. (2004) Calibration of capacitance probe sensors in a saline silty clay soil. *Soil Science Society America Journal* **68**, 770–778.

Kemper, W.D., Nicks A.D., and Corey A.T. (1994) Accumulation of water in soils under sand and gravel mulches. *Soil Science Society of America Journal* **58**, 56–63.

Kirkham, M.B. (2004) *Principles of Soil and Plant Water Relations*. Elsevier Academic Press, Burlington, Massachusetts, p. 500.

Klocke, N.L., Heermann, D.F., and Duke, H.R. (1985) Measurement of evaporation and transpiration with lysimeters. *Transactions of the ASAE* **28**, 183–189.

Klute, A. (1972) The determination of the hydraulic conductivity and diffusivity of unsaturated soils. *Soil Science* **113**, 264–276.

Koorevaar, P., Menelik, G., and Dirksen, C. (1983) Elements of soil physics. *Developments in Soil Science 13*. Elsevier, Amsterdam, the Netherlands. p. 228.

Kostner, B., Granier, A., and Cermak, J. (1998) Sap flow measurements in forest stands-methods and uncertainties. *Annals of Forest Science* **55**, 13–27.

Lal, R. (2003) Soil erosion and the global carbon budget, *Environment International* **29**, 437–450.

Lawson, T. (2009) Guard cell photosynthesis and stomatal function. *New Phytologist* **181**, 13–34.

Lei, Y., Takahashi, H., and Li, W. (2004) Effects of concrete mulch on soil thermal and moisture regimes. *Journal of Agricultural Meteorology* **60**, 17–23.

Leinonen, I. and Jones, H.G. (2004) Combining thermal and visible imagery for estimating canopy temperature and identifying plant stress. *Journal of Experimental Botany* **55**, 1423–1431.

Leuning, R. and Judd, M.J. (1996) The relative merits of open- and closed-path analysers for measurement of eddy fluxes. *Global Change Biology* **2**, 241–253.

Li, X.Y. (2003) Gravel-sand mulch for soil water conservation in the semiarid loess of northwest China. *Catena* **52**, 105–127.

Link, S.O., Thiede, M.E., and van Bavel, M.G. (1998) An improved strain-gauge device for continuous field measurement of stem and fruit diameter. *Journal of Experimental Botany* **49**, 1583–1587.

Malicki, M.A., Plagge, R., and Roth, C.H. (1996) Improving the calibration of dielectric TDR soil moisture determination taking into account the solid soil. *European Journal of Soil Science* **47**, 357–366.

McCready, M.S., Dukes, M.D., and Miller, G.L. (2009) Water conservation potential of smart irrigation controllers on St. Augustinegrass. *Agricultural Water Management* **96**, 1623–1632.

McCurry, J.R., Gray, M.J., and Mercker, D.C. (2010) Early growing season flooding influence on seedlings of three common bottomland hardwood species in western Tennessee. *Journal of Fish and Wildlife Management* **1**, 11–18.

Mellouli, H.J., Van Wesemael, B., Poesen, J., and Hartmann, R. (2000) Evaporation losses from bare soils as influenced by cultivation techniques in semi-arid regions. *Agricultural Water Management* **42**, 355–369.

Meron, M., Hallel, R., Peres, M., Bravdo, B., Wallach, R., and Gieling, T. (2001) Tensiometer actuated automatic micro irrigation of apples. *Acta Horticulturae* **562**, 63–69.

Metselaar, K. and Sinclair, T.R. (2007) Letters to the editor: comment on 'Theoretical analysis of soil and plant traits influencing daily plant water flux on drying soils' by T.R. Sinclair. *Agronomy Journal* **97**,1148–1152 (2005). *Agronomy Journal* **99**, 1188–1189.

Modaihsh, A.S., Horton, R., and Kirkham, D. (1985) Soil water evaporation suppression by sand mulches. *Soil Science* **139**, 357–361.

Moller, M., Alchanatis, V., Cohen, Y., Meron, M., Tsipris, J., Naor, A., Ostrovsky, V., Sprintsin, M., and Cohen, S. (2007) Use of thermal and visible imagery for estimating crop water status of irrigated grapevine. *Journal of Experimental Botany* **58**, 827–838.

Monteith, J.L. (1965) Evaporation and environment. In: *The State and Movement of Water in Living Organisms* (ed. G.E. Fogg), pp. 205–234. XIX Symposium of the Society for Experimental Biology, Cambridge University Press, Cambridge, UK.

Monteith, J. and Szeicz, G. (1962) Radiative temperature in the heat balance of natural surfaces. *Quarterly Journal of the Royal Meteorological Society* **88**, 496–507.

Morison, J.I.L., Baker, N.R., Mullineaux, P.M., and Davies, W.J. (2008) Improving water use in crop production. *Philosophical Transactions of the Royal Society B-Biological Sciences* **363**, 639–658.

Munns, R. (2002) Comparative physiology of salt and water stress. *Plant Cell and Environment* **25**, 239–250.

Munns, R., James, R.A., and Lauchli, A. (2006) Approaches to increasing the salt tolerance of wheat and other cereals. *Journal of Experimental Botany* **57**, 1025–1043.

Muñoz-Carpena, R. (2004) Field devices for monitoring soil water content. *Fact Sheet ABE 343 of the Department of Agricultural and Biological Engineering*, University of Florida. http://edis.ifas.ufl.edu/ae266. Accessed on January 2011

Muñoz-Carpena, R. and Dukes, M.D. (2005) Automatic irrigation based on soil moisture for vegetable crops. *Fact Sheet ABE 356 of the Department of Agricultural and Biological Engineering*, University of Florida. http://edis.ifas.ufl.edu/ae354. Accessed on January 2011

Muñoz-Carpena, R., Dukes, M.D., Li, Y., and Klassen, W. (2008) Design and field evaluation of a new controller for soil-water based irrigation. *Applied Engineering in Agriculture* **24**, 183–191.

Nachabe, M.H. (1998) Refining the interpretation of field capacity in the literature. *American Society of Civil Engineers, Journal of Irrigation and Drainage Engineering* **124**, 230–232.

Nachtergaele, J., Poesen, J., and Wesemael, B.V. (1998) Gravel mulching in vineyards of southern Switzerland. *Soil and Tillage Research* **46**, 51–59.

Nemali, K.S. and van Iersel, M.W. (2006) An automated system for controlling drought stress and irrigation in potted plants. *Scientia Horticulturae* **110**, 292–297.

Nobel, P.S. (1999) *Physicochemical and Environmental Plant Physiology*. 2nd edn., Academic Press, London.

Noorduijn, S.L., Ghadouani, A., Vogwill, R., Smettem, K.R.J., and Legendre, P. (2010) Water table response to an experimental alley farming trial: dissecting the spatial and temporal structure of the data. *Ecological Applications* **20**,1704–1720.

Nortes, P.A., Pérez-Pastor, A., Egea, G., Conejero, W., and Domingo, R. (2005) Comparison of changes in stem diameter and water potential values for detecting water stress in young almond trees. *Agricultural Water Management* **77**, 296–307.

Novaka, M.D., Chenb, W., Orchanskyc, A.L., and Ketler, R. (2000) Turbulent exchange processes within and above a straw mulch. Part II. Thermal and moisture regimes. *Agricultural and Forest Meteorology* **102**, 155–171.

Oades, J. (1984) Soil organic matter and structural stability: mechanisms and implications for management. *Plant and Soil* **76**, 319–337.

O'Leary, G.J. and Connor, D.J. (1997) Stubble retention and tillage in a semi-arid environment. I. Soil water accumulation during fallow. *Field Crops Research* **52**, 209–219.

Oosterbaan, R.J. (1994) Agricultural drainage criteria. In: *Drainage Principles and Applications* (ed. H.P. Ritzema), 2nd edn., International Institute for Land Reclamation and Improvement (ILRI), Publication 16, Wageningen, the Netherlands.

Oosterbaan, R.J. and Nijland, H.J. (1994) Determining the saturated hydraulic conductivity. In: *Drainage Principles and Applications* (ed. H.P. Ritzema), 2nd edn., International Institute for Land Reclamation and Improvement (ILRI), Publication 16, Wageningen, the Netherlands.

Ortuño, M.F., Conejero, W., Moreno, F., Moriana, A., Intrigliolo, D.S., Biel, C., Mellisho, C.D., Perez-Pastor, A., Domingo, R., Ruiz-Sanchez, M.C., Casadesus, J., Bonany, J., and Torrecillas, A. (2010) Could trunk diameter sensors be used in woody crops for irrigation scheduling? A review of current knowledge and future perspectives. *Agricultural Water Management* **97**, 1–11.

Oweis, T., Zhang, H., and Pala, M. (2000) Water use efficiency of rainfed and irrigated bread wheat in a Mediterranean environment. *Agronomy Journal* **92**, 231–238.

Paco, T.A., Ferreira, M.I., and Conceicao, N. (2006) Peach orchard evapotranspiration in a sandy soil: comparison between eddy covariance measurements and estimates by the FAO 56 approach. *Agricultural Water Management* **85**, 305–313.

Pereira, L.S., Oweis, T., and Zairi, A. (2002) Irrigation management under water scarcity. *Agricultural Water Management* **57**, 175–206.

Peretz, J., Evans, R.G., and Proebsting, E.L. (1984) Leaf water potentials for management of high frequency irrigation on apples. *Transactions of the American Society of Agricultural Engineers* **27**, 437–442.

Pérez-Pastor, A., Domingo, R., Torrecillas, A., and Ruiz-Sanchez, M.C. (2009) Response of apricot trees to deficit irrigation strategies. *Irrigation Science* **27**, 231–242.

Price, M. (1996) *Introducing Groundwater*. Chapman and Hall, London, p. 278.

Qureshi, A.S., Eshmuratov, D., and Bezborodov, G. (2010) Determining optimal groundwater table depth for maximizing cotton production in the Sardarya province of Uzbekistan. *Irrigation and Drainage* **60**, 241–252.

Raes, D., Steduto, P., Hsiao, T.C., and Fereres, E. (2009) AquaCrop: the FAO crop model to simulate yield response to water: II. Main algorithms and software description. *Agronomy Journal* **101**, 438–447.

Raupach, M.R. (1994) Simplified expressions for vegetation roughness length and zero-plane displacement as functions of canopy height and area index. *Boundary-Layer Meteorology* **71**, 211–216.

Rengasamy, P. (2006) World salinization with emphasis on Australia. *Journal of Experimental Botany* **57**, 1017–1023.

Roberts, J. (2005) Transpiration. In: *Encyclopedia of Hydrological Sciences* (ed. M.G. Anderson) John Wiley & Sons, Ltd., Chichester, UK.

Robinson, D.A., Gardner, C.M.K., Evans, J., Cooper, J.D., Hodnett, M.G., and Bell, J.P. (1998) The dielectric calibration of capacitance probes for soil hydrology using an oscillation frequency response model. *Hydrology and Earth Systems Science* **2**, 111–120.

Robinson, D.A., Lebron, I., Ryelb, R.J., and Jones, S.B. (2010) Soil water repellency: a method of soil moisture sequestration in pinyon–juniper woodland. *Soil Science Society of America Journal* **74**, 624–634.

Romero, P., Botia, P., and Garcia, F. (2004) Effects of regulated deficit irrigation under subsurface drip irrigation conditions on vegetative development and yield of mature almond trees. *Plant and Soil* **260**, 169–181.

Rosegrant, M.W., Nkonya, E., and Valmonte-Santos, R.A. (2009) Food security and soil water management. In: *Encyclopedia of Soil Science* (ed. R. Lal), 2nd edn. Oxon, UK.

Rowell, D.L. (1994) *Soil Science: Methods and Applications*. Longman Group Ltd., Essex, UK.

Rundle, P.J. and Rundle, B.F. (2002) A case study of farm-based solutions to water logging and secondary salinity in southwestern Australia. *Agricultural Water Management* **53**, 31–38.

Sadras, V.O. (2009) Does partial root-zone drying improve irrigation water productivity in the field? A meta-analysis. *Irrigation Science* **27**, 183–190.

Sauer, T.J. and Horton, R. (2005) Soil heat flux. In: *Micrometeorology in Agricultural Systems* (ed. M.K. Viney), pp. 131–154. American Society of Agronomy, Crop Science Society of America, Soil Science Society of America, Madison, Wisconsin.

Schaap, M.G. and Leij, F.J. (1998) Database related accuracy and uncertainty of pedotransfer functions. *Soil Science* **163**, 765–779.

Scholander, P.F., Hammel, H.T., Bradstreet, E.D., and Hemingsen, E.A. (1965) Sap pressure in vascular plants. *Science* **148**, 339–346.

Sepulcre-Cantó, G., Zarco-Tejada, P.J., Jimenez-Muñoz, J.C., Sobrino, J.A., de Miguel, E., and Villalobos, F.J. (2006) Detection of water stress in an olive orchard with thermal remote sensing imagery. *Agricultural and Forest Meteorology* **136**, 31–44.

Sharma, D.P., Singh, K.N., Rao, K.V.G.K., and Kumbhare, P.S. (1990) Reuse of saline drainage water for irrigation in a sandy loam soil. In: *Symposium on land drainage for salinity control in arid and semi-arid regions*, Vol. 3., pp. 304–312. Drainage Research Institute, Cairo, Egypt.

Shuttleworth, W.J. and Wallace, J.S. (1988) Evaporation from sparse crops—an energy combination theory. *Quarterly Journal of the Royal Meteorological Society* **111**, 839–855.

Siles, P., Vaast, P., Dreyer, E., and Harmand, J.M. (2010) Rainfall partitioning into throughfall, stemflow and interception loss in a coffee (*Coffea arabica* L.) monoculture compared to an agroforestry system with Inga densiflora. *Journal of Hydrology* **395**, 39–48.

Sinclair, T.R. (2005) Theoretical analysis of soil and plant traits influencing daily plant water flux on drying soils. *Agronomy Journal* **97**, 1148–1152.

Smith, D.M. and Allen, S.J. (1996) Measurement of sap flow in plant stems. *Journal of Experimental Botany* **47**, 1833–1844.

Smith, R.J., Raine, S.R., and Minkevich, J. (2005) Irrigation application efficiency and deep drainage potential under surface irrigated cotton. *Agricultural Water Management* **71**, 117–130.

Sobrino, J.A., Jimenez-Munoz, J.C., Zarco-Tejada, P.J., Sepulcre-Canto, G., de Miguel, E., Soria, G., Romaguera, M., Julien, Y., Cuenca, J., Hidalgo, V., Franch, B., Mattar, C., Morales, L., Gillespie, A., Sabol, D., Balick, L., Su, Z., Jia, L., Gieske, A., Timmermans, W., Olioso, A., Nerry, F., Guanter, L., Moreno, J., and Shen, Q. (2009) Thermal remote sensing from Airborne Hyperspectral Scanner data in the framework of the SPARC and SEN2FLEX projects: an overview. *Hydrology and Earth System Sciences* **13**, 2031–2037.

Steduto, P., Katerji, N., Puertos-Molina, H., Unlu, M., Mastrorilli, M., and Rana, G. (1997) Water-use efficiency of sweet sorghum under water stress conditions. Gas-exchange investigations at leaf and canopy scales. *Field Crops Research* **54**, 221–234.

Steduto, P., Hsiao, T.C., Raes, D., and Fereres, E. (2009) AquaCrop: the FAO crop model to simulate yield response to water: I. Concepts and underlying principles. *Agronomy Journal* **101**, 426–437.

Stephens, D. (2002) *National Land and Water Resources Audit. National and Regional Assessments of Crop Yield Trends and Relative Production Efficiency. Theme 5.1. Land Use Change, Productivity and Diversification*. Department of Agriculture of Western Australia, South Perth, Australia.

Stewart, J.B. (1988) Modeling surface conductance of pine forest. *Agricultural and Forest Meteorology* **43**, 19–35.

Stewart, B.A. and Robinson, C.A. (1997) Are agroecosystems sustainable in semiarid regions? *Advances in Agronomy* **60**, 191–228.

Stockle, C.O., Cabelguenne, M., and Debaeke, P. (1997) Comparison of CropSyst performance for water management in southwestern France using submodels of different levels of complexity. *European Journal of Agronomy* **7**, 89–98. www.sciencedirect.com

Stoll, M., Loveys, B., and Dry, P. (2000) Hormonal changes induced by partial rootzone drying of irrigated grapevine. *Journal of Experimental Botany* **51**, 1627–1634.

Stroosnijder, L. and Hoogmoed, W.B. (1984) Crust formation on sandy soils in the Sahel II. Tillage and its effect on the water balance. *Soil and Tillage Research* **4**, 321–337.

Suárez, L., Zarco-Tejada, P.J., Sepulcre-Cantó, G., Perez-Priego, O., Miller, J.R., Jiménez-Muñoz, J.C., and Sobrino, J.A. (2008) Assessing canopy PRI for water stress detection with diurnal airborne imagery. *Remote Sensing of Environment* **112**, 560–575.

Swanson, R.H. (1994) Significant historical development in thermal methods for measuring sap flow in trees. *Agricultural and Forest Meteorology* **72**, 113–132.

Taiz, L. and Zeiger, E. (2002) *Plant Physiology*. 3rd edn., Sinauer Associates, Inc., Sunderland, Massachusetts, p. 623.

Tanji, K.K. and Kielen, N.C. (2002) Agricultural drainage water management in arid and semi-arid areas, FAO Irrigation and Drainage Paper *61*. FAO, Rome, Italy, 188pp.

Tardieu, F. and Simonneau, T. (1998) Variability among species of stomatal control under fluctuating soil water status and evaporative demand: modelling isohydric and anisohydric behaviours. *Journal of Experimental Botany* **49**, 419–432.

Testi, L., Goldhamer, D.A., Iniesta, F., and Salinas, M. (2008) Crop water stress index is a sensitive water stress indicator in pistachio trees. *Irrigation Science* **26**, 395–405.

Todd, D.K. (1980) *Groundwater Hydrology*. 2nd edn., John Wiley & Sons, Inc., New York, p. 535.

Todd, R.W., Evett, S.R., and Howell, T.A. (2000) The Bowen ratio-energy balance method for estimating latent heat flux of irrigated alfalfa evaluated in a semi-arid, advective environment. *Agricultural and Forest Meteorology* **103**, 335–348.

Todorovic, M., Albrizio, R., Zivotic, L., Abi Saab, M.T., Stöckle, C., and Steduto, P. (2009) Assessment of AquaCrop, CropSyst, and WOFOST Models in the simulation of sunflower growth under different water regimes. *Agronomy Journal* **101**, 509–521.

Topp, G.C. and Reynolds, W.D. (1998) Time domain reflectometry: a seminal technique for measuring mass and energy in soil. *Soil and Tillage Research* **47**, 125–132.

Topp G.C. and Ferre, T.P.A. (2005) Measuring soil water content. In: *Encyclopedia of Hydrological Sciences* (eds M.G. Anderson and J.J. McDonnell) John Wiley & Sons, Ltd., Chichester, UK.

Topp, G.C., Davis, J.L., and Annan, A.P. (1980) Electromagnetic determination of soil water content: measurements in coaxial transmission lines. *Water Resources Research* **16**, 574–582.

Topp, G.C., Zegelin, S., and White, I. (2000) Impact of real and imaginary components of relative permittivity on time domain reflectometry measurements in soils. *Soil Science Society of America Journal* **64**, 1244–1252.

Turner, N.C. (2004) Agronomic options for improving rainfall-use efficiency of crops in dryland farming systems. *Journal of Experimental Botany* **55**, 2413–2425.

Turner, N.C. and Ward, P.R. (2002) The role of agroforestry and perennial pasture in mitigating water logging and secondary salinity: summary. *Agricultural Water Management* **53**, 271–275.

Ueda, M. and Shibata, E. (2001) Diurnal changes in branch diameter as indicator of water status of Hinoki cypress (*Chamaecyparis obtuse*). *Trees* **15**, 315–318.

Ueda, M., Yoshikawa, K., and Okitu, J. (1996) Measurement of diurnal changes in stem and branch diameter using strain gauges. *Journal of Forest Research* **1**, 139–142.

Unger, P.W. (1978) Straw-mulch rate effect on soil water storage and sorghum yield. *Soil Science Society of America Journal* **42**, 486–491.

Van Dam, J.C., Stricker, J.N.M., and Droogers, P. (1994) Inverse method to determine soil hydraulic functions from multistep outflow experiments. *Soil Science Society of America Journal* **58**, 647–652.

Van den Honert, T.H. (1948) Water transport in plants as a catenary process. *Discussions of the Faraday Society* **3**, 146–153.

Van den Hurk, B.J.J.M., Verhoef, A., van den Berg, A.R., and de Bruin, H.A.R. (1995) An intercomparison of three vegetation/soil models for a sparse vineyard canopy. *Quarterly Journal of the Royal Meteorological Society* **121**, 1867–1889.

Van Genuchten, M.T. (1980) A closed-form equation for predicting the hydraulic conductivity of unsaturated soils. *Soil Scince Society of America Journal* **44**, 892–898.

Van Schilfgaarde, J. (1974) Non-steady flow to drains. In: *Drainage for Agriculture* (ed. I. Schilfgaarde), pp. 245–307. ASA, Madison, Wisconsin.

Van Wesemael, B., Poesen, J., Kosmas, C.S., Danalatos, N.G., and Nachtergaele, J. (1996) Evaporation from cultivated soils containing rock fragments. *Journal of Hydrology* **182**, 65–82.

Vartapetian, B.B., and Jackson, M.B. (1997) Plant adaptations to anaerobic stress. *Annals of Botany* **79**, 3–20.

Vélez, J.E., Intrigliolo, D.S., and Castel, J.R. (2007) Scheduling deficit irrigation of citrus trees with maximum daily trunk shrinkage. *Agricultural Water Management* **90**, 197–204.

Vereecken, H., Kaiser, R., Dust, M., and Pütz, T. (1997) Evaluation of the multistep outflow method for the determination of unsaturated hydraulic properties of soils. *Soil Science* **162**, 618–631.

Verhoef, A. and Allen, S.J. (2000) A SVAT scheme describing energy and CO_2 fluxes for multi-component vegetation: calibration and test for a Sahelian savannah. *Ecological Modelling* **127**, 245–267.

Verhoef, A. and Campbell, C.L. (2005) Evaporation measurement. In: *Encyclopedia of Hydrological Sciences*, Vol. 1 (ed. M.G. Anderson). John Wiley & Sons, Ltd., Chichester, UK.

Verhoef, A., de Bruin, H.A.R., and van den Hurk, B.J.M. (1997) Some practical notes on parameter kB^{-1} for sparse vegetation. *Journal of Applied Meteorology* **36**, 560–572.

Verhoef, A., Fernández-Gálvez, J., Díaz-Espejo, A., Main, B.E., and El Bishti, M. (2006) The diurnal course of soil moisture as measured by various dielectric sensors: effects of soil temperature and the implications for evaporation estimates. *Journal of Hydrology* **321**, 147–162.

Visser, E.J.W., Voesenek, L.A.C.J., and Vatapetian, B.B. (2003) Flooding and plant growth. *Annals of Botany* **91**, 107–109.

Wallace, J.S. (2000) Increasing agricultural water use efficiency to meet future food production. *Agriculture Ecosystems and Environment* **82**, 105–119.

Wallace, J.S. and Gregory, P.J. (2002) Water resources and their use in food production systems. *Aquatic Science* **64**, 363–375.

Wallace, J.S., Roberts, J.M., and Sivakumar, M.V.K. (1990) The estimation of transpiration from sparse dryland millet using stomatal conductance and vegetation area indexes. *Agricultural and Forest Meteorology* **51**, 35–49.

Wallace, J.S., Jackson, N.A., and Ong, C.K. (1999) Modelling soil evaporation in an agroforestry system in Kenya. *Agricultural and Forest Meteorology* **94**, 189–202.

Wang, L.M., Qiu, G.Y., Zhang, X.Y. et al. (2005) Application of a new method to evaluate crop water stress index. *Irrigation Science* **24**, 49–54.

Wang, F.X., Kang, Y., Liu, S.P., and Chen S.Y. (2007) Effects of soil matric potential on potato growth under drip irrigation in the North China Plain. *Agricultural Water Management* **88**, 34–42.

Whitmore, A.P. and Whalley, W.R. (2009) Physical effects of soil drying on roots and crop growth. *Journal of Experimental Botany* **60**, 2845–2857.

Wilcock, D.N. and Essery, C.I. (1984) Infiltration measurements in a small lowland catchment. *Journal of Hydrology* **74**, 191–204.

Wilson, C.G., Papanicolaou, A.N.T., and Abaci, O. (2009) SOM dynamics and erosion in an agricultural test field. *Hydrology and Earth System Sciences Discuss* **6**, 1581–1619.

Wind, G.P. and Buitendijk, J. (1979) Simulation over 35 years of the moisture content of a topsoil with an electric analog. In: *Proceedings of the International Drainage Workshop* (ed. J. Wesseling), pp. 214–219. ILRI, Publication 25, Wageningen, the Netherlands.

Withers, B. and Vipond, S. (1974) *Irrigation: Design and Practice*. Batsford, London, pp. 306.

Wösten, J.H.M., Lilly, A., Nemes, A. et al. (1999) Development and use of a database of hydraulic properties of European soils. *Geoderma* **90**, 169–185.

Xie, Z.K., Wang, Y.J., Jiang, W.L., Wei, X. H. (2006) Evaporation and evapotranspiration in a watermelon field mulched with gravel of different sizes in northwest China. *Agricultural Water Management* **81**, 173–184.

Yimer, F., Messing, I., Ledin, S., and Abdelkadir, A. (2008) Effects of different land use types on infiltration capacity in a catchment in the highlands of Ethiopia. *Soil Use and Management* **24**, 344–349.

Zarco-Tejada, P.J., Berni, J.A.J., Suárez, L., Sepulcre-Canto, G., Morales, F., and Miller, J.R. (2009) Imaging chlorophyll fluorescence with an airborne narrow-band multispectral camera for vegetation stress detection. *Remote Sensing of Environment* **113**, 1262–1275.

Zhao, X.N., Wu, P.T., Feng, H., Wang, Y.K., and Shao, H.B. (2009) Towards development of eco-agriculture of rainwater-harvesting for supplemental irrigation in the semi-arid loess plateau of China. *Journal of Agronomy and Crop Science* **195**, 399–407.

Zotarelli, L., Dukes, M.D., Scholberg, J.M.S., Munoz-Carpena, R., and Icerman, J. (2009) Tomato nitrogen accumulation and fertilizer use efficiency on a sandy soil, as affected by nitrogen rate and irrigation scheduling. *Agricultural Water Management* **96**, 1247–1258.

Zweifel, R., Steppe, K., and Sterck, F.J. (2007) Stomatal regulation by microclimate and tree water relations: interpreting ecophysiological field data with a hydraulic plant model. *Journal of Experimental Botany* **58**, 2113–2131.

10 Plant-induced changes in soil processes and properties

Philippe Hinsinger

Institut National de la Recherche Agronomique, UMR Eco&Sols, Montpellier, France

10.1 Introduction

Since the early beginning of pedology with the work of Dokuchaev in the nineteenth century, vegetation has been recognised as one of the five major factors explaining the spatial distribution of soil types on planet Earth, together with the climate, parent material, topography and time. However, at such a scale, the biogeography of the vegetation is largely correlated with climatic constraints, so one may wonder whether these two factors are equally needed to explain pedogenesis. While temperature and the amount of water circulating through the system have long been recognised as key drivers of soil formation and biogeochemical cycles, the direct implication of plants has long been overlooked by both soil scientists and geochemists (Hinsinger et al., 2001; Lambers et al., 2009). There is little doubt that soil formation results primarily from a combination of physical and chemical processes leading to the breakdown and weathering of rock particles. The etymology of the word 'weathering' is explicit in this respect, and numerous works in experimental geochemistry (e.g. Bockheim and Gennadiyev, 2009) have demonstrated that purely chemical processes could explain the observed extent of weathering and the formation of typical secondary minerals present in the soil matrix such as clay minerals and metal oxides. However, their nature and amount depend on the combined effects of time, temperature and flux of solute transfer; the latter two being obviously related to the climate. Without plants, though, and especially higher plants with deep root systems, would there be any deep, differentiated soil at all or merely a thin regolith (Lambers et al., 2009)? One feature that distinguishes soil from the loose layer of rock material called regolith is the occurrence of a small amount of organic matter, which primarily originates from plants.

The first section of this chapter describes how higher plants have played a key role as the primary source of organic matter in soils and more generally in the biogeochemical cycle of carbon (C) on planet Earth. Compared with earlier life forms and primary producers such as photosynthetic bacteria, algae, lichens and bryophytes, higher plants have this special feature of deep root systems which have direct interactions with deeper soil horizons and

Soil Conditions and Plant Growth. Edited by Peter J. Gregory and Stephen Nortcliff.
© 2013 Blackwell Publishing Ltd. Published 2013 by Blackwell Publishing Ltd.

can be responsible for a direct input of large amounts of C-rich compounds deep below the soil surface. A special focus of this first section will be on the role of the below-ground processes involved in the biogeochemical cycle of C.

The soil volume surrounding the roots of higher plants is distinct from the bulk soil as a consequence of a number of root activities. It is indeed unique in the intensity and nature of biophysical, biogeochemical/biochemical and biological/ecological processes that occur in the so-called rhizosphere (Darrah, 1993; Hinsinger, 1998; Hiltner, 1904 in Hartmann et al., 2008; Hinsinger et al., 2009). Plant roots are major drivers of such processes but these also largely rely on the indirect effects of microbial communities, which occur and are stimulated in the rhizosphere as a consequence of the rhizodeposition of C-rich (and sometimes N-rich) compounds (Jones et al., 2009). These combined effects of roots and rhizosphere microorganisms ultimately play a key role in the biogeochemical cycles of protons, nutrients and other elements. This will be addressed in two subsequent sections of this chapter, which concentrate on two major processes involved in soil formation and evolution, that is, soil acidification and weathering. While rhizosphere processes are often considered to occur at the microscopic scale of the soil–root interface, various parts of this chapter will attempt to demonstrate that, over time, they have a major impact at the global scale. How these processes affect plant growth either positively or negatively, and their ultimate consequences for ecosystem functions and services, will also be briefly addressed.

10.2 Roles of plants in carbon cycling: deposition, burial and emission

Higher plants are major contributors to photosynthesis among autotrophic organisms on planet Earth, especially in terrestrial ecosystems. These primary producers assimilate atmospheric CO_2 which is thereby incorporated in a whole range of organic compounds that are ultimately supplied to the soil directly as litter or via the loop of primary and secondary consumers, before being recycled back to the atmosphere by decomposers, a process which largely occurs in the soil. This is demonstrated, for example, by the comparison conducted by Johnston et al. (2009) and Hirsch et al. (2009) in the long-term Highfield ley-arable experiment at Rothamsted, United Kingdom. This long-standing grass ley was partly ploughed in 1949 and subsequently either managed by arable rotation or maintained as a bare fallow by regular tillage since 1959. Johnson et al. (2009) reported a significant decline in soil organic C over time, when grass had been converted to arable cultivation, and even more so in the bare fallow. The C content in the topsoil of the grass-treatment soil (54 g C kg^{-1}) was much higher than the arable soil (16 g C kg^{-1}) and the bare-fallow soil (10 g C kg^{-1}). After almost 50 years without significant plant C input, the bare-fallow soil contained ~30 t C ha^{-1}, a value <8% of that in the original grassland soil (Goulding et al., 2010). Hirsch et al. (2009) also showed that the bare-fallow soil supported smaller and less diverse microbial and mesofaunal communities than the planted soil treatments, as a consequence of the absence of an annual input of fresh organic residues of plant origin, compared with the arable and grass treatments which provided a richer C substrate. The changes that occurred were large in a short period of time (50 years) relative to the time needed for pedogenetic development of such soils (thousands or tens of thousands years). It shows that while soil evolution occurs over long time spans, it can also be subject to dramatic changes on a short time scale. The most extreme case of short-term change is that induced by topsoil erosion or landslides, that plant cover tends to restrict (Stokes et al., 2009). Lessons drawn from the analyses of soil chronosequences are of special interest for understanding the long-term patterns of soil development. Richardson et al. (2004) have clearly illustrated with the Franz Joseph chronosequence in New Zealand that the organic C content of the soil increases from close to zero to a few percent in the first hundred of years, then builds further to reach values of up to 15% in the first millennium of soil development and then finally steadily decreases to values between 5% and 10% over the period 5 000 to 120 000 years (Figure 10.1).

10.2.1 Photosynthesis, respiration, litter deposition and C sequestration in plants and soils

Global estimates of components of the C cycle are currently the following (Schlesinger, 1997): photosynthesis in terrestrial ecosystems (gross primary production, GPP) amounts to 120×10^{15} gC year^{-1} while C fixation by primary producers in the ocean amounts to 48×10^{15} gC year^{-1}. In comparison, the sizes of the major reservoirs of terrestrial C are as follows: 550×10^{15} gC in the standing vegetation, 300×10^{15} gC in the litter and 1200×10^{15} gC in the soils, while atmospheric C represents 780×10^{15} gC. This suggests that one-sixth of the atmospheric CO_2 is recycled yearly through photosynthesis by terrestrial vegetation (GPP) (Trumbore, 2006). The net primary production (NPP) of terrestrial ecosystems amounts to 58×10^{15} gC year^{-1} (Field et al., 1998), out of which crops amount to a rather small proportion (8×10^{15} gC year^{-1}; Table 10.1), compared with savannas (17×10^{15} gC year^{-1}) and tropical rainforests (18×10^{15} gC year^{-1}). Such fluxes are worth comparing to that reported for soil respiration, 75×10^{15} gC year^{-1}, which is actually larger than both terrestrial NPP and litterfall (Schlesinger and Andrews, 2000). Soil respiration is greater than NPP because a substantial part of soil respiration is related to respiration of roots and mycorrhizae (deriving from GPP). While above-ground C pools and fluxes are well documented, estimates of below-ground pools and fluxes exhibit greater variability (De Deyn et al., 2008). Distinguishing between root respiration and heterotrophic respiration (due to soil microbial and faunal activity) is especially difficult.

Looking back to geological times, the rise of land plants and particularly the emergence of vascular plants with greater photosynthesis and transpiration capacities (related to their innovative and extensively developed architecture) during the Devonian period is now recognised for its central role in the geochemical history of the Earth and its atmosphere (Beerling and Berner, 2005). The early forms of land plants at the beginning of the Devonian period looked like bryophytes and other plant life forms which had rhizomes but no roots and no large planate leaves (Beerling et al., 2001). Fossil records suggest that the vegetation formed a short chlorophyllous carpet made of short shoots rising from other stems running along the surface (Kenrick and Crane, 1997). In contrast, tall vegetation with megaphyll planate leaves became abundant during the late Devonian/early Carboniferous (Beerling et al., 2001). Interestingly, such evolution coincided with a major drop in atmospheric pCO_2, which was then much above present-day levels. It dropped from 3–4000 ppmv to less than 1000 ppmv, possibly as a result of increased photosynthesis and removal of atmospheric CO_2 by the megaphyll vegetation (Beerling and Berner, 2005). This fixed C was progressively buried as organic matter in soils. Most of the biogeochemical models account for organic C input as related to above-ground parts of the vegetation and implicitly assume that the below-ground portion mostly contributes to an additional input of C that is mostly concentrated in the topsoil. This is derived from the decay of above-ground parts which fall on top of the soil (trunks, branches, shoots of herbaceous plants, leaves and fruits).

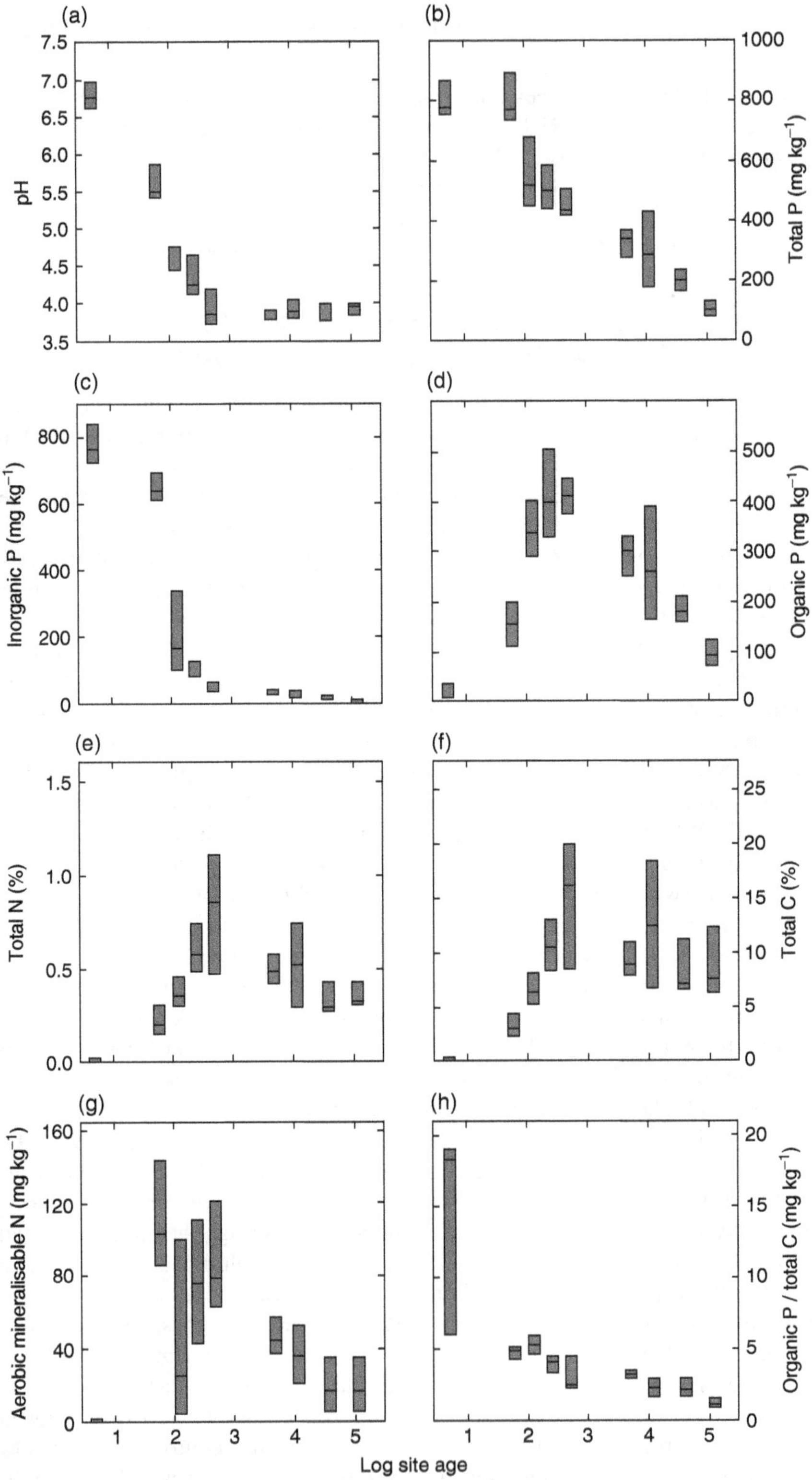

Figure 10.1 Temporal evolution of soil properties along the Franz Josef soil chronosequence. Data are means of five plots at each site with the exception of site 7 where n = 4. Box plot symbols: horizontal lines are the median; shaded bars give 25% and 75% percentiles. From Richardson et al. (2004). Reprinted with kind permission from Springer Science and Business Media (Springer-Verlag)/ International Association for Ecology.

Table 10.1 Net primary production (NPP) in various terrestrial ecosystems.

Tropical rainforests	**17.8**
Broadleaf and needleleaf forests	3.1
Broadleaf deciduous forests	1.5
Needleleaf evergreen forests	3.1
Needleleaf deciduous forest	1.4
Savannas	**16.8**
Perennial grasslands	2.4
Broadleaf shrubs with bare soil	1.0
Tundra	0.8
Desert	0.5
Cultivated agroecosystems	**8.0**

All values are expressed as 10^{15} g C, the total amounting to 56.4 10^{15} g C.
Source: Data compiled from Field et al. (1998). Reprinted with permission from AAAS.

10.2.2 Allocation of C to below-ground parts of the plants

A substantial proportion of the GPP is allocated to below-ground parts of the vegetation. According to Jones (1993), a median value of about 60% of the fixed C is used in the shoots (for shoot biomass production and respiration), while 40% is allocated to below-ground parts (Figure 10.2). For cereals, it has been estimated that over the whole growing season, the total amount of C allocated to below-ground parts ranges from 150 to 220 g C m^{-2} $year^{-1}$ (Kuzyakov and Domanski, 2000). In perennial species, especially when exposed to limitation of below-ground resources (water or nutrient shortage), the allocation of fixed C to the root compartment can exceed 50% of the total C assimilated. In addition, root litter is usually a more recalcitrant C pool than above-ground tissues (shoots), thereby contributing to C sequestration in soils (De Deyn et al., 2008), as will be further discussed in this section. In addition, in contrast to the above-ground litter which is predominantly recycled back to the topsoil (although faunal activities can result in burial of such organic matter at depth), C allocation to roots contributes to a direct C input at depth, that is related to the vertical distribution of roots in the soil profile.

Meta-analyses of the vertical distribution of roots in soils (database of 250 root studies) showed that in most biomes more than 80–90% of the root biomass was found in the top (0–30 cm) soil (Jackson et al., 1996). This was further substantiated by a broader dataset of 475 soil profiles from 209 different geographical locations (Schenk and Jackson, 2002). However, Schenk and Jackson (2002) stressed that many of these studies underestimated root biomass as most authors did not investigate deep soil horizons. They stressed that the proportion of roots found in the topsoil decreased as the depth of soil investigated increased. Canadell et al. (1996) compiled published information about maximum rooting depths and showed that roots commonly grew as deep as 1–2 m in most crops. The biome average value for croplands and boreal forests was similar at about 2 m, while it extended to 2.6 m in temperate grassland, 2.9 m in temperate deciduous forests, 3.7 m in tropical deciduous forest, 7.3 m in tropical evergreen forest and 15 m in tropical grassland/savanna (Canadell et al., 1996). Some detailed studies in deep soil profiles have recently shown that a substantial proportion of root biomass occurs at depth and may thereby contribute a very significant contribution to deep soil C sequestration. Quantification of fine root dynamics down to a

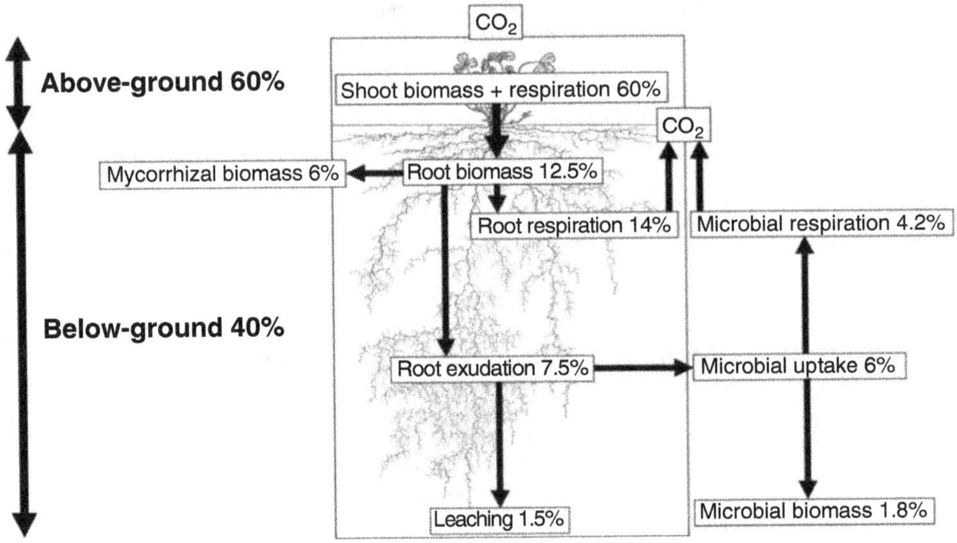

Figure 10.2 Allocation of C to above- and below-ground parts of the whole plant (expressed as percentage of total C assimilated by photosynthesis), including plant biomass formation, respiration and allocation to symbionts or rhizosphere microorganisms. Adapted from Jones (1993), with kind permission from the author.

depth of about 6 m in various Amazonian ecosystems showed a large input of C at depth (Trumbore et al., 2006): out of a total of 91 C m^{-2} $year^{-1}$ in a degraded pasture and 78 g C m^{-2} $year^{-1}$ in a mature forest, about 26% and 35%, respectively, of fine roots were found at 1–6 m depth while 40% and 30%, respectively, were found in the top 0.1 m. Even more impressive are the data obtained in fast-growing eucalypt plantations: in the first year of growth in the Congo, 16% of fine roots were found in the top 0.25 m, while 84% colonised the soil profile almost homogeneously to 3 m depth (Bouillet et al., 2002). In tropical crops such as sugarcane grown on poor, deeply weathered soils in Brazil, Battie Laclau and Laclau (2009) also reported large densities of fine roots to several metres; this was more marked in rainfed than irrigated conditions (Figure 10.3). In temperate crops, much larger root length densities are found in the topsoil for most species, and root length density often decreases exponentially with soil depth, as obtained in cauliflower and wheat in Figure 10.4 (Gregory, 2006a). Some crop species, however, exhibit a different pattern of root distribution with almost constant root length density at depth, as shown for oilseed rape and sugarbeet in Figure 10.4. There is, thus, a need for more quantitative data about root distribution (both in terms of length and biomass) as this is crucial to (1) the exploitation of soil water and nutrient resources by plants and (2) the C sequestration occurring in the subsoil. Indeed, at a global level, only half of the total C sequestered in soils is found in the topsoil. In a temperate pasture, it was shown that this may be linked to greater stability against mineralisation in the subsoil, possibly due to a smaller priming effect occurring at depth because of the smaller supply of fresh material than in the topsoil (Fontaine et al., 2007). This phenomenon might well be related to root litter being more recalcitrant than shoot litter which makes a large proportion of topsoil organic C (De Deyn et al., 2008).

Besides litter deposition (including root turnover), another major process has to be accounted for when considering below-ground processes involved in C cycling. Indeed, a

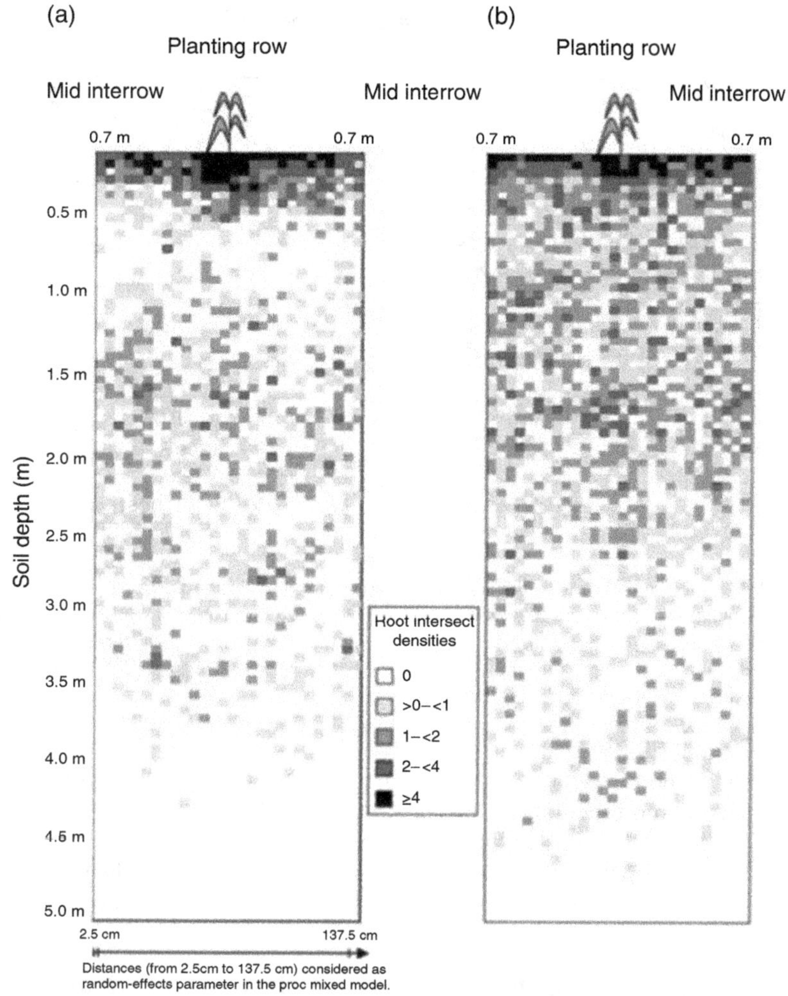

Figure 10.3 Mean number of root intersects (n = 2) in each grid cell of 25 cm^2 delimited in vertical trench walls in the irrigated (a) and rainfed (b) sugarcane crops, at 332 days after planting. Reprinted from Battie-Laclau and Laclau, (2009) with kind permission from Elsevier. For a colour version of this figure, please see Plate 10.1.

substantial portion of the C allocated by plants to their below-ground parts is released during the life of roots via the so-called rhizodeposition process (Nguyen, 2003; Jones et al., 2009). Using ^{14}C labelling experiments, data compiled for a broad range of species (271 datasets comprising 250 datasets of 17 herbaceous species largely dominated by Poaceae, and 21 datasets for 4 woody species) by Jones et al. (2009) showed a median value for C recovered in soil residues of about 5% of fixed C, while the median value for rhizosphere respiration was about 12%. Given that about half of rhizosphere respiration is due to microbial respiration, that is, CO_2 emission as a consequence of the breakdown of rhizodeposits by rhizosphere microorganisms (Kuzyakov, 2006), it can be estimated that rhizodeposition is in about 11% of the total amount of C fixed by plants. This represents about a quarter of the amount of C allocated to below-ground parts of the plants. For cereals, over the whole growing season,

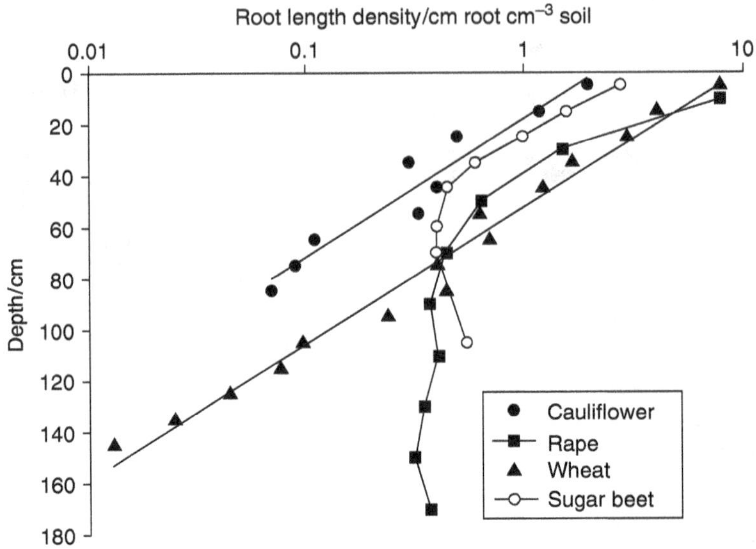

Figure 10.4 Distribution of root length density with soil depth for mature crops of cauliflower, oilseed rape, sugar beet and winter wheat. Linear regressions have been fitted to data for the distributions of cauliflower and winter wheat roots, suggesting an exponential decrease in root density with depth, which is not the case for oilseed rape and sugar beet. Reprinted from Gregory (2006a, b). Reprinted with permission from Blackwell Publishing Ltd.

rhizodeposition would thus amount to 40–60 g C m^{-2} $year^{-1}$ (Jones et al., 2009). This is a significant amount as it represents about 1% of the stock of C stored in the top 30 cm of an arable soil, that is, 4380–5990 g C m^{-2}, assuming a soil organic matter content of 2–3% and a bulk density of 1.2 Mg m^{-3} (Smith et al., 1997). Besides its contribution to delivering C to the soil, rhizodeposition has many functional implications, as described in the following section.

10.2.3 Ecological significance of the rhizodeposition of C compounds

As pointed by Hiltner in 1904 when he first coined the term 'Rhizosphäre', which has a Greek origin (see Hartmann et al. (2008) for an English translation of Hiltner's seminal paper), root exudates have major implications for both plant nutrition and health, via the stimulation of microbial activities (Hartmann et al., 2009). For reviews of these issues refer to Dessaux et al. (2010) and Richardson et al. (2009) for plant nutrition issues and to Raaijmakers et al. (2009) for plant health issues. Instead of root exudates, this chapter will refer to rhizodeposits because this term is broader and includes root secretions, root cap cells and root exudates (Nguyen, 2003; Jones et al., 2009). Rhizodeposits are sources of energy for soil microorganisms whose growth and activities are considerably enhanced in the vicinity of living roots, resulting in the so-called rhizosphere effect (Curl and Truelove, 1986; Lynch, 1990; Hartmann et al., 2009). Soils are a major reservoir of biodiversity and host impressive densities of microbial species and cells, but many of these are only active provided that they are close enough to sources of fresh C substrates. This typically occurs in the rhizosphere as a consequence of rhizodeposition (Hinsinger et al., 2009), as well as in the drilosphere as a consequence of earthworm activities (mucus release for instance) or in the detritusphere, as a consequence of the burial of plant (or animal) residues in the topsoil (Brown et al., 2000). The delivery of fresh C compounds such as rhizodeposits into the soil can enhance the decay of older organic C compounds (i.e. soil organic matter) via the so-called priming effect which occurs as a consequence of the stimulation of microbial activity (Kuzyakov, 2002). Rhizodeposition thus has complex effects on the fate of soil organic matter and the global C cycle (Jones et al., 2009).

Many rhizosphere microbial activities that are stimulated by rhizodeposits have positive consequences on delivering nutrients to plants. This is especially true for the production of enzymes by rhizosphere microorganisms, which contribute to the hydrolysis of organic molecules containing nutrients (e.g. N, P or S) that would otherwise not be bioavailable to higher plants, although these can to some extent also release such enzymes (Richardson et al., 2009). In addition, some microbial activities are also implicated in the fate of inorganic compounds, as illustrated by the well-documented case of P-solubilising bacteria or fungi. The release of inorganic and organic acids by rhizosphere microbes can result in a dissolution of inorganic P compounds such as Ca phosphates, a capacity which seems to be distributed among many microbial genera in the rhizosphere (Rengel and Marschner, 2005; Richardson et al., 2009; Jones and Oburger, 2011). Another important effect is that of plant growth-promoting rhizobacteria (PGPR), which can also improve plant nutrition by indirect effects. For example, the stimulation of PGPR activities as a result of rhizodeposition can have positive feedbacks on root growth via hormonal signals (Faure et al., 2009), which ultimately enables the plant to access a greater pool of soil nutrients (Richardson et al., 2009). In contrast, there is not much evidence of competition for nutrients between plant roots and rhizosphere microorganisms, which would result in negative consequences of rhizodeposition on plant nutrition, although this may well occur when nutrient availability in soils is low (e.g. for iron especially in calcareous environments (Robin et al., 2008) and for ammonium or phosphate). Other major groups of rhizosphere microorganisms that are responsible for improving nutrient acquisition are microsymbionts (e.g. N_2-fixing rhizobacteria and mycorrhizal fungi). However, instead of competing for rhizodeposits with the many other rhizosphere microorganisms, these directly tap C compounds in the root, in the intimacy of symbiotic organs, root nodules and mycorrhizae. This long-lived coevolution of higher plants and soil microorganisms has resulted in highly efficient strategies to acquire nutrients (Brundrett, 2002; Sprent and James, 2007; Franche et al., 2009; Lambers et al., 2009), which is especially documented for N and P. Symbiotic N_2-fixation by Rhizobia legumes is a major component of the terrestrial N cycle (Vitousek et al., 2002; Gruber and Galloway, 2008; Herridge et al., 2008) and an underestimated process contributing to soil formation, which is a unique source of N input as N is almost absent in most rocks and thus in parent materials of soils (Richardson et al., 2004; Parfitt et al., 2005; Figure 10.1). The long soil chronosequences such as the Laupahoehoe site in Hawaii and the Franz Josef chronosequence in New Zealand clearly show that in their early stages of soil development (young soils), terrestrial ecosystems are primarily N limited, while at later stages of development (old soils) they become P limited (Vitousek and Farrington, 1997; Richardson et al., 2004; Parfitt et al., 2005). The role of mycorrhiza in P nutrition is also crucial as most plant species are mycorrhizal and recent reports have shown that even plants which do not positively respond to mycorrhizal infection (i.e. which do not show any improved growth when mycorrhizal) largely rely on mycorrhizal hyphae to acquire soil P (Smith et al., 2009). Indeed, once the mycorrhiza is established, the fungus turns off the P transporters that are located mainly in root epidermal cells and root hairs and turns on specific plant P transporters located in cortical cells where arbuscules are present, thereby forcing the plant to use the mycorrhizal pathway of P uptake, in exchange for C from the host plant (Rausch et al., 2001; Harrison et al., 2002; Smith et al., 2003; Bucher, 2006; Smith et al., 2009). Bypassing rhizodeposition, root symbionts nonetheless represent a significant C cost for the host plants (e.g. mycorrhiza in Figure 10.2 and Jones et al., 2004), and there are several reports of reduced growth as a consequence of mycorrhizal infection (Ryan et al., 2005; Smith et al., 2009), although such symbiosis results in positive effects in many cases.

For plant health, there is an extensive literature on both positive and negative effects of the stimulated microbial populations in the rhizosphere (reviewed by Raaijmakers et al., 2009). Indeed, rhizodeposition can attract both the 'good' (beneficial) and the 'bad' (deleterious, e.g. pathogenic) microorganisms, as they are competing for the same C sources, especially when the availability of fresh C compounds is limited. Understanding how rhizodeposition can alter or even direct the structure and composition of microbial communities has made considerable steps forward with the advent of molecular ecology, and there are many more clues on how plants can select some microbial populations at the expense of others (Faure et al., 2009; Hartmann et al., 2009). There is a sophisticated molecular dialogue operating between microbes and between plants and microbes which suggests that, beyond the amount of C contained in the rhizodeposits or their stoichiometric ratios (e.g. C/N, C/P), minute concentrations of very specific compounds (e.g. antibiotics and signalling molecules) may well be the key driver of rhizosphere microbial communities (Faure et al., 2009). Some of these compounds are volatile and thus may affect a large soil volume, as evidenced by the long-distance recruitment of entomopathogenic nematodes in maize (Rasmann et al., 2005; Turlings and Ton, 2006). Understanding and making use of such processes offers considerable promise for innovative applications in the biocontrol of below-ground pests (Turlings and Ton, 2006) and diseases (Raaijmakers et al., 2009), although the management of their complex ecology remains challenging in agroecosystem conditions.

10.2.4 Root and rhizosphere respiration

Besides organic C, plants also play an important role in the fate of inorganic C in soils, with potential impacts on the global C cycle. The production of CO_2 either directly via root respiration or indirectly via microbial respiration in the rhizosphere as a consequence of rhizodeposition has already been addressed in Section 10.2.3 in terms of contribution to the C budget (Figure 10.2). It will also form part of the next section for its contribution to soil acidification. Plant-induced soil respiration is a major contributor to the build-up of soil pCO_2 and concomitant decrease in pO_2, which have major impacts on soil biology and soil biogeochemistry (Hinsinger et al., 2009). Because of respiration of below-ground organisms, bulk soil pCO_2 is normally 10- to 100-fold higher than that of the atmosphere (currently 360 cm^3 m^{-3}). Values ranging from 7 000 to 24 000 (up to 32 000 under elevated atmospheric pCO_2) cm^3 m^{-3} have been reported in a forest soil (Karberg et al., 2005). It is, however, difficult to quantify how much of this was actually related to higher plant activities. Nevertheless, it has been estimated by various methods that the contribution of root respiration in eucalypt forest amounted to 48–70% of total soil respiration (Epron et al., 2006; Marsden et al., 2008), while the meta-analysis conducted by Hanson et al. (2000) showed a broader range (roughly 10–90%), with mean values consistently high in forest (46%) and non-forest ecosystems including crop-based agroecosystems (60%). Given that roots and rhizosphere microbes are major contributors to soil respiration, it would be expected that the rhizosphere should be a hot spot of elevated pCO_2 and decreased pO_2. However, this is little documented. Gollany et al. (1993) measured pCO_2 values of about 100 000 cm^3 m^{-3} at

1–3 mm from roots but also found only a gentle gradient of rhizosphere pCO_2 consistent with rapid diffusion of respired CO_2 towards the bulk soil. The same should also hold for pO_2 values, which might be expected to decrease in the rhizosphere, at least in aerobic conditions. Again, there are few measurements of pO_2 in the rhizosphere. Bidel et al. (2000) showed that pO_2 values in the rhizosphere of *Prunus persica* (L.) Batsch seedlings were much smaller in the meristematic region of the root as a consequence of intense metabolic activity and respiration. However, these data were obtained in an agar medium, which is known to be hypoxic, and thus hardly mimic the situation of well-aerated soils. In contrast, the case of wetland plants (and, among crops, lowland rice) growing in hypoxic soils has been extensively studied. To ensure adequate respiration of their root cells, those plants have evolved aerenchyma which conducts O_2 from the shoots to the roots (Armstrong, 1979). Rhizosphere pO_2 of such plant species thus increases as a consequence of O_2 leaking from roots (Flessa and Fischer, 1992; Revsbech et al., 1999; Armstrong et al., 2000; Blossfeld and Gansert, 2007). For lowland rice, Revsbech et al. (1999) reported that pO_2 was up to 20% of atmospheric pO_2 at the root surface, while being close to zero at distances greater than 0.4 mm from the root surface. This sharp gradient, which contrasts with the flat gradient of pCO_2, is the consequence of rapid consumption of O_2 in redox reactions occurring at the root surface, especially Fe oxyhydroxide precipitation in the so-called Fe plaque (Bacha and Hossner, 1977; Revsbech et al., 1999; Bravin et al., 2008). *In situ*, non-invasive sensors such as the optodes developed by Blossfeld and Gansert (2007) provide opportunities to acquire quantitative data on rhizosphere pO_2 and possibly pCO_2 as well (and monitor their temporal fate), in both anaerobic and aerobic environments (Blossfeld et al., 2010).

10.2.5 Role of plants in biogenic calcrete formation

Higher plants can leave long-lasting and visible imprints in soil profiles via the precipitation of biogenic $CaCO_3$, predominantly calcite, which ultimately forms calcretes (Freytet et al., 1997; Verboom and Pate, 2006a, b). Plants can thereby make a major and visible contribution to pedogenesis. A surprising case of pedogenic concretions induced by plants was that of calcretes observed by Braissant et al. (2003) and Cailleau et al. (2004, 2005) in tropical, acidic soils of Ivory Coast, that is, under environmental conditions which do not favour abiotic precipitation of Ca carbonate. The iroko (*Milicia excelsa*) trees and soil microorganisms in their vicinity are responsible for the formation of this mineral that would otherwise not occur in such soils (Cailleau et al., 2005). This process results from the large-scale formation of oxalate crystals in the wood tissues in both aerial parts and roots of the iroko trees (Cailleau et al., 2004, 2005), which are thereafter transformed into Ca carbonate by soil bacteria, which make use of only a small proportion of the oxalate for their C requirements (Braissant et al., 2002, 2003). Cailleau et al. (2004) calculated that such a process could be considered significant in the long term as a sink for atmospheric C. Braissant et al. (2002) even concluded that in the plant kingdom, this phenomenon of biologically induced mineralisation was a more efficient and longer term C sink than C sequestration in soil organic matter. Verboom and Pate (2006a, b) also report on the role of higher plants in the genesis of calcretes in acidic soil environments of Western Australia. In contrast with the previous examples, in calcareous environments biogenic precipitation of Ca carbonate as calcite in the rhizosphere has been frequently reported. It may occur first of all during the build-up of Ca which results from the large amount of Ca transferred towards the root surface as a consequence of plant-induced transpiration and subsequent mass flow of water and solutes (Jaillard, 1982; Callot et al., 1983; Hinsinger, 1998; Lambers et al., 2009). In

Figure 10.5 Calcified roots appearing as calcite crystals shaping the root cortex cells while the stele remains free of precipitates (Jaillard, 1987a). Photograph taken in between limestone plates in the topsoil of a natural ecosystem with typical Mediterranean 'garrigues' vegetation cover at St Gély-du-Fesc, close to Montpellier (South of France). The photographed zone is about 2 cm × 2 cm. Photo credit: Philippe Hinsinger. For a colour version of this figure, please see Plate 10.2.

calcareous soils, this process ultimately leads to the precipitation of calcite crystals around roots. If root density is large and the ambient conditions are conducive to such biogenic precipitation in the rhizosphere, the coalescence of the calcite crystals around neighbouring roots can result in continuous calcrete formation (Jaillard, 1982; Callot et al., 1983; Lambers et al., 2009). Another root-mediated process leading to biogenic precipitation of calcite in calcareous soils does not occur in the rhizosphere but inside the root itself (Jaillard, 1982, 1983, 1984, 1985; Jaillard et al., 1991). The resulting, so-called, calcified roots appear as calcite crystals shaping the root cortex cells, while the stele remains free of precipitates (Figure 10.5). This process occurs as a consequence of root activity, which can dissolve abundant amounts of calcium carbonate in the outer rhizosphere as a consequence of respiration and proton release (see Section 10.3), with the Ca ultimately precipitating as calcite in the vacuoles of root cortical cells (Jaillard, 1987a, b). Jaillard (1984) calculated that this biogenic process is responsible for a considerable shift in the particle size of calcareous soils as a consequence of the isometric size of the calcified root cells (about 80 μm diameter). Jaillard (1984) calculated that within about 1000 years, a third of all the Ca carbonate contained in the highly calcareous soil of the studied site (a natural ecosystem with typical Mediterranean 'garrigues' vegetation cover close to Montpellier, South France) could have been dissolved and re-precipitated by the roots. Such an estimate is in line with that of Kuzyakov et al. (2006) who, based on the measurement of ^{14}C incorporation in Ca carbonate derived from ^{14}C-labelled root exudates, calculated that full re-crystallisation of loess carbonate would take 400–2000 years. Contrary to the previous case of calcrete formation in a tropical soil, such dissolution/precipitation of Ca carbonate occurring in calcareous soils would not contribute any net significant C sequestration. They are nonetheless one of the few visible pieces of evidence of the large impact of plants on soil composition and properties (Lambers et al., 2009). Such rhizoconcretions have been observed in some of the oldest paleosoils described on Earth by Retallack (1997). Roots of higher plants can thus leave long-lasting morphological features in soil profiles and be a major contributor to pedogenesis via their direct role in the formation of biogenic Ca carbonates and calcretes (Freytet et al., 1997; Verboom and Pate, 2006a, b). To what extent such processes feedback onto plant growth remains to be studied.

10.3 Roles of plants in soil acidification

Respired CO_2 rapidly forms carbonic acid, H_2CO_3, which is a weak acid with a pK of 6.36. Plant-induced increase in pCO_2 thus results in a decrease in pH, except for those soils that are too acidic ($pH << 6.36$), that is, those soils for which H_2CO_3 remains essentially undissociated. In contrast, in calcareous soils, the actual pH can be quite different from the expected value of 8.3, usually determined in soil-testing procedures which are conducted at ambient atmospheric pCO_2, instead of being conducted at the actual *in situ* pCO_2 of the soil in field conditions. Indeed, soil pH in calcareous soils is constrained by the dissolution/precipitation equilibrium of $CaCO_3$, and thus pCO_2, according to the following equation (Gras, 1974; Hinsinger et al., 2003):

$$pH = -0.67\log(pCO_2) + 0.06 (\text{where } pCO_2 \text{ is expressed in atm}) \tag{10.1}$$

This suggests that for every tenfold increase in pCO_2, the corresponding equilibrium soil pH is expected to decrease by 0.67 pH units in calcareous soils. Based on reported *in situ* values of pCO_2 (Gollany et al., 1993), rhizosphere pH values are expected to be about 6.7–6.8 (Hinsinger et al., 2003), which is lower than the expected value of 8.3 (which would occur if soil pCO_2 was about the same as ambient pCO_2 in the atmosphere). At a global level, this relation explains why the current fast increase in atmospheric pCO_2 is resulting in the acidification of oceans with accompanying detrimental effects on ocean biogeochemistry, global geochemical cycles and aquatic ecosystems such as coral reefs (e.g. Rockström et al., 2009). Equation 10.1 shows that doubling the preindustrial atmospheric pCO_2 should induce a 0.2 pH unit decrease in sea water pH. In spite of the current attention on atmospheric pCO_2, it is surprising that there is so little data available about rhizosphere pCO_2 and its impact on below-ground organisms and biogeochemical cycles. Nevertheless, at the broad scale of ecosystems or watersheds, the contribution of (soil) respiration to acidification and proton budget is quite well documented and consistently shows that acidification rates are greater in soils of neutral to alkaline pH than in acidic soils (Van Breemen et al., 1983, 1984). Van Breemen et al. (1983) reported acidification rates in two oak forests in the Netherlands of up to 14.3 kmol H^+ ha^{-1} $year^{-1}$ in a calcareous environment (soil pH 7–8) but only about 4.8 kmol H^+ ha^{-1} $year^{-1}$ in an acidic environment (soil pH 3.7–3.9). However, the large buffering capacity of the calcareous soil masked this large acidification rate as a consequence of the consumption of protons by the dissolution of $CaCO_3$. As long as all the $CaCO_3$ is not completely dissolved, the soil pH hardly decreases, in spite of a large acidification rate (Van Breemen et al., 1983, 1984).

10.3.1 Role of plants in the proton budget and cycle

Van Breemen et al. (1983, 1984), Lelong et al. (1990) and Bourrié and Lelong (1994) showed that in addition to respiration, plants are largely responsible for the proton budget of terrestrial ecosystems via the control of the cation/anion balance (Table 10.2). In acidic soils, this is by far the major process controlling the acidification rate, with considerable variations according to the composition and productivity of the plant community. In forest ecosystems with acidic soils, cation uptake by the vegetation contributed 37–57% of the overall proton budget for beech, 25–64% for spruce and up to 81% for pine trees (Table 10.2). Lelong et al. (1990) and Bourrié and Lelong (1994) found that compared to tree-based ecosystems, alpine grassland ecosystems yielded smaller acidification rates, as a result of a smaller cation/anion balance (Table 10.2). This is also well documented in agroecosystems. For instance, in tropical pastures of varying composition and relying on various sources of N, Noble et al. (2008) reported over short-term periods (about 3 years) that the maximum acidification rate was obtained when the grass *Andropogon gayanus* was fertilised with ammonium-N, while the minimum rate was obtained when this grass was fertilised with nitrate-N. They also showed that in the absence of any fertiliser supply, the legume *Stylosanthes guianensis* fixing N_2 symbiotically had a much greater acidification rate than the grass (more than twofold larger), although not quite as great as when the grass was fertilised with ammonium-N. It is worth noting that in the latter case, the acidification rate was slightly greater than 16 kmol H^+ ha^{-1} yr^{-1}, which is about 1–3-fold greater than the values reported by Van Breemen et al. (1983) for temperate forest ecosystems and much above those reported in Table 10.2. The ammonium-fertilised grass decreased the bulk soil pH from 4.10 to 3.64 in only about 3 years of growth, while the N_2-fixing legume resulted in a bulk soil pH of 3.82 and the non-fertilised grass produced no significant pH change (Noble et al., 2008). These results of a short-term field experiment in Thailand are in agreement with a number of 'over-the-fence' comparisons made in Australia, which consistently showed that legume-based pastures resulted in greater acidification rates than grass-dominated pastures (Coventry and Slattery, 1991; Noble et al., 1997). This pH decrease in the acidic, poorly buffered soils of Australia caused subsequent problems of Al rhizotoxicity and nutrient deficiencies in crops. Acidification rates ranging from 0.2 to 10.6 kmol H^+ ha^{-1} $year^{-1}$ have been reported to occur (Ridley et al., 1990; Coventry and Slattery, 1991; Noble et al., 1997). Jarvis and Robson (1983a, b) clearly demonstrated that such acidification rates were largely explained by the cation/anion uptake imbalance as a result of N being largely taken up as an uncharged species (N_2) in the N_2-fixing legumes (see Section 10.3.2).

10.3.2 Root-induced changes of pH in the rhizosphere and their ecological relevance

At the rhizosphere scale, it is well established that protons are released by roots (hence rhizosphere acidification occurs) to compensate for an excess of cations over anions being taken up by plants (Nye, 1981; Haynes, 1990; Hinsinger et al., 2003). This is typically what occurs when the dominant form of N taken up is positively charged (ammonium), or to a lesser extent, when a major proportion of the N is taken up as the uncharged species N_2 in N_2-fixing legumes (see Section 10.3.1). In contrast, if the plant takes up more anions than cations, which would typically occur when the dominant form of N taken up is negatively charged (nitrate), rhizosphere alkalisation occurs. However, all of the major nutrient cations and anions actually taken up by the plant must be accounted for in order to predict the resulting

Table 10.2 Proton budget in watersheds corresponding to various ecosystems and dominant vegetation cover.

	Atmospheric deposition of H^+	**CO_2 input = respiration**	**Nitrogen cycle**	**Cation uptake**	**Anion uptake**	**Cation –anion balance**	**Cation release (weathering)**	**Anion fixation**	**Total**
Grass	324	609	25	198	–31	167	–805	–320	1156
Beech	324	250	–48	748	–59	689	–922	–293	1322
Spruce	324	472	62	280	–42	238	–1075	–21	1138

Positive values are proton sources, while negative values represent proton sinks, all values being expressed as mol H^+ ha^{-1} $year^{-1}$.
Source: From Lelong et al. (1990). Reprinted with permission from Elsevier BV/European Geophysical Society.

proton efflux from roots. Moreover, for those plant species responsible for large effluxes of charged exudates such as carboxylates, these should also be accounted for in the balance of ions crossing the plasma membrane of root cells (Hinsinger et al., 2003). Since the early work of Weisenseel et al. (1979) and Marschner and Römheld (1983), the use of pH-sensitive dyes combined with agar gel techniques has led to a large number of reports showing, in addition to the role of N source, (1) the plasticity of roots and their response to both Fe (e.g. Römheld, 1987; Vansuyt et al., 2003) and P deficiencies (e.g. Tang et al., 2004) and (2) the heterogeneity of such responses along the root system. The visual aspect of this technique is appealing, but it is essentially qualitative unless used in soil-less media such as agarose gel, when an adequate treatment of the signal can sometimes be achieved (Jaillard et al., 1996; Rao et al., 2002; Vansuyt et al., 2003). Recently, pH-sensitive optical sensors (the so-called optodes) suitable for noninvasive pH monitoring in soils have been successfully applied to map pH changes over time in plants grown in flat rhizotrons under laboratory conditions (Blossfeld and Gansert, 2007; Blossfeld et al., 2009). This promising technique has revealed the speed with which plant roots respond to changes in photosynthetic activity of the shoots as rhizosphere acidification developed very quickly after the onset of light in the wetland plant *Juncus effusus* (Figure 10.6). In such plants that grow their roots in anaerobic conditions, it has been well established that the root-induced release of O_2 in the rhizosphere (as a consequence of leakage from the aerenchyma), and the subsequent oxidation of the soil, can make a substantial, and sometimes major contribution to the observed rhizosphere acidification, as also shown for lowland rice (Begg et al., 1994; Kirk and Le Van Du, 1997).

Locally, root-induced proton efflux or influx can result in large decreases or increases in rhizosphere pH, respectively, reaching commonly ±0.1 to 1 pH units and in some instances up to ±2 to 3 pH units (Hinsinger et al., 2003). There has been an abundant literature referring to rhizosphere pH changes since the seminal paper by Riley and Barber (1971), which reported on substantial pH variations in the rhizosphere of pot-grown soybean supplied with either nitrate or ammonium, and on their consequences for P bioavailability. Many of the published reports measured such pH changes in rhizobox-grown plants (i.e. more or less artificial settings) with various approaches and more or less spatially resolved and invasive techniques that have been discussed in detail by Jaillard et al. (2003) and more recently by Luster et al. (2009) and Neumann et al. (2009). Several of these works have produced quantitative data on pH gradients occurring in the rhizosphere, and their consequences for the availability of soil P (Gahoonia et al., 1992; Hinsinger and Gilkes, 1995, 1996, 1997; Zoysa et al., 1997). Most of these have reported on root-induced acidification patterns, although Gahoonia et al. (1992) also obtained an upward gradient in rhizosphere pH in ryegrass supplied with nitrate, which interestingly led to less P being depleted from the rhizosphere in the studied luvisol and to more P being depleted in the studied oxisol. The luvisol had a neutral bulk soil pH of 7.4, and the maximum pH increase recorded in the rhizosphere was +0.6 unit, while a 1.5 unit decrease was found when ammonium was supplied. In the oxisol, which had an acidic bulk soil pH of 5.5, similar pH changes were found as the maximum pH increase recorded in the rhizosphere was +0.5 unit for nitrate-supplied ryegrass, while a 1.4 unit decrease was found when ammonium was supplied.

As predicted by the model of Nye (1983), due to the faster diffusion rate of protons compared with phosphate ions, complex patterns of available P can occur in the rhizosphere, with some accumulation of dissolved P occurring at some distance from the roots instead of a continuous depletion gradient (e.g. Hübel and Beck, 1993; Hinsinger and Gilkes, 1996). The rapid diffusion of protons in the rhizosphere can thus lead to more extensive effects than one would first think of for pH-sensitive, poorly mobile nutrients such as P and some

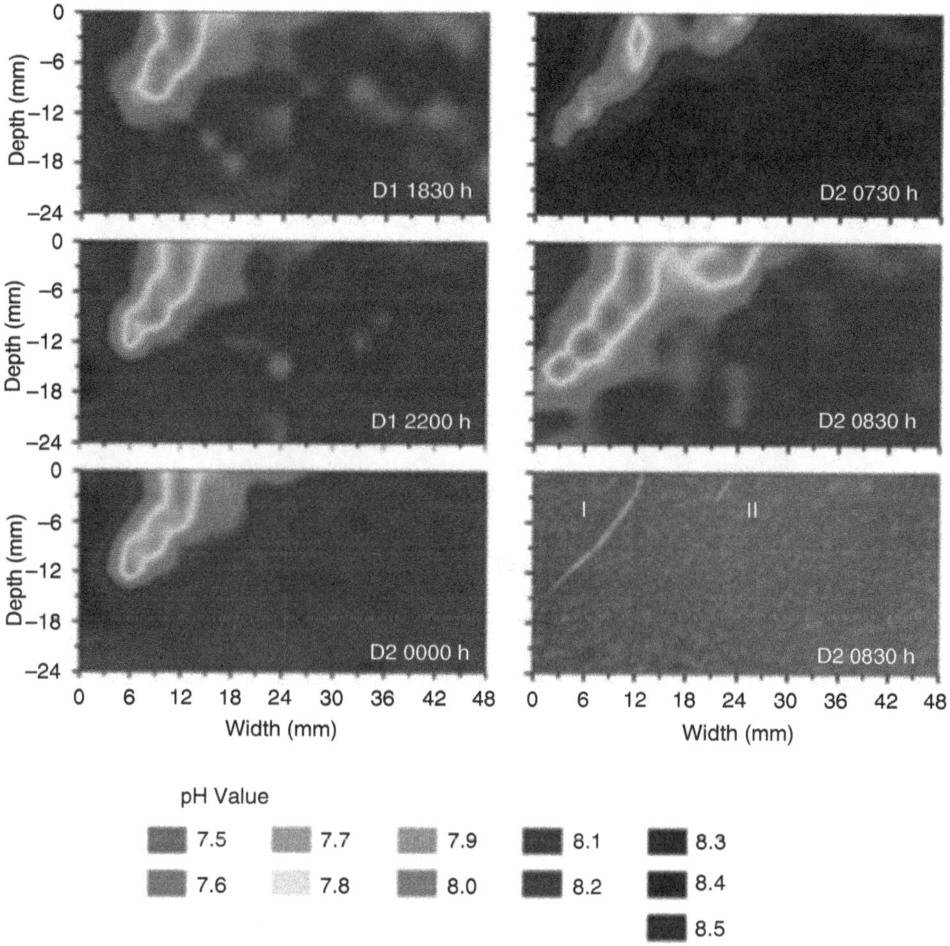

Figure 10.6 Temporal development of root-induced pH changes in the rhizosphere of growing roots of *Juncus effusus*, half an hour before and after the start of daylight, as measured noninvasively with a planar pH optode. The digital photograph (bottom) shows the two investigated roots. Reprinted from Blossfeld and Gansert (2007). With kind permission from Blackwell Publishing Ltd. For a colour version of this figure, please see Plate 10.3.

micronutrients. For example, the work of Bravin et al. (2009a) showed that while Cu uptake models predicted sharp and narrow depletion zones with a computed gradient extending to <0.5 mm in the rhizosphere of durum wheat, the observed Cu gradient extended to about 6 mm from the root mat surface. Bravin et al. (2009a) demonstrated that this decrease in soil solution Cu concentration (and even more so for Cu^{2+}) was a consequence of the root-induced alkalisation of up to 2.8 pH units close to roots and which extended up to 6 mm. In this acidic Cu-contaminated soil, such root-induced pH change enabled the plant to reduce the toxic concentration of Cu found in the bulk soil (Bravin et al., 2009a) and ultimately the bioavailability of Cu (Bravin et al., 2009b; Figure 10.7). This process is likely to be of ecological relevance for the adaptation of plants to acidic soils where Al toxicity may otherwise considerably reduce root and hence plant growth (Bagayoko et al., 2000; Hinsinger et al., 2003, 2009). The artificial setting of rhizobox experiments used to obtain these results may

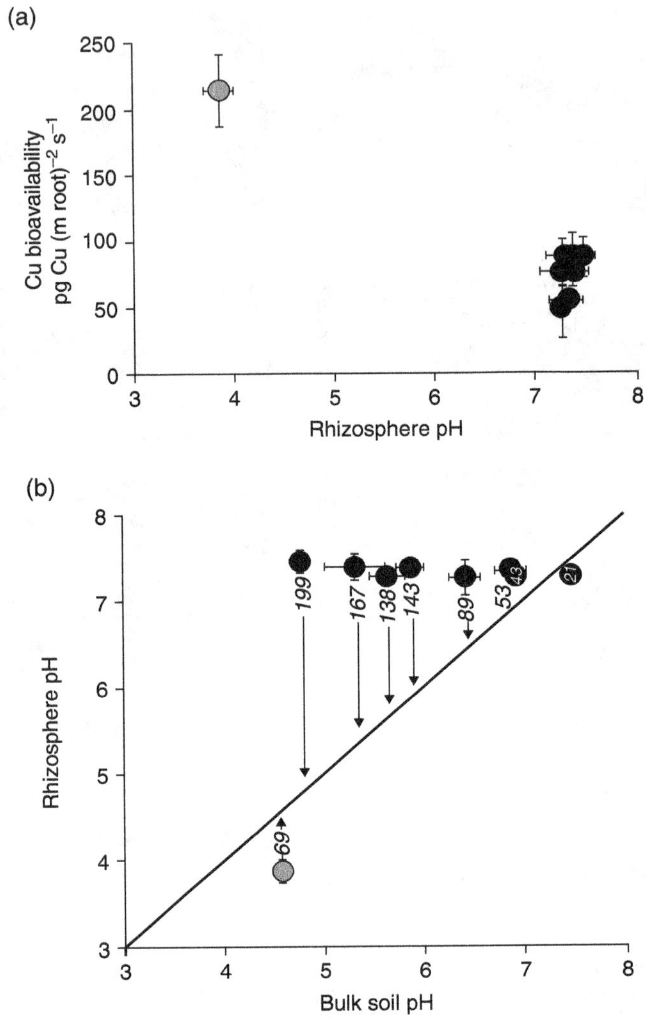

Figure 10.7 (a) Copper bioavailability (influx of Cu into the plant) as a function of rhizosphere pH in durum wheat supplied with either nitrate only (black symbols) or ammonium/nitrate (grey symbol) and (b) rhizosphere pH as a function of bulk soil pH (the range of bulk soil pH values was obtained by lime addition at various rates in an acidic soil for plants grown with nitrate only). The figures stand for OH^- efflux (for points above the y = x axis, corresponding to rhizosphere alkalisation) or H^+ efflux (for points below the y = x axis, corresponding to acidification). Based on unpublished data and data from Bravin et al. (2009b). Reprinted with kind permission from Springer-Verlag.

be thought to lead to large overestimates of the rhizosphere gradients in the vicinity of dense root mats instead of single roots. However, modelling has shown that this is not the case (Bravin et al., 2009a). In addition, pH increases of about one pH unit have also been reported to occur in the rhizosphere of field-grown durum wheat plants (1–5 mm aggregates adhering to roots were sampled for this purpose) in the same acidic soil (Michaud et al., 2007). Sampling field-grown plants, Bagayoko et al. (2000) also reported a substantial pH increase in the rhizosphere of maize in acidic oxisols and suggested that this would have enabled the plants to alleviate Al toxicity and nutrient deficiencies.

Compared with rhizobox experiments in artificial laboratory conditions, there are fewer reports of pH changes in the rhizosphere of field-grown plants, with most such reports on natural or forest ecosystems (Häussling et al., 1985; Marschner et al., 1991; Clegg et al., 1997; Schöttelndreier and Falkengren-Grerup, 1999; Séguin et al., 2004; Turpault et al., 2007). Many of these report slight rhizosphere alkalisation in acidic (forest) soils. In arable soils, there are few reports on changes of pH in the rhizosphere of crops. Yang et al. (1996) reported systematic acidification of the rhizosphere of field-grown soybean, with more acidification associated with lateral roots than primary roots. They showed that such root-induced acidification was positively correlated with the pH of the bulk soil, so that while the pH change was almost nil for a bulk soil pH of 4.0, the pH change increased (in absolute value) with increasing bulk soil pH. This result is consistent with theory which shows that the buffering capacity of soil solution is lower near neutrality (Nye, 1981). It may also reflect the plasticity of roots which enables plants to cope with adverse soil (pH) conditions. For example, Youssef and Chino (1989) showed in rhizobox experiments that barley and soybean were able to increase their rhizosphere pH when grown in an acidic soil and conversely acidify their rhizosphere when grown at higher pH. More convincingly, in a single soil and a broad range of bulk soil pH obtained by various rates of lime addition, Chaignon et al. (2009) showed that oilseed rape was capable of slightly alkalising its rhizosphere when grown at bulk soil pH lower than 4.7 and of acidifying its rhizosphere at greater pH values. Similarly, Bravin et al. (2009b) demonstrated that durum wheat systematically increased its rhizosphere pH to a constant value of about 7.3 whatever the initial pH, unless supplied with ammonium (Figure 10.7). In this case, the plasticity resides in the ability of roots to adjust the proton influx according to the initial pH, as increasing the pH from 4.8 to 7.3 in this soil resulted in a fourfold greater proton influx (or hydroxyl efflux) than for increasing the pH from 6.9 to 7.3. Remarkably, this behaviour of durum wheat roots, which results in increasing alkalisation of the rhizosphere with decreasing pH, was also found in field-grown plants (Michaud et al., 2007).

10.3.3 Role of organic acids in rhizosphere pH changes

There has been a considerable amount of work on the so-called organic acids in the rhizosphere, and excellent reviews have been published on the subject including Jones (1998), Ryan et al. (2001) and Jones et al. (2003). These reviews clearly show that these should more correctly be referred to as organic anions or carboxylates (as most of these are derived from carboxylic acids) because they are exuded as anions and thereby do not result in specific pH changes *per se* (Hinsinger et al., 2003; Jones et al., 2003). They should, however, be accounted for in the anion/cation balance, at least when released in substantial amounts. This release varies considerably among plant species and is generally low in crop species with the exception of white lupin for which citrate and malate (among others) exudation from cluster roots is considerable (as reviewed by Neumann et al., 1999, 2000). The loss of negative charge which results from the release of carboxylates is counterbalanced by the efflux of an equivalent amount of either protons or K-ions (in the latter case, however, this means that protons have been released at a preliminary step to counterbalance for K uptake), hence contributing to rhizosphere acidification unless there is a net excess of anions taken up when accounting for the overall balance of charges crossing the root cell membranes. Indeed, it is well documented for cluster roots of white lupin that they result in a considerable release of both carboxylates (notably citrate) and protons and hence in distinct rhizosphere acidification (e.g. Neumann and Römheld, 1999). This may not be the case for other plants, especially crops which release much smaller amounts of carboxylates, and even rhizosphere alkalisation has sometimes been reported to occur in the rhizosphere of other cluster-rooted plants such as members of the Proteaceae (e.g. Roelofs et al., 2001). These carboxylates play many roles in nutrient acquisition, especially P (and to a lesser extent Fe), and in the alleviation of metal toxicities, as especially documented for the case of Al (Pellet et al., 1995; Ryan et al., 1995a, b, 2001, 2009; Kochian et al., 2004). This is due to their capacity to complex and for some of them chelate metal cations such as Ca-, Fe- or Al-ions (Gerke et al., 2000a, b) and to be involved in ligand exchange processes and thus in the desorption of P-ions (Geelhoed et al., 1999) in the rhizosphere (e.g. Jones, 1998; Hinsinger, 2001). Carboxylates released by roots, such as citrate, or produced by rhizosphere microorganisms and fungi such as some ectomycorrhizal fungal species have thus major implications for plant nutrition and in the weathering of soil minerals, but they play little, and only an indirect, role in root-induced soil pH changes compared with ion uptake and respiration (Hinsinger et al., 2003).

10.4 Roles of plants in the weathering of soil minerals

At the beginning of pedology, Dokuchaev and later Jenny pointed out that, besides climate and rock, vegetation is a major factor involved in soil formation and the diversity of soil types that can be found worldwide. With the development of experimental pedology in the 1950s (Bockheim and Gennadiyev, 2009), several studies showed the role of biochemical compounds and microbially mediated processes in soil formation. Biogeochemistry nevertheless remained essentially confined to geochemistry for a long time, although it was obvious to anyone dealing with the C cycle that higher plants were key players (see Section 10.2.1). At geological temporal scales, the emergence of deep-rooted plants during the Devonian coincided with deep soil profile development and the acceleration of rock weathering and soil formation rates in terrestrial ecosystems (Berner, 1992, 1997; Retallack, 1997; Beerling et al., 2001; Beerling and Berner, 2005; Taylor et al., 2009). Interestingly, Taylor et al. (2009) stressed that the five major plant-induced processes of biological weathering (Table 10.3) were 'driven by plant photosynthesis fuelling plant growth and organic C release into the regolith and soils'. These five key processes, according to Berner et al. (2003) and Taylor et al. (2009) are (1) exudation, (2) respiration, (3) decomposition of plant litter, (4) evapotranspiration and (5) physical stabilisation of soils by root systems and plant cover. Although these authors stress that evapotranspiration plays a role via its impact on the flow of cations to plants, the uptake and translocation of cations and other elements by plants must be considered as another major plant-induced process of rock weathering and soil formation. This will be emphasised in what follows.

The physical effects induced by plants are beyond the scope of the present chapter which will concentrate on plant-mediated chemical weathering. Briefly, the plant-driven transpiration of water by plants directly affects the soil water residence times in numerous ways. Plants favour infiltration at the expense of surface runoff because plant roots create macropore-sized biopores and plant debris creates roughness at the soil surface and decreases the formation of surface crusts. At a local scale, the uptake of water by roots decreases the water content of the soil, which may ultimately reduce the weathering of minerals, given the necessary presence of water for such chemical reactions to occur to any significant extent. At a regional scale, plant communities can significantly affect the amount of rainfall (e.g. Pielke, 2001; Jonko et al., 2010). In addition to and/or combined with water uptake, during

Table 10.3 Major processes of weathering of rocks and minerals induced by plants.

		Chemical processes		
	Physical processes	**Ion concentration changes**	**pH changes**	**Complexation**
Exudation of protons/low molecular weight organic acids			+++	+++
Respiration-induced increase in pCO_2			+++	
Deposition of litter comprising organic matter and cations		+++		+++
Evapotranspiration	+++	+++		
Fragmentation and physical binding of particles/aggregates	+++			

Source: Adapted from Berner et al. (2003) and Taylor et al. (2009).

their growth, and especially their radial growth, roots may contribute to the physical breakdown of rocks via (1) the formation of cracks or expansion of pre-existing cracks where roots proliferate and (2) the promotion of mechanical fractures in mineral grains (April and Keller, 1990; see Chapter 8). At the landscape or slope scale, root systems are key to the stabilisation of thick soil horizons by mechanically preventing processes such as landslides and erosion by surface or subsurface runoff water, as reviewed by Stokes et al. (2009). Such physical processes are linked with the physical properties of the root systems, by which plants contribute to deep soil profile formation and stabilisation. Plant-mediated chemical weathering of rocks and minerals can be subdivided into three major biogeochemical processes, which ultimately shift the chemical equilibria underlying chemical weathering processes, or alter the kinetics of such reactions (Table 10.3), by (1) changes in ionic concentrations as a consequence of water and ion uptake by roots or deposition of cations by throughfall and the decay of organic debris; (2) pH changes as a result of respiration, root-induced efflux or influx of protons or microbially mediated production of organic acids in the rhizosphere and (3) complexation or chelation of metal cations by root exudates or microbial metabolites in the rhizosphere.

10.4.1 Cation uptake by vegetation

Plants are major drivers of changes in ionic concentrations in soil solution which ultimately shift the dissolution/precipitation equilibria and either increase or decrease the kinetics of such processes. The uptake of water and inorganic ions (mineral nutrients, and also non-essential elements such as Si) at high fluxes by plant roots is responsible for important mass exchanges in the rhizosphere, which can drastically modify ionic concentrations in the soil solution and shift the chemical equilibria of reactions involving these ions (Hinsinger, 1998; Hinsinger et al., 2009; Lambers et al., 2009). Hinsinger and Jaillard (1993) demonstrated that the considerable decrease in K concentration occurring in the rhizosphere of ryegrass as a consequence of the sink effect of absorbing roots was directly responsible for the weathering of K-bearing phyllosilicates such as trioctahedral micas. This has been reported by several authors (Mortland et al., 1956; Spyridakis et al., 1967; Hinsinger et al., 1992; Kodama et al., 1994). Hinsinger and Jaillard (1993) and Hinsinger et al. (1992) showed that the root-induced depletion of soil solution K was the driving force for a rapid release of interlayer K

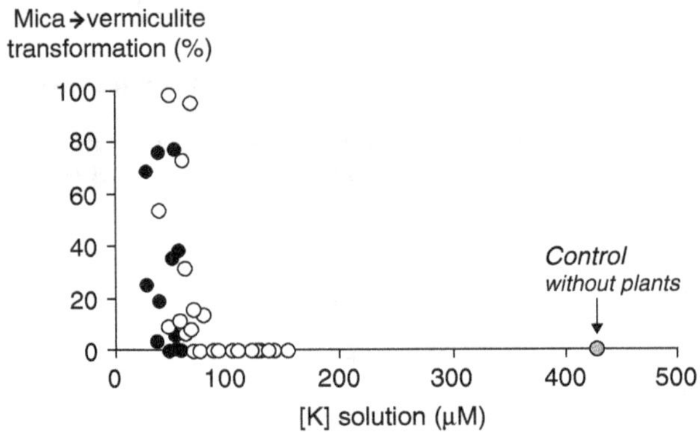

Figure 10.8 Extent of the root-induced transformation of phlogopite mica into vermiculite clay mineral (according to X-ray diffraction) in the rhizosphere of Italian ryegrass (white symbols) and oilseed rape (black symbols) as a function of solution K concentration. Based on unpublished data and data from Hinsinger and Jaillard (1993). Reproduced with permission from John Wiley & Sons.

in the mica and the concurrent formation of the clay mineral vermiculite. Hinsinger and co-workers showed that as soon as the K concentration in the rhizosphere decreased below a threshold value of about 70–80 μM (Figure 10.8), the weathering of the phlogopite mica became detectable by X-ray diffraction (XRD) through the appearance of a typical vermiculite peak forming at the expanse of the mica-characteristic peak. This process took only about two days for ryegrass (Hinsinger and Jaillard, 1993) and three days for oilseed rape (Hinsinger et al., 1993). Hinsinger et al. (2006) later reported that the uptake of K by plant roots led to a five and sixfold increase in the rate of release of interlayer K (Table 10.4) and thus in the rate of weathering of micas such as phlogopite and biotite, respectively, compared to unplanted systems where leaching resulted in only limited weathering over the 14-day experimental period. In the absence of plants, no significant vermiculite formation was detected by XRD, whatever the mica. No significant, root-induced transformation of a dioctahedral mica, muscovite, was found (Table 10.4), although the K budget suggested that some slight weathering had occurred. In contrast, a large proportion of the trioctahedral micas (phlogopite and even more for biotite) had been transformed into vermiculite in the rhizosphere (Hinsinger et al., 2006).

These results were obtained in the laboratory under rather artificial, controlled conditions. They are, however, in agreement with those computed by Taylor and Velbel (1991) and Velbel and Price (2007) who showed that taking account of the uptake of K by the vegetation of forested watersheds led to biotite weathering rates (432 mol K ha^{-1} $year^{-1}$) which were up to 3.5 times greater than when neglecting this term in the K budget (124 mol K ha^{-1} $year^{-1}$). In agroecosystems, accounting for such a rhizosphere process, which is the direct consequence of the uptake activity of plant roots, proved essential to understanding the observed K budgets in long-term fertiliser trials conducted by agronomists (Hinsinger, 2002). Indeed, most of the field trials conducted in Europe have shown that, in the absence of fertiliser K application, the cumulative K offtake by successive crops in the rotation was much greater than the observed decrease in exchangeable K, although only this pool was considered as bioavailable to crops (Hinsinger, 2002). Based on the difference, it can be calculated that the rates of release of nonexchangeable K amounted from 11 to 94 kg K ha^{-1} $year^{-1}$ (Table 10.5; i.e. 281–2404 mol K ha^{-1} $year^{-1}$) as an average value computed over the whole duration of these long-term (several decades) fertiliser trials (Hinsinger, 2002). This

Table 10.4 Rates of weathering of three micas in an experimental weathering experiment in which the micas were leached with a K-free nutrient solution for 14 days in the presence or absence of ryegrass.

		K release (% of initial K) [Log weathering rate] (mol K (g mica)$^{-1}$ s^{-1})	**Vermiculitisation (% – based on CEC increase)**	**Vermiculitisation (% peak surface area in X-ray patterns)**
Muscovite mica	Control (no plants)	2 [−9.93]	1–2	0
	Rhizosphere (ryegrass)	4 [−9.67]	3	0
Phlogopite mica	Control (no plants)	3 [−10.08]	5–6	0
	Rhizosphere (ryegrass)	20 [−9.32]	24–27	≈20
Biotite mica	Control (no plants)	6 [−10.00]	8–9	0
	Rhizosphere (ryegrass)	38 [−9.20]	35–40	≈40

The reactor was designed such that the roots had developed a planar root mat at the upper surface of a dialysis bag containing the particles of mica. The leachate was continuously removed by gravity, so that this device shall be considered as an open-flow system.
Source: From Hinsinger (1990) and Hinsinger et al. (2006).

means that, under such low input conditions, the release of nonexchangeable K contributed a major proportion (up to 90%) of K uptake by crops (Hinsinger, 2002), which is similar to results obtained with short-term experiments (4–8 days) conducted in pots in which K budgets were measured at the rhizosphere scale (Kuchenbuch and Jungk, 1982; Niebes et al., 1993). This major contribution of the release of nonexchangeable K was surprising at first sight as it has often been considered that this process should be negligible in such arable soils, because of bulk soil solution K concentrations typically ranging from 100 to 1000 μM. The release of nonexchangeable K is mostly due to the release of interlayer K and concomitant weathering of soil micas and micaceous clay (illite-like) minerals and is governed by soil solution K concentration (Springob and Richter, 1998). These authors showed that while the rate of K release was considerably enhanced at soil solution K concentrations below 3 μM, it was almost nil at those concentrations commonly found in most soils of arable land in Europe. Claassen and Jungk (1982) estimated that while bulk soil solution concentrations of K were in the order of several hundreds of μM, in the immediate vicinity (about 1 mm) of maize roots, K concentrations could be as low as 2–3 μM; this explains the large rates of release of nonexchangeable K occurring in the rhizosphere (Kuchenbuch and Jungk, 1982; Niebes et al., 1993) in the short term and at the field scale in the long term (Bosc, 1988; Blake et al., 1999; Hinsinger, 2002).

The sink effect of plant roots and the subsequent depletion of soil solution K in the rhizosphere is thus the driving process for the release of interlayer K and concomitant weathering of micaceous minerals in soils (Hinsinger, 2002). Barré et al. (2007a) further proved the causal link between these two processes, based on a quantitative method of assessment of the mineralogical composition of the soil clay fraction derived from a deconvolution of X-ray diffraction patterns. These authors showed that the uptake of K by ryegrass in a pot experiment matched the increased amount of interstratified illite-smectite minerals in the rhizosphere, which formed at the expense of the illite-like clay minerals. These findings confirmed the earlier report of an increased amount of interstratified illite-vermiculite minerals in the rhizosphere of field-grown maize plants (Kodama et al., 1994). Barré et al. (2008) applied

Table 10.5 Potassium budgets and derived estimates of average, annual K release rates in 12 long-term fertiliser trials throughout Europe.

Field trial location (Country)	Duration of trial (years)	Annual K budget = $K_{fert.} - K_{offtake}$ (kgK ha^{-1} $year^{-1}$)	Annual change in exchangeable K (kgK ha^{-1} $year^{-1}$)	Annual K release rate (kgK ha^{-1} $year^{-1}$)
Clermont (France)	30	−105	−11	94
Versailles 2 (France)	18	−87	1	88
Bad Lauchstaedt 2 (Germany)	33	−74	7	81
Bad Lauchstaedt 1 (Germany)	33	−70	4	74
Versailles 1 (France)	32	−66	−4	62
Fjärdingslöv (Sweden)	31	−68	−3	65
Amiens 1 (France)	22	−69	−17	52
Ekebo (Sweden)	31	−54	−3	51
Skierniewice 1 (Poland)	32	−48	0	48
Kungsängen (Sweden)	36	−48	−3	45
Amiens 2 (France)	21	−49	−7	42
Vreta Kloster (Sweden)	30	−58	−18	40
Skierniewice 2 (Poland)	32	−38	0	38
Rothamsted Broadbalk 1 (UK)	26	−28	0	28
Rothamsted Broadbalk 2 (UK)	30	−23	0	23
Toulouse (France)	11	−27	−16	11
Högåsa (Sweden)	30	−25	−18	8

The rate was deduced from the difference between K budget (amount of K applied as fertilisers minus amount of K exported in harvested products) and the change in exchangeable K (negative values indicate a decrease since the start of the trial) cumulated over the whole duration of the trial, divided by this duration in order to get comparable results (average annual rate).

Source: Data compiled from Bosc (1988) for the French trials and Simonsson et al. (2007). Reprinted with permission from Elsevier BV/International Society of Soil Science for the Swedish trials and Blake et al. (1999). Reprinted with permission from Kluwer Academic Publishers, Dordrecht, for other European locations.

the same methodology to soil clay fractions collected in a long-term K fertiliser trial and confirmed that the uptake of K by plants quantitatively matched the release of interlayer nonexchangeable K that was evident by the formation of interstratified illite-smectite minerals at the expense of the illite-like clay minerals. Barré et al. (2007b, 2008) concluded that higher plants have a major role in the biogeochemical cycle of K and in the formation and fate of expandable clay minerals (vermiculites or smectites) in topsoils, with illite-like clay minerals playing the role of a large K reservoir in soils of temperate regions of Europe. As stressed by Hinsinger (2002) and Hinsinger et al. (2006), no such release of interlayer K contained in micaceous minerals would be expected to occur in the absence of plants and the accompanying sink effect of their roots resulting in considerable depletion of soil solution K in the rhizosphere. The corresponding rates of release of about 10–100 kg K ha^{-1} $year^{-1}$ found in long-term fertiliser trials throughout Europe (Table 10.5), either in intensive cropping systems (Bosc, 1988; Blake et al., 1999; Hinsinger, 2002) or in pasture systems (Holmqvist et al., 2003), were sometimes well above those reported by geochemical models based on biotite weathering in forested ecosystems (Taylor and Velbel, 1991; Velbel and Price, 2007). The latter authors computed K release rates of about 4–5 kg K ha^{-1} $year^{-1}$ when neglecting the offtake by the vegetation and up to 17 kg K ha^{-1} $year^{-1}$ when accounting for the offtake by forest trees. In addition to playing a key role in the weathering of micaceous minerals, such root-induced release of nonexchangeable K is of major importance for plant nutrition, covering the requirements of many crops in many soils, except in the most weathered soils of the tropical regions. This should be taken into account when designing K fertilisation strategies for the future, considering that K ores are finite resources (Manning, 2010).

10.4.2 Cation uplift in the soil profile as a consequence of plant activities

The concept of nutrient uplift was first put forward by Jobbagy and Jackson (2001, 2004). One of the most dramatic effect of plant-induced uplift is that reported for Si (Lucas et al., 1993; Lucas, 2001), which is a non-essential element for most plant species, although it is sometimes taken up at rates exceeding those of major nutrients (Epstein, 1999). Lucas et al. (1993) reported that in tropical forest ecosystems of Amazonian Brazil, Si was taken up from depth in soil profiles and subsequently deposited in the topsoil as a consequence of litterfall. This process explained the vertical distribution of Si and clay minerals in soil profiles. While geochemical models normally predict that the most weathered minerals should be found in the topsoil, a number of tropical soils typically exhibit a kaolinite-rich topsoil overlying gibbsite-rich horizons (as, e.g. in the study of Lucas et al., 1993). These authors demonstrated that the annual flux of Si to the topsoil by litterfall amounted to 41 kg Si ha^{-1} $year^{-1}$ while Si leaching at depth was about fourfold smaller, amounting to 11 kg K ha^{-1} $year^{-1}$. This shows that plant-promoted uplift of Si played a prominent role in Si biogeochemical cycling. The accumulated Si in the topsoil most probably occurred as phytoliths which later evolved in Si-rich clay minerals such as kaolinite via dissolution/precipitation processes. Alexandre et al. (1997) observed a considerable increase in Si content in topsoil horizons occurring as phytolites, thus clearly demonstrating the role of higher plants in such biogeochemical uplift of Si. Higher plants are thus now considered to play a key role in the terrestrial biochemical cycle of Si (Basile-Doelsch et al., 2005; Derry et al., 2005) largely via uptake and sometimes via active Si uptake (Gérard et al., 2008).

While the work described in Section 10.4.1 showed that higher plants are major drivers of the weathering of K-bearing minerals and rocks as a consequence of root-induced processes occurring in the rhizosphere, they are also responsible for considerable redistribution of K in soil profiles. Indeed, while K uptake occurs throughout the profile, with substantial contribution of the subsoil horizons (Kuhlmann, 1990; Witter and Johansson, 2001; Buxbaum et al., 2005), a substantial part of K accumulated in the vegetation (all of the K which is not removed with the harvested products) is recovered in the topsoil as a consequence of throughfall and litterfall, as quantified by Barré et al. (2007b, 2009) based on clay mineralogy and K budgets. Barré et al. (2009) and Velde and Barré (2010) stressed that such nutrient uplift can counteract cation leaching from surface soils (i.e. what is often considered as a major driving force of meteoric weathering and soil formation processes). Their measurements and calculations in temperate grassland soils suggest that K uplift by the vegetation may ultimately affect clay mineral stability in upper soil horizons (Barré et al., 2007b, 2009). While weathering is often considered to result in a decrease in soil fertility in the long term, due to nutrient cation leaching from the most weathered horizons, and especially the topsoil, biological activities of plants can conversely improve topsoil fertility through this uplift of nutrients such as K (Barré et al., 2009; Velde and Barré, 2010).

10.4.3 Plant-driven pH shift

Because the pH of interstitial water in rocks and soil solutions is a major driver of the weathering of minerals and rocks, root-induced pH changes in the rhizosphere can have big effects on such soil-forming processes. As reported by Dehérain (1892) in his Treatise of Agricultural Chemistry, an experiment conducted by Sachs in the nineteenth century had already demonstrated such a phenomenon. In this experiment, roots of beans were grown over the surface of a polished marble plate, which ultimately left clearly visible imprints in the rock (see also Trolldenier, 1987). The authors concluded that roots secreted an acid strong enough to dissolve Ca carbonate. More remarkably, such imprints of root systems have also been observed *in situ* (Figure 10.9) for plant roots growing between cracks in dense limestone plates (Jaillard and Hinsinger, 1993; Hinsinger et al., 2003). In the first half of the twentieth century, such acidic root secretions were attributed to carbonic and organic acids produced by rhizosphere microorganisms and roots through respiration and exudation. As explained in Section 10.3.2, it is now well known that such processes are primarily the consequence of excess uptake of cations over anions (possibly accompanied by an exudation of carboxylates), which is counterbalanced by proton efflux and rhizosphere acidification (Hinsinger et al., 2003). In calcareous environments such as in the case depicted in Figure 10.9, the contribution of rhizosphere respiration is another potentially significant cause of such dissolution of Ca carbonates around living roots (Jaillard, 1987a, b). In such soils where the alkaline bulk soil pH would otherwise be a major constraint for plant growth via its negative impact on the bioavailability of a number of nutrients, especially phosphorus (P) and metal micronutrients such as iron and zinc, rhizosphere respiration as well as local release of protons at root tips and the consequent decrease in rhizosphere pH are crucial survival strategies for plants (Hinsinger et al., 2003, 2009).

Poorly soluble minerals and rocks have also been reported to be dissolved as a consequence of rhizosphere acidification (Figure 10.10). The literature is particularly abundant for Ca phosphates and apatitic phosphate rocks because of the significance of their dissolution for plant nutrition and the biogeochemical cycle of P. The root-induced efflux of protons shifts the dissolution/precipitation equilibrium of such minerals and rocks according to the mass action law, thereby enhancing their dissolution (Khasawneh and Doll, 1978; Kirk and Nye, 1986) and ultimately the bioavailability of P (Figure 10.10). The ability of some plant

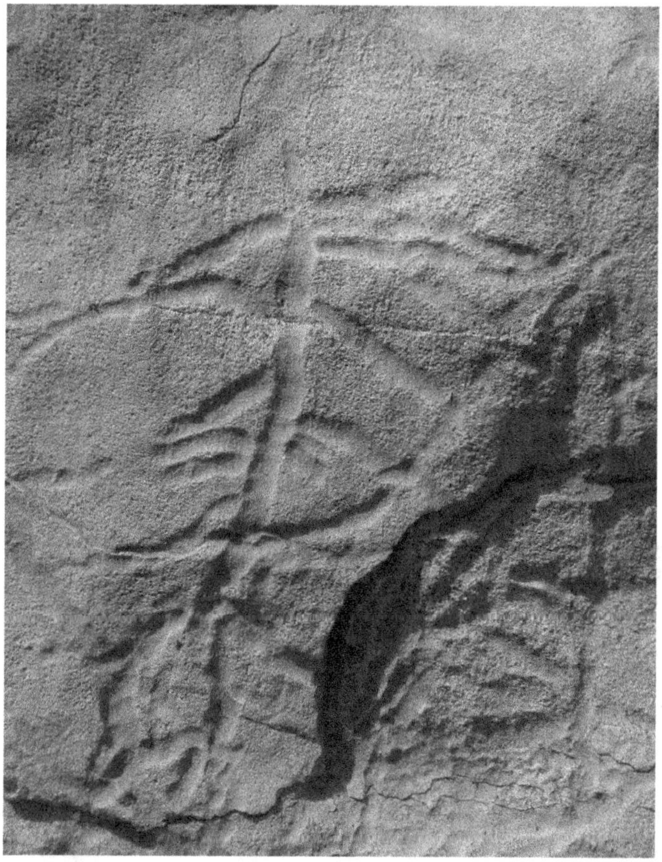

Figure 10.9 Imprints of root systems observed *in situ* for plant roots growing in between cracks in dense limestone plates (Hinsinger et al., 2003). Reproduced with permission from Springer-Verlag. Photograph taken in a sample collected in the topsoil of a natural ecosystem with typical Mediterranean 'garrigues' vegetation cover at St Gély-du-Fesc, close to Montpellier (South of France). The photographed zone is about 5 cm × 7.5 cm. Photo credit: Philippe Hinsinger. From Jaillard and Hinsinger (1993). Reproduced with permission from John Wiley & Sons. For a colour version of this figure, please see Plate 10.4.

species such as buckwheat, oilseed rape and various legumes to utilise P when supplied as a phosphate rock (i.e. francolite-like apatites) has been attributed to their capacity to release protons in their rhizosphere (Aguilar and van Diest, 1981; Bekele et al., 1983; Ruiz, 1992; Hinsinger and Gilkes, 1995, 1997). The impact of the form of N (nitrate vs ammonium) supplied on the extent of phosphate rock dissolution in the rhizosphere of ryegrass was measured by Hinsinger and Gilkes (1996), further substantiating such a causal relationship. In flooded soils, where adapted plants such as lowland rice are expected to rely solely on ammonium because of the ambient reducing conditions, root-induced dissolution of the acid-soluble pool of soil P (presumably Ca phosphates) has been shown to contribute a substantial proportion of the P acquired by rice crops (Kirk and Saleque, 1995; Saleque and Kirk, 1995). These authors showed that, in this particular case, the root-induced dissolution of soil P was only partly due to proton efflux to compensate for excess cation over anion uptake. In addition, the root-induced precipitation of the so-called Fe plaque (which is made of Fe oxyhydroxides) in the rhizosphere of rice and other wetland plants promotes a significant part of the observed rhizosphere acidification (Begg et al., 1994; Kirk and Le Van

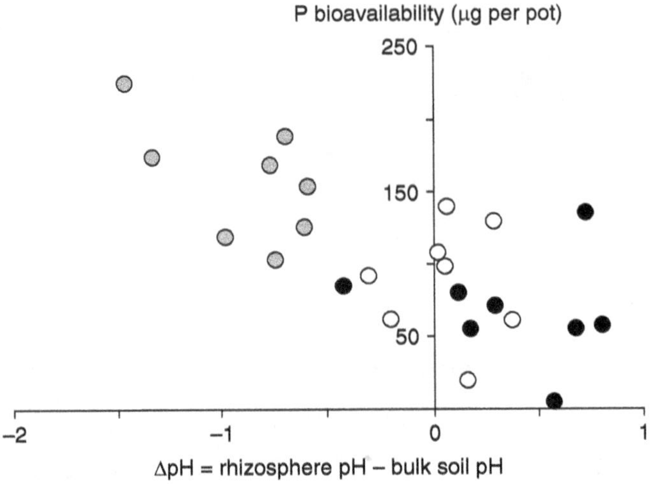

Figure 10.10 Bioavailability of P supplied as phosphate rocks of varying particle size as a function of the root-induced change of pH in the rhizosphere (negative values indicate rhizosphere acidification, positive values indicate rhizosphere alkalisation) of perennial ryegrass (black symbols), oilseed rape (white symbols) and a nitrogen-fixing legume (grey symbols), the annual medic *Medicago truncatula* (Béhi, Arvieu, Benhassaine and Hinsinger, unpublished data).

Du, 1997; Hinsinger et al., 2003, 2009). Besides Ca phosphates, rhizosphere acidification also determines the solubility of silicates and the bioavailability of the cations contained in such minerals and rocks. Compared with ryegrass which did not result in any rhizosphere acidification, oilseed rape led to a substantial root-induced acidification of its rhizosphere and thus to a quite different weathering of a Mg-bearing silicate such as phlogopite mica (Hinsinger et al., 1993). Oilseed rape was responsible for the formation of hydroxy-interlayered vermiculite and for the release of metal cations such as Mg which are part of the octahedral sheet of trioctahedral micas such as phlogopite. These two processes implied that some acid-promoted dissolution of the crystal lattice of phlogopite occurred, contrary to the case of the root-induced formation of vermiculite reported for the rhizosphere of ryegrass (Hinsinger and Jaillard, 1993; Hinsinger et al., 1993).

The impact of pH changes on the weathering of micas and feldspars in the soil around ectomycorrhizae was shown by Arocena et al. (1999) and Arocena and Glowa (2000). Few reports have provided evidence for the *in situ* co-occurrence of such rhizosphere acidification and accelerated weathering rates (Courchesne and Gobran, 1997; Turpault et al., 2005). These works referred to forest stands but there has not been any equivalent report in agroecosystems. None of these studies demonstrated the causal relationship between these two processes but, nevertheless, suggest that rhizosphere acidification should be accounted for when considering soil formation processes (pedogenesis) and plant nutrition.

Beside the release of protons by roots to counterbalance the excess uptake of cations over anions, the production of organic acids by rhizosphere microorganisms has received considerable interest, especially so in the context of ectomycorrhizal fungi associated with forest trees. For instance, Wallander and Wickman (1999) and Bakker et al. (2004) showed the impact of such a process on the weathering of various silicate minerals. Wallander and Wickman (1999) stressed that in such symbioses, the ectomycorrhizal fungal partner was largely responsible for the observed elevated concentrations of citric and oxalic acids and the consequent release of K and Mg from biotite. Ectomycorrhizal fungi, and especially some species and strains, thus play a major role in these processes, rather than the host plant itself (Landeweert et al., 2001; Hoffland et al., 2004). Casarin et al. (2003, 2004) clearly showed that only one of the two tested ectomycorrhizal fungal species was responsible for significant rhizosphere acidification as a consequence of secretion of oxalic acid, while the non-mycorrhizal host plant did not cause any significant pH change. Fungal hyphae of the so-called rock-eating fungi have been shown to dissolve silicate minerals, thereby forming tunnel-like biopores inside mineral grains, presumably as a consequence of the production of organic acids (Jongmans et al., 1997; Landeweert et al., 2001; Hoffland et al., 2004; Van Schöll et al., 2008). More recent works have stressed that when considering the role of mycorrhizae in biological weathering, the potential contribution of the bacteria associated with ectomycorrhizal roots and hyphae should also be taken into account as it may sometimes be large (Calvaruso et al., 2007; Uroz et al., 2007). While Frey-Klett et al. (2005) reported on the so-called mycorrhiza helper bacteria belonging to the group of fluorescent pseudomonads associated with ectomycorrhizae, Calvaruso et al. (2007) showed that different ectomycorrhizal fungal species may exert a selective influence on bacterial communities that are potentially implicated in the production of either acidic or complexing metabolites for promoting the weathering of soil minerals. In the case of arbuscular mycorrhizae, Berthelin and Leyval (1982) and Leyval et al. (1990) showed that the host maize plant was responsible for the largest portion of the biologically mediated release of K and Fe from biotite. Based on current knowledge of the physiology of arbuscular mycorrhizal fungi, these are not expected to release significant amounts of organic acids; this contrasts with ectomycorrhizal species. Biopores of hyphal origin in silicate grains have been reported in boreal forest ecosystems where ectomycorrhizal species are abundant and diverse (Landeweert et al., 2001; Hoffland et al., 2004; Schöll et al., 2007). In contrast, a major proportion of soil bacterial species, the phosphate solubilising bacteria, have been reported to dissolve Ca phosphate minerals and phosphate rocks as a consequence of their ability to produce protons or organic acids (Kucey et al., 1989; Richardson, 2001: Rengel and Marschner, 2005; Richardson et al., 2009). Thus, in the case of non-ectomycorrhizal plants such as crops, roots are expected to be major contributors to acid-promoted weathering of minerals, with a potential contribution of rhizosphere bacteria, but with little or no effect of arbuscular mycorrhizal fungi, which are not to be considered as rock-eating fungi.

10.4.4 Cation complexation or chelation as related to plant activities

Hinsinger et al. (2001) used a microcosm experiment to study the role of different crop species in the weathering of a basalt rock which was leached with a nutrient solution in the presence or absence of plants. In the presence of plants, the dissolution rates of most studied elements (Ca, Mg, Si) increased from two to tenfold, compared with unplanted systems where only leaching was responsible for the observed weathering. The impact of the crops on Fe dissolution rates was much greater, up to 500-fold larger than the minute rate measured in the absence of plants. The latter was expectedly small as a consequence of the neutral pH and oxidising conditions. A significant root-induced acidification was observed which possibly explained the kinetics of release of Ca and Mg from the basalt, but obviously not that of Fe (Hinsinger et al., 2001). Hinsinger et al. (2001) could not elucidate the underlying processes. However, it is likely that the much greater plant-mediated promotion of basalt dissolution measured for Fe compared to other elements was due to rhizosphere processes other than root-induced acidification, such as Fe chelation by root or microbial siderophores.

Bertrand et al. (1999) and Bertrand and Hinsinger (2000) drew similar conclusions in interpreting the results obtained during the experimental dissolution of a poorly soluble Fe oxyhydroxide (goethite) used as a sole source of Fe in the rhizosphere of various crop species. Reichard et al. (2005) showed that phytosiderophores secreted by wheat were implicated in the ligand-promoted dissolution of goethite and that microbial siderophores were also involved (Reichard et al., 2007). These authors also reported a synergetic effect of (phyto)siderophores and carboxylates such as oxalate in the dissolution of goethite.

In the numerous works conducted on the role of carboxylic acids produced by roots and/or rhizosphere microorganisms on the weathering of minerals and rocks, few have attempted to distinguish the direct pH effect from that of the complexing carboxylates. Over the past 50 years, numerous investigations have been conducted on the effects of organic acids on the weathering of rocks and minerals in order to elucidate the possible role of these biochemical agents in soil environments and pedogenesis (e.g. Huang and Keller, 1970; Schnitzer and Kodama, 1976; Razzaghe and Robert, 1979; Tan, 1986; Barman et al., 1992; Eick et al., 1996a, b). In most of these studies, organic acids were implicitly assumed to originate from the decomposition of organic matter by soil microorganisms. Some research showed a more direct implication of soil microorganisms in the weathering of minerals and rocks via the excretion of organic acids or siderophores (e.g. Banfield et al., 1999) and possibly other metabolites which influence pH and redox conditions (see review by Robert and Berthelin, 1986). Until recently, there was little recognition of the potential direct effect of root exudates and secretions of higher plants in the ligand-promoted dissolution of silicate minerals and rocks (see Robert and Berthelin, 1986; Hinsinger et al., 2001). By the end of the twentieth century, the direct implication of land plants in the weathering of rocks and minerals was still a question for debate (Drever, 1994; Jackson, 1996). Mostly this debate focused on lichens which are symbiotic associations of algae and fungi. Lichens are characterised by low growth rates and nutrient requirements that enable them to play the role of pioneer vegetation in the colonisation of fresh rocks (Chapin, 1980). Several studies have shown the role of chelating organic acids released by lichens (e.g. lichenic acid) in the early stages of biological weathering of rocks (Robert and Berthelin, 1986).

The research of Jones and co-workers on carboxylates (e.g. Jones et al., 1996; Jones, 1998) and that of Takagi's group in Japan and Marschner-Römheld's group in Germany (Takagi, 1976; Takagi et al., 1984; Römheld, 1991; Marschner and Römheld, 1994) on phytosiderophores has produced an abundant literature on complexing and chelating agents by plant roots. However, phytosiderophores are only secreted by the Poaceae, which include all the cereals and numerous pasture species (grasses), suggesting that phytosiderophore-promoted dissolution of Fe-bearing minerals and rocks is restricted to these species (Reichard et al., 2005; Robin et al., 2008). In contrast, exudation of carboxylates is a ubiquitous process that all plant species share in common as it is largely the consequence of a passive diffusion of these metabolites which occur at elevated concentration in root cells (Jones, 1998). As stressed, plant species differ considerably in their carboxylate exudation pattern, both qualitatively and quantitatively, and this process is also influenced by the biotic/abiotic environment and nutrient status of the plant (Jones, 1998). Moreover, several of these carboxylates (e.g. the many monocarboxylic anions) do not induce any metal complexation or chelation. Among the most common root exudates, tri-carboxylates such as citrate are by far the most efficient carboxylates in ligand-promoted dissolution of rock-forming or soil-forming minerals, notably via their role in Fe (Robin et al., 2008) and Al chelation. Oxalate is exuded in minor amounts by plant roots except for a few species such as buckwheat (Zheng et al., 1998; Ma et al., 2001), but is of great significance in forest ecosystems as it is produced by many ectomycorrhizal and saprotrophic fungal species (Cromack et al., 1979; Dutton and Evans, 1996; Wallander, 2000; Casarin et al., 2004). For citrate, most plant species exude small amounts but some have been shown to produce large amounts, often combined with other carboxylates such as malate: oilseed rape (e.g. Hoffland et al., 1989) and, among legumes, chickpea (e.g. McLay et al., 1997) and white lupin (e.g. Gardner et al., 1983). The physiology of white lupin has been extensively studied over the past two decades because it releases much larger amounts of citrate than any other crops, thanks to its specialised, so-called cluster roots, which exhibit singular anatomy, architecture and fate (Dinkelaker et al., 1989; Neumann et al., 1999, 2000; Uhde-Stone et al., 2003). This combination of unique morphological and physiological root traits has been described in other cluster-rooted plants, and especially in species of the Proteaceae which are native species occurring predominantly in the highly weathered landscapes of Australia and, to a lesser extent South Africa (Lambers et al., 2006, 2008, 2010). The so-called proteoid roots of these species are capable of exuding citrate and many other carboxylates at rates that are far above those of white lupin and thus orders of magnitude greater than those reported in most crop species (Roelofs et al., 2001). In such environments, Verboom and Pate (2006a) reported on spatial coincidence between the occurrence of ferricretes (pedogenic concretions of Fe oxides in the soil profile) and proteaceous shrub-heathland, while calcretes (pedogenic concretions of Ca carbonates in the soil profile) coincided with myrtaceous woodland. Such differences in soil properties were observed over short (decametric) distances, and thus in conditions of similar climate and parent material in a flat landscape, except for the vegetation cover. This strongly suggests that the vegetation was responsible for such differences. Verboom and Pate (2006b) thereafter developed the 'phytotarium' concept to emphasise the major role of the vegetation on the development of the soil profile. Based on the observation of Fe oxide concretions around root channels in lateritic or deep podzolic soil profiles, they hypothesised that such ferricretes were bioengineered by plants, and especially so Proteaceae, due to the ability of their proteoid roots to exude large amounts of Fe-mobilising carboxylates such as citrate. This mobilisation of Fe favoured the vertical redistribution of Fe in the profile and its re-precipitation along root channels at depth (Verboom and Pate, 2006a). In addition to their role in weathering and mineral formation in soils, complexing and chelating compounds produced by roots and associated microorganisms play a key role in plant growth and nutrition, through their (1) positive impact on the bioavailability of metal micronutrients, especially iron (e.g. Marschner, 1995; Robin et al., 2008), but also major nutrients such as P (e.g. Jones, 1998; Hinsinger, 2001); and (2) negative impact on the bioavailability of toxic metals, as especially documented for Al (e.g. Ryan et al., 2001, 2005; Jones et al., 2003; Kochian et al., 2004).

10.5 Conclusion

The contributions of higher plants to the weathering of primary silicate minerals and rocks and to the formation and fate of secondary minerals (e.g. carbonates, oxides or clay minerals) are manifold and largely based on rhizosphere processes. This weathering and mineral formation also results in the release of plant-available nutrients. Some models of weathering of silicate minerals and rocks now account for the additional effect of the uptake of nutrients by plants (e.g. Taylor and Velbel, 1991; Bormann et al., 1998, Moulton et al., 2000). A number of models of Earth geochemical history have stressed that the considerable drop in atmospheric pCO_2 which started in the early-mid Devonian (410–380 million years ago) coincided with the advent of deep-rooted vascular plants and suggest that such plants were responsible for this dramatic global change (Berner, 1992, 1997; Retallack, 1997; Beerling and Berner, 2005). According to Retallack (1997), 'the early Paleozoic greenhouse may have been curbed by the evolution of rhizospheres with an increased ratio of primary to secondary production and by more effective silicate weathering'. Beerling and Berner (2005) speculated that the decrease in atmospheric pCO_2 was due to enhanced photosynthesis and the subsequent sequestration of C into the soil, burial of C into sediments and enhanced weathering of Ca- and Mg-bearing silicates. However, none of these geochemical models fully accounts for all the other rhizosphere processes that have been reported, in spite of their ability to dramatically influence mineral and rock weathering rates, and thereby the rate of soil formation. Besides the role of roots and rhizosphere microorganisms on elevated soil pCO_2 via respiration which is well acknowledged, the potential impact of protons, carboxylates and siderophores released by roots and rhizosphere microorganisms on the weathering of rocks and pedogenesis is little accounted for. Furthermore, the various plant-mediated processes responsible for (1) the uplift of nutrients such as K or non-essential elements such as Si and subsequent changes of soil mineralogy and (2) the precipitation of calcretes, ferricretes and silcretes need to be better accounted for as they provide explanations for situations where the conventional views of soil geochemistry and pedogenesis accounting only for physically or chemically mediated processes clearly fail (Hinsinger et al., 2001; Verboom and Pate, 2006a; Velde and Barré, 2010).

Plants are thus to be considered as ecosystem engineers (*sensu* Jones et al., 1994), modifying the bioavailability of soil resources for themselves and other living organisms (Lambers et al., 2009). While plants have been long regarded as passive receivers in the acquisition of water and nutrients in ecosystems, compared with animals, this current knowledge about the ecophysiology and ecology of plants shows, conversely, that they are very active. While animals can freely move when encountering adverse conditions and thereby avoid resource depletion, plants have to cope/thrive with the environment where they develop, as a matter of survival. This major constraint probably explains why higher plants have evolved so much plasticity in their root systems and so many strategies to cope with the adverse physical, chemical and biological conditions of soils (Marschner, 1995; Lambers et al., 2006, 2009, 2010). These many processes occur in the rhizosphere, which for some of these represents a very narrow (micrometric-millimetric) layer of soil around roots, while for some other processes it expands centimetres away from roots (Hinsinger et al., 2005; Gregory, 2006b). These rhizosphere processes which are related to the many physiological functions of roots, such as respiration, water and ion uptake, acidification/alkalisation, exudation and rhizodeposition of organic compounds have a number of positive or negative effects on plant growth and nutrition, via the alteration of the distribution and concentration (availability) of resources (water and nutrients) and the alleviation or promotion of toxicities and diseases. Several of these are the consequence of the considerable changes in the structure and activity of microbial communities occurring in the rhizosphere (Dessaux et al., 2010), with rhizodeposition of C-rich compounds being a major driver of such changes (Jones et al., 2009). These rhizosphere processes which all occur, by definition, as a consequence of root activities not only have an impact on the plant itself but thus also on other living biota.

Plant-induced changes in the availability of water and nutrient resources, organic matter and, ultimately, soil fertility are manifold and complex. Through these changes and others (e.g. mechanical effects), plants also alter the habitat of soil biota, or create new habitats for these essential components of soils and ecosystems, and their diversity (Hinsinger et al., 2009). Plants are thereby unique ecosystem engineers, and the challenge for the sustainable management of agroecosystems is to make use of the accumulated knowledge on plant functions and functioning, accounting for their manifold and complex interactions with soils.

References

Aguilar, S.A. and van Diest, A. (1981) Rock phosphate mobilization induced by the alkaline pattern of legumes utilizing symbiotically fixed nitrogen. *Plant and Soil*, **61**, 27–42.

Alexandre, A., Meunier, J.D., Colin, F., and Koud, J.M. (1997) Plant impact on the biogeochemical cycle of silicon and related weathering processes. *Geochimica Cosmochimica Acta*, **61**, 677–682.

April, R. and Keller, D. (1990) Mineralogy of the rhizosphere in forest soils of the eastern United States. Mineralogical studies of the rhizosphere. *Biogeochemistry*, **9**, 1–18.

Armstrong, W. (1979) Aeration in higher plants. *Advances in Botanical Research*, **7**, 225–232.

Armstrong, W., Cousins, D., Armstrong, J., Turner, D.W., and Becket, P.M. (2000) Oxygen distribution in wetland plant roots and permeability barriers to gas-exchange with the rhizosphere: a microelectrode and modelling study with *Phragmites australis*. *Annals of Botany*, **86**, 687–703.

Arocena, J.M. and Glowa, K.R. (2000) Mineral weathering in ectomycorrhizosphere of subalpine fir (*Abies lasiocarpa* (Hook.) Nutt.) as revealed by soil solution composition. *Forest Ecology and Management*, **133**, 61–70.

Arocena, J.M., Glowa, K.R., Massicotte, H.B., and Lavkulich, L. (1999) Chemical and mineral composition of ectomycorrhizosphere soils of subalpine fir (*Abies lasiocarpa* (Hook.) Nutt.) in the Ae horizon of a luvisol. *Canadian Journal of Soil Science*, **79**, 25–35.

Bacha, R.E. and Hossner, L.R. (1977) Characteristics of coatings formed on rice roots as affected by iron and manganese additions. *Soil Science Society of America Journal*, **41**, 931–935.

Bagayoko, M., Alvey, S., Neumann, G., and Buerkert, A. (2000) Root-induced increases in soil pH and nutrient availability to field-grown cereals and legumes on acid sandy soils of Sudano-Sahelian West Africa. *Plant and Soil*, **225**, 117–127.

Bakker, M.R., George, E., Turpault, M.P., Zhang, J.L., and Zeller, B. (2004) Impact of Douglas-fir and Scots pine seedlings on plagioclase weathering under acidic conditions. *Plant and Soil*, **266**, 247–259.

Banfield, J.F., Barker, W.W., Welch, S.A., and Taunton, A. (1999) Biological impact on mineral dissolution: application of the lichen model to understanding mineral weathering in the rhizosphere. *Proceedings of the National Academy of Science USA*, **96**, 3404–3411.

Barman, A.K., Varadachari, C., and Ghosh, K. (1992) Weathering of silicate minerals by organic acids. I. Nature of cation solubilisation. *Geoderma*, **53**, 45–63.

Barré, P., Velde, B., Catel, N., and Abbadie, L. (2007a) Quantification of potassium addition or removal through plant activity on clay minerals by X-ray diffraction. *Plant and Soil*, **292**, 137–146.

Barré, P., Velde, B., and Abbadie, L. (2007b) Dynamic role of 'illite-like' clay minerals in temperate soils: facts and hypotheses. *Biogeochemistry*, **82**, 77–88.

Barré, P., Montagnier, C., Chenu, C., Abbadie, L., and Velde, B. (2008) Clay minerals as a soil potassium reservoir: observation and quantification through X-ray diffraction. *Plant and Soil*, **302**, 213–220.

Barré, P., Berger, G., and Velde, B. (2009) How element translocation by plants may stabilize illitic clays in the surface of temperate soils. *Geoderma*, **151**, 22–30.

Basile-Doelsch, I., Meunier, J.D., and Parron, C. (2005) Another continental pool in the terrestrial silicon cycle. *Nature*, **433**, 399–402.

Battie Laclau, P. and Laclau, J.P. (2009) Growth of the whole root system for a plant crop of sugarcane under rainfed and irrigated environments in Brazil. *Field Crops Research*, **114**, 351–360.

Beerling, D.J. and Berner, R.A. (2005) Feedbacks and the coevolution of plants and atmospheric CO_2. *Proceedings of the National Academy of Science USA*, **102**, 1302–1305.

Beerling, D.J., Osborne, C.P., and Chaloner, W.G. (2001) Evolution of leaf-form in land plants linked to atmospheric CO_2 decline in the Late Palaeozoic era. *Nature*, **410**, 352–354.

Begg, C.B.M., Kirk, G.J.D., MacKenzie, A.F., and Neue, H.-U. (1994) Root-induced iron oxidation and pH changes in the lowland rice rhizosphere. *New Phytologist*, **128**, 469–477.

Bekele, T., Cino, B.J., Ehlert, P.A.I., Van der Maas, A.A., and van Diest, A. (1983) An evaluation of plant-borne factors promoting the solubilization of alkaline rock phosphates. *Plant and Soil*, **75**, 361–378.

Berner, R.A. (1992) Weathering, plants, and the long-term carbon cycle. *Geochimica Cosmochimica Acta*, **56**, 3225–3231.

Berner, R.A. (1997) Paleoclimate – the rise of plants and their effect on weathering and atmospheric CO_2. *Science*, **276**, 544–546.

Berner, E.K., Berner, R.A., and Moulton, K.L. (2003) Plants and mineral weathering: present and past. *Treatise on Geochemistry*, **5**, 169–188.

Berthelin, J. and Leyval, C. (1982) Ability of symbiotic and non-symbiotic rhizospheric microflora of maize (Zea mays) to weather micas and to promote plant growth and plant nutrition. *Plant and Soil*, **68**, 369–377.

Bertrand, I. and Hinsinger, P. (2000) Dissolution of iron oxyhydroxide in the rhizosphere of various crop species. *Journal of Plant Nutrition*, **23**, 1559–1577.

Bertrand, I., Hinsinger, P., Jaillard, B., and Arvieu, J.C. (1999) Dynamics of phosphorus in the rhizosphere of maize and rape grown on synthetic, phosphated calcite and goethite. *Plant and Soil*, **211**, 111–119.

Bidel, L.P.R., Renault, P., Pagès, L., and Rivière, L.M. (2000) Mapping meristem respiration of *Prunus persica* (L.) Batsch seedlings: potential respiration of the meristems, O_2 diffusional constraints and combined effects on root growth. *Journal of Experimental Botany*, **51**, 755–768.

Blake, L., Mercik, S., Koerschens, M., Goulding, K.W.T., Stempen, S., Weigel, A., Poulton, P.R., and Powlson, D.S. (1999) Potassium content in soil, uptake in plants and the potassium balance in three European long-term field experiments. *Plant and Soil*, **216**, 1–14.

Blossfeld, S. and Gansert, D. (2007) A novel non-invasive optical method for quantitative visualization of pH dynamics in the rhizosphere of plants. *Plant, Cell and Environment*, **30**, 176–186.

Blossfeld, S., Perriguey, J., Sterckeman, T., Morel, J.-L., and Lösch, R. (2010) Rhizosphere pH dynamics in trace-metal-contaminated soils, monitored with planar pH optodes. *Plant and Soil*, **330**, 173–184.

Bockheim, J.G. and Gennadiyev, A.N. (2009) The value of controlled experiments in studying soil-forming processes: a review. *Geoderma*, **152**, 208–217.

Bormann, B.T., Wang, D., Bormann, F.H., Benoit, R., April, D., and Snyder, M.C. (1998) Rapid plant induced weathering in an aggrading experimental ecosystem. *Biogeochemistry*, **43**, 129–155.

Bosc, M. (1988) Enseignements fournis par des essais de longue durée sur la fumure phosphatée et potassique. 3. Essais sur la fumure potassique. In: *Phosphore Et Potassium Dans Les Relations Sol-Plante: Consequences Sur La Fertilisation* (ed. L. Gachon), pp. 403–466. INRA, Paris, France.

Bouillet, J.P., Laclau, J.P., Arnaud, M., Thongo M'bou, A., Saint-André, L., Jourdan, C. (2002) Changes with age in the spatial distribution of roots of a Eucalyptus clone in the Congo. Impact on water and nutrient uptake ability. *Forest Ecology and Management*, **171**, 43–57.

Bourrié, G. and Lelong, F. (1994) Les solutions du sol: du profil au bassin versant. In: *Pédologie 2: Constituants et Propriétés du Sol* (eds M. Bonneau and B. Souchier), pp. 239–273. Masson, Paris, France.

Braissant, O., Verrecchia, E.P., and Aragno, M. (2002) Is the contribution of bacteria to terrestrial carbon budget greatly underestimated? *Naturwissenschaften*, **89**, 366–370.

Braissant, O., Cailleau, G., Dupraz, C., and Verrecchia, E.P. (2003) Bacterially induced mineralization of calcium carbonate in terrestrial environments: the role of exopolysaccharides and amino acids. *Journal of Sedimentary Research*, **73**, 485–490.

Bravin, M.N., Travassac, F., Le Floch, M., Hinsinger, P., and Garnier, J.M. (2008) Oxygen input controls the spatial and temporal dynamics of arsenic at the surface of a flooded paddy soil and in the rhizosphere of lowland rice (*Oryza sativa* L.): a microcosm study. *Plant and Soil*, **312**, 207–218.

Bravin, M.N., Tentscher, P., Rose, J., and Hinsinger, P. (2009a) Rhizosphere pH gradient controls copper availability in a strongly acidic soil. *Environmental Science and Technology*, **43**, 5686–5691.

Bravin, M.N., Lara Martí, A., Clairotte, M., and Hinsinger, P. (2009b) Rhizosphere alkalisation – a major driver of copper bioavailability over a broad pH range in an acidic, copper-contaminated soil. *Plant and Soil*, **318**, 257–268.

Brown, G.G., Barois, I., and Lavelle, P. (2000) Regulation of soil organic matter dynamics and microbial activityin the drilosphere and the role of interactions with other edaphic functional domains. *European Journal of Soil Biology*, **36**, 177–198.

Brundrett, M.C. (2002) Coevolution of roots and mycorrhizas of land plants. *New Phytologist*, **154**, 275–304.

Bucher, M. (2006) Functional biology of plant phosphate uptake at root and mycorrhiza interfaces. *New Phytologist*, **173**, 11–26.

Buxbaum, C.A.Z., Nowak, C.A., and White, E.H. (2005) Deep subsoil nutrient uptake in potassium-deficient, aggrading *Pinus resinosa* plantation. *Canadian Journal of Forest Research*, **35**, 1978–1983.

Cailleau, G., Braissant, O., and Verrecchia, E.P. (2004) Biomineralization in plants as a long-term carbon sink. *Naturwissenschaften*, **91**, 191–194.

Cailleau, G., Braissant, O., Dupraz, C., Aragno, M., and Verrecchia, E.P. (2005) Biologically induced accumulations of $CaCO_3$ in orthox soils of Biga, Ivory Coast. *Catena*, **59**, 1–17.

Callot, G., Chamayou, H., Maertens, C., and Salsac, L. (1983) *Mieux comprendre les interactions sol-racine. Incidence sur la nutrition minérale*. INRA, Paris, France.

Calvaruso, C., Turpault, M.P., Leclerc, E., and Frey-Klett, P. (2007) Impact of ectomycorrhizosphere on the functional diversity of soil bacterial and fungal communities from a forest stand in relation to nutrient mobilization processes. *Microbial Ecology*, **54**, 567–577.

Canadell, J., Jackson, R.B., Ehleringer, J.R., Mooney, H.A., Sala, O.E., and Schulze, E.D. (1996) Maximum rooting depth for végétation types at the global scale. *Oecologia*, **108**, 583–595.

Casarin, V., Plassard, C., Souche, G., and Arvieu, J.C. (2003) Quantification of oxalate ions and protons released by ectomycorrhizal fungi in rhizosphere soil. *Agronomie*, **23**, 461–469.

Casarin, V., Plassard, C., Hinsinger, P., and Arvieu, J.C. (2004) Quantification of ectomycorrhizal fungal effects on the bioavailability and mobilisation of soil P in the rhizosphere of *Pinus pinaster*. *New Phytologist*, **163**, 177–185.

Chaignon, V., Quesnoit, M., and Hinsinger, P. (2009) Rhizosphere pH, bioavailability and extractability of Cu in a Cu-contaminated acidic soil as affected by liming. *Environmental Pollution*, **157**, 3363–3369.

Chapin, F.S. III (1980) The mineral nutrition of wild plants. *Annual Review of Ecology, Evolution and Systematics*, **11**, 233–260.

Claassen, N. and Jungk, A. (1982) Kaliumdynamik im wurzelnahen Boden in Beziehung zur Kaliumaufnahme von Maispflanzen. *Zeitschrift für Pflanzenernährung und Bodenkunde*, **145**, 513–525.

Clegg, S., Gobran, G.R., and Guan, X. (1997) Rhizosphere chemistry in an ammonium sulfate and water manipulated Norway spruce [*Picea abies* (L.) Karst.] forest. *Canadian Journal of Soil Science*, **77**, 515–523.

Courchesne, F. and Gobran, G.R. (1997) Mineralogical variation of bulk and rhizosphere soils from a Norway spruce stand. *Soil Science Society of America Journal*, **61**, 1245–1249.

Coventry, D.R. and Slattery, W.J. (1991) Acidification of soil associated with lupins grown in a crop rotation in North-Eastern Victoria. *Australian Journal of Agricultural Research*, **42**, 391–397.

Cromack, K., Sollins, P., Graustein, W.C., Speidel, K., Todd, A.W., Spycher, G., Li, C.Y., and Todd, R.L. (1979) Calcium oxalate accumulation and soil weathering in mats of hypogeous fungus. *Hysterangium crassum. Soil Biology and Biochemistry*, **11**, 463–468.

Curl, E.A. and Truelove, B. (1986) *The Rhizosphere*. Springer-Verlag, Berlin, Germany.

Darrah, P.R. (1993) The rhizosphere and plant nutrition: a quantitative approach. *Plant and Soil*, **155**/156, 1–20.

De Deyn, G.B., Cornelissen J.H.C., and Bardgett R.D. (2008) Plant functional traits and soil carbon sequestration in contrasting biomes. *Ecology Letters*, **11**, 516–531.

Dehérain, P.P. (1892) *Traité de Chimie agricole*, p. 904. Masson, Paris, France.

Derry, L.A., Kurtz, A.C., Ziegler, K., and Chadwick, O.A. (2005) Biological control of terrestrial silica cycling and export fluxes to watersheds. *Nature*, **433**, 728–731.

Dessaux, Y., Hinsinger, P., and Lemanceau, P. (2010) *Rhizosphere: Achievements and Challenges*. Springer, Dordrecht, the Netherlands.

Dinkelaker, B., Römheld, V., and Marschner, H. (1989) Citric acid excretion and precipitation of calcium citrate in the rhizosphere of white lupin (*Lupinus albus* L.). *Plant Cell and Environment*, **12**, 285–292.

Drever, J.I. (1994) The effect of land plants on weathering rates of silicate minerals. *Geochimica Cosmochimica Acta*, **58**, 2325–2332.

Dutton, M.V. and Evans, C.S. (1996) Oxalate production by fungi: its role in pathogenicity and ecology in the soil environment. *Canadian Journal of Microbiology*, **42**, 881–895.

Eick, M.J., Grossl, P.R., Golden, D.C., Sparks, D.L., and Ming, D.W. (1996a) Dissolution of a lunar basalt simulant as affected by pH and organic anions. *Geoderma*, **74**, 139–160.

Eick, M.J., Grossl, P.R., Golden, D.C., Sparks, D.L., and Ming, D.W. (1996b) Dissolution kinetics of a lunar basalt simulant at 298 K. The effect of pH and organic acids. *Geochimica Cosmochimica Acta*, **56**, 157–170.

Epron, D., Nouvellon, Y., Deleporte, P., Ifo, S., Kazotti, G., Thongo, M'Bou A., Mouvondy, W., Saint-André, L., Roupsard, O., Jourdan, C., and Hamel, O. (2006) Soil Carbon balance in a clonal Eucalyptus plantation in Congo: effects of logging on carbon inputs and soil CO_2 efflux. *Global Change Biology*, **12**, 1021–1031.

Epstein, E. (1999) Silicon. *Annual Review of Plant Physiology and Plant Molecular Biology*, **50**, 641–664.

Faure, D., Vereecke, D., and Leveau J.H.J. (2009) Molecular communication in the rhizosphere. *Plant and Soil*, **321**, 279–303.

Field, C.B., Behrenfeld, M.J., Randerson, J.T., and Falkowski, P. (1998) Primary production of the biosphere: integrating terrestrial and oceanic components. *Science*, **281**, 237–240.

Flessa, H. and Fischer, W.R. (1992) Plant-induced changes in the redox potentials of rice rhizospheres. *Plant and Soil*, **143**, 55–60.

Fontaine, S., Barot, S., Barré, P., Bdioui, N., Mary, B., and Rumpel, C. (2007) Stability of organic carbon in deep soil layers controlled by fresh carbon supply. *Nature*, **450**, 277–281.

Franche, C., Lindström, K., and Elmerich, C. (2009) Nitrogen-fixing bacteria associated with leguminous and non-leguminous plants. *Plant and Soil*, **321**, 35–59.

Freytet, P., Plaziat, J.C., and Verrecchia, E.P. (1997) A classification of rhizogenic (root-formed) calcretes, with examples from the upper Jurassic lower Cretaceous of Spain and upper Cretaceous of southern France. *Sediment Geology*, **110**, 299–303.

Frey-Klett, P., Chavatte, M., Clausse, M.L., Courrier, S., Le Roux, C., Raaijmakers, J., Martinotti, M.G., Pierrat, J.C., and Garbaye, J. (2005) Ectomycorrhizal symbiosis affects functional diversity of rhizosphere fluorescent pseudomonads. *New Phytologist*, **165**, 317–328.

Gahoonia, T.S., Claassen, N., and Jungk, A. (1992) Mobilization of phosphate in different soils by ryegrass supplied with ammonium or nitrate. *Plant and Soil*, **140**, 241–248.

Gardner, W.K., Barber, D.A., and Parberry, D.G. (1983) The acquisition of phosphorus by *Lupinus albus* L. III. The probable mechanism by which phosphorus movement in the soil/root interface is enhanced. *Plant and Soil*, **70**, 107–124.

Geelhoed, J.S., Van Riemsdijk, W.H., and Findenegg, G.R. (1999) Simulation of the effect of citrate exudation from roots on the plant availability of phosphate adsorbed on goethite. *European Journal of Soil Science*, **50**, 379–390.

Gérard, F., Mayer, K.U., Hodson, M.J., and Ranger, J. (2008) Modelling the biogeochemical cycle of silicon in soils: application to a temperate forest ecosystem. *Geochimica Cosmochimica Acta*, **72**, 741–758.

Gerke, J., Beissner, L., and Römer, W. (2000a) The quantitative effect of chemical phosphate mobilization by carboxylate anions on P uptake by a single root. I. The basic concept and determination of soil parameters. *Journal of Plant Nutrition and Soil Science*, **163**, 207–212.

Gerke, J., Beissner, L., and Römer, W. (2000b) The quantitative effect of chemical phosphate mobilization by carboxylate anions on P uptake by a single root. II. The importance of soil and plant parameters for uptake of mobilized P. *Journal of Plant Nutrition and Soil Science*, **163**, 213–219.

Gollany, H.T., Schumacher, T.E., Rue, R.R., and Liu, S.-Y. (1993) A carbon dioxide microelectrode for in situ pCO_2 measurement. *Microchemistry Journal*, **48**, 42–49.

Goulding, K.W.T., Murray, P.J., Sohi, S.P., Gilliam, L.M., Williams, J.K., Clark, I.M., and Hirsch, P.R. (2010) Soil without plants: the consequences for microorganisms and mesofauna. In: *Soil Solutions for a Changing World, Proceedings of the 19th World Congress of Soil Science*, Brisbane, Queensland, Australia, IUSS, pp. 134–137.

Gras, F. (1974) Une méthode expérimentale pour la détermination du pH et de la solubilité du carbonate de calcium en sols calcaires. *Bulletin de l'Association Française d'Etude du Sol*, **1**, 33–48.

Gregory, P.J. (2006a) *Plant Roots: Their Growth, Activity and Interactions with Soil.* Blackwell Publishing Ltd, Oxford, UK.

Gregory, P.J. (2006b) Roots, rhizosphere and soil: the route to a better understanding of soil science? *European Journal of Soil Science*, **57**, 2–12.

Gruber, N. and Galloway, J.N. (2008) An Earth-system perspective of the global nitrogen cycle. *Nature*, **451**, 293–296.

Hanson, P.J., Edwards, N.T., Garten, C.T., and Andrews, J.A. (2000) Separating root and soil microbial contributions to soil respiration: a review of methods and observations. *Biogeochemistry*, **48**, 115–146.

Harrison, M.J., Dewbre, G.R., and Liu, J.Y. (2002) A phosphate transporter from *Medicago truncatula* involved in the acquisition of phosphate released by arbuscular mycorrhizal fungi. *The Plant Cell*, **14**, 2413–2429.

Hartmann, A., Rothballer, M., and Schmid, M. (2008) Lorenz Hiltner, a pioneer in rhizosphere microbial ecology and soil bacteriology research. *Plant and Soil*, **312**, 7–14.

Hartmann, A., Schmid, M., van Tuinen, D., and Berg, G. (2009) Plant-driven selection of microbes. *Plant and Soil*, **321**, 235–257.

Häussling, M., Leisen, E., Marschner, H., and Römheld, V. (1985) An improved method for non-destructive measurements of the pH at the root–soil interface (rhizosphere). *Journal of Plant Physiology*, **117**, 371–375.

Haynes, R.J. (1990) Active ion uptake and maintenance of cation–anion balance: a critical examination of their role in regulating rhizosphere pH. *Plant and Soil*, **126**, 247–264.

Herridge, D.F., Peoples, M.B., and Boddey, R.M. (2008) Global inputs of biological nitrogen fixation in agricultural systems. *Plant and Soil*, **311**, 1–18.

Hinsinger, P. (1990) Action des racines sur la libération du potassium et l'altération de minéraux silicatés. PhD thesis, ENSA-M, Montpellier, France, 186pp.

Hinsinger, P. (1998) How do plant roots acquire mineral nutrients? Chemical processes involved in the rhizosphere. *Advances in Agronomy*, **64**, 225–265.

Hinsinger, P. (2001) Bioavailability of soil inorganic P in the rhizosphere as affected by root-induced chemical changes: a review. *Plant and Soil*, **237**, 173–195.

Hinsinger, P. (2002) Potassium. In: *Encyclopedia of Soil Science* (ed. R. Lal). Marcel Dekker, Inc., New York.

Hinsinger, P. and Jaillard, B. (1993) Root-induced release of interlayer potassium and vermiculitization of phlogopite as related to potassium depletion in the rhizosphere of ryegrass. *Journal of Soil Science*, **44**, 525–534.

Hinsinger, P. and Gilkes, R.J. (1995) Root-induced dissolution of a reactive phosphate rock in the rhizosphere of two lupin species grown in alkaline soil. *Australian Journal of Soil Research*, **33**, 477–489.

Hinsinger, P. and Gilkes, R.J. (1996) Mobilization of phosphate from phosphate rock and alumina-sorbed phosphate by the roots of ryegrass and clover as related to rhizosphere pH. *European Journal of Soil Science*, **47**, 533–544.

Hinsinger, P. and Gilkes, R.J. (1997) Dissolution of phosphate rock in the rhizosphere of five plant species grown in an acid, P-fixing mineral substrate. *Geoderma*, **75**, 231–249.

Hinsinger P., Jaillard B., and Dufey J.E. (1992) Rapid weathering of a trioctahedral mica by the roots of ryegrass. *Soil Science Society of America Journal*, **56**, 977–982.

Hinsinger, P., Elsass, F., Jaillard, B., and Robert, M. (1993) Root-induced irreversible transformation of a trioctahedral mica in the rhizosphere of rape. *Journal of Soil Science*, **44**, 535–545.

Hinsinger, P., Fernandes Barros, O.N., Benedetti, M.F., Noack, Y., and Callot, G. (2001) Plant-induced weathering of a basaltic rock: experimental evidence. *Geochimica Cosmochimica Acta*, **65**, 137–152.

Hinsinger, P., Plassard, C., Tang, C., and Jaillard, B. (2003) Origins of root-induced pH changes in the rhizosphere and their responses to environmental constraints: a review. *Plant and Soil*, **248**, 43–59.

Hinsinger, P., Gobran, G.R., Gregory, P.J., and Wenzel, W.W. (2005) Rhizosphere geometry and heterogeneity arising from root-mediated physical and chemical processes. *New Phytologist*, **168**, 293–303.

Hinsinger, P., Plassard, C., and Jaillard, B. (2006) The rhizosphere: a new frontier in soil biogeochemistry. *Journal of Geochemical Exploration*, **88**, 210–213.

Hinsinger, P., Bengough, A.G., Vetterlein, D., and Young, I.M. (2009) Rhizosphere: biophysics, biogeochemistry and ecological relevance. *Plant and Soil*, **321**, 117–152.

Hirsch, P.R., Gilliam, L.M., Sohi, S.P., Williams, J.K., Clark, I.M., and Murray, P.J. (2009) Starving the soil of plant inputs for 50 years reduces abundance but not diversity of soil bacterial communities. *Soil Biology and Biochemistry*, **41**, 2021–2024.

Hoffland, E., Findenegg, G.R., and Nelemans, J.A. (1989) Solubilization of rock phosphate by rape. II. Local root exudation of organic acids as a response to P-starvation. *Plant and Soil*, **113**, 161–165.

Hoffland, E., Kuyper, T.W., Wallander, H., Plassard, C., Gorbushina, A., Haselwandter, K., Holmström, S., Landeweert, R., Lundström, U.S., Rosling, A., Sen, R., Smits, M.M., Van Hees, P.A.W., and Van Breemen, N. (2004) The role of fungi in weathering. *Frontiers in Ecology and Environment*, **2**, 258–264.

Holmqvist, J., Øgaard, A.F., Öborn, I., Edwards, A.C., Mattsson, L., and Sverdrup, H. (2003) Application of the PROFILE model to estimate potassium release from mineral weathering in Northern European agricultural soils. *European Journal of Agronomy*, **20**, 149–163.

Huang, W.H. and Keller, W.D. (1970) Dissolution of rock-forming silicate minerals in organic acids: simulated fist-stage weathering of fresh mineral surfaces. *American Mineralogist*, **57**, 2076–2094.

Hübel, F. and Beck, E. (1993) In-situ determination of the P-relations around the primary root of maize with respect to inorganic and phytate-P. *Plant and Soil*, **157**, 1–9.

Jackson, T.A. (1996) Comment on "The effect of land plants on weathering rates of silicate minerals" by James I. Drever. *Geochimica Cosmochimica Acta*, **60**, 723–724.

Jackson, R.B., Canadell, J., Ehleringer, J.R., Mooney, H.A., Sala, O.E., and Schulze, E.D. (1996) A global analysis of root distributions for terrestrial biomes. *Oecologia*, **108**, 389–411.

Jaillard, B. (1982) Relation entre dynamique de l'eau et organisation morphologique d'un sol calcaire. *Science du Sol*, **20**, 31–52.

Jaillard, B. (1983) Mise en évidence de la calcitisation des cellules corticales de racines de Graminées en milieu carbonaté. *Comptes Rendus de l'Académie des Sciences de Paris*, **297**(série II), 293–296.

Jaillard, B. (1984) Mise en évidence de la néogenèse de sables calcaires sous l'influence des racines: incidence sur la granulométrie du sol. *Agronomie*, **4**, 91–100.

Jaillard, B. (1985) Activité racinaire et rhizostructures en milieu carbonaté. *Pédologie*, **35**, 297–313.

Jaillard, B. (1987a) Les structures rhizomorphes calcaires: modèle de réorganisation des minéraux du sol par les racines. Thèse d'Etat, USTL, Montpellier, France.

Jaillard, B (1987b) Techniques for studying the ionic environment at the soil-root interface. In: *Methodology in Soil-K Research* (ed. International Potassium Institute (IPI)), pp. 231–245. IPI, Bern, Switzerland.

Jaillard, B. and Hinsinger, P. (1993) Alimentation minérale des végétaux dans le sol. In: *Techniques Agricoles, n. 87 (6–93), fasc. 1210*, pp. 1–13. Editions Techniques, Paris, France.

Jaillard, B., Guyon, A., and Maurin, A.F. (1991) Structure and composition of calcified roots, and their identification in calcareous soils. *Geoderma*, **50**, 197–210.

Jaillard, B., Ruiz, L., and Arvieu, J.C. (1996) pH mapping in transparent gel using color indicator videodensitometry. *Plant and Soil*, **183**, 85–95.

Jaillard, B., Plassard, C., and Hinsinger, P. (2003) Measurements of H^+ fluxes and concentrations in the rhizosphere. In: *Handbook of Soil Acidity* (ed. Z. Rengel), pp. 231–266. Marcel Dekker, Inc., New York.

Jarvis, S.C. and Robson, A.D. (1983a) The effects of nitrogen nutrition of plants on the development of acidity in Western Australian soils. I. Effects with subterranean clover grown under leaching conditions. *Australian Journal of Agricultural Research*, **34**, 341–353.

Jarvis, S.C. and Robson, A.D. (1983b) The effects of nitrogen nutrition of plants on the development of acidity in Western Australian soils. II. Effects of differences in cations/anion balance between plant species grown under non-leaching conditions. *Australian Journal of Agricultural Research*, **34**, 355–365.

Jobbagy, E.G. and Jackson, R.B. (2001) The distribution of soil nutrients with depth: global patterns and the imprint of plants. *Biogeochemistry*, **53**, 51–77.

Jobbagy, E.G. and Jackson, R.B. (2004) The uplift of soil nutrients by plants: biogeochemical consequences across scales. *Ecology*, **85**, 2380–2389.

Johnston, A.E., Poulton, P.R., and Coleman, K. (2009) Soil organic matter: its importance in sustainable agriculture and carbon dioxide fluxes. *Advances in Agronomy*, **101**, 1–57.

Jones, D.L. (1993) Rhizosphere carbon flow. PhD thesis, Oxford University, Oxford, UK.

Jones, D.L. (1998) Organic acids in the rhizosphere – a critical review. *Plant and Soil*, **205**, 25–44.

Jones, D.L. and Oburger, E. (eds) (2011) Solubilization of phosphorus by soil microorganisms. In: *Phosphorus in Action – Biological Processes in Soil Phosphorus Cycling*. Springer, Dordrecht, the Netherlands.

Jones, C.G., Lawton, J.H., and Shachak, M. (1994) Organisms as ecosystem engineers. *Oikos*, **69**, 373–386.

Jones, D.L., Darrah, P.R., and Kochian, L.V. (1996) Critical evaluation of organic acid mediated dissolution in the rhizosphere and its potential role in root iron uptake. *Plant and Soil*, **180**, 57–66.

Jones, D.L., Dennis, P.G., Owen, A.G., and Van Hees, P.A.W. (2003) Organic acid behavior in soils – misconceptions and knowledge gaps. *Plant and Soil*, **248**, 31–41.

Jones, D.L., Hodge, A., and Kuzyakov, Y. (2004) Plant and mycorrhizal regulation of rhizodeposition. *New Phytologist*, **163**, 459–480.

Jones, D.L., Nguyen, C., and Finlay, R.D. (2009) Carbon flow in the rhizosphere: carbon trading at the soil-root interface. *Plant and Soil*, **321**, 5–33.

Jongmans, A.G., Van Breemen, N., Lundström, U., Van Hees, P.A.W., Finlay, R.D., Srinivasan, M., Unestam, T., Giesler, R., Melkerud, P.-A., and Olsson, M. (1997) Rock-eating fungi. *Nature*, **389**, 682–683.

Jonko, A.K., Hense, A., and Feddema, J.J. (2010) Effects of land cover change on the tropical circulation in a GCM. *Climate Dynamics*, **35**, 635–649.

Karberg, N.J., Pregitzer, K.S., King, J.S., Friend, A.L., and Wood, J.R. (2005) Soil carbon dioxide partial pressure and dissolved inorganic carbonate chemistry under elevated carbon dioxide and ozone. *Oecologia*, **142**, 296–306.

Kenrick, P. and Crane, P.R. (1997) The origin and early evolution of plants on land. *Nature*, **389**, 33–39.

Khasawneh, F.E. and Doll, E.C. (1978) The use of phosphate rock for direct application to soils. *Advances in Agronomy*, **30**, 159–206.

Kirk, G.J.D. and Nye, P.H. (1986) A simple model for predicting the rates of dissolution of sparingly soluble calcium phosphates in soil. I. The basic model. *Journal of Soil Science*, **37**, 529–540.

Kirk, G.J.D. and Saleque, M.A. (1995) Solubilization of phosphate by rice plants growing in reduced soil: prediction of the amount solubilized and the resultant increase in uptake. *European Journal of Soil Science*, **46**, 247–255.

Kirk, G.J.D. and Le Van Du (1997) Changes in rice root architecture, porosity, and oxygen and proton release under phosphorus deficiency. *New Phytologist*, **135**, 191–200.

Kochian, L.V., Hoekenga, O.A., and Piñeros, M.A. (2004) How do crop plants tolerate acid soils? Mechanisms of aluminum tolerance and phosphorus efficiency. *Annual Review of Plant Biology*, **55**, 459–493.

Kodama, H., Nelson, S., Yang, F., and Kohyama, N. (1994) Mineralogy of rhizospheric and non-rhizospheric soils in corn fields. *Clays and Clay Minerals*, **42**, 755–763.

Kucey, R.M.N., Janzen, H.H., and Leggett, M.E. (1989) Microbially mediated increases in plant-available phosphorus. *Advances in Agronomy*, **42**, 199–228.

Kuchenbuch, R. and Jungk, A. (1982) A method for determining concentration profiles at the soil-root interface by thin slicing rhizospheric soil. *Plant and Soil*, **68**, 391–394.

Kuhlmann, H. (1990) Importance of the subsoil for the K-nutrition of crops. *Plant and Soil*, **127**, 129–136.

Kuzyakov, Y. (2002) Review: factors affecting rhizosphere priming effects. *Journal of Plant Nutrition and Soil Science*, **165**, 382–396.

Kuzyakov, Y. (2006) Sources of CO_2 efflux from soil and review of partitioning methods. *Soil Biology and Biochemistry*, **38**, 425–448.

Kuzyakov, Y. and Domanski, G. (2000) Carbon input by plants into the soil. Review. *Journal of Plant Nutrition and Soil Science*, **163**, 421–431.

Kuzyakov, Y., Shevtzova, T.E., and Pustovoytov, K. (2006) Carbonate re-crystallization in soil revealed by ^{14}C labeling: experiment, model and significance for paleo-environmental reconstructions. *Geoderma*, **131**, 45–58.

Lambers, H., Shane, M.W., Cramer, M.D., Pearse, S.J., and Veneklaas, E.J. (2006) Root structure and functioning for efficient acquisition of phosphorus: matching morphological and physiological traits. *Annals of Botany*, **98**, 693–713.

Lambers, H., Shaver, G., Raven, J.A., and Smith, S.E. (2008) N and P-acquisition change as soils age. *Trends in Ecology and Evolution*, **23**, 95–103.

Lambers, H., Mougel, C., Jaillard, B., and Hinsinger, P. (2009) Plant-microbe-soil interactions in the rhizosphere: an evolutionary perspective. *Plant and Soil*, **321**, 83–115.

Lambers, H., Brundrett, M.C., Raven, J.A., and Hopper, S.D. (2010) Plant mineral nutrition in ancient landscapes: high plant species diversity on infertile soils is linked to functional diversity for nutritional strategies. *Plant and Soil*, **334**, 11–31.

Landeweert, R., Hoffland, E., Finlay, R.D., Kuyper, T.W., and Van Breemen, N. (2001) Linking plants to rocks: ectomycorrhizal fungi mobilize nutrients from minerals. *Trends in Ecology and Evolution*, **16**, 248–254.

Lelong, F., Dupraz, C., Durand, P., and Didonlescot, J.F. (1990) Effects of vegetation type on the biogeochemistry of small catchments (Mont-Lozère, France). *Journal of Hydrology*, **116**, 125–145.

Leyval, C., Laheurte, F., Belgy, G., and Berthelin, J. (1990) Weathering of micas in the rhizospheres of maize, pine and beech seedlings influenced by mycorrhizal and bacterial inoculation. *Symbiosis*, **9**, 105–109.

Lucas, Y. (2001) The role of plants in controlling rates and products of weathering: importance of biological pumping. *Annual Review of Earth and Planetary Sciences*, **29**, 135–163.

Lucas, Y., Luizão, F.J., Chauvel, A., Rouiller, J., and Nahon, D. (1993) The relation between biological activity of the rain forest and mineral composition of soils. *Science*, **260**, 521–523.

Luster, J., Göttlein, A., Nowack, B., and Sarret, G. (2009) Sampling, defining, characterising and modeling the rhizosphere – the soil science tool box. *Plant and Soil*, **321**, 457–482.

Lynch, J.M. (1990) *The Rhizosphere*. John Wiley & Sons, Ltd., Chichester, UK.

Ma, J.F., Ryan, P.R., and Delhaize, E. (2001) Aluminium tolerance in plants and the complexing role of organic acids. *Trends in Plant Science*, **6**, 273–278.

Manning, D.A.C. (2010) Mineral sources of potassium for plant nutrition. A review. *Agronomy for Sustainable Development*, **30**, 281–294.

Marschner, H. (1995) *Mineral Nutrition of Higher Plants*, 2nd edn. Academic Press, London, UK.

Marschner, H. and Römheld, V. (1983) *In vivo* measurement of root-induced pH changes at the soil-root interface: effect of plant species and nitrogen source. *Zeitschrift für Pflanzenernährung und Bodenkunde*, **111**, 241–251.

Marschner, H. and Römheld, V. (1994) Strategies of plants for acquisition of iron. *Plant and Soil*, **165**, 261–274.

Marschner, H., Häussling, M., and George, E. (1991) Ammonium and nitrate uptake rates and rhizosphere pH in non-mycorrhizal roots of Norway spruce (*Picea abies* (L.) Karst.). *Trees Structure and Function*, **5**, 14–21.

Marsden, C., Nouvellon, Y., Thongo M'Bou, A., Saint-André, L., Jourdan, C., Kinana, A., and Epron, D. (2008) Two independent estimations of stand-level root respiration on clonal Eucalyptus stands in Congo: up scaling of direct measurements on roots versus the trenched-plot technique. *New Phytologist*, **177**, 676–687.

McLay, C.D.A., Barton, L., and Tang, C. (1997) Acidification potential of ten grain legume species grown in nutrient solution. *Australian Journal of Agricultural Research*, **48**, 1025–1032.

Michaud, A.M., Bravin, M.N., Galleguillos, M., and Hinsinger, P. (2007) Copper uptake and phytotoxicity as assessed in situ for durum wheat (*Triticum turgidum durum* L.) cultivated in Cu-contaminated, former vineyard soils. *Plant and Soil*, **298**, 99–111.

Mortland, M.M., Lawton, K., and Uehara, G. (1956) Alteration of biotite to vermiculite by plant growth. *Soil Science*, **82**, 477–481.

Moulton, K.L., West, J., and Berner, R.A. (2000) Solute flux and mineral mass balance approaches to the quantification of plant effects on silicate weathering. *American Journal of Science*, **300**, 539–570.

Neumann, G. and Römheld, V. (1999) Root excretion of carboxylic acids and protons in phosphorus-deficient plants. *Plant and Soil*, **211**, 121–130.

Neumann, G., Massonneau, A., Martinoia, E., and Römheld, V. (1999) Physiological adaptations to phosphorus deficiency during proteoid root development in white lupin. *Planta*, **208**, 373–382.

Neumann, G., Massonneau, A., Langlade, N., Dinkelaker, B., Hengeler, C., Römheld, V., and Martinoia, E. (2000) Physiological aspects of cluster root function and development in phosphorus-deficient white lupin (*Lupinus albus* L.). *Annals of Botany*, **85**, 909–919.

Neumann, G., George, T.S., and Plassard, C. (2009) Strategies and methods for studying the rhizosphere – the plant science tool box. *Plant and Soil*, **321**, 431–456.

Nguyen, C. (2003) Rhizodeposition of organic C by plants: mechanisms and controls. *Agronomie*, **23**, 375–396.

Niebes, J.F., Hinsinger, P., Jaillard, B., and Dufey, J.E. (1993) Release of non exchangeable potassium from different size fractions of two highly K-fertilized soils in the rhizosphere of rape (*Brassica napus* cv Drakkar). *Plant and Soil*, **155**/156, 403–406.

Noble, A.D., Cannon, M., and Muller, D. (1997) Evidence of accelerated soil acidification under Stylosanthes-dominated pastures. *Australian Journal of Soil Research*, **35**, 1309–1322.

Noble, A.D., Suzuki, S., Soda, W., Ruaysoongnern, S., and Berthelsen, S. (2008) Soil acidification and carbon storage in fertilized pastures of Northeast Thailand. *Geoderma*, **144**, 248–255.

Nye, P.H. (1981) Changes of pH across the rhizosphere induced by roots. *Plant and Soil*, **61**, 7–26.

Nye, P.H. (1983) The diffusion of two interacting solutes in soil. *Journal of Soil Science*, **34**, 677–691.

Parfitt, R.L., Ross, D.J., Coomes, D.A., Richardson, S.J., Smale, M.C., and Dahlgren, R.A. (2005) N and P in New Zealand soil chronosequences and relationships with foliar N and P. *Biogeochemistry*, **75**, 305–328.

Pellet, D.M., Grunes, D.L., and Kochian, L.V. (1995) Organic acid exudation as an aluminum-tolerance mechanism in maize (*Zea mays* L.). *Planta*, **196**, 788–795.

Pielke, R.A. (2001) Influence of the spatial distribution of vegetation and soils on the prediction of cumulus convective rainfall. *Review of Geophysics*, **39**, 151–177.

Raaijmakers, J.M., Paulitz, T.C., Steinberg, C., Alabouvette, C., and Moënne-Loccoz, Y. (2009) The rhizosphere: a playground and battlefield for soilborne pathogens and beneficial microorganisms. *Plant and Soil*, **321**, 341–361.

Rao, T.R., Yano, K., Iijima, M., Yamauchi, A., and Tatsumi, J. (2002) Regulation of rhizosphere acidification by photosynthetic activity in cowpea (*Vigna unguiculata* L. Walp.) seedlings. *Annals of Botany*, **89**, 213–220.

Rasmann, S., Köllner, T.G., Degenhardt, J., Hiltpold, I., Toepfer, S., Kuhlmann, U., Gershenzon, J., and Turlings, T.C.J. (2005) Recruitment of entomopathogenic nematodes by insect-damaged maize roots. *Nature*, **434**, 732–737.

Rausch, C., Daram, P., Brunner, S., Jansa, J., Lalol, M., Leggewie, G., Amrhein, N., and Bucher, M. (2001) A phosphate transporter expressed in arbuscule-containing cells in potato. *Nature*, **414**, 462–466.

Razzaghe, M.K. and Robert, M. (1979) Géochimie des éléments majeurs des micas en milieu organique: mécanismes de l'altération des silicates. *Annales Agronomiques*, **30**, 493–512.

Reichard, P.U., Kraemer, S.M., Frazier, S.W., and Kretzschmar, R. (2005) Goethite dissolution in the presence of phytosiderophores: rates, mechanisms, and the synergistic effect of oxalate. *Plant and Soil*, **276**, 115–132.

Reichard, P.U., Kretzschmar, R., and Kraemer, S.M. (2007) Dissolution mechanisms of goethite in the presence of siderophores and organic acids. *Geochimica Cosmochimica Acta*, **71**, 5635–5650.

Rengel, Z. and Marschner, P. (2005) Nutrient availability and management in the rhizosphere: exploiting genotypic differences. *New Phytologist*, **168**, 305–312.

Retallack, G.J. (1997) Early forest soils and their role in Devonian global change. *Science*, **276**, 583–585.

Revsbech, N.P., Pedersen, O., Reichardt, W., and Briones, A. (1999) Microsensor analysis of oxygen and pH in the rice rhizosphere under field and laboratory conditions. *Biology and Fertility of Soils*, **29**, 379–385.

Richardson, A.E. (2001) Prospects for using soil microorganisms to improve the acquisition of phosphorus by plants. *Australian Journal of Plant Physiology*, **28**, 897–906.

Richardson, S.J., Peltzer, D.A., Allen, R.B., McGlone, M.S., and Parfitt, R.L. (2004) Rapid development of P limitation in temperate rainforest along the Franz Joseph soil chronosequence. *Oecologia*, **139**, 267–276.

Richardson, A.E., Barea, J.M., McNeill, A.M., and Prigent-Combaret, C. (2009) Acquisition of phosphorus and nitrogen in the rhizosphere and plant growth promotion by microorganisms. *Plant and Soil*, **321**, 305–339.

Ridley, A.M., Helyar, K.R., and Slattery, W.J. (1990) Soil acidification under subterranean clover (*Trifolium subterraneum* L.) pastures in north-eastern Victoria. *Australian Journal of Experimental Agriculture*, **30**, 195–201.

Riley, D. and Barber, S.A. (1971) Effect of ammonium and nitrate fertilization on phosphorus uptake as related to root-induced pH changes at the root-soil interface. *Soil Science Society of America Proceedings*, **35**, 301–306.

Robert, M. and Berthelin, J. (1986) Role of biological and biochemical factors in soil mineral weathering. In: *Interactions of Soil Minerals with Natural Organics and Microbes* (eds P.M. Huang and M. Schnitzer), Special Publication No. 17, pp. 453–495. Soil Science Society of America, Madison, Wisconsin.

Robin, A., Vansuyt, G., Hinsinger, P., Meyer, J.M., Briat, J.F., and Lemanceau, P. (2008) Iron dynamics in the rhizosphere: consequences for plant health and nutrition. *Advances in Agronomy*, **99**, 183–225.

Rockström, J., Steffen, W., Noone, K., Persson, Å., Chapin, F.S. III, Lambin, E.F., Lenton, T.M., Scheffer, M., Folke, C., Schellnhuber, H.J., Nykvist, B., de Wit, C.A., Hughes, T., van der Leeuw, S., Rodhe, H., Sörlin, S., Snyder, P.K., Costanza, R., Svedin, U., Falkenmark, M., Karlberg, L., Corell, R.W., Fabry, V.J., Hansen, J., Walker, B., Liverman, D., Richardson, K., Crutzen, P., and Foley, J.A. (2009) A safe operating space for humanity. *Nature*, **461**, 472–475.

Roelofs, R.F.R., Rengel, Z., Cawthray, G.R., Dixon, K.W., and Lambers, H. (2001) Exudation of carboxylates in Australian Proteaceae: chemical composition. *Plant Cell and Environment*, **24**, 891–904.

Römheld, V. (1987) Different strategies for iron acquisition in higher plants. *Physiologia Plantarum*, **70**, 231–234.

Römheld, V. (1991) The role of phytosiderophores in acquisition of iron and other micronutrients in graminaceous species: an ecological approach. *Plant and Soil*, **130**, 127–134.

Ruiz, L. (1992) Mobilisation du phosphore des apatites dans la rhizosphère. Rôle de l'excrétion de protons par les racines. PhD thesis, Montpellier II University, Montpellier, France, 125pp.

Ryan, P.R., Delhaize, E., and Randall, P.J. (1995a) Characterisation of Al-stimulated efflux of malate from the apices of Al-tolerant wheat roots. *Planta*, **196**, 103–110.

Ryan, P.R., Delhaize, E., and Randall, P.J. (1995b) Malate efflux from root apices and tolerance to aluminium are highly correlated in wheat. *Australian Journal of Plant Physiology*, **22**, 531–536.

Ryan, P.R., Delhaize, E., and Jones, D.L. (2001) Function and mechanism of organic anion exudation from plant roots. *Annual Review of Plant Physiology and Plant Molecular Biology*, **52**, 527–560.

Ryan, M.H., van Herwaarden, A.F., Angus, J.F., and Kirkegaard, J.A. (2005) Reduced growth of autumn-sown wheat in a low-P soil is associated with high colonization by arbuscular mycorrhizal fungi. *Plant and Soil*, **270**, 275–286.

Ryan, P.R., Raman, H., Gupta, S., Horst, W.J., and Delhaize, E. (2009) A second mechanism for aluminum resistance in wheat relies on the constitutive efflux of citrate from roots. *Plant Physiology*, **149**, 340–351.

Saleque, M.A. and Kirk, G.J.D. (1995) Root-induced solubilization of phosphate in the rhizosphere of lowland rice. *New Phytologist*, **129**, 325–336.

Schenk, H.J. and Jackson, R.B. (2002) The global biogeography of roots. *Ecological Monographs*, **72**, 311–328.

Schlesinger, W.H. (1997) *Biogeochemistry: An Analysis of Global Change*, 2nd edn. Academic Press, San Diego, California.

Schlesinger, W.H. and Andrews, J.A. (2000) Soil respiration and the global carbon cycle. *Biogeochemistry*, **48**, 7–20.

Schnitzer, M. and Kodama, H. (1976) The dissolution of micas by fulvic acid. *Geoderma*, **15**, 381–391.

Schötteldreier, M. and Falkengren-Grerup, U. (1999) Plant induced alteration in the rhizosphere and the utilisation of soil heterogeneity. *Plant and Soil*, **209**, 297–309.

Séguin, V., Gagnon, C., and Courchesne, F. (2004) Changes in water extractable metals, pH and organic carbon concentrations at the soil–root interface of forested soils. *Plant and Soil*, **260**, 1–17.

Simonsson, M., Andersson, S., Andrist-Rangel, Y., Hillier, S., Mattsson, L., and Öborn, I. (2007) Potassium release and fixation as a function of fertilizer application rate and soil parent material. *Geoderma*, **140**, 188–198.

Smith, P., Powlson, D.S., Glendining, M.J., and Smith, J.U. (1997) Potential for carbon sequestration in European soils: preliminary estimates for five scenarios using results from long-term experiments. *Global Change Biology*, **3**, 67–79.

Smith, S.E., Smith, F.A., and Jakobsen, I. (2003) Mycorrhizal fungi can dominate phosphate supply to plants irrespective of growth responses. *Plant Physiology*, **133**, 16–20.

Smith, F.A., Grace, E.J., and Smith, S.E. (2009) More than a carbon economy: nutrient trade and ecological sustainability in facultative arbuscular mycorrhizal symbioses. *New Phytologist*, **162**, 347–358.

Sprent, J.I. and James, E.K. (2007) Legume evolution: where do nodules and mycorrhizas fit in? *Plant Physiology*, **144**, 575–581.

Springob, G. and Richter, J. (1998) Measuring interlayer potassium release rates from soil materials. II. A percolation procedure to study the influence of the variable 'solute K' in the <1...10 µM range. *Zeitschrift für Pflanzenernährung und Bodenkunde*, **161**, 323–329.

Spyridakis, D.E., Chesters, G., and Wilde, S.A. (1967) Kaolinization of biotite as a result of coniferous and deciduous seedling growth. *Soil Science Society of America Journal*, **31**, 203–210.

Stokes, A., Atger, C., Bengough, A.G., Fourcaud, T., and Sidle, R.C. (2009) Desirable plant root traits for protecting natural and engineered slopes against landslides. *Plant and Soil*, **324**, 1–30.

Takagi, S. (1976) Naturally occurring iron-chelating compounds in oat- and rice-root washings. I. Activity measurement and preliminary characterization. *Soil Science and Plant Nutrition*, **22**, 423–433.

Takagi, S., Nomoto, K., and Takemoto, T. (1984) Physiological aspect of mugineic acid, a possible phytosiderophore of graminaceous plant. *Journal of Plant Nutrition*, **7**, 469–477.

Tan, K.H. (1986) Degradation of soil minerals by organic acids. In: *Interactions of Soil Minerals with Natural Organics and Microbes* (eds P.M. Huang and M. Schnitzer), Special Publication No. 17, pp. 1–27. Soil Science Society of America, Madison, Wisconsin.

Tang, C., Drevon, J.J., Jaillard, B., Souche, G., and Hinsinger, P. (2004) Proton release of two genotypes of bean (*Phaseolus vulgaris* L.) as affected by N nutrition and P deficiency. *Plant and Soil*, **260**, 59–68.

Taylor, A.B. and Velbel, M.A. (1991) Geochemical mass balances and weathering rates in forested watersheds of the Southern Blue Ridge II. Effects of botanical uptake terms. *Geoderma*, **51**, 29–50.

Taylor, L.L., Leake, J.R., Quirk, J., Hardy, K., Banwarts, S.A., and Beerling, D.J. (2009) Biological weathering and the long-term carbon cycle: integrating mycorrhizal evolution and function into the current paradigm. *Geobiology*, **7**, 171–191.

Trolldenier, G. (1987) Rhizosphere organisms – potassium interactions with emphasis on methodology. In: *Methodology in Soil-K Research* (ed. International Potassium Institute (IPI)), pp. 283–297. IPI, Bern, Switzerland.

Trumbore, S. (2006) Carbon respired by terrestrial ecosystems – recent progress and challenges. *Global Change Biology*, **12**, 141–153.

Trumbore, S., Salazar Da Costa, E., Nepstad, D.C., Barbosa de Camargo, P., Martinelli, L.A., Ray, D., Restom, T., and Silver, W. (2006) Dynamics of fine root carbon in Amazonian tropical ecosystems and the contribution of roots to soil respiration. *Global Change Biology*, **12**, 217–229.

Turlings, T.C.J. and Ton, J. (2006) Exploiting scents of distress: the prospect of manipulating herbivore-induced plant odours to enhance the control of agricultural pests. *Current Opinion in Plant Biology*, **9**, 421–427.

Turpault, M.P., Utérano, C., Boudot, J.P., and Ranger, J. (2005) Influence of mature Douglas fir roots on the solid soil phase of the rhizosphere and its solution chemistry. *Plant and Soil*, **275**, 327–336.

Turpault, M.P., Gobran, G.R., and Bonnaud, P. (2007) Temporal variations of rhizosphere and bulk soil chemistry in a Douglas fir stand. *Geoderma*, **137**, 490–496.

Uhde-Stone, C., Gilbert, G., Johnson, J.M.F., Litjens, R., Zinn, K.E., Temple, S.J., Vance, C.P., and Allan, D.L. (2003) Acclimation of white lupin to phosphorus deficiency involves enhanced expression of genes related to organic acid metabolism. *Plant and Soil*, **248**, 99–116.

Uroz, S., Calvaruso, C., Turpault, M.P., Pierrat, J.C., Mustin, C., and Frey-Klett, P. (2007) Effect of mycorrhizosphere on the genotypic and metabolic diversity of the bacterial communities involved in mineral weathering in a forest soil. *Applied Environmental Microbiology*, **79**, 3019–3027.

Van Breemen, N., Mulder, J., and Driscoll, C.T. (1983) Acidification and alkalinization of soils. *Plant and Soil*, **75**, 283–308.

Van Breemen, N., Driscoll, C.T., and Mulder, J. (1984) Acidic deposition and internal proton sources in acidification of soils and waters. *Nature*, **307**, 599–604.

Van Schöll, L., Kuyper, T.W., Smits, M.M., Landeweert, R., Hoffland, E., and Van Breemen, N. (2008) Rock-eating mycorrhizas: their role in plant nutrition and biogeochemical cycles. *Plant and Soil*, **303**, 35–47.

Vansuyt, G., Souche, G., Strackek, A., Briat, J.F., and Jaillard, B. (2003) Flux of protons released by wild type and ferritin over-expressor tobacco plants: effect of phosphorus and iron nutrition. *Plant Physiology and Biochemistry*, **41**, 27–33.

Velbel, M.A. and Price, J.R. (2007) Solute geochemical mass-balances and mineral weathering rates in small watersheds: methodology, recent advances, and future directions. *Applied Geochemistry*, **22**, 1682–1700.

Velde, B. and Barré, P. (2010) *Soils, Plants and Clay Minerals: Mineral and Biologic Interactions*. Springer, Heidelberg, Germany.

Verboom, W.H. and Pate, J.S. (2003) Relationships between cluster root-bearing taxa and laterite across landscapes in southwest Western Australia: an approach using airborne radiometric and digital elevation models. *Plant and Soil*, **248**, 321–333.

Verboom, W.H. and Pate, J.S. (2006a) Bioengineering of soil profiles in semiarid ecosystems: the 'phytotarium' concept. A review. *Plant and Soil*, **289**, 71–102.

Verboom, W.H. and Pate, J.S. (2006b) Evidence of active biotic influences in pedogenetic processes. Case studies from semiarid ecosystems of south-west Western Australia. *Plant and Soil*, **289**, 103–121.

Vitousek, P.M. and Farrington, H. (1997) Nutrient limitation and soil development: experimental test of a biogeochemical theory. *Biogeochemistry*, **37**, 63–75.

Vitousek, P.M., Cassman, K., Cleveland, C., Crews, T., Field, C.B., Grimm, N.B., Howarth, R.W., Marino, R., Martinelli, L., Rastetter, E.B., and Sprent, J.I. (2002) Towards an ecological understanding of biological nitrogen fixation. *Biogeochemistry*, **57**/58, 1–45.

Wallander, H. (2000) Uptake of P from apatite by *Pinus sylvestris* seedlings colonised by different ectomycorrhizal fungi. *Plant and Soil*, **218**, 249–256.

Wallander, H. and Wickman, T. (1999) Biotite and microcline as potassium sources in ectomycorrhizal and non-mycorrhizal *Pinus sylvestris* seedlings. *Mycorrhiza*, **9**, 25–32.

Weisenseel, M.H., Dorn, A., and Jaffe, L.F. (1979) Natural H^+ currents traverse growing roots of barley (*Hordeum vulgare* L). *Plant Physiology*, **64**, 512–518.

Witter, E. and Johansson, G. (2001) Potassium uptake from the subsoil by green manure crops. *Biological Agriculture and Horticulture*, **19**, 127–141.

Yang, J., Blanchar, R.W., Hammer, R.D., and Thompson, A.L. (1996) Soybean growth and rhizosphere pH as influenced Aa horizon thickness. *Soil Science Society of America Journal*, **60**, 1901–1907.

Youssef, R.A. and Chino, M. (1989) Root-induced changes in the rhizosphere of plants. I. pH changes in relation to the bulk soil. *Soil Science and Plant Nutrition*, **35**, 461–468.

Zheng, S.J., Ma, J.F., and Matsumoto, H. (1998) High aluminum resistance in buckwheat. I. Al-induced specific secretion of oxalic acid from root tips. *Plant Physiology*, **117**, 745–751.

Zoysa, A.K.N., Loganathan, P., and Hedley, M.J. (1997) A technique for studying rhizosphere processes in tree crops: soil phosphorus depletion around camellia (*Camellia japonica* L) roots. *Plant and Soil*, **190**, 253–265.

11 Sustainable management of soil and plant health by optimising soil biological function

Dominic Standing[1] and Ken Killham[2]

[1] *School of Biological Sciences, University of Aberdeen, Aberdeen, UK*
[2] *Honorary Fellow, The James Hutton Institute, Dundee, UK*

11.1 Introduction

11.1.1 Concepts of soil and plant health in relation to management and land use

Soil health and plant health are easily understood concepts and, for plant health, relatively easy to quantify. Is the plant growing well? Do its leaves show signs of chlorosis? Is it likely to produce a full harvest? These are simple questions that any agronomist and agriculturalist will be able to answer without difficulty. Plant health, however, relies on soil health, and soil health is far less easy to quantify as it can be interpreted in numerous ways. Broadly speaking, soil health is the measure of the capacity of that soil to underwrite specified functions. Included here are plant and animal production systems, natural landscape and ecosystem maintenance, regulation of water flow and quality, air quality and nutrient cycling, and the attenuation of many of the negative impacts associated with human habitation and industry. Thus, depending on one's point of view, soil health is primarily an expression of the soil to produce maximum harvests at maximum profit, or cheap and plentiful food, or to support a landscape of diverse plant life and associated wildlife, or provide suitable 'ground' for habitation and industrial activities. However, not all of these are in line with the sustainable use of soil or the maintenance/improvement of soil health.

As national and international policy guidelines tend towards sustainability of resource use, understanding determinants of soil health is essential. Land managers in both the agricultural and industrial sectors can monitor the effects of specific practices on soil health and relate these to its sustainability.

To gain maximum crop yield from their land, farmers have often relied on the input of inorganic chemical fertilizers, particularly NPK, or organic additions such as manures and composts. This results in the alteration of the available and potentially available soil nutrient pools. The pH of the soil is key to determining the level of biological activity and nutrient availability, and management practices must ensure maintenance of soil pH within a range suitable for biological functions which underpin a particular land use.

Soil Conditions and Plant Growth. Edited by Peter J. Gregory and Stephen Nortcliff.
© 2013 Blackwell Publishing Ltd. Published 2013 by Blackwell Publishing Ltd.

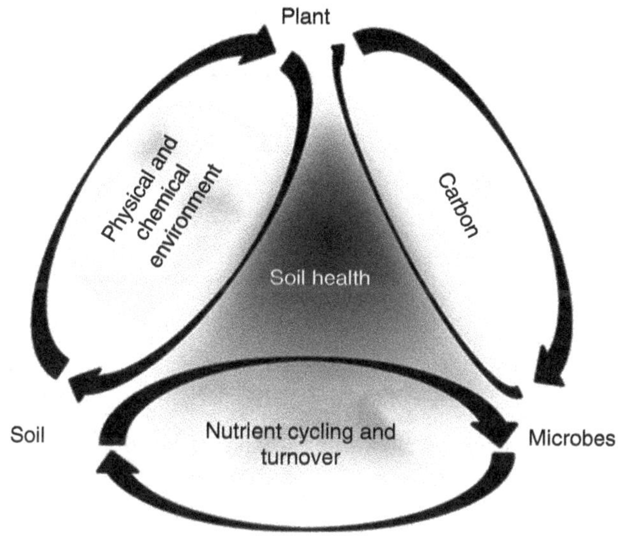

Figure 11.1 Simple schematic of the tripartite interaction between soil, plants and microbes and how this relates to soil health. Plants act as carbon pumps into the soil driving microbial heterotrophic functions and nutrient turnover. This impacts on nonbiological aspects of soil function, in turn impacting on plant growth. Each element of the tripartite interaction can involve multiple feedbacks. Changes in any one of the principal axes will have repercussions on the remaining two with a concomitant change in soil health.

Soil health is a summation of dynamic biotic and abiotic processes and feedbacks. While these are simply presented in Figure 11.1, they can interact in highly complex ways. Consider, for example, that a 60% clay-rich soil may have a specific surface area of nearly 500 m^2 per g of soil. This surface area is packed into an extremely complex system of solids and pore spaces. The pore spaces may be water or gas filled and may be connected or not. The entire surface will have a degree of electro-chemical activity influencing the sorption and desorption of ions and molecules. Superimposed on this will be approximately one billion bacteria, largely attached to pore walls, as well as a large mass of fungal mycelia and a wide range of microfauna (Killham, 1994). Some of the fungal mycelia are mycorrhizal and connected to plant root systems. All of these biotic components interact with the physico-chemical environment. Thus, there is a great level of complexity in terms of the actual and potential interactions between the biota and the soil habitat. However, in terms of land use and management, there is less interest in the microscale of soil (and its biotic interactions) and more in the macroscale (field and landscape), and, specifically, how management practices impact on soil use and health (Table 11.1). The biological (microbial and faunal) components are key to the creation and maintenance of soil organic matter (SOM). SOM plays a vital role in soil structure, water relations, as well as nutrient status and turnover. Soil health is dependent on physical, chemical and biological qualities (Table 11.1). The physical components of the soil often determine water flow and thus the movement of nutrients, contaminants and even micro-organisms. For example, a sandy soil will drain and transport water rapidly in comparison to a silty loam or clay soil (assuming no major cracks). The structure of a soil is a measure of its texture, porosity and bulk density, and while a soil's texture is not easily altered, changes in porosity and bulk density can dramatically impact crop yield.

Table 11.1 Soil health indicators, key soil functions and management strategies to maintain or improve soil health.

Soil health indicators	Soil functions	Management strategies
Soil organic matter (SOM)	Soil structure and stability, water and nutrient retention; reduces soil erosion, C source/sink.	Addition of mulch, plant residues and/or compost
Physical	Water and solute movement and drainage Water stable aggregates Bulk density and root penetration Soil pores as niches for macro- and microfauna	Low/zero tillage
Chemical	pH, extractable soil nutrients NPK and macronutrient base cations Ca, Mg and K Mobility and bioavailability of metals/heavy metals	Maintain or improve SOM (see above)
Biological	Microbial biomass C and N N-fixation Microbial enzyme activity and soil nutrient status	Maintain or improve SOM (see above)

Drivers for managing soil and plant health include political/legislative (soil protection/conservation, watershed protection), economic as well as environmental. Managing soil and plant health is driven by political/legislative, economic and environmental factors. The scope of this chapter is to introduce managing soil and plant health for a variety of services involving soil biological function. Management for biofertilisation is considered. Reference is first made to chemical and organic fertilisation; however, despite success of microbial N_2 fixation and some partial success with P-solubilising micro-organisms, the majority of crop fertilisation for high yields is still achieved through inorganic, chemical applications. The chapter also deals with the restoration of soil after degradation. Here, understanding the role of plants, micro-organisms and animals in maintaining soil structural stability is emphasised. Also highlighted are the bioremediation of soil contamination (increasingly being used as a sustainable solution to pollution problems worldwide) and the many factors determining its success. Finally, crop protection is considered and, despite some successful applications to physically protected (e.g. glasshouses and polytunnels) high-value crops, the field application of biocontrol agents is still in its infancy.

With all of the aspects of managing soil and plant health covered in this chapter, the interested reader is advised to follow-up the references supplied, which provide a more comprehensive picture of current management systems as well as the future potential and ongoing research.

11.2 Managing soil and plant health

11.2.1 Improvement of nutrient supply/status

Chemical/organic fertilisation versus biofertilisation

Globally, management of the soil/plant system to improve nutrient status/supply is largely carried out through application of chemical fertilisers and to some extent by the application of manures and composts. Production of fertiliser nitrogen is currently estimated to be about 170 Tg per year (Galloway, 2003), with the global use of N fertilisers increasing by about 15 Tg per year in order to meet the ever increasing demand for food. Nitrogen fixation in agricultural systems represents about another 40 Tg of reactive nitrogen being added globally. This contribution from fixation is mainly due to legumes and cyanobacteria/rice paddy systems (Killham, 1994). It is hard to estimate the proportion of this contribution from fixation due to biofertilisation as some legumes (e.g. clover in temperate pasture systems) are not inoculated and inoculation of rice paddies is far from routine. However, it does indicate the considerable potential of biofertilisation, a potential likely to be increasingly fulfilled as fertiliser costs escalate and crop/land management is driven towards sustainability.

Fertiliser efficiency in terms of nutrient supply to the target crop is continuing to be increased through innovative management of the plant/soil system such as precision agriculture (Haneklaus and Schnug, 2006) and better transfer of knowledge to farmers, while integrated soil management is increasingly exploiting the combined use of inorganic and organic fertilisation. However, there is a wealth of information on fertiliser and nutrient management regimes. The principles underpinning these systems are well understood and incorporated into existing management practises. Comprehensive references are widely available (e.g. Havlin et al., 2004). Improving nutrient supply using biofertilisers, however, is a very different story and is a challenge that needs serious consideration for sustainable management of the plant/soil system.

Biofertilisers for N_2 fixation, P solubilisation, S oxidation and to introduce plant growth promoting rhizobacteria (PGPRs)

Biofertilisation to improve crop N supply largely revolves around legume inoculation with *Rhizobium* and growth of cyanobacteria with paddy rice, and these are the most important forms of biofertilisation worldwide. Inoculation of various legumes for enhancing N_2 fixation has been highly successful (mainly in tropical latitudes, where gains from fixation tend to be highest). Biofertilisation of rice with cyanobacteria, usually growing in association with the water fern *Azolla*, is also highly successful and supplies fixed N to some 86% of global rice production (Vaishampayan et al., 2001). Other attempts to increase soil fertility through use of free-living nitrogen fixation have been far less successful. The key, fundamental problem is that the energy required to drive the fixation is high and any free living inoculum (other than cyanobacterial systems which are driven by solar energy) must compete for available C with all other heterotrophic micro-organisms in the soil (Killham, 1994). For a temperate cereal crop, there is a rhizosphere carbon flow of approximately 0.5 ton C ha^{-1} y^{-1}. If all of this soil carbon flow were available to a population of the free-living diazotrophic bacterium *Azotobacter*, it would still only be enough to generate about 10 kg N ha^{-1} y^{-1} (Figure 11.2; Killham, 1994). In reality, the *Azotobacter* will only successfully compete for a rather small proportion of this carbon, and hence free living N_2 fixation will not significantly contribute to the needs of a growing crop, and biofertilisation will not be economically viable.

In view of the carbon limitation on free living N_2 fixation illustrated in Figure 11.2, for all cropping systems except the rice paddy, the legume-*Rhizobium* system offers a far more encouraging plant/soil management tool than free living fixation because the rhizobial symbiont carries out N_2 fixation within the plant, and the root nodules formed receive the energy for fixation from translocated, photosynthetically fixed carbon from the plant host leaves (Killham, 1994).

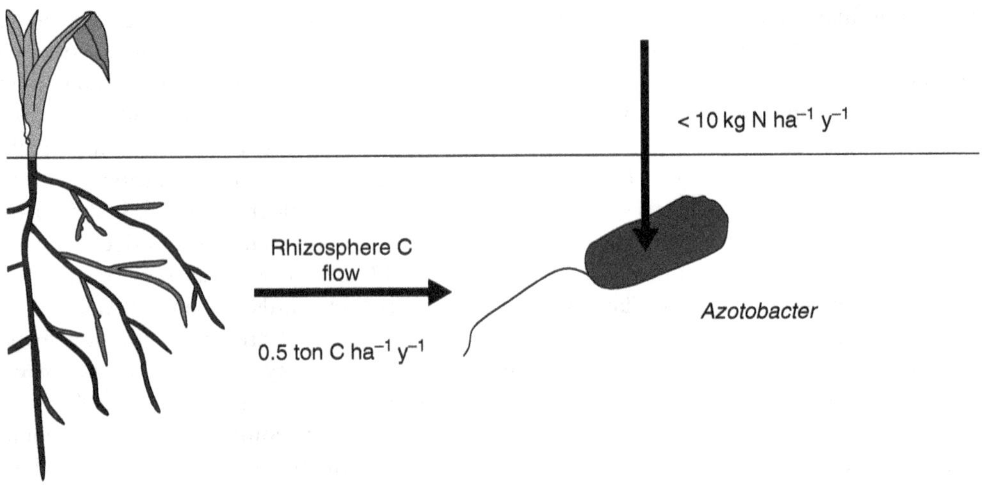

Figure 11.2 The limitation to N_2 fixation by free-living heterotrophic bacteria in soil is the supply of available carbon, which is mainly through rhizosphere flow. The numbers in the figure refer to typical fluxes for a temperate cereal crop. For a colour version of this figure, please see Plate 11.1. From Killham (1994).

Great improvements have been made in terms of management of rhizobial biofertilisation of legumes over recent decades. The soils in tropical latitudes where heavy inoculation is carried out often represent a hostile environment (particular in terms of water stress and adverse pH) when the rhizobial inoculum is introduced, and without protection, loss of viable rhizobial cells can be high. The technology of inoculum delivery to the legume root has therefore become more efficient through the use of effective carriers and seed coats, and better inoculum production and storage (Deaker et al., 2004).

Inoculation programmes with mycorrhizal fungi, phosphate solubilising and S oxidising micro-organisms have encountered much more limited success than the rhizobial/legume and cyanobacterial/rice paddy systems and have only delivered benefits under certain conditions (Siddiqui et al., 2008; Banerjee and Yesmin, 2009; Kahn et al., 2009). Therefore, widespread and routine biofertilisation of field crops/trees with these forms of microbial inoculation has yet to be realised, although it clearly offers considerable potential.

In terms of mycorrhizal systems, inoculation with ectomycorrhizal basidiomycotina has offered promising biofertilisation management systems since the fungi can be readily cultured and bulk inoculum produced and delivered to tree seedling hosts in nurseries (Rincon et al., 2005). However, inoculation has failed to generate a consistent, positive response in the tree host across the wide range of edaphic conditions required/found in the field.

There are numerous examples of trials with P-solubilising rhizobacteria and fungi where successful crop biofertilisation has been achieved (Rai, 2005), and there is also clear evidence of a diverse range of plant growth promoting rhizobacteria (PGPRs), eliciting impressive, increased crop yields in certain soils (Egamberdiyeva, 2007).

In the work by Alogaidi et al. (2011), rice cultivars IAC25 and Azucena were grown in sterile, P-limited subsoil with a biofertiliser inoculum of 5% w/w live topsoil. Controls contained sterile subsoil and sterile topsoil. The photographs of representative root systems (Figure 11.3) clearly show that the presence of a rhizobacterial community is important for IAC25. Interestingly, this does not appear to be the case for the Azucena cultivar. The

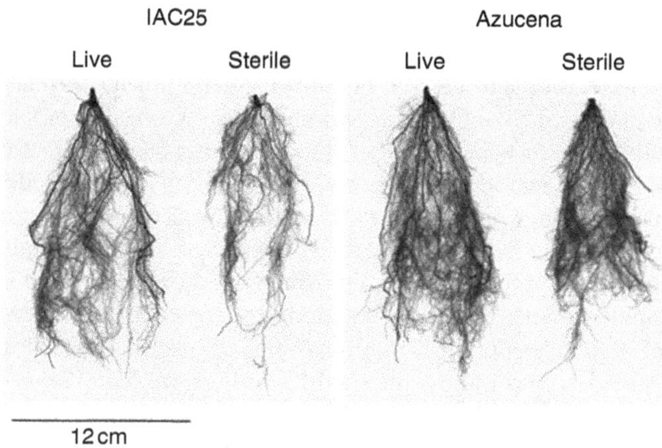

Figure 11.3 Effects on root growth of the addition of a live or sterile soil (1:20 mix with sterile subsoil) as a biofertiliser inoculum source (Alogaidi et al., 2011).

consequences for P-acquisition and growth were that IAC25-sterile was significantly smaller than IAC25-live and took up significantly less P. This was not translated into a lower P-use efficiency. The PGPR response then is by no means universal and this, with appropriate screening, opens up opportunities to match soil conditions to crop cultivars for biofertilisation.

P-solubilising bacteria (PSB) are a widely studied subset of the soil microbiota due to their obvious importance in liberating P for plant uptake. The most efficient PSBs are found in the *Bacillus* and *Pseudomonas* genera (Tilak et al., 2005). Of particular interest are the pseudomonads – an abundant and versatile group of Gram negative, aerobic, chemoheterotrophs. Goldstein and Rogers (1999) demonstrated the solubilising mechanism, extracellular quinoprotein glucose dehydrogenase, oxidises glucose (primary component of seedling exudate-see Yeomans et al., 1999) to gluconic acid. Fungi such as *Aspergillus* and *Penicillium* can solubilise bound phosphates, often through secretion of organic acids and/or phosphatase enzymes. In the case of P solubilisers and other PGPRs, large-scale, consistent, commercial application has yet to be realised, even though, when used as biofertilisers under favourable conditions, microbial inoculants can substitute up to 25% of plant phosphorus requirements (Rai, 2005). As with mycorrhizal systems, the main reason for this is that the soil ecological conditions required vary from one inoculum/crop host/soil combination to another and the rules of engagement are poorly understood. It is widely thought that plant species exert selective pressures on rhizosphere bacterial community structure, resulting in identifiable communities associated with that plant (Berg and Smalla, 2009). Marschner et al. (2001) have demonstrated that rhizobacterial communities change not only with species but also with plant growth and development. In that study, community changes were associated with nutrient cycling and root growth after winter rains. They suggest that the key drivers behind rhizobacterial community change and activity are quantitative and qualitative changes in root exudates.

While it may be expected that, over evolutionary time, different plant species can select for specific rhizobacterial communities, it is fascinating to consider this phenomenon in terms of the plant-breeding history. Traditionally, the key driver for plant breeding has been to maximise yield, and this has most often been achieved through high nutrient input. However, if specific crop cultivars can select for rhizobacterial communities with different functions such as nutrient cycling, then these traits may be exploited for plant growth promotion under the reduced fertiliser input regimes that are becoming increasingly necessary. Germida and Siciliano (2001), in an investigation of modern and ancient wheat cultivars, demonstrated that rhizobacterial community structure was greater on land races than modern cultivars. However, roots of modern wheat cultivars were aggressively colonised by fast growing endophytic bacteria, particularly pseudomonads. It is suggested that either root morphologies or root exudate profiles were responsible for this (Sicilano et al., 1998; Ferrieira et al., 2008), although Briones et al. (2002) found that microscale differences in O_2 availability along rice root cultivars were the key determinant in community profiles of ammonia-oxidising bacteria. In a study that investigated rice root exudate and plant development, Aulakh et al. (2001) confirmed that the chemical composition of exudate is under cultivar-specific control. Intracultivar differences were found as well as a shift from sugars to organic acids as the rice plants developed.

Furthermore, in an experiment that investigated the organic acid profiles of rice cultivars grown with and without access to P fertilisation, Elliot (2011) found the profile response to fertiliser P differed between cultivars (Figure 11.4a and b).

If the chemical signature of root exudate is cultivar specific (as evidenced in Figure 11.5), it can be hypothesised that cultivars will have different effects on rhizosphere chemistry and pH. In turn, this has implications for rhizosphere biological function such as P-mobilisation. The potential for linking cultivar-specific exudates to P-solubilisation and therefore crop nutrition is considerable. A promising way forward is to identify appropriate cultivar genetic markers (Quantitative Trait Loci or QTLs) linked to specific exudate patterns.

Figure 11.5 highlights that for the consistently successful use of microbial inocula, favourable edaphic conditions (and their associated management) are required at the point of inoculum delivery and during microbial inoculum establishment in the soil, and this is challenging to manage at the commercial scale. One of the most important aspects of this relates to the soil physical conditions, in particular, the matric component of soil water potential (Figure 11.6).

Figure 11.6 demonstrates that for most soils, with the exception of periods shortly after rainfall/irrigation or longer periods in soils with impeded drainage, active micro-organisms are confined to water in the soil micropores, and connectivity between these habitats is low. For bacteria introduced into soil, therefore, movement through the soil by chemotactic motility (over short distances) and mass flow (over greater distances) for colonisation of a root system or substrate is not possible without assistance from some other agent. These agents may take the form of motile soil animals (e.g. earthworms, nematodes and protozoa) or the apical tip of a developing plant root.

11.2.2 Restoration of soil after degradation

Restoration of soil structural integrity/soil stability and the role of soil organic matter

Loss of soil structural integrity/soil stability and consequent soil erosion represents one of the greatest challenges to meeting the food and fibre needs of the ever-increasing world population. Salinisation/desertification and soil mismanagement are the main causes of this loss of soil and soil fertility worldwide (Zhao et al., 2009). The loss in crop productivity due to soil degradation (from erosion) is estimated at 18 Tg of food staples per year at the 1990 level of yields for sub-Saharan Africa and 272 Tg for the world at the 1996

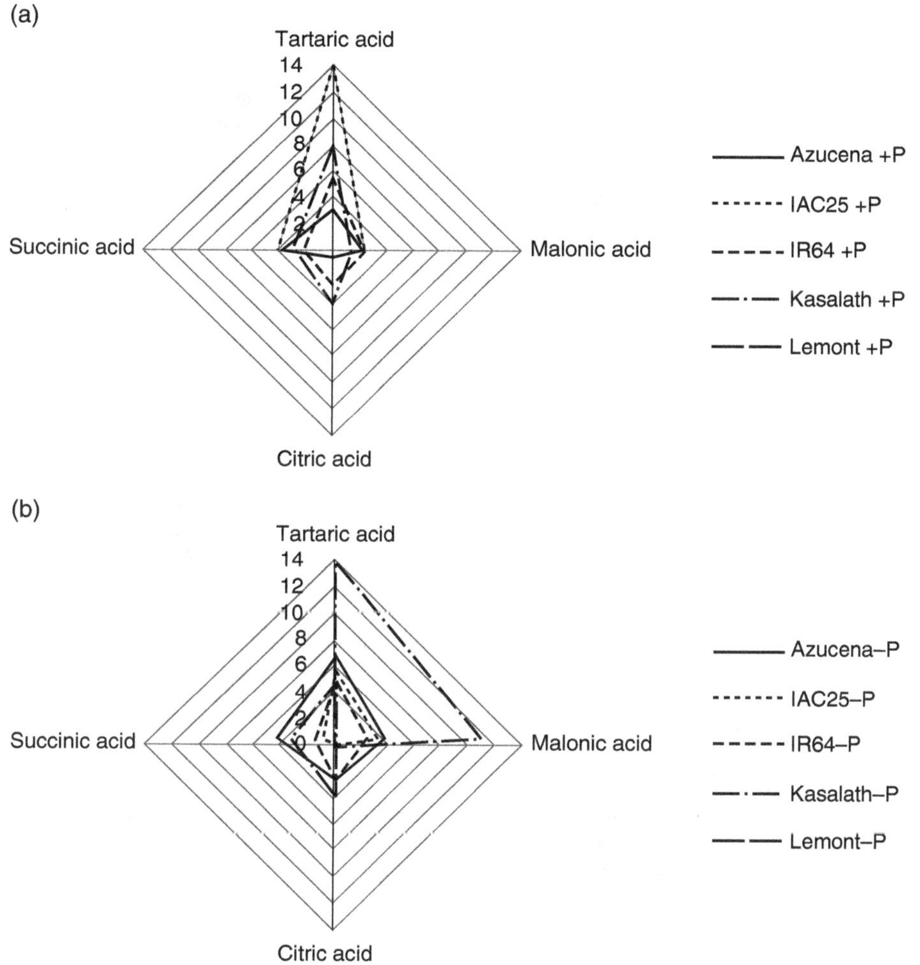

Figure 11.4 Radar plots of organic acid profile of five rice cultivars grown in sterile sand. Yoshida's nutrient solution with P amendment (a) or no P amendment (b), n = 4–5 (Elliot 2011). Note that in Figure 11.4a cultivar IAC25 shows a remarkably different organic acid profile from the other cultivars grown under P amendment. In Figure 11.4b, with limited P conditions, the cultivar Kasalanth shows a different strategy from the other tested cultivars.

level of production (Lal, 2000). This may represent a yield loss at the landscape level or even total crop failure at the farm level. Much of the soil's structural integrity comes from its organic matter, including the living fraction (the soil microbial biomass).

Soil organic matter plays a key role in soil structure and consequently in stability in terms of resistance to erosion (Figure 11.7), with soil structural stability mediated by three main soil organic matter fractions (Cresser et al., 1993). The microbial biomass itself, both living and necromass, contributes through the enmeshing of soil particles. This mainly involves filamentous micro-organisms such as fungi and actinomycetes, which produce networks (mycelia) of tubes called hyphae. Physically protected organic matter, largely polysaccharidic in nature, also plays a vital part in stabilisation as the polysaccharide acts like glue holding fine soil particles together in aggregates. Organic matter moieties which carry a net negative

Figure 11.5 Factors controlling the success of a microbial inoculum in soil. Note that many of these factors can strongly interact – e.g. Predation of microbial inocula by protozoa and nematodes is regulated by matric potential as continuous water films are required for movement of these predators which are considerably larger than their microbial prey.

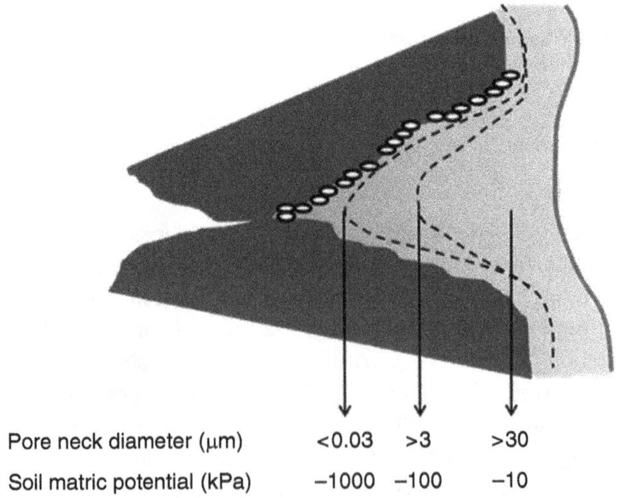

Figure 11.6 The relationship between soil matric potential and the neck diameter of the largest water-filled pores. For a colour version of this figure, please see Plate 11.2.

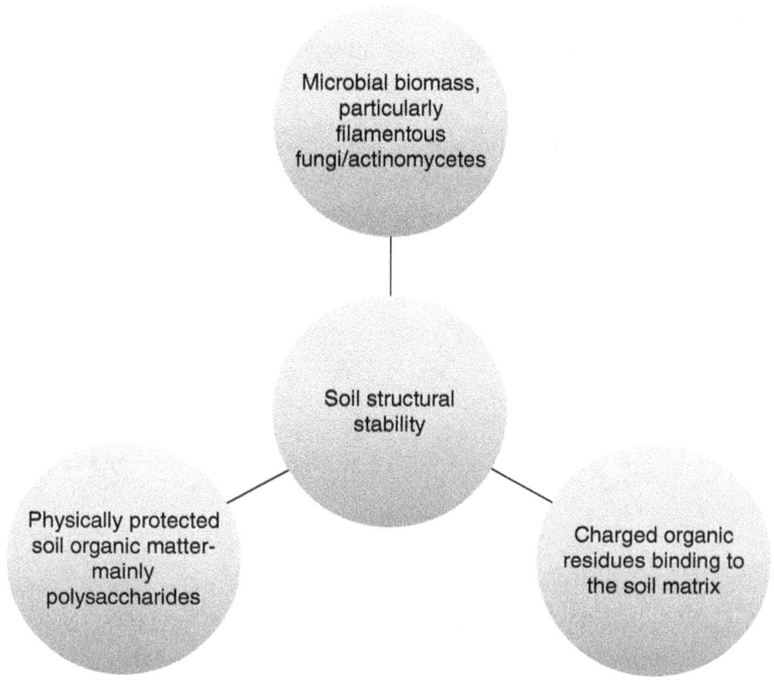

Figure 11.7 The mechanisms by which organic matter contributes to soil structural stability.

charge will bind to other components of the soil matrix which carry a charge. This is often by cationic bridging since much of the mineral and other organic fraction of the soil is also negatively charged (Killham, 1994).

Without the returning of adequate crop and animal residues, continuous agricultural production on the same land reduces the soil organic matter fraction, mainly through the process of mineralisation. Mineralisation without replenishment of stocks represents the 'mining' of soil organic matter. This type of agriculture is common worldwide and complete collapse of some soils can occur when organic matter levels fall below a critical threshold (Russell, 1977), as organic matter serves as both a short-term and more persistent binding agent. However, it should be borne in mind that the soil organic matter threshold value below which structural stability can be lost will vary from one soil type to another mainly because of the complex role of organic matter–clay interactions in the aggregate stabilisation process.

In addition to the problem of declining organic matter, loss of soil structure can also occur through the processes of slaking and dispersion. These processes are often linked to intensive cultivation (particularly ploughing), causing compaction and vital loss of the pore size distribution needed for maintenance of soil fertility (Greenland and Pereira, 1977). Because of these different processes, involving organic matter, soil mineral particles (particularly clays) and water, the mechanisms of soil structural collapse and degradation vary climatically and from one soil type to another, but the impact is generally devastating and the benefits of restoration are therefore considerable. An increase of 1 ton of soil carbon pool to degraded cropland soils may increase crop yield by 20–40 kilograms per hectare (kg/ha) for wheat, 10–20 kg/ha for maize and 0.5–1 kg/ha for cowpeas (Lal, 2004).

Soil management to restore soil structural integrity/soil stability involves the application of a range of technologies and management approaches. Some of these technologies and approaches enhance soil structure and include residue incorporation and mulching, use of appropriate soil conditioners, addition of manures and some forms of agroforestry. A number of additional technologies and approaches are designed to conserve soil and include cover cropping, contour ploughing, riparian zones, minimum tillage and efficient and appropriate irrigation. Álvaro-Fuentes et al. (2009) have demonstrated that in the semiarid Mediterranean agroecosystems, for example, the adoption of zero tillage together with the suppression of long-fallowing period can significantly increase the amount of SOC (particularly the hydrophobic components of humic material) stabilised in the soil surface and this can significantly improve soil structure and aggregation.

Management approaches and technologies to promote/restore structural stability through increasing soil organic matter have to be selected on a soil/site-specific basis but offer the practitioner of plant–soil management tools of soil restoration from physical degradation worldwide.

Plants and algae for soil stabilisation

Because plants are involved in both building and maintaining soil structure, plant management for soil structural stabilisation can be carried out in a number of ways.

Because plant root systems, particularly those with extensive rhizomes, can have a strong influence in stabilising soils, some plants are introduced to stabilise sands and degrade soils. This is a common practice to minimise coastal erosion in Europe and North America where *Spartina* (marram grass) and *Ammophila* (beach grass) have long been planted on unstable sand dunes, which are prone to erosion (Nordstrom, 2008). The mechanisms of stabilisation are threefold. The roots and rhizomes themselves enmesh the sand/soil particles, the carbon flow from the roots drives adhesion of particles both directly and indirectly through microbially produced polysaccharides, and mycorrhizal fungal hyphae associated with the roots also contribute to soil aggregation (Koske and Gemma, 1997).

Photosynthetically driven stabilisation of soils is also mediated by algae, particularly filamentous cyanobacteria such as species of *Microcoleus*, *Phormidium and Scytonema* and *Nostoc*. Threshold wind velocities for erosion of fine desert sands are increased by the presence of these cyanobacteria, particularly those that form thick, cohesive cryptogamic crusts (Hu et al., 2002). With further development of the algal crust, more and more fungi can grow, facilitating the formation of lichens. This further improves soil conditions which can encourage mosses and hence much greater protection against erosion. So, the algae have a key role as primary colonisers and in the early stages of formation of the crusts, which protect desert and other unstable, otherwise bare soils from erosion, as well as provide important inputs of C and N (Orlovsky et al., 2004).

Managing soils to improve structural stability through either encouragement of indigenous algae or the application of algal inocula (algalisation) is a new approach compared to the use of similar algae in rice paddies for biofertilisation. However, if the constraints of adverse water potential and pH can be overcome (most unstable soils in arid and semi-arid zones are too dry and often too acid or alkaline for ready colonisation by algae), then considerable improvements in aggregate stability can be achieved through algalisation (Nisha et al., 2007).

Soil animals – role in soil structure, ecosystem engineering and managing systems lacking soil animals

The soil mesofauna can amount to 2.5 ton of biomass in a single hectare of soil (Killham, 1994), and key members of the soil mesofauna have long been linked with maintenance and improvement of soil structure. This is particularly true of mesofauna such as earthworms and dung beetles. The effects of these can be very considerable, and some soils and their associated organic matter inputs (animal and plant detritus) are heavily processed by these animals. Along with plant roots, as discussed in the previous section, earthworms and other burrowing animals of sufficient size move soil particles to create and then stabilise (with mucilaginous stabilizing agents and faecal materials) cylindrical section pores which are key to good soil structure (Oades, 1993). It is not unusual for some top-soils in faming systems with lower inputs of fertilisers and particularly insecticides to entirely comprise earthworm cast material (Edwards, 2004).

Management of soil structure involving animals is of considerable importance in certain parts of the world. There are two broad approaches. One involves managing soil conditions to control faunal activity. Application of organic residues and manures, for example, encourages earthworm activity in many soils, assuming pesticidal inputs are low. This encourages incorporation of the organic inputs, cycling of the bound nutrients and modification of soil structure through burrowing/soil processing and lining of tunnels. The second management approach involves the actual introduction of fauna and may also comprise managing soil conditions to encourage faunal establishment and activity. In South Australia, for example, earthworms have been reintroduced into wheat belt soils where they had been lost through continuous fertilisation and associated acidification to beyond threshold pH limits for lumbricid worms (Edwards, 2004). Reintroduction involved incorporation of turves of earthworm-rich soil along with liming the soil to enable their successful colonisation/re-establishment. Dung beetles, belonging to the superfamily Scarabacoidca, have also been used as ecosystem engineers and their introduction in pasture soils to cycle nutrients bound in faecal material from grazing cattle and enhance soil structure is well documented in a number of countries (e.g. Bang et al., 2005).

11.2.3 Remediation of soil contamination

Bioremediation versus conventional remediation (in situ and ex situ)

With strong regulatory and other environmental drivers, management of soil and plant health is becoming an increasingly important part of remediation of chemically contaminated land.

Even though estimates of the extent of chemically contaminated (involving organic contaminants and/or metals) land will vary with definition/standards as well as with methods of chemical analysis, and a global figure is unknown, the extent of the problem is very considerable indeed. Estimates of the extent of chemically contaminated land in England and Wales alone range from 50 000 to 200 000 ha (Royal Commission, 1996; EA, 2000), with up to a further 9 000 ha in Scotland (Scottish Government, 1998). The main areas of heaviest chemical contamination are associated with long histories of industrial development. However, the problem is worldwide, with chemical contamination from environmentally persistent pesticides and from diffuse atmospheric pollutants almost ubiquitous.

Bioremediation increasingly provides an attractive alternative (and complement) to the conventional 'dig and dump' remediation of contaminated land. The attraction is often based on lower costs and regulatory drivers particularly tied to reduced dependence on land-filling,

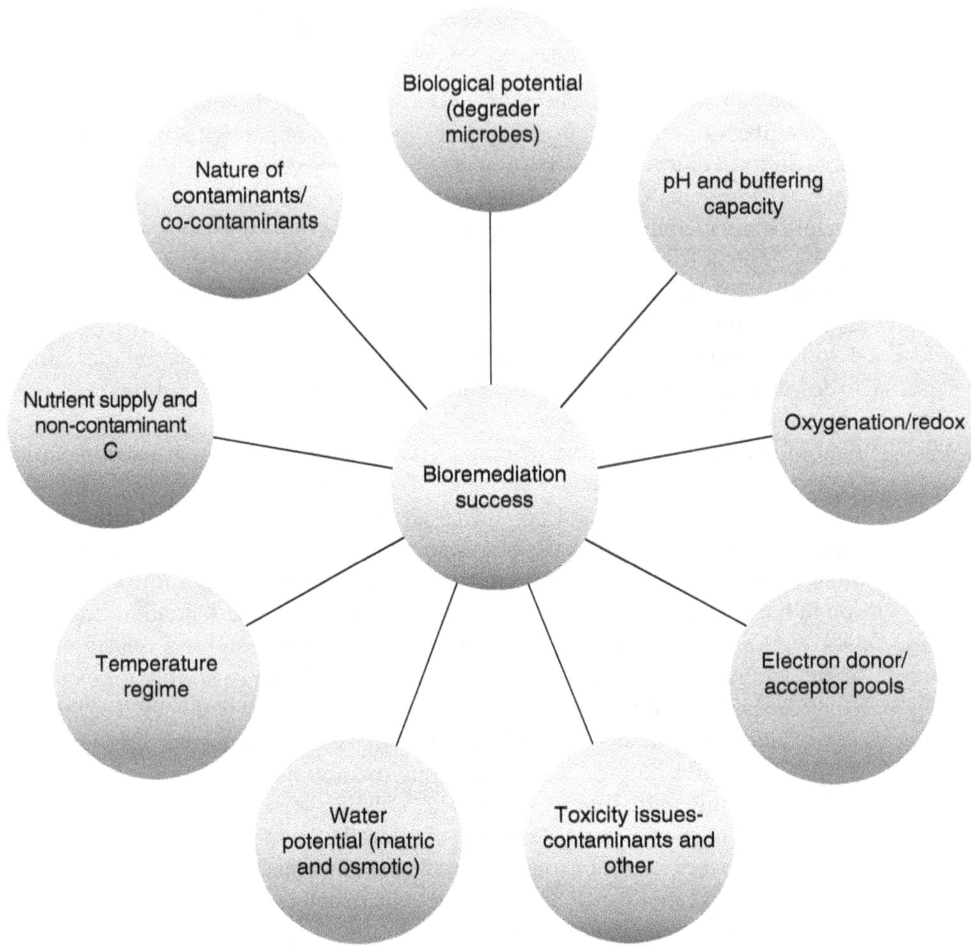

Figure 11.8 Factors controlling the success of bioremediation require careful evaluation on a site/soil-specific basis. Any management strategy requires that any constraints should be alleviated and factors controlling rate of bioremediation optimised. For degradation of petroleum hydrocarbons, this often involves aeration through turning (*ex situ* bioremediation) or sparging (*in situ* bioremediation) to optimise the oxygen supply to the micro-organisms (largely bacteria) driving this highly aerobic process.

but also on the linkage between soil and plant health end points and bioremediation. This linkage is fundamental since bioremediation draws on the activity of the plant/soil system, either for the degradation or immobilisation of many of the target contaminants.

Two broad management approaches to bioremediation can be adopted, depending on whether contaminated material is excavated (*ex situ* bioremediation) or not (*in situ* bioremediation) for biological treatment. Both *in situ* and *ex situ* bioremediation of contaminated land require integrated management of the plant, soil and water system. The key objective is to remove any constraints to the activity of the micro-organisms which carry out bioremediative processes and then optimise the physiochemical conditions, which determine the rate of these processes (Figure 11.8). In managing bioremediation, issues of nutrient status/supply, water potential and oxygenation are of paramount importance to ensure optimal rates of remediation where the aim is degradation of organic pollutants, particularly most petroleum hydrocarbons (Piotrowski et al., 2006). Other issues such as toxicity of contaminants and co-contaminants, varying bioavailability of target contaminants and pH-dependent toxicity (particularly for heavy metals) are more poorly understood and often require careful consideration to ensure successful bioremediation.

The identification and removal of underlying toxic constraints to bioremediation is illustrated in Figure 11.9 and represents a powerful approach to exploiting the full and diverse bioremediative potential of soil micro-organisms in the restoration of contaminated soils.

Bioremediation may aim to restore plant/soil health to its former status, but this is often unrealistic and unnecessary for many types of end use. However, indicators of soil health provide a useful tool for monitoring the bioremediation process. Margesin et al. (2000) showed that hydrocarbon biodegradation could be monitored well by using soil biological parameters, with residual hydrocarbon content correlating positively with soil respiration, biomass-C (substrate-induced respiration) and with activities of soil dehydrogenase, urease and catalase. Robust and ecologically representative indicators of soil health clearly need to be agreed so that relevant end point targets for remediation, whether biologically driven or otherwise, can be set. Useful efforts are now being made towards identifying suites of possible indicators for future monitoring of soil to ensure sustainable management (e.g. Ritz et al., 2009), acceptable to policy makers.

Bioaugmentation versus biostimulation

It is often necessary to carry out some kind of intervention to obtain satisfactory rates of bioremediation. Two key types of bioremediation intervention which offer considerable promise for effective management of plant and soil health are biostimulation and bioaugmentation. Biostimulation is based on the concept that suitable organisms/genes are present for bioremediation but that activity is restricted by a constraint that can be alleviated. It therefore involves the addition of electron acceptors, nutrients or electron donors to increase the numbers or stimulate the activity of indigenous biodegradative micro-organisms (Widada et al., 2002). Bioaugmentation is based on the concept that bioremediation cannot proceed because of a lack of suitable organisms and therefore involves the addition of (indigenous or nonindigenous) laboratory-grown micro-organisms capable of biodegrading the target contaminant (Vogel, 1996; Widada et al., 2002) or serving as donors of catabolic genes (Top et al., 2002). For most contaminated soils which retain reasonable biological activity, organisms for bioremediation are present, but their activity is constrained. So, biostimulation will be commonly applied as a management tool in bioremediation. However, in contaminated sites where suitable organisms for bioremediation are at low population densities or are absent, real benefits can be derived from bioaugementation.

The advantages of bioaugmentation and biostimulation have been combined to remediate pesticide (atrazine) contamination successfully (Silva et al., 2004). Bioaugmenation requires the degrading inoculum to reach the contamination in the soil/fill/stratum as well as carry out biodegradation of the target organoxenobiotic. In the field, this aspect of 'delivering' the degrader micro-organisms to their site of action may well be the most challenging part of a bioaugmentation programme as many organoxenobiotics become bound in micropores of the soil and these are not pathways for the percolating water which can carry the degraders. This constraint to bioaugmentation will often be an important factor in applying a combined approach of bioaugmentation and biostimulation as indigenous degraders may already have colonised these more inaccessible parts of the soil, and the activity of such degraders may be enhanced through biostimulation.

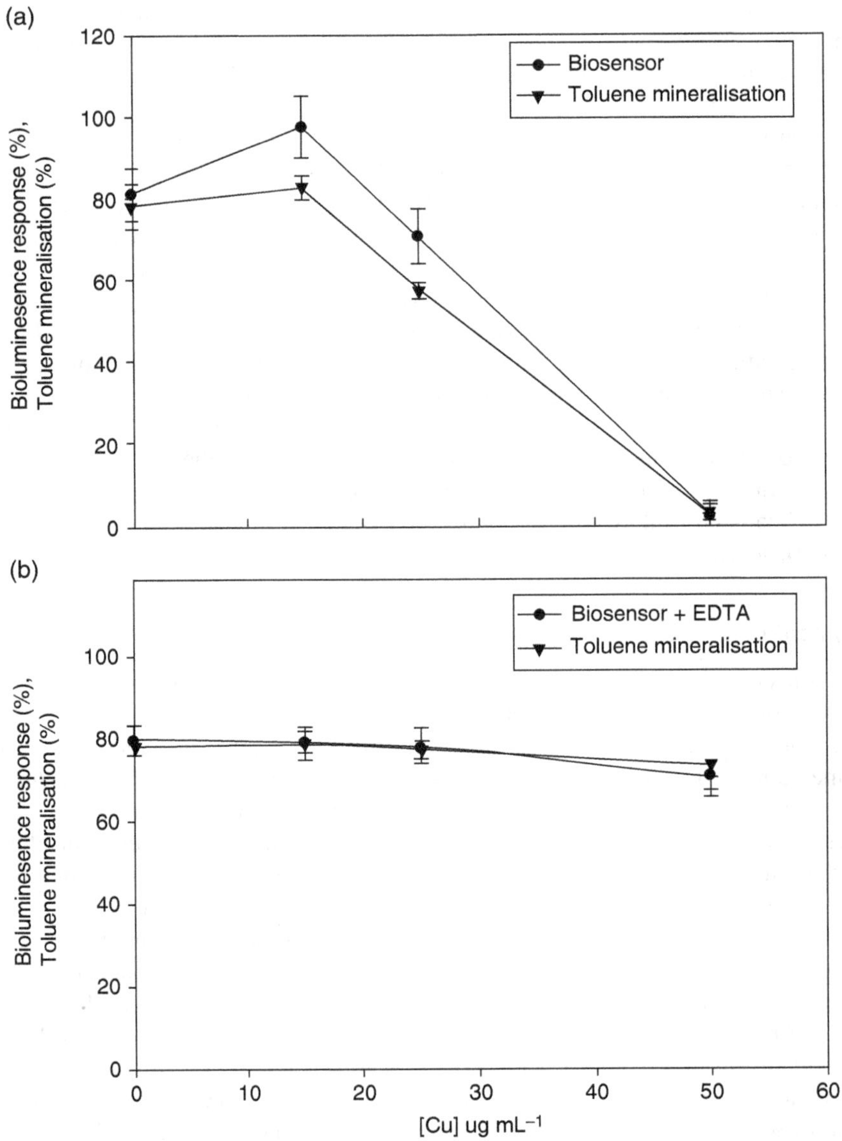

Figure 11.9 (a) The declining bioluminescence response of a *lux* gene marked bacterial biosensor to soil increasingly contaminated with copper (concentrations are given in pore water) demonstrates toxicity and explains why mineralisation of toluene, the target organic contaminant for bioremediation in this case, is increasingly inhibited by high concentrations of copper. (b) The relatively stable bioluminescence and toluene mineralisation across the same range of copper concentrations with the addition of EDTA is because this chelating agent complexes strongly with the copper and makes it unavailable (and hence non toxic) to the toluene mineralising bacteria. Sousa and Killham (2011).

Rhizoremediation

Synergistic enzymatic and chemical activities of plant and microbial metabolism in the rooting zone of contaminated soil can transform and degrade/immobilise contaminants and restore soil health.

In the case of metals, rhizoremediation usually involves their immobilisation in the soil or plant, while in the case of organic pollutants, it usually involves their degradation. Rhizoremediation of metal contaminants is generally most effective when the metals are taken up by the plant, potentially for removal when the plants are harvested, but is constrained by the fact that many of the strongest metal accumulators (particularly the hyperaccumulators) tend to be small plants and so future strategies may involve introducing their key genes involved in transporting and tolerating considerable metal loads (such as the genes encoding the cysteine-rich, metallothioneins metal-binding proteins) into other plants (Milner and Kochian, 2008). In the case of organic pollutants, the aim of rhizoremediation, as introduced earlier, is usually their degradation, either to nontoxic forms or through complete mineralisation to carbon dioxide.

The terms phytoremediation (Harvey et al., 2002; Morikawa and Erkin, 2003) and rhizoremediation (Kuiper et al., 2004) are commonly applied when plant metabolism contributes to the bioremediation process in some way. In certain cases, such as the elevated rates of hydrocarbon degradation in grassed compared to bare soils with similar physico-chemical properties (Gunther et al., 2010), it is the degradation of plant residues which is the main key to driving biodegradation or immobilisation of the target contaminants, while in others, the plant and/or its symbionts are more actively/specifically involved. This is exemplified by deep-rooted poplars which can access mobile, organic contaminants and degrade/detoxify them either by nonmodified plant metabolism, introduced pollutant transformation genes or by the action of microbial endophytes in the tissues of the trees (Doty, 2008).

The mycorrhizal symbiosis involving plant associations with fungal root symbionts (mycobionts) infers a number of properties to the host plant which are key to bioremediation of soil contamination and offer interesting options for plant/soil management. The ectomycorrhizal association, common to most trees, offers considerable metal immobilisation potential, for example, and enables phytoremediation with trees that can colonise sites to greater depths than hyperaccumulators or other metal tolerant plants. There are at least two mechanisms for this – one involves complexation of the metals with oxalates released from the mycorrhizal roots of conifers into the 'mycorrhizosphere' and one involves their uptake via the mycorrhizal fungal hyphae and subsequent immobilisation in the tissue of the tree (Meharg, 2003). Furthermore, because metal contaminated sites select for the mycorrhizal fungal community that can best cope with the presence of potentially toxic metal ions, the associated fungi are able to confer diverse physiological attributes to the host to enable colonisation of a wide range of sites (Meharg, 2003).

The mycorrhizosphere appears to be a key micro-habitat of the plant/soil system – where not only metals can be phytoremediated but organic contaminants also, through a variety of possible mechanisms (Joner and Leyval, 2003) – and offers additional potential for management of the plant/soil system for bioremediation.

There are numerous advantages to rhizoremediation, many of which are associated with the ability of the root system to penetrate otherwise inaccessible contaminants and make them accessible to plant or microbial metabolism. It is also a process ideally suited to the remediation of contaminants such as radioactive (Macek et al., 2009) and explosive (Gerhardt et al., 2009) materials which pose obvious risks to traditional 'dig and dump' approaches. One of the main drawbacks, however, is that it is usually a relatively slow and seasonal process and so it is often necessary to allow rhizoremediation to proceed for a number of years before clean-up soil/groundwater targets are reached.

Indicators for monitoring bioremediation/restoration of soil health

Management of bioremediation to restore soil health after damage from pollution requires robust indicators which can be monitored to assess any restoration of soil quality/health. A great deal of recent research has focussed on identifying which indicators will be most suitable, and although this work is incomplete (due both to the complexity of soils and the wide range of services that soil must supply and hence could be assayed), it is generally accepted that emphasis should be on biological properties of soils to complement existing and well understood physicochemical properties (Lilburne et al., 2004).

A wide suite of indicators was identified in a UK study by Ritz et al. (2009) and consists of molecular fingerprinting techniques as well as more traditional enzymatic and organism-based approaches, addressing the major biological trophic levels which mediate soil function. Reliable measurement of these indicators will require standard operating procedures, and confirmation will be required of their inherent sensitivity. It will also be necessary to confirm their capacity to discriminate between soil and land-use combinations, as well as their ecological/agroecological relevance. The question arises: whether the diversity of soil health indicators available can be integrated into a meaningful index for general application? To attempt to answer this, the approach of using a wide suite of indicators, as Ritz et al. (2009) have suggested, has been demonstrated by the development of a Soil Quality Index (SQI) (Alrumman et al., 2011) using test agricultural soils across a wide textural range (Equations 11.1–11.6). Two key bioassay types were loosely classified as either metabolic assays (soil microbial biomass, soil enzyme activities and respiration) or toxicity assays (higher organisms including plant assays, soil invertebrate assays and genetically modified bacterial sensors representing functional response of soil microbial communities). Bioassay data were transformed into an SQI modified from Canadian soil quality guidelines (www.ccme.ca) using the following formulation and equations:

A test pass/fail is decided, *a priori*, on evidence (or chemical determination) that the test soil is contaminated when compared to the control soil.

$$F1 = \frac{Number of\ failed test}{Total number of\ testes} \times 100 \tag{11.1}$$

Then, the *Excursion* was determined using Equation 11.2 as follows:

$$\delta v = \left| \frac{Mean of\ treated sample}{Mean of\ control} - 1 \right| \tag{11.2}$$

The ***ase*** (average sum of excursion) was calculated using Equation 11.3:

$$as\delta v = \frac{\sum \delta v}{Number of\ failed test} \tag{11.3}$$

Then, *F*2 was calculated using Equation 11.4 was applied to obtain range between 0 and 100

$$F2 = \frac{as\delta v}{0.01 as\delta v + 0.01} \tag{11.4}$$

Finally, the soil quality index (SQI) was calculated using Equation 11.5

$$SQI = 100 - \frac{\sqrt{F1^2 + F2^2}}{1.414} \tag{11.5}$$

The value of the divisor (1.414) was calculated as follows:

$$\frac{\sqrt{100^2 + 100^2}}{100} = 1.414 \tag{11.6}$$

Toxicity assays were found to be more sensitive than metabolic assays. Conversely, metabolic assays were found to be more robust than toxicity assays. The SQI highlighted inter-soil differences according to physical class. Toxicity assays overestimated soil quality values compared to the General SQI. Assay type showed different aspects of soil quality with respect to contamination or treatment with implications for Risk Assessment. Models such as this show promise but should be further validated with a wide range of soil physical classes, history of use and range of contaminants. An enlarged data set will define the specificity of the SQI and highlight key variables contributing to soil health. A recent example of this methodology has been the identification of the recovery of soil following perturbation by hydrocarbon contamination (Alrumman, 2011).

In order for management of the soil/plant system to achieve a more global perspective, indicators for monitoring the effects of soil management should be internationally robust, but it is yet to be seen how readily identification of such a set will be achieved. Soil characterisation has traditionally focussed on distinguishing key physical characteristics. This has helped provide data for soil maps that aid policy decisions for soil use. However, the identification of key soil (bio) indicators for contaminant remediation and post-remediation land use is likely to become increasingly useful. It may be possible to identify a core set of these (bio) indicators that are internationally transferable. Other less transferable indicators may also have to be adopted to meet the specific policy/soils requirements of a particular country or region.

11.2.4 Protection of plants from pests and diseases

Conventional protection versus bioprotection

As with managing nutrient supply, globally, protection of crop plants remains a chemically dominated industry, and this is reflected in the well-documented management systems that prescribe pesticide application. This section will explore management of the plant/soil system involving bioprotection.

With a few exceptions, commercially viable management systems exploiting consistent bioprotection of crops against soil-borne diseases and pests are largely confined to protected crops where soil and other environmental conditions are controlled. There are around 30 commercial biocontrol agents being used worldwide, but with a market value of less than £5M (Yoshihiro, 2006), they are clearly not that widely used. Nonetheless, technical, economical and environmental drivers are forcing agriculture and forestry to consider new and sustainable methods, such as the use of biocontrol, for protection against soil-borne pests and pathogens.

Bioprotection and biocontrol management systems can either involve direct application of protection/control agents or the management of plant-soil conditions to enhance the activity of indigenous agents. The latter can involve suppression of activity of pests or disease causing agents or enhancement of activity of antagonists of those pests or diseases. Field management of potatoes, for example, can involve careful control of soil pH as the potato pathogen *Streptomyces scabies* is most aggressive under near neutral or alkaline conditions (Killham, 1994), while application of stable, mature composts can be highly suppressive to a range of crop pathogens (Hoitink et al., 1996). Management of soil redox conditions through flooding of potato fields (this allows the decomposition of potato haulms to drive down the redox potential and generate anaerobic conditions) between crops can effectively inhibit potato cyst nematodes (McSorley, 1998).

In terms of introducing commercial agents, the most promising results to date for field managed, crop pest and disease control have mainly been derived from antagonistic fungi and bacteria. *Bacillus thuringiensis* is the most widely applied species of bacteria used for commercial, biological control, with at least four subspecies used to control lepidopteran (moth and butterfly caterpillars), coleopteran (beetles) and dipteran (flies) insect pests (Entwistle et al., 1993). *Trichoderma viride* has been used successfully against Dutch Elm disease and to treat the spread of fungal and bacterial pathogen growth on tree wounds, while other species in the genus *Trichoderma* are used to manage some soil-borne plant pathogens (Arora, 2003). However, even these more successful systems have not provided consistent, field-based control.

In terms of managing the plant/soil system for more reliable control of pests and diseases, biocontrol agents can be combined with other chemical products, with physical methods (solarisation or steam sterilisation) or agronomical practices, such as enhancement of suppressive soils, and as part of integrated pest management programs (Spadaro and Gullino, 2005). Because biological control is less consistent, less effective and slower acting than control normally achieved with pesticidal chemicals, successful management may well often depend on integration with other such control measures.

Although biocontrol through manipulation of the soil biota, whether exploiting indigenous populations or introduced inocula, suffers from inconsistent performance across the wide range of edaphic conditions involved, biological control forms a critical part of a 'systems approach' to integrated, sustainable soil pest management to combat insecticide-resistant pests and to minimise the future use of pesticides (Bale et al., 2008). It is becoming clear that successful and consistent biological control beyond the glasshouse requires further research to understand the biology/ecology of the pest/control agent/soil interaction so that soil management can adequately underpin the interaction. One such area where progress is being made is understanding the underlying mechanisms behind 'conducive' and 'suppressive' soils. It has been known for some considerable time that some soils become 'suppressive' over time. In these soils, protection of the crop from soil-borne pathogens is strongly driven by microbial action. This consists of PGPR bacteria and fungi colonising the root system. Much of the protection conferred by many strains of fluorescent pseudomonad PGPRs is attributed to the production of antibiotic secondary metabolites such as 2,4-diacetylphloroglucinol (DAPG) and/or phenazines. There have been many advances in understanding the biochemical and molecular underpinning of antibiotic (particularly DAPG) production. However, our understanding of antibiotic production *in situ* lags behind this. Duffy and Defago (1999) and Shanahan et al. (1992) showed the selective *in vitro* effects of different C sources and minerals on DAPG production. However, how DAPG production is affected by qualitatively changing complex C (i.e. mixtures of sugars, organic acids and amino acids, characteristic of root C flow) has not received much attention. Effects of this root-derived C on the growth and physiology of the soil microbial community is of great interest and a greater understanding of these presents an opportunity for effective bio-control. Figure 11.10 shows the response of a DAPG producer, *Pseudomonas fluorescens* CHA0 to defined C sources. It should be noted that DAPG production is minimal under 100% reducing sugars and that organic acids stimulate antibiotic production (Standing and Killham, 2007).

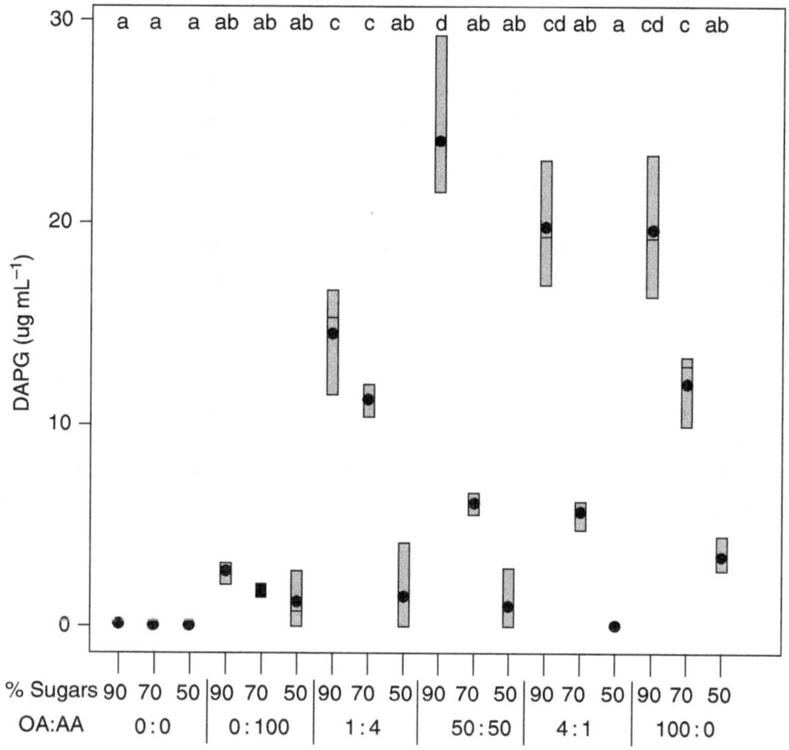

Figure 11.10 Boxplots of DAPG production (μg mL^{-1}) by *P. fluorescens* CHAO in response to reducing sugars (90%, 70%, 50%) amended with different organic acid:amino acid ratios. Significant differences between treatments (n = 3) are indicated by different letters.

While soil microbe–iron relations and the production of siderophores have been implicated in biocontrol, they do not contribute as much as the antibiotics. Having said this, Mazurier et al. (2009) provide experimental validation for the following *Fusarium* wilt disease model: in suppressive soils a high microbial biomass provides a highly competitive environment with respect to C and iron sequestration. This physical lack of resources may suppress the saprophytic growth of the fungus *Fusarium* which leads to reduced pathogenicity and susceptibility of *Fusarium* to fluorescent pseudomonad-produced phenazine. Of the two antibiotics produced by microbial biomass DAPG appeared to be far less effective than phenazine. While the molecular mechanisms underpinning antibiotic production are well characterised in the laboratory the 'field' soil–plant–microbe system is far more complex and many putative signalling systems remain unknown (Haas and Defago, 2005).

Janvier et al. (2007) state that 'Healthy soils are suppressive soils, thus disease suppressiveness can be considered as an indicator of soil health'. Thus, effective biological control and management of soil-borne plant pathogens and pests require a holistic approach – selection, optimisation and regional evaluation of biological control agents and improvement in the understanding of the influence of cultivars, environments and cropping systems on biological control (Janvier et al., 2007). Only full biological control evaluations will provide the critical information on performance of selected control agents under the diverse agroecological conditions of farming and forestry worldwide.

11.3 Conclusion

This chapter has demonstrated that sustainable management of soil and plant health through optimising soil biological function offers an increasingly promising way forward for our land but requires a thorough knowledge of the underpinning factors (many of which were introduced in Figures 11.5 and 11.8) which control this biological function. These factors rarely act singly and generally interact in a highly dynamic way. Despite this complexity, however, we are becoming better able to understand and hence exploit the rules which govern soil–plant–microbe interactions for benefits such as biofertilisation and bioprotection of crops, for restoring and maintaining soil structure, for bioremediation of contaminated land and for minimising climate change.

The rules mentioned in the previous paragraph, which appear to govern soil–plant–microbe interactions, have invariably been deduced from studies carried out at small scales and usually under controlled conditions in the laboratory. The challenge for soil biologists will be to explore how to scale-up and address the temporal and spatial variability of the field and landscape levels. Land managers can then consistently benefit from the increasing knowledge base of soil biology by systematically adopting more sustainable strategies of integrated management of soil and plant health, rather than the 'trial and error' approach often used in the past.

References

Alogaidi, F., Standing, D.B., and Price, A. (2011) Unpublished data.

Alrumman, S. (2011). *Microbial Ecotoxilogical Assessment of Hydrocarbon Impacted Soils Undergone Remediation*. PhD Thesis. University of Aberdeen, Aberdeen, UK.

Alrumman, S., Paton, G.I., and Standing, D.B. (2011) Unpublished data.

Álvaro-Fuentes, J., Cantero-Martínez, J., López, C., Paustian, M.V., Denef, K., Stewart, C.F., and Arrúe, J.L. (2009) Soil aggregation and soil organic carbon stabilization: effects of management in semiarid Mediterranean agroecosystems. *Soil Science Society of America Journal* **73**, 1519–1529.

Arora, D.K. (ed.) (2003) *Fungal Biotechnology in Agricultural, Food and Environmental Applications (Mycology)*. Marcel Dekker, New York, p. 700.

Aulakh, M.S., Wassman, R., Bueno, C., Kreuzwieser, J. and Rennenberg, H. (2001) Characterisation of root exudates at different growth stages of ten rice (*Oryza sativa L.*) cultivars. *Plant Biology* **3**, 139–148.

Bale, J.S., van Lenteren, J.C., and Bigler, F. (2008) Biological control and sustainable food production. *Philosophical Transactions Royal Society London B. Biological Sciences* **363**, 761–776.

Banerjee, M. and Yesmin, L. (2009) Sulfur-oxidizing plant growth promoting rhizobacteria for enhanced canola performance. US Patent 7521236.

Bang, H.S., Lee, J., Kwon, O.S., Na, Y.E., Jang, Y.S., and Kim, W.H. (2005) Effects of paracoprid dung beetles (Coleoptera: Scarabaeidae) on the growth of pasture herbage and on the underlying soil. *Applied Soil Ecology* **29**, 165–171.

Berg, G. and Smalla, K. (2009) Plant species and soil type cooperatively shape the structure and function of microbial communities in the rhizosphere. *FEMS Microbiology Ecology* **68**, 1–13.

Briones A.M., Okabe S., Umemiya Y., Ramsing N.B., Reichardt W., and Okuyama H. (2002) Influence of different cultivars on populations of ammonia-oxidising bacteria in the root environment of rice. *Applied and Environmental Microbiology* **68**, 3067–3075.

Cresser, M.S., Killham, K., and Edwards, A. (1993) *Soil Chemistry and its Applications*. Cambridge University Press, Cambridge, UK, p. 192.

Deaker, R., Roughley, R.J., and Kennedy, I.R. (2004) Legume seed inoculation technology – a review. *Soil Biology and Biochemistry* **36**, 1275–1288.

Doty, S.L. (2008) Tansley review: enhancing phytoremediation through the use of transgenics and endophytes. *New Phytologist* **179**, 318–333.

Duffy, B.K. and Defago, G. (1999) Environmental factors modulating antibiotic and siderophore biosynthesis by *Pseudomonas fluorescens* biocontrol strains. *Applied and Environmental Microbiology* **65** (6), 2429–2438.

Edwards, C.A. (ed.) (2004) *Earthworm Ecology*, 2nd edn. CRC Press, Boca Raton, Florida, p. 456.

Egamberdiyeva, D. (2007) The effect of plant growth promoting bacteria on growth and nutrient uptake of maize in two different soils. *Applied Soil Ecology* **36**, 184–189.

Elliot, M. (2011) Differences in rice cultivar exudates grown in different P conditions. BSc. Hons Thesis. University of Aberdeen, Aberdeen, UK.

Entwistle, P.F., Cory, J.S., Bailey, M.J., and Higgs, S. (eds) (1993) *Bacillus thuringiensis: An Environmental Biopesticide-Theory and Practice*. Wiley-Blackwell, Chichester, UK, p. 330.

Environment Agency (2000) *The State of the Environment of England and Wales: The Land*, HMSO, London, UK.

Ferrieira E.P. de B., Dusi A.N., Xavier G.R., and Rumjanke N.G. (2008) Rhizosphere bacterial communities of potato cultivars evaluated through PCR-DGGE profiles. *Pesquisa Agropecuária Brasileira* **43**, 605–612.

Galloway, J.N. (2003) The global nitrogen cycle. In: *Biogeochemistry* (eds H.D. Holland and K.K. Turekian), *Treatise on Geochemistry*, Vol.8, (ed. W.H. Schlesinger), pp. 557–584. Elsevier-Pergamon, Oxford, UK.

Gerhardt, K.E., Huang, X., Glick, B.R., and Greenberg, B.M. (2009) Phytoremediation and rhizoremediation of organic soil contaminants: potential and challenges. *Plant Science* **176**, 20–30.

Germida J.J. and Siciliano S.D. (2001) Taxonomic diversity of bacteria associated with roots of modern recent and ancient wheat cultivars. *Biology and Fertility of Soils* **33**, 410–415.

Goldstein A.H. and Rogers R.D. (1999) Biomediated continuous release phosphate fertilizer. US Patent 5912398.

Greenland, D.J. and Pereira, H.C. (1977) Soil damage by intensive arable cultivation: temporary or permanent? *Philosophical Transactions of the Royal Society of London. Series B, Biological Sciences* **281**, 193–208.

Gunther T., Dornberger, U., and Fritsche, W. (2010) Rhizoremediation of hydrocarbon contaminated soil using Australian native grasses. *Science of the Total Environment* **408** (17), 3683–3688.

Haas, D. and Defago, G. (2005) Biological control of soil-borne pathogens by fluorescent pseudomonads. *Nature Reviews Microbiology* **3**, 307–319.

Haneklaus, S. and Schnug, E. (2006) Precision agriculture and fertilization. In: *Encyclopedia of Soil Science* (ed. R. Lal), pp. 1364–1366. Taylor & Francis, London, UK.

Harvey, P.J., Campanella, B.F., Castro, P.M.L., Harms, H., Lichtfouse, E., Schaffner, A.R., Smrcek, S., and Werck Reichharts, D. (2002) Phytoremediation of polyaromatic hydrocarbons, anilines and phenols. *Environmental Science and Pollution Research* **9** (1), 29–47.

Havlin, J., Tisdale, S., Nelson, W., and Beaton, J. (2004) *Soil Fertility and Fertilizers. An Introduction to Nutrient Management*, 7th edn. Prentice Hall, New Jersey, p. 528.

Hoitink, H.A.J., Stone, A.J., and Grebus, M.E. (1996) Suppression of plant diseases by composts. In: *The Science of Composting* (ed. E. Epstein), p. 675. CRC Press, Boca Raton, Florida.

Hu, C., Liu, Y., Song, L., and Zhang, D. (2002) Effect of desert soil algae on the stabilization of fine sands. *Journal of Applied Phycology* **14**, 281–292.

Janvier, C., Villeneuve, F., Alabouvette, C., Edel-Hermann, V., Mateille, T., and Steinberg, C. (2007) Soil health through disease suppression: which strategy, from descriptors to indicators? *Soil Biology and Biochemistry* **39**, 1–23.

Joner, E.J. and Leyval, C. (2003) Phytoremediation of organic pollutants using mycorrhizal plants: a new aspect of rhizosphere interactions. *Agronomie* **23**, 495–502.

Khan, M.S., Zaidi, A., and Wani, P.A. (2009) Role of phosphate solubilizing microorganisms in sustainable agriculture – a review. In: *Sustainable Agriculture* (eds E. Lichtfouse, M. Navarrete, P. Debaeke, S. Véronique, and C. Alberola), pp. 551–570. Springer, the Netherlands.

Killham, K. (1994) *Soil Ecology*. Cambridge University Press, Cambridge, UK, p. 242.

Koske, R.E. and Gemma, J.N. (1997) Mycorrhizae and succession in plantings of beachgrass in sand dunes. *American Journal of Botany* **84** (1), 118–130.

Kuiper, I., Lagendijk, E.L., Bloemberg, G.V., and Lugtenberg, B.J.J. (2004) Rhizoremediation: a beneficial plant-microbe interaction. *Molecular Plant–Microbe Interactions* **17**(1), 6–15.

Lal, R. (2000) Soil management in the developing countries. *Soil Science* **165** (1), 57–72.

Lal, R. (2004) Soil carbon sequestration impacts on global climate change and food security. *Science* **304**, 1623–1627.

Lilburne, L., Sparling, G., and Schipper, L. (2004) Soil quality monitoring in New Zealand: development of an interpretative framework. *Agriculture, Ecosystems and Environment* **104**, 535–544.

Macek, T., Uhlik, O., Jecna, K., Novakova, M., Lovecka, P., Rezek, J., Dudkova, V., Stursa, P., Vrchotova, B., Pavlikova, D., Demnerova, K., and Mackova, M. (2009) Advances in phytoremediation and rhizoremediation. In: *Advances in Applied Bioremediation* (eds A. Singh, R.C. Kuhad, and O.P. Ward), pp. 257–277. Springer, Berlin, Germany.

Margesin, R., Zimmerbauer, A., and Schinner, F. (2000) Monitoring of bioremediation by soil biological activities. *Chemosphere* **40**, 339–346.

Marschner, P., Grierson, P.F., and Rengel, Z. (2001) Microbial community composition and functioning in the rhizosphere of three Banksia species in native woodland in Western Australia. *Applied Soil Ecology* **28**, 191–201.

Mazurier, S., Corberand, T., Lemanceau, P., And Raaijmakers, J.M. (2009) Phenazine antibiotics produced by fluorescent pseudomonads contribute to natural soil suppressiveness to *Fusarium* wilt. *The ISME Journal* **3**, 977–991.

McSorley, R. (1998) Alternative practices for managing plant-parasitic nematodes. *American Journal of Alternative Agriculture* **13**, 98–104.

Meharg, A.A. (2003) The mechanistic basis of interactions between mycorrhizal associations and toxic metal cations. *Mycological Research* **107** (11), 1253–1265.

Milner, M.J. and Kochian, L.V. (2008) Investigating heavy-metal hyperaccumulation using *Thlaspi caerulescens*. *Annals of Botany* **102**, 3–13.

Morikawa, H. and Erkin, O.C. (2003) Basic processes in phytoremediation and some applications to air pollution control. *Chemosphere* **52** (9), 1553–1558.

Nisha, R., Kaushik, A., and Kaushik, C.P. (2007) Effect of indigenous cyanobacterial application on structural stability and productivity of an organically poor semi-arid soil. *Geoderma* **138** (1–2), 49–56.

Nordstrom, K.F. (2008) *Beach and Dune Restoration*. Cambridge University Press, Cambridge, UK.

Oades, J.M. (1993) The role of biology in the formation, stabilization and degradation of soil structure, *Geoderma* **56**, 377–400.

Orlovsky, L., Dourikov, M., and Babaev, A. (2004) Temporal dynamics and productivity of biogenic soil crusts in the central Karakum desert, Turkmenistan, *Journal of the Arid Environment* **56** (4), 579–601.

Piotrowski, M.R., Doyle, J.R., and Carraway, J.W. (2006) Integrated bioremediation of soil and groundwater at a superfund site. *Remediation Journal* **2** (3), 293–309.

Rai, M.K. (2005) *Handbook of Microbial Biofertilisers*. CRC Press, Boca Raton, Florida.

Rincón, A., Parladé, J., and Pera, J. (2005) Effects of ectomycorrhizal inoculation and the type of substrate on mycorrhization, growth and nutrition of containerised *Pinus pinea* L. seedlings produced in a commercial nursery. *Annals of Forest Science* **62**, 817–822.

Ritz, K., Black, H.I.J., Campbell, C.D., Harris, J.A., and Wood, C. (2009) Selecting biological indicators for monitoring soils: a framework for balancing scientific opinion to assist policy development. *Ecological Indicators* **9** (6), 1212–1221.

Royal Commission on Environmental Pollution (1996) *Sustainable Use of Soils*. HMSO, London, UK.

Russell, E.W. (1977) The role of organic matter in soil fertility. *Philosophical Transactions of the Royal Society of London. Series B, Biological Sciences* **281**, 209–219.

Scottish Government (1998) *Scottish Vacant and Derelict Land Survey 1998*, ENV/1999/1, Scottish Executive, Edinburgh, UK, www.scotland.gov.uk/library2/doc05/lssb-00,htm. Accessed on 2 September 2011.

Shanahan, P., O'Sullivan, D.J., Simpson, P., Glennon, J.D., and O'Gara, F. (1992) Isolation of 2,4-diacetylphloroglucinol from a fluorescent pseudomonad and investigation of physiological parameters influencing its production. *Applied and Environmental Microbiology* **58**, 353–358.

Sicilano, S.D., Theoret, C.M., Freitas, J.R. de, Hucl, P.J., and Germida, J.J. (1998) Differences in the microbial communities associated with the roots of different cultivars of canola and wheat. *Canadian Journal of Microbiology* **44**, 844–851.

Siddiqui, Z.A., Akhtar, M.S., and Futai, K. (eds) (2008) *Mycorrhizae: Sustainable Agriculture and Forestry*. Springer, New York.

Silva, E., Fialho, A.M., Sá-Correia, I., Burns, R.G., and Shaw, E.J. (2004) Combined bioaugmentation and biostimulation to clean up soil contaminated with high concentrations of atrazine. *Environmental Science and Technology* **38** (2), 632–637.

Sousa, S. and Killham, K. (2011) Unpublished data.

Spadaro, D. and Gullino, M.L. (2005) Improving the efficacy of biocontrol agents against soil-borne pathogens. *Crop Protection* **24**, 601–613.

Standing, D. and Killham, K. (2007) Antibiotic production by a root-associated bacterium – response to simulated plant-root carbon flow. *Laboratory News*, June, **37**.

Tilak, K.V.B.R., Ranganayaki, N., Pal, K.K., Saxena, A.K., Nautiyal, C.S., Mittal, S., Tripathi, A.K., and Johri, B.N. (2005) Diversity of plant growth and soil health supporting bacteria. *Current Science* **89**, 136–150.

Top, E.M., Springael, D., and Boon, N. (2002) Catabolic mobile genetic elements and their potential use in bioaugmentation of polluted soils and waters. *FEMS Microbiology Ecology* **42** (2), 199–208.

Vaishampayan, A., Sinha, R.P., Hader, D.P., Dey, T., Gupta, A.K., Bhan, U., and Rao, A.L. (2001) Cyanobacterial biofertilizers in rice agriculture. *The Botanical Review* **67**, 453–516.

Vogel, T.M. (1996) Bioaugmentation as a soil bioremediation approach. *Current Opinion in Biotechnology* **7**(3), 311–316.

Widada, J., Nojiri, H., and Omori, T. (2002) Recent developments in molecular techniques for identification and monitoring of xenobiotic-degrading bacteria and their catabolic genes in bioremediation. *Applied Microbiology and Biotechnology* **60** (1–2), 45–59.

www.ccme.ca. Website of Canadian Council of Ministers of the Environment. Accessed on 2 September 2011.

Yeomans, C.V., Porteous, F., Paterson, E., Meharg, A.A., and Killham, K. (1999) Assessment of *lux-marked Pseudomonas fluorescens* for reporting on organic carbon compounds. *FEMS Microbiology Letters* **176**, 79–83.

Yoshihiro, T. (2006) Current status and future prospect of commercial biocontrol agents. *Plant Protection* **60** (11), 532–538.

Zhao, H.L., He, Y.H., Zhou, R.L., Su, Y.Z., Li, Y.Q., and Drake, S. (2009) Effects of desertification on soil organic C and N content in sandy farmland and grassland of Inner Mongolia. *Catena* **77**(3), 187–191.

12 Managing the soil–plant system for the delivery of ecosystem services

David W. Hopkins[1] and Edward G. Gregorich[2]

[1] *School of Life Sciences, Heriot-Watt University, Edinburgh, UK*
[2] *Agriculture and Agri-Food Canada, Central Experimental Farm, Ottawa, Ontario, Canada*

12.1 Introduction

Ecosystem services are benefits humans gain from ecosystems either directly or indirectly. The *Millennium Ecosystem Assessment* (MEA, 2005) provides a major synthesis of the ways in which the supporting, provisioning, regulating and cultural services combine to support humanity (Figure 12.1). Soils and plants have explicit, generally fairly obvious and central roles in the supporting services through, for example, soil formation; in the provisioning services through food, wood and fibre production, water supply and in some cases fuel production such as peat for burning; and in the regulating services through climate, flood and disease regulation, and water purification. Soils and plants also contribute to some of the cultural services as outlined in the MEA (2005). Our objective in this chapter is to summarise these roles and indicate how they can be managed with the aim of increasing the sustainability of soil–plant systems. We deliberately extend our coverage beyond the traditional areas of soil and plant science, such as agriculture, ecology and forestry, to include some aspects of cultural ecosystem services (Figure 12.1). Monitoring soil quality is also included because of the ongoing need to be able to assess the ability of the soil–plant system to deliver ecosystem services and the need for early warnings of change and thus potential lack of sustainability (Section 12.3). Sustainability is a widely used term, but for the present purposes we are following the definition in Gregorich et al. (2001): the long-term viability of a particular soil–plant (or land-use) system, which includes steps to ensure natural resource conservation, economic viability (including productivity and security) and social acceptability (including the cultural services). Clearly, for the soil–plant system to be truly sustainable in the sense of being able to continue indefinitely without any resource depletion or degradation is a more exacting and probably unachievable aim because, for example, considering an agricultural system, there is no escape from the thermodynamic principles. To be truly sustainable, all the nutrients and water removed with crops and any soil lost during cultivation and harvesting would need to be replaced by equivalent inputs of regeneration of soil by weathering and recycling. Despite the difficulties of deriving an absolute

Soil Conditions and Plant Growth. Edited by Peter J. Gregory and Stephen Nortcliff.
© 2013 Blackwell Publishing Ltd. Published 2013 by Blackwell Publishing Ltd.

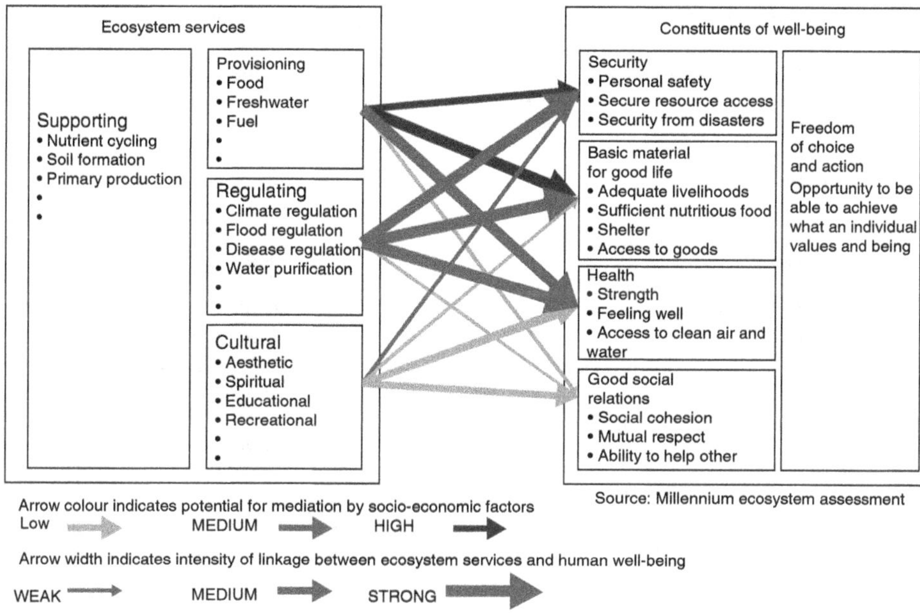

Figure 12.1 Linkages between ecosystem services and human well-being, redrawn from the *Millennium Ecosystem Assessment* (MEA, 2005). For a colour version of this figure, please see Plate 12.1.

definition of sustainability, the term is in general usage and has particular value in the context of the expectations placed on the soil–plant system as the human population expands towards the anticipated plateau of about 9 billion by around 2050. By this time, the amount of land will have declined to 1.6 ha per person, compared with 2.0 ha per person in 2005 and 5.2 ha per person in 1950.

There have been several attempts to place a monetary value on ecosystems goods and services, including those derived from soils. They are united in two different ways: whatever approach is used, the monetary values are astonishingly large (Table 12.1) and the concept that they have a real monetary value is an illusion because ecosystem services such as soil formation cannot be traded because there are no realistic alternative sources. Despite this second criticism, one of the most widely quoted sources (Costanza et al., 1997) places a value of US$17.1 $\times 10^{12}$ on soil formation and substantial values to other processes in which the soil–plant system participates, such as nutrient cycling and water regulation and supply (Table 12.1). The likely development of carbon trading over the next few decades as a response to the need to move to a lower carbon economy will lead to economic values being placed on carbon storage and sequestration in soil. The MEA (2005) suggests that the global value for carbon will be between US$10 billion (for the avoidance of doubt, 1 billion = 1 000 000 000 = 10^9) and US$44 billion in 2010 and that a total of 4.5 billion tonne of carbon (or equivalent) will be traded giving a value of between US$2.2 and US$9.8 per tonne. This places the value of topsoil carbon (assuming 20 cm soil depth, 1.2 g cm^{-3} bulk density and 2% carbon concentration) between about US$100 and US$500 per ha. Pimental et al. (1997) also list a range of biological processes that occur either partly or predominantly in soils as amongst the most valuable benefits of maintaining biodiversity. These include waste disposal, soil formation, biological nitrogen fixation and bioremediation, with global values of US$760 billion, 25 billion, 90 billion

Table 12.1 Estimates of the monetary value of ecosystem services.

Ecosystem service	Value (trillion US$)
Soil formation	17.1
Recreation	3.0
Nutrient cycling	2.3
Water regulation and supply	2.3
Climate regulation	1.8
Habitat	1.4
Flood and storm prevention	1.1
Food and raw materials	0.8
Genetic resources	0.8
Atmospheric gas balance	0.7
Pollination	0.4
All other services	1.6
Total	**33.3**

Note that US$1 trillion = 1×10^{12} and that the total global gross 'national' product was US$18 trillion at the time of the assessment.
Source: Adapted from Costanza et al. (1997).

and 121 billion per annum, respectively. The apparent discrepancy between the value for soil formation and that of Costanza et al. (1997) is in part because Pimental et al. (1997) have considered only the biological contribution, not the geochemical and geophysical processes.

As an alternative to cash values, the concept e*m*ergy (as distinct from e*n*ergy) can be used (Odum, 1996; Odum and Odum, 2000), although it is not without its critics. E*m*ergy (or energy memory) is the amount of energy in total required both now and in the past to produce a particular service. In the context of plant production for food, for example, the e*m*ergy comprises the direct energy inputs harnessed in photosynthesis but also the energy required in cultivation, fertiliser production and distribution, crop harvesting, food processing and distribution, and even the past energy input required for cultivar development, soil formation, biogeochemical and hydrological cycling. Voora and Thrift (2010) estimate the annual e*m*ergy input from soil to global processes is about 1.4×10^{19} J year^{-1}, which is about one-tenth of the amount from oil.

Leaving aside the difficulties of quantifying ecosystem services, the management of the soil–plant system, or any aspect of the ecosystem, for the simultaneous delivery of multiple ecosystem services is extremely challenging because of conflicting effects of different management options and the lack of sufficiently sensitive decision-making tools to account for the multiple interactions. The example in Table 12.2 summarises some of the conflicts for lowland agricultural grassland. Increased grass productivity and therefore milk or meat yield can be at the cost of a range of other ecosystem services, including air and water quality, biodiversity conservation and potential landscape quality (which depends in part on an individual's quality judgement), and can be accompanied by deleterious processes such as unsustainable erosion and losses to groundwater (Table 12.2).

Clearly, assigning a value to different ecosystem services helps prioritise the importance of different ecosystem services in situations where they interact, but because they cannot be traded interchangeably, compromises are still essential, especially with the increasing demands being placed on the soil–plant system by the large and growing human population.

Table 12.2 Relationships between ecosystem services in lowland agricultural grassland with drivers in the rows and responders in the columns compiled from a review of peer-reviewed literature.

	Agricultural production	Climate regulation	Air quality	Water quality	Nutrient cycling	Biodiversity conservation	Landscape quality
Agricultural production		***↕	***↓	***↓	***↓	***↓	***↓
Climate regulation	**↕		***↑	***↑	**↕	**↕	**↕
Air quality	**↕	**↕		***↑	**↑	***↑	***↑
Water quality	↔	**↕	**↕		0	**↑	**↑
Nutrient cycling	***↑	**↑	*↑	***↑		**↑	**↑
Biodiversity conservation	**↕	**↕	↔	**↑	*↑		***↑
Landscape quality	*↕	*↕	*↕	*↑	**↕	***↑	

0, no relationship; ↓, responder declines as driver increases; ↑, responder increases as driver increases; ↕, no conclusive trends (i.e. both increases and decreases in the responder have been observed as the driver increases); ↔, insufficient published information. The intensity of shading indicates the nature of some of the interactions: light grey, win–win and dark grey, lose–lose. The strength of the interaction is indicated by the number of asterisks: ***, high confidence; **, medium confidence; *, low confidence.

Source: Redrawn and adapted from Pilgrim et al. (2010).

12.2 Roles of soils

12.2.1 Food and fibre production

The concept of ecosystem services is well established, but perspectives of the relevance of particular services (Figure 12.1) differ depending on which service is most acutely needed. Where food supply is not secure, the provisioning services, particularly food production, tend to be prioritised. In large parts of the world, some of the other ecosystem services may be regarded as luxuries that can only be afforded when the food supply has been secured. The primary ecosystem service derived from the soil–plant system worldwide is the production of food and fibre. In view of the burgeoning world population and the increasing demand for food, one of the most critical aspects of managing the soil–plant system is the management of soil fertility and the capacity of soils to support crop production. With continued pressures to produce more food, there will be increasing pressures to develop strategies that increase fertiliser efficiency and improve the recycling of organic resources. With the growing public awareness of the need for environmentally sustainable practices in food production, there is increasing recognition of high efficacy and ecologically sound fertiliser practices. Fertilisation of soils can be intensified through agro-ecological methods using biological and technical knowledge in order to make more effective use of locally available resources and processes. This includes the management of soil fertility and innovative methods for recycling of organic matter and nutrients from farms, food processing, household waste and compost, and sewage disposal systems that produce biosolids (Figure 12.2a).

Availability of nutrients is considered a universally important attribute of soil quality and, because of its direct link to food production, it is one of the ecosystem services most studied. A part of soil fertility is the potential of the soil to supply nutrient elements in the quantity, proportion and form needed to support plant growth. In natural ecosystems, the release of these nutrients from organic materials present is often mediated by micro-organisms through decomposition, mineralisation and solubilisation reactions, or as a result of parent material being gradually broken down by weathering. In managed systems, this release of nutrients is supplemented by the addition of fertilisers to offset those removed from the soil–plant system during cropping at a rate that cannot be replaced by the natural processes. In intensively managed systems, this supplementation is achieved by the addition of inorganic fertilisers, manures, slurries, composts and other recycled 'wastes' and by management practices such as acceleration of the decomposition/mineralisation process through tillage as well as the use of plants with nitrogen-fixing symbioses. The fertility of soil can also be diminished by human activities, through use of management practices that promote soil erosion, nutrient mining (harvesting of nutrients at a rate in excess of their rate of replenishment), alteration of the soil biota or structure, salinisation through improperly managed irrigation, acidification through inappropriate fertilisation and deposition of acidifying pollutants.

The usage of inorganic nitrogen, phosphorus and potassium fertilisers (Figure 12.3; FAO, 2003) indicates increasing demands for both nitrogen and phosphorus, and approximately steady demand for potassium. The options for supplying nitrogen include biological fixation that globally contributes about 60% of the nitrogen used in agriculture and industrial fertilisers produced by industrial fixation (the Haber–Bosch process), which contributes about 40%. There are small additions of fixed nitrogen as a result of fixation during lightening strikes. The industrial fixation of nitrogen from the atmosphere by the Haber–Bosch process is an energy-dependent process, requiring high temperature and pressure, which has been possible in the developed world because of a ready supply of relatively inexpensive energy.

Figure 12.2 (a) Compost comprising composted human waste ('nightsoil') and domestic organic waste applied in small-scale vegetable production. (b) Monument to soils in China (soil of five different colours laid in a geometric pattern). (c) Linnaeus terrace in the upper Wright Valley, Antarctica. Redrawn and adapted from Pilgrim et al. (2010). With kind permission from Wiley-Blackwell. For a colour version of this figure, please see Plate 12.2.

However, its long-term sustainability is not certain, but biological fixation remains a viable alternative to industrial fixation. However, the supply of phosphorus for fertilisers is likely to become limiting before restrictions on energy supply limit industrial nitrogen fixation. The global supply of phosphorus from phosphate rock for fertilisers is dwindling and likely to become exhausted in a few decades. This poses a more serious challenge globally and is

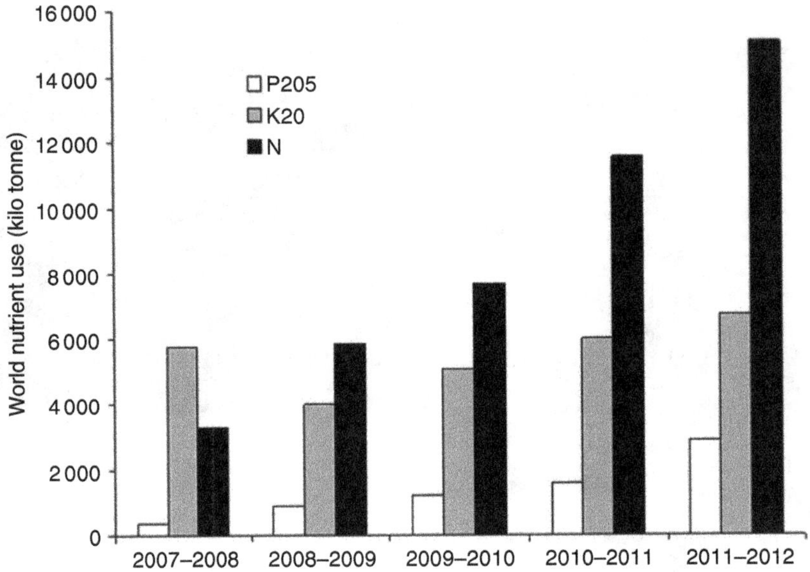

Figure 12.3 World nutrient usage from 2007–2008 to 2011–2012 (FAO, 2010).

leading to considerable attention being focussed on reducing losses and inefficiencies in phosphorus usage, developing new approaches to exploit the reserves of organic phosphorus in soils and recovering phosphorus from sewage waste.

The fact that the supply of nitrogen and phosphorus limits plant productivity and the relative cheapness of fertilisers in some parts of the world has resulted in localised or regional over-application of fertilisers. This has led to subsequent nutrient leakage from agricultural soils with significant environmental implications. Fertiliser applications to agricultural land have more than doubled the quantities of reactive forms of nitrogen and phosphorus entering the biosphere (Galloway et al., 2008). Whilst these inputs have clearly helped to keep world crop production ahead of human population growth, the environmental costs of nutrient pollution from agriculture have been substantial, including the degradation of downstream water quality, eutrophication of aquatic ecosystems, production of photochemical smog and rising global concentrations of nitrous oxide (Vitousek et al., 1997).

An evaluation of nutrient balances in three maize (corn)-based agricultural systems illustrates that nutrient additions in intensive agriculture systems in different world regions range from inadequate to excessive and that both extremes have substantial human and environmental costs. In some developing countries, such as Kenya (Table 12.3), the application of fertilisers often does not replenish soil nutrients removed in the harvest crop, drawing down the nutrient capital of what may once have been high-fertility soils (Table 12.3). By contrast, policy-driven increases in fertiliser use have contributed to rising crop yields as China has strived for increased food security (Table 12.3). Fertiliser additions of nitrogen and phosphorus to many fields far exceed that taken up by the crop and much of the excess is lost to the environment, degrading both air and water quality. In Midwestern USA, increased nitrogen and phosphorus fertilisation has contributed to large increases observed in yields since the 1940s. Consequent hydrologic losses have caused eutrophication of freshwaters and the coastal Gulf of Mexico, which has increased the biological oxygen demand of gulf waters (the so-called hypoxia problem) as a result of the decay of algal and cyanobacterial blooms depleting the dissolved oxygen.

Table 12.3 An evaluation of nitrogen and phosphorus balances in three corn-based agricultural systems (kg ha^{-1} year^{-1}).

	Kenya		China		USA	
Inputs and outputs	**N**	**P**	**N**	**P**	**N**	**P**
Fertiliser	7	8	588	92	93	14
Biological nitrogen fixation					62	
Total agronomic inputs	7	8	588	92	155	14
Removal in grain and/or beans	23	4	361	39	145	23
Removal in other harvested products	36	3				
Total agronomic outputs	59	7	361	39	145	23
Agronomic inputs minus harvest removals	−52	+1	+227	+53	+10	−9

Low input corn in Western Kenya, high input wheat and corn double-cropping systems in Northeast China, and corn–soybean rotations in the upper Midwestern USA.
Source: Derived from Vitousek et al. (2009).

Reducing the risks of fertilisers having detrimental environmental effects will require a combination of better agricultural practices that raise fertiliser efficacy and efficiency and increased efforts to trap agricultural runoff before it leaches into watercourses. Better placement and timing of fertiliser applications including widespread adoption of precision technology (such as GPS-referenced yield mapping and fertiliser placement) can reduce the amount of fertiliser wasted in the field. Other strategies to keep excess fertiliser from contaminating watercourses include establishing vegetation in riparian buffer strips around crop and pasture lands and restoring natural wetlands. Over the long term, controlling water pollution from agriculture will require better coordination between agricultural and environmental objectives.

12.2.2 Water storage and purification

The global hydrological cycle describes the transfer and temporary storage of water in different reservoirs, which include the atmosphere, lakes, streams, oceans, glaciers, vegetation and soil, and the transfers between them by evaporation, transpiration, sublimation, condensation, precipitation, infiltration, runoff and groundwater flow (Table 12.4). Humans have substantially modified the global hydrological cycle in terms of both water quality and water quantity with the development of widespread intensive agriculture necessary to support the global population; 60–70% of the water used by humans is used for agriculture (FAO, 2003). The supplies of surface and groundwater (sometimes referred to as 'blue water') are currently essential inputs to agriculture through irrigation, which is used in systems supplying approximately 40% of the world's food (Gleick, 1993). Water retained in soil and vegetation is sometimes referred to as 'green water' and it is estimated that 80% of agricultural water use comes from this source (Molden, 2007). In the future, explicit consideration of green water is likely to become much more important in the global water budget. For example, some forecasts suggest that 6 billion people will experience water shortages by 2050 if only blue water is taken into account, but this estimate falls to 4 billion people when green water is also accounted for (Rockström et al., 2009), indicating the importance of efficient management of water in the soil–plant system.

Between 15% and 20% of the water used globally goes to producing products that are exported and consumed out of the region/country of production. This water is exported in

Table 12.4 Inventory of water at the Earth's and residence times.

Reservoir	Volume (km³ × 10⁶)	% of total	Average residence time (years)
Oceans	1370	97.25	3 200
Ice caps and glaciers	29	2.05	20–100 (glaciers) 20 000 (Antarctic ice cap)
Groundwater	9.5	0.68	100–200 (shallow) 10 000 (deep)
Lakes	0.125	0.01	50–100
Soil moisture	0.065	0.005	0.1–0.2
Atmosphere	0.013	0.001	0.02
Streams and rivers	0.0017	0.0001	0.1–0.5
Biosphere	0.0006	0.00004	

Source: Derived from Pidwirny (2006).

crops, livestock and industrial products. An increased awareness of how much water is used in production has led to the concept of virtual water (also known as embedded water or hidden water). This refers to the water used in the production of a good or service. In general, crop products have lower virtual water content than livestock products. For example, the global average virtual water contents of maize, wheat and rice are 900, 1300 and 3000 m^3 $tonne^{-1}$, respectively. In contrast, the virtual water contents of roasted coffee, chicken meat, pork and beef are 2100, 3900, 4900 and 15500 m^3 $tonne^{-1}$, respectively (Chapagain and Hoekstra, 2004).

Strategies to increase the amount and residence time of water in the soil involve manipulating the soil and/or crop residue, and the soil organic matter content by tillage or residue/mulch management, both of which increase the water-holding capacity of the soil and reduce evaporation losses. These strategies are relatively easy to implement because they are often part of the overall soil-crop management. Incorporation of organic materials to increase or maintain good structure will enhance the water-holding capacity of soils, particularly in light-textured soils. This incorporation of organic matter supports the soil micro- and macro-biological community. Invertebrates are particularly notable because of their role in water infiltration and runoff, as well as water movement within the soil profile (Swift et al., 2004). Agricultural management practices that focus on altering green water significantly affect the predictions of water shortages. For example, it is estimated that modifying tillage/mulching management can reduce soil evaporation by 35–50% (Rost et al., 2009). Construction of ponds, dykes and subsurface dams to increase on-farm water storage allows farmers to redirect water to crops during droughts and can help recover more than 50% of the water normally lost.

Climate change is likely to induce greater variability in rainfall in some parts of the world which could lead to greater risk of droughts and floods (IPCC, 2007). In addition, higher temperatures will increase water demand (IPCC, 2007). Thus, under a changing climate, more intensive and extensive measures involving water harvesting will be required. Water-harvesting methods involve inducing, collecting and utilising runoff and have been practised since antiquity. These are complemented by modern techniques to increase the total yield of water per unit area and decrease the threshold of rain needed to form runoff. Other indirect methods that can be used to manage water more conservatively and improve the efficiency of water use are to develop crops by breeding or genetics that are more drought tolerant.

Filtration of water through soil provides potable water and water for agricultural (irrigation) and industrial uses, wildlife and aquatic habitat, and recreation. Pollutants such as metals, pathogens, organic contaminants and excess nutrients are adsorbed, modified and filtered out as water moves through soils in agricultural land, wetland areas, forests and riparian zones. Thus, there are human health, environmental, economic and recreational benefits derived from this water purification by soils. There are increased pressures to dispose municipal wastes on soil, and biosolids derived from sewage treatment plants are often spread on land. Liquid municipal biosolids may be applied as a source of nutrients for crops and organic matter to improve soil structure as well as to reduce sewage disposal in landfills. Soils with light texture and with large macropore networks facilitate rapid transport of contaminants from the surface or vadose zone to deep within the soil profile into tile drains and shallow groundwater. This rapid transport often bypasses the soil matrix where the contaminants can be sequestered, stabilised, degraded or available for uptake by plants. In the future, with increased pressures to use farmland as a repository for urban waste, it will be important to develop land application techniques that reduce the risk of contaminants such as pathogens, nutrients and metals from reaching groundwater, adjacent surface water source and/or subsurface artificial drainage networks.

12.2.3 Waste recycling and disposal

In all terrestrial ecosystems, with the exception of some urban settings, soils are the waste-disposal system for the residues, wastes and biomass after death (necromass). Although the estimate of US$760 billion per annum by Pimental et al. (1997) for the total monetary value of organisms involved in waste disposal includes both aquatic and terrestrial organisms, the value soils contribute to this process is inevitably large. Decomposition and recycling of natural organic wastes through the decomposer food chain relies on micro-organisms, which have major catabolic activities, and the soil fauna which contribute to the comminution, mixing, fragmentation and redistribution of organic materials in soils and the recycling by, for example, grazing on the primary microbial decomposer organisms. The decomposer, or detritus, food chain is therefore a complex set of loops in which material and energy are recycled between different groups of soil organisms, each assimilating some of the resources, releasing other components as metabolic wastes such as carbon dioxide, oxygen and ammonium (the process is widely referred to as mineralisation) and leaving some modified residue behind for another group of decomposers. The decomposition process carried out in soils is regulated by a combination of the biophysical conditions in the soil (e.g. temperature, water availability, aeration and acidity/alkalinity), the community of organisms active in decomposition (i.e. which organisms are present, what capacities they have and how active they can be under the prevailing environmental conditions), and the biochemical and physical properties of the material being decomposed (i.e. the nutrient and energy that can be obtained by the decomposer organisms) (Swift et al., 1979).

Environmental conditions

The combined physical and chemical conditions in which decomposition occurs both regulates the rate of decomposition and, in conjunction with the characteristics of the resources, imposes a selection pressure on the decomposer community thereby influencing its composition. Soil temperature, water content, acidity or alkalinity, oxygen supply and texture are amongst the most significant controlling environmental factors.

Organisms

Only 1–5% of soil organic carbon is contained in the living biomass, which predominantly comprises bacteria and fungi. Although probably not strictly correct, it is often assumed that all the micro-organisms necessary to complete the decomposition of any natural compounds (and many anthropogenic compounds) are present in soil and that collectively they are infallible. This assumption holds for many decomposition processes because of the large functional redundancy in decomposer microbial communities (see Section 12.2.5).

Substrate quality

The physiochemical characteristics of organic material – for example its solubility in water or hydrophobicity, water content, nitrogen and other nutrient content, biochemical recalcitrance and toxicity, and physical protection – strongly influence decomposition. Thus, the substrate quality of SOM can be regarded as a suite of combined properties that influence the supply of and energy to heterotrophic soil organisms. Although this is a simple concept, the ability to assess substrate quality is not easy. It has long been recognised that different components of plant litter were decomposed at different rates and that this is a reflection of their resource value to decomposer organisms. The primary source of SOM is plant litter at various stages of degradation. Plant residues usually enter the soil as litter, including leaves, stems, roots and root exudates. Excreta from livestock is another source of organic matter in soils with faeces containing not only the undigested diet but also intestinal-dwelling bacteria and endogenous wastes from animal metabolism.

Nitrogen mineralisation

Nitrogen mineralisation is the result of a set of processes allied to decomposition which lead to the conversion of nitrogen in organic compounds to inorganic nitrogen, usually ammonium, carried out by a wide range of micro-organisms (Harmsen and van Schreven, 1955). This is significant because for many crop plants inorganic nitrogen is the main form in which nitrogen in taken up from the soil. Nitrogen mineralisation operates in effective opposition to nitrogen immobilisation because, depending on the nitrogen supply in the substrate being utilised by decomposer organisms, there may be a surplus of nitrogen in relation to carbon, or a deficit. Jansson and Persson (1982) describe these processes as the mineralisation–immobilisation turnover process in which inorganic nitrogen is released into the soil solution and assimilated from the soil solution. There is more recent evidence that soil micro-organisms assimilate nitrogen from organic sources in the apparently more efficient (from the micro-organisms' perspective), direct immobilisation route with only the nitrogen surplus to the micro-organisms' requirements being released into the soil solution (Figure 12.4).

If the substrate offers a nitrogen surplus to the decomposer organisms, inorganic nitrogen will be released and, if the substrate offers a nitrogen deficit to the decomposer organisms, they will uptake nitrogen from the external (soil) environment, i.e. immobilisation. Thus, the carbon–nitrogen ratio of the substrate strongly influences whether inorganic nitrogen is released during decomposition or whether the decomposer organisms make a demand on the soil nitrogen pool, whereby nitrogen is assimilated into microbial tissues. Typically, a carbon–nitrogen ratio of about 20 is the threshold for nitrogen mineralisation above which it is immobilised, because the substrate offers a nitrogen deficit to the decomposers, and

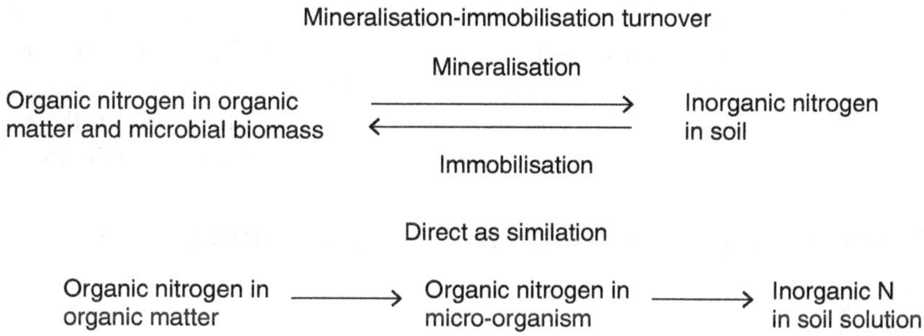

Figure 12.4 Summary of nitrogen mineralisation models showing the mineralisation–immobilisation turnover above and the direct assimilation model below (Barraclough, 1997). Reproduced with kind permission from Elsevier.

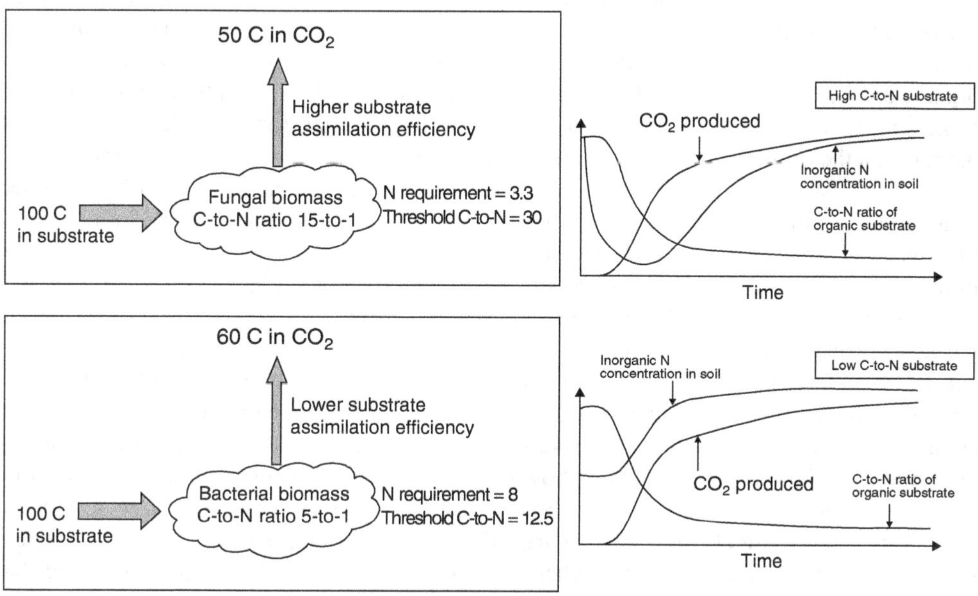

Figure 12.5 Relationships between carbon–nitrogen ratio and the threshold values for net nitrogen mineralisation. Fungi tend to have a larger carbon–nitrogen ratio than bacteria because bacteria contain a larger proportion of structural peptides than bacteria so the threshold for net nitrogen mineralisation for the substrate is greater than for bacteria. It should also be noted that the assimilation efficiency of carbon by fungi is greater than that for bacteria; so the proportion of carbon lost as CO_2 is less for fungi than for bacteria and this affects the amount of carbon the organisms will assimilate and therefore the amount of nitrogen that they will require from the substrate. Redrawn from Killham (1994). Dynamics of CO_2 production, inorganic nitrogen concentration in the soil and the carbon–nitrogen ratio of the organic substrate as the substrate decomposes for two contrasting substrates: one with a high carbon–nitrogen ration (above the threshold for net nitrogen mineralisation) and one with a low carbon–nitrogen ratio (below the threshold for net nitrogen mineralisation).

below which nitrogen is mineralised, because there is a surplus of nitrogen relative to in the substrate. However, the substrate quality and the composition of the decomposer community influence the relationship between carbon–nitrogen ratio of the decomposing substrate and mineralisation and immobilisation (Figure 12.5). Wheat straw has a carbon–nitrogen ratio typically in the range 80–100:1 and will lead to net nitrogen immobilisation, whereas leaf and stem residues from a nitrogen-fixing legume with a carbon–nitrogen ratio in the range 12–15:1 will lead to net mineralisation. This provides an opportunity to manage the synchrony of nitrogen supply to plants, or to remove inorganic nitrogen from the soil pool and thus reduce the chance of nitrogen loss by denitrification and leaching (Figure 12.5).

12.2.4 Managing soils for climate change mitigation

Carbon management

Current efforts to mitigate the effects of human-driven climate change have led to a renewed and extended interest in soil carbon as a reservoir of carbon sequestered for the atmosphere. Soils are the third largest store of carbon globally, storing over twice as much as the atmosphere and almost three times the amount that is in vegetation. Carbon sequestration can be achieved by increasing net photosynthesis (e.g. planting new areas to trees) or by slowing the rate of decomposition of organic matter in soil through a change in land management (e.g. reduce tillage intensity) (Table 12.5). However, not all practices that increase soil organic carbon represent a transfer of additional carbon from the atmosphere to soil, because some simply move carbon from one biosphere pool to another, for example from the plant biomass to the soil, with no immediate implications for the atmospheric carbon dioxide content (Powlson et al., 2011). Assessing whether soil organic carbon content has changed as a result of a management change is itself fraught with difficulties arising from the fact that the rates of increase are slow (decade scale) and the actual amount of additional carbon in the soil may be small relative to the initial amount, the spatial variability of soils which challenges sampling protocols and the fact that in order to determine the soil organic carbon stock, the soil depth, density and carbon concentration need to be determined. With the exception of the unusual situation that occurs in long-term experiments, not all of these data are routinely available. Careful assessment of comparisons of management practices is necessary to determine whether true sequestration is occurring. An assessment of the few available studies with time series data has shown that gains in soil organic carbon were often due to a reduction in or cessation of carbon losses, rather than increases in accumulation (Sanderman and Baldock, 2010). They argued that soil organic carbon data from field trials may have limited predictive power when the state of the soil system is unknown and that current IPCC accounting approaches developed from the trial results may not properly credit these management activities. There are limitations to the extent and effectiveness of sequestration in soil or vegetation for mitigating climate change. First, the amount of soil organic carbon that can be sequestered in soil is finite. This is because the increase in soil usually follows an asymptotic relationship; the rate of increase shortly after initiating a new management practice is greatest in the early years but this rate slows, and eventually ceases, in later years as a new equilibrium level of soil is approached (Johnston et al., 2009). Second, the process of gain is reversible and a management practice such as reduced tillage that may lead to an increase in soil organic carbon content will need to be maintained to avoid subsequent carbon release.

Evaluation of management practices for climate change mitigation needs to be conducted with a full appraisal of all greenhouse gases (Table 12.5). Both nitrous oxide and methane have greater global-warming potential than carbon dioxide, so the effectiveness of management practices that lead to increased vegetation and soil organic carbon stocks have to be evaluated with respect to possible increases in nitrous oxide emission or decreases in methane uptake by soils.

Table 12.5 Comparison of GHG mitigation potential for agricultural land management practices in the USA, summarised from scientific literature.

Activity	Soil carbon(t ha^{-1} yr^{-1})			Land emissions(N_2O and CH_4,t ha^{-1} yr^{-1})			Direct Impact (t ha^{-1} yr^{-1})	Process and upstream emissions(t ha^{-1} yr^{-1})			Average net impact (t ha^{-1} yr^{-1})	maximum area (Mha)
	Mean	Max	Min	Mean	Max	Min		Mean	Max	Min		
Conventional to conservation till	1.09	1.82	0.00	0.08	0.38	0.00	1.18	0.08	0.10	0.03	1.25	72
Conventional to no-till	1.09	2.60	−0.26	−0.18	0.72	−0.91	0.90	0.14	0.18	0.07	1.04	72
Eliminate summer fallow	0.48	2.35	−0.88	−0.03	0.16	−0.30	0.45	−0.13	−0.07	−0.25	0.32	20
Use winter cover crops	0.83	3.24	0.37	0.25	1.05	0.00	1.07	0.61	0.81	0.41	1.69	74
Diversify annual crop rotations	0.58	3.01	−2.50	0.07	0.33	−0.04	0.65	0.00	0.00	0.00	0.65	99
Include perennial crops in rotations	0.57	2.20	−1.75	0.03	0.55	−0.55	0.59	0.17[a]	0.28	0.14	0.77	56
Change from annual to perennial crop	2.26	4.67	0.00	0.12	0.84	−0.55	2.38	0.54	0.70	0.39	2.92	13
Organic soil amendments (esp. manure)	1.53	5.10	0.18	0.05	1.81	−1.35	1.59	0.94	1.42	0.46	2.53	8.5
Manage histosols to reduce GHG emissions	6.17	15.03	2.75	2.66	3.09	2.23	8.83	0.00	0.00	0.00	8.83	0.8
Reduce fertilizer N application rates	0.00	0.00	0.00	0.40	1.32	0.14	0.40	0.06	0.08	0.04	0.46	124
Change fert. N source – slow-release	0.00	0.00	0.00	0.46	1.43	0.00	0.46	0.06	0.31	0.17	0.52	93
Use nitrification inhibitors	0.00	0.00	0.00	0.49	1.65	0.00	0.49		No data		0.49	99
Reduce chemical use (other than N)	0.00	0.00	0.00	0.00	0.00	0.00	0.00	0.07	0.09	0.03	0.07	122
Rice water management for CH_4	0.00	0.00	0.00	1.56	5.22	−0.88	1.56		No data		1.56	1.3
Rice variety development for CH_4	0.00	0.00	0.00	1.17	2.71	0.00	1.17	0.00	0.00	0.00	1.17	1.3
Improved grazing management, rangeland[b]	1.01	4.99	−0.10	0.28	0.31	0.27	1.30		No data		1.30	166
Species composition on grazing lands	2.44	4.84	0.18	−0.94	No range[c]		1.50	0.00	0.00	0.00	1.50	80
Crop to pasture (LUC)	2.87	4.70	0.00	1.00	4.96	0.70	3.87	0.46	0.69	0.24	4.33	Unknown
Reduced rice acreage (LUC)		No data		4.82	No range		4.82	0.00	0.00	0.00	4.82	1.3
Cropland conversion to natural (LUC)[d]	2.49	4.74	−0.15	1.41	5.06	0.36	3.90	1.87	2.63	1.10	5.77	14
Convert Histosol cropland to natural (LUC)[e]	21.17	73.33	2.20	1.50	2.49	0.51	22.67	1.87	2.63	1.10	24.54	0.8

(*continued*)

Table 12.5 *(Cont'd)*

Activity	Soil carbon($t\ ha^{-1}\ yr^{-1}$)			Land emissions(N_2O and CH_4, $t\ ha^{-1}\ yr^{-1}$)			Direct Impact ($t\ ha^{-1}\ yr^{-1}$)	Process and upstream emissions($t\ ha^{-1}\ yr^{-1}$)			Average net impact ($t\ ha^{-1}\ yr^{-1}$)	maximum area (Mha)
	Mean	Max	Min	Mean	Max	Min		Mean	Max	Min		
Wetland restoration	3.11	7.70	0.45	−1.35	0.00	−2.70	1.76	1.87	2.63	1.10	3.62	
Short-rotation woody crops[f]	2.71	10.22	0.00	0.76	1.52	0.00	3.46	1.78	2.26	1.29	5.24	40
Agroforestry (windbreaks, buffers, etc.)	2.72	4.23	0.84	0.76	1.52	0.00	3.47	1.51	1.82	1.19	4.97	10
Switch from dry land to irrigated	1.46	4.77	1.14	−0.42	−0.05	−1.05	1.04	−1.38	−0.41	−3.34	−0.34	n/a[g]
Irrigation improvements (drip, supplemental, etc.)	0.36	0.58	0.18	0.66	0.94	0.14	1.03	0.23	0.27	0.19	1.21	20
Biochar application	3.37	8.92	0.13	1.14	2.93	0.82	4.51	0.70	1.05	0.12	5.22	124
Change fert. N source – between common types	0.00	0.00	0.00	0.42	2.80	−0.48	0.42	0.00	0.00	0.00	0.42	25
Change fert. N timing	0.00	0.00	0.00	0.35	0.52	0.01	0.35	0.00	0.00	0.00	0.35	62
Change fert. N placement	0.00	0.00	0.00	0.33	0.47	0.12	0.33	0.00	0.00	0.00	0.33	99
Improve manure management (N_2O)	No data			0.89	1.22	0.37	0.89	0.00	0.00	0.00	0.89	12
Improved grazing management, pasture	2.94	5.87	0.55	0.28	0.31	0.27	3.22	No data			3.22	48
Rotational grazing, rangeland	Minimal to zero potential											
Rotational grazing, pasture	2.20	2.90	1.50	0.05	0.11	0.02	2.25	0.00	0.00	0.00	2.25	42
Fertilizing grazing lands	1.05	5.87	0.37	−0.75	−0.60	−0.89	1.09	−0.94	−0.59	−1.36	−0.63	n/a
Irrigation on grazing lands[h]	1.04	1.83	0.00	−0.42	−0.05	−1.05	0.62	−1.08	−0.41	−3.34	−0.47	n/a
Pasture to grassland (LUC) – cease grazing	Minimal to negative potential											

All GHG units are in equivalents of carbon dioxide (CO_2) with 100-year time horizon global-warming potential. LUC, land-use change.

[a]Cells that are shaded indicate limited scientific data available (i.e., estimate based on expert opinion or on three or fewer field or laboratory comparisons).

[b]Area for rangeland management does not include any land of federal ownership.

[c]"No range" is indicated where the value comes from one scientific expert opinion or other estimate, with no indication of variability.

[d]Impact of reduced rice acreage depends on subsequent land use. These estimates account for elimination of current CH_4 emissions.

[e]For Histosols, the total area farmed is highly variable in the literature.

[f]For SRWCs and agroforestry, upstream and process emissions do not take into account the end-use for the aboveground biomass,because the GHG impact is highly variable (even though there is a likely GHG benefit). In practice, such lifecycle GHG impacts would need to be assessed.

[g]n/a = not applicable, national total is not calculated because the net GHG benefit is negative.

[h]Land emissions assume same N_2O response as for irrigated cropland

Source: Eagle et al. (2010).

Nitrous oxide

Nitrous oxide is formed as the product/by-product of denitrification and nitrification, both of which are influenced by the aeration status and inorganic nitrogen content of soils, albeit in different ways (nitrification is dependent on ammonium and an aerobic process, whereas nitrate is the main substrate for denitrification, which is carried out by facultatively anaerobic bacteria). Therefore, nitrous oxide emissions from agricultural land are related to the application of nitrogen fertiliser, soil and crop management, and weather conditions. The IPCC (2007) estimated that 70% of the total thermal forcing potential from greenhouse gas emissions from agriculture are attributable to nitrous oxide associated with the use of nitrogen fertiliser; therefore, improvements in nitrogen-use efficiency (i.e. less nitrogen applied for the same crop productivity) have potential to significantly reduce nitrous oxide emissions. Post-harvest residual soil mineral nitrogen concentrations are also positively correlated with leaching of nitrate, which can degrade water quality, as well as nitric oxide emissions and ammonia volatilisation, which also reduce air quality (Mosier et al., 1998). Nitrous oxide fluxes are highly variable temporally and spatially and variations in soil-water content, temperature and concentrations of soil mineral nitrogen are key factors affecting the 'hot-spots and hot-moments' of nitrous oxide emissions. These factors account for the very high variability in nitrous oxide emissions. For example, the majority of nitrous oxide emitted over a growing season is accounted for in short periods, very early in the season or during freeze/thaw cycles in winter/spring (Gregorich et al., 2005). Most fertiliser nitrogen management strategies to reduce nitrous oxide emissions are centred on the '4R' framework described by Roberts (2006) – the right rate, right product, right time and right place.

Methane

In cool temperate regions, fertilisation, tillage and compaction can influence soil methane uptake and emissions. Rice soils emit methane because microbial respiration in flooded conditions reduces oxygen potential, creating anaerobic conditions that lead to methane production. Mineral nitrogen status plays a key role in methane uptake by soil and therefore fertiliser and manure amendment practices play a key role in managing for climate change mitigation.

12.2.5 Conservation of biodiversity

It is well established that organisms in soil contribute to a wide range of biological processes which are fundamental to many ecosystem functions delivered by soils (Section 12.1). The processes include the decomposition of organic matter (notably plant and animal wastes and remains, and organic contaminants); the transformation of nutrient elements, releasing them in plant-available, soluble or volatile forms, which may predispose them to loss from soil (most notably in the case of nitrogen); mixing and formation of channels within the soil matrix by soil fauna; stabilisation of soil structure through the production of extracellular peptides and enmeshing filaments and biocontrol of soil-borne plant pathogens and pests. The diversity of organisms is huge and even with the developments in modern molecular biological techniques, the details of the taxonomy, physiology and ecology of only a small fraction of the organisms is understood (Bardgett et al., 2005). The need to conserve soil biodiversity arises from both the roles of the organisms in delivering ecosystems services and because of the reservoir of the biological potential that may be exploitable in the future, provided it is conserved.

There is a view that the number of micro-organisms in soils, the ability of micro-organisms to be transported from place to place and the massive collective metabolic repertoire of micro-organisms mean steps to conserve specific taxonomic or functional groups are unnecessary. These ideas owe much to the proposition that with respect to micro-organisms in natural environments, 'everything is everywhere, but the environment selects', attributed to Baas Becking and Beijerinck (de Wit and Bouvier, 2006), coupled with the diverse physiological repertoire of both bacteria and fungi, leading to the proposition that microbial communities are infallible with respect to their biodegradation capacity. Although probably not strictly accurate, these propositions are good enough working assumptions because micro-organisms or microbial consortia capable of degrading virtually any natural compound and many xenobiotic compounds can be enriched from soil communities provided environmental conditions are conducive. For many functions carried out by soil organisms, it is usually sufficient to provide appropriate environmental conditions and resources for the functions to occur. Several of the soil processes that operate to benefit crop production are carried out by a wide range of organisms and there is a high degree of functional redundancy in the community (Figure 12.6). These processes may be described as broad processes because they are distributed across a broad range of different organisms which have different environmental tolerance so that the

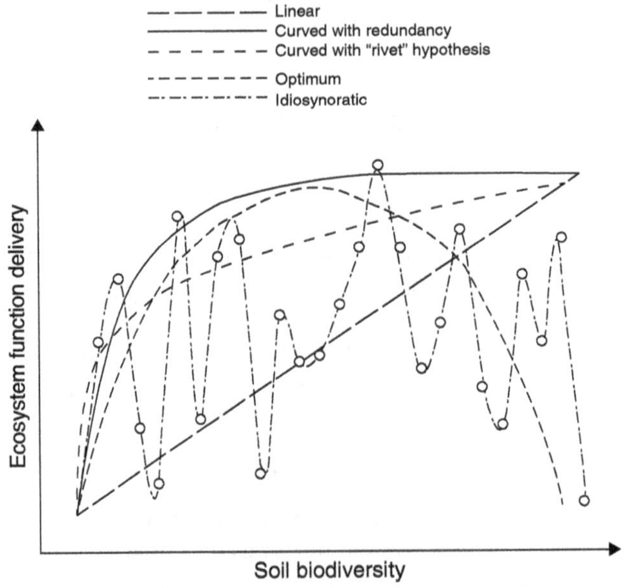

Figure 12.6 Hypothetical models of biodiversity–function relationships. The linear model implies ever increasing functioning with greater biodiversity. The curved model with redundancy (plateau at high biodiversity) implies that eventually no further increase in biodiversity is accompanied with an improvement in function because all the roles necessary for the function are fulfilled and there is spare capacity (redundancy). The corollary of this redundancy is that there is a catastrophic decline in functioning below a certain biodiversity threshold. The curved model with the 'rivet' hypothesis implies there is gradual loss of function with declining biodiversity when biodiversity is high, but a much sharper decline in function when biodiversity is low (the 'rivet' is derived from the analogy with a machine, such as an aircraft, held together with rivets; loss of some rivets is not catastrophic, but with further losses the machine will fail catastrophically). The optimum model implies that function is related to biodiversity at low levels, but that stress due to competition impairs function at high biodiversity. Finally, the idiosyncratic model represents the situation whereby it is not the biodiversity *per se*, but the traits of the individual groups of organisms that determine the functioning.

processes can operate over a wide range of environmental conditions. Such processes include decomposition of organic matter, some of the associated transformation of nutrient elements such as nitrogen mineralisation and biological contributions to the stabilisation of soil structure. Even under harsh environmental conditions (e.g. extremes of coldness or dryness), or when soils have been subject to contamination (e.g. metal-loaded sewage sludge) or have been deprived of significant plant inputs for decades (Hirsch et al., 2009), these processes can be maintained because the functions are distributed amongst a sufficiently large range of soil organisms even though the actual biomass of micro-organisms is severely depressed. These are all examples of the tremendous resilience of soil microbial communities arising from their very large functional diversity.

By contrast to the broad processes, narrow processes are those for which the functions necessary are distributed amongst a limited group of soil organisms or which only operate under a very specific set of environmental conditions. Examples include mycorrhizal associations, symbiotic nitrogen fixation, methane oxidation, nitrification, the decomposition of selected xenobiotic compounds and some antagonist interactions with plant pathogens and pests. These processes are much less resilient and more easily decreased or lost through conditions adverse to the limited groups of organisms performing them (Bardgett et al., 2005). In cases where a function is related to a narrow group of organisms, the absence of the correct species can impair functioning, and function can only be reinstated if the organisms are added and the conditions in the soil managed to ensure that the introduced organisms survive. The reinstatement of mycorrhizal associations and the reintroduction of earthworms to land restored after mining or other civil engineering operations are examples of such practices. Their success in restoring functions relies on both the presence of the organisms and the amelioration of soil conditions to allow their survival.

The second reason for conserving soil biological diversity is that soils are a reservoir of potentially exploitable organisms or capacities. This includes the long-established search, or bioprospecting, for novel antibiotic producers amongst soil micro-organisms that can be developed into therapeutic drugs. Probably the most famous example of pharmacological bioprospecting is the discovery and subsequent production of the antibiotic streptomycin from a soil streptomycete (*actinobacterium*), for which Selman Waksman was awarded the 1952 Nobel Prize for Medicine. Management of this aspect of soil biodiversity falls under the United Nations Convention of Biological Diversity (UNCBD, 1992). The objective of the UNCBD is to ensure conservation of biodiversity, sustainable use of its components and fair and equitable sharing of benefits arising from genetic resources. Implicit in the last of these aims is protection against biopiracy (i.e. the exploitation of indigenous knowledge about nature for commercial gain with no compensation to the indigenous people).

12.2.6 Preservation of soils that deliver cultural ecosystem services

Many systems of soil description regard soils as predominantly natural entities. This may be reasonable in parts of the world which have had low human population densities in the past, but it is not universally valid. Human activities in the past have affected soils by tillage, waste disposal, earth-working (e.g. ditches, dykes, terraces and burial mounds) and the establishment of field systems. Such soils may be regarded as cultural soils, because there is a record of human activity present in the soil structure, composition or morphology. The consequences are that many soils exhibit legacy properties from past land management (Davidson and Smout, 1996).

The terra preta (Portuguese: black earth), or *terra preta do índio*, soils found in pockets in the Amazon basin are an example of a group of cultural soils that are currently receiving considerable attention. There is an argument that the terra preta soils are sufficiently special to merit conservation status. These soils are rich in charcoal, pottery sherds, and plant and animal remains characteristic of midden sites and date back to at least 2500 years before present. They are cultivated by the indigenous people, are particularly productive when compared with the surrounding soils and have a number of desirable characteristics for plant production, including large concentrations of plant nutrients and a low propensity for nutrient loss by leaching. Whether the incorporation of charcoal into the terra preta soils was a deliberate strategy by the original human inhabitants is uncertain, but its current presence is highly likely to contribute to the favourable soil properties for cultivation. The reason for the upsurge in attention being paid the terra preta soils in recent years is because of the suggestion that they provide an analogue for soils amended with biochar (charred biological materials produced under low oxygen conditions), which is being advocated enthusiastically in some quarters as a carbon sequestration treatment to mitigate climate change because of the inherent stability of biochar. In reality, the terra preta soils are probably an example of an anthropogenic soil, the result of long-term management that did not start as a deliberate strategy, but which was maintained because of the empirically observed benefits. Similarly, the plaggen soils of northern Europe are anthropogenic soils created since the twelfth and thirteenth centuries by artificially deepening through the application of turves and livestock bedding enriched with faeces and urine to surrounding fields.

Cultural soils, especially those containing archaeological evidence, can also deliver ecosystem services. Cultural soils serve as a medium for the preservation of buried archaeological remains, including the occurrence of crop marks that indicate the presence beneath the surface of archaeological features. Although the cultural aspects of soils have been investigated on a site-by-site basis, there are few general principles to draw on when considering their management. Management of cultural and archaeological soils is often subordinate to management of the soil–plant system for the provisioning and regulating services, but some aspects of the management have benefits to the delivery of provisioning, regulatory and cultural services, whilst others are in conflict. Examples of management that benefits the provisioning and/or the regulatory ecosystem services include management to reduce erosion and soil organic loss. Minimising erosion helps preserve the stratigraphic context of cultural soils and protects artefacts in archaeological soils, and conserving soil organic matter can preserve both the biomolecular record of human activity and the evidence of past land management. By contrast, tillage, especially deep tillage, disrupts cultural soils, may lead to direct plough damage to archaeological soils and can promote root penetration deep into the soil leading to physical damage to buried archaeological remains. Similarly, drainage may be desirable for food and fibre production, but it also affects moisture regimes and redox conditions threatening the survival of biodegradable organic artefacts and evidence.

Special conservation status is usually extended to ecosystems, landscapes, cultural features (such as monuments and other archaeological remains) and particular species of plants or animals (rarely micro-organisms). In many countries, soils *per se* are not the subject of special conservation measures because of the inherent value or rareness. For example, in the UK soils are not considered within existing geological criteria for Sites of Special Scientific Interest (SSSI), but the plant community they support may be. The review of geological conservation criteria for the designation of SSSIs (JNCC, 1996) states 'the importance of soils, somewhat neglected hitherto in nature conservation, is now gaining increasing prominence and their study within the concept of sustainability is likely to become an important task in the future'. The importance attached to soils does, however, differ between cultures. The 'red centre' of Australia is a clear reference to the soils of a harsh landscape in which Australian aborigine people can find sustenance; there is a strong notion of 'geospirituality' amongst many indigenous people including First Nations people in Canada and in the Middle and Far East. Figure 12.2b is a monument to the different soils in China, illustrating the importance of soils in that country and the recognition of their importance for food production. Indeed, the importance of indigenous peoples' knowledge of soils is an important dimension of agricultural development (Pawluk et al., 1992). The soils of the environmentally extreme Antarctic dry valleys support no vascular plants because of the extreme cold and dry conditions (there has been no rainfall for millennia), but they do support a small active soil biological community (Figure 12.2c). They are considered as having inherent conservation value because of their rareness, extremeness and fragility, and they fall within the designation of either Antarctic Specially Protected Areas or an Antarctic Specially Managed Area. At the other end of a productivity scale, it might be argued that the most productive agricultural soils merit protection from development and degradation, and therefore remain available for food production.

12.3 Sustainable soil use and soil quality

Assessing the capacity of the soil–plant system to deliver ecosystem services is critical to assessing the sustainability. Rapid depletion of soil resources or soil degradation will impair functioning and thereby impair ecosystem service delivery. The concept of soil quality has been developed to address the need for a framework to assess soils. Soil quality is a long-established concept (e.g. Storie, 1932; Jenny, 1961), but the actual term was first introduced in the 1970s (Warkentin and Fletcher, 1977). Since then it has been variously defined and subtly reinterpreted (Table 12.6). However, the ability to perform functions and supply services is common to many of the definitions. Table 12.6 is not intended as a definitive summary of the definitions and interpretations, but it illustrates that the concept of value is common to many definitions, and that the soil properties linked to crop production predominate. This is because of the dominance of agricultural science in the development of soil and plant science during the twentieth and twenty-first centuries. Soil quality is closely allied to the soil health concept, which is an approach to assessing soil quality analogous to assessing human health in which the condition of a soil's properties are assessed relative to some optimum or the ability of a soil to perform particular functions is assessed against goals placed on it (Gregorich et al., 2001). The important difference between soil quality and soil health is that a soil may be of low inherent quality, but still be in good health if the expectations placed on it are either suitably low or tailored to the environment. A comparison of this interpretation of soil health against the definitions of soil quality in Table 12.6 reveals why the terms are often used synonymously.

Because soil quality is multivariate, frameworks for assessing soil properties from measurable indicators are needed. The development of such systems has been a major activity by many agencies in recent years. Some examples of indicator sets are summarised in Table 12.7. The general principle is that indicators need collectively to cover the biological, chemical and physical attributes as well as often including geophysical attributes (e.g. climatic) and management (e.g. workability) of soils. In some cases, a minimum dataset has been proposed for soil quality assessment (Doran and Parkin, 1996), in others sophisticated

Table 12.6 Examples of definitions and interpretations of soil quality.

Definitions and interpretations	Country or region of origin	Source
... key soil qualities important for crop production have been derived, namely nutrient availability and retention capacity, oxygen availability for roots, excess salts, toxicities and workability See Table 13.6 for specific details of the specific 'qualities'	Global	FAO (2010)
... an account of the soil's ability to provide ecosystem services and social services through its capacities to perform its functions under changing conditions See Table 13.6 for specific details of the specific 'qualities'	European Union	European Commission Joint Research Centre Tóth et al. (2008)
... the capacity of a specific kind of soil to function, within natural or managed boundaries, to sustain plant and animal productivity, maintain or enhance water and air quality and support human health	USA	USDA Natural Resources Conservation Service NRCS (2010)
... value placed on a soil with respect to its fitness for a specific use and ... categorisation of the soil for certain use-based ecological aspects	Canada	Gregorich et al. (2001)
.... a soil that is productive and easy to manage under the intended land use; it has biological, chemical and physical properties that promote the health of plants, animals (including humans) ... whilst also maintaining environmental quality	Australia	Soil Quality and Health Information for Australia (2010)
... soil quality is not a single concept, but encompasses aspects of soil physical structure, chemical fertility, nutrient storage, organic matter resources and the biological life in the soil See Table 13.6 for specific details of the specific 'qualities'	New Zealand	SINDI (2010)

'calculators', sometimes web based, have been developed to assist land managers, farmers, policymakers and environmental regulators (e.g. SINDI, 2010). In some other schemes that include a large number of indicators, the importance and desirability of different indicators are ranked (e.g. Aalders et al., 2009). The main challenges in developing indicator sets for soil quality assessment has been to include indicators that are meaningful, readily measurable, cost-effective and which can be compared with data in existing databases to assess change. Scientific innovation has led to a wide range of research-orientated techniques being available, such as molecular biological approaches to assess the soil biological community, but these rarely find a place in routine soil quality assessment because they are often expensive to measure, rely on sophisticated equipment and highly specialised personnel, and because it is rarely possible to interpret the data unambiguously. Such approaches may find their way into future assessment procedures in the same way indicators now regarded as routine were previously cutting-edge research techniques. For this reason, soil archives and sample archives (e.g. –80°C stores of DNA extracted from soil) have an important role in long-term soil

Table 12.7 Examples of indicators used in soil quality assessment.

System	Indicator	Soil quality information	Source
SINDI	CHEMICAL INDICATORS		SINDI (2010)
	Olsen P	Plant-available phosphorus indicative of fertility	
	pH	Acidity or alkalinity	
	Total carbon	Organic matter reserves with implications for soil structural stability and water retention	
	Total N	Organic nitrogen reserves	
	BIOLOGICAL/ BIOCHEMICAL INDICATORS		
	Anaerobically mineralisable N	Availability of nitrogen to plants from the reserve	
	PHYSICAL INDICATORS		
	Bulk density	Soil compaction, physical environment for roots and soil organisms	
	Macroporosity	Availability of water and air, water retention and drainage properties	
System	**Soil qualities**	**Soil characteristics**	**Source**
FAO Soil quality for crop production	NUTRIENT AVAILABILITY	Soil texture Soil organic carbon Soil pH Total exchangeable bases	FAO (2010)
	NUTRIENT RETENTION CAPACITY	Soil organic carbon Soil texture Base saturation Cation exchange capacity of soil overall Cation exchange capacity of the clay fraction	
	ROOTING CONDITIONS	Soil texture Bulk density Coarse fragments Vertic properties and soil phases affecting root penetration Soil depth Soil volume	
	OXYGEN AVAILABILITY TO ROOTS	Soil drainage and phases influencing soil drainage	
	EXCESS SALTS TOXICITY	Soil salinity, sodicity and factors affecting salt conditions Calcium carbonate and gypsum (in excess)	
	WORKABILITY (constraining field management)	Soil texture Effective soil depth/volume Phases constraining soil management (such as shallowness, rock outcrops, excessive stoniness, gravel, concretions and hardpans)	

(continued)

Table 12.7 (*cont'd*).

Threat	Indicator (referred to as 'criterion' in the source document)	Source
Loss of soil organic matter	Soil type Soil texture Soil organic concentration Soil organic content Climate (mean annual precipitation and temperature) Slope, aspect and position in relief (top or bottom of slope) Land use or land cover	European Commission Joint Research Centre Tóth et al. (2008)
Erosion	Soil type Soil texture Soil (bulk) density and hydraulic properties (water retention) Topography (slope) Land use or land cover Climate (annual precipitation and temperature, number of rain days, storm events and potential evapotranspiration) Hydrological conditions (catchment information and digital elevation model to give slopes) Agro-ecological zone (based on soil, climate and landscape)	
Compaction	Soil mapping unit/soil typological unit Top and subsoil texture Climate (rain fall and evapotranspiration) Land use and land cover Slope	
Salinisation	Soil mapping unit/soil typological unit Soil texture Soil hydraulic properties (conductivity, water retention and drainage) Irrigation area and intensity, salt content, sodicity and alkalinity of irrigation water Groundwater information (depth, salt contents, sodicity and alkalinity) Climate (annual rainfall and annual potential evapotranspiration)	
Landslides	Record of previous landslides Bedrock Soil texture, structure and permeability Slope class (<10°, 10–30° >30°) Land use and land cover (including infrastructure, cultivation and mining) Climate (storm events) Seismic risks	

The New Zealand SINDI and the FAO soil quality for crop production are included to compare the relatively narrow range of indicators needed for a small country with a relatively narrow range of land management and land-use options, compared with the system developed for global coverage. The European Commission Joint Research Centre system is included because it is focussed on identifying the risk of major threats involving soil that would affect its ability to deliver ecosystem functions. The European example also illustrates the widespread applicability of some key soil indicators, such as texture and climate, which provide valuable information for several contexts, as well as the subtle differences in common indicators (e.g. the climate indicator includes annual precipitation and/or number of rain days and/or number of storm events, and potential evapotranspiration depending on the threat).

quality assessment as they provide material that can be accessed when new techniques become more routine and when the purpose of soil monitoring switches between applications, such as changing emphasis from agricultural production to carbon sequestration for climate amelioration. When assessing a soil indicator against a target, or when interpreting any change in an indicator value either with time or in response to a treatment, it is important to realise that different indicators will respond differently; an increase in the indicator does not necessarily imply improved functioning or ability to deliver an ecosystem service. Therefore, the interpretation relies on either expert judgement at the time or on the expert judgement and evidence used to develop soil quality calculators.

Monitoring the condition of soil and how it changes over time is the major reason for collecting indicator data. This requires repeat measurements to judge both changes in soil quality and the efficacy of soil protection strategies (Arrouays et al., 2008). Whilst there are some examples of repeat measurements in the context of long-term experiments (e.g. Johnston et al., 2009), widespread re-sampling and measurement of indicators at a national or regional scale is relatively rare. For example, across Europe, soil at the majority of designated environmental monitoring sites has only been sampled once and therefore most soil monitoring is still at the pre-monitoring, inventory or baseline stage (Morvan et al., 2008). Re-sampling is not as straightforward as it might at first appear. To be useful in monitoring, a soil indicator should not be over-reactive and fluctuate or oscillate too rapidly; the most useful indicators are ones which capture the slow rates of change over years and decades but sensitive enough to initial subtle changes that may provide an early warning of a trend. This means that sampling has to be done on a scale measured over decades. As techniques, personnel and skills change, because records may be lost, and because the requirements of monitoring programmes change over these timescales, particular care and foresight are needed. Without wanting to diminish the efforts of major national scale monitoring and re-sampling exercises, the fact that one of the demands on soil monitoring changed over time from determination of soil organic carbon concentration (i.e. percent carbon by weight) to the need for estimates of total carbon stocks (i.e. quantity of carbon per unit area) has highlighted the difficulties of missing indicators from earlier surveys (e.g. bulk density and depth profiling), changing methods of sample collection and analysis and relocation of precise sampling sites from the map-based era to the satellite-positioning era, especially after land-use change. However, with increasing demands for soil monitoring information by regulators, policymakers and land managers, there will be a continued need for the collection of soil indicator data and the evolution of soil monitoring frameworks.

12.4 Concluding remarks

Managing the soil–plant system for the delivery of the range of ecosystem services humans expect is complex because of the competing demands and the opposing effects of different services on the properties of the soil–plant system. The soil–plant system contributes to all the major groups of ecosystems services (supporting, provisioning, regulating and cultural services), and rarely are any straightforward choices available when all the systems have to be managed to deliver all four different types of services. As the human population continues to increase, the demands on the soil–plant system will increase in proportion. Inevitably, this leads to choices and different services being prioritised by different land owners or societies. It has been traditional for the management of the soil–plant system to be addressed from the point of view of delivering provisioning services, and sometimes regulating services. But the soil–plant system also contributes to other services and prioritisation is more complex than simply choosing, for example, food and fibre production over water quality. The choices and prioritisation will become more intense with increasing human population; yet unless it is possible to reconcile these demands, the soil–plant system will be unable to continue to deliver the full range of ecosystem services. As we explained at the outset, assigning either a monetary or an energetic value to ecosystems services delivered by soils provides an approach to prioritising different attributes of soils, but such attribution is only an analogy and the value cannot be realised. Managing the soil–plant system for the delivery of ecosystem goods and services will always be complex because of the many competing expectations placed on soils. Good knowledge of the physics, chemistry, biology and environmental science dimensions of soils will be essential, but so will a sound appreciation of soil and plant (crop) management, planning, economics and even the social sciences.

References

Aalders, I., Hough, R., Towers, W., Black, H., Ball, B., Griffiths, B., Hopkins, D.W., Lilly, A., McKenzie, B., Rees, R., Sinclair, A., Watson, C., and Campbell, C. (2009) Considerations for Scottish soil monitoring in the European context. *European Journal of Soil Science*, **60**, 833–843.

Arrouays, D., Richer de Forges, A., Morvan, X., Saby, N., Jones, A., and Le Bas, C. (2008) *Environmental Assessment of Soil for Monitoring: Volume IIb Survey of National Networks*. In: EUR 23490 EN/ 2B, pp. 188. Office for the Official Publications of the European Communities, Luxembourg, Europe.

Bardgett, R.D., Bowman, W.D., Kaufmann, R., and Schmidt, S.K. (2005) A temporal approach to linking aboveground and belowground ecology. *Trends in Ecology and Evolution*, **20**, 634–641.

Barraclough, D. (1997) The direct or MIT route for nitrogen immobilization: a ^{15}N mirror image study with leucine and glycine. *Soil Biology and Biochemistry*, **29**, 101–108.

Chapagain A.K. and Hoekstra, A.Y. (2004) Water footprints of nations. In: *Value of Water Research Report Series No. 16*. UNESCO-IHE, Institute for Water Education, Delft, the Netherlands, 80pp. Available at http://www.waterfootprint.org/Reports/Report16Vol1.pdf

Costanza, R., d'Arge, R., de Groot, R., Farber, S., Grasso, M., Hannon, B., Naeem, S., Limburg, K., Paruelo, J., O'Neill, R.V., Raskin, R., Sutton, P., and van den Belt, M. (1997) The value of the world's ecosystem services and natural capital. *Nature*, **387**, 253–260.

Davidson, D.A. and Smout, B. (1996) Soil change in Scotland: the legacy of past land improvement processes. In: *Soils, Sustainability and the Natural Heritage* (eds A.G. Taylor, J.E. Gordon, and M.B. Usher), pp. 44–54. Edinburgh, Scotland, UK.

Doran, J.W. and Parkin, T.B. (1996) Quantitative indicators of soil quality: a minimum dataset. In: *Methods for Assessing Soil Quality* (eds J.W. Doran and A.J. Jones), pp. 25–38. SSSA Special Publication Number 49. Madison, Wisconsin.

Eagle, A.J., Henry, L.R., Olander, L.P., Haugen-Kozyra, K., Millar, N., and Robertson, G.P. (2010) *Greenhouse Gas Mitigation Potential of Agricultural Land Management in the United States: A Synthesis of the Literature*. Report NI R 10–04, Nicholas Institute for Environmental Policy Solutions, Duke University, Durham, North Carolina.

FAO (2003) *World Agriculture: Towards 2015/2030*. Interim report. FAO, Rome, Italy.

FAO (Food and Agriculture Organization of the United Nations). (2010) *Soil Quality for Crop Production*. http://www.fao.org/nr/land/soils/harmonized-world-soil-database/soil-quality-for-crop-production/en/ accessed 27 April 2011.

Galloway, J.N., Townsend, A.R., Erisman, J.W., Bekunda, M., Cai, Z., FreneyJ.F., Martinelli, L.A., Seitzinger, S.P., and Sutton, M.A. (2008) Transformation of the nitrogen cycle: recent trends, questions, and potential solutions. *Science*, **320**, 889–892.

Gleick, P.H. (ed.) (1993) *Water in Crisis*. Oxford University Press, New York.

Gregorich, E.G., Turchenek, L.W., Carter, M.R., and Angers, D.A. (Eds) (2001) *Soil and Environmental Science Dictionary*. CRC Press, Boca Raton, Florida.

Gregorich, E.G., Rochette, P., Van den Bygaart, A., and Angers, D.A. (2005). Greenhouse gas contributions of agricultural soils and potential mitigation practices in Eastern Canada. *Soil and Tillage Research*, **83**, 53–72.

Harmsen, G.W. and van Schreven, D.A. (1955) Mineralization of organic nitrogen in soil. *Advances in Agronomy*, **7**, 299–398.

Hirsch, P.R., Gilliam, L.M., Sohi, S.P., Williams, J.K., Clark, I.M., and Murray, P.J. (2009) Starving the soil of plant inputs for 50 years reduces abundance but not diversity of soil bacterial communities. *Soil Biology and Biochemistry*, **41**, 2021–2024.

IPCC Intergovernmental Panel on Climate Change. (2007) *Changes in Atmospheric Constituents and in Radiative Forcing*. Fourth Assessment Report (AR4), Working Group 1, Chapter 2., pp. 131–234.

Jansson, S.L. and Persson, J. (1982) Mineralization and immobilization of soil nitrogen. In *Nitrogen in Agricultural Soils* (ed F.J. Stevenson), pp. 229–252. America Society of Agronomy, Crop Science Society of America, Soil Science Society of America, Madison, Wisconsin.

Jenny, H. (1961) Derivation of state factor equations of soils and ecosystems. *Soil Science Society of America, Proceedings*, **25**, 385–388.

JNCC (1996) Joint Nature Conservation Committee; http://jncc.defra.gov.uk/page-2303. Accessed on 2 October 2011.

Johnston, A.E., Poulton, P.R., and Coleman, K. (2009) Soil organic matter: its importance in sustainable agriculture and carbon dioxide fluxes. *Advances in Agronomy*, **101**, 1–57.

Killham, K. (1994) *Soil Ecology*. Cambridge University Press, Cambridge, UK.

MEA (2005) *Millennium Ecosystem Assessment, Ecosystems and Human Well-Being: synthesis*. Island Press, Washington, D.C.

Mosier, A., Kroeze, C., Nevison, C., Oenema, O, Seyzinger, S. and van Cleemput, O. (1998). Closing the global N_2O budget: nitrous oxide emissions through the agricultural nitrogen cycle – OECD/IPSS/IEA phas II development of IPCC guidelines for national greenhouse gas inventory methodology. *Nutrient Cycling in Agroecosystems*, **52**, 225–248.

Molden, D. (ed.) (2007) *Water for Food, Water for Life*. Earthscan, London.

Morvan, X.P.P., Saby, N.P.A., Arrouays, D., Le Bas, C., Jones, R.J.A., Verheijen, F.G.A., Bellamy, P.H., Stephens, M., and Kibblewhite, M.G. (2008) Soil monitoring in Europe: a review of existing systems and requirements for harmonisation. *Science of the Total Environment*, **391**, 1–12.

NRCS Natural Resources Conservation Service. (2010) *Soil Quality: definition and Purpose*. http://soils.usda.gov/technical/handbook/contents/part624.html accessed Apr 27, 2011.

Odum, H.T. (1996) *Environmental Accounting: EMERGY and Environmental Decision Making*. John Wiley & Sons, Inc., New York.

Odum, H. T. and Odum, E.P. (2000) The energetic basis for valuation of ecosystem services. *Ecosystems*, **3**, 21–23.

Pawluk, R.R., Sandor, J.A., and Tabor, J.A. (1992) The role of indigenous soil knowledge in agricultural development. *Journal of Soil and Water Conservation*, **47**, 298–302.

Pidwirny, M. (2006) *The Hydrologic Cycle. Fundamentals of Physical Geography*, 2nd edn. *University of British Colombia*, Okanagan. http://www.physicalgeography.net/fundamentals/8b.html. Accessed on 2 October 2011.

Pilgrim, E.S., Macleod, C.J.A., Blackwell, M.S.A., Bol, R., Hogan, D.V., Chadwick, D.R., Cardenas, L., Misselbrook, T.H., Haygarth, P.M., Hobbs, P., Jarvis, S.C., Hodgson, C., Murray, P.J., and Firbank, L.G. (2010) Interactions among agricultural production and other ecosystem services delivered from European temperate grassland systems. *Advances in Agronomy*, **109**, 117–0154.

Pimental, D., Wilson, C., McCullum, C., Huang, R., Dwen, P., Flack, J., Tran, Q., Saltman, T., and Cliff, B. (1997) Economic and environmental benefits of biodiversity. *BioScience*, **47**, 747–758.

Powlson, D.S., Whitmore, A.P., and Goulding, K.W.T. (2011) Soil carbon sequestration to mitigate climate change: a critical re-examination to identify the true and the false. *European Journal of Soil Science*, **62**, 42–55.

Roberts, T.L. (2006) Improving nutrient use efficiency. In: *Proceedings of the IFA Agriculture Conference Optimizing Resource Use Efficiency for Sustainable Intensification of Agriculture*, Kunming, China. International Fertilizer Industry Association, Paris, France. February 27–March 2.

Rockström, J., Falkenmark, M., Karlberg, L., Hoff, H., Rost, R., and Gerten, D. (2009) Future water availability for global food production: the potential of green water for increasing resilience to global change. *Water Resources Research*, **45**, 16–26.

Rost, S., Gerten, D., Hoff, H., Lucht, W., Falkenmark, M., and Rockström, J. (2009) Global potential to increase crop production through water management in rainfed agriculture. *Environmental Research Letters*, **4**, 044002.

Sanderman, J. and Baldock, J.A. (2010) Accounting for soil carbon sequestration in national inventories: a soil scientist's perspective. *Environmental Research Letters*, **5**, 1–6.

SINDI: soil quality indicators. (2010) *Soil Indicators on the Web*. Landcare Research – Manaaki Whenua, Dunedin, New Zealand. http://sindi.landcare.cri.nz accessed Apr 27, 2011.

Soil Quality and Health Information for Australia. (2010) http://soilquality.org.au accessed December 1, 2010.

Storie, R. (1932) *An Index for Rating the Agricultural Values of Soils*. Bulletin 556. California Agricultural Experiment Station, Berkeley, California.

Swift, M.J., Heal. O.W., and Anderson, J.M. (1979) *Decomposition in Terrestrial Ecosystems*. Blackwell Scientific Publications, Oxford, UK.

Swift, M.J., Izac, A.M.N., and van Noordwijk, M. (2004) Biodiversity and ecosystem services in agricultural landscapes – are we asking the right questions? *Agriculture, Ecosystems and Environment*, **104**, 113–134.

Tóth, G., Montanarella, L., and Rusco, E. (eds) (2008) *Threats to Soil Quality in Europe*. European Commission, Joint Research Centre, Institute for Environment and Sustainability, Ispra, Italy.

Vitousek, P.M., Aber, J.D., Howarth, R.W., Likens, G.E., Matson, P.A., Schindler, D.W., Schlesinger, W.H., and Tilman, D.G. (1997) Human alteration of the global nitrogen cycle: sources and consequences. *Ecological Applications*, **7**, 737–750.

Vitousek, P.M., Naylor, R., Crews, T., David, M.B., Drinkwater, L.E., Holland, E., Johnes, P.J., Katzenberger, J., Martinelli, L.A., Matson, P.A., Nziguheba, G., Ojima, D., Palm, C.A., Robertson, G.P., Sanchez, P.A., Townsend, A.R., and Zhang, F.S. (2009) Nutrient imbalances in agricultural development. *Science*, **324**, 1519–1520.

Voora, V. and Thrift, C. (2010) *Using Emergy to Value Ecosystem Goods and Services*. International Institute for Sustainable Development, Winnipeg, Manitoba.

Warkentin, B.P. and Fletcher, H.R. (1977) Soil quality for intensive agriculture. In: *Proceedings of the International Seminar on Soil Environment and Fertility Management in Intensive Agriculture*, pp. 594–598. Society of Science of Soil and Manure, Tokyo, Japan.

de Wit, R. and Bouvier, T. (2006) 'Everything is everywhere, but, the environment selects'; what did Baas Becking and Beijerinck really say? *Environmental Microbiology*, **8**, 755–758.

UNCB (1992) *United Nations Convention on Biological Diversity. 1992*. Text and Annexes, the Interim Secretariat for the UNCBD, Geneva Executive Center, Chatelaine, Switzerland.

13 The new challenge – sustainable production in a changing environment

Peter J. Gregory[1,2] and Stephen Nortcliff[3]

[1]Centre for Food Security, School of Agriculture, Policy and Development, University of Reading, Reading, UK
[2]East Malling Research, Reading, UK
[3]Department of Geography and Environmental Science, Soil Research Centre, University of Reading, Reading, UK

13.1 Introduction

The last 60 years have seen a large increase in the world's population from about 2.2 billion in 1950 to about 7 billion in 2011. Accompanying this increase has been a substantial increase in food production and use of resources such as energy, water and land not only to produce food but also to satisfy changed, generally more affluent and resource-demanding, lifestyles. In the last 50 years, production of cereal grains such as maize, rice, wheat and barley has almost tripled, while production of root crops such as potato and cassava has increased by around 40% and that of the coarse grains millet and sorghum by about 20%. Simultaneously, diets have changed to consume more meat so that numbers of chickens have increased 4.5-fold, pigs have increased 2-fold and cattle and sheep have increased by about 30% (Godfray et al., 2010).

Much of this increased production has come about by intensifying production systems and increasing yields (see Section 13.2.1 for more details), but it has often been achieved at some cost to other aspects of the environment. For example, excessive nutrient inputs, especially of nitrogen and phosphorus, have resulted in coastal eutrophication and reduced the quality of water in reservoirs used for drinking water (Vitousek et al., 2009). Similarly, cultivating soils for crop production has often increased the frequency of substantial soil erosion by either water, tillage or wind so that the current mobilisation of soil globally is 35 (±10) Pg year^{-1} or about 5 t year^{-1} for every person on the planet (Quinton et al., 2010).

This increasingly common degradation of soils, land and water as a result of agricultural intensification has led to the exploration of notions of land being needed to provide multiple functions simultaneously. These are commonly referred to as ecological or environmental goods and services and include elements of provisioning (e.g. food and fibre production), regulating (e.g. climate regulation and water purification) and cultural and non-material benefits (e.g. recreation and aesthetic and cultural benefits), which are all underpinned by the supporting services of soil formation, nutrient cycling and primary production (Millennium Ecosystem Assessment (MEA, 2005); see Chapters 1 and 12 for details). Such thinking has led to demands for changed agricultural practices that integrate biological and

Soil Conditions and Plant Growth. Edited by Peter J. Gregory and Stephen Nortcliff.
© 2013 Blackwell Publishing Ltd. Published 2013 by Blackwell Publishing Ltd.

ecological processes into food production so that agricultural systems become more sustainable and deliver the other essential services (Pretty, 2008; Powlson et al., 2011).

The purpose of this chapter is to examine the multiple environmental changes that humans are inducing and to reflect on the potential contributions that better understanding of soil and plant interactions might have on sustaining the requisite intensive crop production needed in future.

13.2 Human-induced environmental change

13.2.1 Land use change – extensification and intensification

Increased crop production has occurred as a result of both extensification (altering natural ecosystems to produce products) and intensification (producing more of the desired products per unit area of land already used for agriculture or forestry). Only about 3 billion hectares of the world's 13.4 billion hectare land surface is suitable for crop production and about one-half of this is already cultivated (1.4 billion ha in 2008; Greenland et al., 1998). The remaining potentially cultivatable land is currently beneath tropical forests and some grassland areas, but it would be undesirable to convert this to arable land because of the effects on conserving biodiversity, greenhouse gas emissions, regional climate and hydrological changes, and because of the high costs of providing the requisite infrastructure. As in the recent past, future increases in crop production are generally projected to come from increased yield (68%) and higher cropping intensity (12%) with only a small proportion coming from the cultivation of new land (about 20%; Bruinsma, 2003; Smith et al., 2010b). Table 13.1 shows that while intensification will dominate, extensification will contribute significantly to crop production in sub-Saharan Africa (27%) and Latin America and the Caribbean (33%). There is almost no land available for expansion of agriculture in South and East Asia and the Near East/North Africa (there may, indeed, be loss of agricultural land to urban development) so that intensification will be the main means of increasing production (Gregory et al., 2002; Bruinsma, 2003).

Intensifying crop production through increased yields per unit area will continue to be a major means of delivering the increasing human demand for food and fibre. As cereal production (wheat, maize and rice) has increased from 877 million tons in 1961 to 2342 million tonnes in 2007, the world average cereal yield has increased from 1.35 to 3.35 t ha^{-1} over the same period and is projected to be about 4.8 t ha^{-1} in 2040. Simultaneously, *per capita* arable

Table 13.1 Projected contributions (%) to increased crop production between 1997/1999 and 2030.

	Land area expansion	Increase in cropping intensity	Yield increase
All developing countries	21	12	67
Sub-Saharan Africa	27	12	61
Near East/North Africa	13	19	68
Latin America and Caribbean	33	21	46
South Asia	6	13	81
East Asia	5	14	81

Source: Derived from Bruinsma (2003).

land area has decreased from 0.415 ha in 1961 to 0.214 ha in 2007 (Smith et al., 2010b). Put another way, had the increases in yield of the last 60–70 years not been achieved, almost three times more land would have been required to produce crops to sustain the present population, land that, as indicated, does not exist except by using some that is unsuitable for cropping. Without the increased yields resulting from intensified practices, competition for land for different purposes would have been greater (Smith et al., 2010b). As demands for land to satisfy the required multiple ecosystem services increase, so competition for land will intensify, although there are major uncertainties in such projections (Smith et al., 2010b).

While the immediate loss of biodiversity as a consequence of clearing land for crop production has been well documented; the higher level consequences of increased crop yields on sparing land for biodiversity conservation is a matter of considerable debate. This is because (1) the on-farm losses of biodiversity due to practices giving high yields may outweigh the benefits of sparing biodiverse habitats; (2) high-yielding crops may have negative effects on offsite biodiversity and (3) land-sparing does not occur or is imperfect. The complexity of the factors involved is evident in the findings of Ewers et al. (2009), who analysed the changes in yields of 23 staple crops for 124 countries between 1979 and 1999. While per capita area of the 23 staple crops decreased in developing countries where large yield increases occurred, this was counteracted by a tendency for an increased area of non-staple crops leading to only a weak tendency for land-sparing overall. In developed countries, there was no evidence that higher yields reduced per capita cropped area possibly because of the role of agricultural subsidies in promoting production thereby overriding any land-sparing effects. Ewers et al. (2009) concluded that land-sparing is a weak process, but that improved agricultural technology may have contributed to the maintenance of natural vegetation cover in the past and that future conservation benefits, while debatable, are potentially available if land-use policies are also modified.

Another factor influencing the future use of land will be the changing climate. For example, warming of the climate in Scotland as documented by Barnett et al. (2006) and Gregory and Marshall (2012) could have substantial beneficial effects on the land-use potential for agriculture in Scotland resulting in potential expansion of areas of 'prime' agricultural land especially in the east and south of the country (Brown et al., 2008). However, while warming is well established, a crucial and presently largely unknown factor influencing future land use is precipitation. In Scotland, wetter winters are already being experienced leading to waterlogging of some autumn-sown crops such as wheat and rape and to delays in land preparation for spring-sown crops. However, should summer rainfall decline (there is no significant decrease currently identified using long-term data), then this could act in future to reduce land capability (Brown et al., 2011). In parts of the world in which water already limits crop production, the combined effects of increased temperature and decreased rainfall could have severe detrimental effects on land capability and yields (Fischer et al., 2005).

13.2.2 Greenhouse gases and biogeochemical cycles

Changes in the terrestrial carbon cycle as a consequence of the burning of fossil fuels are well documented (see Chapter 4), and the contribution of changes in land use to soil organic carbon and emissions of carbon dioxide (CO_2) to the atmosphere have been detailed in Chapters 4 and 12. This section focuses on the role of soils and plants in the production of the greenhouse gases (GHGs) methane (CH_4) and nitrous oxide (N_2O).

GHGs comprise CO_2, CH_4, N_2O, hydrofluorocarbons, perfluorocarbons and sulphur hexafluoride, and all have different degrees of effectiveness in retaining heat in the earth's

Table 13.2 The contributions of methane and nitrous oxide from agricultural sources to total national emissions and the contributions of the major processes for these gases to the agricultural emissions (2005 data; AEA, 2007).

	Methane		Nitrous oxide	
Country	**% Total emissions**	**Enteric fermentation (% agric. emissions)**	**% total emissions**	**Fertilisers and manures (% agric. emissions)**
England	28	84	61	64
Scotland			77	63
Wales	60	90	82	71
N. Ireland	78	86	79	69

atmosphere. Several surveys have indicated that the direct emissions of GHGs from agriculture contribute about 7% of a country's total GHG emissions (calculated as mass of CO_2 equivalent; AEA, 2007; EEA, 2007). While this is small in comparison with emissions from the energy sector (85.6% of the total), it is comparable with those from waste and industrial processes combined (7.7%), and attention will continue to be focused on attempting to reduce agricultural emissions. The methodology used to calculate these figures does not take account of the emissions resulting from the use of energy on farms for heating and fuel, nor the energy required to transport and process farm products for markets. When the whole food chain is considered, the contribution to GHGs increases to about 18% (EEA, 2007).

Net GHG emissions from agriculture consist entirely of CH_4 and N_2O because CO_2 uptake in photosynthesis is normally greater than any losses of CO_2 from, for example, soil respiration, unless major clearance of trees occurs (AEA, 2007). Emissions of these gases are very dependent on the form of agriculture practised. For example, in the UK, agriculture is the major contributor to emissions of N_2O in all parts of the UK but to emissions of CH_4 in only Scotland, Wales and Northern Ireland, where animals constitute a relatively larger component of agricultural activity (Table 13.2). Enteric fermentation is the major process contributing to CH_4 emissions (typically 85–90% of the release from agriculture), so stocking levels play a major role in determining the amounts of CH_4 released. This picture in the UK is broadly similar across Europe as a whole where agricultural emissions of GHGs comprise direct soil emissions (26%), cattle (CH_4, 26%), indirect emissions (N_2O, 18%), animal production (N_2O, 7%), cattle (CH_4, 5%), pigs (CH_4, 5%), sheep (CH_4, 5%), solid stores and drylots (N_2O, 5%) and others (4%; EEA, 2007).

Methane

The concentration of CH_4 in the atmosphere has increased rapidly from its pre-industrial concentration of about 750 ppbv to over 1750 ppbv today. This concentration is greater than that at any time in the last 650,000 years and is estimated to result in an additional radiative forcing of about 0.5 Wm^{-2} compared with that in 1750 (IPCC, 2007). There are many natural sources of CH_4 including termites, wild animals and oceans but wetlands (174 Tg $year^{-1}$) dominate the estimated 238 Tg $year^{-1}$ emitted (Reay et al., 2010). Anthropogenic sources of CH_4 (336 Tg $year^{-1}$) are made up of many components but with land-based activities dominant including emissions from ruminants (84 Tg $year^{-1}$), rice agriculture, landfills and waste (54 Tg $year^{-1}$ each), and biomass burning (47 Tg $year^{-1}$; Reay et al., 2010).

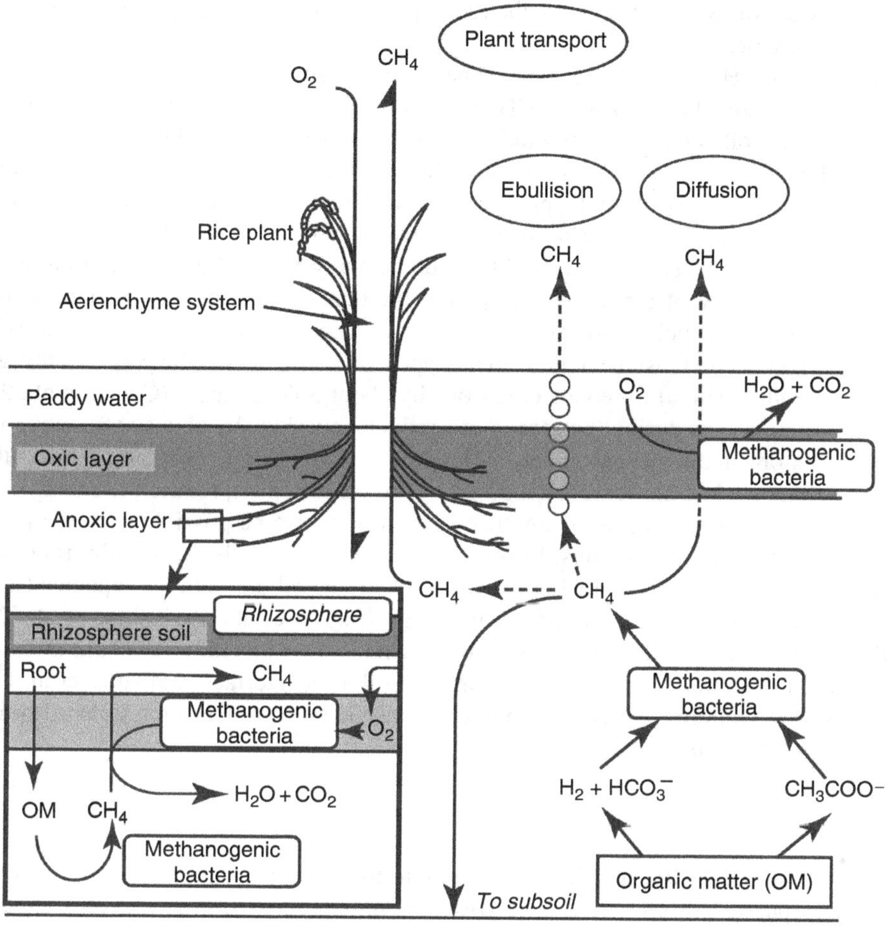

Figure 13.1 The production, oxidation and emission of methane in paddy rice fields. Reproduced from Yagi et al. (1997). With permission from Springer Science and Business Media.

More than one-third of CH_4 emissions come from soils as a result of the microbial breakdown of organic matter in anaerobic conditions. This process of methanogenesis is commonplace in flooded soils used to grow rice, in landfill sites and in the gut of some species of soil-dwelling termites. Rice is produced on about 160 Mha worldwide on a variety of soils in a range of production systems (Greenland, 1997). Flooded, paddy soils used to produce rice are complex systems comprising both aerobic (the soil–floodwater interface and zones around plant roots) and anaerobic (the bulk soil) regions (Figure 13.1). Respiration by plant roots and micro-organisms depletes the oxygen in the soil–water system so that other chemical species (nitrate, manganese (IV), iron (III), sulphate and CO_2) then act in turn as electron acceptors in the respiratory process and are reduced (Ponnamperuma, 1972). Following their reduction, methanogenesis occurs with both hydrogen and acetate acting as major drivers of CH_4 production. Both hydrogen and acetate are produced from the degradation of organic materials, but hydrogen is rapidly consumed (in minutes) as an electron donor in many redox reactions, while acetate turnover is typically slower and occurs in hours or days (Conrad, 1989; Yao et al., 1999). Methanogenic bacteria become active as the partial pressure of oxygen decreases, the redox potential declines and the partial pressure of hydrogen increases.

Emissions of CH_4 occur through several pathways (Figure 13.1; Yagi et al., 1997). In temperate regions, more than 90% of the CH_4 can be lost directly from plants following transport of the gas from soil via the aerenchyma, whereas in tropical paddy fields, significant quantities of CH_4 are evolved through bubbles (ebullision; Schütz et al., 1989; Denier van der Gon and Neue, 1995). Some of the CH_4 produced is consumed by methanotrophs inhabiting oxic regions (such as the soil–floodwater interface), which oxidise the CH_4 to CO_2.

The production of CH_4 from rice fields is substantially affected by the duration of flooding (i.e. the intensity of the anaerobicity) and the quantity and quality of organic matter added; lesser factors include the chemical fertility of the soil and the rice cultivar (Conen et al., 2010). Water and residue management appear as key factors in the potential mitigation of this gas with potential to reduce emissions by about 30% globally (Conen et al., 2010). Draining at least once during the season can reduce emissions by about 15%, while incorporation of straw in the off season before flooding, or composting it off-site, can simultaneously retain most plant nutrients (including most of the nitrogen), reduce the energy content and thereby cut emissions also by another 15%. It may also be possible in future to select rice cultivars to reduce emissions. Dissolved organic compounds released by roots are an energy source for methanogens, but the quantities released are cultivar dependent so that scope may exist to reduce this source of energy (Sigren et al., 1997; Lou et al., 2008). Furthermore, increasing the rate of oxygen diffusion from roots into soil could inhibit CH_4 production and stimulate CH_4 oxidation at the soil–root interface. Satpathy et al. (1998) found a fourfold difference between ten rice cultivars in seasonal CH_4 emissions with similar yield in nine of the cultivars.

Nitrous oxide

For most of the last 4000 years, N_2O concentrations in dry air were close to 270 ppbv, but since 1850, atmospheric concentrations have risen and now exceed 320 ppbv. This is significant because N_2O in the atmosphere has two distinct environmental consequences (Smith, 2010). First, its capacity to absorb infrared radiation is about 300 times greater than that of CO_2 so it is a significant contributor to the greenhouse effect (about 6% of the current greenhouse effect; IPCC, 2007) even though its concentration is 1000 times less than that of CO_2. Second, when it reaches the stratosphere, it contributes to the loss of ozone and thereby to the penetration of ultraviolet radiation to the Earth's surface with consequent direct effects on human health.

N_2O is a natural product of microbial processes occurring in soils principally via the pathways of nitrification (ammonia oxidation) and nitrate dissimilation (denitrification and ammonification of nitrate), and its atmospheric concentration was kept in check by natural destruction mostly in the stratosphere. Although nitrification requires aerobic conditions and denitrification occurs in anaerobic conditions, there are many circumstances in which both processes can occur simultaneously because of the heterogeneity of soil aeration status (see, for example, Figure 13.1 for the locally variable conditions around rice roots). Emissions of N_2O from soils have increased since 1860 mainly because of the increased use of nitrogen to promote crop productivity. Most nitrogen is in the form of unreactive N_2 gas, but human activity has generated increasingly large amounts of reactive nitrogen through combustion of fossil fuels, the promotion of biological nitrogen fixation in crop rotations and the manufacture of fertilisers using the Haber–Bosch process (Galloway et al., 2004; Figure 13.2). In 1860, natural nitrogen fixation amounted to between 100 and 200 Tg N $year^{-1}$

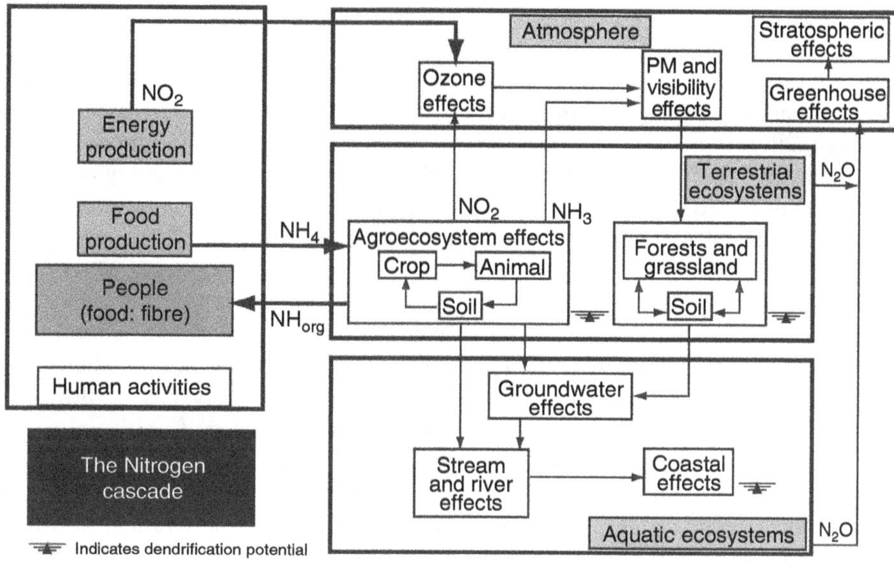

Figure 13.2 The nitrogen cascade showing the movement of human-produced reactive nitrogen through the atmosphere, terrestrial and aquatic ecosystems. From Galloway and Cowling, 2002; copyright The Royal Swedish Academy of Sciences.

and synthetic nitrogen fertilisers about 15 Tg N year^{-1}; by 2005, these values had increased to about 350 and 187 Tg N year^{-1} respectively (Galloway et al., 2008). Without this introduced reactive nitrogen, it is estimated that over 40% of the world's present population would not be fed (Galloway et al., 2004, 2008), but Crutzen et al. (2008) estimated that 3–5% of this reactive N input is converted to N_2O and contributes to the greenhouse effect. Smith et al. (2010a) conducted a top-down analysis and concluded that $4 \pm 1\%$ of all anthropogenically released reactive N is transformed into N_2O; this analysis demonstrates that agriculture is the dominant pathway for reactive N entering the biosphere.

Nitrogen from animal manures is also a source of atmospheric N_2O. Bouwman et al. (2010) estimated that in 2000 about 65% of the 112 Tg N year^{-1} in manure was generated by mixed and landless systems and 26% from pastoral systems. Manure collected in barns and storage systems was mainly spread on cropland (33.3 Tg N year^{-1}) with a smaller amount (5.3 Tg N year^{-1}) on grassland. Again for 2000, Bouwman et al. (2010) estimated that the combined N_2O emission from the 79.1 Tg N year^{-1} applied as fertiliser and 33.3 Tg N year^{-1} applied as manure was 4 Tg N_2O–N year^{-1}, a value that is similar to the top-down estimates of Smith et al. (2010a).

Reducing these N_2O emissions is a difficult problem. Except where over-application of N fertilisers is an issue (e.g. some parts of China), simply reducing N fertiliser application is not an option because this would reduce yields. Substituting N fertiliser with biologically fixed N will also not reduce emissions (Mosier et al., 1998) and may lead to an increase as, for example, when legumes were introduced into a pasture resulting in N_2O emissions that were two to three times greater (Duxbury et al., 1982). Improved timing and application rates of fertilisers and use of nitrification inhibitors with fertilisers can reduce emissions from arable crops (Smith et al., 1997), but it is also important to recognise that a reduction in emissions for one part of the system may result in an increase in another part of the nitrogen cascade (Figure 13.2).

13.2.3 Urbanisation

In 2008, the world's urban population exceeded its rural population for the first time. This process of urbanisation has being going on for some time but accelerated significantly during the twentieth century. In 1900, there were, globally, an average of 6.7 rural dwellers for each urban dweller, but by 2025, there will be three urban dwellers for every rural dweller (Satterthwaite et al., 2010). Along with this change has been a decrease in the ratio of food producers to food consumers brought about by rapid growth in the world's economy and the shift from primary industries (agriculture, forestry, mining and fishing) to manufacturing and services.

Expansion of towns and cities often takes place at the expense of agricultural land not least because most cities have been there for some time and were located where they are because their hinterlands could supply food from fertile soils. However, Satterthwaite et al. (2010) suggest that the loss of agricultural land is often overstated, with only Western Europe having > 1% of its land area as urban among all of the world's regions. Moreover, while land is clearly used for buildings, roads, etc., the remaining land may be used for more intensive production (for example, vegetable and fruit crops) or for a variety of urban and peri-urban agricultural schemes (Smit et al., 1996). Many people living in urban areas rely on urban or peri-urban agriculture for part of their food. For example, in urban areas of East Africa in the 1990s, 17–36% of the population grew crops (often maize) or kept livestock (Lee-Smith, 2010).

In many urban centres, waste materials are efficiently recycled through livestock and composting. For example, in Nairobi, municipal solid waste is used for composting by a range of community-based organisations, and animal manure is imported into the city from the arid and semiarid livestock-producing areas from up to 300 km away (Njenga et al., 2007). These represent business opportunities for some urban dwellers. However, the quality of the solid wastes used to produce food is crucial as heavy metals may accumulate causing toxicities in leafy vegetable crops. In Vietnam, Khai et al. (2007) measured the flows of N, P, K, Cu and Zn for intensive small-scale aquatic and terrestrial vegetable systems in two peri-urban areas of Hanoi City. They found that high inputs of chemical fertilisers, chicken manure and irrigation water were used resulting in surpluses of all nutrients (85–882 kg N ha^{-1} $year^{-1}$, 109–196 kg P ha^{-1} $year^{-1}$, 20–306 kg K ha^{-1} $year^{-1}$, 0.2–2.7 kg Cu ha^{-1} $year^{-1}$ and 0.6 – 7.7 kg Zn ha^{-1} $year^{-1}$). They concluded that irrigation with wastewater contributed to the high inputs and that inputs at these levels constituted a major threat to the soil and water environment.

One major consequence of urbanisation that is little reported is the major shift of nutrients from rural to urban areas. Nutrient budgets for different production systems have been conducted at a range of scales in many countries to demonstrate either gains or losses of nutrients from soils (e.g. Smaling et al., 1997; Vitousek et al., 2009). With the increasing movement of the nutrients present in crops from rural to urban areas, nutrients in the resulting wastes are likely to become concentrated in towns and cities unless effective methods of recycling can be found.

13.2.4 Climate change

There is now little doubt that human activity since industrialisation has led to warming of the globe and that future change is inevitable (IPCC, 2007). In the last 50 years or so, global mean air temperature has increased by 0.13 K $decade^{-1}$ (IPCC, 2007), with warming particularly noticeable in northern latitudes. Gregory and Ingram (2008) reviewed literature on climate change and yield and found that many studies noted an increase in mean temperature over the last 30–40 years in northern temperate latitudes of about 1.0–1.4 K (Chmielewski et al., 2004 in Germany; Jaggard et al., 2007 in the UK). They also noted that in some regions, the change in minimum temperature was greater than that in maximum temperature (e.g. Cutforth et al., 2007, in the Great Northern Plains of the USA) but that none of the studies detected any trend in precipitation. Warmer temperatures have resulted in plants developing and flowering earlier in Europe (Menzel et al., 2006), earlier harvesting of cereal crops in the UK (Sparks et al., 2005) and some reports of earlier outbreaks of pests and pathogens (Gregory et al., 2009). However, while past warming can be linked to simple phenological changes in plants, attributing effects to agricultural production is much more difficult because of the many changes in production practices (including new varieties, better weed control, increased fertiliser applications, etc.) that have occurred over the same period. In a German study, the phenology of 78 developmental stages (including sowing, emergence, flowering and maturity) for a wide range of agricultural and horticultural crops was advanced by an average of 1.1–1.3 d decade^{-1} (or about 4 d K^{-1}) between 1951 and 2004 (Estrella et al., 2007). However, while technological advances or altered farming practices such as new cultivars can influence phenology, de-trending of the data suggested that such effects were only 13% of the response observed (Estrella et al., 2007).

Global circulation models based on well-established physical principles have been developed that reproduce observed changes in climate and project future changes (IPCC, 2007; Jarvis et al., 2010). Current knowledge suggests increases in mean annual air temperature of an additional 1–3 °C by 2050 with some complex, and spatially variable, changes in rainfall (Jarvis et al., 2010). There are considerable uncertainties in these projections arising from uncertainties in future GHG emissions, modelled responses to emission scenarios and misrepresentation of physical processes by the models. Nevertheless, some future warming appears inevitable. Much less certain, and in many agricultural contexts more important, are projections of future rainfall, which influence the likelihood of increased soil erosion by water, the oxidation of soil organic matter (see Chapters 4 and 12) and the yields of crops. Projecting change in rainfall is more difficult than that for temperature because more atmospheric processes are involved and the outcome is therefore more dependent on the representation of interacting processes within the model (Jarvis et al., 2010). This means that the consequences of climate change for rainfed agriculture are presently uncertain. Generally, the effects of climate change on production range from being very negative in areas that are already water limited to more favourable in regions that are temperature limited (Fuhrer, 2006). Typically, warming is projected to have an overall negative effect on yields of many crops because the duration of grain-filling is adversely affected by high temperatures (e.g. Ewert et al., 2005). For example, Lobell and Field (2007) showed a negative response of global yields of wheat, maize and barley to increased temperatures since 1980 equivalent to about 2–3% of production in 2002. The production foregone (about 19 Mt year^{-1} of wheat, 12 Mt year^{-1} of maize and 8 Mt year^{-1} of barley) would have been worth about $US 5 billion per year and the loss of wheat and maize together was equivalent to that of Argentina (Lobell and Field, 2007).

In some specific regions and for particular crops, though, in which temperature (but not rainfall) currently restricts growth, warming may be beneficial. In the UK, Jaggard et al. (2007) estimated that about 66% of the increased yield of sugar beet over the period 1976–2004 was attributable to changes in weather (mainly warming). About 10% was due to the ability to sow earlier in the year, but 56% was due to within-growing-season changes in weather. Similarly, in eastern Scotland, Gregory and Marshall (2012) demonstrated that warming over the last 50 years could have contributed up to 25% of the increased yields of potato through earlier emergence and canopy closure allowing a longer period of photosynthesis at maximum rate.

Mitigation of climate change essentially involves the adoption of practices that either reduce emissions of GHGs to the atmosphere and/or remove GHGs from the atmosphere into long-term stores (see Section 4.9.2). Farmers can contribute to the mitigation of climate change by (1) decreasing the use of fossil fuel for farm operations or in inputs having large emissions from fossil fuels during their manufacture (e.g. N fertiliser; see Section 13.2.2); (2) growing bioenergy crops or utilising agricultural wastes to partially replace fossil fuels for electricity production, heating or transport; (3) adopting management practices that decrease net emissions of N_2O and CH_4 (see Section 13.2.2) and (4) adopting management practices that permit sequestration of carbon (from CO_2) in soil or in long-lived plants.

There is already some evidence that farmers in many parts of the world are adapting their management to cope with the warmer temperatures and variable rainfall. For example, the date by which 50% of the sugar beet crop in eastern England is sown has advanced by about 15 d since the 1970s because of warmer spring temperatures (Jaggard et al., 2007, 2010). In sub-Saharan Africa, farmers have experienced variable rainfall for many years and are used to varying sowing dates and other practices accordingly; however, these strategies may not be as effective if climatic variability and the incidence of extreme weather events become more pronounced (Haile, 2005).

13.3 Food security and food production

13.3.1 Food security

There are several definitions of food security but the one most commonly used is that it represents the state when 'all people, at all times, have physical and economic access to sufficient, safe, and nutritious food to meet their dietary needs and food preferences for an active and healthy life' (FAO, 1996). In the UK, this definition has been modified to give a greater emphasis on consumer preferences and the required safety of food: 'ensuring the availability of, and access to, affordable, safe and nutritious food sufficient for an active lifestyle, for all, at all times' (Defra, 2009). Food security is, then, a social construct so that, while production is clearly a critical element contributing to its achievement, access to food and the utilisation of food are also crucial and, in many cases, of overriding importance (Gregory et al., 2005; Ericksen, 2008). Food security is underpinned by effective food systems, which are a set of dynamic interactions between and within biogeophysical and human environments. They include a number of activities (producing food; processing, packaging and distributing food; and retailing and consuming food) that lead to a number of associated outcomes some of which contribute to food security (i.e. food availability, access to food and food utilisation) and others that relate to environmental and other social welfare concerns (Ericksen, 2008). Because food security is diminished when food systems are disrupted or stressed, food security policy must address the whole food system.

Achieving food security for all is no easy matter to achieve, not least because there are distinct, and contradictory, views on the understandings that different communities have about food and the relations between food, society and the environment. Eakin et al. (2010) explore three different perspectives on food, which highlight these different modes of discourse and result in different policies and courses of action. One view perceives food as a global commodity in which a diverse range of products are bought and sold competitively and markets operate to reward efficiencies in production, distribution and processing, and favour product differentiation. McMichael (1994) refers to this as the 'exchange-value' of food in which there is demand without differentiation across a market. Maize is maize, food is a commodity to be traded and ideas of 'terroir' or soils contributing to product quality have little resonance. Classical soil science, with its emphasis on soil fertility and increasing crop yields, has arguably, made substantial contributions to this view of food. This perspective, in which the economic value of food is paramount, has contributed to policies relating to failures of entitlement as major drivers of food insecurity and to regulations to make globalised food commodity markets work for all.

As human-induced changes to the environment have accelerated in the last 60 years, a second discourse has emerged in which food is viewed as an environmental service. This view was formally stated in 2005 in the MEA (2005) and perceives food as one of several provisioning, regulating and supporting services provided by natural ecosystems. This view has rapidly gained credence in scientific discourse (see, for example, the UK's ecosystem assessment; UK National Ecosystem Assessment, 2011) and has been referred to several times in this book (Chapters 3, 4 and 12). However, this discourse is often in conflict with the commoditisation and globalisation perspective, which seeks to secure production by controlling the environmental conditions in which food is produced through technological advances. Resolving these conflicts by the development of systems of food production that can utilise ecological perspectives and achieve the yields required from food provisioning is a substantial challenge for the future.

Finally, the globalisation of food as a commodity and the increasing pressures brought about by global environmental change has given renewed life to a third view of food as a basic human right. Maslow (1943) proposed a hierarchy of human needs in which food, clothing and shelter were the underpinning needs for life. Communities to whom this right is denied are likely to rise up and protest as happened in many countries during the 'food crisis' of 2008. Associated with this basic right are also ideas of food sovereignty and the right to choose food of particular types, produced in specific ways. Again, this view of food is at odds with food as a commodity. Eakin et al. (2010) conclude that future food security requires governance systems that can reconcile these contrasting views and achieve, in addition to freedoms of availability, access and utilisation, the freedom to make food choices that support individual, communal (and environmental) preferences.

13.3.2 Food production and yields

While food production alone does not guarantee food security, as stated in Section 13.1, the growth in the world's human population from about 2.2 billion in 1950 to about 7 billion in 2011 has been sustained by increased production (Godfray et al., 2010). Simultaneously, there has been increased demand for fuel and fibre, and this increased production will need to be maintained if the projected population of 9 billion by about 2040 is to be sustained; demand for forest products is, similarly, projected to increase to 2030 by 1.4% per annum for sawnwood and about 3% per annum for paper and wood-based panels (FAO, 2009). Increases in crop yields have been the major factor supporting increased food production (Section 13.2.1), and there is much interest in how past success can be sustained.

Yield is a consequence of the interaction of three factors: genotype (G) × environment (E) × management (M; Hay and Porter, 2006). Evans (1993, 1998) highlights how the synergistic effects of these interactions linked to innovative technologies have contributed to past increases in yield. Among the important contributors to these have been (1) improved germ plasm able to grow vigorously (e.g. hybrids), resist pathogens and respond to fertilisers without lodging (in particular, the use of dwarfing and semi-dwarfing genes in rice and wheat); (2) the application of fertilisers and particularly the availability of affordable nitrogen fertiliser; (3) the development of chemicals to control weeds, pests and diseases and (4) improved irrigation systems especially in rice-producing countries and for some previously rainfed crops. Together, these technological innovations have modified G, E and M to greatly increase yields. The consequence of these innovations has been that since about 1960, yields of many crops have increased year on year (Figure 13.3). Global yield increases for a number of crops have typically been linear (Figure 13.3a) with values of 53 kg ha^{-1} $year^{-1}$ for rice, 41 kg ha^{-1} $year^{-1}$ for wheat and 63 kg ha^{-1} $year^{-1}$ for maize over the period 1961–2004. Increases in yield have also been linear with time in many individual countries (e.g. Scotland in Figure 13.3b and see Ewert et al. (2005) for wheat yields in several European countries) although in a few instances technological innovations have produced more rapid, stepwise increases in yield (e.g. Australia, Angus (2001); Figure 13.3c). An arithmetic consequence of the linear yield increases with time is that the proportionate increase in yield obtained through breeding programmes has decreased. Of greater concern, though, is that in some countries and regions, absolute increases in yields are no longer occurring at the same rate despite yields being less than the attainable value. For example, rice yields in SE Asia and wheat yields in the UK over almost the last decade have appeared to be approaching an asymptote (Figure 13.3d). The reasons for this are complex and probably include almost constant inputs of nitrogen fertiliser to wheat crops in the case of the UK from 2000 onwards (average N fertiliser application was 188 kg ha^{-1} in 2000 and 190 kg ha^{-1} in 2007; Defra 2010) and concerns about the long-term availability of mineralised soil N in paddy crops because of the formation phenolic lignin residues that stabilise soil organic nitrogen (Olk et al., 2006).

Although crop yields have increased globally and throughout North and South America, Europe, Australia and much of Asia, a notable exception has been that of Africa where, for example, per capita food production decreased by about 5–10% between 1980 and 1995 (Sanchez et al., 1997). FAO data analysed by Greenland et al. (1998) demonstrated significant decreases in crop yields in many countries in sub-Saharan Africa, and there are still many countries where yields have improved little. The reasons for the poor performance in Africa relative to other countries are many and include social unrest and war, poor institutions and governance, climatic variability making reliable irrigation difficult and weathered soils that are deficient in nutrients. Greenland et al. (1998) concluded that a major factor behind many of the observed decreases in yield in African countries was the decline of soil fertility accompanied by the lack of fertiliser application. Much has been written about the need to 'recapitalise' the soils of Africa, especially with regard to P status, but progress has been limited (Sanchez et al., 1997).

Because the area of cropped land is likely to increase proportionately less than the future demand for food, reducing the gap between current yields and potential yields is a major goal for the future (Jaggard et al., 2010). Potential yield is a theoretical upper limit to yield imposed by solar radiation (affecting growth), temperature (affecting development and growth) and water supply (affecting mainly growth but also development; see Chapter 2 for explanations). Lobell et al. (2009) reviewed data from crops of maize, rice and wheat grown in a range of countries and compared potential yields with measured experimental, maximum farmer and average farmer yields. The gap between potential and the actual yields ranged from about 20% to 80%. In many irrigated cereal systems, yield appeared to plateau

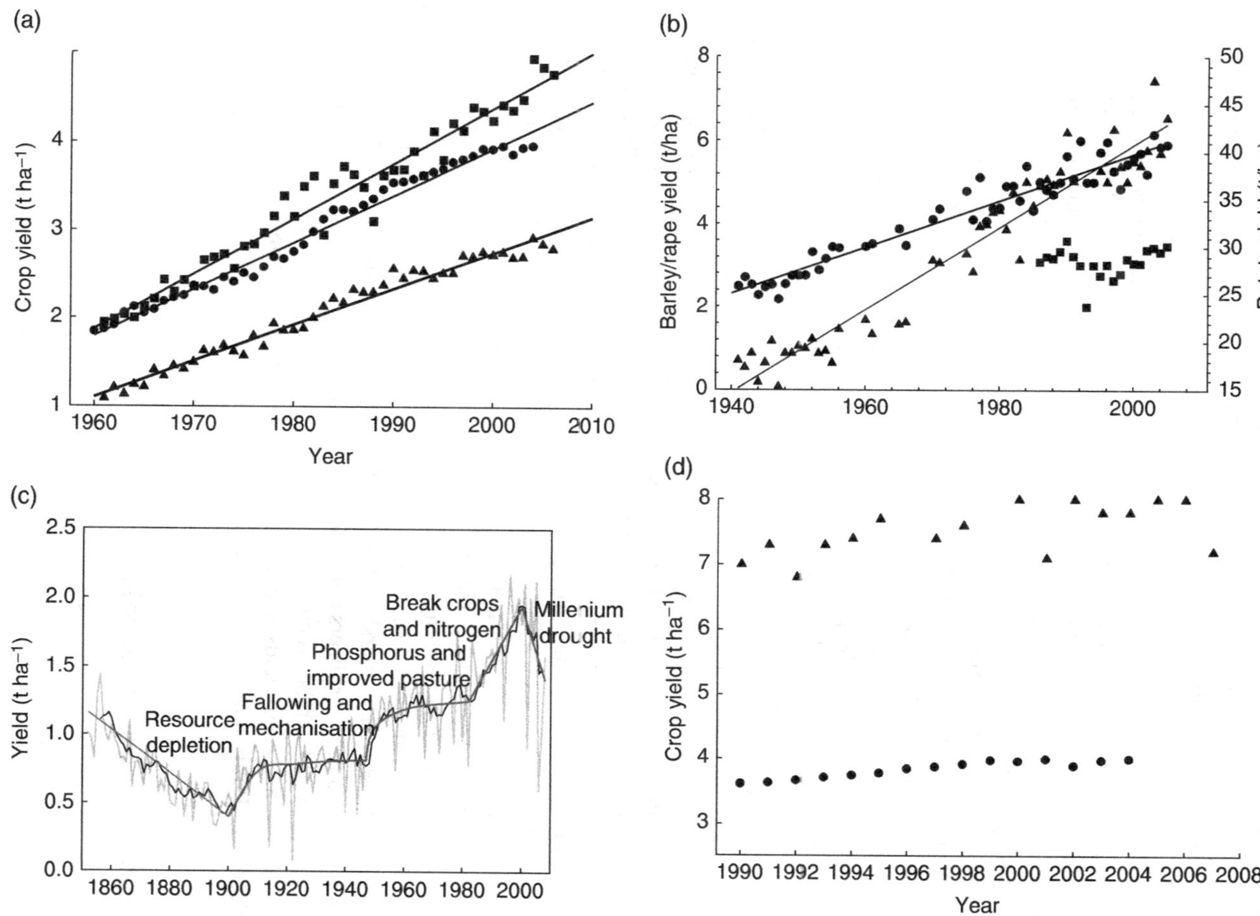

Figure 13.3 Changes in yields with time: (a) global average annual yields (data from USDA) for maize (squares), rice (circles) and wheat (triangles); the lines are linear regressions: (b) national average annual yields in Scotland for barley (circles), rape (squares) and potato (triangles); the lines are linear regressions (data from annual farm surveys conducted by Scottish agencies responsible for agriculture): (c) national average annual yield of wheat in Australia (from Angus 2001 and personal communication); the dashed line joins annual values, the thin solid ine is a 5-year running average and the solid black line summarises major trends in yield with respect to changes in crop husbandry: and (d) regional average annual rice yields in Asia (circles – data from USDA) and wheat in the UK (triangles – data from Defra). Figure reproduced from Gregory and George (2011). With kind permission from Wiley-Blackwell.

at or about 80% of potential yield while in rainfed systems, average yields were commonly 50% or less of potential (Lobell et al., 2009). While part of the yield gap is inevitable because of crop losses during harvest, storage and transport and the way that land areas are reported (Jaggard et al., 2010), there are still large differences in performance between adjoining farms. Lobell et al. (2009) concluded that a fundamental constraint in irrigated systems was uncertainty in growing season weather; this is also a factor in rainfed systems where interactions between water and nutrient availability are complex (Cooper et al., 1987). Raising yields above 80% of yield potential is possible, but only if technologies can be developed and adopted that reduce the uncertainties faced by farmers in assessing soil and climatic conditions or respond dynamically to these conditions or both (for example, installation of nutrient and water sensors; Lobell et al., 2009). Such technologies may have the added benefits of increasing the efficiency of use of inputs and reducing losses off-site as well as increasing yields.

In a global analysis of wheat, maize and rice production, Neumann et al. (2010) found that yield gaps were significantly correlated with irrigation, market accessibility and influence, availability of agricultural labour and slope; the contribution of these factors varied substantially between regions and generalisations as to the best means of reducing yield gaps were not possible. In China, the potential to reduce yield gaps of maize was demonstrated on 66 on-farm experimental plots, raising yields to 13 t ha^{-1} on average (nearly twice the typical farmer yield) without any increase in N fertiliser application (Chen et al., 2011). This was achieved using simulation models to identify appropriate combinations of planting date, crop density and cultivar at each site based on long-term weather data and then changing variety, sowing date and spacing as appropriate. This was combined with an in-season root-zone N management strategy, which matched supply to demand; in particular, a greater proportion of the N currently applied was applied later in the growing season. This integrated agronomic approach combined with cultivars of appropriate duration increased yields and reduced off-site nutrient movement (Chen et al., 2011). In an analysis of yield gaps for rice in South-east Asia, Laborte et al. (2012) examined farmers' yields in four intensively cropped regions in Indonesia, the Philippines, Thailand and Vietnam and compared them with yields simulated in a rice model. Three yield gaps were identified based on climatic potential (Van Ittersum and Rabbinge, 1997), economic and best farmers' yields (Figure 13.4). Average yields varied between 43% and 75% of the climatic yield potential and 61% and 83% of the best farmers' yields. The farmers with the best yields were typically more educated and used fertilisers and labour more efficiently than others. This analysis showed that if the yield gap between the average and the best farmers were closed, then this would cover the increased yields needed for 2050 (assuming no change in diet) in all countries except the Philippines, where other institutional changes will also be required (Laborte et al., 2012).

13.3.3 Biomass and biofuels

Growing crops for biomass and biofuels has been highlighted as a potential competitor for land with food crops and, thereby, a potential contributor to food insecurity (see, for example, Karp and Richter, 2011). It is noteworthy that the area occupied by biofuels and their by-products in 2004 was only 14 Mha compared with 1500 Mha of crops (i.e. about 1% of the total cropped area) and 4,500 Mha of pastures worldwide (IEA, 2006). However, there has been a substantial increase in the area of land used to grown biofuels since 2004, especially in the USA, with one-third of the maize grown in 2007 used for bioethanol. While the

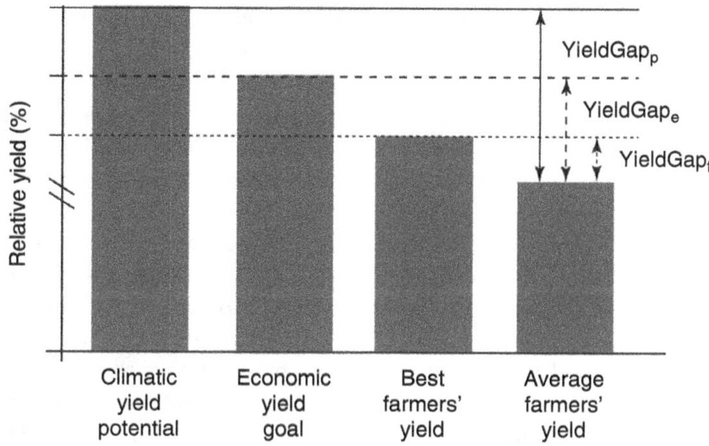

Figure 13.4 A conceptual expression of the relative yield determined by climatic yield potential, economic yield goal and farmers' yield. Three different yield gaps are identified based on the differences between average farmers' yield and the climatic yield potential (YieldGapp), the economic yield goal (YieldGape) and the best farmers' yield (YieldGapf). Reproduced from Laborte et al. (2012). With kind permission from Elsevier.

reasons for growing crops for biofuels are complex (including increased energy security, increased supplies of transportation fuels and a desire to decrease net emissions of GHGs), the use of land for them is likely to increase (FAO, 2008 gives some projections), and the emergence of biofuels as a new source of demand for agricultural commodities may assist in the revitalisation of agriculture in developing countries with positive benefits for personal livelihoods and economic growth (Pingali et al., 2008).

Biomass for energy can be obtained from various sources including specific energy crops (e.g. willow and perennial grasses; Karp and Shield, 2008) and agricultural residues. The potential energy production from these residues is large and globally estimated at over 12×10^{18} J (Hall et al., 1993). However, many agricultural residues are reused as feed for livestock so that a waste stream unsuitable for human consumption is upgraded into valuable food products. Nonhebel (2007) demonstrated that, from the perspective of land use, it was better to produce energy from dedicated energy crops and to use the agricultural residues for livestock food rather than use the residues for energy. This was because when dedicated energy crops were grown, less land was required to produce bioenergy than the additional land area required to compensate for protein losses in the human food.

Field et al. (2007) suggest that the interaction of four major factors will determine the success or otherwise of biomass energy in future: (1) the intrinsic productive capacity of land and ocean ecosystems; (2) the alternative uses for the land and water resources used for biomass energy production; (3) the effects of biomass energy technologies on off-site factors such as pollution and invasive species and (4) the conversion processes for increasing the energy yield per unit of biomass and from each unit of land or water. All four factors have implications for the production of both food and GHGs but especially factor two. The obligations imposed by the Renewable Transport Fuel Obligation (RTFO) pose considerable challenges for land use for individual countries if the fuels are to be produced locally. For example, in the UK, the 1.2 billion litres of bioethanol and 1.35 billion litres of bio-diesel required to meet the UK RTFO targets would require about 500,000 ha of wheat (about 20–30% of the current area) and 340,000 ha of oilseed rape (compared with the 670,000 ha grown in 2007), respectively (Cottrill et al., 2007). So about 15–20% of the UK arable area would be required to meet the RTFO targets (Karp and Richter, 2011).

Several studies have attempted to calculate the net effect of growing biofuel crops on GHG emissions after allowing for energy inputs such as fertiliser applications, cultivation and other inputs in the production process (e.g. Hill et al., 2006). In the USA, it has been estimated that, relative to the energy used in their production, ethanol from maize yields 25% more and bio-diesel from soya beans 93% more (Hill et al., 2006). Furthermore, GHGs are reduced by 12% by the production and burning of bioethanol and 41% by diesel relative to the fossil fuels that they replace. However, studies in both the UK and the USA have demonstrated that home-grown crops for biofuels will meet only a small proportion of current fuel needs. Powlson et al. (2005) calculated that making a variety of reasonable assumptions about the land that might be used to produce biomass or biofuels, biomass crops might provide up to 12% of UK electricity demand and sugar beet or oilseed rape crops about 1.2–1.8% of UK diesel or petrol requirements. Similarly, in the USA, using all maize and soya bean production for biofuels would contribute only 12% of petrol and 6% of diesel demand (Hill et al., 2006).

As in the past, land will be used to grow both food and fuel in the future (it is worth remembering that until relatively recently, much land was used to grow forage for horses), so the bigger question is not fuel vs food but how land might best be used to contribute to both food and energy security (Karp and Richter, 2011). Several studies have indicated that the worst option is to clear new land for the production or biomass or biofuel crops (Field et al., 2007; Fargione et al., 2008; Searchinger et al., 2008). Clearing can result in GHG emissions that are much greater than any benefits provided by biofuels because soils and plant biomass are biologically active stores of carbon that are decomposed by microbes when disturbed releasing large quantities of CO_2. For example, replacing a lowland tropical rainforest in Indonesia or Malaysia with oil palm to be used for bio-diesel would produce about 610 Mg CO_2 ha^{-1}, which would take about 86 years to recoup from the use of bio-diesel instead of fossil fuel, while soya bean diesel replacing Amazonian rainforest producing >280 Mg CO_2 ha^{-1} would take about 320 years (Figure 13.5; Righelato and Spracklen, 2007). Such considerations have led to the conclusion that production of bioenergy crops should be focused either on previously cleared and abandoned agricultural lands or on lower grade, marginal land (Field et al., 2007; Campbell et al., 2008; Karp and Richter, 2011). However, much remains to be done to identify crops and production systems that can yield an economic return from such land.

13.3.4 Use of energy in crop production

Concerns about the amount of fossil fuel energy expended in crop production and food processing are not new (Pimental et al., 1973; Spedding and Walsingham, 1976), but have come to the fore again as energy costs have increased and concerns about CO_2 emissions and the need to develop low carbon cropping practices have emerged (Woods et al., 2010). Pimental and Pimental (2008) provide a variety of examples to illustrate the poor energy returns of many crop production practices ranging from maize production in Mexico using human power and an axe and hoe returning 10.7 times as much energy as consumed in production to a return of 2.2 times for rice production in the USA (Table 13.3). In the Mexican example, the only fossil fuel used was in the production of the axe and hoe; unfortunately, the yield (1.94 t ha^{-1}) is well below that required to sustain the future global population. These figures omit the energy required to convert the grains

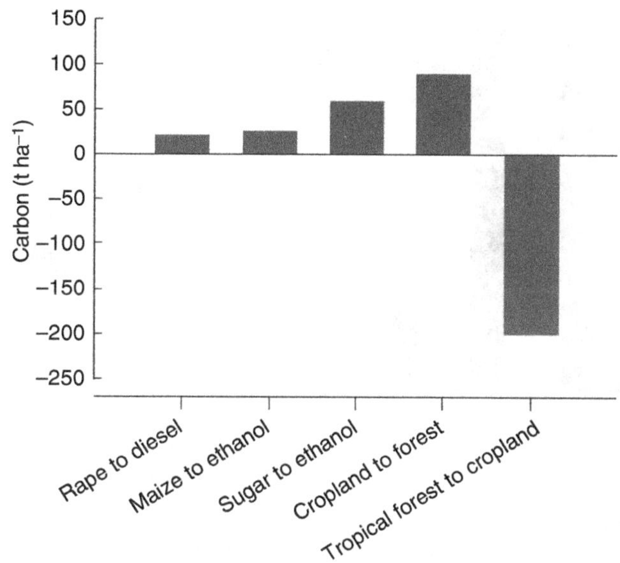

Figure 13.5 Cumulative avoided emissions of carbon over 30 years from several biofuels compared with the loss of carbon to the atmosphere following the conversion of tropical forest to cropland. Based on Righelato and Spracklen (2007).

Table 13.3 Some examples of energy use in grain and legume production.

Crop	Country	Tillage	Yield ($t\ ha^{-1}$)	Inputs ($kJ \times 10^3\ ha^{-1}$)	Output ($kJ \times 10^3\ ha^{-1}$)	Energy ratio
Groundnut	Thailand	Buffalo	1.28	8,048	20,892	2.60
Groundnut	USA	Mechanised	3.72	45,817	64,051	1.40
Maize	Mexico	Human	1.94	2,687	28,881	10.7
Maize	Mexico	Oxen	0.94	3,222	13,982	4.34
Maize	USA	Mechanised	8.66	33,961	130,396	3.84
Rice	Borneo	Human	2.02	4,327	30,626	7.08
Rice	Philippines	Animal	1.65	7,638	25,126	3.29
Rice	Japan	Mechanised	6.33	34,405	96,163	2.80
Rice	USA	Mechanised	7.37	49,542	110,995	2.24
Soybean	USA	Mechanised	2.67	12,609	40,197	3.19
Wheat	USA	Mechanised	2.67	17,740	35,354	2.13

Source: Derived from Pimental and Pimental (2008).

into human food – negligible in the Mexican example but substantial enough in the case of the USA to render the overall energy return close to unity.

Although agriculture accounts for only about 2% of total energy usage, Table 13.3 demonstrates that agriculture has become very dependent on fossil fuels to produce the current levels of yield. Fossil fuel is used directly for tillage and crop management and indirectly through the application of energy-intensive inputs such as fertilisers and pesticides. Moreover, the use of energy embedded in buildings, tractors and other infrastructure continues to increase. For the UK, Woods et al. (2010) showed that primary energy inputs ranged from about 1 GJ t^{-1} for potato to 6 GJ t^{-1} for oilseed rape; in part, this range reflects the

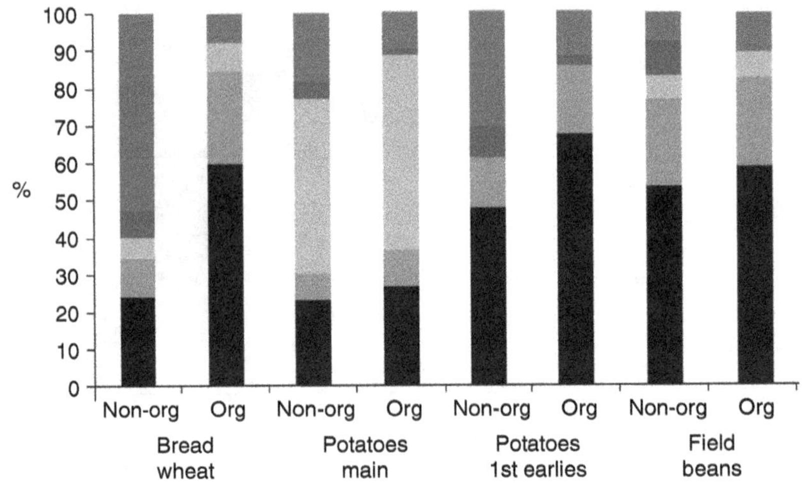

Figure 13.6 A breakdown of the energy used in the production of wheat, potatoes and field beans in the UK in conventional and organic production systems. Green, fertiliser manufacture; red, pesticide manufacture; blue, post-harvest; purple, machinery manufacture; black, field diesel. Reproduced from Woods et al. (2010). With permission from the Royal Society, London. For a colour version of this figure, please see Plate 13.1.

different energy contents of the harvested organs in addition to the level of inputs required. Their analysis showed that while organic production systems were often apparently more energy efficient than conventional systems, human energy inputs were frequently higher for organic systems. The principal energy inputs into crops are for diesel, fertiliser and pesticides, but the proportions of the energy used for different elements of the production process differ substantially between crops (Figure 13.6). For wheat, fertiliser (mostly nitrogen) is over 50% on the energy input into the conventionally grown crop, while for field beans that receive little N fertiliser and have lower yields than wheat, more diesel is used per tonne of yield produced. Woods et al. (2010) show that maincrop potatoes in the UK are very energy intensive because they are stored in cool rooms for a long time to provide a year-round supply. This contrasts with early potatoes that are generally not stored on farms.

The energy required to produce fertilisers is still substantial but has decreased significantly as a result of innovation by manufacturers. The main components of nitrogen fertilisers differ in the quantities of energy required to manufacture them (typically 32 MJ kg^{-1} for ammonium, 22 MJ kg^{-1} for urea and 13 MJ kg^{-1} for ammonium nitrate) and are considerably higher than the 5 MJ kg^{-1} required for P and K fertilisers. These energy inputs are much greater than those required to apply the products via spreaders in the field (Woods et al., 2010).

13.4 Routes to sustainable intensification

13.4.1 Elements of sustainability

The increasing global food demand coupled with the acknowledged environmental change resulting from changed production practices has led many to challenge the technologies that have resulted in today's intensified agriculture and to call for the development of sustainable production practices (Conway, 1997; Gregory et al., 2002; Pretty, 2008; Foresight, 2011).

While there is general agreement that agricultural sustainability includes elements of profitable production, environmental stewardship and social responsibility, there is much less agreement as to how sustainability is to be achieved in practice beyond the need to integrate biological and ecological insights into the production process. Pretty (2008) reviewed the literature to establish four key principles of sustainability that 'i) integrate biological and ecological processes such as nutrient cycling, nitrogen fixation, soil regeneration, allelopathy, competition, predation and parasitism into food production processes; ii) minimise the use of those non-renewable inputs that cause harm to the environment or to the health of farmers and consumers; iii) make productive use of the knowledge and skills of farmers, thus improving their self-reliance and substituting human capital for costly external inputs; and iv) make productive use of people's collective capacities to work together to solve common agricultural and natural resource problems, such as for pest, watershed, irrigation, forest and credit management'. Such principles go well beyond the need for technical innovations including improvements to genotypes and ecological and agronomic management systems and embrace the need to develop important capital assets for agricultural systems including natural, social, human, physical and financial capital. A corollary of this analysis is that many disciplines and ways of thinking will be required to develop sustainable systems and that there is unlikely to be a single solution appropriate to all soils and production systems.

As outlined in Section 13.3.2, the general expectation is that intensive production systems will be the main route by which future food supplies are secured. Cassman et al. (2003) concluded that in order to avoid severe degradation of natural resources and to reduce GHG emissions, such systems would need to produce higher yields on existing cropland, limit expansion of the cultivated area, achieve a substantial increase in N fertiliser efficiency and improve soil quality through increasing soil organic matter. There is little doubt that for most soils, sustainable production is inextricably linked to the maintenance of soil organic matter contents through appropriate additions to offset the losses caused by cultivation and nutrient depletion (Greenland et al., 1998; Lal, 2004).

In a wide-ranging assessment of future food and farming, the UK Foresight (2011) concluded that sustainable intensification was required and that all technological means of achieving this should be assessed and appropriately utilised. How might this sustainable intensification be achieved? Tilman et al. (2011) approached this problem by first defining the global demand for crop production in 2050 and then examining the environmental impacts of different ways of achieving this. Their analysis, based on measured relations between per capita energy and protein demands with per capita gross domestic product and allowing for different economic groups, indicates a 100–110% increase in global crop production; this is larger than the 70% estimated by Alexandratos (2009) using expert opinion only. Using past N fertiliser rates as a surrogate for soil fertility enhancement (though recognising that soil fertility can be enhanced in other ways), alternative pathways to increasing yields to the required levels were investigated. The options examined were (1) current technology in which each economic group retained its 2005 N-dependent yield function; (2) technological improvement in which technological advances continue along existing temporal trends to 2050; (3) technology transfer in which low-yielding countries adopt and adapt the existing high-yielding technologies of high-yielding countries and (4) technology improvement and transfer in which all countries achieve soil- and climate-adjusted yields. Figure 13.7 shows the consequences for yields, land clearance and GHG emissions from adopting these different routes together with trajectories based on minimising land clearance or increasing the efficiency of N use. In summary, with present trends of intensification

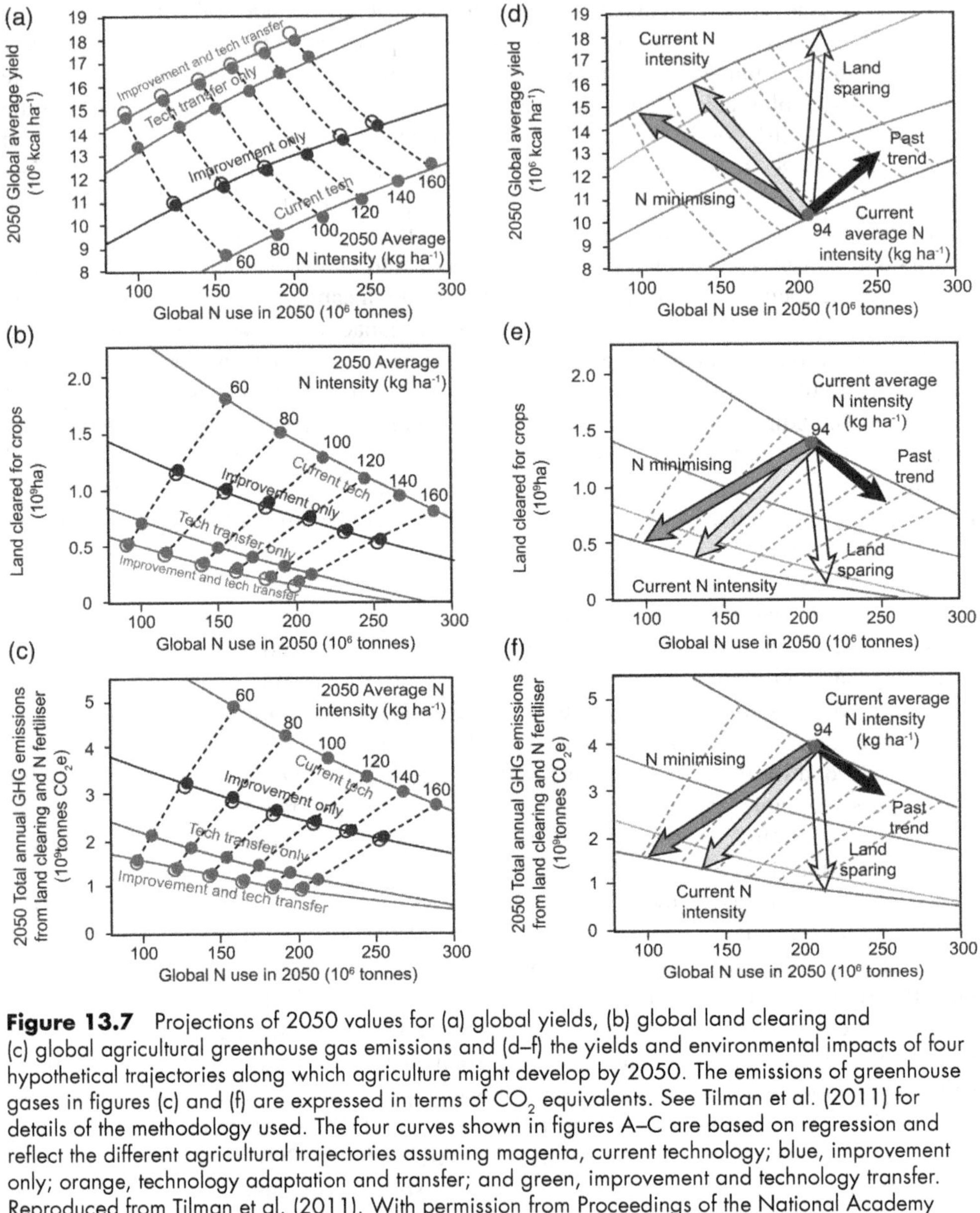

Figure 13.7 Projections of 2050 values for (a) global yields, (b) global land clearing and (c) global agricultural greenhouse gas emissions and (d–f) the yields and environmental impacts of four hypothetical trajectories along which agriculture might develop by 2050. The emissions of greenhouse gases in figures (c) and (f) are expressed in terms of CO_2 equivalents. See Tilman et al. (2011) for details of the methodology used. The four curves shown in figures A–C are based on regression and reflect the different agricultural trajectories assuming magenta, current technology; blue, improvement only; orange, technology adaptation and transfer; and green, improvement and technology transfer. Reproduced from Tilman et al. (2011). With permission from Proceedings of the National Academy of Science. For a colour version of this figure, please see Plate 13.2.

in rich nations and extensification in poor nations, by 2050, an additional 1 billion ha would be cleared with GHG emissions of 3 Gt year^{-1} and N use of 250 Mt year^{-1}. However, if intensification were concentrated on existing cropland and transfer and adoption of high-yielding technologies were successful, then only 0.2 billion ha would be cleared, GHG emissions would be reduced to one-third (1 Gt year^{-1}) and global N use would be 225 Mt year^{-1}. Although this analysis omits any effects of future climate change, it indicates what might be possible with investment in innovative technologies and infrastructure.

13.4.2 Improved efficiency of resource use

Several chapters (Chapters 1, 3, 6, 9 and 12) have already mentioned to need to improve the efficiency with which resources such as water and nutrients are used in systems of agricultural production. The efficient application of water to crops through sprinklers and drip systems has been dealt with in Section 9.5, so this section focuses on nutrients.

A primary requirement for the future is to produce higher yields with inputs that do not lead to environmental problems either on- or off-site. Nutrient additions that are inadequate relative to crop offtake degrade land through nutrient mining while additions that are excessive degrade land, water and air through leaching, eutrophication and gaseous emissions (Vitousek et al., 2009). Ideally, nutrient additions (whether as mineral fertilisers or manures) and soil biota should be managed to deliver nutrients to crops synchronously with demand (Myers et al., 1994), but this has proved difficult to achieve in practice because applications must normally be made before the demand exists and large canopies do not permit application of solid sources to soils.

Precision agriculture has evolved rapidly in the last decade so that it is now possible to identify agronomic issues, and apply inputs, at the sub-field scale (McCormick et al., 2009; Lawes and Robertson, 2011). Harvesters with weighing facilities and global positioning systems have allowed the production of yield maps on an almost routine basis and these, together with grid and transect sampling of soils to produce maps of nutrients, are allowing farmers to contemplate site-specific nutrient management (Robertson et al., 2008). Such management is knowledge intensive and requires multiple forms of knowledge to be integrated in a way that can be practically managed. For example, should fertiliser be applied according to past yield performance or soil fertility analysis? Khosla et al. (2008) compared the yields and monetary returns from irrigated maize crops grown in northeastern Colorado managed according to management zones based either on site-specific soil properties such as bulk density, texture and organic matter or on yields. They found that, overall, management of N fertiliser based on soil properties produced higher grain yields with net monetary returns ranging from \$US 188 to \$US 679 ha^{-1} and concluded that site-specific management zones were an inexpensive and pragmatic approach to N management.

The best approach to adopt, though, is itself likely to be site specific. For example, from an analysis of rainfed yields of 20 fields in Western Australia, Lawes and Robertson (2011) concluded that while applying variable rates of inputs added some value to many fields, the gains were small so that the technology needed to be cheap for it to be implemented by farmers. In their study (a rainfed system with much lower yields and inputs than that of Khosla et al., 2008), the monetary returns were greater for fields with soils of higher fertility and variable yields than where soil fertility was low and uniform across the field.

It has been appreciated for a long time that there is an interaction between the efficiencies which water and nutrients are used, not least because the availability and mobility of nutrients in soils depends on water (see Chapter 6). In rainfed production systems, this interaction is a major constraint to the efficient use of inputs because rainfall is not known in advance. Where soil nutrients are deficient for maximum growth of crops, application of fertilisers and manures may result not only in increased growth but also in increased water use efficiency. This effect of modest applications of fertiliser has been well documented in several studies (Cooper et al., 1987; Latiri-Souki et al., 1998; Bolger and Turner, 1999; Ogola et al., 2002) and is illustrated in Table 13.4. Fertiliser use may increase slightly the total amount of water used (e.g. Cooper et al., 1987 for barley in Syria; Ogola et al. 2002 for maize in the UK), but the principal effect is to increase early canopy growth so that it shades

Table 13.4 Effects of modest applications of fertiliser on shoot dry matter, water use and water use efficiency (WUE) for crops of barley at Breda, Syria (from Cooper et al. 1987) and pearl millet at Sadore and Dosso, Niger (ICRISAT).

Crop	Season	Rainfall (mm)	Fertiliser	Dry matter ($t\ ha^{-1}$)	Water use (mm)	*WUE* ($kg\ ha^{-1}\ mm^{-1}$)
Barley	1981/82	324	+	6100	231	26.4
			–	4540	231	19.7
	1983/84	204	+	2880	176	16.3
			–	1340	171	7.8
Millet	1984	260	+	4750	165	28.8
			–	2417	163	14.8
	1985	380	+	5000	247	20.2
			–	3100	270	11.5
	1986	440	+	3850	268	14.4
			–	1140	211	5.4

Source: From Gregory et al. (1997).

the surface and thereby reduces evaporation from the soil surface as a proportion of the total water that is evaporated. However, the beneficial effect of fertiliser in increasing growth and reducing E_s is not universal and is dependent on the wetness of the soil surface and the evaporative demand (Gregory et al., 1997).

In semi-arid production systems, the efficiency of N and P fertilisers depends on the amount of water available to the crop so that the response to N, in particular, is variable and limited in dry years. Typically, crop response to N increases with increasing rainfall while response to P decreases on P-deficient soils (Jones and Wahbi, 1992). Studies in the Sahel have also concluded that soil fertility is often a more important factor in rangeland and crop productivity than rainfall, so that effective management of water cannot be achieved without also managing soil nutrient constraints (Payne, 2000). The limiting factors to crop growth at different times during any particular season could be either water or nutrient availability or both. A practical problem to be resolved in many semi-arid regions is how to afford and apply the optimum amount of fertiliser to produce an economically viable yield in a given season. For example, Sadras (2002) analysed financial returns over a 40-year period from fertiliser applications in an area with low and erratic rainfall in south-eastern Australia and found that a very low input of N fertiliser (5 kg N ha^{-1}) ensured the greatest economic stability at all sites examined. Most scientific analysis is based on hindsight, but farmers must operate without this benefit so that conservative practices tend to dominate.

13.4.3 Improved tillage and conservation agriculture

In addition to managing chemical and biological elements of soil fertility, there is increasing emphasis on soil physical fertility and health as a means of improving both root growth and function, and soil ecosystem services. The practical means by which these ends can be achieved often relies on tillage. In much conventional agriculture, soil is often inverted (with, for example, a mouldboard plough or discs) with crop residues and weeds buried. There are many reasons why such tillage practices arose including (1) the preparation of a seedbed allowing uniform germination; (2) removal of weed competition early in the crop's growing cycle; (3) release of nutrients through mineralisation after oxidation of soil organic matter; (4) burial of crop residues reducing carry-over of diseases and facilitating easier sowing; (5) easier incorporation of nutrients; (6) relief from soil compaction and (7) control of some soil-borne diseases and insects (Hobbs et al., 2008). However, by creating a bare soil surface, such tillage also has the undesirable consequences of substantially increasing rates of water and wind erosion and of the oxidation of soil carbon.

Since the 1930s, following the dust storms in the midwest of the USA, there has been a growing movement to reduced tillage systems (Unger et al., 1988); this was given added impetus by the development of herbicides to control weeds. Conservation tillage is a wide-ranging term used to describe the many practices that have evolved, initially to retain at least 30% surface coverage by crop residues, but to varying extents to also conserve time, fuel, earthworms, soil water, soil structure and nutrients (Baker et al., 2006). Several studies have shown that minimum tillage coupled with residue retention increased infiltration and profile moisture storage (e.g. Gicheru, 1994, in Kenya), reduced soil erosion by water by 10–100 fold because of greater aggregate stability and reduced run-off (Rhoton et al., 2002; Montgomery, 2007) and increased soil organic matter content (e.g. Govaerts et al., 2009).

Conservation tillage has been widely adopted in the USA, Brazil, Argentina, Canada and Australia (about 89 Mha in 2004; Derpsch, 2005, cited by Hobbs et al., 2008) and has now been included within a system of production referred to as 'conservation agriculture'. This system removes the emphasis from tillage alone and combines three principles: (1) reduced tillage – ideally zero tillage but may involve controlled tillage during seeding to disturb no more than 20–25% of the soil surface; (2) retention of crop residues and surface cover – variable rates of retention to reduce erosion and run-off and enhance soil properties associated with long-term production and (3) use of crop rotations – moderation of weed, disease and pest problems coupled with utilisation of beneficial effects of some crops on soil conditions (Verhulst et al., 2010). These principles are applicable to a wide range of production systems from low-yielding, rainfed conditions to high-yielding, irrigated conditions. In practice, though, the techniques used will vary with the specific situation depending on farmer circumstances and biophysical and management conditions. Table 13.5 compares the effects of conventional tillage, conservation tillage and conservation agriculture on a range of soil and crop factors and demonstrates the advantages to sustainable production and other ecosystem services from adopting such an approach.

13.4.4 Managing roots and the rhizosphere

The rhizosphere is a complex zone of soil both influenced by and influencing roots. Some elements of the interactions taking place in this zone have already been referred to in Chapters 3, 6, 8, 9, 10, 11 and 12, but there is increasing interest in the possibilities of managing this interface for the improved acquisition of water and nutrients and prevention of soil-borne diseases in cropping systems (Ryan et al., 2009; Richardson et al., 2011). The architecture of root systems differs markedly between species and within genotypes and has been progress in exploiting this variation to benefit yields (Gregory, 2006; Lynch, 2007). For example, genotypes of common bean with shallow root architecture have been shown to grow and yield better in soils of low P status than genotypes with deep architecture (Rubio et al., 2001; Ho et al., 2004; Henry et al., 2010). Such genotypes have the added benefit that their increased growth results in greater ground cover and less P loss by erosion of topsoil (Henry et al., 2010). In soya bean, too, the most P-efficient genotypes had longer and larger root systems with a greater proportion of the root system in the topsoil (Ao et al., 2010).

Table 13.5 A comparison of tillage, conservation tillage (CT) and conservation agriculture (CA) for various issues.

Issues	Traditional tillage (TT)	Conservation tillage (CT)	Conservation agriculture (CA)
Practice	Disturbs the soil and leaves a bare surface	Reduces the soil disturbance in TT and keeps the soil covered	Minimal soil disturbance and soil surface permanently covered
Erosion	Wind and soil erosion: maximum	Wind and soil erosion: reduced significantly	Wind and soil erosion: the least of the three
Soil physical health	The lowest of the three	Significantly improved	The best practice of the three
Compaction	Used to reduce compaction and can also induce it by destroying biological pores	Reduced tillage is used to reduce compaction	Compaction can be a problem but use of mulch and promotion of biological tillage helps reduce this problem
Soil biological health	The lowest of the three owing to frequent disturbance	Moderately better soil biological health	More diverse and healthy biological properties and populations
Water infiltration	Lowest after soil pores clogged	Good water infiltration	Best water infiltration
Soil organic matter	Oxidises soil organic matter and causes its loss	Soil organic build-up possible in the surface layers	Soil organic build-up in the surface layers even better than CT
Weeds	Controls weeds and also causes more weed seeds to germinate	Reduced tillage controls weeds and also exposes other weed seeds for germination	Weeds are a problem especially in the early stages of adoption, but problems are reduced with time and residues can help suppress weed growth
Soil temperature	Surface soil temperature: more variable	Surface soil temperature: intermediate in variability	Surface soil temperature: moderated the most
Diesel use and costs	Diesel use: high	Diesel use: intermediate	Diesel use: much reduced
Production costs	Highest costs	Intermediate costs	Lowest costs
Timeliness	Operations can be delayed	Intermediate timeliness of operations	Timeliness of operations more optimal
Yield	Can be lower where planting delayed	Yields same as TT	Yields same as TT but can be higher if planting done more timely

Source: From Hobbs et al. (2008)

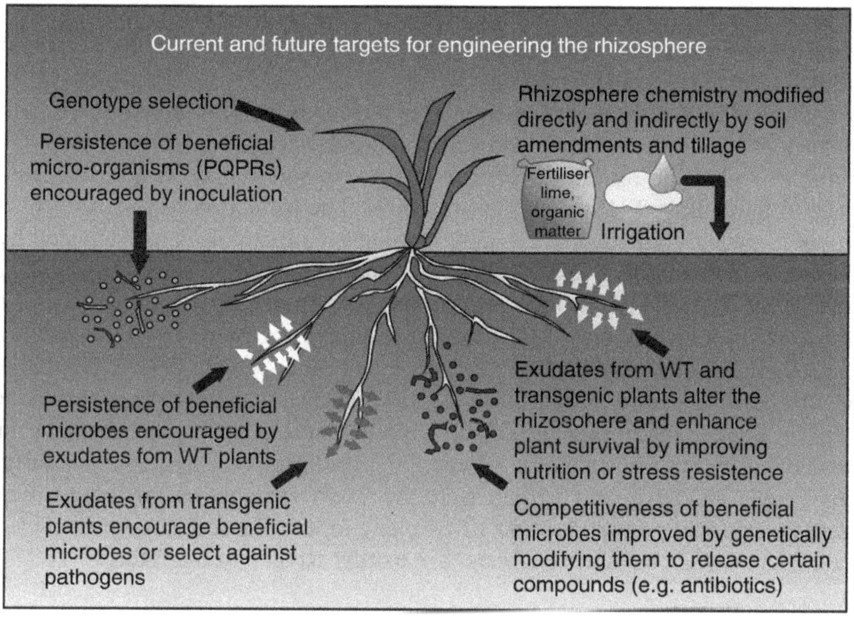

Figure 13.8 Current and future targets for engineering the rhizosphere to improve crop performance. Reproduced from Ryan et al. (2009). With kind permission from Springer Science and Business Media. For a colour version of this figure, please see Plate 13.3.

In addition, roots release a wide variety of compounds that modify their immediate environment to increase the availability of nutrients and to keep pathogens at bay (Gregory, 2006; Hiltpold et al., 2010, 2011). Figure 13.8 shows some current and future targets for rhizosphere engineering to increase the efficiency of nutrient recovery and to provide defences against pests and pathogens. Ryan et al. (2009) detail many possible routes to effect these changes, but these are still predominantly in the research stage with development only recently starting. One example of chemical modification with the potential for large-scale benefits is the release of nitrification inhibitors that biologically inhibit nitrification and reduce emissions of N_2O (see Section 13.2.2; Subbarao et al., 2009). Engineering plants to exude organic anions such as malate and citrate have been shown in the laboratory to confer some tolerance to aluminium toxicity (Delhaize et al., 2004; Magalhaes et al., 2007), but the realisation of this potential in field soils has yet to be achieved. In the field, the increased exudation of organic acids may also influence the availability of nutrients and change the populations of soil microbes to metabolise these carbon sources. The net effect of these changes may be to negate the tolerance effect, but more work in different soils is needed to develop reliable systems.

Similarly, the modification of phosphatase activity has been investigated as a means of improving the availability of soil organic phosphorus to plants (Asmar et al., 1995; George et al., 2007; Richardson et al., 2011). Many plants exude phosphatase enzymes from their roots naturally, and this can be associated with a depletion of soil organic phosphorus (e.g. George et al., 2002). Plants can be genetically modified to express an extracellular microbial phytase enabling them to use P from inositol phosphate on simple growing substrates (Richardson et al., 2001; George et al., 2004). However, when grown in soil, adsorption of the enzyme by clay minerals substantially reduces its activity, and the enzyme is a poor competitor with existing microbial communities (George et al., 2007; Giaveno et al., 2010). Achieving greater hydrolysis of organic P by plants would be beneficial on many soils, but, again, this is a topic for further research.

The biological and physical interactions occurring in the rhizosphere especially as a consequence of root growth, rhizodeposition, microbial activity, and the repeated wetting and drying of soil at the root–soil interface result in a soil matrix with different physical properties to the bulk soil (Hinsinger et al., 2009). The formation of rhizosheaths around some roots (especially grasses) is a widely recognised structural feature induced by roots and relies on immature xylem vessels, root hairs and the release of water and mucilage from roots (McCully, 1999). Mycorrhizal hyphae are also implicated in the adhesion of soil particles to roots together with the exopolysaccharides produced by rhizosphere micro-organisms (Amellal et al., 1998). The release of root mucilage may also change the water relations of the rhizosphere through the release of surfactants that reduce surface tension (Read et al., 2003). Whalley et al. (2005) also found that rhizosphere soil of maize and barley tended to be drier at a given matric potential than bulk soil but suggested that differences in wetting angle and pore connectivity were the likely explanation for these differences. Managing the physical properties of the rhizosphere to stabilise soils, improve soil structure and enable plants to access deep soil water are worthy targets for the future.

References

AEA Technology plc (2007). *Greenhouse Gas Inventories for England, Scotland, Wales and Northern Ireland: 1990–2005*. AEA Energy & Environment, Didcot, UK.

Alexandratos, N. (2009) How to feed the world in 2050. *Proceedings of a Technical Meeting of Experts*. FAO, Rome, Italy, pp. 1–32.

Amellal, N., Burtin, G., Bartoli, F., and T. Heulin, T. (1998) Colonization of wheat roots by an exopolysaccharide-producing *Pantoea agglomerans* strain and its effect on rhizosphere soil aggregation. *Applied and Environmental Microbiology* **64**, 3740–3747.

Angus, J.F. (2001) Nitrogen supply and demand in Australian agriculture. *Australian Journal of Experimental Agriculture* **41**, 277–288.

Ao, J., Fu, J., Tian, J., Yan, X., and Liao, H. (2010) Genetic variability for root morph-architecture traits and root growth dynamics as related to phosphorus efficiency in soybean. *Functional Plant Biology* **37**, 304–312.

Asmar, F., Gahoonia, T.S., and Nielsen, N.E. (1995) Barley genotypes differ in activity of soluble extracellular phosphatase and depletion of organic phosphorus in the rhizosphere soil. *Plant and Soil* **172**,117–122.

Baker, C.J., Saxton, K.E., and Ritchie, W.R. (2006) *No-Tillage Seeding: Science and Practice*, 2nd edn., CAB International, Wallingford, UK.

Barnett, C., Hossell, J., Perry, M., Procter, C., and Hughes, G. (2006) A handbook of climate trends across Scotland. SNIFFER project CC03, Scotland & Northern Ireland Forum for Environmental Research, 62pp.

Bolger, T.P. and Turner, N.C. (1999) Water use efficiency and water use of Mediterranean annual pastures in southern Australia. *Australian Journal of Agricultural Research* **50**, 1035–1046.

Bouwman, L., Stehfest, E., and van Kessel, C. (2010) Nitrous oxide emissions from the nitrogen cycle in arable agriculture: estimation and mitigation. In: *Nitrous Oxide and Climate Change* (ed. K. Smith), pp. 85–106. Earthscan, London.

Brown, I., Towers, W., Rivington, M., and Black, H.I.J. (2008) Influence of climate change on agricultural land-use potential: adapting and updating the land capability system for Scotland. *Climate Research*, **37**, 43–57.

Brown, I., Poggio, L., Gimona, A., and Castellazzi, M. (2011) Climate change, drought risk and land capability for agriculture: implications for land use in Scotland. *Regional Environmental Change* **11**, 503–518.

Bruinsma, J. (2003) *World Agriculture: Towards 2015/2030, an FAO Perspective*. Earthscan Publications, London.

Campbell, J.E., Lobell, D.B., Genova, R.C., and Field, C.B. (2008) The global potential of bioenergy on abandoned agriculture lands. *Environmental Science and Technology* **42**, 5791–5794.

Cassman, K.G., Dobermann, A., Walters, D.T., and Yang, H. (2003) Meeting cereal demand while protecting natural resources and improving environmental quality. *Annual Review of Environmental Resources* **28**, 315–358.

Chen, X.-P., Cui, Z.-L., Vitousek, P.M., Cassman, K.G., Matson, P.A., Bai, J.-S., Meng, Q,-F., Hou, P., Yue, S.-C., Römheld, V., and Zhang, F.-S. (2011) Integrated soil-crop system management for food security. *Proceedings of the National Academy of Science* **108**, 6399–6404.

Chmielewski, F.-M., Müller, A., and Bruns, E. (2004) Climate changes and trends in phenology of fruit trees and field crops in Germany, 1961–2000. *Agricultural and Forest Meteorology*, **121**, 69–78.

Conen, F., Smith, K.A., and Yagi K. (2010) Rice cultivation. In: *Methane and Climate Change* (eds D. Reay, P. Smith, and A. van Amstel), pp. 115–135. Earthscan, London.

Conrad, R. (1989) Temporal change of gas metabolism by hydrogen syntrophic methanogenic bacterial associations in anoxic paddy soil. *FEMS Microbiology Ecology* **62**, 265–274.

Conway, G.R. (1997) *The Doubly Green Revolution*. Penguin, London.

Cooper, P.J.M., Gregory, P.J., Tully, D., and Harris, H.C. (1987) Improving water use efficiency of annual crops in the rainfed farming systems of West Asia and North Africa. *Experimental Agriculture* **23**, 113–158.

Cottrill, B., Smith, C., Berry, P., Weightman, R., Wiseman, J., White, G., and Temple, M. (2007) Opportunities and implications of using the co-products from biofuel production as feeds for livestock. *HGCA Research Review* 66. HGCA, London.

Crutzen, P.J., Mosier, A.R., Smith, K.A., and Winiwarter, W. (2008) N_2O release from agro-biofuel production negates global warming reduction by replacing fossil fuels. *Atmospheric Chemistry and Physics* **8**, 389–395.

Cutforth, H.W., McGinn, S.M., McPhee, K.E., and Miller, P.R. (2007) Adaptation of pulse crops to the changing climate of the Northern Great Plains. *Agronomy Journal* **99**, 1684–1699.

Defra (2009) *UK Food Security Assessment: Our Approach*. Department for Environment, Food and Rural Affairs, London.

Defra (2010) *The British Survey of Fertiliser Practice: Fertiliser Use on Farm Crops for Crop Year 2009*. Department for Environment, Food and Rural Affairs, London.

Delhaize, E., Ryan, P.R., Hebb, D.M., Yamamoto, Y., Sasaki, T., and Matsumoto, H. (2004) Engineering high-level aluminium tolerance in barley with the ALMT1 gene. *Proceedings of the National Academy of Science* **101**, 15249–15254.

Denier van der Gon, H.A.C. and Neue, H.-U. (1995) Influence of organic matter incorporation on the CH_4 emission from a wetland rice field. *Global Biogeochemical Cycles* **9**, 11–22.

Duxbury, J.M., Bouldin, D.R., Terry, R.E., and Tate, R.L. (1982) Emissions of nitrous oxide from soils. *Nature* **298**, 462–464.

Eakin, H., Bohle, H.-G., Izac, A.-M., Reenberg, A., Gregory, P., and Pereira, L. (2010) Food, violence and human rights. In: *Food Security and Global Environmental Change* (eds J. Ingram, P. Ericksen, and D. Liverman), pp. 245–271. Earthscan, London.

EEA (2007) Annual European Community greenhouse gas inventory 1990–2005 and inventory report 2007. European Environment Agency, Technical report No 7/2007. EEA, Copenhagen, Denmark.

Ericksen, P.J. (2008) Conceptualizing food systems for global environmental change research. *Global Environmental Change* **18**, 234–245.

Estrella, N., Sparks, T.H., and Menzel, A. (2007) Trends and temperature response in the phenology of crops in Germany. *Global Change Biology* **13**, 1737–1747.

Evans, L.T. (1993) *Crop Evolution, Adaptation and Yield*. Cambridge University Press, Cambridge, UK.

Evans, L.T. (1998) *Feeding the Ten Billion: Plants and Population Growth*. Cambridge University Press, Cambridge, UK.

Ewers, R.M., Scharlemann, J.P.W., Balmford, A., and Green, R.E. (2009) Do increases in agricultural yield spare land for nature? *Global Change Biology* **15**, 1716–1726.

Ewert, F., Rounsevell, M.D.A., Reginster, I., Metzger, M.J., and Leemans R. (2005) Future scenarios of European agricultural land use. I. Estimating changes in crop productivity. Agriculture, *Ecosystems and Environment* **107**, 101–116.

FAO (1996) *Report of the World Food Summit*. FAO, Rome, Italy.

FAO (2008) *The State of Food and Agriculture. Biofuels: Prospects, Risks and Opportunities*. FAO, Rome, Italy.

FAO (2009) *State of the World's Forests 2009*. FAO, Rome, Italy.

Fargione, J., Hill, J., Tilman, D., Polasky, S., and Hawthorne P. (2008) Land clearing and the carbon debt. *Science* **319**, 1235–1238.

Field, C.B., Campbell, J.E., and Lobell D.B. (2007) Biomass energy: the scale of the potential resource. *Trends in Ecology and Evolution* **23**, 65–72.

Fischer, G., Shah, M., Tubiello, F.N., and van Velhuizen, H. (2005) Socio-economic and climate change impacts on agriculture: an integrated assessment, 1980–2080. *Philosophical Transactions of the Royal Society* B **360**, 2067–2083.

Foresight (2011) *The Future of Food and Farming*. Government Office of Science, London, UK.

Fuhrer, J. (2006) Agricultural systems: sensitivity to climate change. *CAB Reviews: Perspectives in Agriculture, Veterinary Science, Nutrition and Natural Resources* **1**, No. 052.

Galloway, J.N. and Cowling, E.B. (2002) Reactive nitrogen and the world: 200 years of change. *Ambio* **31**, 64–71.

Galloway, J.N., Dentener, F.J., Capone, D.G., Boyer, E.W., Howarth, R.W., Seitzinger, S.P., Asner, G.P., Cleveland, C.C., Green, P.A., Holland, E.A., Karl, D.M., Michaels, A.F., Porter, J.H., Townsend, A.R., and Vörösmarty, C.J. (2004) Nitrogen cycles: past, present and future. *Biogeochemistry* **70**, 153–226.

Galloway, J.N., Townsend, A.R., Erisman, J.W., Bekunda, M., Cai, Z., Freney, J.R., Martinelli, L.A., Seitzinger, S.P., and Sutton, M.A. (2008) Transformation of the nitrogen cycle: recent trends, questions, and potential solutions. *Science* **320**, 889–892.

George, T.S., Gregory, P.J., Robinson, J.S., Buresh, R.J., and Jama, B. (2002) Utilisation of soil organic P by agroforestry and crop species in the field, western Kenya. *Plant and Soil* **246**, 53–63.

George, T.S., Richardson, A.E., Hadobas, P.A., and Simpson, R.J. (2004) Characterization of transgenic *Trifolium subterraneum* L. which expresses *phyA* and releases extracellular phytase: growth and P nutrition in laboratory media and soil. *Plant, Cell and Environment* **27**, 1351–1361.

George, T.S., Simpson, R.J., Gregory, P.J., and Richardson, A.E. (2007) Differential interaction of *Aspergillus niger* and *Peniophora lycii* phytases with soil particles affects the hydrolysis of inositol phosphates. *Soil Biology and Biochemistry* **39**, 793–803.

Giaveno, C., Celi, L., Richardson, A.E., Simpson, R.J., and Barberis, E. (2010) Interaction of phytases with soil minerals and availability of substrate affects the hydrolysis of inositol phosphates. *Soil Biology and Biochemistry* **42**, 491–498.

Gicheru, P.T. (1994) Effects of residue mulch and tillage on soil-moisture conservation. *Soil Technology* **7**, 209–220.

Godfray, H.C.J., Beddington, J.R., Crute, I.R., Haddad, L., Lawrence, D., Muir, J.F., Pretty, J., Robinson, S., Thomas, S.M., and Toulmin, C. (2010) Food security: the challenge of feeding 9 billion people. *Science* **327**, 812–818.

Govaerts, B., Verhulst, N., Castellanos-Navarrete, A., Sayre, K.D., Dixon, J., and Dendooven, L. (2009) Conservation agriculture and soil carbon sequestration: between myth and farmer reality. *Critical Reviews in Plant Science* **28**, 97–122.

Greenland, D.J. (1997) *The Sustainability of Rice Farming*. CAB International, Wallingford, UK.

Greenland, D.J., Gregory, P.J., and Nye, P.H. (1998) Land resources and constraints to crop production. In: *Feeding a World Population of More Than Eight Billion People: A Challenge to Science* (eds J.C. Waterlow, D.G. Armstrong, L. Fowden, and R. Riley), pp. 39–55. Oxford University Press, Oxford, UK.

Gregory, P.J. (2006) *Plant Roots: Growth, Activity and Interactions with Soil*. Blackwell Publishing Ltd., Oxford, UK.

Gregory, P.J. and Ingram, J.S.I. (2008) Climate change and the current 'food crisis'. *CAB Reviews: Perspectives in Agriculture, Veterinary Science, Nutrition and Natural Resources* **3**(99), 1–10.

Gregory, P.J. and George, T.S. (2011) Feeding nine billion: the challenge to sustainable crop production. *Journal of Experimental Botany* **62**, 5233–5239.

Gregory, P.J. and Marshall, B. (2012) Attribution of climate change: a methodology to estimate the potential contribution to increases in potato yield in Scotland since 1960. *Global Change Biology* **18**, 1372–1388.

Gregory, P.J., Simmonds, L.P., and Warren, G.P. (1997) Interactions between plant nutrients, water and carbon dioxide as factors limiting crop yields. *Philosophical Transactions of the Royal Society London B* **352**, 987–996.

Gregory, P.J., Ingram, J.S.I., Andersson, R., Betts, R.A., Brovkin, V., Chase, T.N., Grace, P.R., Gray, A.J., Hamilton, N., Hardy, T.B., Howden, S.M., Jenkins, A., Meybeck, M., Olsson, M., Ortiz-Monasterio, I., Palm, C.A., Payn, T.W., Rummukainen, M., Schulze, R.E., Thiem, M., Valentin, C. and Wilkinson, M.J. (2002) Environmental consequences of alternative practices for intensifying crop production. *Agriculture, Ecosystems and Environment* **88**, 279–290.

Gregory, P.J., Ingram, J.S.I., and Brklacich, M. (2005) Climate change and food security. *Philosophical Transactions of the Royal Society B* **360**, 2139–2148.

Gregory, P.J., Johnson, S.N., Newton, A.C., and Ingram, J.S.I. (2009) Integrating pests and pathogens into the climate change/food security debate. *Journal of Experimental Botany* **60**, 2827–2838.

Haile, M. (2005) Weather patterns, food security and humanitarian response in sub-Saharan Africa. *Philosophical Transactions of the Royal Society London B* **360**, 2169–2182.

Hall, D.O., Rosillo-Calle, F., Williams, R.H., and Woods, J. (1993) Biomass for energy: supply prospects. In: *Renewable Energy: Sources for Fuels and Electricity* (eds T.B. Johansson, H. Kelly, A.K.N. Reddy, and R.H. Williams), pp. 593–651. Island Press, Washington, D.C.

Hay, R.K.M. and Porter, J.R. (2006) *The Physiology of Crop Yield*, 2nd edn. Blackwell Publishing Ltd., Oxford, UK.

Henry, A., Chaves, N.F., Kleinman, P.J.A., and Lynch, J.P. (2010) Will nutrient-efficient genotypes mine the soil? Effects of genetic differences in root architecture in common bean (*Phaseolus vulgaris* L.) on soil phosphorus depletion in a low-input agro-ecosystem in Central America. *Field Crops Research* **115**, 67–78.

Hill, J., Nelson, E., Tilman, D., Polasky, S., and Tiffany, D. (2006) Environmental, economic and energetic costs and benefits of biodiesel and ethanol biofuels. *Proceedings of the National Academy of Sciences* **103**, 11206–11210.

Hiltpold, I., Baroni, M., Toepfer, S., Kuhlmann, U., and Turlings, T.C.J. (2010) Selection of entomopathogenic nematodes for enhanced responsiveness to a volatile root signal helps to control a major root pest. *Journal of Experimental Biology* **213**, 2417–2423.

Hiltpold, I., Erb, M., Robert, C.A.M., and Turlings, T.C.J. (2011) Systemic root signalling in a belowground, volatile-mediated tritrophic interaction. *Plant, Cell and Environment* **34**, 1267–1275.

Hinsinger, P., Bengough, A.G., Vetterlein, D., and Young, I.M. (2009) Rhizosphere: biophysics, biogeochemistry and ecological relevance. *Plant and Soil* **321**, 117–152.

Ho, M.D., McCannon, B.C., and Lynch, J.P. (2004) Optimization modeling of plant root architecture for water and phosphorus acquisition. *Journal of Theoretical Biology* **226**, 331–340.

Hobbs, P.R., Sayre, K., and Gupta, R. (2008) The role of conservation agriculture in sustainable agriculture. *Philosophical Transactions of the Royal Society B* **363**, 543–555.

IEA (2006) International Energy Agency. *World Energy Outlook 2006*. International Energy Agency, Paris, France.

IPCC (2007) Climate change 2007: The physical science basis. *Fourth Assessment Report of the Intergovernmental Panel on Climate Change*, Geneva, Switzerland.

Jaggard, K.W., Qi, A., and Semenov, M.A. (2007) The impact of climate change on sugarbeet yield in the UK: 1976–2004. *Journal of Agricultural Science* **14**, 367–75.

Jaggard, K.W., Qi, A., and Ober, E.S. (2010) Possible changes to arable crop yields by 2050. *Philosophical Transactions of the Royal Society B, London* **365**, 2835–2851.

Jarvis, A., Ramirez, J., Anderson, B., Leibing, C., and Aggarwal, P. (2010) Scenarios of climate change within the context of agriculture. In: *Climate Change & Crop Production* (ed. M.P. Reynolds), pp. 9–37. CAB International, Wallingford, UK.

Jones, M.J. and Wahbi, A. (1992) Site-factor influence on barley response to fertilizer in on-farm trials in northern Syria: descriptive and predictive models. *Experimental Agriculture* **28**, 63–87.

Karp, A. and Shield, I. (2008) Bioenergy from plants and the sustainable yield challenge. *New Phytologist* **179**, 15–32.

Karp, A. and Richter, G.M. (2011) Meeting the challenge of food and energy security. *Journal of Experimental Botany* **62**, 3263–3271.

Khai, N.M., Ha, P.Q., and Öborn, I. (2007) Nutrient flows in small-scale peri-urban vegetable farming systems in Southeast Asia – a case study in Hanoi. *Agriculture, Ecosystems and Environment* **122**, 192–202.

Khosla, R., Inman, D., Westfall, D.G., Frasier, M., Mzuka, M., Koch, B., and Hornung, A. (2008) A synthesis of multi-disciplinary research in precision agriculture: site-specific management zones in the semi-arid western Great Plains of the USA. *Precision Agriculture* **9**, 85–100.

Laborte, A.G., de Bie, K.(C.A.J.M.), Smaling, E.M.A., Moya, P.F., Boling, A.A., and Van Ittersum, M.K. (2012) Rice yields and yield gaps in Southeast Asia: past trends and future outlook. *European Journal of Agronomy* **36**, 9–20.

Lal, R. (2004) Soil carbon sequestration impacts on global climate change and food security. *Science* **304**, 1623–1627.

Latiri-Souki, K., Nortcliff, S., and Lawlor, D.W. (1998) Nitrogen fertilizer can increase dry matter, grain production and radiation and water use efficiencies for durum wheat under semi-arid conditions. *European Journal of Agronomy* **9**, 21–34.

Lawes, R.A. and Robertson, M.J. (2011) Whole farm implications on the application of variable rate technology to every cropped field. *Field Crops Research* **124**, 142–148.

Lee-Smith, D. (2010) Cities feeding people: an up-date on urban agriculture in equatorial Africa. *Environment and Urbanization* **22**, 483–499.

Lobell, D.B. and Field, C.B. (2007) Global scale climate-crop yield relationships and the impacts of recent warming. *Environmental Research Letters* **2**, 1–7.

Lobell, D.B., Cassman, K.G., and Field, C.B. (2009) Crop yield gaps: their importance, magnitudes, and causes. *Annual Review of Environmental Resources* **34**, 179–204.

Lou, Y.S., Inubushi, K., Mizuno, T., Hasegawa, T., Lin, Y., Sakai, H., Cheng, W., and Kobayashi, K. (2008) CH_4 emissions with differences in atmospheric CO_2 enrichment and rice cultivars in a Japanese paddy soil. *Global Change Biology* **14**, 2678–2687.

Lynch, J.P. (2007) Roots of the second green revolution. *Australian Journal of Botany* **55**, 1–20.

Magalhaes, J.V., Liu, J., Guimaraes, C.T., Lana, U.G.P., Alves, V.M.C., Wang, Y.H., Schaffert, R.E., Hoekenga, O.A., Pineros, M.A., Shaff, J.E., Klein, P.E., Carneiro, N.P., Coelho, C.M., Trick, H.N., and Kochian, L.V. (2007) A gene in the multidrug and toxic compound extrusion (MATE) family confers aluminium tolerance in sorghum. *Nature Genetics* **39**, 1156–1161.

Maslow, A. (1943) A theory of human motivation. *Psychological Review* **50**, 370–396.

McCormick, S., Jordan, C., and Bailey, J.S. (2009) Within and between-field spatial variation in soil phosphorus in permanent grassland. *Precision Agriculture* **3**, 262–276.

McCully, M.E. (1999) Roots in soil: unearthing the complexities of roots and their rhizospheres. *Annual Review of Plant Physiology and Plant Molecular Biology* **50**, 695–718.

McMichael, P. (1994) *The Global Restructuring of Agro-Food Systems.* Cornell University Press, Ithaca, New York.

MEA (2005) *Millennium Ecosystem Assessment*, Island Press, Washington, D.C.

Menzel, A., Sparks, T.H., Estrella, N., Koch, E., Aasa, A., Ahas, R., Alm-Kübler, K., Bissolli, P., Braslavská, O., Briede, A., Chmielwski, F.M., Crepinsek, Z., Curnel, Y., Dahl, Å., Defila, C., Donnelly, A., Filella, Y., Jatczak, K., Måge, F., Mestre, A., Nordli, Ø., Peñuelas, J., Pirinen, P., Remišová, V., Scheifinger, H., Striz, M., Susnik, A., van Vliet, A.J.H., Wielgolaski, F.-E., Zach, S., and Zust, A. (2006) European phenological response to climate change matches the warming pattern. *Global Change Biology* **12**, 1969–1976.

Mosier, A.R., Kroeze, C., Nevison, C., Oenema, O., Seitzinger, S., and Cleemput, O.V. (1998) Closing the global atmospheric N_2O budget: nitrous oxide emissions through the agricultural nitrogen cycle. *Nutrient Cycling in Agroecosystems* **52**, 225–248.

Montgomery, D.R. (2007) Soil erosion and agricultural sustainability. *Proceedings of the National Academy of Science, USA* **104**, 13268–13272.

Myers, R.J.K., Palm, C.A., Cuevas, E., Gunatilleke, I.U.N., and Brossard M. (1994) The synchronization of nutrient mineralization and plant nutrient demand. In: *The Biological Management of Tropical Soil Fertility* (eds P.L. Woomer and M.J. Swift), pp. 81–116. John Wiley & Sons, Ltd., Chichester, UK.

Neumann, K., Verburg, P.H., Stehfest, E., and Mǘller, C. (2010) The yield gap of global grain production: a spatial analysis. *Agricultural Systems* **103**, 316–326.

Njenga, M., Kimani, S., Romney, D., and Karanja, N. (2007) Nutrient recovery from solid waste and linkage to urban and peri-urban agriculture in Nairobi, Kenya. In: *Advances in Integrated Soil Fertility Management in Sub-Saharan Africa: Challenges and Opportunities* (eds A. Bationo, B. Waswa, J. Kihara, and J. Kimetu), pp. 487–491. Springer, Berlin, Germany.

Nonhebel, S. (2007) Energy from agricultural residues and consequences for land requirements for food production. *Agricultural Systems* **94**, 586–592.

Ogola, J.B.O., Wheeler, T.R., and Harris, P.M. (2002) Effects of nitrogen and irrigation on water use of maize crops. *Field Crops Research* **78**, 105–117.

Olk, D.C., Cassman, K.G., Schmidt-Rohr, K., Anders, M.M., Mao, J.-D., and Deenik, J.L. (2006) Chemical stabilization of soil organic nitrogen by phenolic lignin residues in anaerobic agroecosystems. *Soil Biology and Biochemistry* **38**, 3303–3312.

Payne, W.A. (2000) Optimizing crop water use in sparse stands of pearl millet. *Agronomy Journal* **92**, 808–814.

Pimental, D. and Pimental, M.H. (2008) *Food, Energy and Society*, 3rd edn. CRC Press, Boca Raton, Florida.

Pimental, D., Jurd, L.E., Bellotti, A.C., Forster, M.J., Oka, I.N., Sholes, O.D., and Whitman, R.J. (1973) Food production and the energy crisis. *Science* **182**, 443–449.

Pingali, P., Raney, T., and Wiebe, K. (2008) Biofuels and food security: missing the point. *Review of Agricultural Economics* **30**, 506–516.

Ponnamperuma, F.N. (1972) The chemistry of submerged soils. *Advances in Agronomy* **24**, 29–96.

Powlson, D.S., Riche, A.B., and Shield, I. (2005) Biofuels and other approaches for decreasing fossil fuel emissions from agriculture. *Annals of Applied Biology* **146**, 193–201.

Powlson, D.S., Gregory, P.J., Whalley, W.R., Quinton, J.N., Hopkins, D.W., Whitmore, A.P., Hirsch, P.R., and Goulding, K.W.T. (2011) Soil management in relation to sustainable agriculture and ecosystem services. *Food Policy* **36**, S72–S87.

Pretty, J. (2008) Agricultural sustainability: concepts, principles and evidence. *Philosophical Transactions of the Royal Society B* **363**, 447–465.

Quinton, J.N., Govers, G., Van Oost, K., and Bardgett, R.D. (2010) The impact of agricultural soil erosion on biogeochemical cycling. *Nature Geoscience* **3**, 311–314.

Read, D.B., Bengough, A.G., Gregory, P.J., Crawford, J.W., Robinson, D., Scrimgeour, C.M., Youmg, I.M., Zhang, K., and Zhang, X. (2003) Plant roots release phospholipid surfactants that modify the physical and chemical properties of soil. *New Phytologist* **157**, 315–326.

Reay, D., Smith, P., and van Amstel, A. (2010) Methane sources and the global methane budget. In: *Methane and Climate Change* (eds D. Reay, P. Smith, and A. van Amstel), pp. 1–13. Earthscan, London.

Richardson, A.E., Hadobas, P.A., and Hayes, J.E. (2001) Extracellular secretion of Aspergillus phytase from *Arabidopsis* roots enables plants to obtain phosphorus from phytate. *The Plant Journal* **25**, 641–649.

Richardson, A.E., Lynch, J.P., Ryan, P.R., Delhaize, E., Smith, F.A., Smith, S.E., Harvey, P.R., Ryan, M.H., Veneklaas, E.J., Lambers, H., Oberson, A., Culvenor, R.A., and Simpson, R.J. (2011) Plant and microbial strategies to improve the phosphorus efficiency of agriculture. *Plant and Soil* **349**, 121–156.

Righelato, R. and Spracklen, D.V. (2007) Carbon mitigation by biofuels or by saving and restoring forests? *Science* **317**, 902.

Rhoton, F.E., Shipitalo, M.J., and Lindbo, D.L. (2002) Runoff and soil loss from Midwestern, and southeastern US silt loam soils as affected by tillage practice and soil organic matter content. *Soil & Tillage Research* **66**, 1–11.

Robertson, M.J., Lyle, G., and Bowden, J.W. (2008) Within-field variability of wheat yield and economic implications for spatially variable nutrient management. *Field Crops Research* **105**, 211–220.

Rubio, G., Walk, T., Ge, Z.Y., Yan, X.L., Liao, H., and Lynch, J.P (2001) Root gravitropism and below-ground competition among neighbouring plants: a modelling approach. *Annals of Botany* **88**, 929–940.

Ryan, P.R., Dessaux, Y., Thomashow, L.S., and Weller, D.M. (2009) Rhizosphere engineering and management for sustainable agriculture. *Plant and Soil* **321**, 363–383.

Sadras, V. (2002) Interaction between rainfall and nitrogen fertilization of wheat in environments prone to terminal drought: economic and environmental risk analysis. *Field Crops Research* **77**, 201–215.

Sanchez, P.A., Shepherd, K.D., Soule, M.J., Place, F.M., Buresh, R.J., Izac, A.-M.N., Mokwunye, A.U., Fwesiga, F.R., Ndiritu, C.G., and Woomer, P.L. (1997) Soil fertility replenishment in Africa: an investment in natural resource capital. In: *Replenishing Soil Fertility in Africa* (eds R.K. Buresh, P.A. Sanchez, and F. Calhoun), pp. 1–46. Soil Science Society of America, Madison, Wisconsin.

Satpathy, S.N., Mishra, S., Adhya, T.K., Ramakrishnan, B., Rao, V.R., and Sethunathan, N. (1998) Cultivar variation in methane efflux from tropical rice. *Plant and Soil* **202**, 223–229.

Satterthwaite, D., McGranahan, G., and Tacoli, C. (2010) Urbanization and its implications for food and farming. *Philosophical Transactions of the Royal Society B* **365**, 2809–2820.

Schütz, H., Seiler, W., and Conrad, R. (1989) Processes involved in formation and emission of methane in rice paddies. *Biogeochemistry* **7**, 33–53.

Searchinger, T., Heimlich, R., Houghton, R.A., Dong, F., Elobeid, A., Fabiosa, J., Tokgoz, S., Hayes, D., and Yu, T.-H. (2008) Use of U.S. croplands for biofuels increases greenhouse gases through emissions from land-use change. *Science* **319**, 1238–1240.

Sigren, L.K., Byrd, G.T., Fisher, F.M., and Sass, R.L. (1997) Comparison of soil acetate concentrations and methane production, transport, and emission in two rice cultivars. *Global Biogeochemical Cycles* **11**, 1–14.

Smaling, E.M.A., Nandwa, S.M., and Janssen, B.H. (1997) Soil fertility in Africa is at stake. In: *Replenishing Soil Fertility in Africa* (eds R.J. Buresh, P.A. Sanchez, and F. Calhoun), pp. 47–61. SSSA Special Publication No 51, Soil Science Society of America, Madison, Wisconsin.

Smith, K. (2010) Introduction. In: *Nitrous Oxide and Climate Change* (ed. K. Smith), pp. 1–3. Earthscan, London.

Smit, J., Ratta, A., and Nasr, J. (1996) *Urban Agriculture: Food, Jobs and Sustainable Cities*. UNDP, New York.

Smith K.A., McTaggart I.P., and Tsuruta H. (1997) Emissions of N_2O and NO associated with nitrogen fertilization in intensive agriculture, and the potential for mitigation. *Soil Use and Management* **13**, 296–304.

Smith, K., Crutzen, P., Mosier, A., and Winiwarter, W. (2010a) The global nitrous oxide budget: a reassessment. In: *Nitrous Oxide and Climate Change* (ed. K. Smith), pp. 63–84. Earthscan, London.

Smith, P., Gregory, P.J., van Vuuren, D., Obersteiner, M., Havlik, P., Rounsevell, M., Woods, J., Stehfest, E., and Bellarby, J. (2010b) Competition for land. *Philosophical Transactions of the Royal Society B* **365**, 2941–2957.

Sparks, T.H., Croxton, P.J., Collinson, N., and Taylor, P.W. (2005) Examples of phenological change, past and present, in UK farming. *Annals of Applied Biology* **146**, 531–537.

Spedding, C.R.W. and Walsingham, J.M. (1976) The production and use of energy in agriculture. *Journal of Agricultural Economics* **27**, 19–30.

Subbarao, G.V., Kishii, M., Nakahara, K., Ishikawa, T., Ban, T., Tsujimoto, H., George, T.S., Berry, W.L., Hash, C.T., and Ito, O. (2009) Biological nitrification inhibition (BNI) – is there potential for genetic interventions in the Triticeae? *Breeding Science* **59**, 529–545.

Tilman, D., Balzer, C., Hill, J., and Befort, B.L. (2011) Global food demand and the sustainable intensification of agriculture. *Proceedings of the National Academy of Science* **108**, 20260–20264.

UK National Ecosystem Assessment (2011) *UK National Ecosystem Assessment: Understanding Nature's Value to Society. Synthesis of the Key Findings*. National Ecosystem Assessment, UK.

Unger, P.W., Langdale, D.W., and Papendick, R.I. (1988) Role of crop residues – improving water conservation and use. In: *Cropping Strategies for Efficient Use of Water and Nitrogen* (ed. W.K. Hargrove), pp. 69–100. American Society of Agronomy, Madison, Wisconsin.

Van Ittersum, M.K. and Rabbinge, R. (1997) Concepts in production ecology for analysis and quantification of agricultural input-output combinations. *Field Crops Research* **52**, 197–208.

Verhulst, N., Govaerts, B., Verachtert, E., Castellanos-Navarrete, A., Mezzalama, M., Wall, P.C., Chocobar, A., Deckers, J., and Sayre, K.D. (2010) Conservation agriculture, improving soil quality for sustainable production systems? In: *Food Security and Soil Quality* (eds R. Lal and B.A. Stewart). CRC Press, Boca Raton, Florida.

Vitousek, P.M., Naylor, R., Crews, T., David, M.B., Drinkwater, L.E., Holland, E., Johnes, P.J., Katzenberger, J., Martinelli, L.A., Matson, P.A., Nziguheba, G., Ojima, D., Palm, C.A., Robertson, G.P., Sanchez, P.A., Townsend, A.R., and Zhang, F.S. (2009) Nutrient imbalances in agricultural development. *Science* **324**, 1519–1520.

Whalley,W.R., Riseley, B., Leeds-Harrison, P.B., Bird, N.R.A., Leech, P.K., and Adderley, W.P. (2005) Structural differences between bulk and rhizosphere soil. *European Journal of Soil Science* **56**, 353–360.

Woods, J., Williams, A., Hughes, J.K., Black, M., and Murphy, R. (2010) Energy and the food System. *Philosophical Transactions of the Royal Society B* **365**, 2991–3006.

Yagi, K., Tsuruta, H., and Minami, K. (1997) Possible options for mitigating methane emissions from rice cultivation. *Nutrient Cycling in Agroecosystems* **49**, 213–220.

Yao, H., Conrad, R., Wassmann, R., and Neue, H.U. (1999) Effect of soil characteristics on sequential reduction and methane production in sixteen rice paddy soils from China, the Philippines, and Italy. *Biogeochemistry* **47**, 269–295.

Index

Note: Page numbers in *italics* refer to Figures; those in **bold** to Tables

Soil Conditions and Plant Growth. Edited by Peter J. Gregory and Stephen Nortcliff.
© 2013 Blackwell Publishing Ltd. Published 2013 by Blackwell Publishing Ltd.

WILEY-BLACKWELL

More research in Soil Science from Wiley-Blackwell

Sustaining Soil Productivity in Response to Global Climate Change: Science, Policy, and Ethics

Edited by Thomas J. Sauer, John Norman and Mannava V. K. Sivakumar

ISBN: 978-0-470-95857-5

A multi-disciplinary volume exploring the ethical, political and social issues surrounding the stewardship of our vital soil resources.

Ecology and Management of Forest Soils, 4th Edition

Edited by Dan Binkley and Richard Fisher

ISBN: 978-0-470-97946-4

A completely revised and updated Fourth Edition of this classic textbook in the field. Clearly structured throughout with new material on approaches to the management of forest soils and geostatistics, GIS and taxonomies.

Root Genomics and Soil Interactions

Edited by Martin Crespi

ISBN: 978-0-470-96043-1

Examines the use of genome-based technologies to understand root development and adaptability to biotic and abiotic stresses and changes in the soil environment.

Get your copy today!

Order online at www.wiley.com/go/soil

Find more at **www.wiley.com/go/soil**

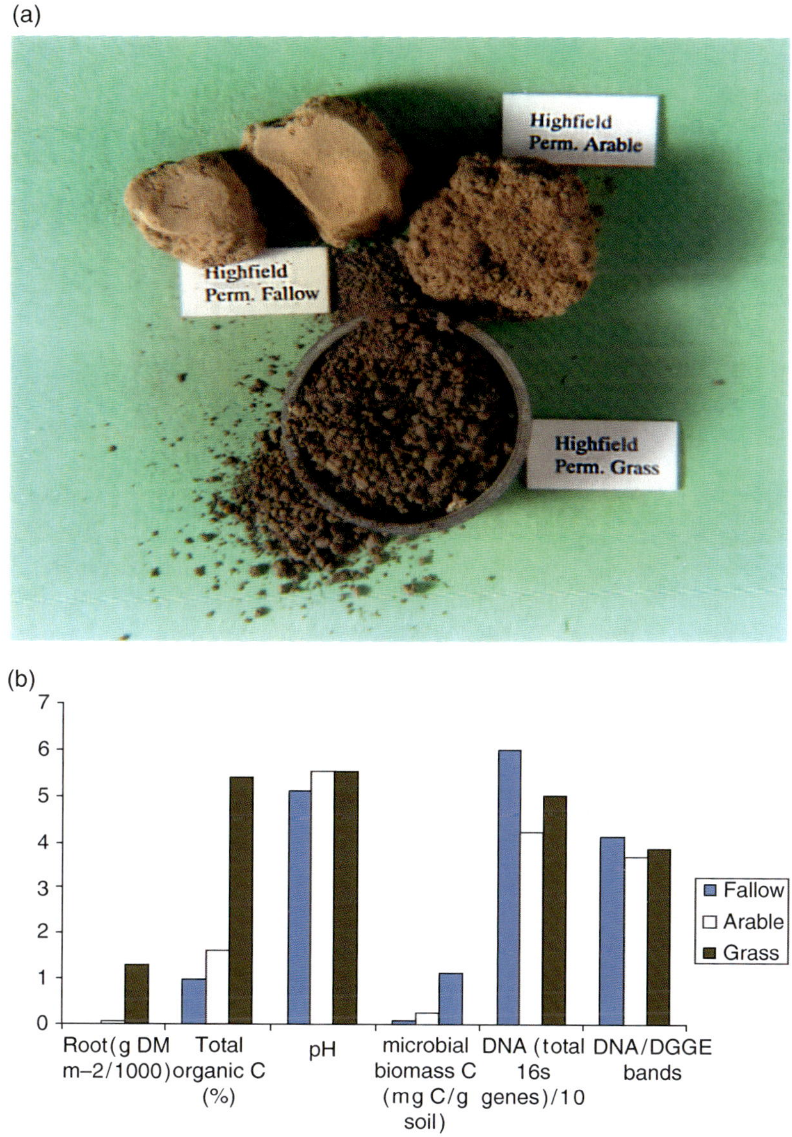

Plate 3.1 Data from the long-term Highfield experiment at Rothamsted, England, shows the impacts of long-term fallowing with intensive cultivation on (a) visible soil physical properties and (b) on some chemical and biological properties. The almost zero inputs of organic carbon via plants and tillage in the fallow plots has very significantly reduced soil organic carbon levels and soil microbial biomass but had no impact on pH in this well-buffered clay soil. Soil microbial diversity indicated by the number of bands in the DNA/DGGE is not reduced by fallowing; in fact, underlying soil microbial diversity is apparently revealed when the dominance of plant-associated bacteria is reduced. From Hirsch et al. (2009). With kind permission from Elsevier.

Soil Conditions and Plant Growth. Edited by Peter J. Gregory and Stephen Nortcliff.
© 2013 Blackwell Publishing Ltd. Published 2013 by Blackwell Publishing Ltd.

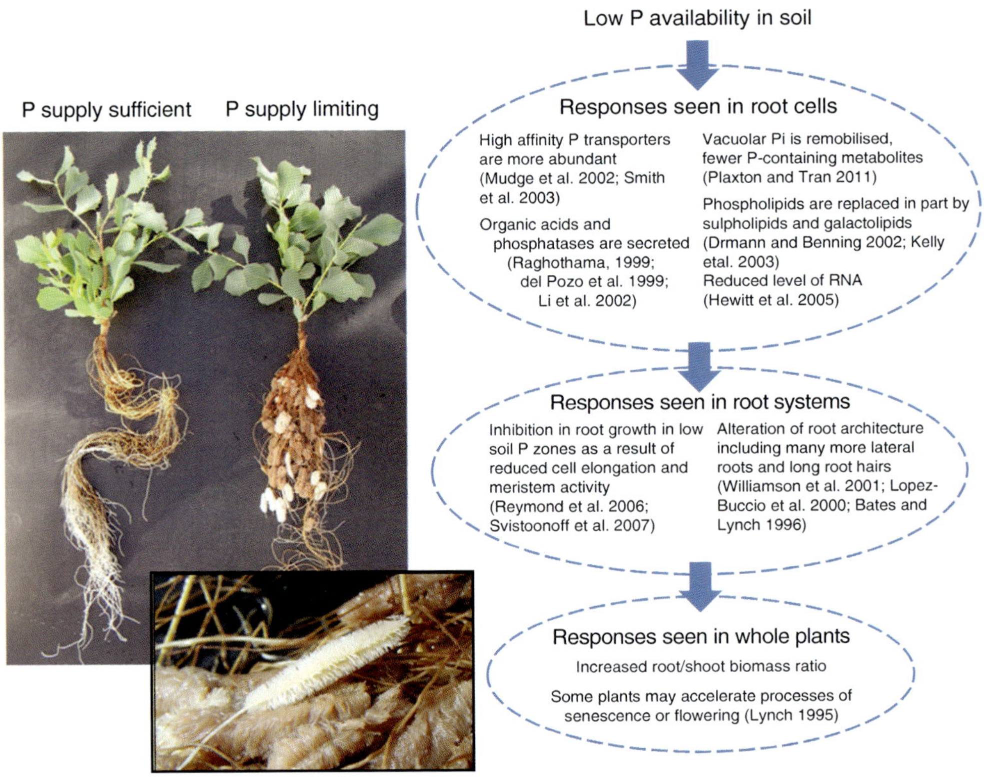

Plate 3.2 Complex and interlinked changes in plant physiology at root cell, root system and whole plant scales occur in response to low P availability in soil as the result of up-regulation of many genes. Photographs show *Hakea prostrata* R. Br. grown under sufficient and limiting phosphate supply. Cluster roots develop in some species as an adaptation for P acquisition in low P environments (details in lower image). Thanks to Stuart Pearse and Michael Shane from the University of Western Australia for access to greenhouse trials which provided the photographs.

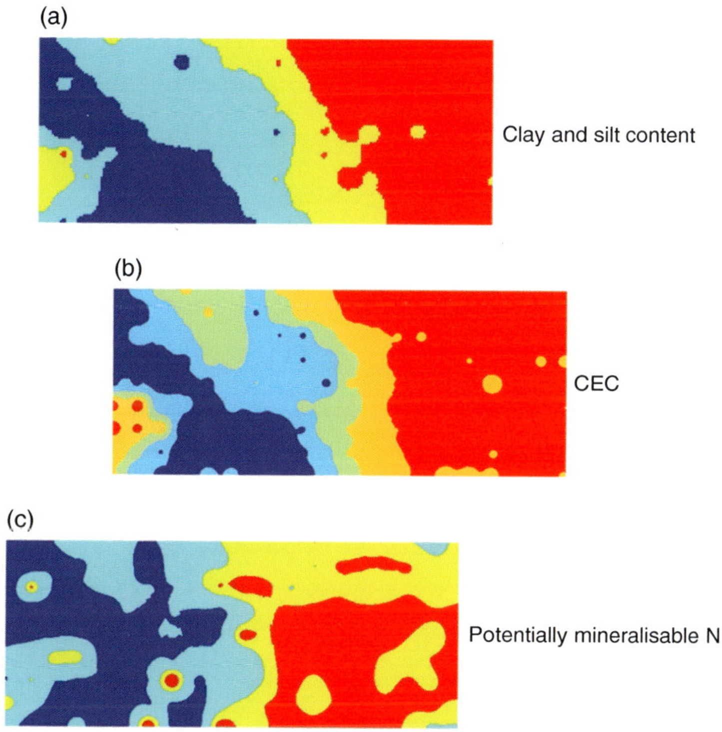

Plate 3.3 Spatial maps of (a) clay and silt content, (b) cation exchange capacity and (c) potentially mineralisable N (where the colours show quartiles ranges as an illustration) over a 10 ha area (200 × 500 m; 180 sampling points on a 25 × 25 m grid) where the properties were determined using mid-infrared sensors. The spatial patterns were very strongly correlated with the same properties determined by traditional laboratory-based methods (r^2 = 0.91, 0.91 and 0.82, respectively). Unpublished data from the central wheatbelt, Western Australia provided by D. V. Murphy.

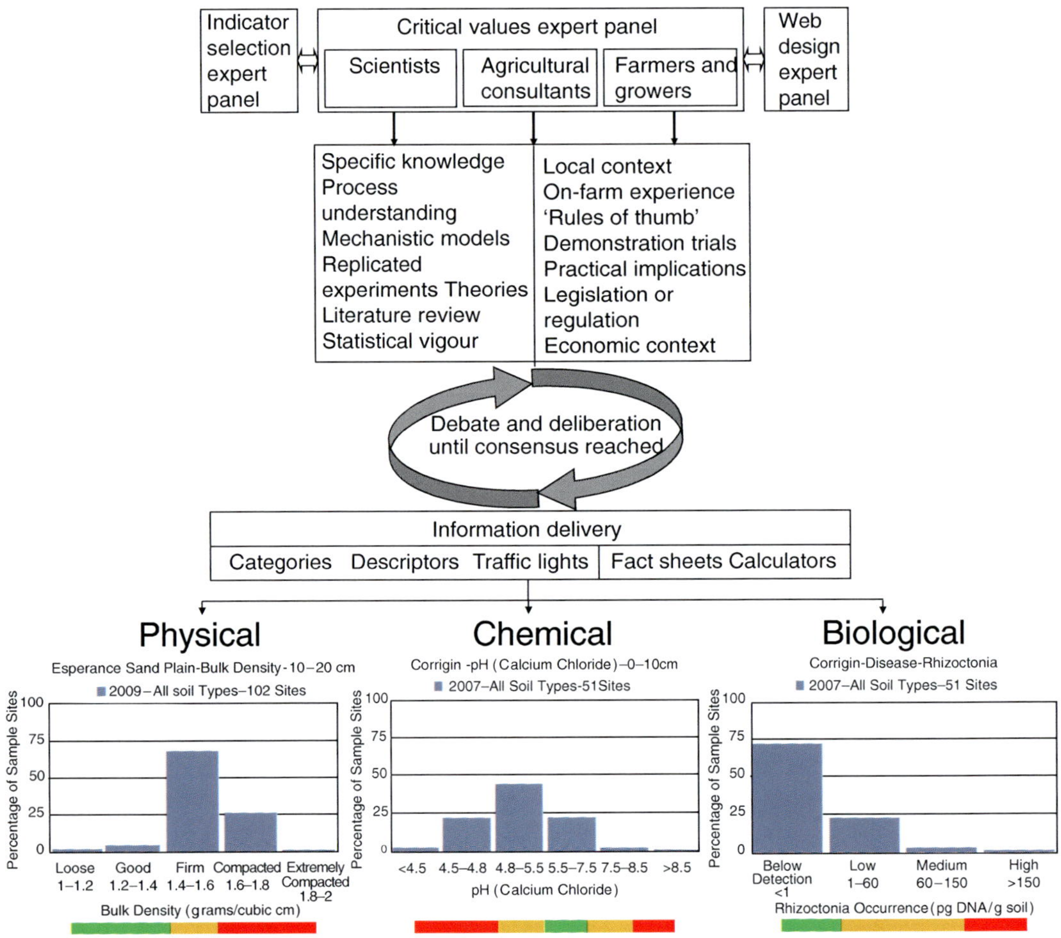

Plate 3.4 Example of the approaches used to develop a soil quality indicator package, highlighting the role of expert groups. These groups drew together a wide range of stakeholders with relevant information and experience including literature review and practical experience. Through facilitated discussion, the groups derived the relationships between yield response and a range of soil properties to inform the interpretation of the indicators used within the soil quality monitoring framework (www.soilquality.org.au) which focuses on soil quality for crop production. Data presentation using the derived categories, descriptors and traffic lights shown for three soil indicators as examples.

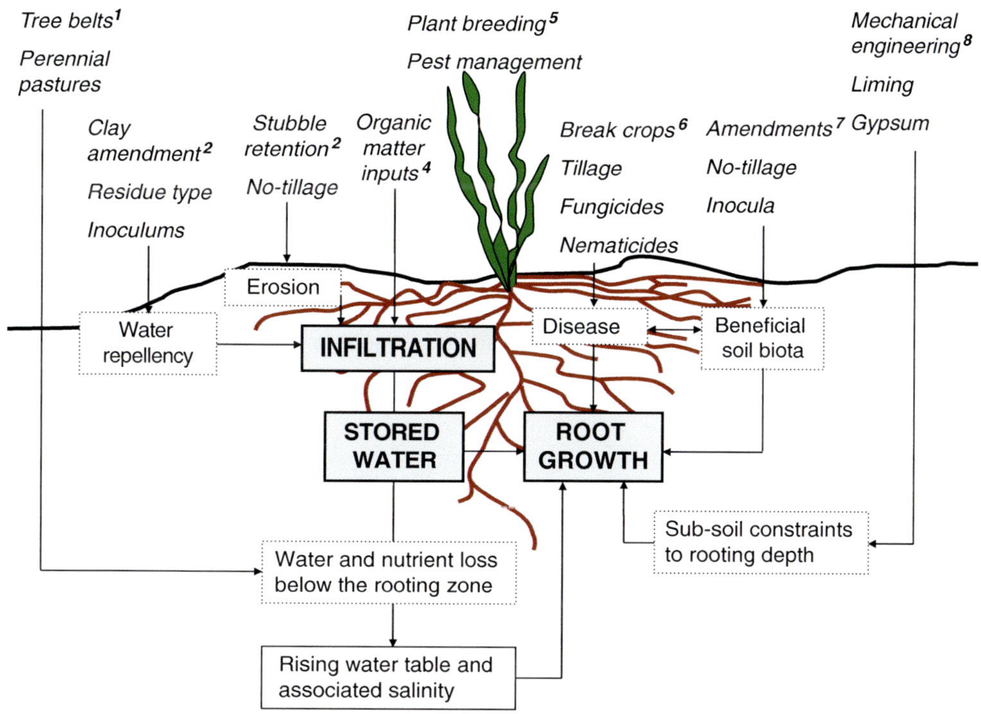

Plate 3.5 Developing integrated strategies for effective capture and utilisation of rainfall together with management of soil water is critical to achieve crop growth in dryland cropping systems (Tow *et al.*, 2012). In South Australia, a range of soil management approaches are integrated within farming systems so that the potential yields (defined by available water in any season) can be attained. These include (1) introduction of tree belts (Robinson *et al.*, 2006) and deep rooting perennial pastures to slow a rising (often saline) water table. (2) While wax degrading micro-organisms have been isolated (Roper, 2004), their success as inocula under field conditions has been limited. Instead clay amendment is an economically viable option (Hall *et al.*, 2010) to overcome water repellence where it can be sourced locally (e.g. from subsoil on the same farm). (3) No-tillage and stubble retention practices are now common place to manage erosion and ensure maintenance of soil carbon levels and associated soil biological function (Hoyle and Murphy, 2006). (4) Organic matter stabilises soil aggregates and improves water infiltration into soil by contributing to the development of a more porous soil structure (Hoyle *et al.*, 2012). (5) Plant breeding for improved water use efficiency (more grain per mm plant available water) and tolerance to toxicity (e.g. Al; Tang *et al.*, 2003a). Early vigour lines (Ludwig and Asseng, 2010), which enable roots to track early season water and nutrients in leaching soils, also have the benefit of growing the root tip through surface soil constraints (e.g. diseases) quicker. (6) Development of disease-suppressive soils has occurred through crop management and rotation (Wiseman *et al.*, 1996). (7) Amendments to soil can have either positive or negative effects on the soil biota (Bunemann *et al.*, 2006) which can impact on the level of micro-organisms involved in carbon and nutrient cycling. (8) Deep ripping to remove compaction and at same time including gypsum or lime (Tang *et al.*, 2003b) to depth to ameliorate subsoil constraints. Use of controlled traffic (Kingwell and Fuchsbichler, 2011) to stop vehicle movement over the majority of the soil surface and tractors with tank tracks instead of wheels to minimise compaction.

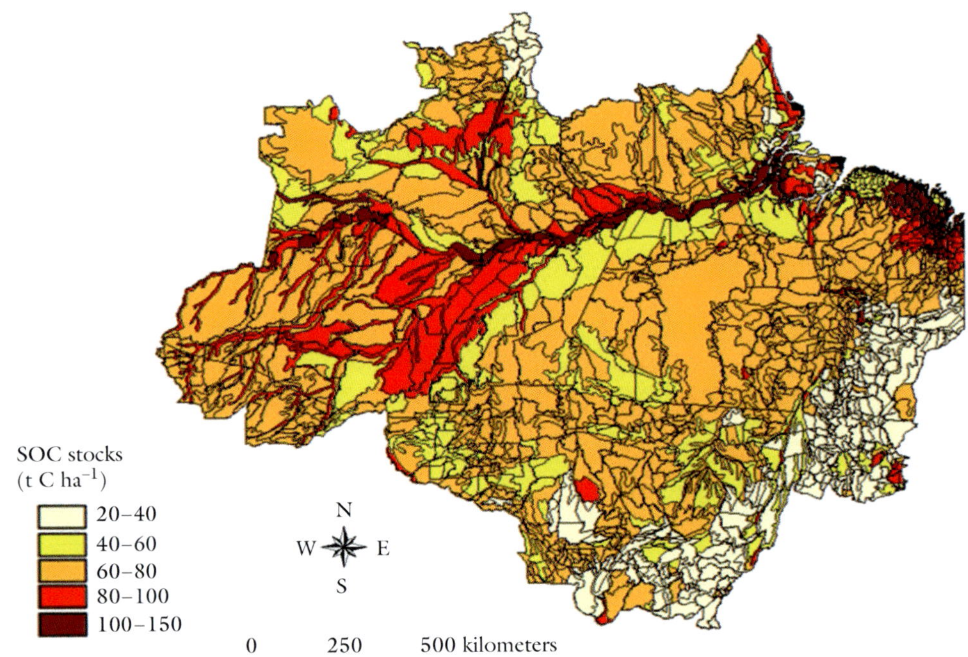

Plate 4.1 Map of soil organic carbon stocks (carbon density) in the Brazilian Amazon. Reprinted from Cerri et al. (2007). With kind permission from Elsevier.

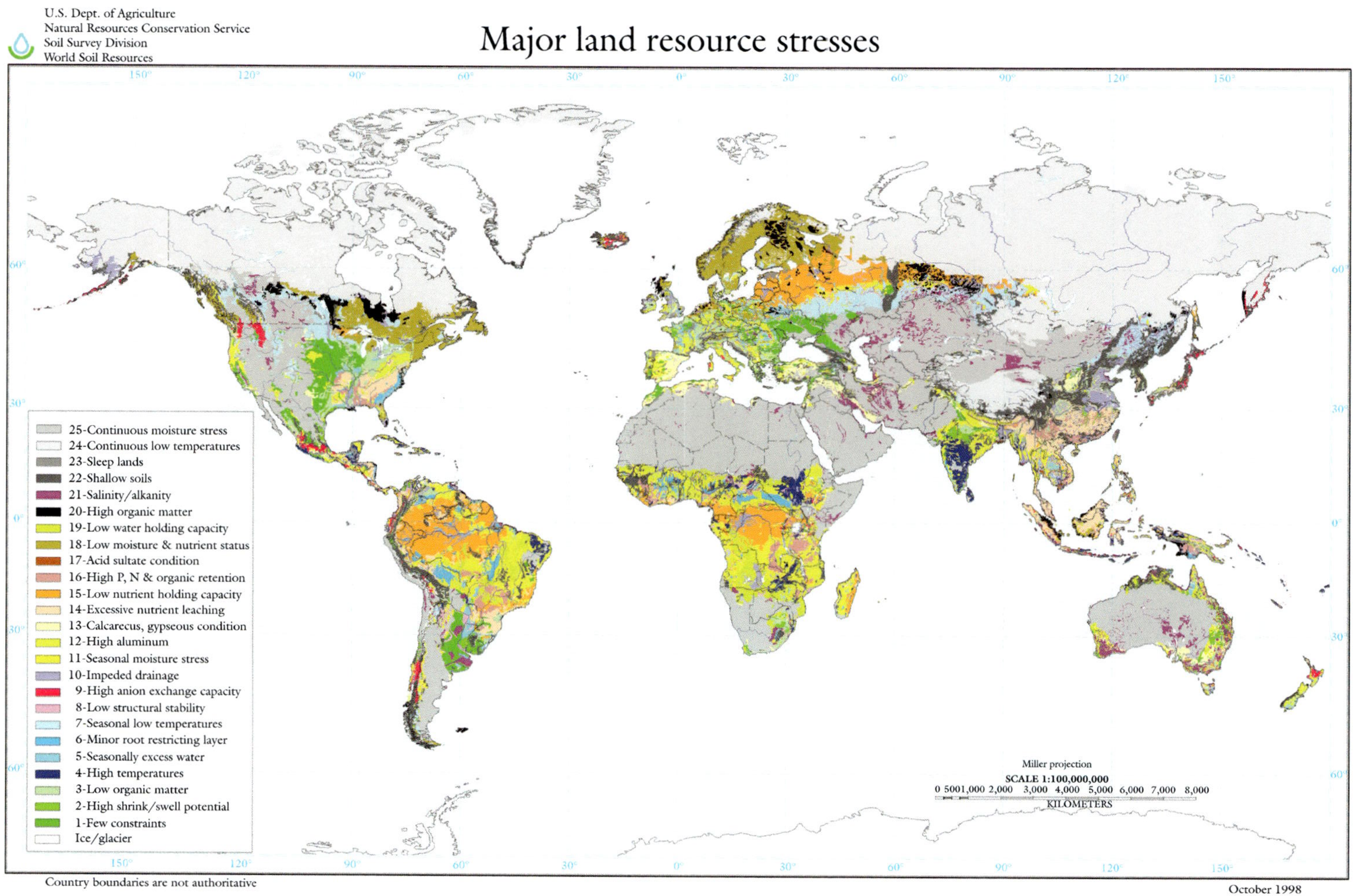

Plate 7.1 Global distribution of saline/sodic soils (these soils would fall within the category of salinity/alkalinity according to this map key). http://soils.usda.gov/use/worldsoils/mapindex/stresses.html. With permission from USDA-NRCS, Soil Survey Division, World Soil Resources.

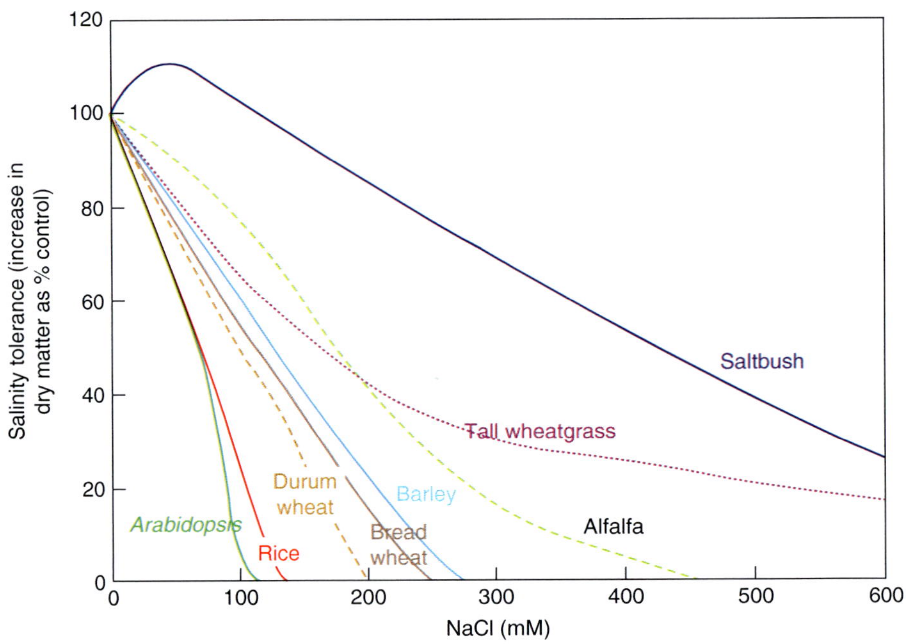

Plate 7.2 Diversity in the salt tolerance of different plant species, shown as increases in shoot dry matter after growth in solution or sand culture containing NaCl for at least 3 weeks, relative to plant growth in the absence of NaCl.

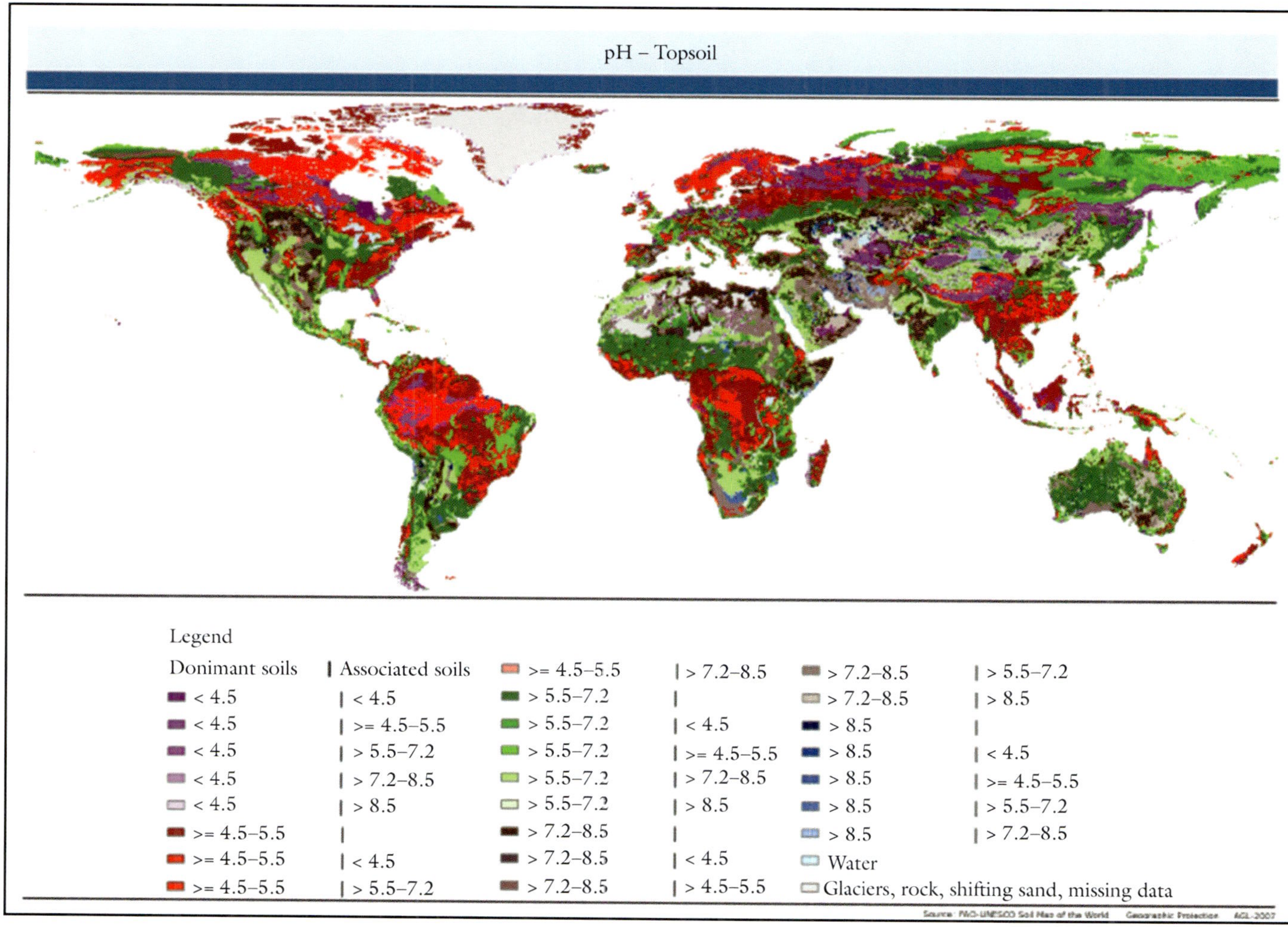

Plate 7.3 Global distribution of topsoil pH. Used with permission. FAO-UNESCO Soil Map of the World 2007. (http://www.fao.org/fileadmin/templates/nr/images/resources/images/Maps/geonetwork/ph_t.png).

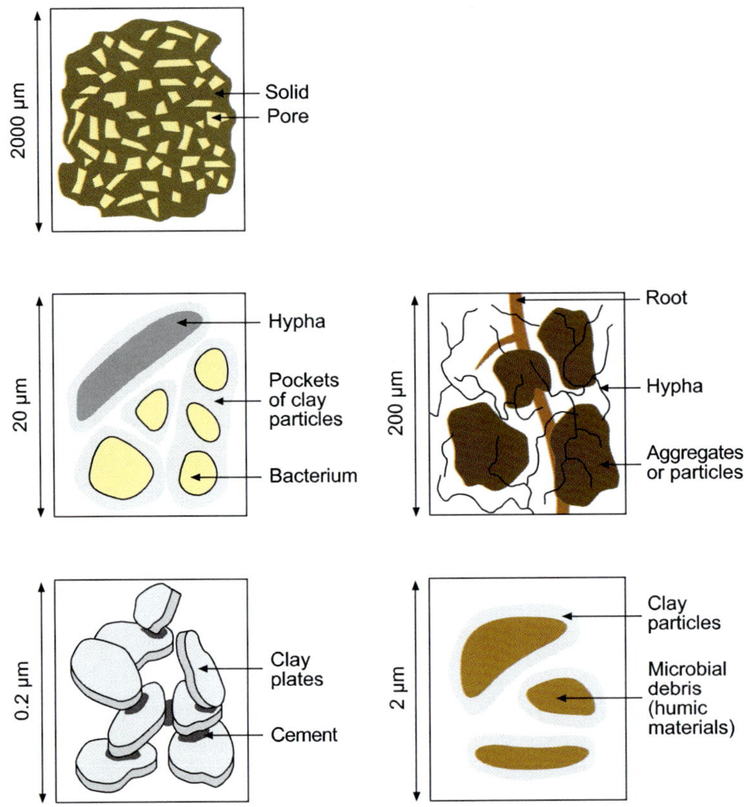

Plate 8.1 Soil physical structure across a wide range of scales. Redrawn from Tisdall and Oades (1982). With kind permission from John Wiley & Sons.

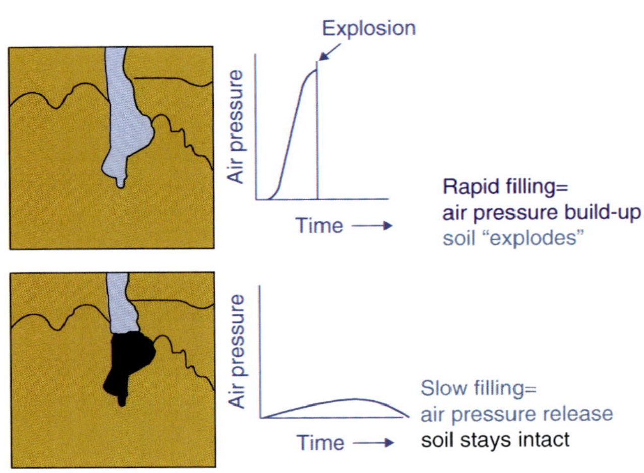

Plate 8.2 Build-up and dissipation of air pressure of soil is a major process leading to slaking of soil aggregates.

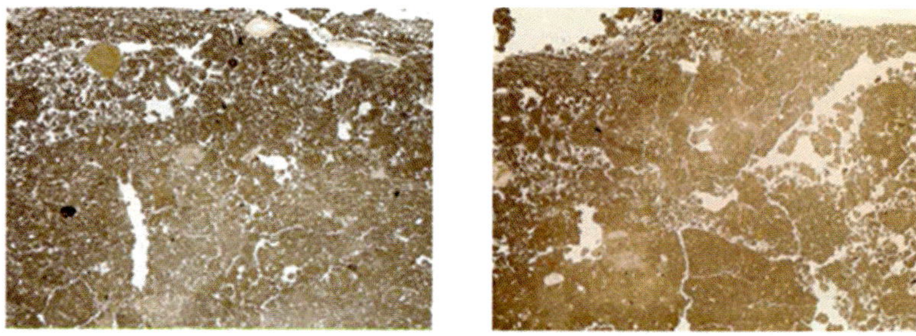

Plate 8.3 Thin sections of topsoil from plots after 11 years of zero tillage (left) and conventional tillage by ploughing and discing (right). From VandenBygaart et al. (1999). With permission from Elsevier.

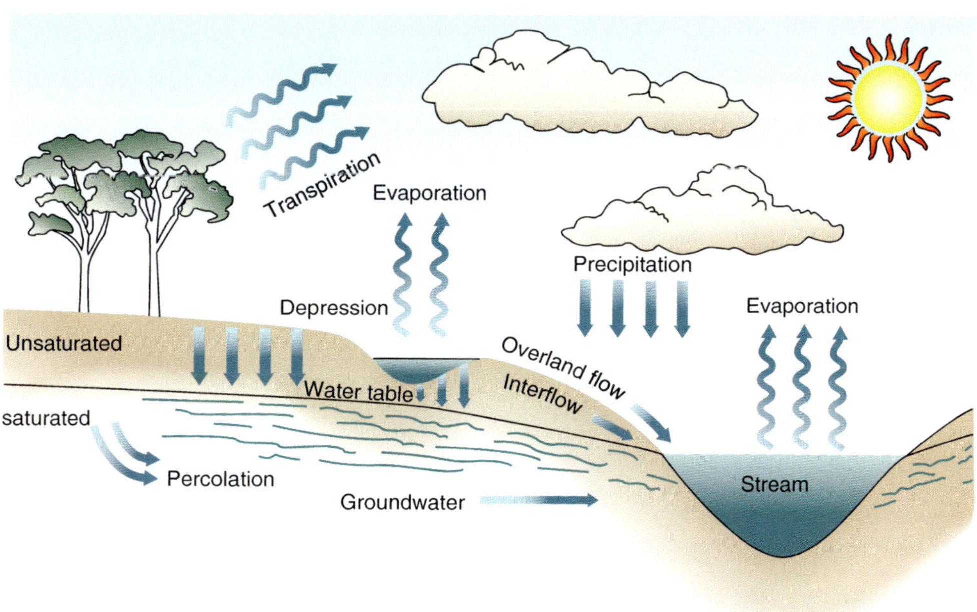

Plate 9.1 The hydrological cycle; focus is on key fluxes related to the soil water balance. From http://soer.justice.tas.gov.au/2003/image/267/index.php; image owner: Government of Tasmania.

Plate 9.2 Lysimeter set-up based on loadcells in an olive plantation near Seville; changes in weight (with a resolution of 0.01 mm) were measured using three 50 kg capacity load cells per lysimeter from which evaporation loss was derived. (Top left) Perspex lysimeter box placed on loadcells bolted on to aluminium baseplate, (top right) holes to place inner (near tree trunk) and outer lysimeters, (bottom left), placement of lysimeters, now reinforced around edges and equipped with tubes to feed through instrument cables, (bottom right) a photo of the finished installation. Also visible are some meteorological sensors. Photos courtesy of Dr Antonio Diaz-Espejo, CSIC, Seville, Spain.

Plate 9.3 Detail of eddy covariance (EC) equipment, comprising a three-axis sonic anemometer and the adjacent Infra Red Gas Analyser (IRGA). Picture taken at the experimental station of the University of Cartagena (Spain).

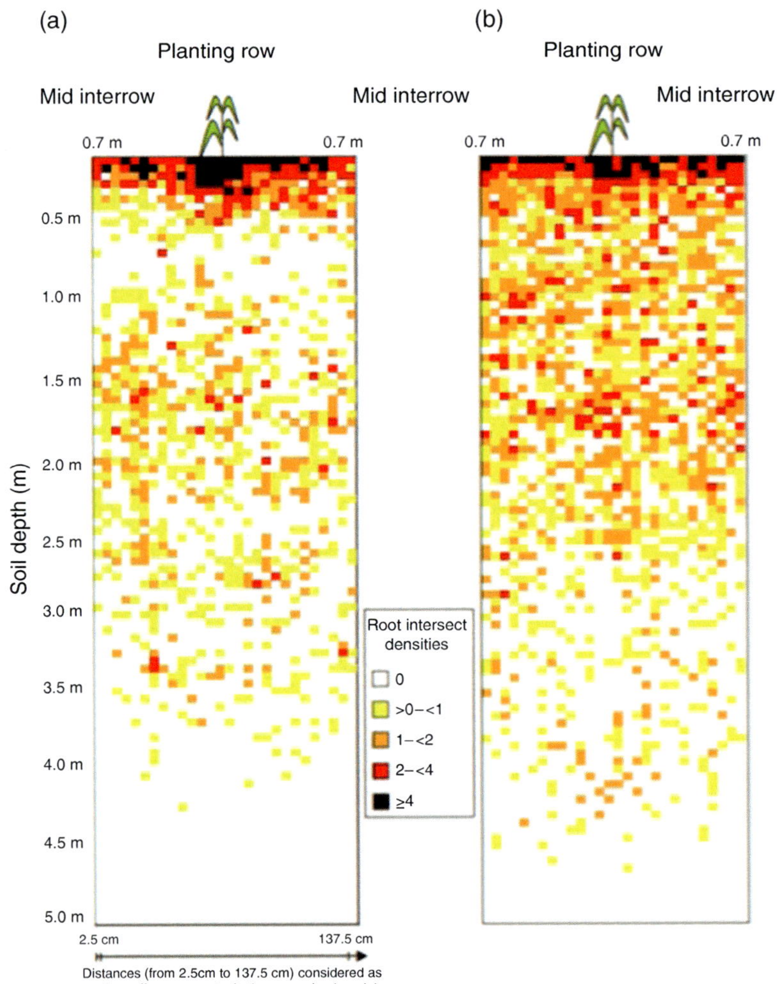

Plate 10.1 Mean number of root intersects (n = 2) in each grid cell of 25 cm^2 delimited in vertical trench walls in the irrigated (a) and rainfed (b) sugarcane crops, at 332 days after planting. Reprinted from Battie-Laclau and Laclau, (2009) with kind permission from Elsevier.

Plate 10.2 Calcified roots appearing as calcite crystals shaping the root cortex cells while the stele remains free of precipitates (Jaillard, 1987a). Photograph taken in between limestone plates in the topsoil of a natural ecosystem with typical Mediterranean 'garrigues' vegetation cover at St Gély-du-Fesc, close to Montpellier (South of France). The photographed zone is about 2 cm × 2 cm. Photo credit: Philippe Hinsinger.

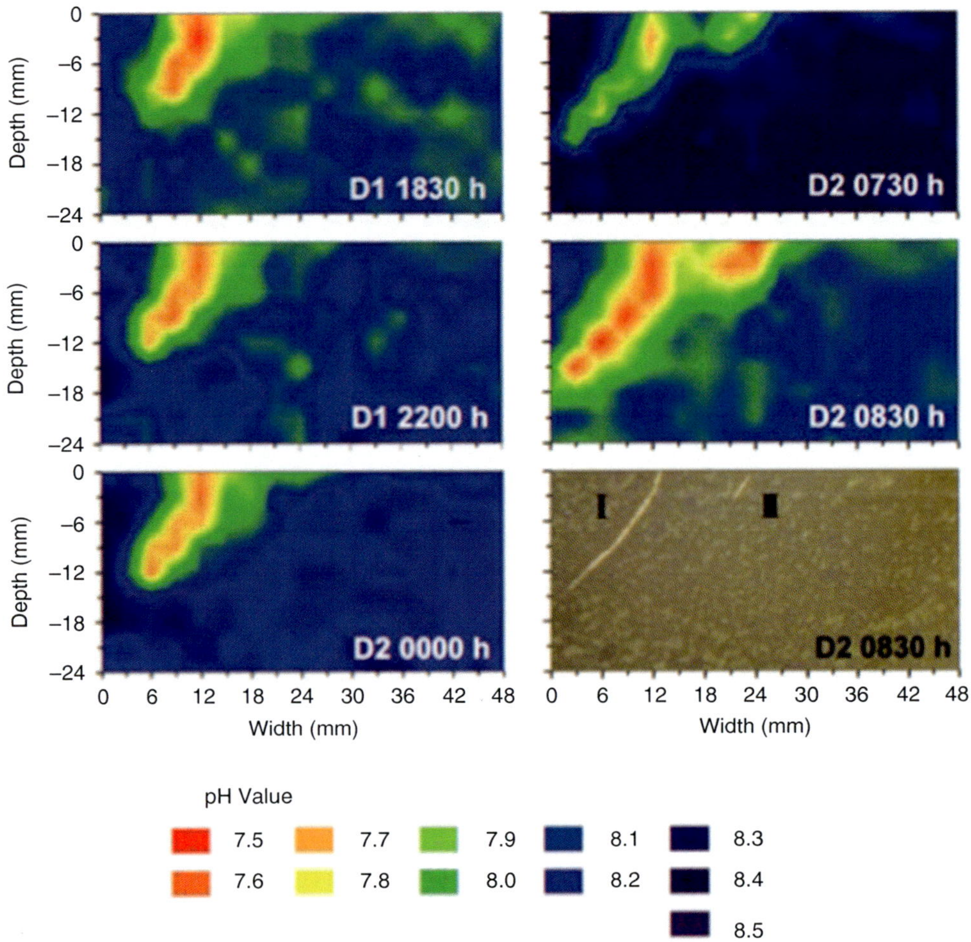

Plate 10.3 Temporal development of root-induced pH changes in the rhizosphere of growing roots of *Juncus effusus*, half an hour before and after the start of daylight, as measured noninvasively with a planar pH optode. The digital photograph (bottom) shows the two investigated roots. Reprinted from Blossfeld and Gansert (2007). With kind permission from Blackwell Publishing Ltd.

Plate 10.4 Imprints of root systems observed *in situ* for plant roots growing in between cracks in dense limestone plates (Hinsinger et al., 2003). Reproduced with permission from Springer-Verlag. Photograph taken in a sample collected in the topsoil of a natural ecosystem with typical Mediterranean 'garrigues' vegetation cover at St Gély-du-Fesc, close to Montpellier (South of France). The photographed zone is about 5 cm × 7.5 cm. Photo credit: Philippe Hinsinger. From Jaillard and Hinsinger (1993). Reproduced with permission from John Wiley & Sons.

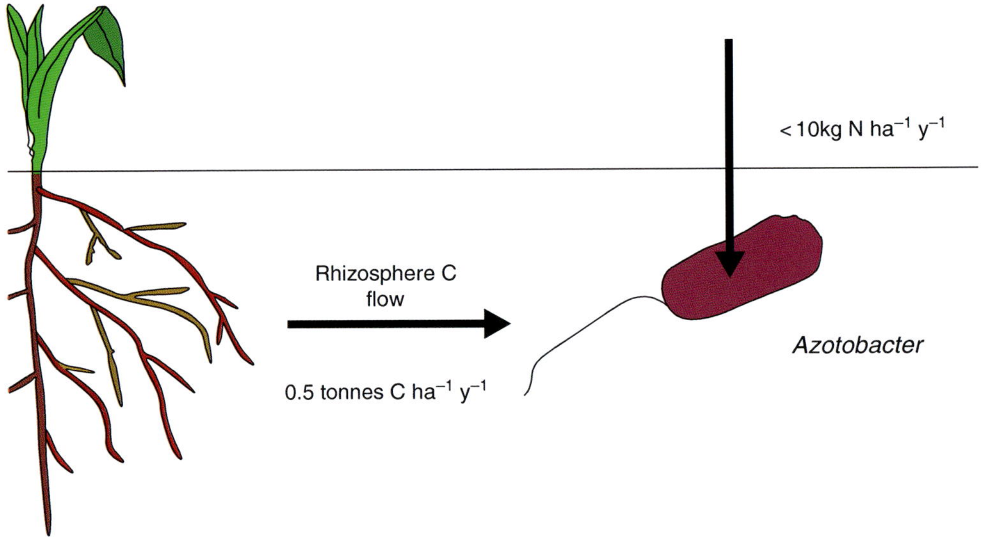

Plate 11.1 The limitation to N_2 fixation by free-living heterotrophic bacteria in soil is the supply of available carbon, which is mainly through rhizosphere flow. The numbers in the figure refer to typical fluxes for a temperate cereal crop. From Killham (1994).

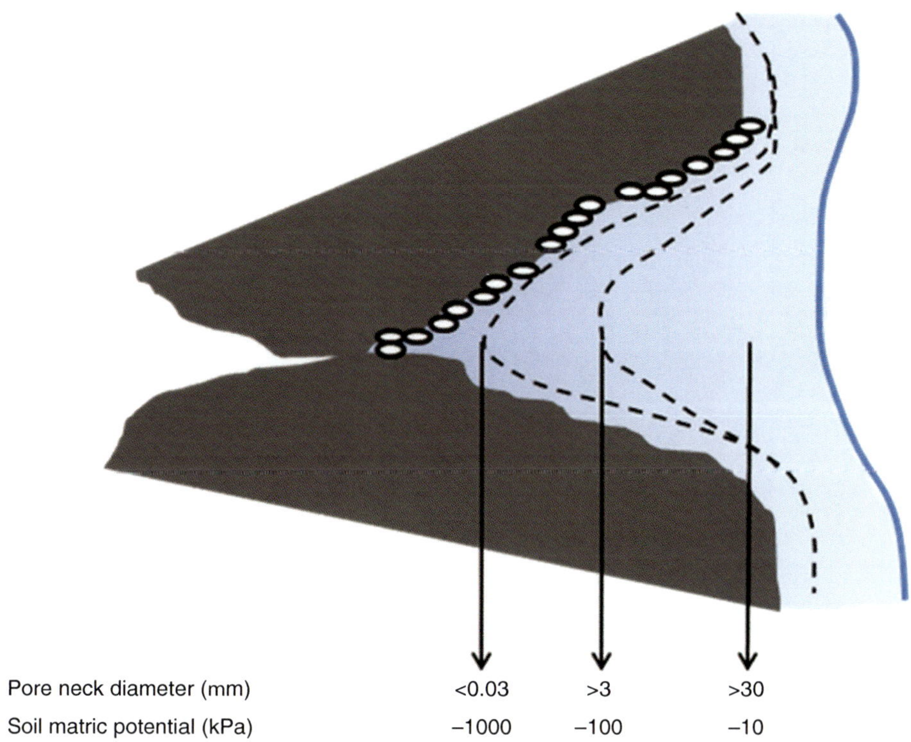

Plate 11.2 The relationship between soil matric potential and the neck diameter of the largest water-filled pores.

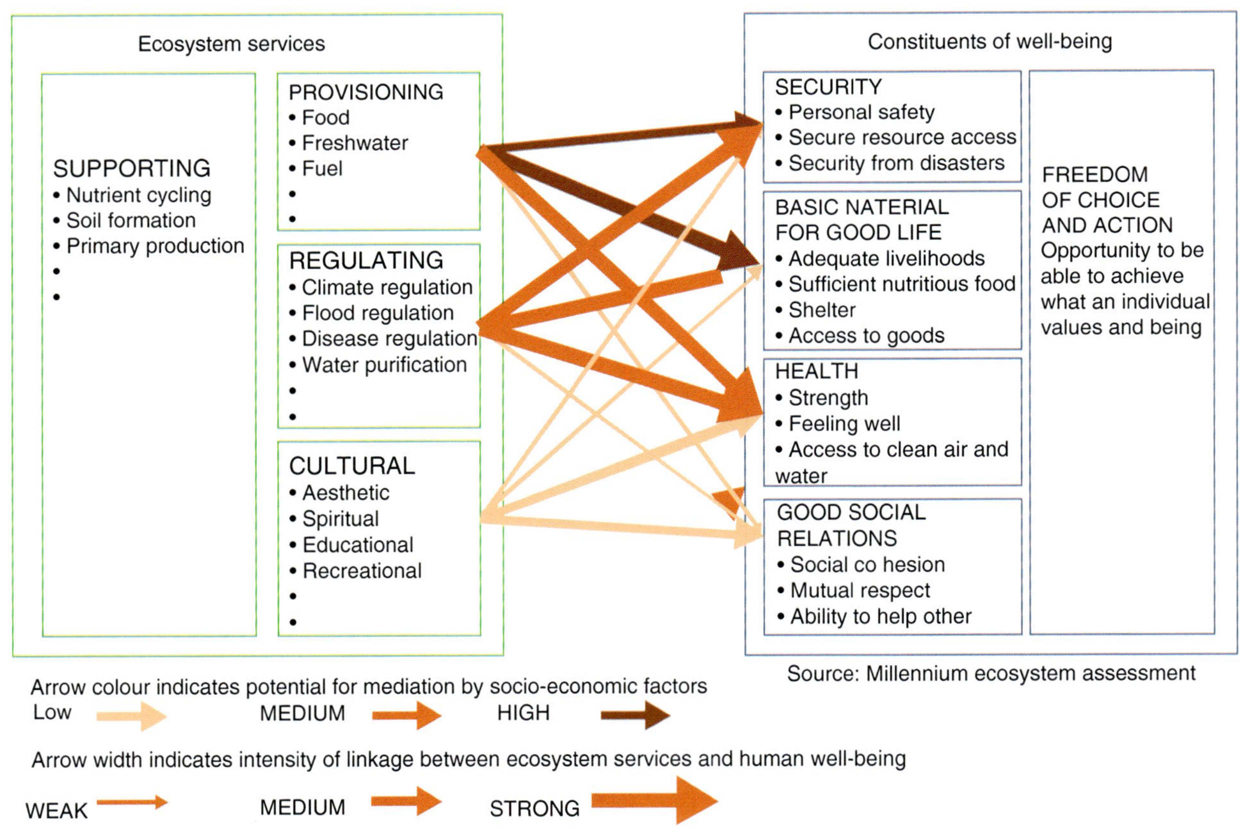

Plate 12.1 Linkages between ecosystem services and human well-being, redrawn from the *Millennium Ecosystem Assessment* (MEA, 2005).

Plate 12.2 (a) Compost comprising composted human ('nightsoil') and domestic organic waste applied in small-scale vegetable production. (b) Monument to soils in China (soil of five different colours laid in a geometric pattern). (c) Linnaeus terrace in the upper Wright Valley, Antarctica. Redrawn and adapted from Pilgrim et al. (2010). With kind permission from Wiley-Blackwell.

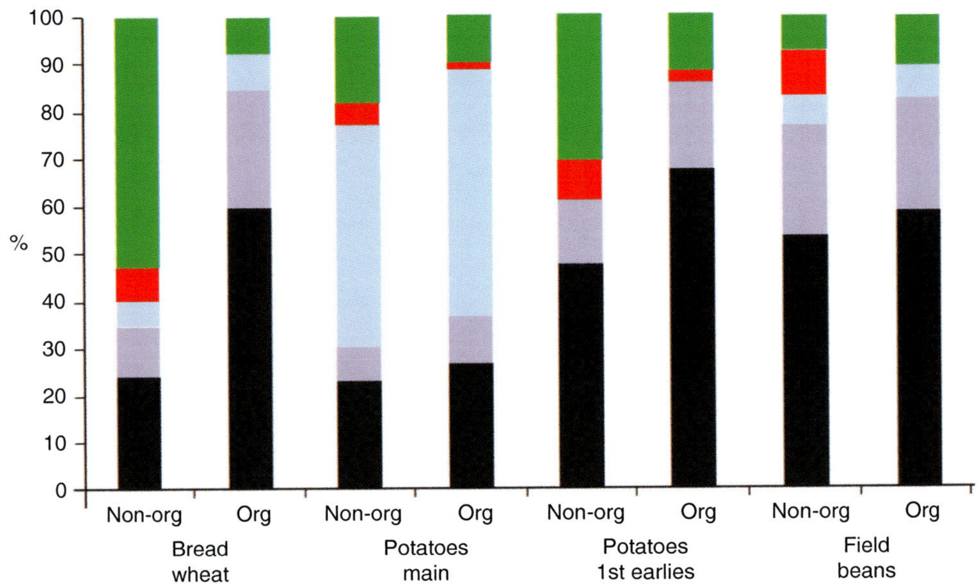

Plate 13.1 A breakdown of the energy used in the production of wheat, potatoes and field beans in the UK in conventional and organic production systems. Green, fertiliser manufacture; red, pesticide manufacture; blue, post-harvest; purple, machinery manufacture; black, field diesel. Reproduced from Woods et al. (2010) with permission from the Royal Society, London.

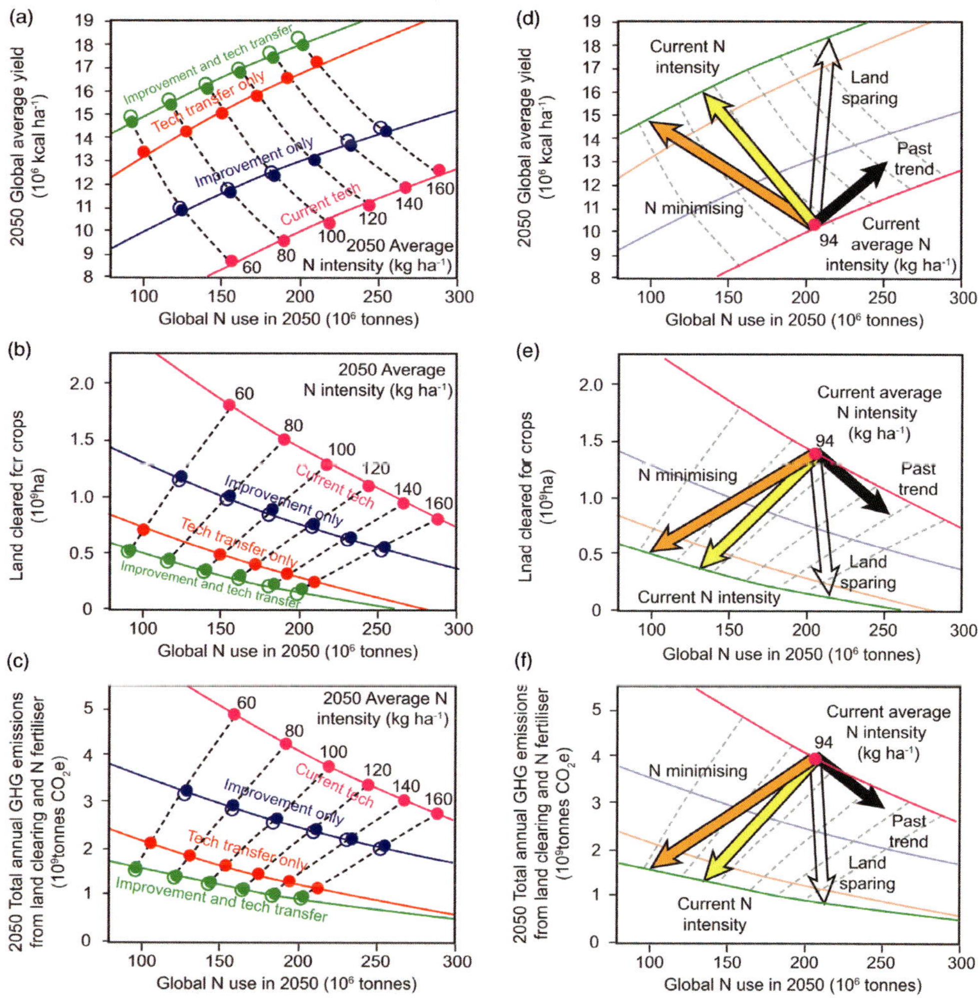

Plate 13.2 Projections of 2050 values for (a) global yields, (b) global land clearing and (c) global agricultural greenhouse gas emissions and (d–f) the yields and environmental impacts of four hypothetical trajectories along which agriculture might develop by 2050. The emissions of greenhouse gases in figures (c) and (f) are expressed in terms of CO_2 equivalents. See Tilman et al. (2011) for details of the methodology used. The four curves shown in figures A–C are based on regression and reflect the different agricultural trajectories assuming magenta, current technology; blue, improvement only; orange, technology adaptation and transfer; and green, improvement and technology transfer. Reproduced from Tilman et al. (2011). With permission from Proceedings of the National Academy of Science.

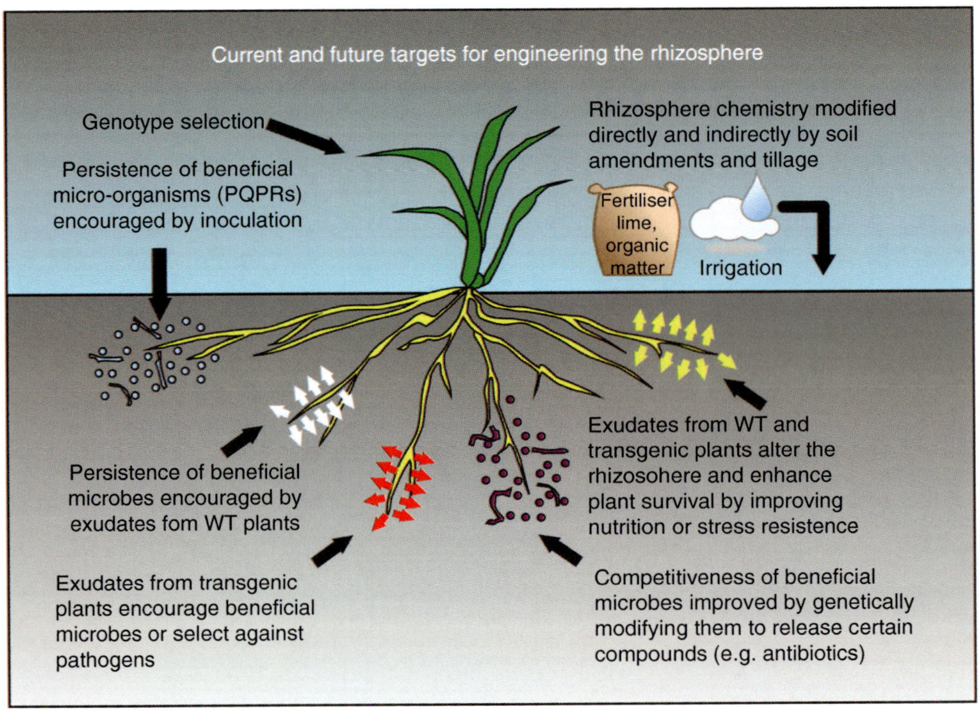

Plate 13.3 Current and future targets for engineering the rhizosphere to improve crop performance. Reproduced from Ryan et al. (2009). With kind permission from Springer Science and Business Media.

Printed in the USA/Agawam, MA
May 29, 2019

704034.003